Principles of Human Anatomy

Principles of Human Anatomy

SECOND EDITION

Gerard J. Tortora
BERGEN COMMUNITY COLLEGE

HARPER & ROW, PUBLISHERS, New York
Cambridge, Hagerstown, Philadelphia, San Francisco,
London, Mexico City, São Paulo, Sydney

1817

Special Projects Editor: Ann Torbert
Project Editor: David Nickol
Designer: Emily Harste
Senior Production Manager: Kewal K. Sharma
Compositor: The Clarinda Company
Printer and Binder: Kingsport Press
Art Studio: J & R Services Inc.
Medical Illustrators: George Weisbrod, Nelva B. Richardson,
    Marsha J. Dohrmann, Helen Gee Jeung

Principles of Human Anatomy, Second Edition

Library of Congress Cataloging in Publication Data

Tortora, Gerard J
    Principles of human anatomy.

    Bibliography: p.
    Includes index.
    1. Human physiology. 2. Anatomy, Human. I. Title.
[DNLM: 1. Anatomy. 2. Physiology. QS4 T712pa]
QP34.5.T68 1980   612   79-29669
ISBN 0-06-046637-5

Contents

CHAPTER 4
THE INTEGUMENTARY SYSTEM 79

CHAPTER 5
OSSEOUS TISSUE 91

CHAPTER 6
THE SKELETAL SYSTEM: THE AXIAL SKELETON 108

CHAPTER 7
THE SKELETAL SYSTEM: THE APPENDICULAR SKELETON 143

CHAPTER 8
ARTICULATIONS 173

CHAPTER 17
THE ENDOCRINE SYSTEM 471

CHAPTER 18
THE REPRODUCTIVE SYSTEMS 491

Preface

Audience

Designed for the introductory course in human anatomy, the second edition of *Principles of Human Anatomy* assumes no previous study of the human body. The text is geared to students in biological, medical, and health-oriented programs. Among the students specifically served by this volume are those aiming for careers as nurses, medical assistants, physician's assistants, medical laboratory technologists, radiologic technologists, respiratory therapists, dental hygienists, physical therapists, morticians, and medical record keepers. However, because of the scope of the text, *Principles of Human Anatomy,* Second Edition, is also useful for students in the biological sciences, premedical and predental programs, science technology, liberal arts, and physical education.

Objectives

The objectives of the second edition of *Principles of Human Anatomy* remain unchanged. Because the subject matter of human anatomy is an exceedingly large and complex body of knowledge to present in an introductory course, an attempt has been made to present unified concepts and data considered useful to a basic understanding of the structure of the human body. Data unessential to this objective has been minimized. The second principal objective has been to present the material at a reading level that can be handled by the average student; however, technical vocabulary and vital, but difficult, concepts have not been avoided. Instead, step-by-step, easy-to-comprehend explanations of each concept have been developed to meet this objective.

Themes

The second edition of this textbook departs from the approaches of other anatomy texts in that somewhat more emphasis is given to physiology and applications to health. This approach is premised on the idea that it is motivationally and conceptually more effective to present basic anatomy with some reference to function. This framework is especially valuable in discussion of certain systems of the body, such as the nervous system. It makes little sense to present students with anatomical detail without relating anatomy to function because structure determines function. With function included, this text will give students a better understanding of anatomical concepts. Moreover, it is important that a study of anatomy include reference to clinical situations. For this reason, disorders are discussed that can be understood once the context of normal anatomy is established. These disorders are treated under the special heading of **Applications to Health.** The basic content of this book is anatomy, but its presentation within the context of physiology and disorders will improve student comprehension of the course material.

Organization

The book is organized by systems rather than by regions. The first chapter introduces the levels of structural organization, body cavities, anatomical and directional terms, planes of the body, surface anatomy, and units of measurement. A new section on radiographical anatomy comparing roentgenograms and computed tomography has been added. A generalized cell is used to demonstrate the basic features of cells. Here new sections on microtubules, cells and aging, cells and cancer, and medical terminology have been added. Tissue organization is presented through descriptions of the structure, functions, and locations of the principal kinds of epithelium and connective tissue. The histology of bone, muscle, nervous tissue, and blood is considered along with the relevant organ systems. The chapter on tissues includes a new section on the growth of cartilage. The skin and its accessory structures are utilized as an example of the organ and system levels of organization. This chapter has also been enhanced by the inclusion of a new section on the development and distribution of hair and a revised section on the structure of hair.

The first body system studied in detail is the skeletal system. This is accomplished by examining the principal features of osseous tissue, the axial skeleton, the appendicular skeleton, and articulations. Several changes have been made in these chapters. In the osseous tissue chapter, a new section has been added on the structure and growth of the epiphyseal plate. Fractures and fracture repair have been moved from Chapter 7 into Chapter 5. The chapter on the axial skeleton now contains a new section on thoracic vertebrae, a revised section on ribs, and several new bone markings have been added. Some new markings have also been added to the chapter on the appendicular skeleton.

One of the most extensively revised chapters is the one on articulations. It contains new sections on fibrous joints, synovial joints, and the movements of synovial joints, and new sections on the principal joints of the body. For each joint discussed in exhibit format, there is a definition, classification by type, a list of anatomical components, and a labeled diagram. Added also is a new discussion of dislocation and sprains.

The muscular system is next analyzed through a study of muscle tissue and the locations and actions of the principal muscles of the body. The chapter on muscular tissue now contains revised sections on the motor unit and mechanism of contraction. New to the chapter on the muscles of the body are sections on the arrangement of fasciculi and their relationship to movement. Surface anatomy photos are now adjacent to muscle diagrams.

The student is next introduced to the cardiovascular and lymphatic systems. The chapter includes a study of blood, interstitial fluid, lymph, the heart, blood vessels, circulatory routes, and lymphatic structures and contains new material on the development of blood cells and a revised section on atherosclerosis. A major change here is the inclusion of the location and drainage of the principal groups of lymph nodes in exhibit format as well as in labeled diagrams.

The next major area of emphasis is the nervous system. Students are introduced to the structure and physiology of nervous tissue, the spinal cord and spinal nerves, the brain and cranial nerves, and the autonomic nervous system. The chapter on nervous tissue now presents a revised organization of the nervous system in a flowchart. The chapter on the spinal cord includes a revised discussion of Wallerian degeneration and new sections on sciatica and neuritis. In the chapter on the brain, the discussion of the cerebrum now comes before the discussion of the cerebellum. Disorders have also been moved from the chapter on the autonomic nervous system into this chapter and sections on dyslexia and Tay-Sachs disease have been added. The chapter on the autonomic nervous system now includes sections on its control by higher centers, biofeedback, and meditation. The structure and function of the visual, auditory, gustatory, and olfactory receptors are also considered. New to the chapter on receptors is discussion of sensory pathways (pain and temperature; crude touch and pressure; fine touch, proprioception, and vibrations; and cerebellar tracts) and motor pathways (pyramidal and extrapyramidal).

Attention is then turned to a discussion of the anatomy and physiology of the endocrine and reproductive systems. Here the interrelationships between hormones and reproduction are established. The endocrine system chapter contains revised sections on the role of cyclic AMP and the thymus gland and a new section on the blood and nerve supply of the parathyroids. The reproductive systems chapter contains new sections on spermatogenesis, oogenesis, mammography and thermography.

The text concludes with descriptions of the respiratory, digestive, and urinary systems. The respiratory system chapter has revised sections on bronchography and hyaline membrane disease and new sections on spirometry and the Heimlich maneuver. New to the digestive system chapter are sections on intrinsic muscles of the tongue, the nerve supply to the salivary glands, directional terms for the teeth, and the blood and nerve supply of the teeth. The chapter on the urinary system contains a revised section on the ultrastructure of a nephron and a new section on abnormal constituents in urine.

Discussions of all major body systems include some reference to physiology, disorders, and a listing of pertinent medical terms.

Special Features

The text contains a number of special learning aids for students, including the following:

1. Student Objectives appear at the beginning of each chapter. Each objective describes a knowledge or skill the student should acquire while studying the chapter. (See **Note to the Student** for an explanation of how the objectives can be used.) End-of-chapter **Review Questions** are designed specifically to help meet the stated objectives. In addition, each **Study Outline** provides a checklist of major topics the student should learn.

2. The health-science student is generally expected to learn a great deal about the anatomy of certain organ systems—specifically, about bones, articulations, skeletal muscles, blood vessels, lymph nodes, and nerves. In these areas, many anatomical details have been removed from the narrative and placed in exhibits, most of which are closely tied to illustrations. This method organizes the data and deemphasizes rote learning of concepts presented in the narrative.

3. An unusually large number of disorders are described in the section entitled **Applications to Health** in many chapters. The topics provide review of normal anatomy and physiology and allow the student to understand why the study of anatomy is fundamental to a career in any of the health fields.

4. Glossaries of selected **Key Medical Terms** appear at the end of each chapter that deals with a major body system.

5. The line drawings are unusually large, so that details are clearly labeled and easily discernible. Color is used functionally to differentiate structures and regions. In the second edition we have introduced yellow to denote nerves, blue to distinguish venous blood from arterial blood (shown in red), and a combination of all four colors to delineate the various bones of the skull. New to this edition is a tone-rendering technique for skeletal muscles.

6. Photomicrographs and electron micrographs are frequently accompanied by adjacent labeled diagrams that amplify and aid observation.

7. Scanning electron micrographs are new to this edition as are surface anatomy photographs of skeletal muscles. Selected computed tomography (CT) scans have also been incorporated.

8. Photographs of specimens have also been added to enhance pedagogy.

9. There are numerous roentgenograms, especially of bones. These are labeled and designed to provide students with an opportunity to transfer anatomical knowledge to clinical situations.

10. Many students find muscle identification an onerous task. To help, the following learning aids have been provided: Illustrations of muscles are shown with duplicates of drawings used for bone identification. In this way, the student is given consistent points of reference. A brief section on the criteria for naming skeletal muscles has been included. Each exhibit dealing with muscles also contains a listing of prefixes, suffixes, roots, and definitions for each muscle discussed. These will help the student understand why a particular skeletal muscle is so named.

11. Another distinctive feature of this textbook is the inclusion of a very large number of photomicrographs of various tissues of the body, designed to help the student understand anatomy at the microscopic level.

12. A fairly extensive list of **Selected Readings** at the end of the book provides numerous references for instructors and students.

13. A comprehensive **Glossary** of important terms used in the textbook, with a pronunciation key, concludes the book.

Supplementary Material

A complimentary Instructor's Manual accompanies *Principles of Human Anatomy,* Second Edition. For each chapter of the textbook, the manual contains a listing of key instructional concepts, selected audiovisual materials, and twenty multiple-choice questions. A directory of the distributors of the audiovisual materials is also provided. The questions are carefully designed to evaluate student understanding of data, concepts, clinical situations, and applications of this knowledge.

Acknowledgments

Since the inception of this textbook, Harper & Row has provided me with the services of several individuals who reviewed the manuscript and provided me with invaluable assistance in preparing it. Among those to whom I wish to express my deepest gratitude are Alphonse R. Burdi, of the University of Michigan Medical School; Warren W. Burggren, of the University of Massachusetts; Edward Carlson, of the University of California, Davis; David J. Garvey, of Wright State University, School of Medicine; Royce L. Montgomery, of the University of North Carolina at Chapel Hill; and Tom Swain, of the University of Colorado, at Boulder. All of the reviewers have helped me to develop an accurate, logical, and pedagogically sound presentation of human anatomy.

I wish to particularly acknowledge Victor B. Eichler, of Wichita State University, and Steve Harper, for providing me with many excellent photographs and photomicrographs, John C. Bennett, of St. Mary's Hospital in San Francisco, for supplying many high-quality roentgenograms, the General Electric Company for the outstanding CT scans, and the Fisher Scientific Company and S.T.E.M. Laboratories, Inc. for the extraordinary scanning electron micrographs. Gratitude is also extended to the many other individuals, publishers, and companies that provided photographs, photomicrographs, and electron micrographs.

The editorial assistance provided by Harper & Row for the development of this project has been outstanding. I wish to express my appreciation to Kyle Wallace, Editor, who offered me all the resources of Harper & Row and personally provided the continuous guidance and encouragement to complete the project; David Nickol, Project Editor, who coordinated the various phases of the project; Emily Harste, Designer; and Ann Torbert, Special Projects Editor.

All drafts of the manuscript were typed by Geraldine C. Tortora, my wife. She also handled all secretarial duties related to the project, for which I am deeply grateful.

I would like to invite readers of this book to send their reactions and suggestions concerning the book to me at the address given below. These responses will be helpful to me in formulating plans for subsequent editions.

Gerard J. Tortora
Biology Department
Bergen Community College
400 Paramus Road
Paramus, N. J. 07652

Note to the Student

At the beginning of each chapter is a listing of **Student Objectives.** Before you read the chapter, please read the objectives carefully. Each objective is a statement of a skill or knowledge that you should acquire. To meet these objectives, you will have to perform several activities. Obviously, you must read the chapter very carefully and if there are sections of the chapter that you do not understand after one reading, you should reread those sections before continuing. In conjunction with your reading, pay particular attention to the figures and exhibits; they have been carefully coordinated to the textual narrative. At the end of each chapter are two and sometimes three other guides that you may find useful. The first, **Study Outline,** is a concise summary of important topics discussed in the chapter. This section is designed to consolidate the essential points covered in the chapter, so that you may recall and relate them to one another. The second guide, **Review Questions,** is a series of questions designed specifically to help you meet your objectives. Sometimes a third aid, **Key Medical Terms,** can be found directly following the chapter. After you have answered the review questions, you should return to the beginning of the chapter and reread the objectives to determine whether or not you have achieved the goals.

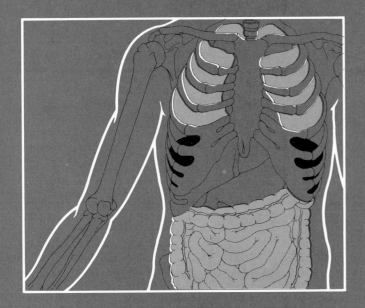

1

An Introduction to the Human Body

STUDENT OBJECTIVES

- Define anatomy, with its subdivisions and physiology.

- Determine the relationship between structure and function.

- Compare the levels of structural organization that make up the human body.

- Define a cell, a tissue, an organ, a system, and an organism.

- Define directional terms used in association with the body.

- List by name and location the principal body cavities and their major organs.

- Describe the subdivisions of the abdominopelvic cavity into nine regions and four quadrants.

- Define the anatomical position.

- Compare common and anatomical terms used to describe the external features of the body.

- Identify by visual inspection or palpation various surface features of the head, neck, trunk, upper extremity, and lower extremity.

- Explain the principle of computed tomography (CT) scanning in radiographical anatomy.

- Define the common metric units of length, mass, and volume, and their U.S. equivalents, that are used in measuring the human body.

You are about to begin a study of the human body in order to learn not only how your body is organized but also, in many instances, how it functions. The study of the human body involves many branches of science. Each contributes to a comprehensive understanding of how your body normally works and what happens when it is injured, diseased, or placed under stress.

Anatomy and Physiology Defined

Two branches of science that will help you understand your body parts and functions are anatomy and physiology. **Anatomy** (or **morphology**) refers to the study of *structure* and the relationships among structures. Anatomy is a broad science, and the study of structure becomes more meaningful when specific aspects of the science are considered. **Gross anatomy** deals with structures that can be studied without a microscope. Another kind of anatomy, **systemic anatomy,** covers specific systems of the body, such as the system of nerves, spinal cord, and brain or the system of heart, blood vessels, and blood. **Regional anatomy** deals with a specific region of the body, such as the head, neck, chest, or abdomen. **Developmental anatomy** is the study of development from fertilized egg to adult form. **Embryology** is generally restricted to the study of development from the fertilized egg to the end of the eighth week in utero. Other branches of anatomy are **pathological anatomy,** the study of structural changes caused by disease, **histology,** the microscopic study of the structure of tissues, and **cytology,** the study of cells.

Whereas anatomy and its branches deal with structures of the body, **physiology** deals with *functions* of the body parts—that is, how the body parts work. As you will see in later chapters, physiology cannot be completely separated from anatomy. Thus you will learn about the human body by studying its structures with references to functions. Each structure of the body is custom-modeled to carry out a particular set of functions. For instance, bones function as rigid supports for the body because they are constructed of hard minerals. Thus the structure of a part often determines the functions it will perform. In turn, body functions often influence the size, shape, and health of the structures. Glands perform the function of manufacturing chemicals, for example, some of which stimulate bones to build up minerals so they become hard and strong. Other chemicals cause the bones to give up minerals so they do not become too thick or too heavy.

Levels of Structural Organization

The human body consists of several levels of structural organization that are associated with one another in several ways. The lowest level of organization, the chemical level, includes all chemical substances essential for maintaining life. All these chemicals are made up of atoms joined together in various ways (Figure 1-1). The chemicals, in turn, are put together to form the next higher level of organization: the cellular level. Cells are the basic structural and functional units of the organism. Among the many kinds of cells in your body are muscle cells, nerve cells, and blood cells. Figure 1-1 shows several isolated cells from the lining of the stomach. Each has a different structure, and each performs a different job.

The next higher level of structural organization is the tissue level. **Tissues** are made up of groups of similar cells and the intercellular material that perform a specific function. When the isolated cells shown in Figure 1-1 are joined together, they form a tissue called epithelium, which lines the stomach. Each cell in the tissue has a specific function. Mucous cells produce mucus, a secretion that lubricates food as it passes through the stomach. Parietal cells produce acid in the stomach. Chief cells produce enzymes needed to digest proteins. Other examples of tissues in your body are muscle tissue, connective tissue, and nervous tissue.

In many places of the body, different kinds of tissues are joined together to form an even higher level of organization: the organ level. **Organs** consist of two or more different tissues that perform a specific function. Organs usually have a recognizable shape. Examples of organs are the heart, liver, lungs, brain, and stomach. In Figure 1-1 you see that the stomach is an organ since it consists of two or more kinds of tissues. Three of the tissues that make up the stomach are shown here. The serous layer (also called the serosa) protects the stomach and reduces friction when the stomach moves and rubs against other organs. The muscle tissue layers of the stomach contract to mix food and pass it on to the next digestive organ. The epithelial tissue layer produces mucus, acid, and enzymes.

The next higher level of structural organization in the body is the system level. A **system** consists of an association of organs that have a common function. The digestive system, which functions in the breakdown of food, is composed of the mouth, saliva-producing glands called salivary glands, pharynx (throat), esophagus (gullet), stomach, small intestine, large intestine, rectum, liver, gallbladder, and pancreas. All the parts of the body functioning with each

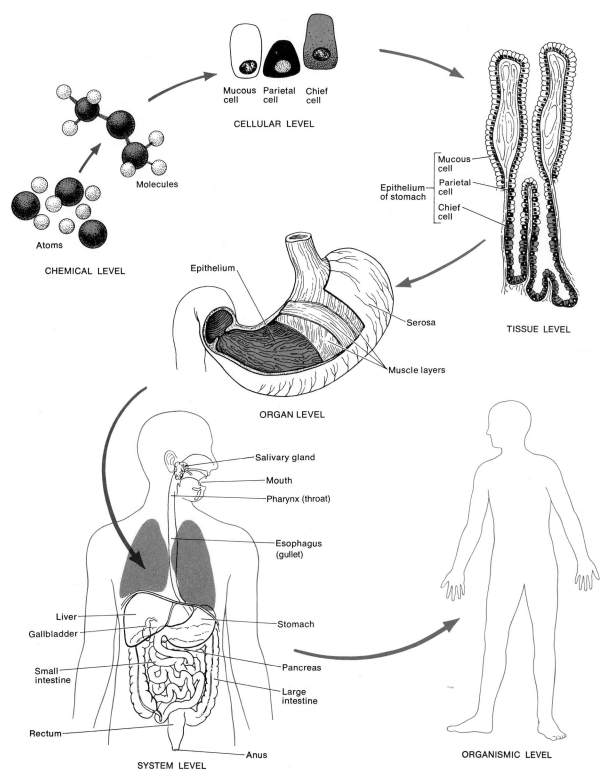

Molecules

Atoms

CHEMICAL LEVEL

Mucous cell · Parietal cell · Chief cell

CELLULAR LEVEL

Mucous cell
Epithelium of stomach — Parietal cell
Chief cell

TISSUE LEVEL

Epithelium

Serosa

Muscle layers

ORGAN LEVEL

Salivary gland
Mouth
Pharynx (throat)
Esophagus (gullet)
Liver
Gallbladder
Stomach
Small intestine
Pancreas
Large intestine
Rectum
Anus

SYSTEM LEVEL

ORGANISMIC LEVEL

FIGURE 1-1 Levels of structural organization that compose the human body.

other constitute the total **organism**—one living individual.

In the chapters that follow, you will examine the anatomy and selected physiology of the major body systems. Exhibit 1-1 illustrates these systems, their representative organs, and their general functions. The systems are presented in the exhibit in the order in which they are discussed in later chapters.

EXHIBIT 1-1

PRINCIPAL SYSTEMS OF HUMAN BODY, REPRESENTATIVE ORGANS, AND FUNCTIONS

1. Integumentary

Definition: The skin and structures derived from it, such as hair, nails, and sweat and oil glands.

Function: Regulates body temperature, protects the body, eliminates wastes, and receives certain stimuli such as temperature, pressure, and pain.

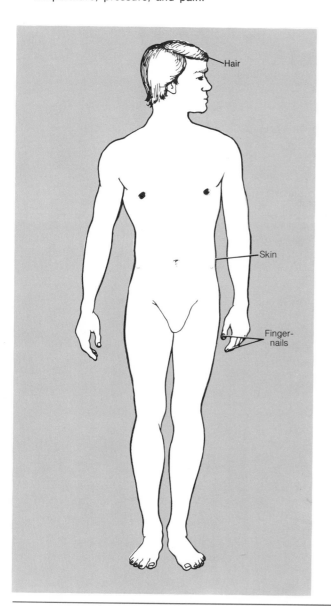

2. Skeletal

Definition: All the bones of the body, their associated cartilages, and the joints of the body.

Function: Supports and protects the body, gives leverage, produces blood cells, and stores minerals.

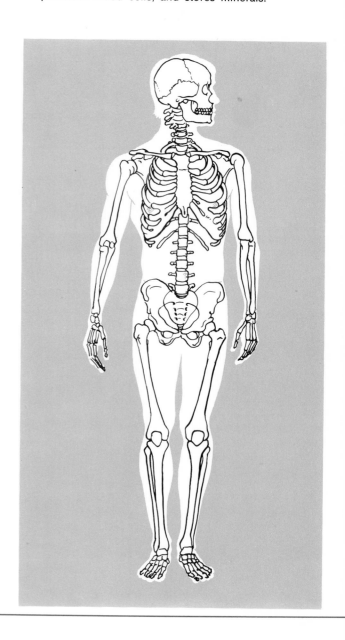

3. Muscular

Definition: All the muscle tissue of the body, including skeletal, visceral, and cardiac.

Function: Allows movement, maintains posture, and produces heat.

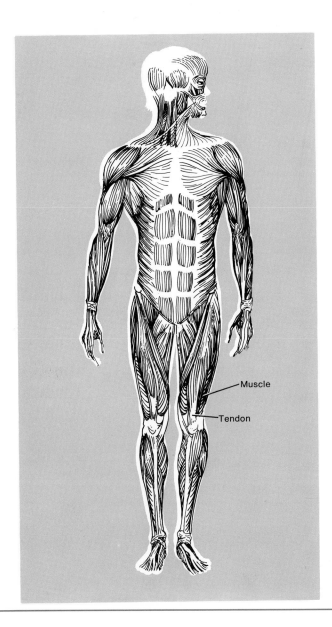

4. Cardiovascular

Definition: Blood, heart, and blood vessels.

Function: Distributes oxygen and nutrients to cells, carries carbon dioxide and wastes from cells, maintains the acid-base balance of the body, protects against disease, prevents hemorrhage by forming blood clots, and helps regulate body temperature.

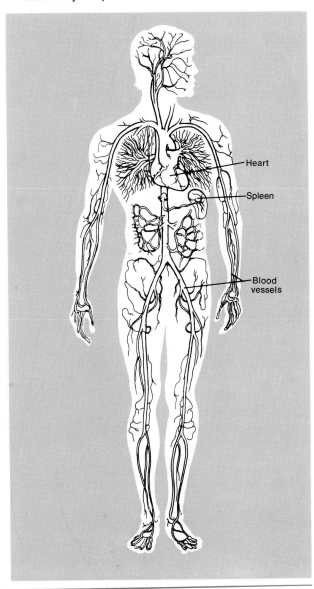

EXHIBIT 1-1
(Continued)

5. Lymphatic

Definition: Lymph, lymph nodes, lymph vessels and lymph glands, such as the spleen, thymus gland, and tonsils.

Function: Returns proteins to the cardiovascular system, filters the blood, produces blood cells, and protects against disease.

6. Nervous

Definition: Brain, spinal cord, nerves, and sense organs, such as the eye and ear.

Function: Regulates body activities through nerve impulses.

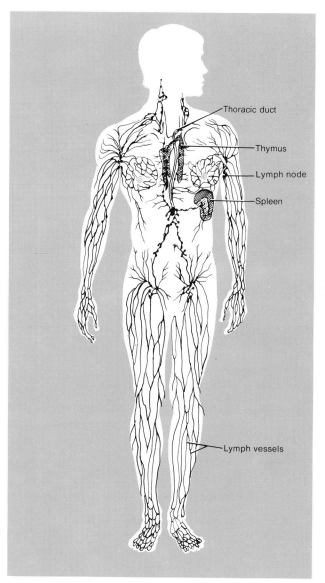

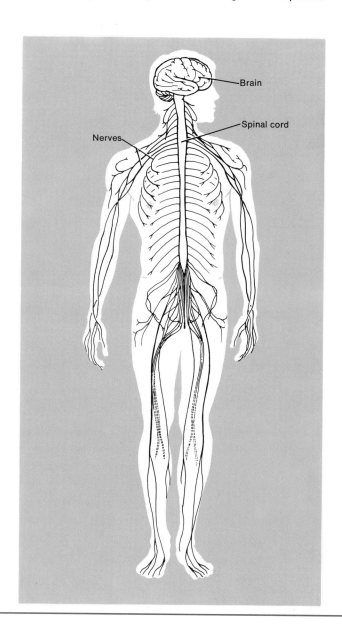

7. Endocrine

Definition: All glands that produce hormones.
Function: Regulates body activities through hormones transported by the cardiovascular system.

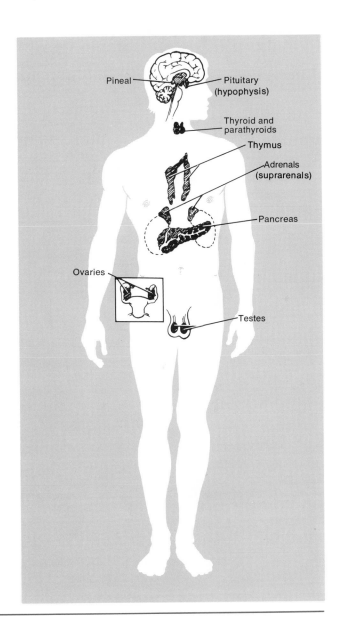

EXHIBIT 1-1
(Continued)

8. Reproductive

Definition: Organs (testes and ovaries) that produce reproductive cells (sperm and ova) and organs that transport and store reproductive cells.

Function: Reproduces the organism.

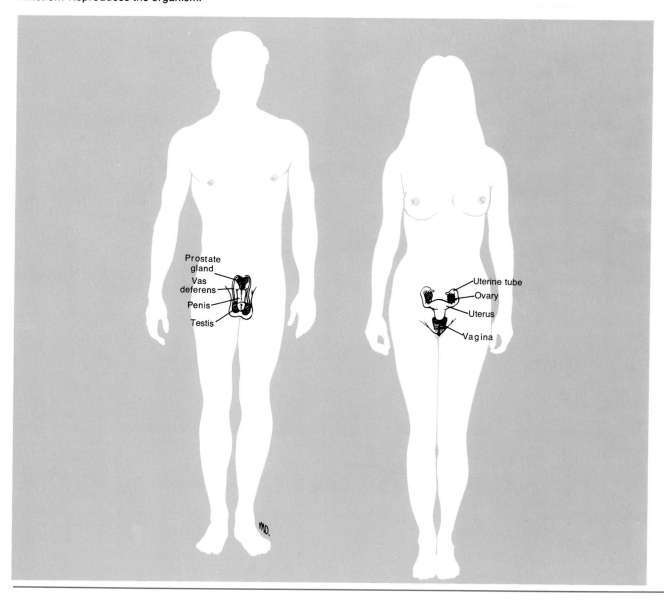

9. Respiratory

Definition: The lungs and a series of passageways leading into and out of them.
Function: Supplies oxygen, eliminates carbon dioxide, and regulates the acid-base balance of the body.

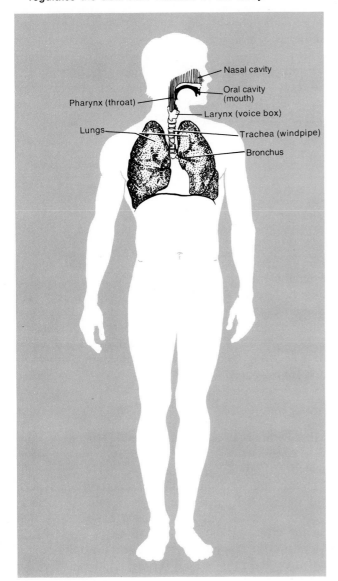

10. Digestive

Definition: A long canal and associated organs such as the salivary glands, liver, gallbladder, and pancreas.
Function: Performs the physical and chemical breakdown of food for use by cells and eliminates solid wastes.

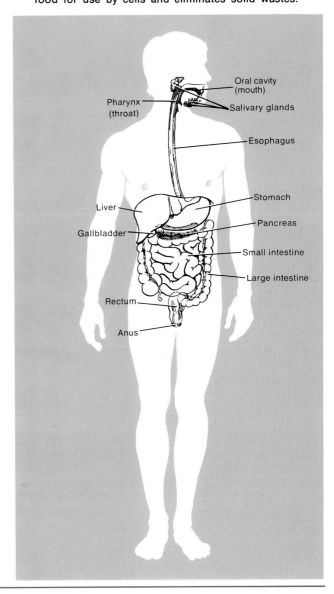

EXHIBIT 1-1
(Continued)

11. Urinary

Definition: Organs that produce, collect, and eliminate urine.

Function: Regulates the chemical composition of blood, eliminates wastes, regulates fluid and electrolyte balance and volume, and maintains the acid-base balance of the body.

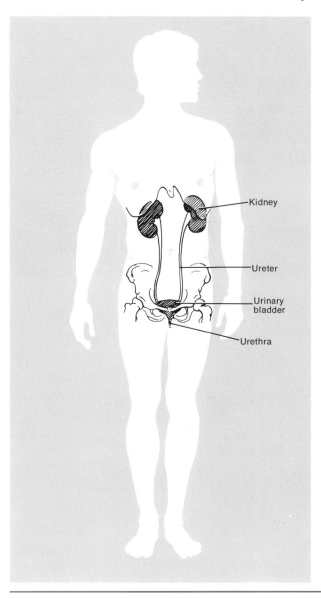

Kidney

Ureter

Urinary bladder

Urethra

Structural Plan

The human organism has certain **anatomical characteristics** that are easily identifiable and can, therefore, serve as landmarks. For example, humans have a _backbone_, a characteristic that places them in a large group of organisms called vertebrates. Moreover, humans are for the most part _bilaterally symmetrical_—the left and right sides of the body are mirror images. Another characteristic of the body's organization is that it resembles a *tube within a tube*. The outer tube is formed by the body wall; the inner tube is the digestive tract. You will need to know such general characteristics as well as the terms used to describe positions and directions in the body.

DIRECTIONAL TERMS

In order to explain exactly where a structure of the body is located, anatomists must use certain directional terms. If you want to point out the sternum (breastbone) to someone who knows where the clavicle (collarbone) is, you can say that the sternum is **inferior** (farther away from the head) and **medial** (toward the middle of the body) to the clavicle. As you can see, using the terms inferior and medial avoids a great deal of complicated description. Many of the directional terms defined in Exhibit 1-2 may be understood by referring to Figure 1-2. Essential parts of the figure are labeled so that you can see the directional relationships among parts.

BODY CAVITIES

Spaces within the body that contain internal organs are called **body cavities.** Specific cavities may be distinguished if the body is viewed after making a **median,** or **midsagittal,** section—that is, after cutting it into right and left halves. Figure 1-3 shows the two principal body cavities. The **dorsal body cavity** is located near the dorsal surface of the body. It is further subdivided into a **cranial cavity,** which is a bony cavity formed by the cranial (skull) bones and contains the brain, and a **vertebral** or **spinal canal,** which is a bony cavity formed by the vertebrae of the backbone and contains the spinal cord and the beginnings of spinal nerves.

The other principal body cavity is the **ventral body cavity.** This cavity, also known as the *coelom,* is located on the ventral aspect of the body. The organs inside the ventral body cavity are called the **viscera.** The ventral cavity walls are composed of skin, connective tissue, bone, muscles, and serous membrane. Like the dorsal body cavity, the ventral body cavity has two principal subdivisions—an upper portion, called the **thoracic cavity** (or chest cavity), and a lower

EXHIBIT 1-2
DIRECTIONAL TERMS

TERM	DEFINITION	EXAMPLE
Superior (**cephalad** or **cranial**)	Toward the head or the upper part of a structure; generally refers to structures in the trunk.	The heart is superior to the liver.
Inferior (**caudad**)	Away from the head or toward the lower part of a structure; generally refers to structures in the trunk.	The stomach is inferior to the lungs.
Anterior (**ventral**)	Nearer to or at the front of the body.	The sternum is anterior to the heart.
Posterior (**dorsal**)	Nearer to or at the back of the body.	The esophagus is posterior to the trachea.
Medial	Nearer the midline of the body or a structure.	The ulna is on the medial side of the forearm.
Lateral	Farther from the midline of the body or a structure.	The ascending colon is lateral to the urinary bladder.
Proximal	Nearer the attachment of an extremity to the trunk.	The humerus is proximal to the radius.
Distal	Farther from the attachment of an extremity to the trunk.	The phalanges are distal to the carpals (wrist bones).
Superficial	Toward or on the surface of the body.	The muscles of the thoracic wall are superficial to the viscera in the thoracic cavity. (See Figure 1-4b.)
Deep (**internal**)	Away from the surface of the body.	The muscles of the arm are deep to the skin of the arm.
Parietal	Pertaining to the outer wall of a body cavity.	The parietal pleura forms the outer layer of the pleural sacs that surround the lungs. (See Figure 1-4b.)
Visceral	Pertaining to the covering of an organ (viscus).	The visceral pleura forms the inner layer of the pleural sacs and covers the external surface of the lungs. (See Figure 1-4b.)

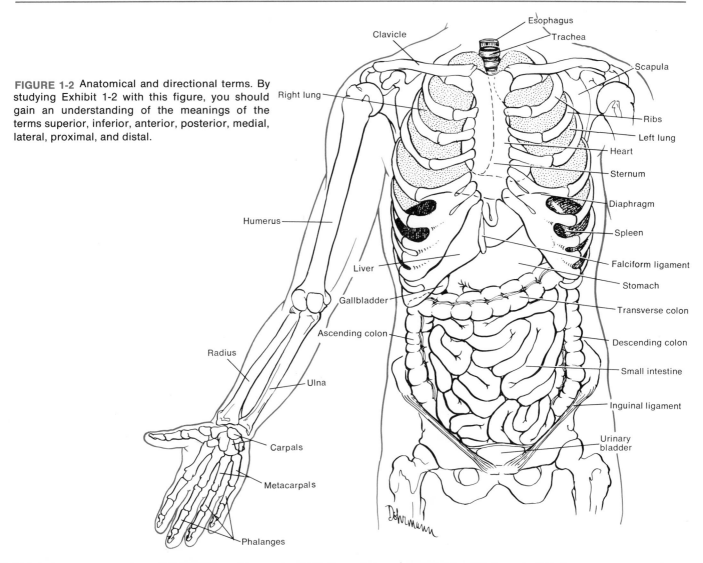

FIGURE 1-2 Anatomical and directional terms. By studying Exhibit 1-2 with this figure, you should gain an understanding of the meanings of the terms superior, inferior, anterior, posterior, medial, lateral, proximal, and distal.

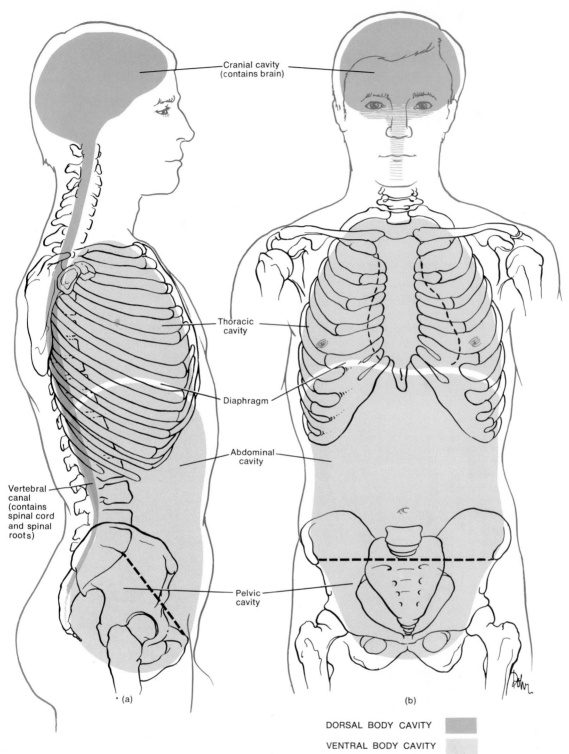

DORSAL BODY CAVITY

VENTRAL BODY CAVITY

FIGURE 1-3 Body cavities. (a) Median section through the human body to indicate the location of the dorsal and ventral body cavities. (b) Subdivisions of the ventral body cavity.

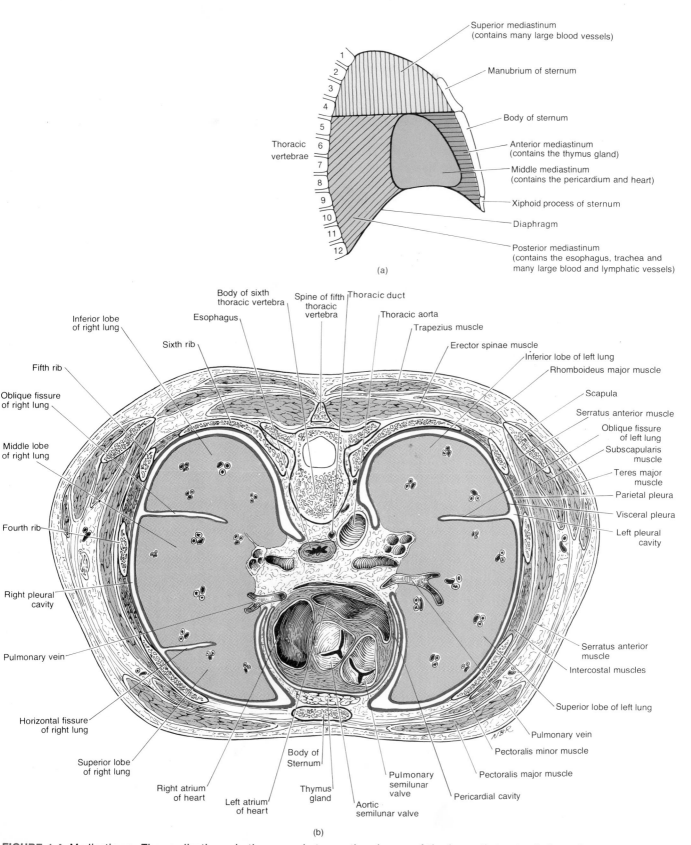

FIGURE 1-4 Mediastinum. The mediastinum is the space between the pleurae of the lungs that extends from the sternum to the vertebral column. (a) Subdivisions of the mediastinum seen in right lateral view. (b) Mediastinum seen in a cross section of the thorax. Many of the structures shown and labeled are unfamiliar to you now. They are discussed in detail in later chapters.

portion, called the **abdominopelvic cavity.** The anatomical landmark that divides the ventral body cavity into the thoracic and abdominopelvic cavities is the muscular diaphragm. The thoracic cavity, in turn, contains several divisions.

There are two **pleural cavities** (Figure 1-4b), one around each lung, and a **mediastinum,** a mass of tissue between the pleurae of the lungs that extends from the sternum to the vertebral column (Figure 1-4a, b). The **pericardial cavity** is located around the heart (Figure 1-4b). The abdominopelvic cavity, as the name suggests, is divided into two portions, although no wall lies between them (Figure 1-3). The upper portion, the abdominal cavity, contains the stomach, spleen, liver, gallbladder, pancreas, small intestine, most of the large intestine, the kidneys, and the ureters. The lower portion, the pelvic cavity, contains the urinary bladder, sigmoid colon, rectum, and the internal male or female reproductive organs. One way to demarcate the abdominal and pelvic cavities is by drawing an imaginary line from the symphysis pubis (anterior joint be-

tween hipbones) to the superior border of the sacrum (sacral promontory).

ABDOMINOPELVIC REGIONS

To describe the location of organs easily, the abdominopelvic cavity may be divided into the nine regions shown in Figure 1-5. Although some unfamiliar terms are used in describing the nine regions and their contents, follow the description as best as you can. When the organs are studied in detail in later chapters, they will have more meaning. The *epigastric region* contains the left lobe and medial part of the right lobe of the liver, the pyloric part and lesser curvature of the stomach, the superior and descending portions of the duodenum, the body and upper part of the head of the pancreas, and the two adrenal (suprarenal) glands. The *right hypochondriac region* contains the right lobe of the liver, the gallbladder, and the upper third of the right kidney. The *left hypochondriac region* contains the body and fundus of the stomach, the spleen, the

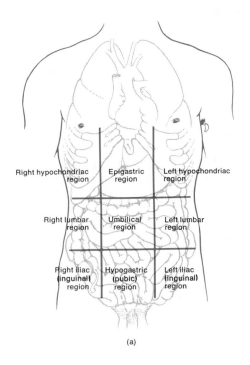

(a)

FIGURE 1-5 Abdominopelvic cavity. (a) The nine regions. The top horizontal line is drawn just below the bottom of the rib cage, and the bottom horizontal line is drawn just below the tops of the hipbones. The two vertical lines are drawn just medial to the nipples. The horizontal and vertical lines divide the area into a larger middle section and smaller left and right sections. (b) Anterior view of the abdomen and pelvis showing the most superficial organs. The greater omentum has been removed.

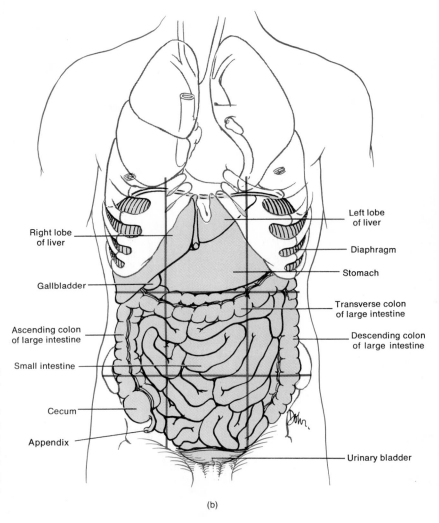

(b)

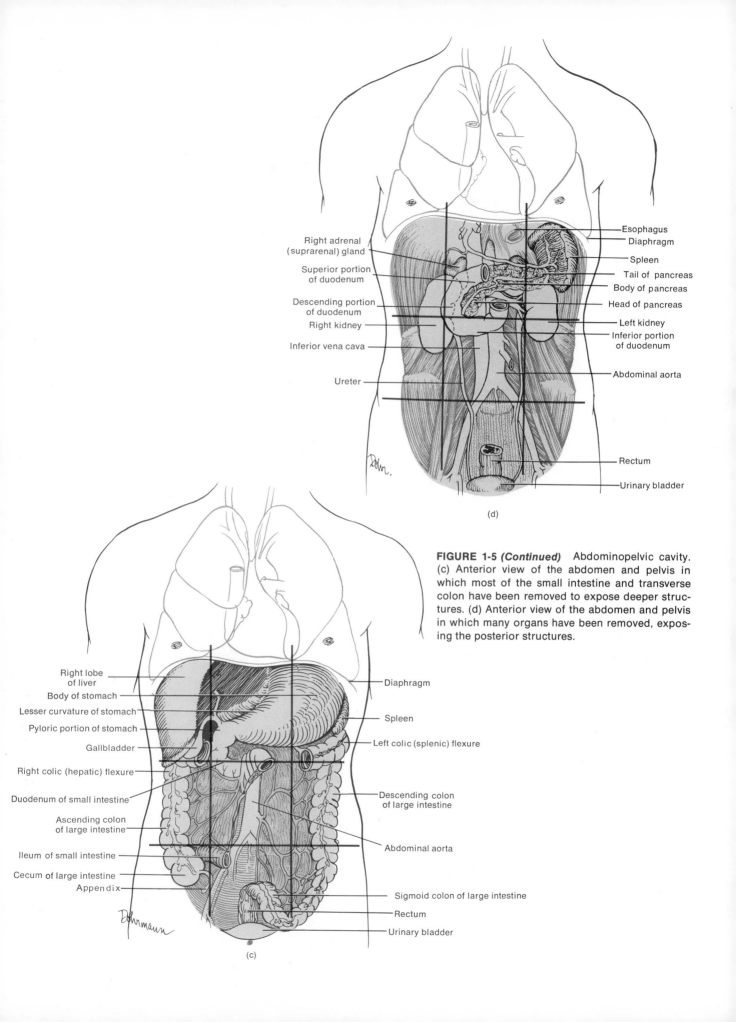

Right adrenal (suprarenal) gland

Superior portion of duodenum

Descending portion of duodenum

Right kidney

Inferior vena cava

Ureter

Esophagus

Diaphragm

Spleen

Tail of pancreas

Body of pancreas

Head of pancreas

Left kidney

Inferior portion of duodenum

Abdominal aorta

Rectum

Urinary bladder

(d)

FIGURE 1-5 (Continued) Abdominopelvic cavity. (c) Anterior view of the abdomen and pelvis in which most of the small intestine and transverse colon have been removed to expose deeper structures. (d) Anterior view of the abdomen and pelvis in which many organs have been removed, exposing the posterior structures.

Right lobe of liver

Body of stomach

Lesser curvature of stomach

Pyloric portion of stomach

Gallbladder

Right colic (hepatic) flexure

Duodenum of small intestine

Ascending colon of large intestine

Ileum of small intestine

Cecum of large intestine

Appendix

Diaphragm

Spleen

Left colic (splenic) flexure

Descending colon of large intestine

Abdominal aorta

Sigmoid colon of large intestine

Rectum

Urinary bladder

(c)

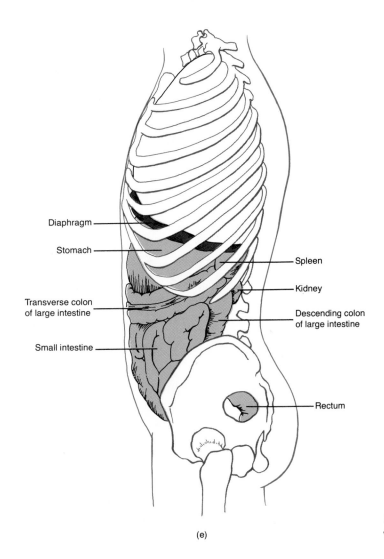

Diaphragm

Stomach

Spleen

Kidney

Transverse colon
of large intestine

Descending colon
of large intestine

Small intestine

Rectum

(e)

FIGURE 1-5 *(Continued)* Abdominopelvic cavity. (e) Left lateral view of the abdomen and pelvis.

left colic (splenic) flexure, the upper two-thirds of the left kidney, and the tail of the pancreas. The *umbilical region* contains the middle of the transverse colon, the inferior part of the duodenum, the jejunum, the ileum, the hilar regions of the kidneys, and the bifurcations (branching) of the abdominal aorta and inferior vena cava. The *right lumbar region* contains the superior part of the cecum, the ascending colon, the right colic (hepatic) flexure, the lower portion of the right kidney, and the small intestine. The *left lumbar region* contains the descending colon, the lower third of the left kidney, and the small intestine. The *hypogastric* (or *pubic*) *region* contains the urinary bladder when full, the small intestine, and part of the sigmoid colon. The *right iliac* (or *right inguinal*) *region* contains the lower end of the cecum, the appendix, and the small intestine. The *left iliac* (or *left inguinal*) *region* contains the junction of the descending and sigmoid parts of the colon and the small intestine.

An easier way to divide the abdominopelvic cavity is into four quadrants (Figure 1-6). In this method, fre-

quently used by clinicians, a horizontal plane is passed through the umbilicus together with a midsagittal plane. These two planes divide the abdomen into a *right superior (upper) quadrant, left superior (upper) quadrant, right inferior (lower) quadrant,* and *left inferior (lower) quadrant.* Whereas the nine-region designation is more widely used for anatomical studies, the four-quadrant designation is better suited for locating the site of an abdominopelvic pain, tumor, or other abnormality.

ANATOMICAL POSITION AND REGIONAL NAMES

When a region of the body is described in an anatomical text or chart, we assume that the body is in the **anatomical position**—erect and facing the observer. The arms are at the sides, and the palms of the hands are turned forward, as in Figure 1-7. The common terms and the anatomical terms, in parentheses, of certain body regions also are presented in Figure 1-7.

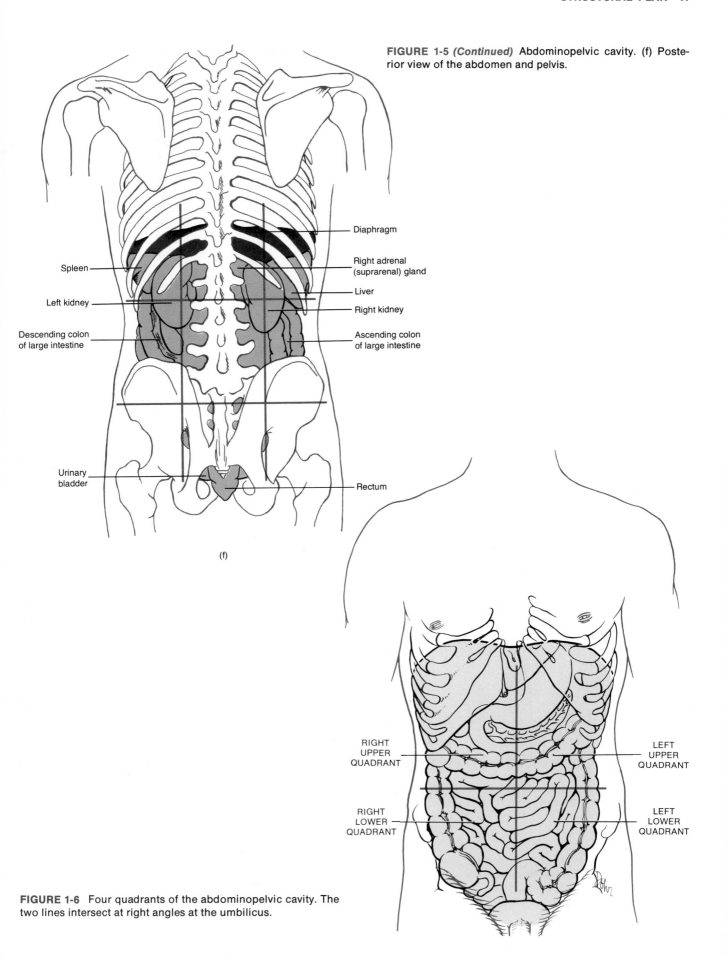

FIGURE 1-5 (Continued) Abdominopelvic cavity. (f) Posterior view of the abdomen and pelvis.

Diaphragm

Right adrenal (suprarenal) gland

Liver

Right kidney

Ascending colon of large intestine

Rectum

Spleen

Left kidney

Descending colon of large intestine

Urinary bladder

(f)

FIGURE 1-6 Four quadrants of the abdominopelvic cavity. The two lines intersect at right angles at the umbilicus.

RIGHT UPPER QUADRANT

LEFT UPPER QUADRANT

RIGHT LOWER QUADRANT

LEFT LOWER QUADRANT

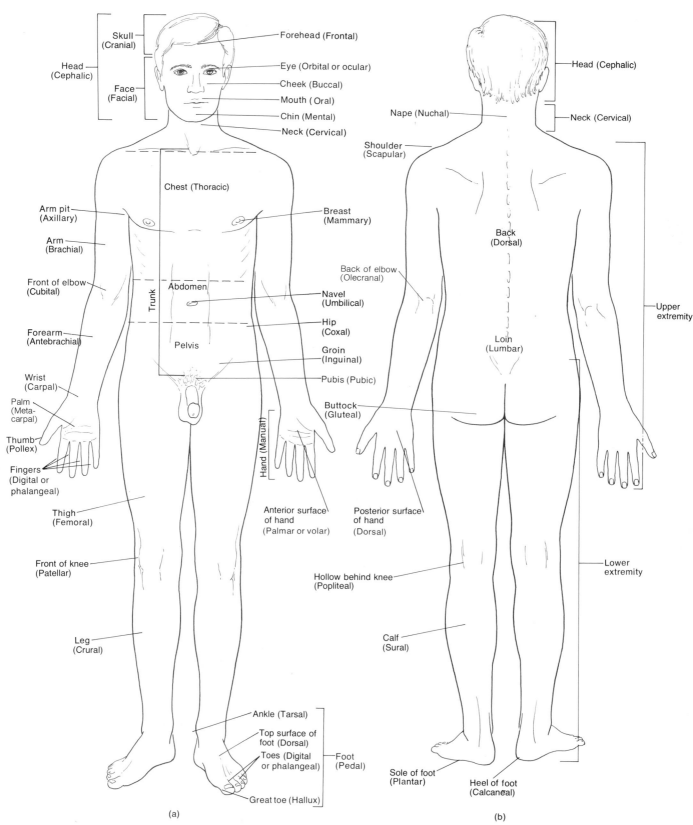

FIGURE 1-7 Anatomical position. (a) Anterior view. (b) Posterior view. Both common terms and anatomical terms, in parentheses, are indicated for many of the regions of the body.

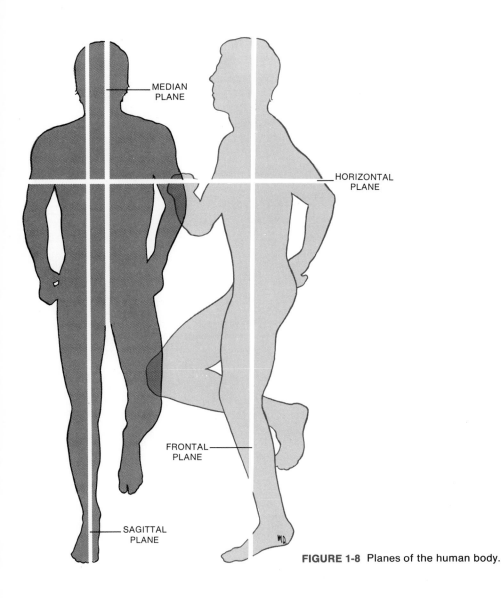

MEDIAN
PLANE

HORIZONTAL
PLANE

FRONTAL
PLANE

SAGITTAL
PLANE

FIGURE 1-8 Planes of the human body.

PLANES

The structural plan of the human body may also be discussed with respect to **planes** (flat surfaces) that pass through it. Several of the commonly used planes are illustrated in Figure 1-8. A *midsagittal* (or *median*) *plane* through the midline of the body runs vertically and divides the body into equal right and left sides. A *sagittal* (or *parasagittal*) *plane* also runs vertically, but it divides the body into unequal left and right portions. A *frontal* (or *coronal*) *plane* runs vertically and divides the body into anterior and posterior portions. Finally, a *horizontal* (or *transverse*) *plane* runs parallel to the ground and divides the body into superior and inferior portions.

When you examine sections of organs, it is important to know the plane of the section so that you can understand the anatomical relationship of one part to another. Figure 1-9 indicates how three different sections—a *cross section,* an *oblique section,* and *longitudinal sections*—are made through a simple tube and through the spinal cord.

Introduction to Surface Anatomy

In the beginning of this chapter, several branches of anatomy were defined and their importance to an understanding of the structure of the body was noted. Now that you have a fairly good idea as to how the body is organized, we can take a look at another branch of anatomy called surface anatomy.

Very simply, **surface anatomy** is the study of the form and markings of the surface of the body. A knowledge of surface anatomy will help you to identify certain superficial structures through visual inspection or palpation through the skin. **Palpation** means to feel with the hand.

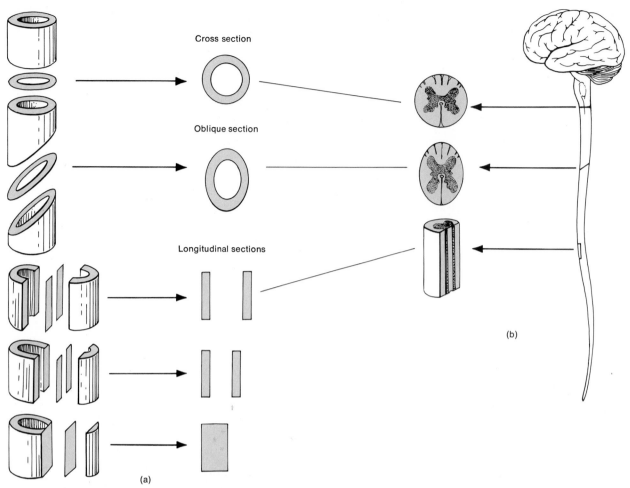

Cross section

Oblique section

Longitudinal sections

(a)

(b)

FIGURE 1-9 How sections are made. (a) Three sections through a tube. (b) The same three sections through the spinal cord.

To introduce you to surface anatomy, the body will first be divided into its primary divisions. These divisions are outlined as follows and may be reviewed in Figure 1-7.

I. Head
  A. Cranium
  B. Face

II. Neck

III. Trunk
  A. Back
  B. Thorax
  C. Abdomen
  D. Pelvis

IV. Extremities
  A. Upper extremity
    1. Axilla
    2. Shoulder

    3. Arm
    4. Elbow
    5. Forearm
    6. Hand
      a. Wrist
      b. Palm
      c. Dorsum
  B. Lower extremity
    1. Buttocks
    2. Thigh
    3. Knee
    4. Leg
    5. Foot
      a. Heel
      b. Dorsum
      c. Sole

THE HEAD

Our discussion of surface anatomy will first consider the head. The **head** *(caput)* is divided into the cranium and face. The **cranium** *(brain case)* consists of a crown *(vertex);* front *(sinciput),* including the forehead *(frons);* back *(occiput);* sides *(tempora);* and ears *(aures).* The **face** *(facies)* is subdivided into the ocular region which includes the eyes *(oculi),* eyebrows *(supercilia),* and eyeballs *(bulbi oculorum);* nose *(nasus);* mouth *(os),* including the lips *(labia)* and oral cavity *(cavum oris);* cheek *(bucca* of *mala);* and chin *(mentum).*

Examine Exhibit 1-3. It contains illustrations of various surface features of the head with an accompanying description of each of the features. Although a large number of surface features are included in Exhibit 1-3 and other exhibits dealing with surface anatomy, do not become alarmed. The exhibits are presented to give you an understanding of the importance of surface anatomy. In subsequent chapters, when you have a better understanding of anatomical terms, we

EXHIBIT 1-3
SURFACE ANATOMY OF THE HEAD

EYE

1. **Pupil.** Opening of center of iris of eyeball for light transmission.

2. **Iris.** Circular pigmented muscular membrane behind cornea.

3. **Sclera.** "White" of eye, a coat of fibrous tissue that covers entire eyeball except for cornea.

4. **Conjunctiva.** Membrane that covers exposed surface of eyeball and lines eyelids.

5. **Eyelids.** Folds of skin and muscle lined by conjunctiva.

6. **Palpebral fissure.** Space between eyelids when they are open.

7. **Medial canthus.** Site of union of upper and lower eyelids near nose.

8. **Lateral canthus.** Site of union of upper and lower eyelids away from nose.

9. **Lacrimal caruncle.** Fleshy, yellowish projection of medial commissure that contains modified sweat and sebaceous glands.

10. **Eyelashes.** Hairs on margins of eyelids, usually arranged in two or three rows.

11. **Eyebrows.** Several rows of hairs superior to upper eyelids.

EAR

1. **Auricle.** Portion of external ear not contained in head, also called pinna or trumpet.

2. **Tragus.** Cartilaginous projection anterior to external opening to ear.

3. **Antitragus.** Cartilaginous projection opposite tragus.

4. **Concha.** Hollow of auricle.

5. **Helix.** Superior and posterior free margin of auricle.

6. **Antihelix.** Semicircular ridge posterior and superior to concha.

7. **Triangular fossa.** Depression in superior portion of antihelix.

8. **Lobule.** Interior portion of auricle devoid of cartilage.

9. **External auditory meatus.** Canal extending from external ear to eardrum.

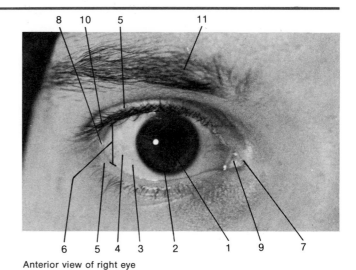

Anterior view of right eye

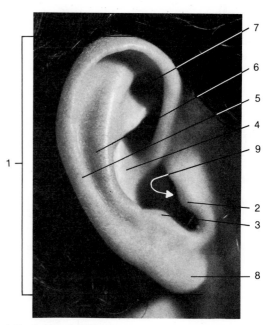

Lateral view of right ear

Photographs courtesy of Victor B. Eichler, of Wichita State University, and Steve Harper.

---

will refer back to the exhibits. They will have even more meaning then.

THE NECK

The **neck** (collum) is divided into an anterior region called the *cervix,* two lateral regions, and a posterior region referred to as the *nucha.* The most prominent structure in the midline of the anterior region is the thyroid cartilage or Adam's apple. Just above it, the hyoid bone can be palpated. Below the Adam's apple, the cricoid cartilage of the larynx can be felt. A major portion of the lateral regions of the neck is formed by the sternocleidomastoid muscles. Each muscle extends from the mastoid process of the temporal bone, felt as a bump behind the auricle of the ear, to the sternum and clavicle. A muscle that extends downward and outward from the base of the skull and occupying a portion of the lateral region of the neck is the trapezius muscle. You might be interested to know that a

EXHIBIT 1-3
(Continued)

NOSE AND LIPS

1. **Root.** Superior attachment of nose at forehead located between eyes.

2. **Apex.** Tip of nose.

3. **Dorsum nasi.** Rounded anterior border connecting root and apex; in profile, may be straight, convex, concave, or wavy.

4. **Nasofacial angle.** Point at which side of nose blends with tissues of face.

5. **Ala.** Convex flared portion of inferior lateral surface; unites with upper lip.

6. **External nares.** External openings into nose.

7. **Bridge.** Superior portion of dorsum nasi, superficial to nasal bones.

8. **Philtrum.** Vertical groove in medial portion of upper lip.

9. **Lips.** Upper and lower fleshy border of the oral cavity.

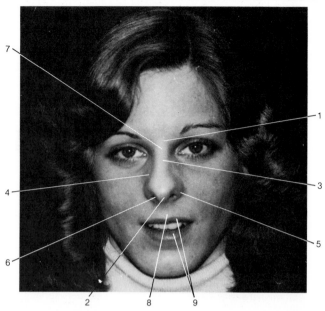

Anterior view of nose and lips

Photograph courtesy of Donald Castellaro and Deborah Massimi.

EXHIBIT 1-4
SURFACE ANATOMY OF THE NECK

NECK

1. **Thyroid cartilage.** Anterior portion of larynx (voice box).

2. **Sternocleidomastoid muscle.** Forms major portion of lateral surface of neck; bends head on neck and turns it to opposite side.

3. **Trapezius muscle.** Occupies a portion of lateral surfaces of neck; used in shrugging shoulders.

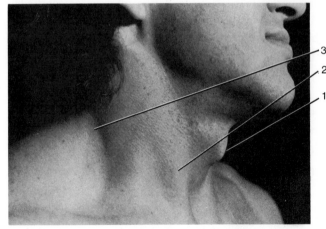

Lateral view of neck

Photograph courtesy of Victor B. Eichler, of Wichita State University, and Steve Harper.

"stiff neck" is frequently associated with an inflammation of this muscle. A very prominent vein that runs along the lateral surface of the neck is the external jugular vein. It is readily seen if you are angry of if your collar is too tight. The locations of a few surface features of the neck are shown in Exhibit 1-4.

THE TRUNK

The **trunk** is divided into the back *(dorsum)*, chest *(thorax)*, abdomen *(venter)*, and pelvis. One of the most striking surface features of the back is the vertebral spines, the dorsally pointed projections of the vertebrae. A very prominent vertebral spine is the vertebra prominens of the seventh cervical vertebra. When the head is bent forward, it is easily seen. Another easily identifiable surface landmark of the back is the scapula. In fact, several parts of the scapula may also be seen or palpated. In lean individuals the ribs may also be seen. Among the superficial muscles of the back that can be seen are the latissimus dorsi, erector spinae, infraspinatus, trapezius, and teres major. These, as well as the other surface features of the back, are described in Exhibit 1-5.

The chest presents a number of anatomical landmarks. At its superior region are the clavicles. The sternum lies in the midline of the chest. Its superior end attaches to the clavicles. Between the medial ends of the clavicles, there is a depression on the superior surface of the sternum called the jugular notch. The sternal angle is formed by a junction line between the manubrium and body of the sternum and is palpable under the skin. It locates the sternal end of the second rib and is the most reliable surface landmark of the chest. The inferior portion of the sternum, the xiphoid

EXHIBIT 1-5
SURFACE ANATOMY OF THE TRUNK

BACK

1. **Vertebral spines.** Dorsally pointed projections of vertebrae (backbones).

2. **Scapula.** Shoulder blade.

3. **Latissimus dorsi muscle.** Broad muscle of back that helps draw shoulders backward.

4. **Erector spinae muscle.** Parallel to vertebral column; moves backbone.

5. **Infraspinatus muscle.** Located inferior to spine of scapula; helps rotate humerus (armbone) laterally.

6. **Trapezius muscle.** See Exhibit 1-4 for description.

7. **Teres major muscle.** Located inferior to infraspinatus; helps move humerus.

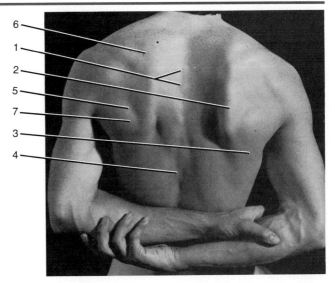

Posterior view of back

CHEST

1. **Clavicle.** Collarbone.

2. **Sternum.** Breastbone.

3. **Pectoralis major muscle.** Principal upper chest muscle; helps move humerus.

4. **Serratus anterior muscle.** Below and lateral to pectoralis major; helps move scapula.

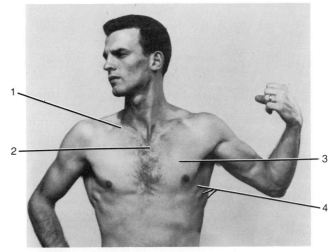

Anterior view of chest

Photographs courtesy of Victor B. Eichler, of Wichita State University, and Steve Harper (top) and Vincent P. Destro, Mayo Foundation (bottom).

EXHIBIT 1-5
(Continued)

ABDOMEN

1. **Umbilicus.** Also called navel; previous site of attachment of umbilical cord in fetus.

2. **External oblique muscle.** Located inferior to serratus anterior; helps compress abdomen.

3. **Rectus abdominis muscle.** Located just lateral to midline of abdomen; helps compress abdomen.

4. **Linea alba.** Flat, tendinous raphe between rectus abdominis muscles.

5. **Tendinous intersections.** Fibrous bands that run transversely or obliquely across the rectus abdominis muscles.

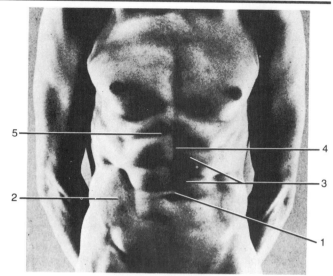

Anterior view of abdomen

Photograph courtesy of R. D. Lockhart, *Living Anatomy*, Faber & Faber, 1974.

process, may be palpated. Also visible or palpable are the ribs. Among the more prominent superficial chest muscles are the pectoralis major and serratus anterior. These, and other surface features of the chest, are described in Exhibit 1-5.

The abdomen and pelvis have already been discussed in terms of their nine regions or four quadrants (see Figures 1-5 and 1-6). External features of the abdomen include the umbilicus, serratus anterior muscle, external oblique muscle, and rectus abdominis muscle, all described in Exhibit 1-5.

THE UPPER EXTREMITY

The **upper extremity** consists of the armpit *(axilla)*, shoulder *(omos)*, arm *(brachium)*, elbow *(cubitus)*, forearm *(antebrachium)*, and hand *(manus)*. The hand, in turn, is subdivided into the wrist *(carpus)*, palm *(metacarpus)*, and fingers *(digits* or *phalanges)*. See Exhibit 1-6.

Moving laterally along the top of the clavicle, it is possible to palpate a slight elevation at the lateral end of the clavicle, the acromioclavicular joint. Less than 2.5 cm (1 in.) from this joint, one can also feel the acromion of the scapula which forms the tip of the shoulder. The rounded prominence of the shoulder is formed by the deltoid muscle, a frequent site for intramuscular injections.

Most of the anterior surface of the upper extremity is occupied by the biceps brachii muscle, whereas most

of the posterior surface is occupied by the triceps brachii muscle. At the elbow it is possible to locate three bony protuberences. The medial and lateral epicondyles of the humerus form visible eminences on the dorsum of the elbow. The olecranon of the ulna forms the large eminence in the middle of the dorsum of the elbow and lies between and slightly superior to the epicondyles when the forearm is extended. The ulnar nerve can be palpated in a groove behind the medial epicondyle. The triangular space of the anterior region of the elbow is the cubital fossa. The median cubital vein usually crosses the cubital fossa obliquely. This vein is the one frequently selected for removal of blood for diagnosis, transfusions, and intravenous therapy.

The most prominent landmarks of the forearm are the olecranon and the styloid process of the ulna. The styloid process of the ulna may be seen as a protuberance on the medial (little finger) side of the wrist. The ulna is the medial bone of the forearm, and the radius is the lateral bone of the forearm. On the lateral side of the upper forearm is the brachioradialis muscle. Next to it is the flexor carpi radialis muscle. On the medial side is the flexor carpi ulnaris muscle.

At the wrist, several structures may be felt. On the anterior surface of the wrist, it is possible to see the tendon of the palmaris longus muscle by making a fist. Next to this tendon, as you move toward the thumb, you can feel the tendon of the flexor carpi radialis muscle. If you continue toward the thumb, you can palpate the radial artery just medial to the styloid process of

EXHIBIT 1-6
SURFACE ANATOMY OF THE UPPER EXTREMITY

SHOULDER

1. **Acromion.** Expanded end of spine of scapula; forms tip of shoulder.

2. **Deltoid muscle.** Triangular-shaped muscle that forms rounded prominence of shoulder.

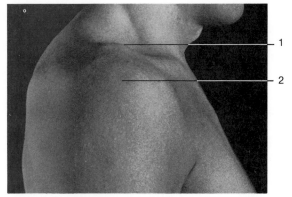

Lateral view of shoulder

ARM

1. **Biceps brachii muscle.** Forms bulk of anterior surface of arm; helps move forearm.

2. **Triceps brachii muscle.** Forms bulk of posterior surface of arm; helps move forearm.

3. **Medial epicondyle.** Medial projection at distal end of humerus.

4. **Lateral epicondyle.** Lateral projection at distal end of humerus.

5. **Olecranon.** Projection of proximal end of ulna (medial bone of forearm); forms elbow.

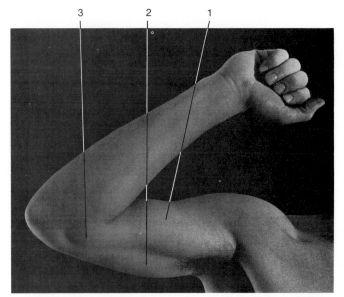

Anterior view of upper extremity

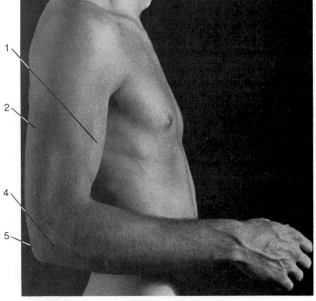

Lateral view of upper extremity

EXHIBIT 1-6
(Continued)

FOREARM

1. **Styloid process of ulna.** Projection of distal end of ulna at medial side of wrist (little-finger side).

2. **Brachioradialis muscle.** Located at superior and lateral aspect of forearm; helps move forearm.

3. **Flexor carpi radialis muscle.** Located along midportion of forearm; helps move wrist.

4. **Flexor carpi ulnaris muscle.** Located at medial aspect of forearm; helps move wrist.

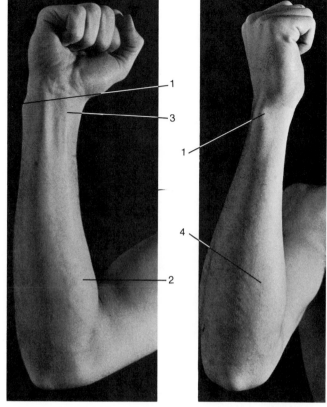

Medial view of forearm   Anterior view of forearm

WRIST AND HAND

1. **Tendon of palmaris longus muscle.** Tendon on anterior surface of wrist nearer ulna.

2. **Tendon of flexor carpi radialis muscle.** Tendon on anterior surface of wrist lateral to palmaris longus tendon.

3. **"Knuckles."** Commonly refers to the distal ends of second through fifth metacarpal (palm) bones; also includes the dorsal aspects of the metacarpophalangeal and interphalangeal joints.

4. **Bracelet flexure lines.** Creases in wrist.

5. **Proximal transverse flexure line.** Crease in palm running in an oblique transverse direction; closer to wrist.

6. **Distal transverse flexure line.** Crease in palm running in an oblique transverse direction; farther from wrist.

7. **Phalanges.** Bones of fingers.

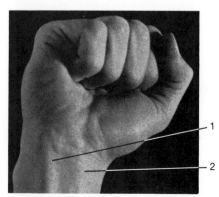

Anterior view of wrist

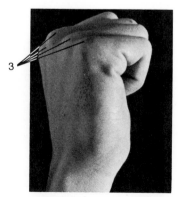

Medial view of wrist

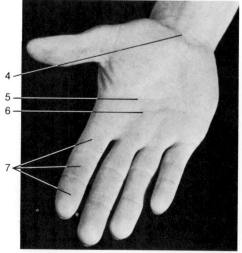

Anterior view of hand

Photographs courtesy of Victor B: Eichler, of Wichita State University, and Steve Harper.

the radius. This artery is frequently used to take the pulse. By bending the thumb backward, two prominent tendons may be located along the posterior surface of the wrist. The one closer to the styloid process of the radius is the tendon of the extensor pollicus brevis muscle. The one closer to the styloid process of the ulna is the tendon of the extensor pollicus longus muscle. The depression between these two tendons is known as the "anatomical snuffbox." By palpating the depression, you can feel the radial artery.

The distal ends of the second through fifth metacarpal (palm) bones are commonly referred to as the "knuckles." By examining the palmar surface of the hand, it is possible to see a number of creases in the skin, as well as the location of the phalanges between joints of the fingers.

THE LOWER EXTREMITY

The **lower extremity** consists of the buttocks *(gluteal region)*, thigh *(femoral region)*, knee *(genu)*, leg *(crus)*, and foot *(pes)*. The foot includes the ankle *(tarsus)* and toes *(digitis* or *phalanges)*. See Exhibit 1-7.

The iliac crest forms the outline of the superior border of the bottock. About 20 cm (8 in.) below the highest portion of the iliac crest, the greater trochanter of the femur can be felt on the lateral side of the thigh. As you will see later, the iliac crest and greater trochanter are useful landmarks when giving an intramuscular injection in the gluteal muscle. Most of the prominence of the buttocks is formed by the gluteal muscles. The bony prominence in each buttock is the

EXHIBIT 1-7
SURFACE ANATOMY OF THE LOWER EXTREMITY

BUTTOCKS AND THIGH

1. **Iliac crest.** Superior portion of hipbone.

2. **Greater trochanter.** Projection of proximal end of femur (thighbone).

3. **Gluteus maximus muscle.** Forms major portion of prominence of buttock; helps move femur.

4. **Gluteus medius muscle.** Located above gluteus maximus; helps move femur.

5. **Adductor longus muscle.** Located on medial side of thigh; helps move femur.

6. **Rectus femoris muscle.** Located along midportion of thigh; helps move tibia (shinbone).

7. **Vastus medialis muscle.** Located at medial, inferior portion of thigh; helps move tibia.

8. **Vastus lateralis muscle.** Located along anterolateral surface of thigh; helps move tibia.

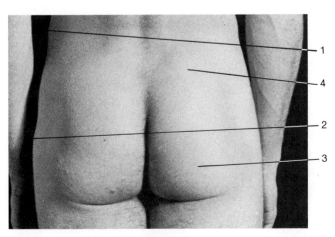

Posterior view of buttocks and thigh

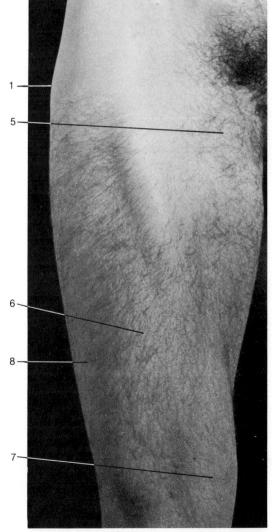

Anterior view of thigh

EXHIBIT 1-7
(Continued)

KNEE

1. **Patella.** Also called kneecap; located on anterior surface of knee along midline.

2. **Patellar ligament.** Located below patella.

3. **Popliteal fossa.** Diamond-shaped space on posterior surface of knee.

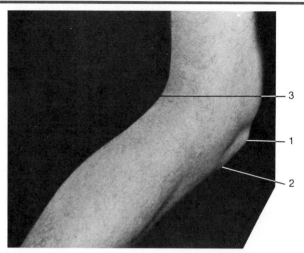

Posteromedial view of knee

LEG

1. **Medial condyles of femur and tibia.** Medial projections just below patella; upper part of projection belongs to distal end of femur; lower part of projection belongs to proximal end of tibia.

2. **Lateral condyles of femur and tibia.** Lateral projections just below patella; upper part of projection belongs to distal end of femur; lower part of projection belongs to proximal end of tibia.

3. **Tibial tuberosity.** Bony prominence below patella.

4. **Tibialis anterior muscle.** Located on anterior surface of leg along midportion; helps move foot.

5. **Gastrocnemius muscle.** Forms bulk of mid and upper portion of posterior surface of leg; helps move foot.

6. **Soleus muscle.** Located deep to the gastrocnemius; helps move foot. See also posterior view of foot.

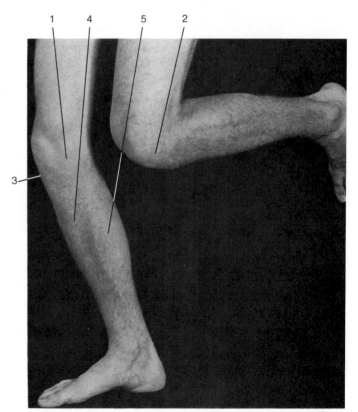

Medial view of right leg and lateral view of left leg

ischial tuberosity of the hipbone. This structure bears the weight of the body when you are seated. Among the prominent superficial muscles on the anterior surface of the thigh are the adductor longus, vastus lateralis, vastus medialis, and rectus femoris. The vastus lateralis is frequently used as an injection site by diabetics when administering insulin.

On the anterior surface of the knee, the patella, or kneecap, is observable. Below it is the patellar ligament. On the posterior surface of the knee is a diamond-shaped space, the popliteal fossa.

Just below the patella, on either side of the patellar ligament, the medial and lateral condyles of the femur and tibia can be felt. The bony prominence below the

FOOT

1. **Medial malleolus.** Projection of distal end of tibia (medial bone of leg) that forms medial prominence of ankle.

2. **Lateral malleolus.** Projection of distal end of fibula (lateral bone of leg) that forms lateral prominence of ankle.

3. **Calcaneus.** Heel bone.

4. **Calcaneal tendon.** Also called Achilles tendon; conspicuous tendon attached to calcaneus.

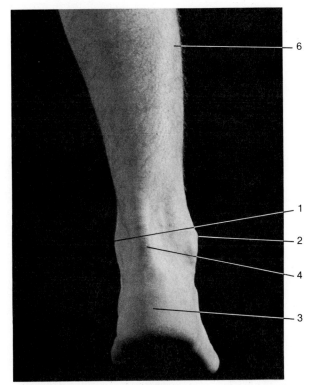

Posterior view of foot

Photographs courtesy of Victor B. Eichler, of Wichita State University, and Steve Harper.

patella in the middle of the leg is the tibial tuberosity. The tibia is the medial bone of the leg, and the fibula is the lateral bone of the leg. Prominent superficial muscles of the leg include the tibialis anterior, gastrocnemius, and soleus.

At the ankle, the medial malleolus of the tibia and the lateral malleolus of the fibula can be noted as two prominent eminences. Arising from the heel bone (calcaneus) is the calcaneal (Achilles) tendon.

Radiographical Anatomy

A very specialized branch of anatomy essential for the diagnosis of many disorders is called **radiographic**

anatomy, which includes the use of x-rays. One type of radiographic anatomy employs the use of a single barrage of x-rays. In this conventional usage, the x-rays pass through the body and expose an x-ray film, producing a photographic image called a *roentgenogram* (Figure 1-10a). A roentgenogram provides a two-dimensional shadow image of the interior of the body. As valuable as they are in diagnosis, x-rays compress the body image onto a flat sheet of film, often resulting in an overlap of organs and tissues that can make diagnosis difficult. Moreover, x-rays do not always dif-

ferentiate between subtle differences in tissue density, again resulting in difficulties in diagnosis.

These difficulties have been virtually eliminated by the use of a recent x-ray technique called **computed tomography (CT) scanning.** First introduced in 1971, CT scanning combines the principle of x-ray radiography with advanced computer technology. An x-ray source moves in an arc around the part of the body being scanned and repeatedly sends out x-ray beams. The beams are then converted into electronic impulses that produce more than 46,000 readings on the density

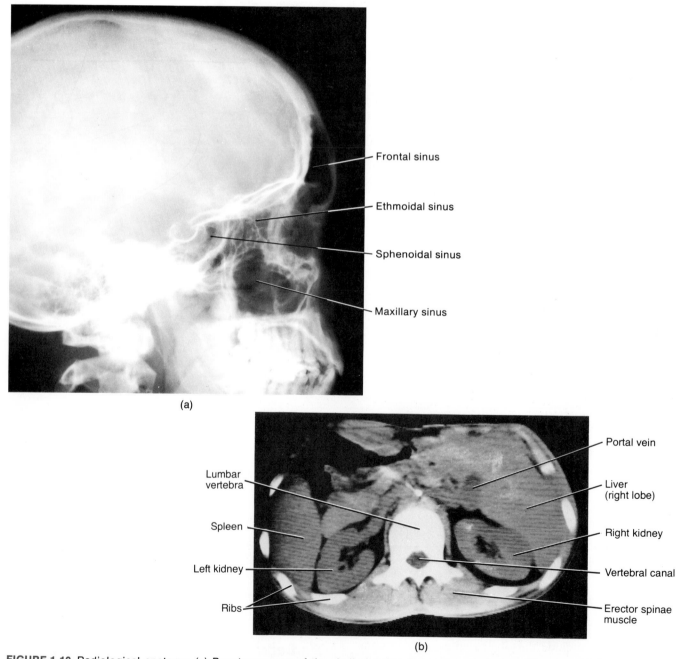

(a)

(b)

FIGURE 1-10 Radiological anatomy. (a) Roentgenogram of the skull showing the paranasal sinuses. (Courtesy of Eastman Kodak Company). (b) CT scan of the abdomen. (Courtesy of General Electric Co.)

of the tissue in a one centimeter slice of body tissue. Next, a computer produces a graphic, detailed representation from these readings, called a *CT scan,* which can be viewed on a TV screen called the physician's console (Figure 1-11). The CT scan provides a very accurate cross-sectional picture of any area of the body in which the organ or tissue is reproduced exactly as it exists (Figure 1-10b). The CT scan permits a significant differentiation of body parts never before possible. The whole CT scanning process takes only seconds, is completely painless, and the x-ray dose is equal to or less than that of many other diagnostic procedures.

Measuring the Human Body

One of the most important concepts you should understand about your body is the concept of measurement. You will constantly be exposed to various kinds of measurement in order to understand how your body works, how big various organs are, how much various organs weigh, and how much of a given medication should be administered, to mention a few. Measurements involving time, weight, temperature, size, length, and volume are a routine part of your studies in a medical science program.

Whenever you come across a measurement in the

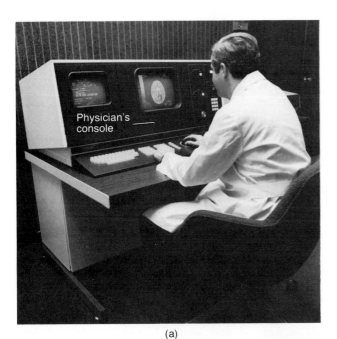

(a)

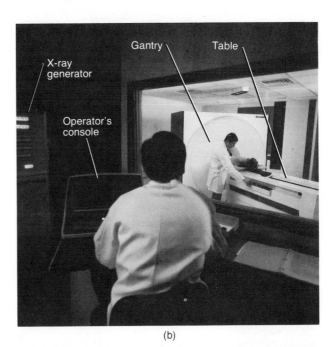

(b)

FIGURE 1-11 General Electric CT scanner system. (a) Physician's console. Here, images can be displayed, adjusted for contrast and density, magnified, and measured. In addition, previous scans stored in the computer may be instantly recalled and displayed. (b) Operator's console (foreground) controls the scan procedure. The number of slices, slice thickness, scan speed, and slice location are determined here. The CT table and gantry (background) perform the scans. After each slice, the table automatically moves to the next slice location. The gantry can angulate to scan from a variety of positions. The x-ray generator, shown to the left, supplies power to the x-ray tube located in the gantry. (c) The computer (left) processes thousands of bits of information to create each image. It also helps set up exams, controls scan procedures, displays images, and stores and recalls previous studies. The multiformat camera (right) can record CT scans on film for physician reference. (Courtesy of General Electric Co.)

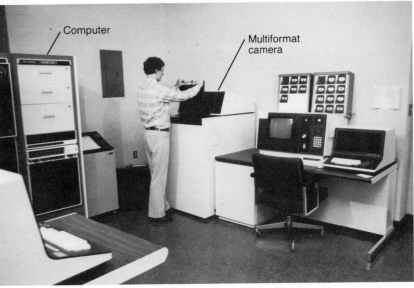

(c)

EXHIBIT 1-8
EXHIBIT 1-8
METRIC UNITS OF LENGTH AND SOME U.S. EQUIVALENTS

METRIC UNIT (ABBREVIATION, SYMBOL)	MEANING OF PREFIX	METRIC EQUIVALENT	U.S. EQUIVALENT
1 kilometer (km)	kilo = 1,000	1,000 m	3,280.84 ft or 0.62 mi; 1 mi = 1.61 km
1 hectometer (hm)	hecto = 100	100 m	328. ft
1 dekameter (dam)	deka = 10	10 m	3.28 ft
1 meter (m)	Standard unit of length		39.37 in. or 3.28 ft or 1.09 yd
1 decimeter (dm)	deci = $1/10$	0.1 m	3.94 in.
1 centimeter (cm)	centi = $1/100$	0.01 m	0.394 in.; 1 in. = 2.54 cm
1 millimeter (mm)	milli = $1/1,000$	0.001 m	0.0394 in.
1 micrometer (μm) [formerly micron (μ)]	micro = $1/1,000,000$	0.000,001 m	3.94×10^{-5} in.
1 nanometer (nm) [formerly millimicron (mμ)]	nano = $1/1,000,000,000$	0.000,000,001 m	3.94×10^{-8} in.
1 angstrom (Å)		0.000,000,000,1 m	3.94×10^{-9} in.

EXHIBIT 1-9
METRIC UNITS OF MASS AND SOME U.S. EQUIVALENTS

METRIC UNIT (ABBREVIATION, SYMBOL)	METRIC EQUIVALENT	U.S. EQUIVALENT
1 kilogram (kg)	1,000 g	2.205 lb
1 hectogram (hg)	100 g	
1 dekagram (dag)	10 g	
1 gram (g)	1 g	1 lb = 453.6 g 1 oz = 28.35 g
1 decigram (dg)	0.1 g	
1 centigram (cg)	0.01 g	
1 milligram (mg)	0.001 g	
1 microgram (gmg)	0.000,001 g	

text, the measurement will be given in metric units. To help you compare the metric unit to a familiar unit, the approximate U.S. equivalent will also be given in parentheses directly after the metric unit. For example, you might be told that the length of a particular part of the body is 2.54 cm (1 inch).

As a first step in helping you understand the relationships of the metric to U.S. systems of measurement, three exhibits have been prepared. Exhibit 1-8 contains metric units of length and some U.S. equivalents. Exhibit 1-9 contains metric units of mass and some U.S. equivalents. Exhibit 1-10 contains metric units of volume and some U.S. equivalents. Before you begin reading Chapter 2, carefully examine the exhibits. Even if you do not learn all the metric units and their equivalents at this point, you can still refer back to the exhibits later.

EXHIBIT 1-10
METRIC UNITS OF VOLUME AND SOME U.S. EQUIVALENTS

METRIC UNIT (ABBREVIATION, SYMBOL)	METRIC EQUIVALENT	U.S. EQUIVALENT
1 liter (l)	1,000 ml	33.81 fl oz or 1.057 qt 946 ml = 1 qt
1 milliliter (ml)	0.001 liter	0.0338 fl oz; 30 ml = 1 fl oz 5 ml = 1 teaspoon
1 cubic centimeter (cm^3)	0.999972 ml	0.0338 fl oz

STUDY OUTLINE

Anatomy and Physiology Defined

1. Anatomy is the study of structure and how structures are related to each other.
2. Subdivisions of anatomy include gross anatomy (macroscopic), systemic anatomy (systems), regional anatomy (regions), developmental anatomy (development from fertilization to adulthood), embryology (development before eighth week), pathological anatomy (disease), and histology (microscopic study of tissues and cells).
3. Physiology is the study of how structures function.

Levels of Structural Organization

1. The human body consists of levels of structural organization from the chemical level to the organismic level.
2. The chemical level is represented by all the atoms and molecules in the body. The cellular level consists of cells. The tissue level is represented by tissues. The organ level consists of body organs, and the system level is represented by organs that work together to perform a general function.
3. The human organism is a collection of structurally and functionally integrated systems.

Structural Plan

Directional Terms

1. Directional terms indicate the relationship of one part of the body to another.
2. Examples of directional terms are superior (toward the head), anterior (near the front), medial (nearer the midline), distal (farther from the attachment of a limb), and external (toward the surface).

Body Cavities

1. Spaces in the body that contain internal organs are called cavities.
2. Dorsal and ventral cavities are the two principal body cavities. The dorsal cavity contains the brain and spinal cord. The organs of the ventral cavity (coelom) are collectively called the viscera.
3. The dorsal cavity is subdivided into the cranial cavity and vertebral canal.
4. The ventral body cavity is subdivided by the diaphragm into an upper thoracic cavity and a lower abdominopelvic cavity.
5. The thoracic cavity contains two pleural cavities and a mediastinum, which includes the pericardial cavity.

Abdominopelvic Regions

1. The abdominopelvic cavity, actually an upper abdominal cavity and a lower pelvic cavity, is divided into nine anatomical regions. It may also be divided into four quadrants.

Anatomical Position and Regional Names

1. The position in which the body is studied and described is the anatomical position. The subject stands erect and faces the observer with arms at sides and palms turned forward.
2. Regional names are terms given to specific regions of the body for reference. Examples of regional names include cranial (skull), thoracic (chest), brachial (arm), patellar (knee), cephalic (head), and gluteal (buttock).

Planes

1. Planes of the body are flat surfaces that divide the body into definite areas. The midsagittal or median plane divides the body into equal right and left sides; the sagittal (parasagittal) plane, into unequal right and left sides, the frontal (coronal) plane, into anterior and posterior portions; the horizontal (transverse) plane, into superior and inferior portions.
2. Sections of organs include cross sections, oblique sections, and longitudinal sections.

Introduction to Surface Anatomy

1. Surface anatomy is the study of the form and markings of the surface of the body.
2. Surface anatomy features may be noted by visual inspection or palpation.
3. The primary divisions of the body used to study surface anatomy are the head, neck, trunk, upper extremity, and lower extremity.
4. A review of surface anatomy is presented in Exhibits 1-3 through 1-7.

Radiographical Anatomy

1. This branch of anatomical study includes the use of x-rays.
2. In conventional usage, x-rays produce a two-dimensional shadow image of the interior of the body called a roentgenogram.
3. CT scanning employs the use of x-rays and advanced computer technology which produces an accurate cross-sectional image of the body called a CT scan.

Measuring the Human Body

1. Measurements involving length, mass, and volume are integral components of a knowledge of the human body.
2. Metric units of length may be reviewed in Exhibit 1-8, metric units of mass may be reviewed in Exhibit 1-9, and metric units of volume may be reviewed in Exhibit 1-10.

REVIEW QUESTIONS

1. Define anatomy. How does each subdivision of anatomy help you understand the structure of the human body? Define physiology.

2. Construct a diagram to illustrate the levels of structural organization that characterize the body. Be sure to define each level.

3. Outline the function of each system of the body, and list several organs that compose each system.
4. What does bilateral symmetry mean? Why is the body considered to be a tube within a tube?
5. What is a directional term? Why are these terms important? Can you use each of the directional terms listed in Exhibit 1-2 in a complete sentence?
6. Define a body cavity. List the body cavities discussed, and tell which major organs are located in each. What landmarks separate the various body cavities from each other?
7. Discuss how the abdominopelvic area is subdivided into nine regions. Name and locate each region and list the organs, or parts of organs, in each. Describe how the abdominopelvic cavity is divided into four quadrants and name each quadrant.
8. When is the body in the anatomical position? Why is the anatomical position used?
9. Review Figure 1-7. See if you can locate each region on your own body, and name each by its common and anatomical term.
10. Describe the various planes that may be passed through the body. Explain how each plane divides the body.
11. What is meant by the phrase "a part of the body has been sectioned"? Given an orange, can you make a cross section, oblique section, and longitudinal section with a knife?
12. Using Exhibits 1-3 through 1-7 as an outline, locate as many of the surface features as you can on your partner's body, wall charts, models, photographs, and skeletons.
13. Explain the principle of CT scanning. Contrast it with simple x-ray procedures.

14. Convert the following lengths:
    a. If a bacterial cell measures 100 μm in length, how many nanometers is this?
    b. How many meters are in 1 mi?
    c. If a road sign reads 35 km/hour, what would your speedometer have to read to obey the sign?
    d. How many millimeters are in 1 km? In 1 in.?
    e. A person's arm measures 2 ft in length. How many centimeters is this?
    f. Convert 0.40 m to millimeters.
    g. How many millimeters are in 5 in.?
    h. If you ran 295.2 ft, how many meters would you have run?
    i. If the distance to the moon is 239,000 mi, what is it in meters?
15. Solve the following conversions of mass:
    a. Calculate the milligrams in 0.4 kg and in 1 lb.
    b. If a bottle contains 1.42 g, how many centigrams does it contain?
    c. The indicated dosage of a certain drug is 50 μg. How many milligrams is this?
    d. If you weigh 110 lb, how many kilograms do you weigh?
    e. How many centigrams are in 1 g?
16. Convert the following volumes:
    a. If you excrete 1,200 ml of urine in a day, how many liters is this?
    b. How many milliliters are in 2 liters?
    c. Convert 2 pt to milliliters.
    d. If you remove 15 cm^3 of blood from a patient, how many milliliters have you removed?

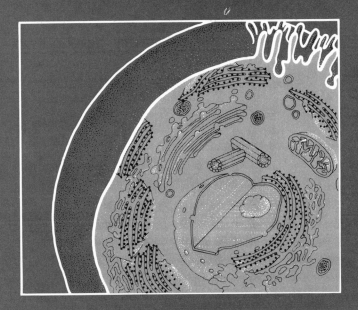

2

Cells: Structural Units of the Body

STUDENT OBJECTIVES

- Define and list a cell's generalized parts.
- Describe the structure and molecular organization of the plasma membrane.
- List the factors related to semipermeability of the plasma membrane.
- Define diffusion, facilitated diffusion, osmosis, filtration, dialysis, active transport, phagocytosis, and pinocytosis.
- Describe the structure and function of several modified plasma membranes.
- Describe the chemical composition and list the functions of cytoplasm.
- Describe two general functions of a cell nucleus.
- Distinguish between agranular and granular endoplasmic reticulum.
- Define the function of ribosomes.
- Describe the role of the Golgi complex in the synthesis, storage, and secretion of glycoproteins.
- Describe the function of mitochondria as "powerhouses of the cell."
- Explain why a lysosome in a cell is called a "suicide packet," and describe microtubules.
- Describe the structure and function of centrioles in cellular reproduction.
- Differentiate between cilia and flagella.
- Define cell inclusion and give several examples.
- Define extracellular material and give several examples.
- Describe the stages and events involved in cell division.
- Discuss the significance of cell division.
- Describe the relationship of cancer and aging to cells.
- Define key medical terms associated with cells.

A study of the body at the cellular level of organization is important to a total understanding of the structure and function of the body because many activities essential to life occur in cells and many disease processes originate in cells. A **cell** may be defined as the basic, living, structural and functional unit of the body and, in fact, of all organisms. **Cytology** is the specialized branch of science concerned with the study of cells. In this chapter, we shall concentrate on the structure of cells, the functions of cells, and the reproduction of cells. A series of illustrations accompanies each cell structure that you study. The first illustration shows the location of the structure within the cell. The second is an *electron micrograph,* which is a photograph taken with a high-powered electron microscope.* The third illustration that accompanies each structure is a drawing of the electron micrograph. The drawing will clarify some of the small details by exaggerating their outlines.

*The electron microscope can magnify an object more than 200,000 times. In comparison, the light microscope that you probably use in your laboratory magnifies objects 1000 to 2000 times their size (1000–2000×).

Generalized Animal Cell

A generalized animal cell is a composite of many different cells in the body. Examine the generalized cell illustrated in Figure 2-1, but keep in mind that no such single cell actually exists.

For convenience, we can divide the generalized cell into four principal parts:

1. Plasma (or **cell**) **membrane.** The outer, limiting membrane separating the cell's internal parts from the extracellular fluid and external environment.

2. Cytoplasm. The substance between the nucleus and cell surface.

3. Organelles. The cellular components that are highly specialized for specific cellular activities.

4. Inclusions. The secretions and storage areas of cells.

Extracellular materials, which are substances external to the cell surface, will also be examined in connection with cells.

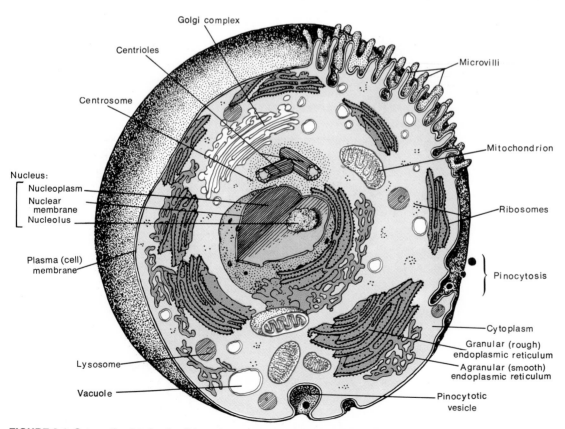

FIGURE 2-1 Generalized animal cell based on electron microscope studies.

Plasma (Cell) Membrane

The exceedingly thin structure that separates one cell from other cells and from the external environment is called the **plasma membrane,** or **cell membrane.** Electron microscopy studies have shown that the plasma membrane ranges from 65 to 100 angstroms (Å) in thickness, a dimension below light-microscope limits.* Scientists have known for a long time that plasma membranes are composed of phospholipid and protein molecules. Recent studies of membrane structure sug-

*One **angstrom,** usually written as 1Å, = 0.000,000,000,1 m ($^{1}/_{250,000,000}$ in.). Another microscopic unit of measurement is the **micrometer (μm),** which is equal to 0.000,001 m ($^{1}/_{25,000}$ in.).

gest a new concept regarding the arrangement of phospholipid and protein molecules. This new concept of membrane structure is called the *fluid mosaic hypothesis*. These studies seem to indicate that the phospholipid molecules, which account for about half the mass of the membrane, are arranged in two parallel rows. This double row of phospholipid molecules is termed a lipid bilayer (Figure 2-2). The protein molecules are arranged somewhat differently. Some lie at or near the inner and outer surface of the membrane. Others penetrate the membrane partway, completely, singly, or in pairs. Such an arrangement suggests that membranes are not static but that the proteins and phospholipids have a considerable degree of movement. It

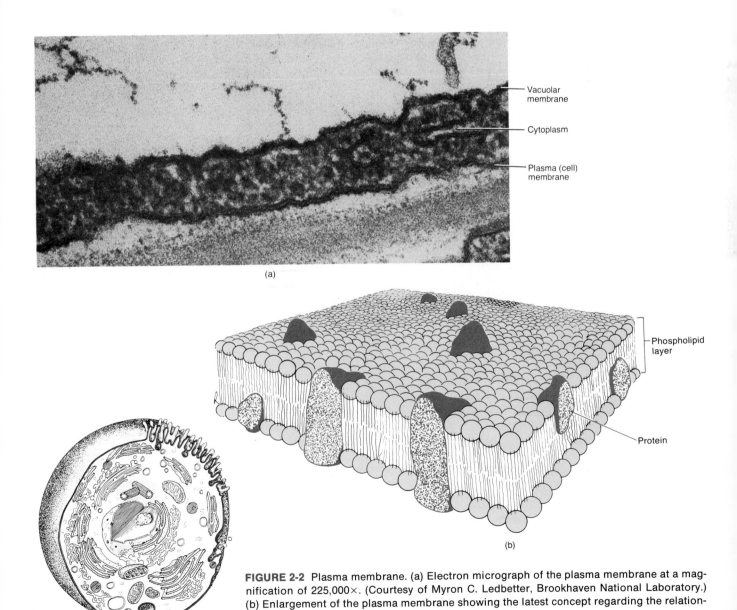

(a)

(b)

FIGURE 2-2 Plasma membrane. (a) Electron micrograph of the plasma membrane at a magnification of 225,000×. (Courtesy of Myron C. Ledbetter, Brookhaven National Laboratory.) (b) Enlargement of the plasma membrane showing the latest concept regarding the relationship of the phospholipid layer and protein molecules.

is quite possible that many key functions of membranes will be fully explained once the reasons for protein and phospholipid movements are understood. The fluid mosaic hypothesis may explain how specific receptor sites on membranes attach to hormones (for example, insulin), to transmitter substances produced by axons, and to antigens on the surfaces of red blood cells. Other features of the membrane structure are areas that appear as breaks along the membrane surface. These breaks occur at intervals and range in size from 7 to 10 Å in diameter. Researchers suspect they may be pores.

The basic functions of the plasma membrane are to enclose the components of the cell and to serve as a boundary through which substances must pass to enter or exit the cell. One important characteristic of the plasma membrane is that it permits certain ions and molecules to enter or exit the cell but restricts the passage of others. For this reason, plasma membranes are described as *semipermeable, differentially permeable,* or *selectively permeable.* In general, plasma membranes are freely permeable to water. In other words, they let water into and out of the cell. However, they act as partial barriers to the movement of almost all other substances. The ease with which a substance passes through a membrane is called the membrane's *permeability* to that substance. The permeability of a plasma membrane appears to be a function of several factors:

1. Size of molecules. Large molecules cannot pass through the plasma membrane. Water and amino acids are small molecules and can enter and exit the cell easily. However, most proteins, which consist of many amino acids linked together, seem to be too large to pass through the membrane. Many scientists believe that the giant-sized molecules do not enter the cell because they are larger than the diameters of the suspected membrane pores.

2. Solubility in lipids. Substances that dissolve easily in lipids pass through the membrane more readily than other substances since a major part of the plasma membrane consists of lipid molecules.

3. Charge on ions. The charge of an ion attempting to cross the plasma membrane can determine how easily the ion enters or leaves the cell. The protein portion of the membrane is capable of ionization. If an ion has a charge opposite that of the membrane, it is attracted to the membrane and passes through more readily. If the ion attempting to cross the membrane has the same charge as the membrane, it is repelled by the membrane and its passage is restricted. This phenomenon conforms to the rule of physics that opposite charges attract, whereas like charges repel each other.

4. Presence of carrier molecules. Plasma membranes contain special molecules called carriers that

are capable of attracting and transporting substances across the membrane regardless of size, ability to dissolve in lipids, or membrane charge.

MOVEMENT OF MATERIALS ACROSS PLASMA MEMBRANES

The mechanisms whereby substances move across the plasma membrane are essential to the life of the cell. Certain substances, for example, must move into the cell to support life, whereas waste materials or harmful substances must be moved out. The processes involved in these movements may be classed as either passive or active. In *passive processes,* substances move across plasma membranes without assistance from the cell. The substances move, on their own, from an area where their concentration is greater to an area where their concentration is less. The substances could also be forced across the plasma membrane by pressure from an area where the pressure is greater to an area where it is less. In *active processes,* the cell contributes energy in moving the substance across the membrane.

Passive Processes

● *Diffusion* A passive process called **diffusion** occurs when there is a *net* or greater movement of molecules or ions from a region of high concentration to a region of low concentration. The movement from high to low concentration continues until the molecules are evenly distributed. At this point, they move in both directions at an equal rate. This point of even distribution is called *equilibrium.* The difference between high and low concentrations is called the *concentration gradient.* Molecules moving from the high-concentration area to the low-concentration area are said to move *down* or *with* the concentration gradient. If a dye pellet is placed in a beaker filled with water, the color of the dye is seen immediately around the pellet. At increasing distances from the pellet, the color becomes lighter (Figure 2-3). Later, however, the water solution will be a uniform color. The dye molecules possess kinetic energy, which causes them to move about at random throughout the entire area. The dye molecules move down the concentration gradient from an area of high concentration to an area of low concentration. The water molecules also move from a high-concentration to a low-concentration area. When dye molecules and water molecules are evenly distributed among themselves, equilibrium is reached and diffusion ceases, even though molecular movements continue.

In the example cited, no membranes were involved. Diffusion may occur, however, through semipermeable membranes in the body. A good example of this

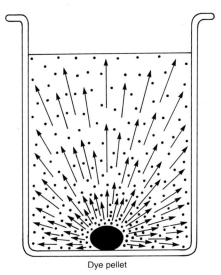

FIGURE 2-3 Principle of diffusion.

kind of diffusion in the human body is the movement of oxygen from the blood into the cells and the movement of carbon dioxide from the cells back into the blood.

● *Facilitated Diffusion* Another example of diffusion through a semipermeable membrane occurs by a process called **facilitated diffusion.** Although some chemical substances are insoluble in lipids, they can still pass through the plasma membrane. Among these are different sugars, especially glucose. In the process of facilitated diffusion, glucose combines with a carrier substance. The combined glucose-carrier is soluble in the lipid layer of the membrane, and the carrier transports the glucose to the inside of the membrane. At this point, the glucose separates from the carrier and enters the cell. The carrier then returns to the outside

of the membrane to pick up more glucose and transport it inside. The carrier makes the glucose soluble in the lipid portion of the membrane so it can pass through the membrane. By itself, glucose is insoluble and cannot penetrate the membrane. In the process of facilitated diffusion, the cell does not expend energy, and the movement of the substance is from a region of higher to lower concentration.

The rate of facilitated diffusion depends on (1) the difference in concentration of the substance on either side of the membrane, (2) the amount of carrier available to transport the substance, and (3) how quickly the carrier and substance combine. The process is greatly accelerated by insulin, a hormone produced by the pancreas. One of insulin's functions is to lower the blood-glucose level by accelerating the transportation of glucose from the blood into body cells. This transportation, as we have just seen, is by facilitated diffusion.

● *Osmosis* Another passive process by which materials move across membranes is **osmosis.** Unlike diffusion, this process specifically refers to the net movement of water molecules through a semipermeable membrane from an area of high water concentration to an area of lower water concentration. Once again, a simple apparatus may be used to demonstrate the process. The apparatus shown in Figure 2-4 consists of a tube constructed from cellophane, a semipermeable membrane. The cellophane tube is filled with a colored, 20 percent sugar (sucrose) solution. The upper portion of the cellophane tube is plugged with a rubber stopper through which a glass tubing is fitted. The cellophane tube is placed into a beaker containing pure water. Initially, the concentrations of water on either side of the semipermeable membrane are different. There is a lower concentration of water

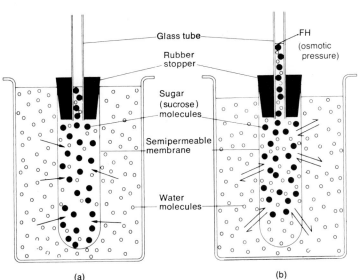

(a)   (b)

FIGURE 2-4 Principle of osmosis. (a) Apparatus at the start of the experiment. (b) Apparatus at equilibrium. In (a), the cellophane tube contains a 20 percent sugar solution and is immersed in a beaker of distilled water. The arrows indicate that water molecules can pass freely into the tube, but that sugar (sucrose) molecules are held back by the semipermeable membrane. As water moves into the tube by osmosis, the sugar solution is diluted and the volume of the solution in the cellophane tube increases. This increased volume is shown in (b), with the sugar solution moving up the glass tubing. The final height reached (FH) occurs at equilibrium and represents the osmotic pressure. At this point, the number of water molecules leaving the cellophane tube is equal to the number of water molecules entering the tube.

inside the cellophane tube than outside. Because of this difference, water moves from the beaker into the cellophane tube. The force with which the water moves is called osmotic pressure. Very simply, osmotic pressure is the force under which a solvent moves from a solution of lower solute concentration to a solution of higher solute concentration when the solutions are separated by a semipermeable membrane. There is no movement of sugar from the cellophane tube inside the beaker, however, since the cellophane is impermeable to molecules of sugar— sugar molecules are too large to go through the pores of the membrane. As water moves into the cellophane tube, the sugar solution becomes increasingly diluted and begins to move up the glass tubing. In time, the water that has accumulated in the cellophane tube and the glass tubing exerts a downward pressure that forces water molecules back out of the cellophane tube and into the beaker. When water molecules leave the cellophane tube and enter the tube at the same rate, equilibrium is reached.

● *Isotonic, Hypotonic, and Hypertonic Solutions* Osmosis may also be understood by considering the effects of different water concentrations on red blood cells. If the normal shape of a red blood cell is to be maintained, the cell must be placed in an **isotonic solution** (Figure 2-5a). This is a solution in which the concentrations of water molecules and solute molecules are the same on both sides of the semipermeable cell membrane. The concentrations of water and solute in the extracellular fluid outside the red blood cell must be the same as the concentration of the intracellular fluid. Under ordinary circumstances, a 0.85 percent NaCl solution is isotonic for red blood cells. In this condition, water molecules enter and exit the cell at the same rate, allowing the cell to maintain its normal shape. A different situation results if red blood cells are placed in a solution that has a lower concentration of solutes and, therefore, a higher concentration of water. This is called a **hypotonic solution.** In this condition, water molecules enter the cells faster than they can leave—causing the red blood cells to swell and eventually burst (Figure 2-5b). The rupture of red blood cells in this manner is called *hemolysis* or *laking*. Distilled water is a strongly hypotonic solution.

On the other hand, a **hypertonic solution** has a higher concentration of solutes and a lower concentration of water than the red blood cells. One example of a hypertonic solution is a 10 percent NaCl solution. In such a solution, water molecules move out of the cells faster than they can enter. This situation causes the cells to shrink (Figure 2-5c). The shrinkage of red blood cells in this manner is called *crenation*. Red blood cells may be greatly impaired or destroyed if placed in solutions that deviate significantly from the isotonic state.

● *Filtration* A third passive process involved in moving materials in and out of cells is **filtration.** This process involves the movement of solvents such as water and dissolved substances such as sugar across a semipermeable membrane by mechanical pressure, usually hydrostatic pressure. Such a movement is always from an area of higher pressure to an area of lower pressure and continues as long as a pressure difference exists. Most small to medium-sized molecules can be forced through a cell membrane. An example of filtration occurs in the kidneys, where the blood pressure supplied by the heart forces water and urea through thin cell membranes of tiny blood vessels and into the kidney tubule cells. In this basic process, protein molecules are retained by the body since they are too large to be forced through the cell membranes of the blood vessels. Harmful substances such as urea are small enough to be forced through and eliminated, however.

● *Dialysis* The final passive process to be considered is **dialysis.** Essentially, dialysis involves the separation of small molecules from large molecules by diffusion of the smaller molecules through a semipermeable membrane. Suppose a solution containing molecules of various sizes is placed in a tube that is permeable only to the smaller molecules. The tube is then placed in a beaker of distilled water. Eventually, the smaller molecules will move from the tube into the water in the beaker and the larger molecules will be left behind. This principle of dialysis is employed in artificial kidneys. The patient's blood is passed into a dialysis tube outside the body. The dialysis tube takes the place of the kidneys. As the blood moves through the tube, waste products pass from the blood

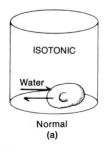

ISOTONIC

Water

Normal
(a)

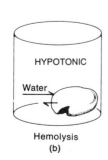

HYPOTONIC

Water

Hemolysis
(b)

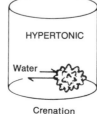

HYPERTONIC

Water

Crenation
(c)

FIGURE 2-5 Principle of osmosis applied to red blood cells. Shown here are the effects on red blood cells when placed in (a) an isotonic solution, in which they maintain normal shape; (b) a hypotonic solution, in which they undergo hemolysis; and (c) a hypertonic solution, in which they undergo crenation.

into a solution surrounding the dialysis tube. At the same time, certain nutrients are passed from the solution into the blood. The blood is then returned to the body.

Active Processes

When cells actively participate in moving substances across membranes, they must expend energy. Cells can even move substances against a concentration gradient. The active processes considered here are active transport, phagocytosis, and pinocytosis.

● *Active Transport* The process by which substances, usually ions, are transported across plasma membranes from an area of lower concentration to an area of higher concentration is called **active transport** (Figure 2-6a). Although the exact mechanism is not known, the following sequence is believed to occur:

1. An ion in the extracellular fluid outside the plasma membrane is attached to an enzymelike carrier molecule located in or on the plasma membrane.

2. The ion-carrier complex is soluble in the lipid portion of the membrane.

3. The compound moves toward the interior of the membrane where it is split by enzymes.

4. The ion is then transported into the intracellular fluid of the cell, and the carrier returns to the outer surface of the membrane to pick up another ion.

The energy for the attachment and release of the carrier molecule is supplied by adenosine triphosphate (ATP).

● *Phagocytosis* Another active process by which cells take in substances across the plasma membrane is called **phagocytosis,** or "cell eating" (Figure 2-6b). In this process, projections of cytoplasm, called *pseudopodia,* engulf solid particles external to the cell. Once the particle is surrounded, the membrane folds inwardly, forming a membrane sac around the particle. This newly formed sac, called a *digestive vacuole,* breaks off from the outer cell membrane, and the solid material inside the vacuole is digested. Indigestible particles and cell products are removed from the cell by a reverse phagocytosis. This process is important because molecules and particles of material that would normally be restricted from crossing the plasma membrane can be brought into or removed from the cell. The phagocytic white blood cells of the body constitute a vital defense mechanism. Through phagocytosis, the white blood cells destroy bacteria and other foreign substances.

● *Pinocytosis* In **pinocytosis,** or "cell drinking," the engulfed material consists of a liquid rather than a solid (Figure 2-6c). Moreover, no cytoplasmic projections are formed. Instead, the liquid is attracted to the surface of the membrane. The membrane folds inwardly, surrounds the liquid, and detaches from the

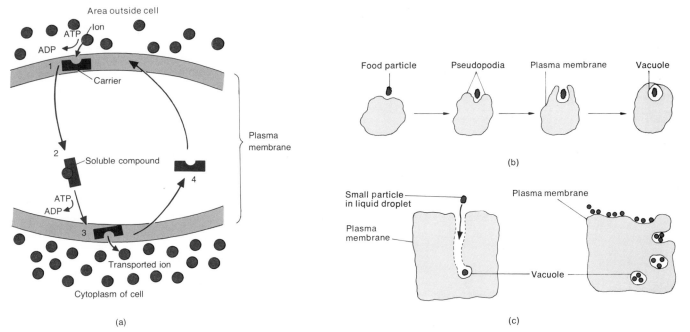

FIGURE 2-6 Active processes. (a) Mechanism of active transport. (b) Phagocytosis. (c) Two variations of pinocytosis. In the variation on the left, the ingested substance enters a channel formed by the plasma membrane and becomes enclosed in a vacuole at the base of the channel. In the variation on the right, the ingested substance becomes enclosed in a vacuole that forms and detaches at the surface of the cell.

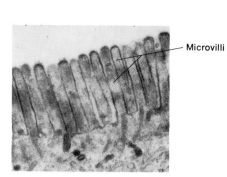

Microvilli

(a)

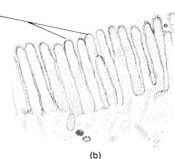

Microvilli

(b)

FIGURE 2-7 Modified plasma membranes. (a) Microvilli: electron micrograph of a portion of the small intestine at a magnification of 20,000×. (b) Microvilli: labeled diagram of the electron micrograph. (c) Rod sacs: electron micrograph of a portion of a rod of the eye at a magnification of 2000×. (d) Rod sacs: labeled diagram of the electron micrograph. (e) Stereocilia: photomicrograph of stereocilia projecting from the lining cells of the ductus epididymis at a magnification of 200×. (f) Stereocilia: labeled diagram of the photomicrograph. (g) Myelin sheath: electron micrograph of a myelin sheath in cross section at a magnification of 20,000×. (h) Myelin sheath: labeled diagram of the electron micrograph. (Electron micrographs courtesy of E. B. Sandburn, University of Montreal. Photomicrograph courtesy of Victor B. Eichler, Wichita State University.)

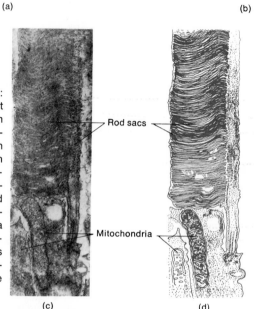

Rod sacs

Mitochondria

(c)

Rod sacs

Mitochondria

(d)

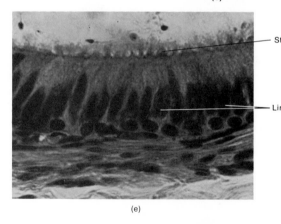

Stereocilia

Lining cells

(e)

Stereocilia

Lining cells

(f)

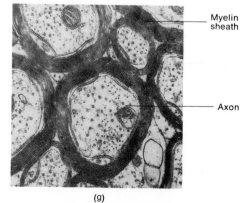

Myelin sheath

Axon

(g)

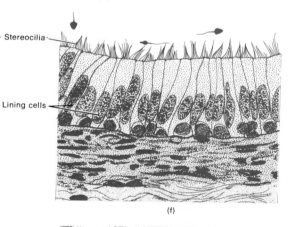

Myelin sheath

Axon

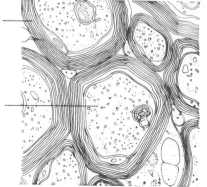

(h)

rest of the intact membrane. Few cells are capable of phagocytosis, but many cells may carry on pinocytosis. Examples include cells in the kidneys and urinary bladder.

When both phagocytosis and pinocytosis involve the inward movement of materials, they are together referred to as *endocytosis*. However, since phagocytosis and pinocytosis can also work in reverse by exporting materials from cells in vacuoles, they are also referred to as *exocytosis*.

MODIFIED PLASMA MEMBRANES

Electron microscope studies have revealed that plasma membranes of certain cells contain a number of modifications. That is, they have different structures for very specific purposes. For example, the membranes of some cells lining the small intestine have small, cylindrical projections called **microvilli** (Figure 2-7a, b). These fingerlike projections enormously increase the absorbing area of the cell surface. A single cell may have as many as 3000 microvilli, and a one sq mm (0.0394 sq in.) area of small intestine may contain as many as 200 million microvilli.

Another membrane modification is found in the rods and cones of the eye. They serve as photoreceptors, or light-receiving cells. The upper portion of each rod contains two-layered, disc-shaped membranes called **sacs** that contain the pigments involved in vision (Figure 2-7c, d). Another example of a membrane modification is the **stereocilia.** They are found only in cells lining a duct (epididymis) of the male reproductive system. They appear by light microscopy as long, slender, branching processes at the free surfaces of the lining cells (Figure 2-7e, f). Electron micrographs show stereocilia to be microvilli. A final example of a membrane modification is the **myelin sheath** that surrounds portions of certain nerve cells (Figure 2-7g, h). It is thought that the myelin sheath increases the velocity of impulse conduction, protects the portion of the nerve cell it surrounds, and is related to the nutrition of the nerve cell.

Cytoplasm

The living matter inside the cell's plasma membrane and external to the nucleus is called **cytoplasm** (Figure 2-8a, b). It is the matrix or ground substance in which various cellular components are found. Physically, cytoplasm may be described as a thick, semitransparent,

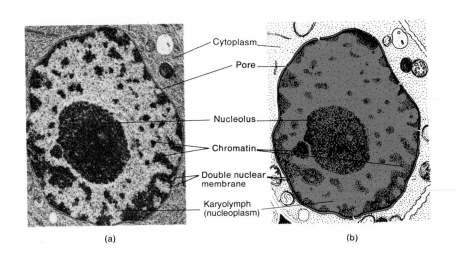

(a)

Cytoplasm
Pore
Nucleolus
Chromatin
Double nuclear membrane
Karyolymph (nucleoplasm)

(b)

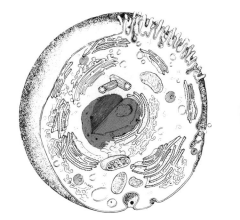

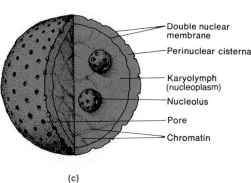

Double nuclear membrane
Perinuclear cisterna
Karyolymph (nucleoplasm)
Nucleolus
Pore
Chromatin

(c)

FIGURE 2-8 Cytoplasm and nucleus. (a) Electron micrograph of cytoplasm and the nucleus at a magnification of 31,600×. (Courtesy of Myron C. Ledbetter, Brookhaven National Laboratory.) (b) Diagram of the electron micrograph. (c) Diagram of a nucleus with two nucleoli.

elastic fluid containing suspended particles. Chemically, cytoplasm is 75 to 90 percent water plus solid components. Proteins, carbohydrates, lipids, and inorganic substances compose the bulk of the solid components. The inorganic substances and most carbohydrates are soluble in water and are present as a true solution. The majority of organic compounds, however, are found as colloids—particles that remain suspended in the surrounding ground substance. Since the particles of a colloid bear electrical charges that repel each other, they remain suspended and separated from each other.

Functionally, cytoplasm is the substance in which chemical reactions occur. The cytoplasm receives raw materials from the external environment and converts them into usable energy by decomposition reactions. Cytoplasm is also the site where new substances are synthesized for cellular use. It packages chemicals for transport to other parts of the cell or other cells of the body and facilitates the excretion of waste materials.

Organelles

Despite the myriad chemical activities occurring simultaneously in the cell, there is little interference of one reaction with another. The cell has a system of compartmentalization provided by structures called **organelles.** These structures are specialized portions of the cell that assume specific roles in growth, maintenance, repair, and control.

NUCLEUS

The **nucleus** is generally a spherical or oval organelle (Figure 2-8a, b). In addition to being the largest structure in the cell, the nucleus contains hereditary factors of the cell, called genes, which control cellular structure and direct many cellular activities. Certain cells, such as mature red blood cells, do not have nuclei. These cells carry on only limited types of chemical activity and are not capable of growth or reproduction.

The nucleus is separated from the cytoplasm by a double membrane called the *nuclear membrane* or *envelope* (Figure 2-8c). Between the two layers of the nuclear membrane is a space called the *perinuclear cisterna*. This arrangement of the nuclear membrane resembles the structure of the plasma membrane. Minute pores in the nuclear membrane allow the nucleus to communicate with a membranous network in the cytoplasm called the endoplasmic reticulum. Substances entering and exiting the nucleus are believed to pass through the tiny pores. Three prominent structures are visible within the nuclear membrane. The

first of these is a gel-like fluid that fills the nucleus called *karyolymph (nucleoplasm)*. Spherical bodies called the *nucleoli* are also present. These structures are composed primarily of RNA and assume a function in protein synthesis. Finally, there is the *genetic material* consisting principally of DNA. When the cell is not reproducing, the genetic material appears as a threadlike mass called *chromatin*. Prior to cellular reproduction the chromatin shortens and thickens into rod-shaped bodies called *chromosomes*.

ENDOPLASMIC RETICULUM AND RIBOSOMES

Within the cytoplasm, there is a system consisting of pairs of parallel membranes enclosing narrow cavities of varying shapes. This system is known as the **endoplasmic reticulum,** or **ER** (Figure 2-9). The ER, in other words, is a network of channels *(cisternae)* running through the entire cytoplasm. These channels are continuous with both the plasma membrane and the nuclear membrane. It is thought that the ER provides a surface area for chemical reactions, a pathway for transporting molecules within the cell, and a storage area for synthesized molecules. And, together with the Golgi complex, the ER is thought to secrete certain chemicals. Attached to the outer surfaces of some of the ER are exceedingly small, dense, spherical bodies called **ribosomes.** The ER in these areas is referred to as *granular,* or *rough,* reticulum. Portions of the ER that lack ribosomes are called *agranular,* or *smooth,* reticulum. Ribosomes are thought to serve as the sites of protein synthesis in the cell.

GOLGI COMPLEX

Another structure found in the cytoplasm is the **Golgi complex.** This structure usually consists of four to eight flattened channels, stacked upon each other with expanded areas at their ends. Like those of the ER, the stacked elements are called *cisternae,* and the expanded, terminal areas are *vesicles* (Figure 2-10). Generally, the Golgi complex is located near the nucleus and is directly connected, in parts, to the ER. One function of the Golgi complex is the secretion of proteins. *Secretion* is the production and release from the cell of a fluid that usually contains a variety of substances. Proteins synthesized by the ribosomes associated with granular ER are transported into the ER cisternae. They migrate along the ER cisternae until they reach the Golgi complex. As proteins accumulate in the cisternae of the Golgi complex, the cisternae expand to form vesicles. After a certain size is reached, the vesicles pinch off from the cisternae. The protein

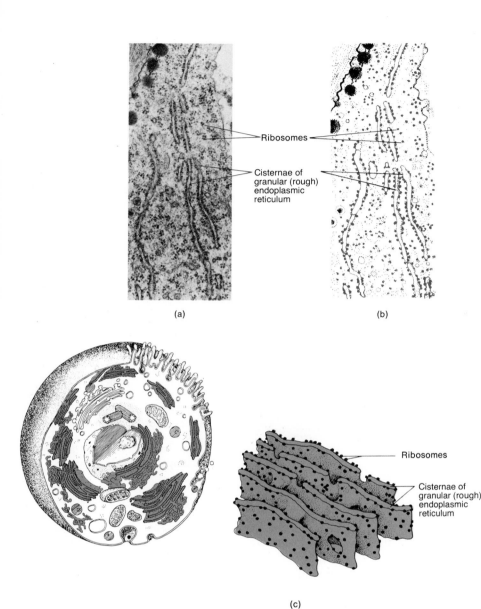

Ribosomes

Cisternae of
granular (rough)
endoplasmic
reticulum

(a)

(b)

Ribosomes

Cisternae of
granular (rough)
endoplasmic
reticulum

(c)

FIGURE 2-9 Endoplasmic reticulum and ribosomes. (a) Electron micrograph of the endoplasmic reticulum and ribosomes at a magnification of 76,000×. (Courtesy of Myron C. Ledbetter, Brookhaven National Laboratory.) (b) Diagram of the electron micrograph. (c) Diagram of the endoplasmic reticulum and ribosomes. See if you can find the agranular (smooth) endoplasmic reticulum in Figure 2-10.

and its associated vesicle is referred to as a *secretory granule*. The secretory granule then moves toward the surface of the cell where the protein is secreted. Cells of the digestive tract that secrete protein enzymes utilize this mechanism. The vacuole prevents "digestion" of the cytoplasm of the cell as it moves toward the cell surface.

Another function of the Golgi complex is associated with lipid secretion. It occurs in essentially the same way as protein secretion, except the lipids are synthesized by the agranular ER. The lipids pass through the ER into the Golgi complex. As in the mechanism just described, the lipids migrate into the cisternae and vesicles and are discharged at the surface of the cell. In the course of moving through the cytoplasm, the vesicle may release lipids into the cytoplasm before

being discharged from the cell. These appear in the cytoplasm as lipid droplets. Among the lipids secreted in this manner are the steroids.

The Golgi complex also functions in the synthesis of carbohydrates. Recent evidence indicates that carbohydrates synthesized by the Golgi complex are combined with proteins synthesized by ribosomes associated with granular ER to form carbohydrate-protein complexes. These complexes of carbohydrate and protein are called *glycoproteins*. As glycoproteins are assembled, they accumulate in the flattened channels of the Golgi complex. The channels expand and form vesicles. After a critical size is reached, the vesicles pinch off from the channel, migrate through the cytoplasm, and pass out of the cell through the plasma membrane. Outside the plasma membrane, the vesi-

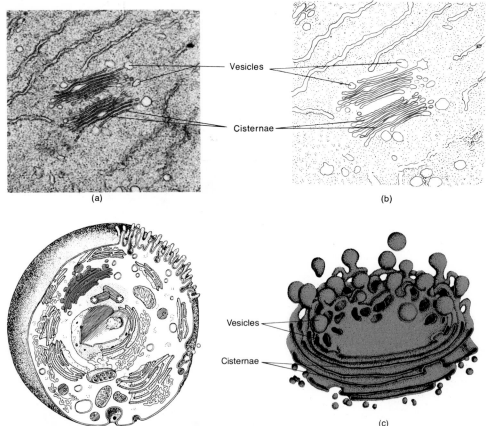

Vesicles

Cisternae

(a)

(b)

Vesicles

Cisternae

(c)

FIGURE 2-10 Golgi complex. (a) Electron micrograph of two Golgi complexes at a magnification of 78,000×. (Courtesy of Myron C. Ledbetter, Brookhaven National Laboratory.) (b) Diagram of the electron micrograph. (c) Diagram of the Golgi complex.

cles rupture and release their contents. Essentially, the Golgi complex synthesizes carbohydrates and combines them with proteins. It then packages the resulting glycoprotein and secretes it from the cell. The Golgi complex is well developed and highly active in secretory cells such as those found in the pancreas and the salivary glands.

MITOCHONDRIA

Small, spherical, rod-shaped, or filamentous structures called **mitochondria** appear throughout the cytoplasm. When sectioned and viewed under an electron microscope, each reveals an elaborate internal organization (Figure 2-11). A mitochondrion consists of a double membrane each of which is similar in structure to the plasma membrane. The outer mitochondrial membrane is smooth, but the inner membrane is arranged in a series of folds called *cristae*. The center of the mitochondrion is called the *matrix*. Because of the nature and arrangement of the cristae, the inner membrane provides an enormous surface area for chemical reactions. Enzymes involved in energy-releasing reactions that form ATP are located on the cristate. Mitochondria are frequently called the "powerhouses of

the cell" because of their central role in the production of ATP. Active cells such as muscle cells have a large number of mitochondria because of their high energy expenditure.

LYSOSOMES

When viewed under the electron microscope, **lysosomes** appear as membrane-enclosed spheres. They are formed from Golgi complexes (Figure 2-12). Unlike mitochondria, lysosomes have only a single membrane and lack detailed structure. Moreover, they contain powerful digestive enzymes capable of breaking down many kinds of molecules. These enzymes are also capable of digesting bacteria that enter the cell. White blood cells, which ingest bacteria by phagocytosis, contain large numbers of lysosomes. Scientists have wondered why these powerful enzymes do not also destroy their own cells. Perhaps the lysosome membrane in a healthy cell is impermeable to enzymes so they cannot move out into the cytoplasm. However, when a cell is injured, the lysosomes release their enzymes. The enzymes then promote reactions that break the cell down into its chemical constituents. The chemical remains are either reused by the body or

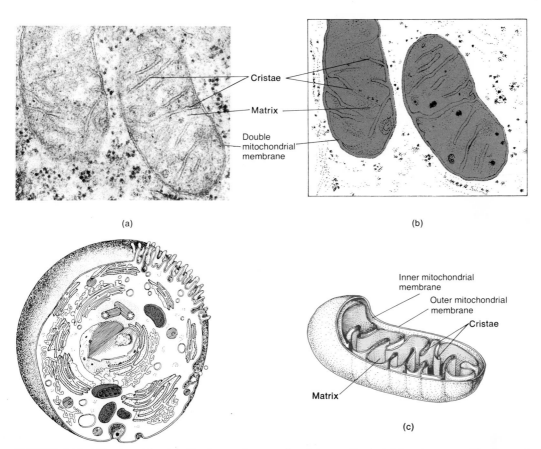

FIGURE 2-11 Mitochondria. (a) Electron micrograph of two mitochondria at a magnification of 20,000×. (Courtesy of Myron C. Ledbetter, Brookhaven National Laboratory.) (b) Diagram of the electron micrograph. (c) Diagram of a mitochondrion.

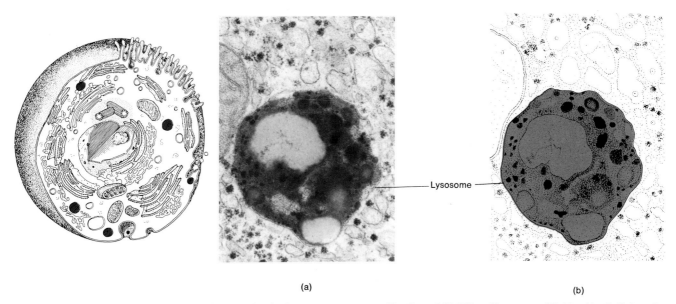

FIGURE 2-12 Lysosome. (a) Electron micrograph of a lysosome at a magnification of 55,000×. (Courtesy of F. Van Hoof, Universite Catholique de Louvain.) (b) Diagram of the electron micrograph.

excreted. Because of this function, lysosomes have been called "suicide packets."

MICROTUBULES

Small tubules made of protein and called **microtubules** are found in most cells. They range from 200 to 270 Å in diameter and are remarkably uniform in size. They do not branch. Microtubules are thought to be internal "skeletons" that afford cell shape and stiffness to the regions they occupy. They are also believed to function as intracellular channels along which substances move from place to place. Microtubules make up the internal structure of cilia and flagella. They also make up the spindle fibers and centrioles (see Figure 2-13) required for cell division.

CENTROSOME AND CENTRIOLES

A dense area of cytoplasm, generally spherical and located near the nucleus, is called the **centrosome** or **centrosphere.** Within the centrosome is a pair of cylindrical structures: the **centrioles** (Figure 2-13). Each centriole is composed of a ring of nine evenly spaced bundles surrounding two central tubules. Each bundle, in turn, consists of three microtubules. The two centrioles are situated so that the long axis of one is at right angles to the long axis of the other. Centrioles assume a role in cell reproduction. Certain cells, such as most mature nerve cells, do not have a centrosome and so do not reproduce. This is why they cannot be replaced if destroyed.

FLAGELLA AND CILIA

Some body cells possess projections for moving the entire cell or for moving substances along the surface of the cell. These projections contain cytoplasm and are bounded by the plasma membrane. If the projec-

tions are few and long in proportion to the size of the cell, they are called **flagella.** An example of a flagellum is the tail of a sperm cell, used for locomotion. If the projections are numerous and short, resembling many hairs, they are called cilia. In humans, ciliated cells of the respiratory tract move lubricating fluids over the surface of the tissue and trap foreign particles. Electron microscopy has revealed no fundamental structural difference between cilia and flagella (Figure 2-14). Both consist of nine pairs of microtubules that form a ring around two microtubules in the center.

Cell Inclusions

The **cell inclusions** are a large and diverse group of chemical substances. These products are principally organic and may appear or disappear at various times in the life of the cell. *Hemoglobin* lies inside red blood cells. It performs the function of attaching to oxygen molecules and carrying the oxygen to other cells. *Melanin* is a pigment stored in the cells of the skin, hair, and eyes. It protects the body by screening out harmful ultraviolet rays from the sun. *Glycogen* is a polysaccharide that is stored in liver and skeletal muscle cells. When the body requires quick energy, liver cells can break down the glycogen into glucose and release it. *Lipids,* which are stored in fat cells, may be decomposed when the body runs out of carbohydrates for producing energy. A final example of an inclusion is *mucus,* which is produced by cells that line organs. Its function is to provide lubrication.

The major parts of the cell and their functions are summarized in Exhibit 2-1.

Extracellular Materials

The substances that lie outside cells are called **extracellular materials.** They include the body fluids, which provide a medium for dissolving, mixing, and trans-

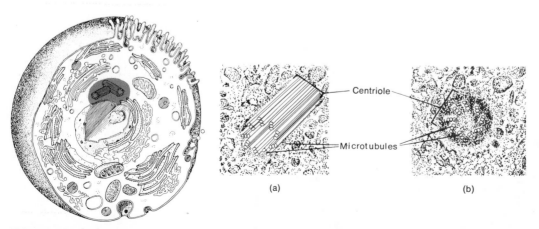

FIGURE 2-13 Centrosome and centrioles. (a) Centriole in longitudinal section. (b) Centriole in cross section.

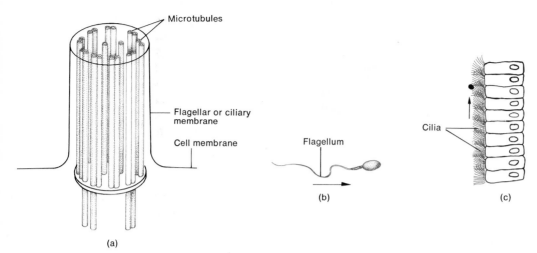

FIGURE 2-14 Flagella and cilia. (a) Structure of a flagellum or cilium. (b) Flagellum of a sperm cell. (c) Cilia of the respiratory tract moving a particle upward toward the mouth.

EXHIBIT 2-1
CELL PARTS AND THEIR FUNCTIONS

PART	FUNCTIONS
Plasma membrane	Protects and allows substances to enter or exit the cell through diffusion, facilitated diffusion, osmosis, filtration, dialysis, active transport, phagocytosis, and pinocytosis.
Cytoplasm	Serves as the ground substance in which chemical reactions occur.
Organelles Nucleus	Contains genes and controls cellular activities.
Endoplasmic reticulum	Provides a surface area for chemical reactions; provides a pathway for transporting chemicals; serves as a storage area.
Ribosomes	Act as sites of protein synthesis.
Golgi complex	Synthesizes carbohydrates, combines carbohydrates with proteins, packages materials for secretion, and secretes lipids and glycoproteins.
Mitochondria	Produce ATP.
Lysosomes	Digest chemicals and foreign microbes.
Microtubules	Serve as intracellular "skeleton" and passageway and component of cilia, flagella, and spindle fibers.
Centrioles	Help organize spindle fibers during cell division.
Flagella and cilia	Allow movement of cell or movement of particles along surface of cell.
Inclusions	Involved in overall body functions. Include materials retained in cell (hemoglobin), reserve materials (glycogen, lipids), and secretions (mucus).

porting substances. Among the body fluids are interstitial fluid, the fluid that fills the microscopic spaces (interstitial spaces) between cells and plasma; and plasma, the fluid in blood vessels. Extracellular materials also include secreted inclusions like mucus and special substances that form the matrix in which some cells are embedded.

The matrix materials are produced by certain cells and deposited outside their plasma membranes. The matrix supports the cells, binds them together, and gives strength and elasticity to the tissue. Some matrix materials are *amorphous*—they have no specific shape. These include hyaluronic acid and chondroitin sulfate. *Hyaluronic acid* is a viscous, fluidlike substance that binds cells together, lubricates joints, and maintains the shape of the eyeballs. *Chondroitin sulfate* is a jellylike substance that provides support and adhesiveness in cartilage, bone, heart valves, the cornea of the eye, and the umbilical cord. Other matrix materials are *fibrous,* or threadlike. Fibrous materials provide strength and support for tissues. Among these are **collagen,** or *collagenous fibers.* Collagen is found in all types of connective tissue, especially in bones, cartilage, tendons, and ligaments. **Reticulin,** also called *reticular fibers,* is a matrix material that forms a network around fat cells, nerve fibers, muscle cells, and blood vessels. It forms the framework or stroma for many soft organs of the body such as the spleen. **Elastin,** found in *elastic fibers,* gives elasticity to skin and to tissues forming the walls of blood vessels.

Cell Division

Most of the cell activities mentioned thus far maintain the life of the cell on a day-to-day basis. However, cells become damaged, diseased, or worn out and then die. Thus new cells must be produced for growth.

Cell division is the process by which cells reproduce themselves. Cell division or, more appropriately, nuclear division, may be of two kinds. In the first kind a single parent cell duplicates itself. This process consists of mitosis (nuclear division) and cytokinesis (cystoplasmic division). The process ensures that each new daughter cell has the same *number* and *kind* of chromosomes as the original parent cell. After the process is complete, the two daughter cells have the same hereditary material and genetic potential as the parent cell. This kind of cell division results in an increase in the number of body cells. Mitosis and cytokinesis are the means by which dead or injured cells are replaced and new cells are added for body growth. The second kind of division is a mechanism by which sperm and egg cells are produced. This process, meiosis, is the mechanism that enables the reproduction of an entirely new organism (Chapter 18).

MITOSIS

In a 24-hour period, the average adult loses about 500 million cells from different parts of the body. Obviously, these cells must be replaced. Cells that have a short life span—the cells of the outer layer of skin, the cornea of the eye, the digestive tract—are continually being replaced. The succession of events that takes place during mitosis and cytokinesis is plainly visible under a microscope after the cells have been stained in the laboratory.

When a cell reproduces, it must replicate its chromosomes so its hereditary traits may be passed on to succeeding generations of cells. A **chromosome** is a highly coiled **deoxyribonucleic acid (DNA)** molecule. This is the molecule that contains your hereditary information. Each human chromosome consists of about 20,000 genes.

Before taking a look at the relationship between chromosomes and cell division, it will first be necessary to briefly examine the structure of DNA, the basic component of chromosomes.

A molecule of DNA is a chain composed of repeating units called *nucleotides.* Each nucleotide of DNA consists of three basic parts (Figure 2-15a). (1) It contains one of four possible *nitrogen bases,* which are ring-shaped structures containing atoms of C, H, O, and N. The nitrogen bases found in DNA are named adenine, thymine, cytosine, and guanine. (2) It contains a sugar called *deoxyribose.* (3) It has a phosphoric acid called the *phosphate group.* The nucleotides are named according to the nitrogen base that is present. Thus, a nucleotide containing thymine is called a *thymine nucleotide.* One containing adenine is called an *adenine nucleotide,* and so on.

The chemical composition of the DNA molecule was known before 1900, but it was not until 1953 that a model of the organization of the chemicals was constructed. This model was proposed by J. D. Watson and F. H. C. Crick on the basis of data from many investigations. Figure 2-15b shows the following structural characteristics of the DNA molecule: (1) The molecule consists of two strands with crossbars. The strands twist about each other in the form of a *double helix* so that the shape resembles a twisted ladder. (2) The uprights of the DNA ''ladder'' consist of alternating phosphate groups and the deoxyribose portions of the nucleotides. (3) The rungs of the ladder contain paired nitrogen bases. As shown, adenine always pairs off with thymine, and cytosine always pairs off with guanine.

The process called **mitosis** is the replication of chromosomes and the distribution of the two sets of chromosomes into two separate and equal nuclei. For convenience, biologists divide the process into four stages:

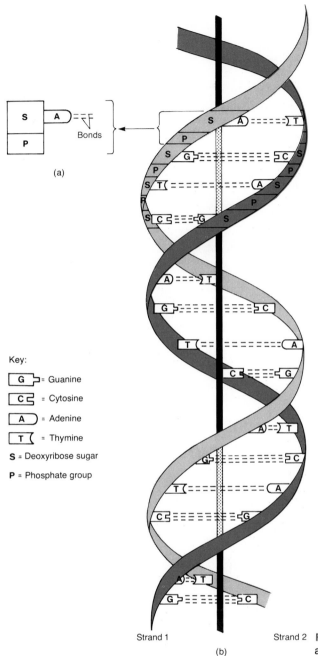

Key:

G ⊐ = Guanine

C ⊏ = Cytosine

A ◯ = Adenine

T ◖ = Thymine

S = Deoxyribose sugar

P = Phosphate group

Strand 1   Strand 2   **FIGURE 2-15** DNA molecule. (a) Adenine nucleotide. (b) Portion of an
                        (b)   assembled DNA molecule.

prophase, metaphase, anaphase, and telophase. These are arbitrary classifications. Mitosis is actually a continuous process, one stage merging imperceptibly into the next. Interphase is the stage that occurs between consecutive cell divisions.

Interphase

When a cell is carrying on every life process except division, it is said to be in **interphase** (Figure 2-16a). One of the principal events of interphase is the repli-

cation of DNA. When DNA replicates, its helical structure partially uncoils (Figure 2-17). Those portions of DNA that remain coiled stain darker than the uncoiled portions. This unequal distribution of stain causes the DNA to appear as a granular mass called **chromatin.** (See Figure 2-16a). During uncoiling, DNA separates at the points where the nitrogen bases are connected. Each exposed nitrogen base then picks up a complementary nitrogen base (with associated sugar and phosphate group) from the cytoplasm of the cell. This uncoiling and complementary base pairing con-

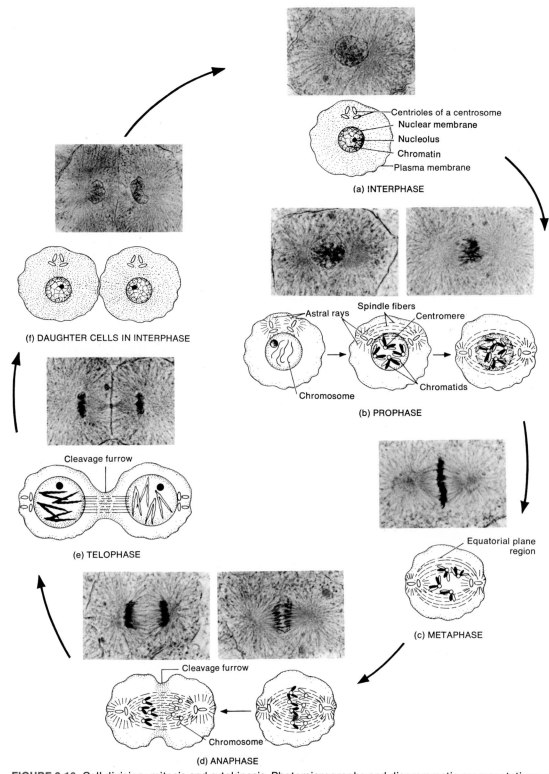

FIGURE 2-16 Cell division: mitosis and cytokinesis. Photomicrographs and diagrammatic representations of the various stages of division in whitefish eggs. Read the sequence starting at (a), and move clockwise until you complete the cycle. (Courtesy of Carolina Biological Supply Company.)

tinues until each of the two original DNA strands is matched and joined with two newly formed DNA strands. The original DNA molecule has become two DNA molecules.

During interphase the cell is also synthesizing most of its RNA (ribonucleic acid) and proteins. It is producing chemicals so that all cellular components can be doubled during division. A microscopic view of a cell during interphase shows a clearly defined membrane, nucleoli, karyolymph, and chromatin. As interphase progresses, a pair of centrioles appears. The centrioles duplicate and the resulting two pairs of centrioles separate. Once a cell completes its activities during interphase, mitosis begins.

Prophase

During **prophase** (Figure 2-16b) the centrioles move apart and project a series of radiating fibers called *astral rays*. The centrioles move to opposite poles of the cell and become connected by another system of fibers called *spindle fibers*. The astral rays and spindle fibers consist of microtubules. Together, the centrioles, astral rays, and spindle fibers are referred to as the *mitotic apparatus*. Simultaneously, the chromatin has been shortening and thickening into chromosomes. The nucleoli have become less distinct, and the nuclear membrane has disappeared. Each prophase ''chromosome'' is actually composed of a pair of separate structures called *chromatids*. Each chromatid is a complete chromosome consisting of a double-stranded DNA molecule. Each chromatid is attached to its chromatid pair by a small spherical body called the *centromere*. During prophase, the chromatid pairs move toward the equatorial plane region or equator of the cell.

Metaphase

During **metaphase** (Figure 2-16c), the second stage of mitosis, the chromatid pairs line up on the equatorial plane of the spindle fibers. The centromere of each chromatid pair attaches itself to a spindle fiber.

Anaphase

Anaphase (Figure 2-16d), the third stage of mitosis, is characterized by the division of the centromeres and the movement of complete identical sets of chromatids, now called chromosomes, to opposite poles of the cell. During this movement, the centromeres attached to the spindle fibers seem to drag the trailing parts of the chromosomes toward opposite poles.

Telophase

Telophase (Figure 2-16e), the final stage of mitosis, consists of a series of events nearly the reverse of prophase. By this time, two identical sets of chromosomes have reached opposite poles. New nuclear membranes begin to enclose them. The chromosomes start to assume their chromatin form. Nucleoli reappear, and the spindle fibers and astral rays disappear. The centrioles also replicate so that each cell has two centriole pairs. The formation of two nuclei identical to those of cells in interphase terminates telophase. A mitotic cycle has thus been completed (Figure 2-16f).

CYTOKINESIS

Division of the cytoplasm, called **cytokinesis,** often begins during late anaphase and terminates at the same time as telophase. Cytokinesis begins with the formation of a *cleavage furrow* that runs around the cell's equator. The furrow progresses inward, resembling a constricting ring, and cuts completely through the cell to form two separate portions of cytoplasm (Figure 2-16d to f).

Applications to Health

CELLS AND AGING

As Frédéric Verzar, the Swiss dean of gerontologists, once said: ''Old age is not an illness; it is a continuation of life with decreasing capacities for adaptation.'' Only recently has his view of aging as a progressive failure of the body's homeostatic adaptive responses gained wide acceptance. There has been a strong tendency to confuse aging with many diseases frequently associated with it—especially cancer and atherosclerosis. Each, in fact, probably accelerates the other.

The obvious characteristics of aging are well known: graying and loss of hair, loss of teeth, wrinkling of skin, decreased muscle mass, and increased fat deposits. The physiological signs of aging are gradual deterioration in function and capacity to respond to environmental stress. Thus basic kidney and digestive metabolic rates decrease, as does the ability to maintain a constant internal environment despite changes in temperature, diet, and oxygen supply. These manifestations of aging are related to a net decrease in the number of cells in the body and to the disordered functioning of the cells that remain.

The extracellular components of tissues also change with age. *Collagen* fibers, responsible for the strength in tendons, increase in number and change in quality with aging. These changes in arterial walls are as much responsible for the loss of elasticity as those in atherosclerosis. *Elastin,* another constituent of the intercel-

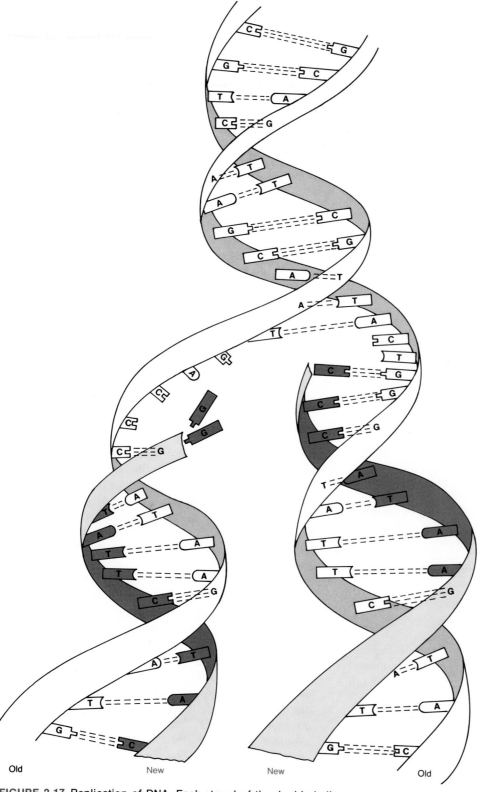

FIGURE 2-17 Replication of DNA. Each strand of the double helix separates by breaking the bonds between nucleotides. New nucleotides attach at the proper sites, and two new strands of DNA are paired off with the two old strands. After replication, the two DNA molecules, each consisting of a new and an old strand, return to their helical structure.

lular matrix, is reponsible for the elasticity of blood vessels and skin. It thickens, fragments, and acquires a greater affinity for calcium with age—changes that may be associated with the development of arteriosclerosis.

Several kinds of cells in the body—heart cells, skeletal muscle cells, neurons—are incapable of replacement. Recent experiments have proved that certain other cell types are limited when it comes to cell division. Cells grown outside the body divided only a certain number of times and then stopped. The number of divisions correlated with the donor's age. The number of divisions also correlated with the normal lifespan of the different species from which the cells were obtained—strong evidence for the hypothesis that cessation of mitosis is a normal, genetically programed event.

Just as the factors that limit the life of an individual cell are unknown, so are those that restrict the growth or life of a tissue or organ. At menopause, the ovary ceases to function. Its cells die long before the rest of the female body. Perhaps similar mechanisms determine longevity.

Some investigators have studied aging from the standpoint of immunology. The ability to develop antibodies is said to diminish with age. Senescence, according to researchers, results in the older person's immunological system having a "shotgun," rather than a specific, response to foreign protein. This shotgun response may include an autoimmune reaction that attacks and gradually destroys the individual's own normal tissue and organs.

The following generalizations on aging can be made:

1. Aging is a general process that produces observable changes in structure and function.

2. Aging produces increased vulnerability to environmental stress and disease.

3. Evidence suggests that the life span is about 110 years but maximum life expectancy is no greater than 85 years.

4. The mechanism behind aging is not known. Improvements in life expectancy may be due to an overall reduction in the number of life-threatening situations or to modification of the aging process.

5. Eventually the aging process may be modified and life span and life expectancy lengthened.

CELLS AND CANCER

Cancer is not a single disease but many. The human body contains more than a hundred different types of cells, each of which can malfunction in its own distinctive way to cause cancer. When cells in some area of the body duplicate unusually quickly, the excess of tissue that develops is called a growth, or **tumor.** Tumors may be cancerous and fatal or quite harmless. A cancerous growth is called a *malignant tumor,* or **malignancy.** A noncancerous growth is called a *benign growth.*

Cells of malignant growths all have one thing in common: they duplicate continuously and often very quickly. This growth continues until the victim dies or until every malignant cell is removed or destroyed. As the cancer grows, it expands and begins to compete with normal tissues for space and nutrients. Eventually, the normal tissues are crowded out, regress, and die. The organ functions less and less efficiently until finally it ceases to function altogether. Cancer cells may spread from the original, or *primary,* growth. Cancer of the breast, for instance, has a tendency to spread to the lungs. The spread of cancer to other regions of the body is called **metastasis.** Metastasis occurs when a malignant cell breaks away from the growth, enters the bloodstream, and is carried through the body. Wherever the cell comes to rest, it establishes another tumor: a *secondary* growth. Usually death is caused when a vital organ regresses because of competition with the cancer cells for room and nutrients. Pain develops when the growth impinges on nerves or blocks a passageway so that secretions build up pressure.

Cancer cells multiply without control. Individual cells vary in size and shape, and the orderly orientation of normal cells is replaced by disorganization which may be so extensive that no recognizable structures remain.

At present, cancers are classified by their microscopic appearance and the body site from which they arise. At least 100 different cancers have been identified in this way. If finer details of appearance are taken into consideration, the number can be increased to 200 or more. The name of the cancer is derived from the type of tissue in which it develops. **Sarcoma** is a general term for any cancer arising from connective tissue. *Osteogenic sarcomas* (*osteo* = bone; *genic* = origin), the most frequent type of childhood cancer, destroy normal bone tissue and eventually spread to other areas of the body. *Myelomas* (*myelos* = marrow) are malignant tumors, occurring in middle-aged and older people, that interfere with the blood-cell-producing function of bone marrow and cause anemia. *Chondrosarcoma* is a cancerous growth of the cartilage (*chondro* = cartilage).

Cancer has been observed in all species of vertebrates. In fact, cellular abnormalities that resemble cancer—such as crown gall of tomatoes—have been observed in plants as well. Cell masses that resemble cancers of higher animals have also been produced and studied in insects.

What triggers a perfectly normal cell to lose control and become abnormal? Scientists are uncertain. First there are environmental agents: substances in the air we breathe, the water we drink, the food we eat. The World Health Organization estimates that these agents—called *carcinogens*—may be associated with 60 to 90 percent of all human cancer. Examples of carcinogens are the hydrocarbons found in cigarette tar. Ninety percent of all lung cancer patients are smokers. Another environmental factor is radiation. Ultraviolet light from the sun, for example, may cause genetic mutations in exposed skin cells and lead to cancer, especially among light-skinned people.

Viruses are a second cause of cancer, at least in animals. These agents are tiny packages of nucleic acids that are capable of infecting cells and converting them to virus-producers. Virologists have linked tumor viruses with cancer in many species of birds and mammals, including primates. Since these experiments have not been performed on humans, there is no absolute proof that viruses cause human cancer. Nevertheless, with over 100 separate viruses identified as carcinogens in many species and tissues of animals, it is also probably that at least some cancers in humans are due to virus.

Benign tumors, unlike malignant tumors, are composed of cells that do not metastasize—that is, the growth does not spread to other organs. Removing all or part of a tumor to determine whether it is benign or malignant is called a **biopsy.** A benign tumor may be removed if it impairs a normal body function or causes disfiguration.

A reminder: As mentioned in the preface, each chapter in this text that discusses a major system of the body is followed by a glossary of **key medical terms.** Both normal and pathological conditions of the system are included in these glossaries. You should familiarize yourself with the terms since they will play an essential role in your medical vocabulary. Some of these disorders, as well as disorders discussed in the text, are referred to as local or systemic. A **local disease** is one that affects one part or a limited area of the body. A **systemic disease** affects either the entire body or several parts.

KEY MEDICAL TERMS ASSOCIATED WITH CELLS

Atrophy (*a* = without; *tropho* = nourish)   A decrease in the size of cells with subsequent decrease in the size of the affected tissue or organ; wasting away.

Biopsy   The removal and microscopic examination of tissue from the living body for diagnosis.

Carcinogen (*carc* = cancer)   A chemical or other environmental agent that produces cancer.

Carcinoma (*oma* = tumor)   A malignant tumor or cancer made up of epithelial cells.

Hyperplasia (*hyper* = over; *plas* = grow)   Increase in the number of cells due to an increase in the frequency of cell division.

Hypertrophy   Increase in the size of cells without cell division.

Metaplasia (*meta* = change)   The transformation of one cell into another.

Necrosis (*necros* = death; *osis* = condition)   Death of a group of cells.

Neoplasm (*neo* = new)   Any abnormal formation or growth, usually a malignant tumor.

Progeny   Offspring or descendants.

Senescence   The process of growing old.

STUDY OUTLINE

Generalized Animal Cell

1. A cell is the basic, living, structural and functional unit of the body.
2. A generalized cell is a composite that represents various cells of the body.
3. Cytology is the science concerned with the study of cells.
4. The principal parts of a cell are the plasma membrane, cytoplasm, organelles, and inclusions. Extracellular materials are manufactured by the cell and deposited outside the plasma membrane.

Plasma (Cell) Membrane

1. The plasma membrane, or cell membrane, surrounds the cell and separates it from other cells and the external environment.
2. The plasma membrane is composed of proteins and a bilayer of lipids. It is believed that the membrane possesses pores.
3. The membrane's semipermeable nature restricts the passage of certain substances. Substances can pass through the membrane depending on their molecular size, lipid solubility, electrical charges, and the presence of carriers.

Movement of Materials Across Plasma Membranes

1. Passive processes involve the innate energy of individual molecules.
2. Diffusion is the net movement of molecules or ions from an area of higher concentration to an area of lower concentration until an equilibrium is reached.
3. In facilitated diffusion, certain molecules like glucose

combine with a carrier to become soluble in the lipid portion of the membrane.

4. Osmosis is the movement of water through a semipermeable membrane from an area of higher water concentration to an area of lower water concentration.

5. Osmotic pressure is the force under which a solvent moves from a solution of lower solute concentration to a solution of higher solute concentration when the solutions are separated by a semipermeable membrane.

6. Filtration is the movement of water and dissolved substances across a semipermeable membrane by pressure.

7. Dialysis is the separation of small molecules from large molecules by diffusion through a semipermeable membrane.

8. Active processes involve the use of ATP by the cell.

9. Active transport is the movement of ions across a cell membrane from lower to higher concentration. This process relies on the participation of carriers.

10. Phagocytosis is the ingestion of solid particles by pseudopodia. It is an important process used by white blood cells to destroy bacteria that enter the body.

11. Pinocytosis is the ingestion of a liquid by the plasma membrane. In this process, the liquid becomes surrounded by a vacuole.

Modified Plasma Membranes

1. The membranes of certain cells are structured for specific functions.

2. Microvilli are microscopic fingerlike projections of the plasma membrane that increase the surface area for absorption.

3. Rod and cone cells of the eye contain sacs of light-sensitive pigments.

4. The myelin sheath of nerve cells protects, aids impulse conduction, and provides nutrition.

5. Stereocilia are long, slender, branching cells which line the epididymis.

Cytoplasm

1. The cytoplasm is the living matter inside the cell that contains organelles and inclusions.

2. It is composed mostly of water plus proteins, carbohydrates, lipids, and inorganic substances. The chemicals in cytoplasm are either in solution or in a colloid (suspended) form.

3. Functionally, cytoplasm is the medium in which chemical reactions occur.

Organelles

Organelles are specialized structures that carry on specific activities.

Nucleus

1. Usually the largest organelle, the nucleus controls cellular activities.

2. Cells without nuclei, such as mature red blood cells, do not grow or reproduce.

3. The parts of the nucleus include the nuclear membrane, nucleoplasm, nucleoli, and genetic material (DNA), comprising the chromosomes.

Endoplasmic Reticulum and Ribosomes

1. The ER is a network of parallel membranes continuous with the plasma membrane and nuclear membrane.

2. It functions in chemical reactions, transportation, and storage.

3. Granular or rough ER has ribosomes attached to it. Agranular or smooth ER does not contain ribosomes. Ribosomes are small spherical bodies that serve as sites of protein synthesis.

Golgi Complex

1. This structure consists of four to eight flattened channels vertically stacked on each other.

2. In conjunction with the ER, the Golgi complex synthesizes glycoproteins and secretes lipids.

3. It is particularly prominent in secretory cells such as those in the pancreas or salivary glands.

Mitochondria

1. These structures consist of a smooth outer membrane and a folded inner membrane. The inner folds are called cristae.

2. The mitochondria are called "powerhouses of the cell" because ATP is produced in them.

Lysosomes

1. Lysosomes are spherical structures containing digestive enzymes.

2. They are found in large numbers in white blood cells, which carry on phagocytosis.

3. If the cell is injured, lysosomes release enzymes and digest the cell. Thus they are called "suicide packets."

Microtubules

1. Microtubules consist of small protein tubules.

2. They form intracellular "skeletons" and intracellular channels and compose cilia, flagella, and spindle fibers.

Centrosome and Centrioles

1. The dense area of cytoplasm containing the centrioles is called a centrosome.

2. Centrioles are paired cylinders arranged at right angles to one another. They assume an important role in cell reproduction.

Flagella and Cilia

1. These cell projections have the same basic structure and are used in movement.

2. If projections are few and long, they are called flagella. If they are numerous and hairlike, they are called cilia.

3. The flagellum on a sperm cell moves the entire cell. The cilia on cells of the respiratory tract move foreign matter along the cell surfaces toward the throat for elimination.

Cell Inclusions

1. These chemical substances are produced by cells. They may be stored, may participate in chemical reactions, and may have recognizable shapes.

2. Examples of cell inclusions are glycogen, hemoglobin, mucus, and melanin.

Extracellular Materials

1. These are all the substances that lie outside the cell membrane.
2. They provide support and a medium for the diffusion of nutrients and wastes.
3. Some, like hyaluronic acid, are amorphous or have no shape. Others, like collagen, are fibrous or threadlike.

Cell Division

1. Cell division that results in the formation of new cells is called mitosis and cytokinesis. Nuclear division that results in the production of sperm and egg cells is termed meiosis.
2. Mitosis and cytokinesis replace and add body cells. Prior to mitosis and cytokinesis, the DNA molecules, or chromosomes, replicate themselves so the same chromosomal complement can be passed on to future generations of cells.

Mitosis

1. Mitosis—division of the nucleus—consists of prophase, metaphase, anaphase, and telophase.

2. A cell carrying on every life process except division is said to be in interphase.

Cytokinesis

1. Cytokinesis—division of the cytoplasm—begins in late anaphase and terminates in telophase.

Applications to Health

Cells and Aging

1. Aging is a progressive failure of the body's homeostatic adaptive responses.
2. Many theories of aging have been proposed, including genetically programed cessation of cell division and excessive immune responses, but none successfully answers all the experimental objections.

Cells and Cancer

1. Cancerous tumors are referred to as malignancies; noncancerous tumors are called benign growths.
2. The spread of cancer from its primary site is called metastasis.
3. Carcinogens include environmental agents and viruses.

REVIEW QUESTIONS

1. Define a cell. What are the four principal portions of a cell? What is meant by a generalized cell?
2. Discuss the structure of the plasma membrane. What determines the permeability of the plasma membrane? How are plasma membranes modified for various functions?
3. What are the major differences between active processes and passive processes in moving substances across plasma membranes?
4. Define and give an example of each of the following: diffusion, facilitated diffusion, osmosis, filtration, active transport, phagocytosis, pinocytosis.
5. Compare the effect on red blood cells of an isotonic, hypertonic, and hypotonic solution. What is osmotic pressure?
6. Describe the structure and function of microvilli, rod sacs, myelin sheaths, and stereocilia as membrane modifications.
7. Discuss the chemical composition and physical nature of cytoplasm. What is its function?
8. What is an organelle? By means of a diagram, indicate the structure and describe the function of the following organelles: nucleus, endoplasmic reticulum, ribosomes, Golgi complex, mitochondria, lysosomes, microtubules, centrosome, centrioles, cilia, flagella.
9. Define a cell inclusion. Provide examples and indicate their functions.
10. What is an extracellular material? Give examples and the functions of each.
11. How does DNA replicate itself?
12. Discuss mitosis and cytokinesis with regard to stages. What are the characteristics of each stage, the relative duration, and the importance?
13. List the five generalizations on aging that can be made based on current information.
14. What is a tumor? Distinguish between malignant and benign tumors.
15. Define metastasis. Why is it clinically important?
16. Discuss how certain carcinogens may be related to cancer.
17. Refer to the glossary of key medical terms at the end of the chapter and be sure that you can define each term.

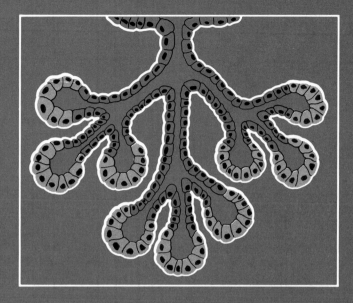

3

Tissues

STUDENT OBJECTIVES

- Define a tissue.

- Classify the tissues of the body into four major types.

- Describe the distinguishing characteristics of epithelial tissue.

- Contrast the structural and functional differences of covering, lining, and glandular epithelium.

- Compare the shape of cells and the layering arrangements of covering and lining epithelium.

- List the structure, function, and location of simple, stratified, and pseudostratified epithelium.

- Define a gland.

- Distinguish between exocrine and endocrine glands.

- Classify exocrine glands according to structural complexity and physiology.

- Describe the distinguishing characteristics of connective tissue.

- Contrast the structural and functional differences between embryonic and adult connective tissues.

- Describe the ground substance, fibers, and cells that constitute connective tissue.

- List the structure, function, and location of loose connective tissue, adipose tissue, and dense, elastic and reticular connective tissue.

- List the structure, function, and location of the three types of cartilage.

- Distinguish between the interstitial and appositional growth of cartilage.

- Define an epithelial membrane.

- List the location and function of mucous, serous, synovial, and cutaneous membranes.

The cells discussed in the preceding chapter are highly organized units, but they do not function as isolated units. Instead, they work together in a group of somewhat similarly constructed cells, called a tissue. We shall now examine how the body operates at the tissue level of organization.

A **tissue** is an aggregation of similar cells working together to perform a specialized activity. For example, some tissues of the body function to move body parts. Others move food through body organs. Some tissues protect and support the body. And still others function to produce chemicals such as enzymes and hormones. Depending on their functions and structure, the various tissues of the body are classified into four principal types: (1) **epithelial tissue,** which covers body or tissue surfaces, lines body cavities, and forms glands; (2) **connective tissue,** which protects and supports the body and its organs and binds organs together; (3) **muscular tissue,** which is responsible for movement; and (4) **nervous tissue,** which initiates and transmits nerve impulses that coordinate body activities. Some of these tissues will be discussed in later chapters as parts of a particular system. Others, such as epithelial tissue and most connective tissues, will be treated in detail in this chapter.

Epithelial Tissue

The tissues falling into this main category perform many activities in the body, ranging from protection to secretion. **Epithelial tissue,** or more simply **epithelium,** may be divided into two subtypes (1) *covering and lining epithelium* and (2) *glandular epithelium.* (See Exhibit 3-1.) Covering and lining epithelium forms the outer covering of external body surfaces and the outer covering of some internal organs. It lines the body cavities and the interiors of the respiratory and digestive tracts, blood vessels, and ducts. It comprises, along with nervous tissue, the parts of the sense organs that are sensitive to stimuli such as light and sound. Glandular epithelium constitutes the secreting portion of glands.

Both types of epithelium consist largely or entirely of closely packed cells with little or no intercellular material between adjacent cells. Such materials are collectively called the matrix. In addition, the epithelial cells are arranged in continuous sheets that may be either single or multilayered. Nerves may run through these sheets, but blood vessels do not. Thus they are *avascular*. The vessels that supply nutrients and remove wastes are located in underlying connective tissue. Epithelium overlies and adheres firmly to the con-

nective tissue, which holds the epithelium in position and prevents it from being torn. The surface of attachment between the epithelium and connective tissue is a thin extracellular layer of modified epithelial and connective tissue called the **basement membrane.** Since all epithelium is subjected to a certain degree of wear, tear, and injury, its cells must divide and produce new cells to replace those that are destroyed. These general characteristics are found in both types of epithelial tissue.

COVERING AND LINING EPITHELIUM

Arrangement of Layers

Covering and lining epithelium is arranged in several different ways related to location and function. If the epithelium is specialized for absorption or filtration and is in an area that has minimal wear and tear, the cells of the tissue are arranged in a single layer. Such an arrangement is called **simple epithelium.** If the epithelium is not specialized for absorption of filtration and is found in an area with a high degree of wear and tear, then the cells are stacked in several layers. This tissue is referred to as **stratified epithelium.** A third, less common, arrangement of epithelium is called **pseudostratified.** Like simple epithelium, pseudostratified epithelium has only one layer of cells. However, some of the cells do not reach the surface—an arrangement that gives the tissue a multilayered, or stratified, appearance. The pseudostratified cells that do reach the surface either secrete mucus or contain cilia that move mucus and foreign particles for eventual elimination from the body.

Cell Shapes

In addition to classifying covering epithelium according to the number of its layers, it may also be categorized by cell shape. The cells may be flat, cubelike, or columnar or may resemble a cross between shapes. **Squamous** cells are flattened, scalelike, and attached to each other like a mosaic. **Cuboidal** cells are usually cube-shaped in cross section. They sometimes appear as hexagons. **Columnar** cells are tall and cylindrical, appearing as rectangles set on end. **Transitional** cells often have a combination of shapes and are found where there is a great degree of distention or expansion in the body. Transitional cells on the bottom layer of an epithelial tissue may range in shape from cuboidal to columnar. In the intermediate layer, they may be cuboidal or polyhedral. Transitional cells in the superficial layer may range from cuboidal to squamous, depending on how much they are pulled out of shape during certain body functions.

Classification

Considering layers and cell type in combination, we may classify covering and lining epithelium as follows:

Simple
1. *Squamous*
2. *Cuboidal*
3. *Columnar*

Stratified
1. *Squamous*
2. *Cuboidal*
3. *Columnar*
4. *Transitional*

Psuedostratified

Each of the epithelial tissue described in the following section is illustrated in Exhibit 3-1.

Simple Epithelium

● *Simple Squamous Epithelium* This type of simple epithelium consists of a single layer of flat, scalelike cells. The surface of this epithelium resembles a tiled floor. The nucleus of each cell is centrally located and oval or spherical. Since simple squamous epithelium has only one layer of cells, it is highly adapted to diffusion, osmosis, and filtration. Thus it lines the air sacs of the lungs, where oxygen is exchanged with carbon dioxide. It is present in the part of the kidney

---

EXHIBIT 3-1
EPITHELIAL TISSUE

COVERING AND LINING EPITHELIUM

Simple Squamous (130×)

Description: Single layer of flat, scalelike cells; large, centrally located nucleus.
Location: Lines air sacs of lungs, glomerular capsule of kidneys, crystalline lens of eyes, and eardrum. Called endothelium when it lines heart, blood, and lymphatic vessels, and forms capillaries. Called mesothelium when it lines the ventral body cavity and covers viscera as part of a serous membrane.
Function: Filtration, absorption, and secretion in serous membranes.

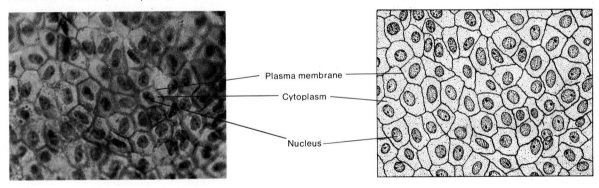

Plasma membrane
Cytoplasm
Nucleus

Simple Cuboidal (130×)

Description: Single layer of cube-shaped cells; centrally located nucleus.
Location: Covers surface of ovary; lines inner surface of cornea and lens of eye, kidney tubules, and smaller ducts of many glands.
Function: Secretion and absorption.

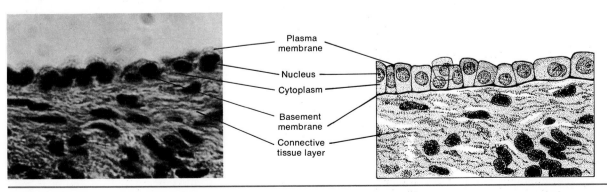

Plasma membrane
Nucleus
Cytoplasm
Basement membrane
Connective tissue layer

EXHIBIT 3-1
(*Continued*)

Simple Columnar (Nonciliated) (130×)

Description: Single layer of nonciliated rectangular cells; contains goblet cells; nuclei at bases of cells.
Location: Lines stomach, small and large intestines, digestive glands, and gallbladder.
Function: Secretion and absorption.

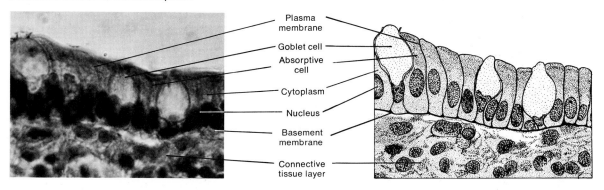

Simple Columnar (Ciliated) (130×)

Description: Single layer of ciliated rectangular cells; contains goblet cells; nuclei at bases of cells.
Location: Lines some portions of upper respiratory tract, uterine (fallopian) tubes, and uterus.
Function: Moves mucus by ciliary action.

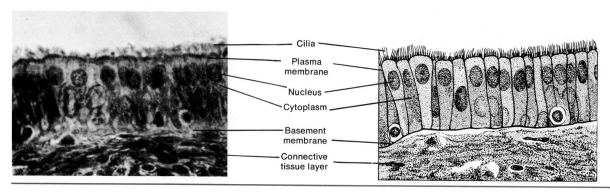

that filters the blood. It is also found in delicate structures such as the crystalline lens of the eye and the lining of the eardrum. Simple squamous epithelium is found in body parts that have little wear or tear. A tissue similar to simple squamous epithelium is endothelium. **Endothelium** lines the heart, the blood vessels, and the lymph vessels and forms the walls of capillaries. The term **mesothelium** is applied to another simple squamous epitheliumlike tissue that lines the thoracic, abdominal, and pelvic cavities and covers the viscera within them.

● *Simple Cuboidal Epithelium* Viewed from above, the cells of simple cuboidal epithelium appear as closely fitted polygons. The cuboidal nature of the cells is obvious only when the tissue is sectioned at right angles. Like simple squamous epithelium, these cells possess a central nucleus. Simple cuboidal epithelium covers the surfaces of the ovaries and lines the inner surfaces of the cornea. In the kidneys, where it forms the kidney tubules and contains microvilli, it functions in water reabsorption. It also lines the smaller ducts of some glands and the secreting units of glands, such as the thyroid. This tissue performs the functions of secretion and absorption. *Secretion,* usually a function of epithelium, is the production and release by cells of a fluid that may contain a variety of substances, such as mucus, per-

Stratified Squamous (65×)

Description: Several layers of cells; deeper layers are cuboidal to columnar; superficial layers are flat and scalelike; basal cells replace surface cells as they are lost.

Location: Nonkeratinizing variety lines wet surfaces such as mouth, esophagus, part of epiglottis, and vagina. Keratinizing variety forms outer layer of skin.

Function: Protection.

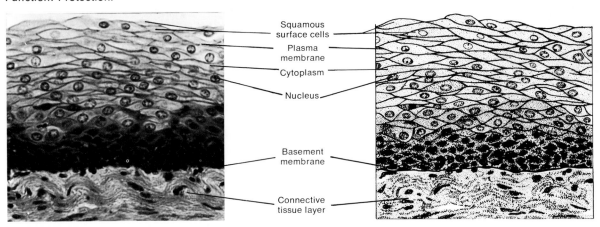

Stratified Cuboidal (80×)

Description: Two or more layers of cube-shaped cells.
Location: Ducts of adult sweat glands.
Function: Protection.

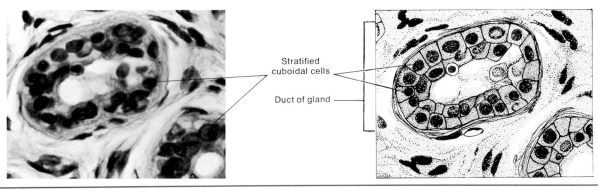

spiration, or enzymes. *Absorption* is the intake of fluids or other substances by cells of the skin or mucous membranes.

● *Simple Columnar Epithelium* The surface view of simple columnar epithelium is similar to that of simple cuboidal tissue. When sectioned at right angles to the surface, however, the cells appear somewhat rectangular. The nuclei are located near the bases of the cells. Simple columnar epithelium is modified in several ways, depending on location and function. Simple columnar epithelium lines the stomach, the small and large intestines, the digestive glands, and the gallbladder. In such sites, the cells protect the underlying tis-

sues. Many of them are also modified to aid in food-related activities. In the small intestine especially, the plasma membranes of the cells are folded into *micro-villi*. (See Figure 2-7a.) The microvilli arrangement increases the surface area of the plasma membrane and thereby allows larger amounts of digested nutrients and fluids to diffuse into the body. Interspersed among the typical columnar cells of the intestine are other modified columnar cells called *goblet cells*. These cells, which secrete mucus, are so named because the mucus accumulates in the upper half of the cell, causing the area to bulge out. The whole cell resembles a goblet or wine glass. The secreted mucus serves as a lubricant between the food and the walls of the diges-

EXHIBIT 3-1
(*Continued*)

Stratified Columnar (130×)

Description: Several layers of polyhedral cells; only superficial layer is columnar.
Location: Lines part of male urethra and some larger excretory ducts.
Function: Protection and secretion.

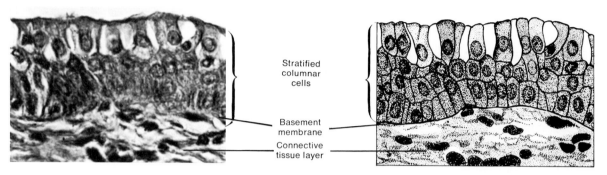

Stratified Transitional (110×)

Description: Resembles stratified squamous nonkeratinizing tissue, except superficial cells are larger
and more rounded.
Location: Lines urinary tract.
Function: Permits distention.

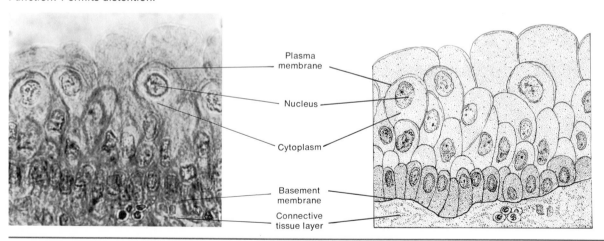

tive tract. A third modification of columnar epithelium is found in cells with hairlike processes called *cilia*. In some portions of the upper respiratory tract, ciliated columnar cells are interspersed with goblet cells. Mucus secreted by the goblet cells forms a film over the respiratory surface. This film traps foreign particles that are inhaled. The cilia, which wave in unison, move the mucus and foreign particles toward the throat, where it can be swallowed or eliminated. Thus air is filtered by this process before entering the lungs. Ciliated columnar epithelium, combined with goblet cells, is also found in the uterus and uterine tubes of the female reproductive system.

Stratified Epithelium

In contrast to simple epithelium, stratified epithelium consists of at least two layers of cells. Thus it is dura-

Pseudostratified (170×)

Description: Not a true stratified issue; nuclei of cells are present at different levels; some cells do not reach surface, but all sit on the basement membrane.
Location: Lines larger excretory ducts of many large glands and male urethra; ciliated variety with goblet cells lines most of the upper respiratory passageways and some ducts of male reproductive system.
Function: Secretion and movement of mucus by ciliary action.

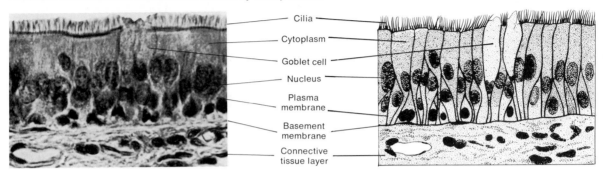

Cilia
Cytoplasm
Goblet cell
Nucleus
Plasma membrane
Basement membrane
Connective tissue layer

GLANDULAR EPITHELIUM

Exocrine Gland (110×)

Description: Secretes products into ducts.
Location: Sweat, oil, and wax glands of the skin; digestive glands such as salivary glands which secrete into mouth cavity.
Function: Produce perspiration, oil, wax, and digestive enzymes.

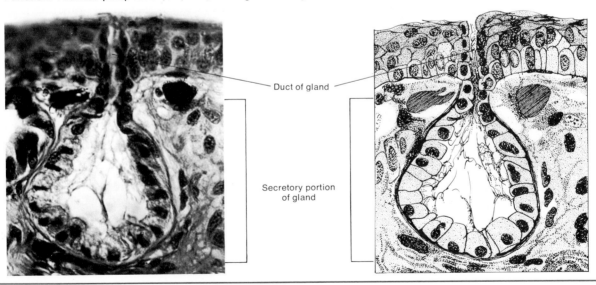

Duct of gland

Secretory portion of gland

ble and can protect underlying tissues from the external environment and from wear and tear. Some stratified epithelium cells also produce secretions. The name of the specific kind of stratified epithelium depends on the shape of the surface cells.

● *Stratified Squamous Epithelium* In the more superficial layers of this type of epithelium, the cells are flat, whereas the deep cells vary in shape from cuboidal to columnar. The basal, or bottom, cells are continually multiplying by cell division. As new cells grow, they compress the cells on the surface and push them outward. According to this growth pattern, basal cells continually shift upward and outward. As they move farther from the deep layer and their blood supply, they become dehydrated, shrink, and grow harder. At the surface, the cells rub off. New cells continually emerge, are sloughed off, and replaced.

EXHIBIT 3-1
(*Continued*)

Endocrine Gland (45×)

Description: Secretes hormones into blood.
Location: Pituitary gland at base of brain; thyroid gland near larynx; adrenal (suprarenal) glands above kidneys.
Function: Produce hormones that regulate various body activities.

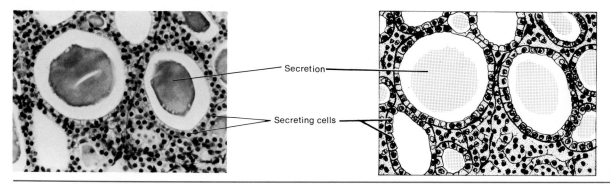

Photomicrographs courtesy of Victor B. Eichler, Wichita State University, except for pseudostratified tissue, which is courtesy of Donald I. Patt, from *Comparative Vertebrate Histology,* by Donald I. Patt and Gail R. Patt, Harper & Row, Publishers, Inc., New York, 1969.

One form of stratified squamous epithelium is called *nonkeratinized stratified squamous epithelium.* This tissue is found on wet surfaces that are subjected to considerable wear and tear and do not perform the function of absorption—the linings of the mouth, the esophagus, and the vagina. Another form of stratified squamous epithelium is called *keratinized stratified squamous epithelium.* The surface cells of this type are modified into a tough layer of material containing keratin. **Keratin** is a protein that is waterproof, resistant to friction, and impervious to bacterial invasion. The outer layer of skin consists of keratinized tissue.

● *Stratified Cuboidal Epithelium*   This rare type of epithelium is found in the ducts of the sweat glands of adults. It sometimes consists of more than two layers of cells. Its function is mainly protective.

● *Stratified Columnar Epithelium*   Like stratified cuboidal epithelium, this type of tissue is also uncommon in the body. Usually the basal layer or layers consist of shortened, irregularly polyhedral cells. Only the superficial cells are columnar in form. This kind of epithelium lines part of the male urethra and some larger excretory ducts such as lactiferous ducts in the mammary glands. It functions in protection and secretion.

● *Transitional Epithelium*   This kind of epithelium is very much like nonkeratinized stratified squamous epithelium. The distinction is that the outer layer of cells in transitional epithelium tend to be large and rounded rather than flat. This feature allows the tissue to be stretched without the outer cells breaking apart from one another. When stretched, they are drawn out into squamouslike cells. Because of this arrangement, transitional epithelium lines hollow structures that are subjected to expansion from within, such as the urinary bladder. Its function is to help prevent a rupture of the organ.

Pseudostratified Epithelium

The third category of covering and lining epithelium is called pseudostratified epithelium. The nuclei of the cells in this kind of tissue are at varying depths. Even though all the cells are attached to the basement membrane in a single layer, some do not reach the surface. This feature gives the impression of a multilayered tissue, the reason for the designation *pseudo*stratified epithelium. It lines the larger excretory ducts of many glands and parts of the male urethra. Pseudostratified epithelium may be ciliated and may contain goblet cells. In this form, it lines most of the upper respiratory passages and certain ducts of the male reproductive system.

GLANDULAR EPITHELIUM

The function of glandular epithelium is secretion, accomplished by glandular cells that lie in clusters deep to the covering and lining epithelium. A **gland** may consist of one cell or a group of highly specialized epithelial cells that secrete substances into ducts or into the blood. The production of such substances always requires active work by the glandular cells and results in an expenditure of energy.

On the basis of this distinction, all glands of the body are classified as exocrine or endocrine (Exhibit 3-1). **Exocrine glands** secrete their products into ducts or tubes that empty at the surface of the covering and lining epithelium. The product of an exocrine gland may be released at the skin surface or into the lumen of a hollow organ. The secretions of exocrine glands include enzymes, oil, and sweat. Examples of exocrine glands are sweat glands, which eliminate perspiration to cool the skin, and salivary glands, which secrete a digestive enzyme. **Endocrine glands,** by contrast, are ductless and, consequently, must secrete their products into the blood. The secretions of endocrine glands are always hormones. The pituitary, thyroid, and adrenal glands are endocrine glands.

Structural Classification of Exocrine Glands

Exocrine glands are classified into two structural types: unicellular and multicellular. **Unicellular glands** are single-celled. A good example of a unicellular gland is a goblet cell (Exhibit 3-1). Goblet cells are found in the epithelial lining of the digestive, respiratory, urinary, and reproductive systems. They produce mucus to lubricate the free surfaces of these membranes. **Multicellular glands** are many-celled glands and occur in several different forms (Figure 3-1). If the secretory portions of the gland are tubular, they are referred to as *tubular glands*. If flasklike, they are called *acinar glands*. If the gland contains both tubular and flasklike secretory portions, it is called a *tubuloacinar gland*.

Further, if the duct of the gland does not branch, it is referred to as a *simple gland;* if the duct does branch, it is called a *compound gland*. By combining the shape of the secretory portion with the degree of branching of the duct, we arrive at the following structural classification for exocrine glands:

Unicellular. One-celled gland that secretes mucus. Example: goblet cell of the digestive system.

Multicellular. Many-celled glands.
  Simple. Single, nonbranched duct.
  1. **Tubular.** The secretory portion is straight and tubular. Example: crypts of Lieberkuhn of intestines.
  2. **Branched tubular.** The secretory portion is branched and tubular. Examples: gastric and uterine glands.
  3. **Coiled tubular.** The secretory portion is coiled. Examples: sudoriferous (sweat) glands.
  4. **Acinar.** The secretory portion is flasklike. Example: seminal vesicle glands.
  5. **Branched acinar.** The secretory portion is branched and flasklike. Example: sebaceous (oil) glands.
  Compound. Branched duct.
  1. **Tubular.** The secretory portion is tubular. Examples: bulbourethral glands, testes, and liver.
  2. **Acinar.** The secretory portion is flasklike. Examples: mammary and salivary glands (sublingual and submandibular).

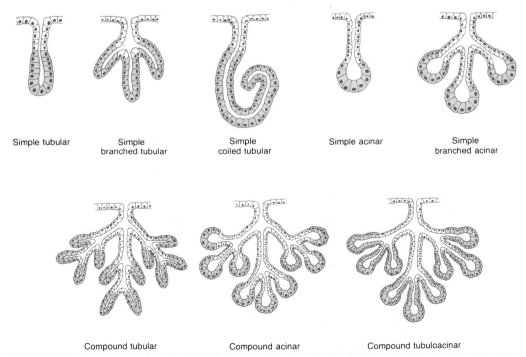

Simple tubular   Simple branched tubular   Simple coiled tubular   Simple acinar   Simple branched acinar

Compound tubular   Compound acinar   Compound tubuloacinar

FIGURE 3-1 Structural types of multicellular exocrine glands. The secretory portions of the glands are indicated in color. The uncolored areas represent the ducts of the glands.

3. Tubuloacinar. The secretory portion is both tubular and flasklike. Examples: salivary glands (parotid) and pancreas.

Functional Classification of Exocrine Glands

The functional classification of exocrine glands is based on how the gland releases its secretion. The three recognized categories are holocrine, merocrine, and apocrine glands. **Holocrine glands** accumulate a secretory product in their cytoplasm. The cell then dies and is discharged with its contents as the glandular secretion (Figure 3-2a). The discharged cell is replaced by a new cell. One example of a holocrine gland is a sebaceous gland of the skin. **Merocrine glands** are those that produce secretions that do not contain any portion of the secretory cell (Figure 3-2b). The secretion is simply formed and discharged by the cell. An example of a merocrine gland is the pancreas. Another is the salivary glands. **Apocrine glands** are those whose secretory product accumulates at the apical (outer) margin of the secreting cell. The apical region of the cell and its secretory contents pinch off from the rest of the cell to form the secretion (Figure 3-2c). The remaining part of the cell repairs itself and repeats the process. An example of an apocrine gland is the mammary gland.

Connective Tissue

The most abundant tissue in the body is **connective tissue.** This binding and supporting tissue usually has a rich blood supply. Thus it is *vascular*. The cells are widely scattered, rather than closely packed, and there is considerable intercellular material comprising the matrix. In contrast to epithelium, connective tissues do not occur on free surfaces, such as the surfaces of a body cavity or the external surface of the body. The general functions of connective tissues are protection, support, and the binding together of various organs.

The intercellular substance in a connective tissue largely determines the tissue's qualities. These substances are nonliving and may consist of fluid, semifluid, or mucoid (mucuslike) material. In cartilage, the intercellular material is firm. In bone, it is quite rigid. The living parts of connective tissue are the cells, which produce the intercellular substances. The cells may also store fat, ingest bacteria and cell debris, form anticoagulants, or give rise to antibodies that protect against disease.

LOOSE AND DENSE CONNECTIVE TISSUE

Before classifying and studying connective tissue, it will be helpful to distinguish between loose and dense connective tissues. **Loose connective tissue** refers to the arrangement of intercellular substance. That is, the fibers in the intercellular substance are neither abundant nor arranged to prevent stretching. In addition, the intercellular substance is soft or jellylike in consistency. By contrast, **dense connective tissue** is characterized by the close packing of fibers and less intercellular substance. In areas of the body where tensions are exerted in various directions, the fiber bundles are interwoven and without regular orientation. Such a dense connective tissue is referred to as *irregularly arranged*. Dense, irregularly arranged connective tissue occurs in sheets and forms most fasciae, the dermis of the skin, the fibrous capsules of some organs (testes, liver, lymph nodes), the periosteum of bone, and the perichondrium of cartilage. In other areas of the body, dense connective tissue is adapted for tension in one direction and the fibers have an orderly, parallel arrangement. Such a connective tissue is known as *regularly arranged*. Dense, regularly arranged connective tissue comprises tendons, ligaments, and aponeuroses.

Loose connective tissue contains each of the three types of fibers—collagenous, elastic, and reticular. Dense, irregularly arranged connective tissue also contains each of these three types, but collagenous fi-

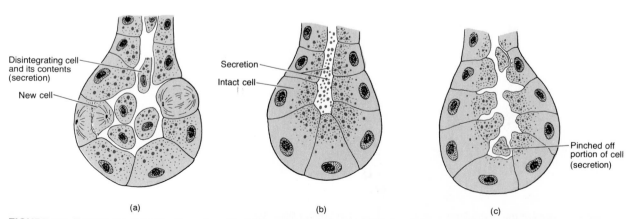

Disintegrating cell
and its contents
(secretion)

New cell

Secretion

Intact cell

Pinched off
portion of cell
(secretion)

(a)　　　　　　　　　(b)　　　　　　　　　(c)

FIGURE 3-2 Functional classification of multicellular exocrine glands. (a) Holocrine gland. (b) Merocrine gland. (c) Apocrine gland.

bers predominate. The dense, regularly arranged connective tissue of many ligaments and aponeuroses is composed of collagenous fibers. Such ligaments are known as collagenous ligaments. However, some ligaments are composed of elastic fibers and are known as yellow elastic ligaments. Examples of these include the ligamenta flava of the vertebrae, the suspensory ligament of the penis, and the true vocal cords.

CLASSIFICATION

Connective tissue may be classified in several ways. We will classify them as follows:

Embryonic connective tissue
Adult connective tissue
Connective tissues proper
   1. Loose connective (areolar) tissue
   2. Adipose tissue
   3. Dense connective (collagenous) tissue
   4. Elastic connective tissue
   5. Reticular connective tissue
Cartilage
   1. Hyaline cartilage
   2. Fibrocartilage
   3. Elastic cartilage
Bone (osseous) tissue
Vascular (blood) tissue

Each of the connective tissue described in the following sections is illustrated in Exhibit 3-2.

EMBRYONIC CONNECTIVE TISSUE

Connective tissue that is present primarily in the embryo or fetus is called **embryonic connective tissue.** Whereas the term *embryo* refers to a developing human from fertilization through the first two months of

EXHIBIT 3-2
CONNECTIVE TISSUES

EMBRYONAL

Mesenchymal (130×)

Description: Consists of highly branched mesenchymal cells embedded in a fluid substance.
Location: Under skin and along developing bones of embryo; some mesenchymal cells are found in adult connective tissue, especially along blood vessels.
Function: Forms all other kinds of connective tissue.

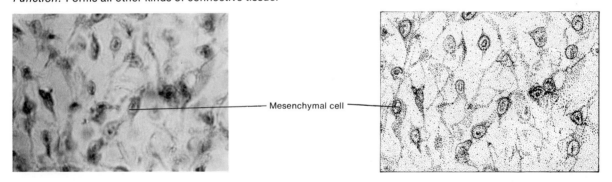

Mesenchymal cell

Mucous (130×)

Description: Consists of flattened or spindle-shaped cells embedded in a mucuslike substance containing fine collagenous fibers.
Location: Umbilical cord of fetus.
Function: Support.

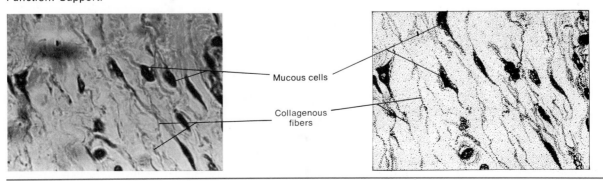

Mucous cells

Collagenous fibers

EXHIBIT 3-2
(*Continued*)

ADULT

Loose or Areolar (65×)

Description: Consists of fibers (collagenous and elastic) and several kinds of cells (fibroblasts, macrophages, plasma cells, and mast cells) embedded in a semifluid ground substance.
Location: Subcutaneous layer of skin, mucous membranes, blood vessels, nerves, and body organs.
Function: Strength, elasticity, support, phagocytosis, produces antibodies, and produces an anticoagulant.

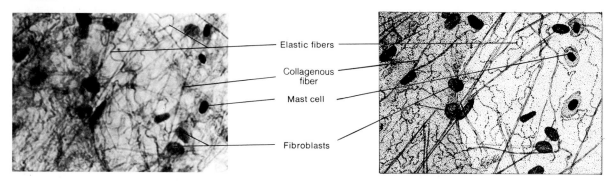

Elastic fibers
Collagenous fiber
Mast cell
Fibroblasts

Adipose (90×)

Description: Contains fibroblasts specialized for fat storage; cells have a "signet-ring shape."
Location: Subcutaneous layer of skin, around heart and kidneys, marrow of long bones, padding around joints.
Function: Reduces heat loss through skin, serves as a food reserve, supports, and protects.

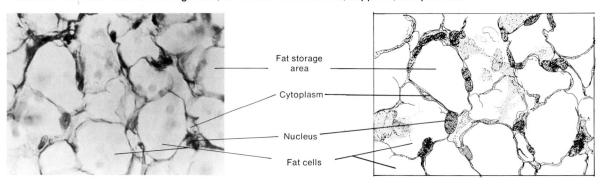

Fat storage area
Cytoplasm
Nucleus
Fat cells

pregnancy, a developing human from the third month of pregnancy to birth is regarded as a *fetus*. One example of embryonic connective tissue found almost exclusively in the embryo is <u>mesenchyme</u>—the tissue from which all other connective tissues eventually arise. Mesenchyme may be observed beneath the skin and along the developing bones of the embryo. Some mesenchymal cells are scattered irregularly throughout adult connective tissue, most frequently around blood vessels. Here mesenchymal cells differentiate into fibroblasts that assist in wound healing. Another kind of embryonic connective tissue is *mucous connective tissue,* found only in the fetus. This tissue, also called *Wharton's jelly,* is located in the umbilical cord of the fetus where it supports the wall of the cord.

ADULT CONNECTIVE TISSUE

Adult connective tissue is connective tissue that exists in the newborn and that does not change after birth. It is subdivided into several kinds.

Connective Tissue Proper

Connective tissue that has a more or less fluid intercellular material and a fibroblast as the typical cell is termed *connective tissue proper*. Five examples of such tissues may be distinguished.

● *Loose Connective (Areolar) Tissue* This type of tissue is one of the most widely distributed connective tissues in the body. Structurally, it consists of fibers

Dense or Collagenous (65×)

Description: Collagenous, or white, fibers predominate and are arranged in bundles; fibroblasts are in rows between bundles.
Location: Forms tendons, ligaments, aponeuroses, membranes around various organs, and fasciae.
Function: Provides strong attachment between various structures.

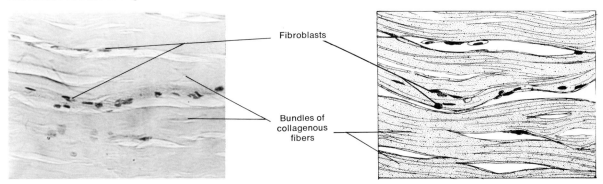

Fibroblasts

Bundles of collagenous fibers

Elastic (65×)

Description: Elastic, or yellow, fibers predominate and branch freely; fibroblasts present in spaces between fibers.
Location: Lung tissue, cartilage of larynx, walls of arteries, trachea, bronchial tubes, true vocal cords, and ligamenta flava of vertebrae.
Function: Allows stretching of various organs.

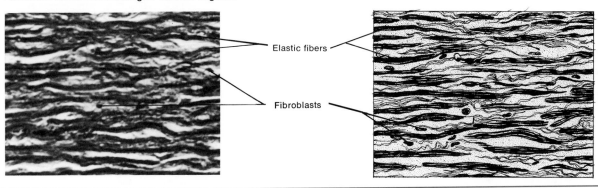

Elastic fibers

Fibroblasts

and several kinds of cells embedded in a semifluid intercellular substance. This intercellular substance consists of a viscous material called **hyaluronic acid,** which normally facilitates the passage of nutrients from the blood vessels of the connective tissue into adjacent cells and tissue. However, the thick consistency of this material may impede the movement of substances through the tissue. But, if an enzyme called hyaluronidase is injected into the tissue, hyaluronic acid changes to a water consistency. This feature is of clinical importance because the reduced viscosity hastens the absorption and diffusion of injected drugs and fluids through the tissue and thus can lessen tension and pain.

The three types of fibers embedded between the

cells of loose connective tissue are collagenous fibers, elastic fibers, and reticular fibers. **Collagenous,** or **white, fibers** are very tough and resistant to a pulling force, yet are somewhat flexible because they are usually wavy. These fibers often occur in bundles. They are composed of many minute fibers called fibrils lying parallel to one another. The bundle arrangement affords a great deal of strength. Chemically, collagenous fibers consist of the protein **collagen. Elastic,** or **yellow, fibers,** by contrast, are smaller than collagenous fibers and freely branch and rejoin one another. Elastic fibers consist of a protein called **elastin.** These fibers also provide strength and have great elasticity, up to 50 percent of their length. **Reticular fibers** also consist of collagen (plus some carbohydrate) but they are very

EXHIBIT 3-2
(Continued)

Reticular (65×)

Description: Consists of a network of interlacing fibers; cells are thin and flat and wrapped around fibers.
Location: Liver, spleen, and lymph nodes.
Function: Forms stroma, or framework, of organs; binds together smooth muscle tissue cells.

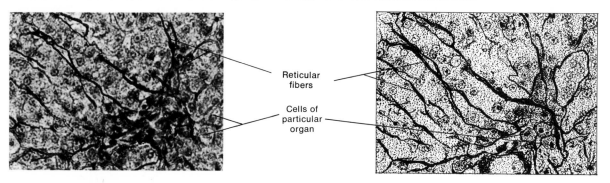

Reticular fibers

Cells of particular organ

Hyaline Cartilage (30×)

Description: Also called gristle; appears as a bluish white, glossy mass; contains numerous chondrocytes; is the most abundant type of cartilage.
Location: Ends of long bones, ends of ribs, nose, parts of larynx, trachea, bronchi, bronchial tubes, and embryonic skeleton.
Function: Provides movement at joints, flexibility, and support.

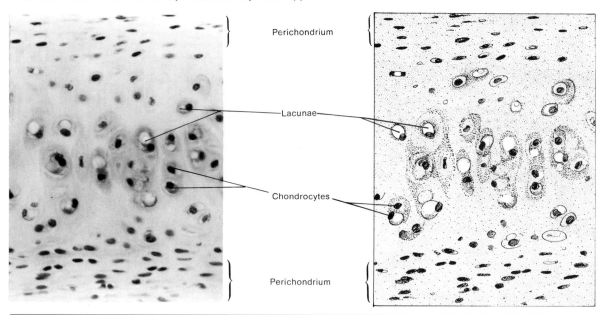

Perichondrium

Lacunae

Chondrocytes

Perichondrium

thin fibers that branch extensively. Some authorities believe that reticular fibers are immature collagenous fibers. Like collagenous fibers, reticular fibers provide support and strength and form the *stroma* (framework) of many soft organs.

The cells in loose connective tissue are numerous and varied. Most are **fibroblasts**—large, flat cells with branching processes. If the tissue is injured, the fibro-

blasts are believed to form collagenous fibers. Some evidence suggests that fibroblasts also form elastic fibers and the viscous ground substance. Mature fibroblasts are referred to as **fibrocytes.** The basic distinction between the two is that fibroblasts (or *"blasts"* of any form of cell) are involved in the formation of immature tissue or repair of mature tissue, and fibrocytes (or *"cytes"* of any form of cell) are involved in main-

Fibrocartilage (65×)

Description: Consists of chondrocytes scattered among bundles of collagenous fibers.
Location: Symphysis pubis and intervertebral discs.
Function: Support and fusion.

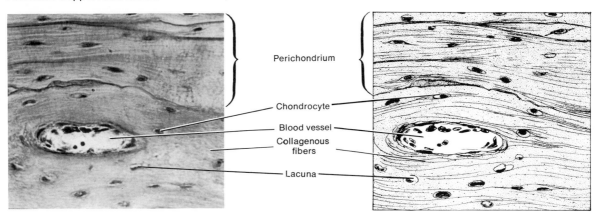

Perichondrium

Chondrocyte
Blood vessel
Collagenous fibers
Lacuna

Elastic Cartilage (65×)

Description: Consists of chondrocytes located in a threadlike network of elastic fibers.
Location: Epiglottis, parts of larynx, external ear; and eustachian (auditory) tubes.
Function: Gives support and maintains shape.

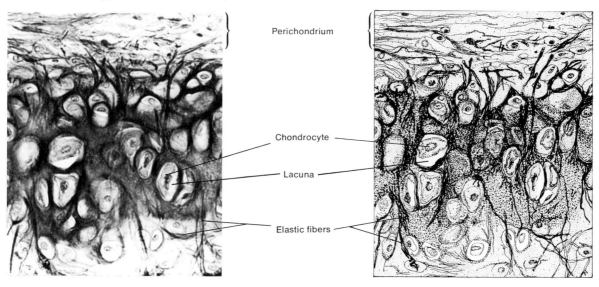

Perichondrium

Chondrocyte

Lacuna

Elastic fibers

Photomicrographs courtesy of Victor B. Eichler, Wichita State University.

taining health of mature tissue. Other cells found in loose connective tissue are called **macrophages,** or **histiocytes.** They are irregular in form with short branching processes. These cells are derived from undifferentiated mesenchymal cells called hemocytoblasts and are capable of engulfing bacteria and cellular debris by the process of phagocytosis. Thus they provide a vital defense for the body. A third kind of cell in loose connective tissue is the **plasma cell.** These cells are small and either round or irregular. Plasma cells probably develop from a type of white blood cell called a lymphocyte. They give rise to antibodies and, accordingly, provide a defensive mechanism through immunity. Plasma cells are found in many places of the body, but most are found in connective tissue, especially that of the digestive tract and the mammary glands. Another cell in loose connective tissue is the **mast cell.** It may develop from another type of white blood cell called a basophil. The mast cell is somewhat larger than a basophil and is found in abundance along blood vessels. It forms heparin, an anticoagulant that prevents blood from clotting in the vessels. Mast cells are also believed to produce histamine, a chemical that dilates, or enlarges, small blood vessels. Other cells in loose connective tissue include **melanocytes,** or pigment cells, fat cells, and white blood cells.

Loose connective tissue is continuous throughout the body. It is present in all mucous membranes and around all blood vessels and nerves. And it occurs around body organs. Combined with adipose tissue, it forms the *subcutaneous layer*—the layer of tissue that attaches the skin to underlying tissues and organs. The subcutaneous layer is also referred to as the *superficial fascia*.

● *Adipose Tissue* This kind of tissue is basically a form of loose connective tissue in which the fibroblasts are specialized for fat storage. Adipose cells have the shape of a "signet ring" because the cytoplasm and nucleus are pushed to the edge of the cell by a large droplet of fat. Adipose tissue is found wherever loose connective tissue is located. Specifically, it is in the subcutaneous layer below the skin, around the kidneys, at the base and on the surface of the heart, in the marrow of long bones, as a padding around joints, and behind the eyeball in the orbit. Adipose tissue is a poor conductor of heat and therefore reduces heat loss through the skin. It is also a major food reserve and generally supports and protects various organs.

● *Dense Connective (Collagenous) Tissue* This tissue has a predominance of collagenous fibers, or white fibers, arranged in bundles. The cells found in collagenous connective tissue are fibroblasts, which are placed in rows between the bundles. The tissue is silvery white, tough, yet somewhat pliable. Because of the great strength of this dense, regularly arranged tissue, it is the principal component of *tendons,* which attach muscles to bones; many *ligaments,* which hold bones together at joints; and *aponeuroses,* which are flat bands connecting one muscle with another or with a bone. Dense, irregularly arranged collagenous connective tissue forms *membrane capsules* around the kidneys, heart, testes, liver, and lymph nodes. It also forms the *deep fascia*—sheets of connective tissue wrapped around muscles to hold them in place

● *Elastic Connective Tissue* Unlike collagenous connective tissue, elastic connective tissue has a pre-

dominance of freely branching elastic fibers. These fibers give the tissue a yellowish color. Fibroblasts are present only in the spaces between the fibers. Elastic connective tissue can be stretched and will snap back into shape. It is a component of the cartilages of the larynx, the walls of elastic arteries, the trachea, bronchial tubes to the lungs, and the lungs themselves. Elastic connective tissue provides stretch and strength, allowing structures to perform their functions efficiently. Yellow elastic ligaments, composed mostly of elastic fibers, comprise the ligamenta flava of the vertebrae and the true vocal cords.

● *Reticular Connective Tissue* This kind of connective tissue consists of interlacing reticular fibers. It helps to form the framework, or stroma, of many organs, including the liver, spleen, and lymph nodes. Reticular connective tissue also helps to bind together the cells of smooth muscle tissue. It is especially adapted to providing strength and support.

Cartilage

One type of connective tissue is capable of enduring considerably more stress than the tissues just discussed. **Cartilage** consists of a dense network of collagenous fibers and elastic fibers firmly embedded in a gel-like substance. The cells of mature cartilage, called *chondrocytes,* occur singly or in groups within spaces called *lacunae* in the intercellular substance. The surface of cartilage is surrounded by irregularly arranged dense connective tissue called the *perichondrium* (*chondro* = cartilage; *peri* = around). Three kinds of cartilage are recognized: hyaline cartilage, fibrocartilage, and elastic cartilage (Exhibit 3-2).

● *Hyaline Cartilage* This cartilage, also called gristle, appears as a bluish white, glossy, homogeneous mass. The collagenous fibers, although present, are not visible, and the prominent chondrocytes are found in lacunae. Hyaline cartilage is the most abundant kind of cartilage in the body. It is found at joints over the ends of long bones (where it is called *articular cartilage*) and forms the *costal cartilages* at the ventral ends of the ribs. Hyaline cartilage also helps to form the nose, larynx, trachea, bronchi, and bronchial tubes leading to the lungs. Most of the embryonic skeleton consists of hyaline cartilage. It affords flexibility and support.

● *Fibrocartilage* Chondrocytes scattered through many bundles of visible collagenous fibers are found in this type of cartilage. Fibrocartilage is found at the symphysis pubis, the point where the coxal (hip) bones fuse anteriorly at the midline. It is also found in the discs between the vertebrae. This tissue combines strength and rigidity.

● *Elastic Cartilage* In this tissue, chondrocytes are located in a threadlike network of elastic fibers. Elastic cartilage provides strength and maintains the shape of certain organs—the larynx, the external part of the ear (the pinna), and the eustachian (auditory) tubes (the internal connection between the middle ear cavity and the upper throat).

● *Growth of Cartilage* The growth of cartilage follows two basic routes. In *interstitial (endogenous) growth,* the cartilage increases rapidly in size through the cell division of existing chondrocytes and a continuous deposition of increasing amounts of intercellular matrix by the chondrocytes. The formation of new chondrocytes and their production of new intercellular matrix causes the cartilage to expand from within; thus the name interstitial growth. This growth pattern occurs as long as the cartilage is young and pliable, such as during childhood and adolescence.

In *appositional (exogenous) growth,* the growth of cartilage occurs because of the activity of the inner chondrogenic layer of the perichondrium. The deeper cells of the perichondrium, the fibroblasts, divide. Some differentiate into chondroblasts and then chondrocytes. As differentiation occurs, the chondroblasts become surrounded with intercellular matrix and become chondrocytes. As a result, the matrix is deposited on the surface of the cartilage, thus increasing its size. The new layer of cartilage is added under the perichondrium on the surface of the cartilage causing it to grow in width. Appositional growth starts later in life, and unlike interstitial growth, it continues throughout life.

Bone Tissue

The details of bone, or osseous, tissue, another kind of connective tissue, are discussed in Chapter 5 as part of the skeletal system.

Vascular Tissue

This kind of connective tissue, also known as blood, is treated in Chapter 11 as a component of the circulatory system.

Muscle Tissue and Nervous Tissue

Epithelial and connective tissue can take a variety of forms to provide a variety of body functions. They are all-purpose tissues. By contrast, the third major type

of tissue, **muscle tissue,** consists of highly modified cells that perform one basic function: contraction (see Chapter 9). The fourth major type, **nervous tissue,** is specialized to conduct electrical impulses (see Chapter 12).

Membranes

The combination of an epithelial layer and an underlying connective tissue layer constitutes an **epithelial membrane.** The principal epithelial membranes of the body are mucous membranes, serous membranes, and the cutaneous membrane or skin. Another kind of membrane, a synovial membrane, is also significant.

MUCOUS MEMBRANES

Mucous membranes line the body cavities that open to the exterior. Examples include the membranes lining the entire digestive, respiratory, excretory, and reproductive tracts. The surface tissue of a mucous membrane may vary in type—it is stratified squamous epithelium in the esophagus and simple columnar epithelium in the intestine. The epithelial layer of a mucous membrane secretes mucus, which prevents the cavities from drying out. It also traps dust in the respiratory passageways and lubricates food as it moves through the digestive tract. In addition, the epithelial layer is responsible for the secretion of digestive enzymes and the absorption of food. The connective tissue layer of a mucous membrane binds the epithelium to the underlying structures. It is referred to as the *lamina propria* and allows some flexibility of the membrane. It holds the blood vessels in place, protects underlying muscles from abrasion or puncture, provides the epithelium covering it with oxygen and nutrients, and removes wastes.

SEROUS MEMBRANES

Serous membranes line the body cavities that do not open to the exterior, and they cover the organs that lie within those cavities. They consist of thin layers of loose connective tissue covered by a layer of mesothelium. Serous membranes are in the form of double-walled sacs. The part attached to the cavity wall is called the *parietal* portion. The part that covers the organs inside these cavities is the *visceral* portion. The serous membrane lining the thoracic cavity and covering the lungs is called the pleura. The membrane lining the heart cavity and covering the heart is the pericardium (*cardio* = heart). The serous membrane lining the abdominal cavity and covering the abdominal organs and some pelvic organs is called the peritoneum.

The epithelial layer of a serous membrane secretes a lubricating fluid that allows the organs to glide easily against each other or the wall of the cavities. The connective tissue layer of the serous membrane consists of a relatively thin layer of loose connective tissue.

CUTANEOUS MEMBRANE

The **cutaneous membrane,** or skin, constitutes an organ of the integumentary system and is discussed in the next chapter.

SYNOVIAL MEMBRANES

Synovial membranes line the cavities of the joints. Like serous membranes, they line structures that do not open to the exterior. Unlike mucous and serous membranes, synovial membranes do not contain epithelium. They are composed of loose connective tissue with elastic fibers and varying amounts of fat. Synovial membranes, therefore, are not epithelial membranes. Synovial membranes secrete *synovial fluid,* which lubricates the ends of bones as they move at joints.

STUDY OUTLINE

Types of Tissues
1. A tissue is a group of similar cells and their intercellular substance specialized for a particular function.
2. Certain tissues of the body function in moving body parts, while others move food through organs.
3. Some tissues protect and support the body, and still others produce chemicals such as enzymes and hormones.
4. Depending on their function and structure, the various tissues of the body are classified into four principal types: epithelial, connective, muscular, and nervous.

Epithelial Tissue
1. Epithelium covers and lines body surfaces and forms glands.
2. Epithelium has many cells, little intercellular material, and no blood vessels. It is attached to connective tissue by a basement membrane. It can replace itself.

Covering and Lining Epithelium
1. Simple epithelium is a single layer of cells adapted for absorption or filtration.
2. Stratified epithelium has several layers of cells and is adapted for protection.
3. Epithelial cell shapes include squamous (flat), cuboidal (cubelike), columnar (rectangular), and transitional (variable).
4. Simple squamous epithelium is adapted for diffusion and filtration and is found in lungs and kidneys. Endothelium lines the heart and blood vessels. Mesothelium lines body cavities and covers internal organs.
5. Simple cuboidal epithelium is adapted for secretion and absorption in kidneys and glands.
6. Simple columnar epithelium lines the digestive tract. Specialized cells containing microvilli perform absorption. Goblet cells perform secretion. In some portions of the respiratory tract, the cells are ciliated to move foreign particles out of the body.
7. Stratified squamous epithelium is protective. It lines the upper digestive tract and forms the outer layer of skin.
8. Stratified cuboidal epithelium is found in adult sweat glands.
9. Stratified columnar epithelium protects and secretes. It is found in the male urethra and excretory ducts.
10. Transitional epithelium lines the urinary bladder and is capable of stretching.
11. Pseudostratified epithelium has only one layer but gives the appearance of many. It lines excretory and upper respiratory structures where it protects and secretes.

Glandular Epithelium
1. A gland is a single cell or a mass of epithelial cells adapted for secretion.
2. Exocrine glands (sweat, oil, and digestive glands) secrete into ducts.
3. The structural classification includes unicellular and multicellular glands.
4. The functional classification includes holocrine, merocrine, and apocrine glands.
5. Endocrine glands secrete hormones into the blood.

Connective Tissue
1. Connective tissue is the most abundant body tissue. It has few cells, an extensive matrix, and a rich blood supply.
2. General types include loose and dense connective tissues.
3. It protects, supports, and binds organs together.

Loose and Dense Connective Tissue
1. Loose connective (areolar) tissue is widely distributed. It contains three kinds of fibers (collagenous, elastic, and reticular). It also has several kinds of cells (fibroblasts, macrophages, plasma cells, mast cells, and white blood cells). Loose connective tissue forms the subcutaneous layer. It is present in mucous membranes and around blood vessels, nerves, and body organs. When the fibroblasts in loose connective tissue become infiltrated with fat, the tissue is then known as adipose tissue.
2. Dense connective (collagenous) tissue forms tendons, ligaments, and deep fasciae. All three are usually primarily composed of closely packed collagenous fibers. In tendons and ligaments collagenous fibers are arranged in parallel bundles. In fasciae, these fibers are inter-

woven at various angles with each other. A few ligaments are composed of closely packed elastic fibers.

Classification

Connective tissue may be classified in several ways. We classify them as follows:

Embryonic connective tissue
Adult connective tissue
    Connective tissue proper
        1. Loose connective (areolar) tissue
        2. Adipose tissue
        3. Dense connective (collagenous) tissue
        4. Elastic connective tissue
        5. Reticular connective tissue
    Cartilage
        1. Hyaline cartilage
        2. Fibrocartilage
        3. Elastic cartilage
    Bone (osseous) tissue
    Vascular (blood) tissue

Embryonic Connective Tissue

1. Mesenchyme forms all other connective tissue.
2. Mucous connective tissue is found only in the umbilical cord of the fetus, where it gives support.

Adult Connective Tissue

1. Adult connective tissue is connective tissue that exists in the newborn and that does not change after birth. It is subdivided into several kinds: connective tissue proper, cartilage, bone tissue, and vascular tissue.
2. Connective tissue proper has a more or less fluid intercellular material, and a typical cell is the fibroblast. Five examples of such tissues may be distinguished.
3. Loose connective (areolar) is one of the most widely distributed connective tissues in the body.
4. Adipose tissue is a form of loose connective tissue in which the fibroblasts are specialized for fat storage.

5. Dense connective (collagenous) tissue has a predominance of collagenous or white fibers arranged in bundles.
6. Elastic connective tissue has a predominance of freely branching elastic fibers that give it a yellow color.
7. Reticular connective tissue consists of interlacing reticular fibers.
8. Cartilage has a gel-like matrix of collagenous and elastic fibers that contains chondrocytes. Cartilage grows by interstitial growth (expansion from within) and appositional growth (surface growth).
9. Hyaline cartilage is found at the ends of bones, in the nose, and in respiratory structures. It is flexible, allows movement, and provides support.
10. Fibrocartilage connects the pelvic bones and the vertebrae. It provides strength.
11. Elastic cartilage maintains the shape of organs such as the external ear.

Muscle Tissue and Nervous Tissue

1. Muscle tissue performs one major function: contraction.
2. Nervous tissue is specialized to conduct electrical impulses.

Membranes

1. An epithelial membrane is an epithelial layer overlying a connective tissue layer. Examples are mucous, serous, and cutaneous.
2. Mucous membranes line cavities that open to the exterior, such as the digestive tract.
3. Serous membranes (pleura, pericardium, peritoneum) line closed cavities and cover the organs in the cavities. These membranes consist of parietal and visceral portions.
4. The cutaneous membrane is the skin.
5. Synovial membranes line joint cavities and do not contain epithelium.

REVIEW QUESTIONS

1. Define a tissue. What are the four basic kinds of human tissue?
2. What characteristics are common to all epithelium? Distinguish covering and lining epithelium from glandular epithelium.
3. Discuss the classification of epithelium based on layering and cell type.
4. For each of the following kinds of epithelium, briefly describe the microscopic appearance, location in the body, and functions: simple squamous, simple cuboidal, simple columnar, stratified squamous, stratified cuboidal, stratified columnar, transitional, and pseudostratified.
5. What is a gland? Distinguish between endocrine and exocrine glands. Describe the classification of exocrine glands according to structural complexity and function and give at least one example of each.

6. Enumerate the way in which connective tissue differs from epithelium. Distinguish between loose and dense connective tissues. How are connective tissues classified?
7. Compare embryonal connective tissue with adult connective tissue.
8. Describe the following connective tissues with regard to microscopic appearance, location in the body, and function: loose (areolar), adipose, dense (collagenous), elastic, reticular, hyaline cartilage, fibrocartilage, and elastic cartilage.
9. Distinguish between interstitial and appositional growth as it applies to cartilage.
10. Define the following kinds of membranes: mucous, serous, cutaneous, and synovial. Where is each located in the body? What are their functions?

11. Below are some descriptive statements for various tissues. For each statement, name the tissue described:

A stratified epithelium that permits distention

A single layer of flat cells concerned with filtration and absorption

Forms all other kinds of connective tissue

Specialized for fat storage

An epithelium with waterproofing qualities

Forms the framework of many organs

Produces perspiration, wax, oil, and digestive enzymes

Cartilage that shapes the external ear

Contains goblet cells and lines the intestine

Most widely distributed connective tissue

Forms tendons, ligaments, and aponeuroses

Specialized for the secretion of hormones

Provides support in the umbilical cord

Lines kidney tubules and is specialized for absorption and secretion

Permits extensibility of lung tissue

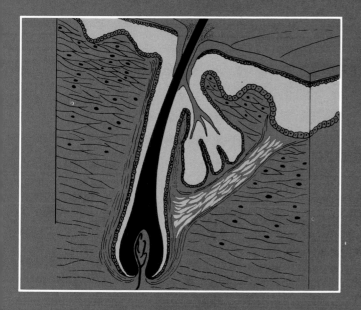

4

The Integumentary System

STUDENT OBJECTIVES

- Define the skin as the organ of the integumentary system.

- Explain how the skin is structurally divided into epidermis and dermis.

- List the structural layers of the epidermis and describe their functions.

- Explain the composition and function of the dermis.

- Contrast the structure and functions of derivatives of the epidermis such as hair, glands, and nails.

- Describe the effects of a burn.

- Classify burns into first, second, and third degrees.

- Define the "rule of nines" for estimating the extent of a burn.

- Define key medical terms associated with the integumentary system.

An aggregation of tissues that performs a specific function is an **organ.** The next higher level of organization is a **system**—a group of organs operating together to perform specialized functions. The skin and its derivatives, such as hair, nails, glands, and several specialized receptors, constitute the **integumentary system** of the body. First, let us consider the skin as an organ.

Skin

The **skin** or **cutis** is an organ because it consists of tissues structurally joined together to perform specific activities. Not just a simple thin covering that keeps the body together and give it protection, the skin is quite complex in structure and performs several vital functions. In fact, this organ is essential for survival.

The skin is the largest organ of the body. For the average adult, the skin occupies a surface area of approximately 19,344 sq cm (3,000 sq inches). It varies in thickness from 0.05 to 3 mm (0.0197–0.1182 inch) and is somewhat thicker on the dorsal and extensor surfaces than on the ventral and flexor surfaces. It covers the body and protects the underlying tissues, not only from bacterial invasion but also from drying out and from harmful light rays. Moreover, the skin helps to control body temperature, prevents excessive loss of inorganic and organic materials, receives stimuli from the environment, stores chemical compounds, excretes water and salts, and synthesizes several important compounds, including vitamin D.

STRUCTURE

Structurally, the skin consists of two principal parts (Figure 4-1). The outer, thinner portion, which is com-

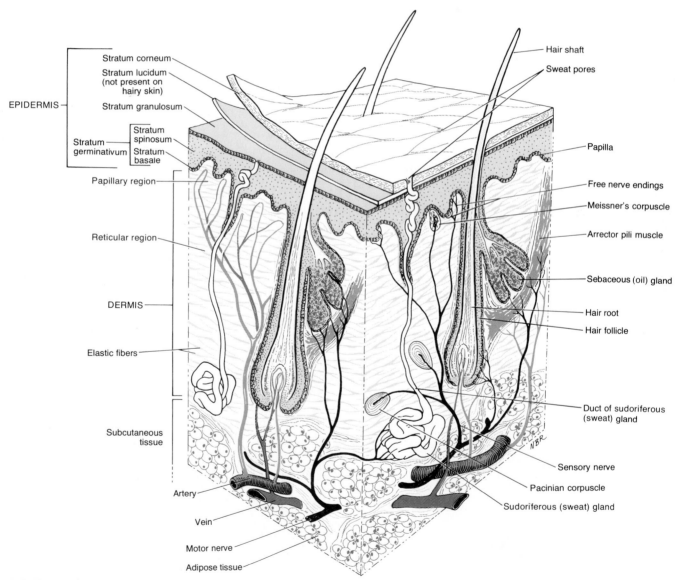

FIGURE 4-1 Structure of the skin and underlying subcutaneous layer.

posed of epithelium, is called the **epidermis.** The epidermis is cemented to the inner, thicker, connective tissue part called the **dermis.** Beneath the dermis is a *subcutaneous layer* of tissues (*sub* = under). This layer, also called the *superficial fascia,* consists of areolar and adipose tissues. Fibers from the dermis extend down into the superficial fascia and anchor the skin to the subcutaneous layer. The superficial fascia, in turn, is firmly attached to underlying tissues and organs.

EPIDERMIS

The **epidermis** is composed of stratified epithelium in four or five cell layers, depending on its location in the body (see Figure 4-1). Where exposure to friction is greatest, such as the palms and soles, the epidermis has five layers. In all other parts it has four layers. The names of the layers from the inside outward are as follows:

1. Stratum basale. This single layer of columnar cells is capable of continued cell division. As these cells multiply, they push up toward the surface. Their nuclei degenerate, and the cells die. Eventually, the cells are shed in the top layer of the epidermis.

2. Stratum spinosum. This layer of the epidermis contains eight to ten rows of polygonal (many-sided) cells that fit closely together. The surfaces of these cells may assume a prickly appearance when prepared for microscope examination (*spinosum* = prickly). The stratum basale and stratum spinosum are some-

times collectively referred to as the *stratum germinativum* to indicate that they are the layers where new cells are germinated.

3. Stratum granulosum. This third layer of the epidermis consists of two or three rows of flattened cells that contain darkly staining granules of a substance called *keratohyaline.* This compound is involved in the first step of keratin formation. **Keratin** is a waterproofing protein found in the top layer of the epidermis.

4. Stratum lucidum. This layer is quite pronounced in the thick skin of the palms and soles. It consists of three to four rows of clear, flat, dead cells that contain droplets of a translucent substance called *eleidin.* Eleidin is formed from keratohyaline and is eventually transformed to keratin.

5. Stratum corneum. This layer consists of 25 to 30 rows of flat, dead cells containing keratin. These cells are continuously shed and replaced. The stratum corneum serves as an effective barrier against light and heat waves, bacteria, and many chemicals. Be sure to examine the photomicrograph of the skin in Figure 4-2.

The color of the skin is due to a pigment called **melanin.** The amount of melanin varies the skin color from pale yellow to black. This pigment is found throughout the basale and spinosum layers and in the granulosum of all Caucasian people. In Negroes melanin is found in all epidermal layers. Melanin is synthesized in cells called **melanocytes.** They produce the pigment and pass it on to the epidermal cells. When the skin is exposed to ultraviolet radiation, both the amount and the

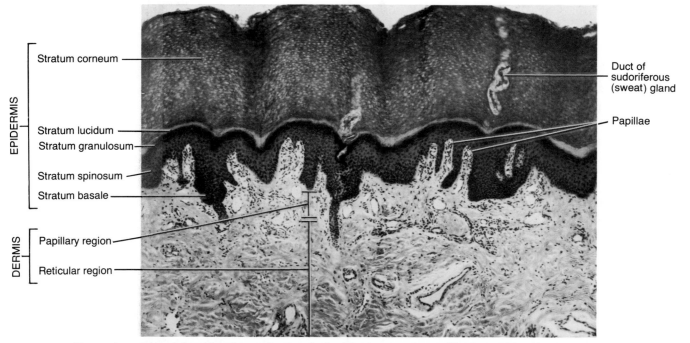

FIGURE 4-2 Photomicrograph of the skin of the palmar surface of the hand at a magnification of 65×. (Courtesy of Edward J. Reith, from *Atlas of Descriptive Histology,* by Edward J. Reith and Michael H. Ross, Harper & Row, Publishers, Inc., New York, 1970.)

darkness of melanin increase, tanning and further protecting the body against radiation. Thus melanin serves a vital protective function. In mammals, the melanocyte-stimulating hormone (MSH) produced by the anterior pituitary gland causes increased melanin synthesis. The exact role of MSH in humans is not clear. Another pigment called **carotene** is found in the corneum and fatty areas of the dermis in Oriental people. Together carotene and melanin account for the yellowish hue of their skin. The pink color of Caucasian skin is due to blood vessels in the dermis. The redness of the vessels is not heavily masked by pigment. The epidermis has no blood vessels, a characteristic of all epithelia.

DERMIS

The second principal part of the skin, the **dermis,** is composed of connective tissue containing collagenous and elastic fibers. (See Figure 4-1.) It is about 0.5 to 2.5 mm (0.0197–0.1085 inch) thick. Numerous blood vessels, nerves, glands, and hair follicles are also embedded in the dermis. The upper region of the dermis, about one-fifth of the thickness of the total layer, is named the *papillary region*—its surface area is greatly increased by small, fingerlike projections called **papillae.** These structures project into the epidermis and contain loops of capillaries. Some papillae contain *Meissner's corpuscles*—nerve endings sensitive to light touch. The dermis also contains nerve endings called *Pacinian corpuscles,* which are sensitive to deep pressure.

The ridges marking the external surface of the epidermis are caused by the size and arrangement of the papillae in the dermis. Some ridges cross at various angles and can be seen on the back of your hand. Other ridges on your palms and fingertips prevent slipping. The ridge patterns on the fingertips and thumbs are different in each individual.

The remaining portion of the dermis is called the *reticular region.* It is a dense, irregular, collagenous connective tissue. The irregular arrangement of fibers permits flexibility and strength in all directions. This area of the dermis contains many blood vessels and also collagenous and elastic fibers. The spaces between the interlacing fibers are occupied by adipose tissue and sweat glands. The reticular zone is attached to the organs beneath it, such as bone and muscle, by the subcutaneous layer.

Epidermal Derivatives

Organs derived from the skin—hair, glands, nails—perform functions that are necessary and sometimes vital. Hair and nails protect the body. The sweat glands help regulate body temperature.

HAIR

Growths of the epidermis variously distributed over the body are **hairs** or **pili.** The primary function of hair is protection. Though the protection is limited, hair guards the scalp from injury and the sun's rays. Eyebrows and eyelashes protect the eyes from foreign particles. Hair in the nostrils and external ear canal protects these structures from insects and dust.

Development and Distribution

Around the beginning of the third month of fetal life, the epidermis develops downgrowths into the dermis called hair follicles. By the fifth or sixth month, the follicles produce very delicate hairs called *lanugo* (*lana* = wool) that covers the fetus. The lanugo is shed prior to birth, except in the regions of the eyebrows, eyelids, and scalp. Here they persist and become stronger. Several months after birth, these hairs are shed and are replaced by still even coarser ones while the remainder of the body develops a new hair growth called the *vellus* (fleece). At puberty, coarse hairs develop in the axilla and pubic regions of both sexes and on the face and to a lesser extent on other parts of the body in the male. The coarse hairs that develop at puberty, plus the hairs of the scalp and eyebrows, are referred to as *terminal hairs.* The life span of a hair is from two to five years in most parts of the body. Eyelashes are shed between three to five months. If hairs are not replaced when they are shed, baldness results. Hairs are distributed on nearly all parts of the body. They are absent from the palms of the hands, soles of the feet, dorsal surfaces of the distal phalanges, lips, nipples, clitoris, glans penis, inner surface of the prepuce, inner surfaces of the labia majora and minora, and outer surfaces of the labia minora. Straight hairs are oval or cylindrical in cross section, whereas curly hairs are flattened. Straight hairs are stronger than curly ones.

Structure

Each hair consists of a shaft and a root (Figure 4-3). The **shaft** is the superficial portion, most of which projects above the surface of the skin. The shaft consists of three principal parts. The inner *medulla* is composed of rows of polyhedral cells containing granules of eleidin and air spaces. The medulla is poorly developed or not developed at all in most hairs. The second principal part of the shaft is the middle *cortex.* It forms the major part of the shaft and consists of elongated cells which contain pigment granules in dark hair, but only air in white hair. The outermost *cuticle* consists of a single layer of thin, flat, scalelike cells that are arranged like shingles on the side of a house.

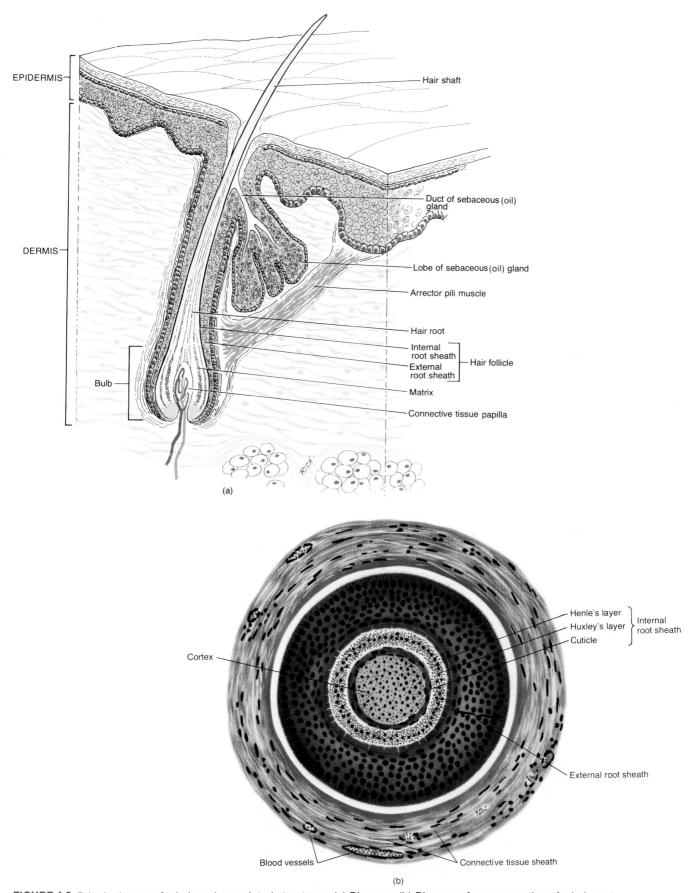

FIGURE 4-3 Principal parts of a hair and associated structures. (a) Diagram. (b) Diagram of a cross section of a hair root.

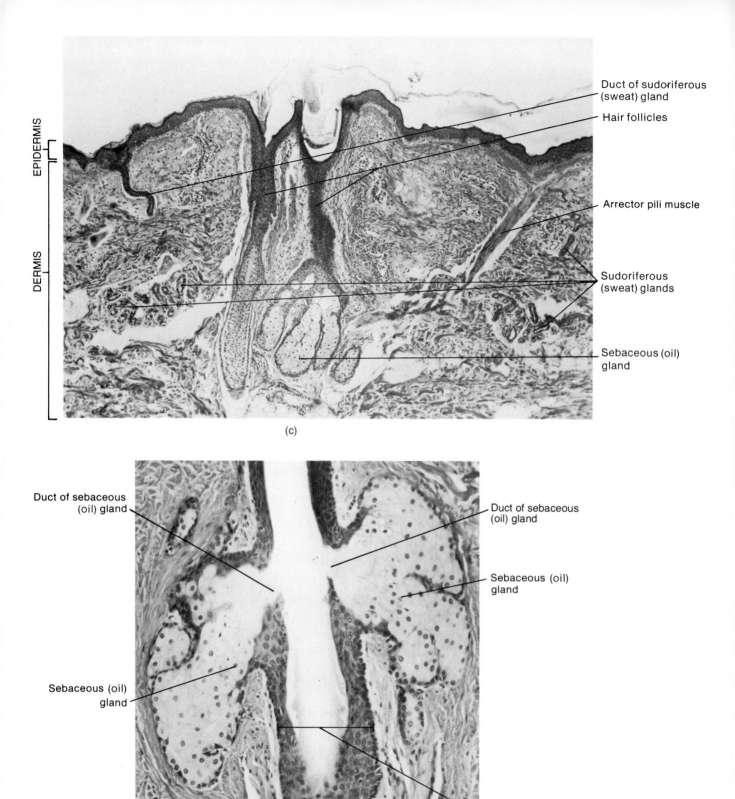

(c)

(d)

FIGURE 4-3 *(Continued)* Principal parts of a hair and associated structures. (c) Photomicrograph of a section from the skin of the face at a magnification of 45×. (d) Photomicrograph of two sebaceous glands opening into a hair follicle at a magnification of 160×. (Courtesy of Edward J. Reith, from *Atlas of Descriptive Histology,* by Edward J. Reith and Michael H. Ross, Harper & Row, Publishers, Inc., New York, 1970.)

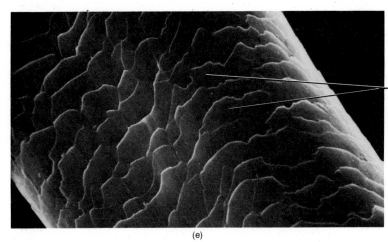

Cuticular
scales

(e)

FIGURE 4-3 *(Continued)* Principal parts of a hair and associated structures. (e) Scanning electron micrograph of the surface of a hair shaft showing the shinglelike cuticular scales at a magnification of 1000×. (Courtesy of Fisher Scientific Company and S.T.E.M. Laboratories, Inc., Copyright 1975.)

But the free edges of the cuticle cells point upward rather than downward like shingles (Figure 4-3e). The **root** is the portion below the surface that penetrates deep into the dermis and, like the shaft, also contains a medulla, cortex, and curticle (Figure 4-3b). Surrounding the root is the *hair follicle,* which is made up of an external zone of epithelium (the external root sheath) and an internal zone of epithelium (the internal root sheath). The *external root sheath* is a downward continuation of the epidermis. Near the surface it contains all the epidermal layers. As it descends, it does not exhibit the superficial epidermal layers. At the bottom of the hair follicle, the external root sheath contains only the stratum basale. The *internal root sheath* is formed from proliferating cells of the matrix (described shortly) and takes the form of a cellular tubular sheath that separates the hair from the external root sheath. It extends only part way up the follicle. The internal root sheath consists of (1) an inner layer of *cuticle,* which is a single layer of flattened cells with atrophied nuclei, (2) a middle *Huxley's layer,* which is one or two layers of flattened nucleated cells, and (3) an outer *Henle's layer,* which is a single layer of cubical cells with flattened nuclei.

The base of each follicle is enlarged into an onion-shaped structure, the *bulb.* This structure contains an indentation, the *connective tissue papilla,* filled with loose connective tissue. The papilla contains many blood vessels and provides nourishment for the growing hair. The base of the bulb also contains a region of cells called the *matrix,* a germinal layer. The cells of the matrix produce new hairs by cell division when older hairs are shed. This replacement occurs within the same follicle. Hair grows about 1 mm (0.039 inch) every three days. Hair loss in an adult is about 70 to 100 hairs a day. But the rate of growth may be altered by illness.

Sebaceous glands and bundles of smooth muscle are also associated with hair. These smooth muscles, called *arrector pili* extend from the dermis of the skin

to the side of the hair follicle. In its normal position hair is arranged at an angle to the surface of the skin. The arrector pili muscles contract under stresses of fright and cold and pull the hairs into a vertical position. This contraction results in ''goosebumps'' or ''gooseflesh'' because the skin around the shaft forms slight elevations.

GLANDS

Two kinds of glands associated with the skin are sebaceous glands and sweat glands.

Sebaceous (Oil) Glands

Sebaceous (oil) glands, with few exceptions, are connected to hair follicles. (See Figure 4-3). They are simple branched acinar glands connected directly to the follicle by a short duct. Absent in the palms and soles, they vary in size and shape in other regions of the body. For example, they are small in most areas of the trunk and extremities but large in the skin of the breasts, face, neck, and upper chest. The sebaceous glands secrete an oily substance called *sebum,* a mixture of fats, cholesterol, proteins, and inorganic salts. These glands keep the hair from drying and becoming brittle and also form a protective film that prevents excessive evaporation of water from the skin. The sebum keeps the skin soft and pliable, too. When sebaceous glands of the face become enlarged, because of accumulated sebum, blackheads develop. Since sebum is nutritive to certain bacteria, pimples or boils often result. The color of blackheads is due to oxidized oil, not dirt.

Sudoriferous (Sweat) Glands

Sudoriferous (sweat) glands (*sudor* = sweat; *ferre* = to bear) are distributed throughout the skin except on the nail beds of the fingers and toes, margins of the lips, eardrums and tips of the penis and clitoris. In contrast to sebaceous glands, sudoriferous glands are most numerous in the skin of the palms and the soles. They are also found in abundance in the armpits and forehead. Each gland consists of a coiled end embedded in the subcutaneous tissue and a single tube that projects upward through the dermis and epidermis. This tube, the excretory duct, terminates in a pore at the surface of the epidermis. (See Figure 4-1). The base of each sudoriferous gland is surrounded by a network of small blood vessels. In the axillary region, sudoriferous glands are of the simple branched tubular type. Elsewhere they are simple coiled tubular glands.

Perspiration, or *sweat,* is the substance produced by sudoriferous glands. It is a mixture of water, salts (mostly NaCl), urea, uric acid, amino acids, ammonia, sugar, lactic acid, and ascorbic acid. Its principal function is to help regulate body temperature. It also helps to eliminate wastes.

NAILS

Modified horny cells of the epidermis are referred to as **nails.** The cells form a clear, solid covering over the dorsal surfaces of the terminal portions of the fingers and toes. Each nail (Figure 4-4) consists of a *nail body,* a *free edge,* and the *nail root.* Most of the nail body is pink because of the underlying vascular tissue. The whitish semilunar area at the proximal end of the body is called the *lunula.* It appears whitish because the vascular tissue underneath does not show through.

The *eponychium* (''cuticle'') is a narrow band of stratum corneum on the proximal border of the nail extending from the margins of the nail wall (lateral borders). The thickened stratum corneum below the free edge of the nail is the *hyponychium.*

The epithelium of the proximal part of the nail bed is known as the *nail matrix.* Its function is to bring about the growth of nails. Essentially, growth occurs by the transformation of superficial cells of the matrix into nail cells. In the process, the outer, harder layer is pushed forward over the stratum germinativum. The average growth in the length of fingernails is about 1 mm (0.039 inch) a week. The growth rate is somewhat slower in toenails.

Applications to Health

BURNS

Tissues may be damaged by thermal (heat), electrical, radioactive, or chemical agents. These agents can destroy the proteins in the exposed cells and cause cell injury or death. Such damage results in a **burn.** The tissues that are directly or indirectly in contact with the environment, such as the skin or the linings of the respiratory and digestive tracts, are affected. Generally, however, the systemic effects of a burn are a greater threat to life than the local effects. *Systemic* effects occur throughout the body. *Local* effects occur in one area of the body. The systemic effects of burn include:

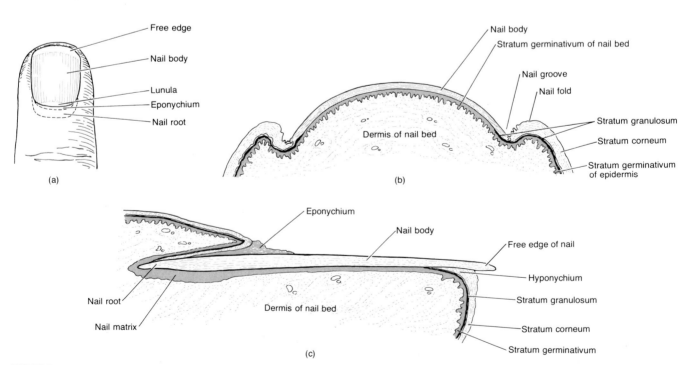

FIGURE 4-4 Structure of nails (a) Fingernail viewed from above. (b) Cross section of a fingernail and the nail bed. (c) Longitudinal section of a fingernail.

1. A large loss of water, plasma, and plasma proteins, which causes shock
2. Bacterial infection
3. Reduced circulation of blood
4. A decrease in urine production

Classification

Burns are classified into three types: first degree, second degree, and third degree. In *first-degree* burns, the damage is restricted to the epidermal layers of the skin. Symptoms are limited to local effects such as redness, tenderness, pain, and edema—the cardinal signs of inflammation. In *second-degree* burns, the epidermal and portions of the dermal layers of the skin are damaged, but rapid regeneration of epithelium is still possible. Blisters containing elements of blood and lymph form on the skin surface or beneath the epidermis. Blisters beneath or within the epidermis are called *bullae*. In *third-degree* burns, both the epidermis and the dermis are destroyed. The skin surface may be charred or white or both. It is lifeless and insensitive to touch. The regeneration of epithelium

originates from the wound edges. Regeneration is slow, and much granulation tissue forms before being covered by epithelium. Even if skin grafting is quickly begun, these wounds frequently contract and produce disfiguring or disabling scars.

"Rule of Nines"

A fairly accurate method for estimating the extent of a burn is to apply the *"rule of nines"* (Figure 4-5):

1. If the anterior and posterior surfaces of the head and neck are affected, the burn covers 9 percent of the body surface.
2. The anterior and posterior surfaces of each shoulder, arm, forearm, and hand also comprise 9 percent of the body surface.
3. The anterior and posterior surfaces of the trunk, including the buttocks, constitute 36 percent.
4. The anterior and posterior surfaces of each foot, leg, and thigh as far up as the buttocks total 18 percent.
5. The perineum consists of 1 percent. The perineum includes the anal and urogenital regions.

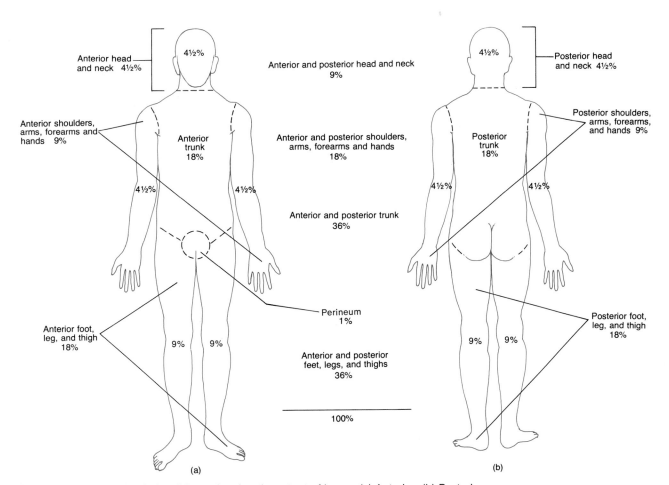

FIGURE 4-5 The "rule of nines" for estimating the extent of burns. (a) Anterior. (b) Posterior.

Treatment

A severely burned individual should be moved as quickly as possible to a hospital. Treatment may then include:

1. Cleansing the burn wounds thoroughly

2. Removing all necrotic (dead) tissue so antibacterial agents can directly contact the wound surface and thereby prevent infection
3. Replacing lost body fluids
4. Covering wounds with grafts as soon as possible.

KEY MEDICAL TERMS ASSOCIATED WITH THE INTEGUMENTARY SYSTEM

Albinism (*alb* = white; *ism* = condition) Congenital (existing at birth) absence of pigment from the skin, hair, and parts of the eye.

Anhidrosis (*an* = without; *hidr* = sweating; *osis* = condition) Inability to sweat.

Callus An area of hardened and thickened skin that is usually seen in palms and soles and is due to pressure and friction.

Carbuncle A hard, round, deep, painful inflammation of the subcutaneous tissue that causes necrosis (deadness) and pus formation (abscess).

Carcinogenesis (*carc* = cancer; *gen* = origin) Production of cancer.

Comedo A collection of sebaceous material and dead cells in the hair follicle and excretory duct of the sebaceous gland. Usually found over the face, chest, and back, and more commonly during adolescence. Also called *blackhead* or *whitehead*.

Cyst (*cyst* = sac containing fluid) A sac with a distinct connective tissue wall, containing a fluid or other material.

Decubitus ulcer A bedsore. An ulcer formed due to continual pressure over the skin.

Dermatome (*derm* = skin) An instrument for excising areas of skin to be used for grafting.

Epidermophytosis (*epi* = upon; *derm* = skin; *phyto* = pertaining to plants; *osis* = condition) Any fungus infection of the skin producing scaliness with itching. Called athlete's foot when it affects the feet.

Erythema (*eryth* = red) Congestive or exudative redness of the skin caused by engorgement of capillaries in lower layers of the skin. It occurs with any skin injury, infection, or inflammation.

Furuncle A boil; an abscess resulting from infection of a hair follicle.

Hypodermic (*hypo* = under; *derm* = skin) The area beneath the skin.

Impetigo An inflammatory skin disease.

Intradermal (*intra* = within) Within the skin. Also called intracutaneous.

Maceration Softening of a solid by soaking; wasting away.

Melanoma (*melano* = dark-colored; *oma* = tumor) A cancerous tumor consisting of melanocytes that produce skin pigment.

Nevi Round, pigmented, flat, or raised skin areas that may be present at birth or develop later. Varying in color from yellow-brown to black. Also called *moles* or *birthmarks*.

Nodule A large cluster of cells raised above the skin but extending deep into the tissues.

Papilloma Tumors, such as warts, in the skin or the lining of internal organs.

Papule A small, round skin elevation varying in size from a pinpoint to that of a split pea. One example is a pimple.

Pathogenesis (*path* = disease) Origination and development of a disease.

Polyp A tumor on a stem found especially on mucous membranes.

Pustule A small, round elevation of the skin containing pus.

Subcutaneous (*sub* = under; *cutis* = skin) Beneath the skin.

Topical Pertaining to a definite area; local.

Wart A common, contagious, noncancerous epithelial tumor caused by a virus.

STUDY OUTLINE

Skin
1. The skin and its derivatives of hair, glands, and nails constitute the integumentary system.
2. The skin is the largest body organ. It performs the functions of protection, maintaining body temperature, picking up stimuli, and excretion.
3. The principal parts of the skin are the outer epidermis and inner dermis. The dermis overlies the subcutaneous layer.
4. The epidermal layers, from the inside outward, are the stratum basale, spinosum, granulosum, lucidum, and cor-

neum. The basale and spinosum undergo continuous cell division and produce all other layers.
5. The dermis consists of a papillary region and a reticular region. The papillary region is connective tissue containing blood vessels, nerves, oil glands, hair follicles, and papillae. The reticular region is connective tissue containing fat and sweat glands.

Epidermal Derivatives
1. The epidermal derivatives of the skin are hair, sebaceous glands, sudoriferous glands, and nails.

2. Hairs consist of a shaft above the surface, a root anchored in the dermis, and a hair follicle.
3. Sebaceous glands are connected to hair follicles. They secrete sebum, which moistens hair and waterproofs the skin.
4. Sudoriferous glands produce perspiration, which carries wastes to the surface and assists temperature regulation.
5. Nails are modified epidermal cells. The principal parts of a nail are the body, free edge, root, eponychium, and matrix.

Applications to Health

1. Tissue damage that destroys protein is called a burn.
2. Depending on the extent of damage, burns are classified as first-degree, second-degree, and third-degree.
3. One method employed for determining the extent of a burn is to apply the ''rule-of-nines.''
4. Burn treatment may include cleansing the wound, removing dead tissue, replacing lost body fluids, and covering wounds with grafts.

REVIEW QUESTIONS

1. Define an organ. In what respect is the skin an organ? What is the integumentary system?
2. List the principal functions of the skin.
3. Compare the structures of epidermis and dermis. What is the subcutaneous layer?
4. List and describe the epidermal layers from the inside outward. What is the importance of each layer?
5. How is the dermis adapted to receive stimuli for touch, pressure, or pain?
6. Describe the structure of a hair. How are hairs moistened? What produces ''goosebumps'' or ''gooseflesh''?
7. Contrast the locations and functions of sebaceous glands and sudoriferous glands. What are the name and chemical components of the secretions of each?
8. From what layer of the skin do nails form? Describe the principal parts of a nail.
9. Define a burn. Classify burns according to degree. How is the ''rule of nines'' used? What steps may be employed in treating burns?
10. Refer to the glossary of key medical terms associated with the integumentary system. Be sure you can define each term.

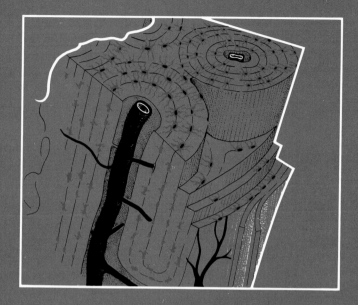

5

Osseous
Tissue

STUDENT OBJECTIVES

- Describe the components of the skeletal system.

- Describe the functions of the skeletal system.

- Describe the gross features of a long bone.

- Describe the histological features of dense bone tissue.

- Compare the histological characteristics of spongy and dense bone.

- Define ossification.

- Contrast the steps involved in intramembranous and endochondral ossification.

- Identify the parts and growth pattern of the epiphyseal plate.

- Interpret roentgenograms of normal ossification.

- Describe the processes of bone construction and destruction involved in bone replacement.

- Describe the conditions necessary for normal bone growth.

- Define rickets and osteomalacia as vitamin deficiency disorders.

- Contrast the causes and clinical symptoms associated with osteoporosis, Pagets disease, and osteomyelitis.

- Define a fracture and list 13 kinds of fractures.

- Describe the sequence of events involved in fracture repair.

- Define key medical terms associated with the skeletal system.

Without the skeletal system we would be unable to perform movements, such as walking or grasping. The slightest jar to the head or chest could damage the brain or heart. It would even be impossible to chew food. The framework of bones that protects our organs and allows us to move is called the **skeletal system.** Besides bone, the skeletal system consist of cartilage in the nose, larynx, outer ear, and at bone attachments. The points where bones attach to each other are called *joints* or *articulations*.

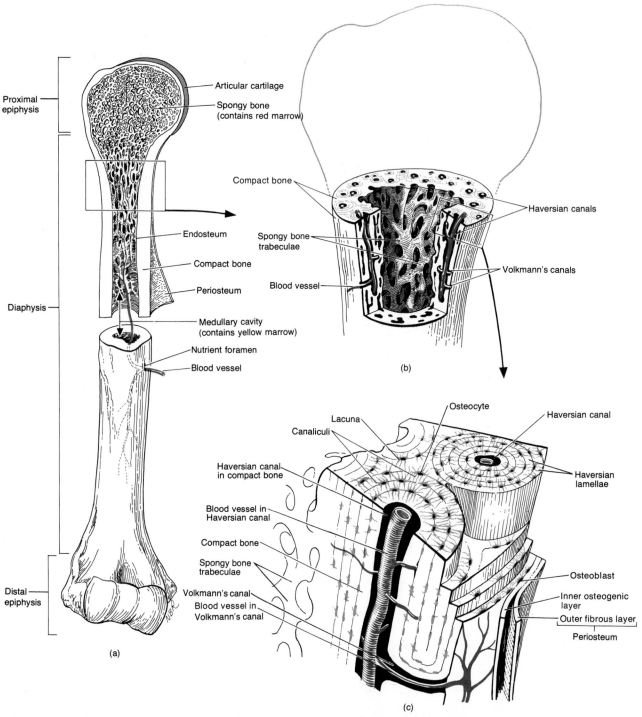

FIGURE 5-1 Osseous tissue. (a) Macroscopic appearance of a long bone that has been partially sectioned lengthwise. (b) Histological structure of bone. (c) Enlarged aspect of Haversian systems in compact bone.

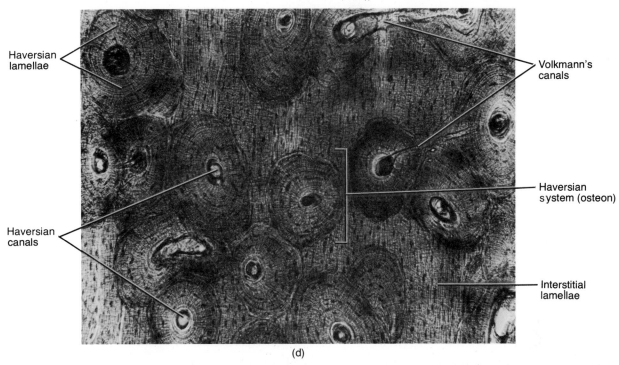

Haversian lamellae

Volkmann's canals

Haversian canals

Haversian system (osteon)

Interstitial lamellae

(d)

FIGURE 5-1 (Continued) Osseous tissue. (d) Photomicrograph of several Haversian systems with interstitial lamellae between systems at a magnification of 150×. If you look carefully, you can see the canaliculi as fine lines running radially, like the spokes of a wheel, and in circles within the Haversian systems. (Courtesy of Carolina Biological Supply Company.)

Functions

The skeletal system performs several basic functions. First, it *supports* the soft tissues of the body so that the form of the body and an erect posture can be maintained. Second, the system *protects* delicate structures—the brain, the spinal cord, the lungs, the heart, the major blood vessels in the thoracic cavity. Third, the bones serve as *levers* to which the muscles of the body are attached. When the muscles contract, the bones acting as levers produce *movement*. Fourth, the bones serve as *storage areas* for mineral salts—especially calcium and phosphorus. A fifth major feature of the skeletal system is *blood-cell production,* which occurs in the red marrow of the bones. This process is referred to as *hematopoiesis.*

Histology

Structurally, the skeletal system consists of two types of connective tissue: cartilage and bone. In Chapter 3, we discussed the microscopic structure of cartilage. Here our attention will be directed to discussing the microscopic structure of bone tissue.

Like other connective tissues, **bone,** or **osseous tissue,** contains a great deal of intercellular substance surrounding widely separated cells. Unlike other connective tissues, the intercellular substance of bone contains abundant mineral salts, primarily calcium

phosphate and calcium carbonate. These salts are responsible for the hardness of bone, which is thus said to be ossified. Embedded in the intercellular substance are collagenous fibers that reinforce the tissue.

The microscopic structure of bone may be analyzed by considering the anatomy of a long bone such as the humerus (arm bone). As shown in Figure 5-1a, a typical long bone consists of the following parts:

1. Diaphysis, which is the shaft or long, main portion

2. Epiphyses, which are the extremities or ends of the bone

3. Articular cartilage, a thin layer of hyaline cartilage covering the epiphysis where the bone forms a joint with another bone

4. Periosteum, a dense, white, fibrous membrane covering the remaining surface of the bone. The periosteum (*peri* = around; *osteo* = bone) consists of two layers. The outer, *fibrous layer* is composed of connective tissue containing blood vessels, lymphatic vessels, and nerves that pass into the bone. The inner *osteogenic layer* contains elastic fibers, blood vessels, and **osteoblasts**—cells responsible for forming new bone during growth and repair (Figure 5-1c). The word *blast* means a germ or bud. It denotes an immature cell or tissue that later develops into a specialized form. The periosteum is essential for bone growth, repair, and nutrition. It also serves as a point of attach-

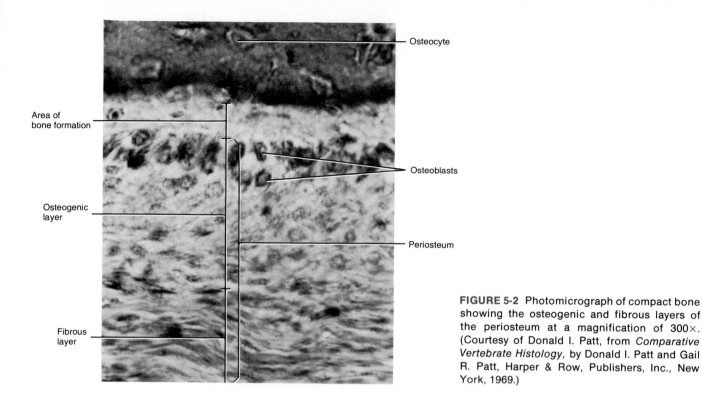

Osteocyte

Area of
bone formation

Osteoblasts

Osteogenic
layer

Periosteum

Fibrous
layer

FIGURE 5-2 Photomicrograph of compact bone showing the osteogenic and fibrous layers of the periosteum at a magnification of 300×. (Courtesy of Donald I. Patt, from *Comparative Vertebrate Histology*, by Donald I. Patt and Gail R. Patt, Harper & Row, Publishers, Inc., New York, 1969.)

ment for ligaments and tendons. A photomicrograph of the periosteum is shown in Figure 5-2.

5. Medullary (or **marrow**) **cavity,** which is the space within the diaphysis that contains the fatty *yellow marrow* in adults

6. Endosteum, a layer of osteoblasts that lines the medullary cavity and contains scattered **osteoclasts**—cells that may assume a role in the removal of bone.

Bone is not a solid, homogeneous substance. In fact, all bone is porous. The pores contain living cells and blood vessels that supply the cells with nutrients. The pores also make bones lighter. Depending on the degree of porosity, the regions of a bone may be categorized as spongy or compact (Figure 5-1b). **Spongy,** or **cancellous,** bone tissue contains many large spaces filled with marrow. It makes up most of the bone tissue of short, flat, and irregularly shaped bones and most of the epiphyses of long bones. Spongy bone tissue also provides a storage area for marrow. **Compact,** or **dense,** bone tissue, by contrast, contains few spaces. It is deposited in a thin layer over the spongy bone tissue. The layer of compact bone is thicker in the diaphyses than the epiphyses. Compact bone tissue provides protection and support and helps the long bones resist the stress of weight placed on them (Figure 5-3).

We can compare the differences between spongy and compact bone tissue by looking at the highly magnified, transverse sections in Figure 5-1. One main difference is that adult compact bone has a concentric-ring structure, whereas spongy bone does not. Blood vessels and nerves from the periosteum penetrate the compact bone through *Volkmann's canals* (Figure 5-1c). The blood vessels of these canals connect with

blood vessels and nerves of the medullary cavity and those of the *Haversian canals.* The Haversian canals run longitudinally through the bone. Around them are *lamellae*—concentric rings of hard, calcified, intercellular substance. Between the lamellae are small spaces called *lacunae,* where osteocytes are found. **Osteocytes** are mature osteoblasts that have lost their ability to produce new bone tissue. Radiating in all directions from the lacunae are minute canals called *canaliculi,* which connect with other lacunae and, eventually, with the Haversian canals. Thus an intricate network is formed throughout the bone. This branching network of canaliculi provides numerous routes for blood vessels so that nutrients can reach the osteocytes and wastes can be removed. Each Haversian canal, with its surrounding lamellae, lacunae, osteocytes, and canaliculi, is called a **Haversian system** or **osteon.** Haversian systems are characteristic of adult bone. The areas between Haversian systems contain *interstitial lamellae.* These also possess lacunae with osteocytes and canaliculi, but their lamellae are usually not connected to the Haversian systems.

In contrast to compact bone, spongy bone does not contain true Haversian systems. It consist of an irregular latticework of thin plates of bone called *trabeculae.* The spaces between the trabeculae of some bones are filled with *red marrow.* Within the trabeculae lie the small spaces called lacunae, which contain the osteocytes. Blood vessels from the periosteum penetrate the spongy bone, and the osteocytes in the trabeculae are nourished directly from the blood circulating through the marrow cavities. The cells of red marrow are responsible for producing new blood cells.

Most people think of bone as a very hard, white ma-

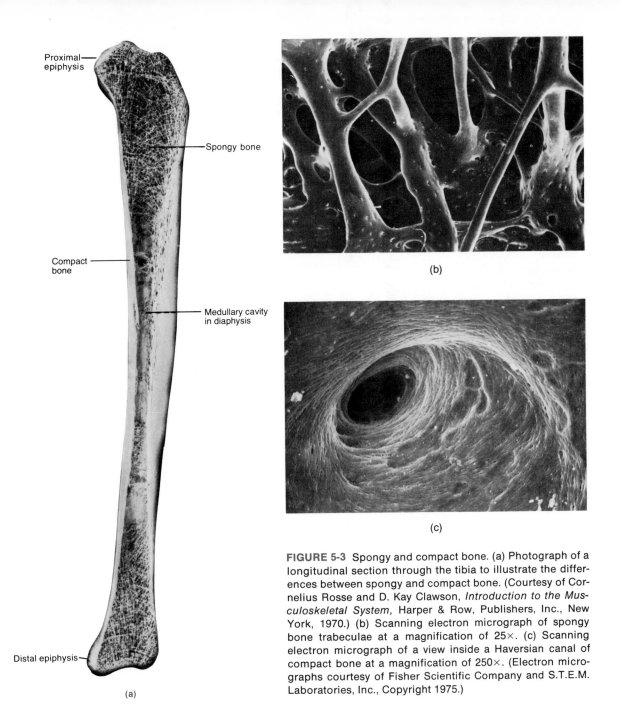

Proximal epiphysis

Spongy bone

Compact bone

Medullary cavity in diaphysis

Distal epiphysis

(a)

(b)

(c)

FIGURE 5-3 Spongy and compact bone. (a) Photograph of a longitudinal section through the tibia to illustrate the differences between spongy and compact bone. (Courtesy of Cornelius Rosse and D. Kay Clawson, *Introduction to the Musculoskeletal System,* Harper & Row, Publishers, Inc., New York, 1970.) (b) Scanning electron micrograph of spongy bone trabeculae at a magnification of 25×. (c) Scanning electron micrograph of a view inside a Haversian canal of compact bone at a magnification of 250×. (Electron micrographs courtesy of Fisher Scientific Company and S.T.E.M. Laboratories, Inc., Copyright 1975.)

terial. Yet the bones of an infant are not hard at all. It is common knowledge that it is dangerous to drop an infant, especially on its head. Its bones are "soft," and the fall may change the shape of its head or damage its brain. Moreover, most of us know that a child's bones are generally more pliable than those of an adult. The final shape and hardness of adult bones require many years to develop and depend on a complex series of chemical changes. Let us now see how bones are formed and how they grow.

Ossification

The process by which bone forms in the body is called **ossification** or **osteogenesis.** The "skeleton" of a human

embryo is composed of fibrous membranes and hyaline cartilage. Both are shaped like bones and provide the medium for ossification (Figure 5-4). Ossification begins around the sixth week of embryonic life and continues until adulthood. Two kinds of bone formation are recognized. The first is called intramembranous ossification. This term refers to the formation of bone directly on or within the fibrous membranes (*intra* = within; *membranous* = membrane). The second kind, endochondral (intracartilaginous) ossification, refers to the formation of bone in cartilage (*endo* = within; *chondro* = cartilage). These two kinds of ossification do *not* lead to differences in the structure of mature bones. They simply indicate different methods of bone formation.

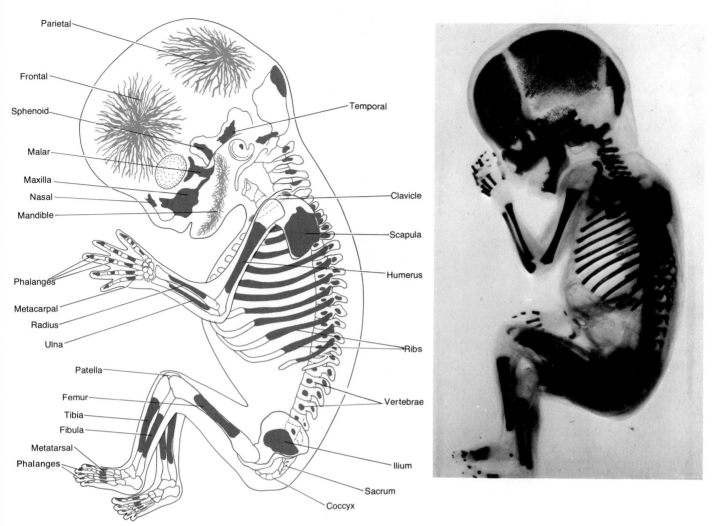

Parietal
Frontal
Sphenoid
Malar
Maxilla
Nasal
Mandible
Phalanges
Metacarpal
Radius
Ulna
Patella
Femur
Tibia
Fibula
Metatarsal
Phalanges
Temporal
Clavicle
Scapula
Humerus
Ribs
Vertebrae
Ilium
Sacrum
Coccyx

FIGURE 5-4 Diagrammatic representation of the ten-week-old human embryo shown in the photograph. Colored bones are those that develop by endochondral ossification. Gray bones are those that develop by intramembranous ossification. (Photograph courtesy of Carolina Biological Supply Company.)

The first stage in the development of bone is the migration of embryonic connective tissue cells (mesenchymal cells) into the area where bone formation is about to begin. Soon these cells increase in number and size. In some skeletal structures they become chondroblasts; in others some become osteoblasts. The chondroblasts will be responsible for cartilage formation. The osteoblasts will form bone tissue by intramembranous or endochondral ossification.

INTRAMEMBRANOUS OSSIFICATION

Of the two types of bone formation, the simpler and more direct is **intramembranous ossification.** The flat bones of the roof of the skull, mandible (lower jawbone), and probably part of the clavicles are formed in this way. The essentials of this process are as follows.

Osteoblasts formed from mesenchymal cells cluster in the fibrous membrane. The site of such a cluster is called a *center of ossification*. The osteoblasts then secrete intercellular substances. These substances are partly composed of collagenous fibers that form a framework, or matrix, in which calcium salts are quickly deposited. The deposition of calcium salts is called *calcification*. When a cluster of osteoblasts is completely surrounded by a calcified matrix, it is called a *trabecula*. As trabeculae form in nearby ossification centers, they fuse into the open latticework characteristic of spongy bone. With the formation of successive layers of bone, some osteoblasts become entrapped in the minute spaces called lacunae. The entrapped osteoblasts lose their ability to form bone and are called osteocytes. The spaces between the trabeculae fill with red marrow. The original connective tissue that surrounds the growing mass of bone then becomes the periosteum. The ossified area has now become true spongy bone. Eventually, the surface layers of the spongy bone will be reconstructed into compact bone. Much of this newly formed bone will be destroyed and reformed so the bone may reach its final adult size and shape.

ENDOCHONDRAL OSSIFICATION

The replacement of cartilage by bone is called **endochondral ossification.** Most bones of the body, includ-ing the skull, are formed in this way. Since this type of ossification is best observed in a long bone, we will investigate the tibia, or shin bone (Figure 5-5).

Early in embryonic life, a cartilage model or tem-plate of the future bone is laid down. This model is covered by a membrane called the *perichondrium.* Midway along the shaft of this model a blood vessel penetrates the perichondrium, stimulating the cells in

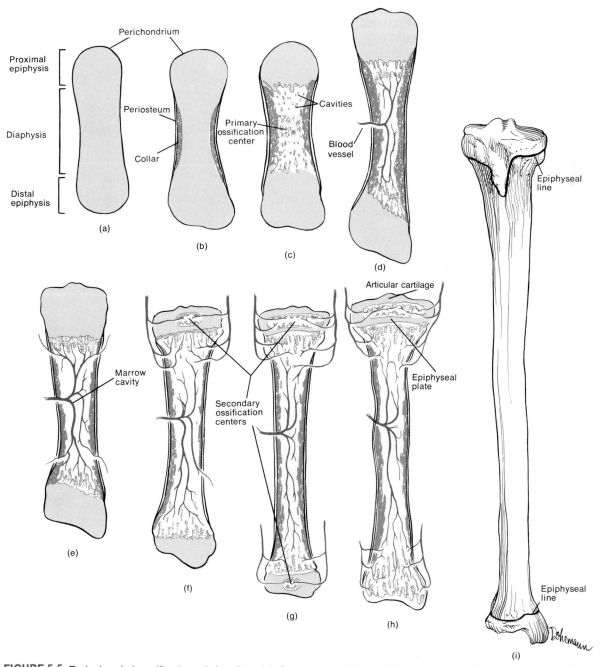

FIGURE 5-5 Endochondral ossification of the tibia. (a) Cartilage model. (b) Collar formation. (c) Development of primary ossification center. (d) Entrance of blood vessels. (e) Marrow-cavity formation. (f) Thickening and lengthening of the collar. (g) Formation of secondary ossification centers. (h) Remains of cartilage as the articular cartilage and epiphyseal plate. (i) Formation of the epiphyseal line.

the internal layer of the perichondrium to enlarge and become osteoblasts. The cells begin to form a collar of spongy bone around the middle of the diaphysis of the cartilage model. Once the perichondrium starts to form bone, it is called the periosteum. Simultaneously with the appearance of the bone collar and the penetration of blood vessels, changes occur in the cartilage in the center of the diaphysis. In this area, the *primary ossification center,* cartilage cells hypertrophy (increase in size)—probably because they accumulate glycogen for energy and produce enzymes to catalyze future chemical reactions. When the hypertrophied cells burst, there is a change in extracellular pH to a more alkaline pH causing the intercellular substance

to become *calcified*—that is, minerals are deposited within it. Once the cartilage becomes calcified, nutritive materials required by the cartilage cells can no longer diffuse through the intercellular substance; this may cause the cartilage cells to die. Then the intercellular substance begins to degenerate, leaving large cavities in the cartilage model. The blood vessels grow along the spaces where cartilage cells were previously located and enlarge the cavities further. Gradually, these spaces in the middle of the shaft join with each other, and the marrow cavity is formed.

As these developmental changes are occurring, the osteoblasts of the periosteum deposit successive layers of bone on the outer surface so that the collar

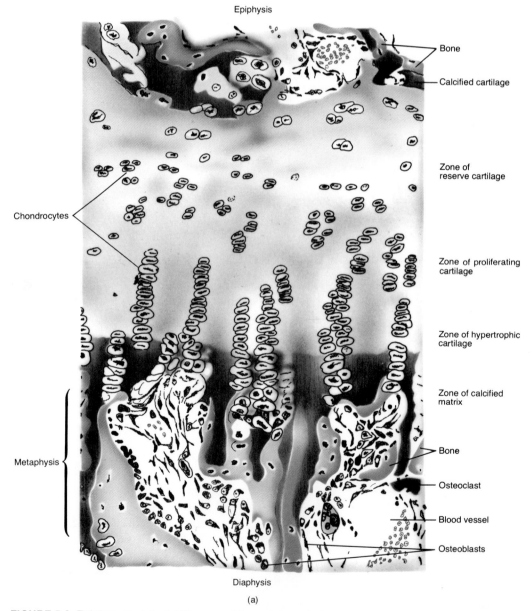

FIGURE 5-6 Epiphyseal plate. (a) Diagram of the epiphyseal plate.

thickens, becoming thickest at the diaphysis. The cartilage model continues to grow at its ends, steadily increasing in length. Eventually, blood vessels enter the epiphyses and secondary ossification centers appear in the epiphyses and also lay down spongy bone. In the tibia, one secondary ossification center develops in the proximal epiphysis soon after birth. The other center develops in the distal epiphysis during the child's second year.

After the two secondary ossification centers have formed, bone tissue has completely replaced cartilage, except in two regions. Cartilage continues to cover the articular surfaces of the epiphyses, where it is called articular cartilage. It also remains as a plate between the epiphysis and diaphysis, in which case it is called the **epiphyseal plate.** The epiphyseal plate, from epiphysis to diaphysis, consists of four zones (Figure 5-6). The *zone of reserve cartilage* is adjacent to the epiphysis of the bone. It consists of small chondrocytes that are scattered irregularly throughout the intercellular matrix. The cells of this zone do not function in bone growth. The zone of reserve cartilage

functions in anchoring the epiphyseal plate to the bone of the epiphysis. Its blood vessels also provide nutrients for the various zones of the epiphyseal plate. The second zone, the *zone of proliferating cartilage,* consists of slightly larger chondrocytes that are arranged like stacks of coins. The function of this zone is to make new chondrocytes (by cell division) to replace those that die at the diaphyseal surface of the epiphyseal plate. The third zone, the *zone of hypertrophic cartilage,* consists of even larger chondrocytes also arranged in columns. The cells are in various stages of maturation, with the more mature cells closer to the diaphysis. The lengthwise expansion of the epiphyseal plate is the result of cellular proliferation of the zone of proliferating cartilage and maturation of the cells in the zone of hypertrophic cartilage. Near the diaphyseal end of the bone, the intercellular matrix of the cells of the zone of hypertrophic cartilage become calcified and die. The fourth zone, the *zone of calcified matrix,* is only a few cells thick and consists of mostly dead cells because the intercellular matrix around them has calcified. The calcified matrix breaks up and is invaded by osteoblasts and capillaries from the bone of the diaphysis. These cells lay down bone on the calcified cartilage that persists. As a result, the diaphyseal border of the epiphyseal plate is firmly cemented to the bone of the diaphysis. The region between the diaphysis and epiphysis of a bone where the calcified matrix is replaced by bone is called the *metaphysis.* The activity of the epiphyseal plate is the only mechanism by which the diaphysis can increase in length. Unlike cartilage which can grow by both interstitial and appositional growth, bone can grow only by appositional growth.

The epiphyseal plate allows the bone to increase in length until early adulthood. As the child grows, cartilage cells are produced by mitosis on the epiphyseal side of the plate. Cartilage cells are then destroyed and the cartilage is replaced by bone on the diaphyseal side of the plate. In this way, the thickness of the epiphyseal plate remains fairly constant but the bone on the diaphyseal side increases in length. In this process, the bone lining the marrow cavity is destroyed so that the cavity increases in diameter. At the same time, osteoblasts from the periosteum add new osseous tissue around the outer surface of the bone. Initially, diaphyseal and epiphyseal ossification produce only spongy bone. Later, by reconstruction, the outer region of spongy bone is reorganized into compact bone.

Around the age of 17, the epiphyseal cartilage cells stop duplicating and the entire cartilage is slowly replaced by bone. Bone growth stops. The remnant of the epiphyseal plate is called the **epiphyseal line.** Ossification of all bones is usually completed by age 25. Figure 5-7 consists of a series of **roentgenograms** (pho-

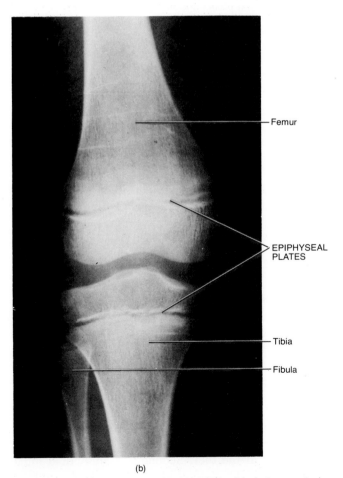

Femur

EPIPHYSEAL
PLATES

Tibia

Fibula

(b)

Figure 5-6 (Continued) Epiphyseal plate. (b) Anteroposterior projection showing epiphyseal plates in the knee of a ten-year-old child. (Courtesy of Arthur Provost, R. T.)

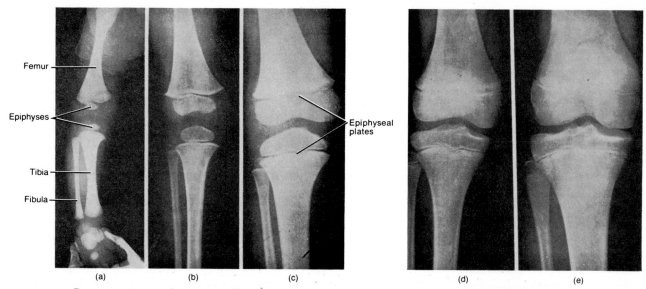

FIGURE 5-7 Roentgenograms of normal ossification at the knee. (a) One-month-old infant. The epiphyses at the knee are mostly cartilage. Ossification centers have formed for the femur and tibia. The space between the ends of the bones is occupied by epiphyseal cartilage. (b) Two-year-old child. Centers of ossification have grown. The white transverse zones marking the ends of the shafts are areas where mineral salts are deposited temporarily around the degenerating cartilage cells. (c) Five-year-old child. The epiphyses have assumed the shape of adult bones. The epiphyseal plates are clearly visible between the epiphyses and diaphyses of all three bones. (d) Eight-year-old child. The epiphyseal plates are still distinct as ossification continues. (e) Twelve-year-old child. The epiphyses have ossified almost completely. The epiphyseal plates are assuming the character of epiphyseal lines. (Courtesy of Lester W. Paul and John H. Juhl, *The Essentials of Roentgen Interpretation,* 3d ed., Harper & Row, Publishers, Inc., New York, 1972.)

tographs taken with x-rays) that show ossification in the epiphyses of two long bones at the knee. Bones undergoing either intramembranous or endochondral ossification are continually remodeling from the time that initial calcification occurs until the final structure appears. Compact bone is formed by the transformation of spongy bone. The diameter of a long bone is increased by the destruction of the bone closest to the marrow cavity and the construction of new bone around the outside of the diaphysis. However, even after bones have reached their adult shapes and sizes, old bone is perpetually destroyed and new osseous tissue is formed in its place.

Bone Replacement

Bone shares with skin the unique feature of replacing itself throughout adult life. This **remodeling** takes place at different rates in various body regions. The distal portion of the femur (thighbone) is replaced about every four months. By contrast, bone in certain areas of the shaft will not be completely replaced during the individual's life. Remodeling allows worn or injured bone to be removed and replaced with new tissue. It also allows bone to serve as the body's storage area for calcium. Many other tissues in the body need small amounts of calcium in order to perform their functions. For example, muscle needs calcium in order

to contract. The muscle cells take their calcium from the blood. However, the blood itself needs calcium in order to clot. The blood continually trades off calcium with the bones removing calcium when other tissues are not receiving enough of this element and resupplying the bones with dietary calcium when available to keep them from losing too much bone mass.

The cells believed to be responsible for the destruction of bone tissue are the **osteoclasts** (*clast* = break). In the healthy adult, a delicate homeostasis is maintained between the action of the osteoclasts and the removal of calcium on one hand, and the action of the bone-making osteoblasts and the deposition of calcium on the other. Should too much new tissue be formed, the bones become abnormally thick and heavy. If too much calcium is deposited in the bone, the surplus may form thick bumps, or spurs, on the bone that interfere with movement at joints. A loss of too much tissue or calcium weakens the bones and allows them to break easily.

Normal bone growth in the young and bone replacement in the adult depend on several factors. First, sufficient quantities of calcium and phosphorus, components of the primary salt that makes bone hard, must be included in the diet. Second, the individual must obtain sufficient amounts of vitamins A, C, and D. These substances are particularly responsible for the proper utilization of calcium and phosphorus by the body. Third, the body must manufacture the proper amounts of the hormones responsible for bone tissue activity.

Pituitary growth hormones are responsible for the general growth of bones. Too much or too little of these hormones during childhood makes the adult abnormally tall or short. Other hormones specialize in regulating the osteoclasts. And still others, especially the sex hormones, aid osteoblastic activity and thus promote the growth of new bone. The sex hormones act as a double-edged sword. They aid in the growth of new bone, but they also bring about the degeneration of all the cartilage cells in the epiphyseal plates. Because of the sex hormones, the typical adolescent experiences a spurt of growth during puberty, when sex hormone levels start to increase. The individual then quickly completes the growth process as the epiphyseal cartilage disappears. Premature puberty can actually prevent one from reaching an average adult height because of the simultaneous premature degeneration of the plates. Still another kind of hormone, produced by the parathyroid glands in the neck, determines whether the blood will deposit calcium and phosphorus in osseous tissue or whether it will remove these elements from the bones.

Applications to Health

Many bone disorders result from deficiencies in vitamins or minerals or from too much or too little of the hormones that regulate bone growth and development. Infections and tumors are also responsible for certain bone disorders.

VITAMIN DEFICIENCIES

Vitamin D is important to normal bone growth and maintenance. It is essential for the synthesis of a protein that transports the calcium obtained from foods across the lining of the intestine and into the extracellular fluid. When the body lacks this vitamin, it is unable to absorb calcium and phosphorus. A deficiency of vitamin D produces rickets in children and osteomalacia in adults.

Rickets

In the condition called **rickets,** epiphyseal cartilage cells cease to degenerate and new cartilage continues to be produced. Epiphyseal cartilage thus becomes wider than normal. At the same time, the soft matrix laid down by the osteoblasts in the diaphysis fails to calcify. As a result, the bones stay soft. When the child walks, the weight of the body causes the bones

in the legs to bow. Malformations of the head, chest, and pelvis also occur.

Osteomalacia

A deficiency of vitamin D in the adult causes the bones to give up excessive amounts of calcium and phosphorus. This loss, called *demineralization,* is especially heavy in the bones of the pelvis, legs, and spine. Demineralization caused by vitamin D deficiency is called **osteomalacia** (*malacia* = softness). After the bones demineralize, the weight of the body produces a bowing of the leg bones, a shortening of the backbone, and a flattening of the pelvic bones. Osteomalacia mainly affects women who live on poor cereal diets devoid of milk, are seldom exposed to the sun, and have repeated pregnancies that deplete the body of calcium.

OSTEOPOROSIS

Osteoporosis or **osteopenia** is a bone disorder affecting the middle-aged and elderly. Between puberty and the middle years, the sex hormones maintain osseous tissue by stimulating the osteoblasts to form new bone. After menopause, however, women produce smaller amounts of sex hormones. During old age, both men and women produce smaller amounts. As a result, the osteoblasts become less active and there is a decrease in bone mass (Figure 5-8). Osteoporosis affects the entire skeletal system, especially the spine, legs, and feet. As the spine collapses and curves, the breasts sag and the ribs fall on the pelvic rim. This condition leads to gastrointestinal distension and an overall decrease in muscle tone.

Among the factors implicated in bone loss is a decrease in estrogen, calcium deficiency and malabsorption, vitamin D deficiency, loss of muscle mass, inactivity, and high-protein diets. Osteoporosis seems to be an inevitable accompaniment of aging.

PAGET'S DISEASE

Another disorder, called **Paget's disease,** is characterized by an irregular thickening and softening of the bones. It is rarely seen in people under the age of 50. The cause, or *etiology,* of the disease is unknown. But the bone-producing osteoblasts and the bone-destroying osteoclasts apparently become uncoordinated. In some areas, too much bone is produced, whereas in other areas too much old bone is removed. The balance between bone formation and bone destruction is altered. Paget's disease affects the skull, the pelvis, and the bones of the extremities. Very little can be done to alter the course of the disease.

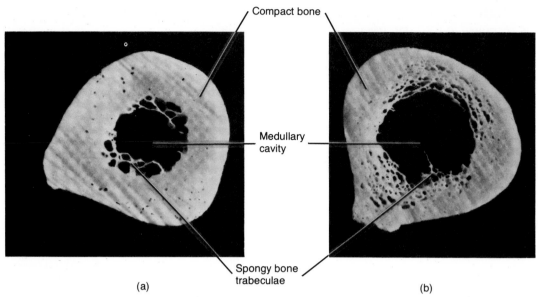

Compact bone

Medullary cavity

Spongy bone trabeculae

(a)   (b)

FIGURE 5-8 Osteoporosis. Photographs of a cross section of the femur from (a) a normal female aged 25 and (b) a female with osteoporosis aged 82. In (b) note the decreased thickness of the compact bone of the diaphysis, the increase in the diameter of the medullary cavity, and the thinning of the spongy bone trabeculae. (Photographs courtesy of D. P. Jenkins, from *Introduction to the Musculoskeletal System* by Cornelius Rosse and D. Kay Clawson, Harper & Row, Publishers, Inc., New York, 1970.)

OSTEOMYELITIS

The term **osteomyelitis** includes all the infectious diseases of bone. These diseases may be localized, or they may affect many bones. The infections may also involve the periosteum, marrow, and cartilage. Quite a few microorganisms may give rise to bone infection, but the most frequent are bacteria known as *Staphylococcus aureus,* commonly called "staph." These bacteria may reach the bone by various means: the bloodstream, an injury such as a fracture, or an infection, such as a soft-tissue abscess, a sinus infection, or an abscess of the tooth.

Before antibiotic treatment became available, osteomyelitis often became a long-lasting condition. It can destroy extensive areas of bone, spread to nearby joints, and, in rare cases, lead to death by producing abscesses in many parts of the body. Penicillin and other antibiotics have been effective in treating the disease and in preventing it from spreading through extensive areas of the bone.

FRACTURES

In simplest terms, a **fracture** is any sudden break in a bone. Sometimes the fracture is restored to normal position by manipulation without surgery. This procedure of setting a fracture is called *closed reduction.* In other cases, the fracture must be exposed by surgery before the break is rejoined. This procedure is known as *open reduction.* Although fractures of bones of the extremities may be classified in several different ways, the following scheme is useful (see Figure 5-9):

1. Partial. The break across the bone is incomplete.

2. Complete. The break occurs across the entire bone. The bone is completely broken into two pieces.

3. Simple or **closed.** The fractured bone does not break through the skin.

4. Compound or **open.** The broken ends of the fractured bone protrude through the skin.

5. Comminuted. The bone is splintered at the site of impact and smaller fragments of bone are found between the two main fragments.

6. Greenstick. A partial fracture in which one side of the bone is broken and the other side bends.

7. Spiral. The bone is usually twisted apart.

8. Transverse. A fracture at right angles to the long axis of the bone.

9. Impacted. A fracture in which one fragment is firmly driven into the other.

10. Pott's. A fracture of the distal end of the fibula, with serious injury of the distal tibial articulation.

11. Colles'. A fracture of the distal end of the radius in which the distal fragment is displaced posteriorly.

12. Displaced. A fracture in which the anatomical alignment of the bone fragments is not preserved.

13. Nondisplaced. A fracture in which the anatomical alignment of the bone fragments has not been disrupted.

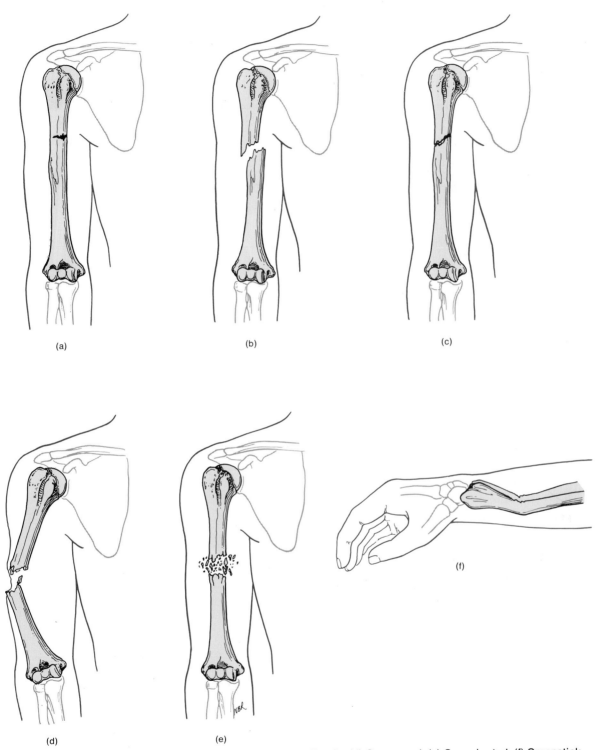

FIGURE 5-9 Types of fractures. (a) Partial. (b) Complete. (c) Simple. (d) Compound. (e) Comminuted. (f) Greenstick.

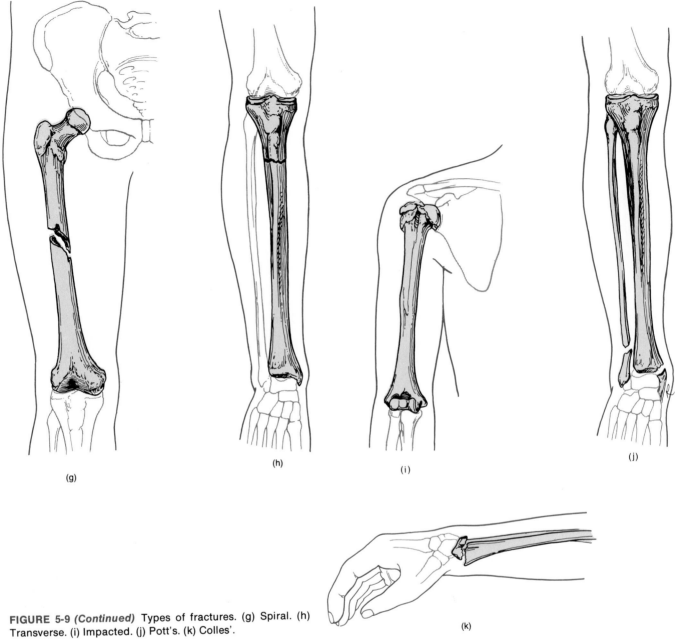

(g)

(h)

(i)

(j)

(k)

FIGURE 5-9 *(Continued)* Types of fractures. (g) Spiral. (h) Transverse. (i) Impacted. (j) Pott's. (k) Colles'.

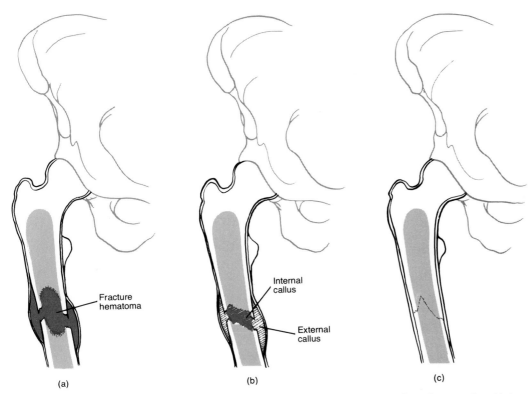

FIGURE 5-10 Fracture repair. (a) Formation of fracture hematoma. (b) Formation of external and internal calli. (c) Completely healed fracture.

Unlike the skin, which may repair itself within days, or muscle, which may mend in weeks, a bone sometimes requires months to heal. A fractured femur, for example, may take six months to heal because sufficient calcium to strengthen and harden new bone is deposited only gradually. Bone cells also grow and reproduce slowly. Moreover, the blood supply to bone is poor, which helps to explain the difficulty in the healing of an infected bone.

The following steps occur in the repair of a fracture (Figure 5-10):

1. As a result of the fracture, blood vessels crossing the fracture line are broken. These vessels are found in the periosteum, Haversian systems, and marrow cavity. As blood pours from the torn ends of the vessels, it coagulates and forms a clot in and about the site of the fracture. This clot, called a **fracture hematoma,** usually occurs six to eight hours after the injury. Since the circulation of blood ceases when the fracture hematoma forms, bone cells and periosteal cells at the fracture line die.

2. A growth of new bone tissue—a **callus**—develops in and around the fractured area. It forms a bridge between separated areas of bone. The callus that forms from the osteogenic cells of the torn periosteum and develops around the outside of the fracture is called an *external callus*. The callus that forms from the osteogenic cells of the endosteum and develops between the two ends of bone fragments and between the two marrow cavities is called the *internal callus*.

Approximately 48 hours after a fracture occurs, the cells that ultimately repair the fracture become actively mitotic. These cells come from the osteogenic layer of the periosteum, the endosteum of the marrow cavity, and the bone marrow. As a result of their accelerated mitotic activity, the cells of the three regions grow toward the fracture. During the first week following the fracture, the cells of the endosteum and bone marrow form new trabeculae in the marrow cavity near the line of fracture. This is the internal callus. During the next few days, osteogenic cells of the periosteum form a collar around each bone fragment. The collar, or external callus, is replaced by trabeculae. The trabeculae of the calli are joined to living and dead portions of the original bone fragments.

3. The final phase of fracture repair is the **remodeling** of the calli. Dead portions of the original fragments are gradually resorbed. Compact bone replaces spongy bone around the periphery of the fracture. In some cases, the healing is so complete that the fracture line is undetectable, even by x-ray. However, a thickened area on the surface of the bone usually remains as evidence of the fracture site. The healing of a fracture of the femur is illustrated by sequential roentgenograms in Figure 5-11.

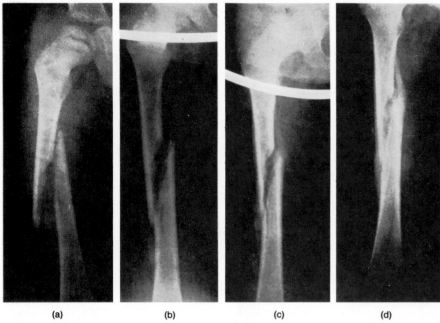

(a) (b) (c) (d)

FIGURE 5-11 Roentgenograms of the repair of a fractured femur. (a) Immediately after fracture. (b) Two weeks later, a hazy external callus is visible around the margins of the fracture. (c) About 3½ weeks after the fracture, the internal and external calli begin to bridge the separated fragments. (d) Almost 9 weeks after the fracture, the bridge between fragments is fairly well developed. In time, remodeling will occur and the fracture will be repaired. (Courtesy of Ralph C. Frank, Eau Claire, Wisconsin, from Lester W. Paul and John H. Juhl, *The Essentials of Roentgen Interpretation*, 3d ed., Harper & Row, Publishers, Inc., New York, 1972.)

KEY MEDICAL TERMS ASSOCIATED WITH THE SKELETAL SYSTEM

Achondroplasia (*a* = without; *chondro* = cartilage; *plasia* = growth) Imperfect ossification within cartilage of long bones during fetal life; also called *fetal rickets*.

Brodie's abscess Infection in the spongy tissue of a long bone, with a small inflammatory area.

Craniotomy (*cranium* = skull; *tome* = a cutting) Any surgery that requires cutting through the bones surrounding the brain.

Necrosis (*necros* = death; *osis* = condition) Death of tissues or organs; in the case of bone, necrosis results from deprivation of blood supply; could result from fracture, extensive removal of periosteum in surgery, exposure to radioactive substances, and other causes.

Osteitis (*osteo* = bone) Inflammation of infection of bone.

Osteoarthritis (*arthro* = joint) A degenerative condition of bone and also the joint.

Osteoblastoma (*oma* = tumor) A benign tumor of the osteoblasts.

Osteochondroma (*chondro* = cartilage) A benign tumor of the bone and cartilage.

Osteoma A benign bone tumor.

Osteomyelitis (*myel* = marrow) Infection that involves bone marrow.

Osteosarcoma (*sarcoma* = connective tissue tumor) A malignant tumor composed of osseous tissue.

Pott's disease Inflammation of the backbone, caused by the microorganism that produces tuberculosis.

STUDY OUTLINE

Functions

1. The skeletal system consists of all bones attached at joints, cartilage between joints, and cartilage found elsewhere (nose, larynx, and outer ear).
2. The functions of the skeletal system include support, protection, leverage, mineral storage, and blood-cell formation.

Histology

1. Parts of a typical long bone are the diaphysis (shaft), epiphyses (ends), articular cartilage, periosteum, medullary (marrow) cavity, and endosteum.
2. Cancellous, or spongy, bone has many marrow-filled pores and does not contain Haversian systems. It consists of trabeculae containing osteocytes and lacunae.

3. Dense, or compact, bone has fewer pores and fewer networks of passageways called Haversian systems. Dense bone lies over spongy bone and composes most of the bone tissue of the diaphyses.

Ossification

1. Bone forms by a process called ossification or osteogenesis, which begins when mesenchymal cells become transformed into osteoblasts.
2. Intramembranous ossification occurs within fibrous membranes of the embryo and the adult.
3. Endochondral ossification occurs within a cartilage model.
4. The primary ossification center of a long bone is in the diaphysis. Cartilage degenerates, leaving cavities that merge to form the marrow cavity. Osteoblasts lay down bone. Next ossification occurs in the epiphyses, where bone replaces cartilage, except for the epiphyseal plate.
5. In both types of ossification, spongy bone is laid down first. Dense bone is later reconstructed from spongy bone.

Bone Replacement

1. The growth and development of bone depends on a balance between bone formation and destruction.
2. Old bone is constantly destroyed by osteoclasts while new bone is constructed by osteoblasts. This process is called remodeling.
3. Normal growth depends on calcium, phosphorus, and vitamins (A, C, and D) and is controlled by hormones that are responsible for bone mineralization and reabsorption.

Applications to Health

1. Rickets is a vitamin D deficiency in children in which the body does not absorb calcium and phosphorus. The bones soften and bend under the body's weight.
2. Osteomalacia is a vitamin D deficiency in adults that leads to demineralization.
3. With osteoporosis, the amount and strength of bone tissue decrease due to decreases in hormone output.
4. Paget's disease is the irregular thickening and softening of bones in which osteoclast and osteoblast activities are imbalanced.
5. Osteomyelitis is a term for the infectious diseases of bones, marrow, and periosteum. It is frequently caused by "staph" bacteria.
6. A fracture is any break in a bone.
7. The types of fractures include: partial, complete, simple, compound, comminuted, greenstick, spiral, transverse, impacted, Pott's, Colles', displaced, and nondisplaced.
8. Fracture repair consists of forming a fracture hematoma, forming a callus, and remodeling.

REVIEW QUESTIONS

1. Define the skeletal system. What are its five principal functions?
2. Diagram the parts of a long bone, and list the functions of each part. What is the difference between compact and spongy bone tissue? Diagram the microscopic structure of bone.
3. What is meant by ossification? When does the process begin and end?
4. Distinguish between the two principal kinds of ossification.
5. Outline the major events involved in intramembranous and endochondral ossification.
6. Describe the various zones of the epiphyseal plate. How does the plate grow?
7. What is a roentgenogram?
8. List the primary factors involved in bone growth.
9. How does osteoblast activity in balance with osteoclast activity control the replacement of bone?
10. Define rickets and osteomalacia in terms of symptoms, cause, and treatment. What do these two diseases have in common?
11. What are the principal symptoms of osteoporosis, osteomyelitis, and Paget's disease? What is the etiology of each?
12. What is a fracture? Distinguish 13 principal kinds.
13. Outline the steps involved in fracture repair.
14. Refer to the glossary of key medical terms associated with the skeletal system. Be sure that you can define each term.

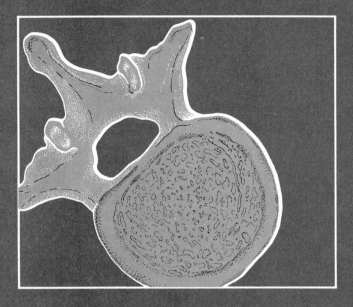

6

The Skeletal System: The Axial Skeleton

STUDENT OBJECTIVES

- Define the four principal types of bones in the skeleton.
- Explain the relationship between bone structure and function.
- Describe the various markings on the surfaces of bones.
- Relate the structure of the marking to its function.
- Describe the components of the axial and appendicular skeleton.
- Identify the bones of the skull and the major markings associated with each.
- Identify the sutures and fontanels of the skull.
- Identify the paranasal sinuses of the skull in projection diagrams and roentgenograms.
- Identify the principal foramina of the skull.
- Identify the bones of the vertebral column and their principal markings.
- List the defining characteristics and curves of each region of the vertebral column.
- Identify the bones of the thorax and their principal markings.

The skeletal system forms the framework of the body. For this reason, a familiarity with the names, shapes, and positions of individual bones will help you to understand some of the other organ systems. For example, movements such as throwing a ball, typing, and walking require the coordinated use of bones and muscles. To understand how muscles produce different movements, you need to learn the parts of the bones to which the muscles attach. The respiratory system is also highly dependent on bone structure. The bones in the nasal cavity form a series of passageways that help clean, moisten, and warm inhaled air. Furthermore, the bones of the thorax are specially shaped and positioned so the chest can expand during inhalation. Many bones also serve as landmarks to students of anatomy as well as to surgeons. Blood vessels and nerves often run parallel to bones. These structures can be located more easily if the bone is identified first.

We shall study bones by examining the various regions of the body. For instance, we shall look at the skull first and see how the bones of the skull relate to each other. Then we shall move on to the chest. This regional approach will allow you to see how all the many bones of the body relate to each other.

Types of Bones

The bones of the body may be classified into four principal types: long, short, flat, and irregular. **Long bones** have greater length than width and consist of a diaphysis and two epiphyses. They are slightly curved for strength. A curved bone is structurally designed to absorb the stress of the body weight at several different points so the stress is evenly distributed. If such bones were straight, the weight of the body would be unevenly distributed and the bone would easily fracture. Examples of long bones include bones of the thighs, legs, toes, arms, forearms, and fingers. Figure 5-1a shows the parts of a long bone. **Short bones** are somewhat cube-shaped and nearly equal in length and width. Their texture is spongy throughout, except at the surface, where there is a thin layer of compact bone. Examples of short bones are the wrist and ankle bones. **Flat bones** are generally thin and composed of two more or less parallel plates of compact bone enclosing a layer of spongy bone. The term *diploe* is applied to the spongy bone of the cranial bones. Flat bones afford considerable protection and provide extensive areas for muscle attachment. Examples of flat bones include the cranial bones, which protect the brain, the sternum and ribs, which protect organs in the thorax, and the scapulas. **Irregular bones** have complex shapes and cannot be grouped into any of the three categories just described. They also vary in the amount of spongy and compact bone present. Such bones are the vertebrae and certain facial bones.

Besides these four principal types of bones, two other kinds are recognized. **Wormian,** or **sutural, bones** are small clusters of bones between the joints of certain cranial bones. Their number varies greatly from person to person. **Sesamoid bones** are small bones in tendons where considerable pressure develops—for instance, in the wrist. These bones, like the Wormian bones, are also variable in number. Two sesamoid bones, are patellas, or kneecaps, are present in all individuals.

Surface Markings

The surfaces of bones reveal various **markings.** The structure of many of these markings indicates their functions. Long bones that bear a great deal of weight have large, rounded ends that can form sturdy joints. Other bones have depressions that receive the rounded ends. Rough areas serve for the attachment of muscles, tendons, and ligaments. Grooves in the surfaces of bones provide for the passage of blood vessels, and openings occur where blood vessels and nerves pass through the bone. Exhibit 6-1 describes the different markings and their functions.

Divisions of the Skeletal System

The adult human skeleton usually consists of 206 bones grouped in two principal divisions: the **axial** and the **appendicular.** The longitudinal *axis,* or center, of the human body is a straight line that runs vertically along the body's center of gravity. This imaginary line runs through the head and down to the space between the feet. The midsagittal section is drawn through this line. The axial division of the skeleton consists of the bones that lie around the axis; ribs, breastbone, the bones of the skull, and backbone. The appendicular division contains the bones of the free *appendages,* which are the upper and lower extremities, plus the bones called *girdles,* which connect the free appendages to the axial skeleton. The 80 bones of the axial division and the 126 bones of the appendicular division are typically grouped as follows:

Axial skeleton
Skull
   1. Cranium   8
   2. Face   14
Hyoid (above the larynx)   1
*Auditory ossicles,** 3 in each ear   6
Vertebral column   26

*Although the auditory ossicles are not considered part of the axial or appendicular skeleton, but rather as a separate group of bones, they are placed with the axial skeleton for convenience.

Thorax
   1. Sternum   1
   2. Ribs   24
                              80

Appendicular skeleton
  Shoulder girdles
   1. Clavicle   2
   2. Scapula   2
  Upper extremities
   1. Humerus   2
   2. Ulna   2
   3. Radius   2
   4. Carpals   16

   5. Metacarpals   10
   6. Phalanges   28
  Pelvic girdle
   1. Coxal, hip, or pelvic bone   2
  Lower extremities
   1. Femur   2
   2. Fibula   2
   3. Tibia   2
   4. Patella   2
   5. Tarsals   14
   6. Metatarsals   10
   7. Phalanges   28
                       126

EXHIBIT 6-1
BONE MARKINGS

MARKING	DESCRIPTION	EXAMPLE
DEPRESSIONS AND OPENINGS		
Fissure	A narrow, cleftlike opening between adjacent parts of bones through which blood vessels or nerves pass.	Superior orbital fissure of the sphenoid bone (Figure 6-2).
Foramen (*foramen* = hole)	A rounded opening through which blood vessels, nerves, or ligaments pass.	Infraorbital foramen of the maxilla (Figure 6-2).
Meatus (canal)	A tubelike passageway running within a bone.	External auditory meatus of the temporal bone (Figure 6-2).
Paranasal sinus (*sin* = cavity)	An air-filled cavity within a bone connected to the nasal cavity.	Frontal sinus of the frontal bone (Figure 6-8).
Groove or sulcus (*sulcus* = ditchlike groove)	A furrow or groove that accommodates a soft structure such as a blood vessel, nerve, or tendon.	Intertubercular sulcus of the humerus (Figure 7-4).
Fossa (*fossa* = basinlike depression)	A depression in or on a bone.	Mandibular fossa of the temporal bone (Figure 6-4).
PROCESS	Any prominent projection.	Mastoid process of the temporal bone (Figure 6-2).
Processes that form joints		
Condyle (*condylus* = knucklelike process)	A large, convex or concave articular prominence.	Medial condyle of the femur (Figure 7-10).
Head	A rounded articular projection supported on the constricted portion (neck) of a bone.	Head of the femur (Figure 7-10).
Facet	A smooth, flat surface.	Articular facet for the tubercle of rib on a vertebra (Figure 6-18).
Processes to which tendons, ligaments, and other connective tissues attach		
Tubercle (*tuber* = knob)	A small, rounded process.	Greater tubercle of the humerus (Figure 7-4).
Tuberosity	A large, rounded, usually roughened process.	Ischial tuberosity of the hipbone (Figure 7-8).
Trochanter	A large, blunt projection found only on the femur.	Greater trochanter of the femur (Figure 7-10).
Crest	A prominent border or ridge on a bone.	Iliac crest of the hipbone (Figure 7-7).
Line	A less prominent ridge.	Linea aspera of the femur (Figure 7-10).
Spinous process (spine)	A sharp, slender process.	Spinous process of a vertebra (Figure 6-12).
Epicondyle (*epi* = above)	A prominence above a condyle.	Medial epicondyle of the femur (Figure 7-10).

Now that you understand how the skeleton is organized into axial and appendicular divisions, refer to Figure 6-1 to see how the two divisions are joined to form the skeleton. The bones of the axial skeleton are shown in gray. Be certain to locate the following regions of the skeleton: skull, cranium, face, hyoid bone, vertebral column, thorax, shoulder girdle, upper extremity, pelvic girdle, and lower extremity.

Skull

The **skull,** which contains 22 bones, rests on the superior end of the vertebral column and is composed of two sets of bones: cranial bones and facial bones. The **cranial bones** enclose and protect the brain and the organs of sight, hearing, and balance. The eight cranial bones are the frontal bone, parietal bones (two), temporal bones (two), the occipital bone, sphenoid, and ethmoid. There are 14 **facial bones:** the nasal bones (two), maxillae (two), zygomatic bones (two), mandible, lacrimal bones (two), palatine bones (two), inferior nasal conchae (two), and vomer. Be sure you can locate all the skull bones in the anterior, lateral, median, and posterior views of the skull (Figure 6-2).

SUTURES

A **suture,** meaning seam or stitch, is an immovable joint found only between skull bones. Very little connective tissue is found between the bones of the suture. Four prominent skull sutures include the:

1. Coronal suture between the frontal bone and the two parietal bones
2. Sagittal suture between the two parietal bones
3. Lambdoidal suture between the parietal bones and the occipital bone
4. Squamosal suture between the parietal bones and the temporal bones

Refer to Figures 6-2 and 6-3 for the locations of these sutures. Several other sutures are also shown. Their names are descriptive of the bones they connect. For example, the frontonasal suture is between the frontal bone and the nasal bones. These sutures are indicated in Figures 6-2 to 6-5.

FONTANELS

The "skeleton" of a newly formed embryo consists of cartilage or fibrous membrane structures shaped like bones. Gradually the cartilage or fibrous membrane is replaced by bone. At birth, membrane-filled spaces called **fontanels,** meaning fountains, are found between cranial bones (Figure 6-3). These "soft spots" are areas where the bone-making process is not yet complete. They allow the skull to be compressed during birth. Physicians find the fontanels helpful in determining the position of the infant's head prior to delivery. Although an infant may have many fontanels at birth, the form and location of six are fairly constant.

The **anterior (frontal) fontanel** is located between the angles of the two parietal bones and the two segments of the frontal bone. This fontanel is roughly diamond-shaped, and it is the largest of the six fontanels. It usually closes in 18 to 24 months.

The **posterior (occipital) fontanel** is situated between the two parietal bones and the occipital bone. This fontanel is considerably smaller than the anterior fontanel. It is diamond-shaped and generally closes about two months after birth.

The **anterolateral (sphenoidal) fontanels** are paired. One is located on each side of the skull at the junction of the frontal, parietal, temporal, and sphenoid bones. These fontanels are quite small and irregular in shape. They normally close by the third month after birth.

The **posterolateral (mastoid) fontanels** are also paired. One is situated on each side of the skull at the junction of the parietal, occipital, and temporal bones. These fontanels are irregularly shaped. They begin to close one or two months after birth, but closure is not generally complete until the age of one year.

FRONTAL BONE

The **frontal bone** forms the forehead, the anterior part of the cranium; the superior portion of the *orbits* (eye sockets); and most of the anterior part of the cranial floor. Soon after birth the left and right parts of the frontal bone are united by a suture. The suture usually disappears by age six. If, however, the suture persists throughout life, it is referred to as the **metopic suture.**

If you examine the anterior and lateral views of the skull in Figure 6-2, you will note the **frontal squama,** or vertical plate (*squam* = scale). This scalelike plate, which corresponds to the forehead, gradually slopes down from the coronal suture, then turns abruptly downward. It projects slightly above its lower edge on either side of the midline to form the **frontal eminences.** Inferior to each eminence is a horizontal ridge, the **superciliary arch,** caused by the projection of the frontal sinuses posterior to the eyebrow. Between the eminences and the arches just superior to the nose is a flattened area, the **glabella.** A thickening of the frontal bone inferior to the superciliary arches is called the **supraorbital margin.** From this margin the frontal bone extends posteriorly to form the roof of the orbit and part of the floor of the cranial cavity. Within the supraorbital margin, slightly medial to its midpoint, is a hole called the **supraorbital foramen** (*foramen* = hole). The supraorbital nerve and artery pass through this foramen. The **frontal sinuses** lie deep to the superciliary arches. These mucus-lined cavities act as sound chambers which give the voice resonance.

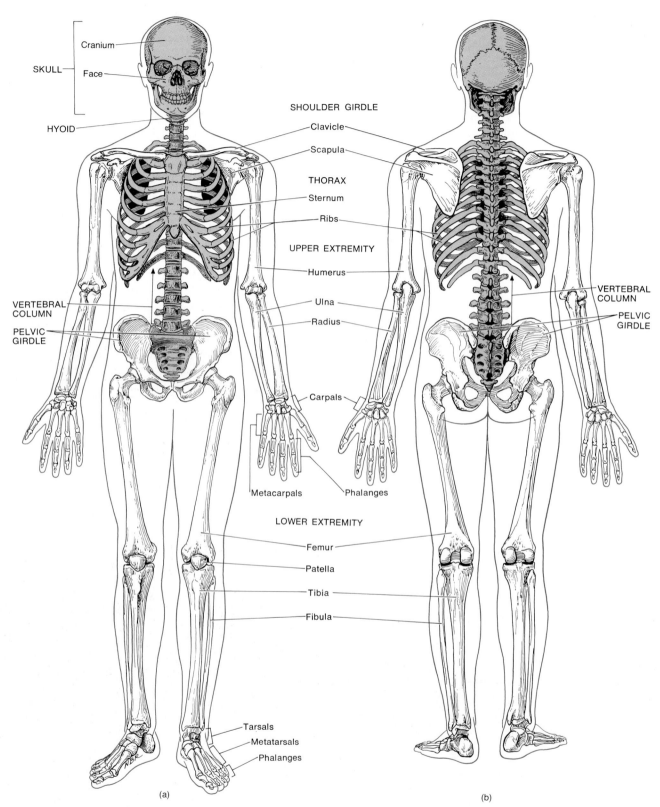

FIGURE 6-1 Divisions of the skeletal system. (a) Anterior view. (b) Posterior view.

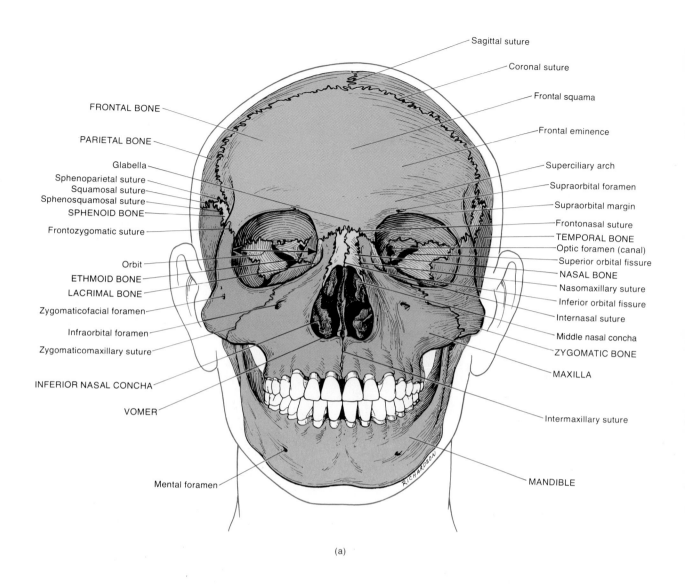

Sagittal suture

Coronal suture

Frontal squama

Frontal eminence

Superciliary arch

Supraorbital foramen

Supraorbital margin

Frontonasal suture

TEMPORAL BONE

Optic foramen (canal)

Superior orbital fissure

NASAL BONE

Nasomaxillary suture

Inferior orbital fissure

Internasal suture

Middle nasal concha

ZYGOMATIC BONE

MAXILLA

Intermaxillary suture

MANDIBLE

FRONTAL BONE

PARIETAL BONE

Glabella

Sphenoparietal suture

Squamosal suture

Sphenosquamosal suture

SPHENOID BONE

Frontozygomatic suture

Orbit

ETHMOID BONE

LACRIMAL BONE

Zygomaticofacial foramen

Infraorbital foramen

Zygomaticomaxillary suture

INFERIOR NASAL CONCHA

VOMER

Mental foramen

(a)

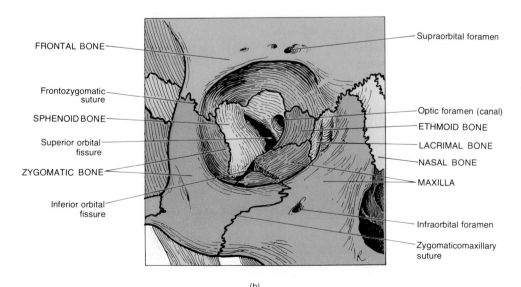

FRONTAL BONE

Frontozygomatic suture

SPHENOID BONE

Superior orbital fissure

ZYGOMATIC BONE

Inferior orbital fissure

Supraorbital foramen

Optic foramen (canal)

ETHMOID BONE

LACRIMAL BONE

NASAL BONE

MAXILLA

Infraorbital foramen

Zygomaticomaxillary suture

(b)

FIGURE 6-2 Skull. (a) Anterior view. (b) Detail of the right orbit in anterior view.

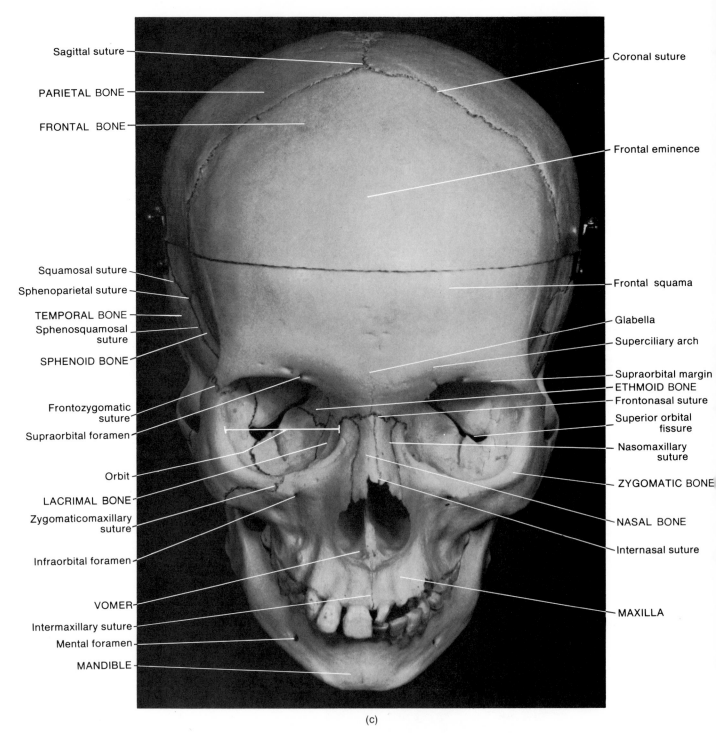

Sagittal suture

PARIETAL BONE

FRONTAL BONE

Squamosal suture

Sphenoparietal suture

TEMPORAL BONE

Sphenosquamosal suture

SPHENOID BONE

Frontozygomatic suture

Supraorbital foramen

Orbit

LACRIMAL BONE

Zygomaticomaxillary suture

Infraorbital foramen

VOMER

Intermaxillary suture

Mental foramen

MANDIBLE

Coronal suture

Frontal eminence

Frontal squama

Glabella

Superciliary arch

Supraorbital margin

ETHMOID BONE

Frontonasal suture

Superior orbital fissure

Nasomaxillary suture

ZYGOMATIC BONE

NASAL BONE

Internasal suture

MAXILLA

(c)

FIGURE 6-2 *(Continued)* Skull. (c) Photograph of the skull in anterior view. (Courtesy of Lenny Patti.)

PARIETAL BONES

The two **parietal bones** (*paries* = wall) form the greater portion of the sides and roof of the cranial cavity.

The external surface contains two slight ridges that may be observed by looking at the lateral view of the skull in Figure 6-2. These are the **superior temporal line** and a less conspicuous **inferior temporal line** below it. The internal surface has many eminences and depressions that accommodate the blood vessels supplying the outer meninx (covering) of the brain called the *dura mater*.

TEMPORAL BONES

The two **temporal bones** form the inferior sides of the cranium and part of the cranial floor. The term *tempora* pertains to the temples.

In the lateral view of the skull in Figure 6-2, notice the **squama** or **squamous portion**—a thin, large, expanded area that forms the anterior and superior part

of the temple. Projecting from the inferior portion of the squama is the **zygomatic process,** which articulates with the temporal process of the zygomatic bone. The zygomatic process of the temporal bone together with the temporal process of the zygomatic bone constitutes the **zygomatic arch.** At the floor of the cranial cavity, shown in Figure 6-5, is the **petrous portion** of the temporal bone. This portion is triangular and located at the base of the skull between the sphenoid and occipital bones. The petrous portion contains the internal ear, the essential part of the organ of hearing. It also contains the **carotid foramen (canal)** through which the internal carotid artery passes (see Figure 6-4). Posterior to the carotid foramen and anterior to the occipital bone is the **jugular foramen (fossa)** through which the internal jugular vein and the glossopharyngeal nerve (IX), vagus nerve (X), and accessory nerve (XI) pass. (As you will see later, the Roman numerals associated with cranial nerves indicate the order in which the nerves arise from the brain, from front to back.) Between the squamous and petrous portions is

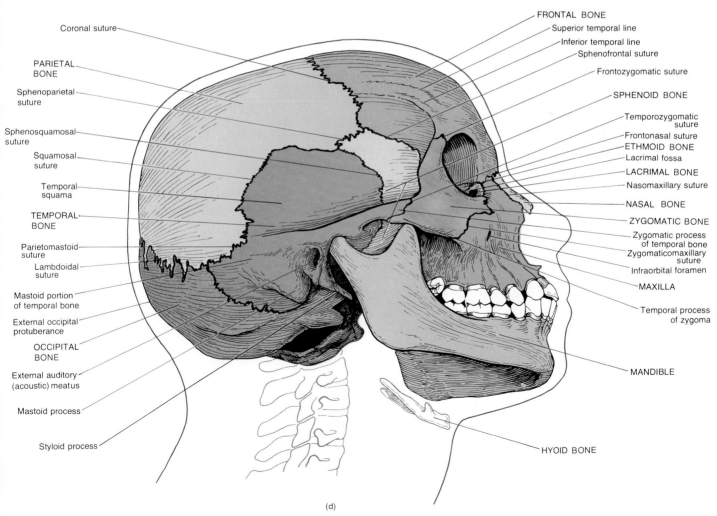

(d)

FIGURE 6-2 *(Continued)* Skull. (d) Right lateral view.

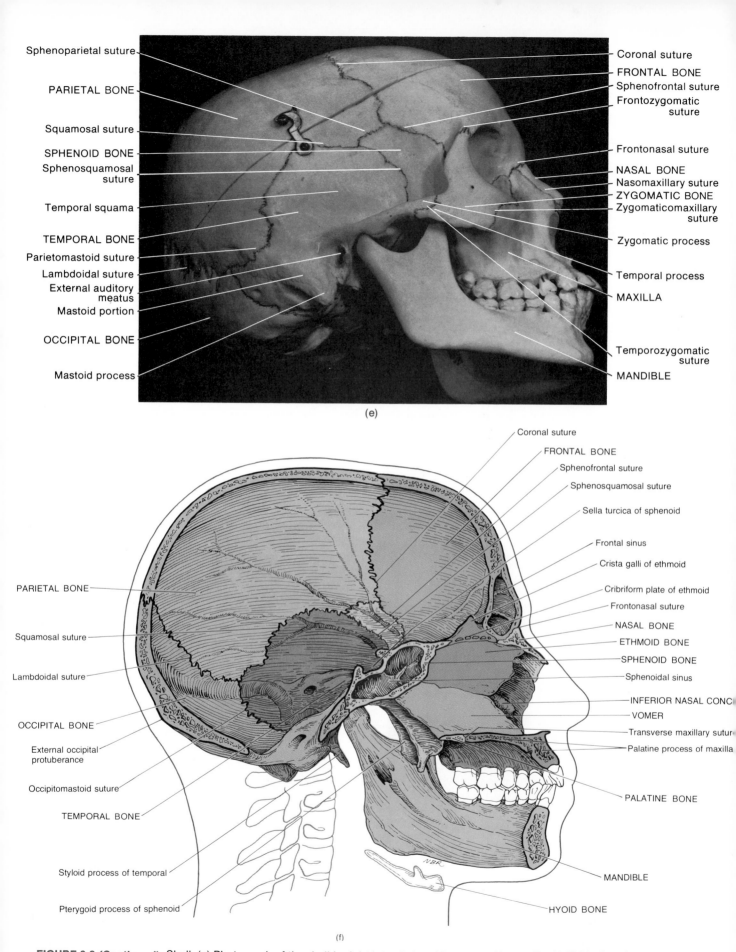

FIGURE 6-2 (Continued) Skull. (e) Photograph of the skull in right lateral view. (Courtesy of Lenny Patti.) (f) Median view.

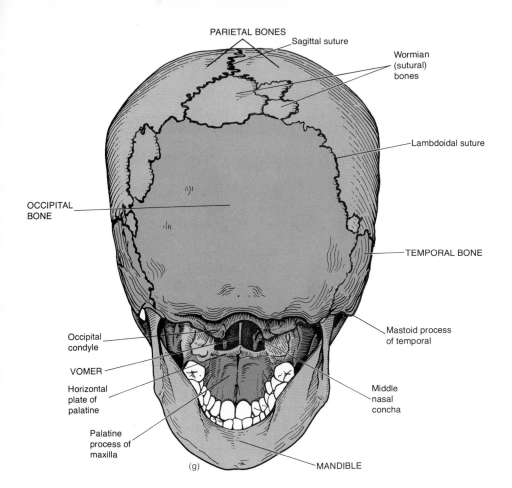

PARIETAL BONES
Sagittal suture
Wormian
(sutural)
bones
Lambdoidal suture
OCCIPITAL
BONE
TEMPORAL BONE
Mastoid process
of temporal
Occipital
condyle
VOMER
Middle
nasal
concha
Horizontal
plate of
palatine
Palatine
process of
maxilla
(g)
MANDIBLE

FIGURE 6-2 *(Continued)* Skull. (g)
Posterior view showing sutural
(Wormian) bones.

a socket called the **mandibular fossa.** Anterior to the mandibular fossa is a rounded eminence, the **articular tubercle.** The mandibular fossa and articular tubercle articulate with the condylar process of the mandible (lower jaw bone) to form the temporomandibular joint. The mandibular fossa and articular tubercle are seen best in Figure 6-4. In the lateral view of the skull in Figure 6-2, you will see the **mastoid portion** of the temporal bone, located posterior and inferior to the external auditory meatus, or ear canal. In the adult, this portion of the bone contains a number of **mastoid air "cells,"** These air spaces are separated from the brain only by thin bony partitions. If *mastoiditis,* inflammation of these bony cells, occurs, the infection may spread to the brain or its outer covering. The mastoid air cells do not drain as do the paranasal sinuses. The **mastoid process** is a rounded projection of the temporal bone posterior to the external auditory meatus. It serves as a point of attachment for several neck muscles. Near the posterior border of the mastoid process is the **mastoid foramen** through which a vein to the transverse sinus and a small branch of the occipital artery to the dura mater pass. The **external auditory meatus** is the canal in the temporal bone that leads to the middle ear. The **internal acoustic meatus** is superior to the jugular foramen. It transmits the facial and acoustic nerves and the internal auditory artery. The **styloid process** projects downward from the undersurface of the tem-

poral bone and serves as a point of attachment for muscles and ligaments of the tongue and neck. Between the styloid process and the mastoid process is the **stylomastoid foramen,** which transmits the facial nerve (VII) and stylomastoid artery (see Figure 6-4).

OCCIPITAL BONE

The **occipital bone** forms the posterior part and a prominent portion of the base of the cranium (Figure 6-4)

The **foramen magnum** is a large hole in the inferior part of the bone through which the medulla oblongata and its membranes, the accessory nerve (XI), and the vertebral and spinal arteries pass. The **occipital condyles** are oval processes with convex surfaces, one on either side of the foramen magnum, which articulate with depressions on the first cervical vertebra. Extending laterally from the posterior portion of the condyles are quadrilateral plates of bone called the **jugular processes.** At the base of the condyles is the **hypoglossal canal (fossa)** through which the hypoglossal nerve (XII) passes (see Figure 6-5). The **external occipital protuberance** is a prominent projection on the posterior surface of the bone just superior to the foramen magnum. You can feel this structure as a definite bump on the back of your head, just above your neck. The protuberance is also visible in Figure 6-2g.

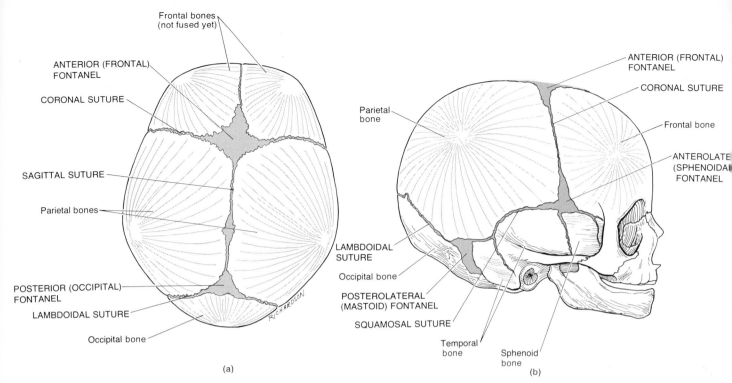

FIGURE 6-3 Fontanels of the skull at birth. (a) Superior view. (b) Right lateral view.

FIGURE 6-4 Skull. (a) Inferior view.

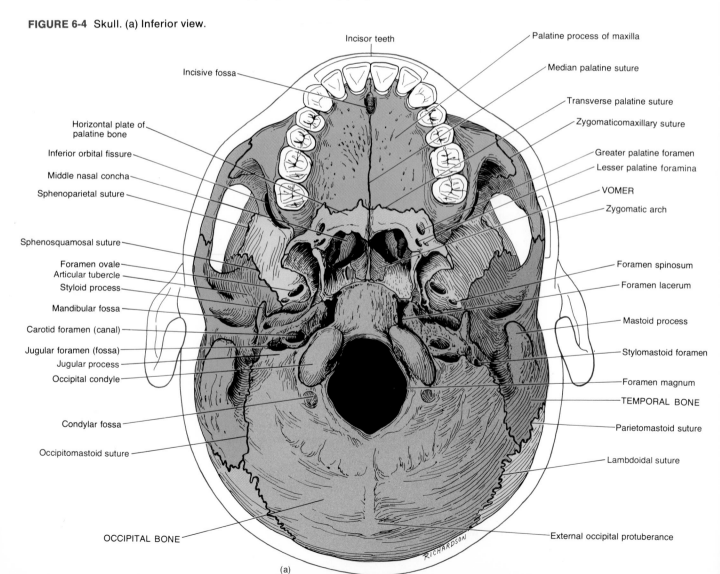

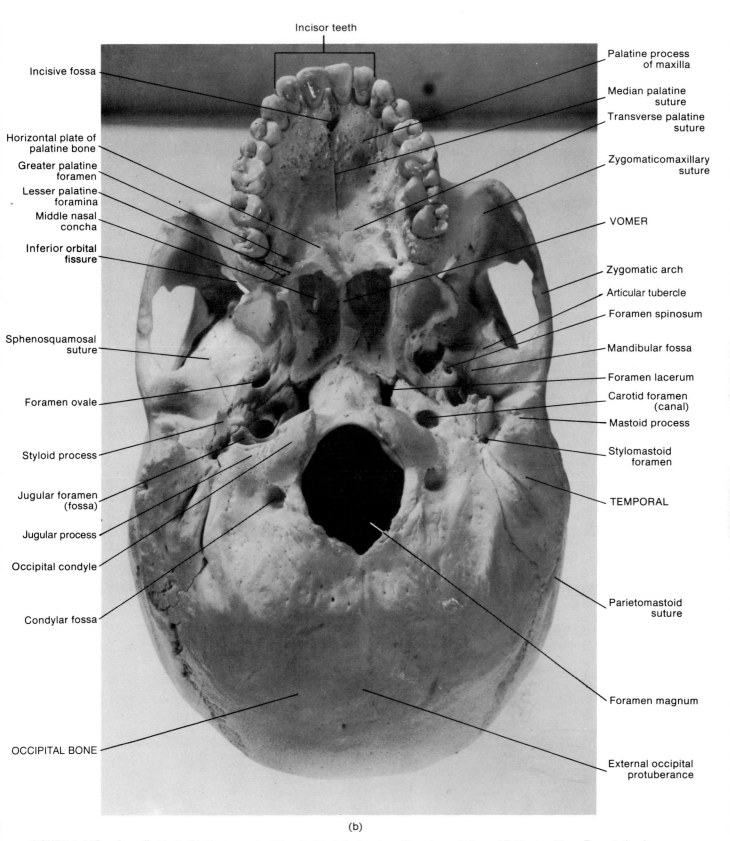

Incisor teeth

Incisive fossa

Horizontal plate of palatine bone

Greater palatine foramen

Lesser palatine foramina

Middle nasal concha

Inferior orbital fissure

Sphenosquamosal suture

Foramen ovale

Styloid process

Jugular foramen (fossa)

Jugular process

Occipital condyle

Condylar fossa

OCCIPITAL BONE

Palatine process of maxilla

Median palatine suture

Transverse palatine suture

Zygomaticomaxillary suture

VOMER

Zygomatic arch

Articular tubercle

Foramen spinosum

Mandibular fossa

Foramen lacerum

Carotid foramen (canal)

Mastoid process

Stylomastoid foramen

TEMPORAL

Parietomastoid suture

Foramen magnum

External occipital protuberance

(b)

FIGURE 6-4 *(Continued)* Skull. (b) Photograph of the skull in inferior view. (Courtesy of Vincent P. Destro, Mayo Foundation.)

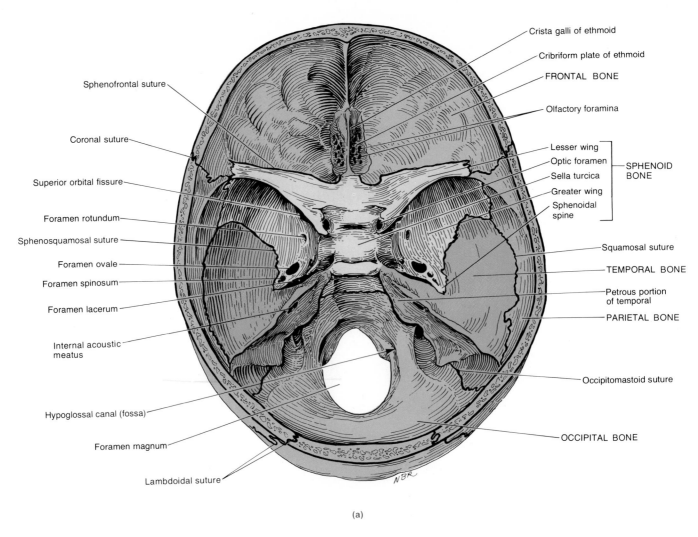

(a)

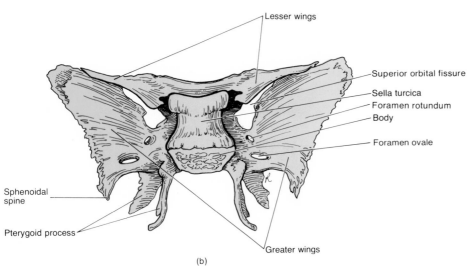

(b)

FIGURE 6-5 Sphenoid bone. (a) Viewed in the floor of the cranium from above. (b) Posterior view.

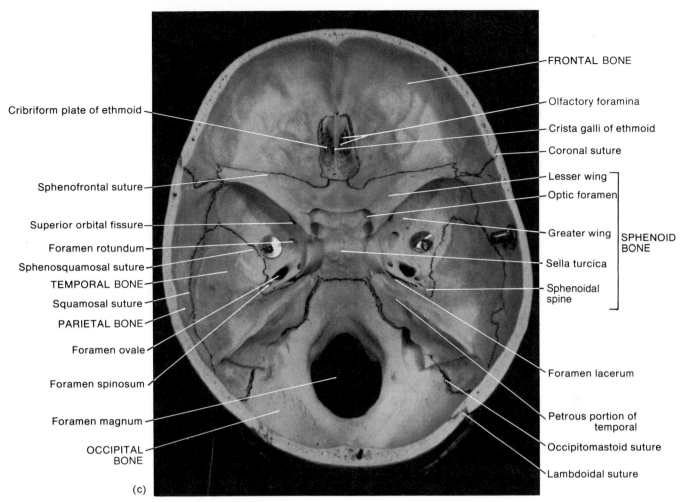

Cribriform plate of ethmoid

Sphenofrontal suture

Superior orbital fissure

Foramen rotundum

Sphenosquamosal suture

TEMPORAL BONE

Squamosal suture

PARIETAL BONE

Foramen ovale

Foramen spinosum

Foramen magnum

OCCIPITAL BONE

FRONTAL BONE

Olfactory foramina

Crista galli of ethmoid

Coronal suture

Lesser wing

Optic foramen

Greater wing

Sella turcica

Sphenoidal spine

SPHENOID BONE

Foramen lacerum

Petrous portion of temporal

Occipitomastoid suture

Lambdoidal suture

(c)

FIGURE 6-5 *(Continued)* Sphenoid bone. (c) Photograph of the sphenoid bone viewed in the floor of the cranium from above. (Courtesy of Lenny Patti.)

SPHENOID BONE

The **sphenoid bone** is situated at the anterior part of the base of the skull (Figure 6-5). The combining form *spheno* means wedge. This bone is referred to as the keystone of the cranial floor because it binds the other cranial bones together. If you view the floor of the cranium from above, you will note that the sphenoid articulates with the temporal bones anteriorly, and the occipital bone posteriorly. It lies posterior and slightly superior to the nasal cavities and forms part of the floor and sidewalls of the eye socket. The shape of the sphenoid is frequently described as a bat with outstretched wings.

The **body** of the sphenoid is the cubelike central portion between the ethmoid and occipital bones. It contains a large air space, the **sphenoidal sinus,** which drains into the nasal cavity (see Figure 6-8). On the superior surface of the sphenoid body is a depression called the **sella turcica,** meaning Turk's Saddle. This depression houses the pituitary gland. The **greater wings** of the sphenoid are lateral projections from the body and form the anterolateral floor of the cranium. The greater wings also form part of the lateral wall of the skull just anterior to the temporal bone. The pos-

terior portion of each greater wing contains a triangular projection, the **sphenoidal spine,** that fits into the angle between the squama and petrous portion of the temporal bone. The **lesser wings** are anterior and superior to the greater wings. They form part of the floor of the cranium and the posterior part of the **orbit,** or eye socket. Between the body and lesser wing, you can locate the **optic foramen** through which the optic nerve (II) and ophthalmic artery pass. Lateral to the body between the greater and lesser wings is a somewhat triangular slit called the **superior orbital fissure.** It is an opening for the oculomotor nerve (III), trochlear nerve (IV), ophthalmic branch of the trigeminal nerve (V), and abducens nerve (VI). This fissure may also be seen in the anterior view of the skull in Figure 6-2. On the inferior part of the sphenoid bone you can see the **pterygoid processes.** These structures project inferiorly from the points where the body and greater wings unite. The pterygoid processes form part of the lateral walls of the nasal cavity. At the base of the lateral pterygoid process in the greater wing is the **foramen ovale** through which the mandibular branch of the trigeminal nerve (V) passes. Another foramen, the **foramen spinosum,** lies at the posterior angle of the sphenoid and transmits the middle meningeal vessels.

The **foramen lacerum** is bounded anteriorly by the sphenoid bone, and medially by the sphenoid and occipital bones. Although the foramen is covered in part by a layer of fibrocartilage in living subjects, it transmits the internal carotid artery and the meningeal branch of the ascending pharyngeal artery. A final foramen associated with the sphenoid bone is the **foramen rotundum** through which the maxillary branch of the trigeminal nerve (V) passes. It is located at the junction of the anterior and medial parts of the sphenoid bone.

ETHMOID BONE

The **ethmoid bone** is a light, spongy bone located in the anterior part of the floor of the cranium between the orbits. It is anterior to the sphenoid and posterior to the nasal bones (Figure 6-6). The combining form *ethmos* means sieve. This bone forms part of the anterior portion of the cranial floor, the medial wall of the orbits, the superior portions of the nasal septum, or partition, and most of the sidewalls of the nasal roof. The ethmoid is the principal supporting structure of the nasal cavity.

Its **lateral masses** or **labyrinths** compose most of the wall between the nasal cavity and the orbits. They contain several air spaces, or "cells." which together form the **ethmoidal sinuses.** The sinuses are shown in Figure 6-8. The **perpendicular plate** (Figure 6-7) forms the superior portion of the nasal septum. The **cribriform plate,** or **horizontal plate,** lies in the anterior floor of the cranium and forms the roof of the nasal cavity. The cribriform plate contains the **olfactory foramina** through which the olfactory nerves (I) pass. These nerves function in smell (see Figure 6-6). Projecting upward from the horizontal plate is a triangular process called the **crista galli,** which means cock's comb. This structure serves as a point of attachment for the membranes that cover the brain. The labyrinths contain two thin, scroll-shaped bones on either side of the nasal septum. These are called the **superior nasal concha** and the **middle nasal concha.** The conchae allow for the efficient circulation and filtration of inhaled air before it passes into the trachea, the bronchi, and the lungs.

NASAL BONES

The paired **nasal bones** are small, oblong bones that meet at the middle and superior part of the face. Their fusion forms the superior part of the bridge of the nose. The inferior portion of the nose, indeed the major portion, consists of cartilage. See Figures 6-2 and 6-7.

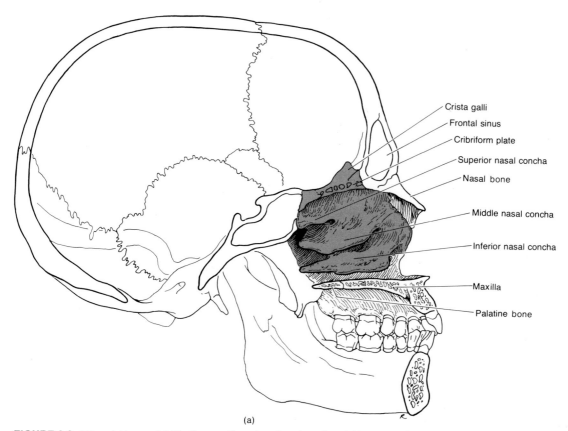

(a)

FIGURE 6-6 Ethmoid bone. (a) Median section showing the ethmoid bone on the inner aspect of the left part of the skull.

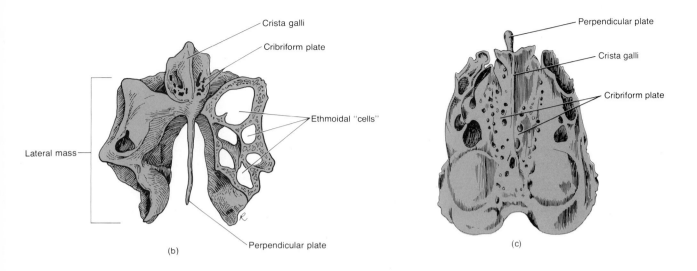

(b)

(c)

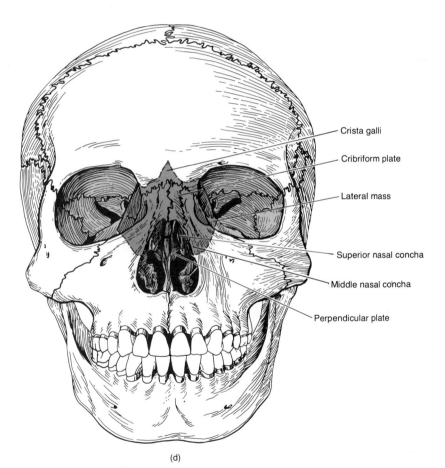

(d)

FIGURE 6-6 *(Continued)* Ethmoid bone. (b) Anterior view. A frontal section has been made through the left side to expose the ethmoidal "cells." (c) Superior view. (d) Frontal section of the head showing the ethmoid bone in relation to surrounding structures.

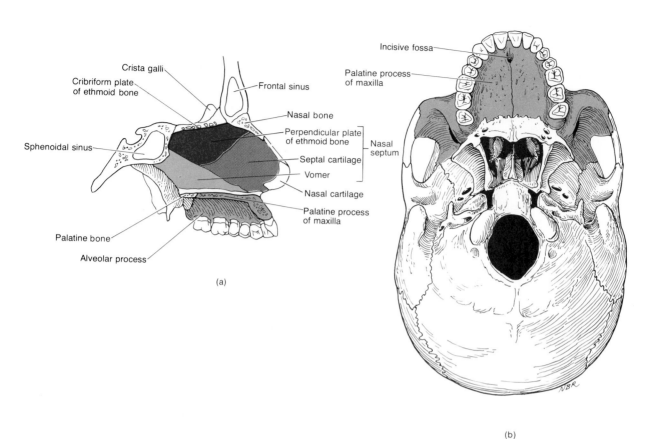

FIGURE 6-7 Maxillae. (a) Median view of the left maxilla. (b) Inferior view of the skull, showing the maxillae.

MAXILLAE

The paired maxillary bones unite to form the upper jawbone (Figure 6-7). The **maxillae** articulate with every bone of the face except the mandible, or lower jawbone. They form part of the floor of the orbits, part of the roof of the mouth, and part of the lateral walls and floor of the nasal cavities. The two portions of the maxillary bones unite, and the fusion is normally completed before birth. If the palatine processes of the maxillary bones do not unite before birth, a condition called **cleft palate** results. Another form of this condision, called **cleft lip,** involves a split in the upper lip. Cleft lip is often associated with cleft palate. Depending on the extent and position of the cleft, speech and swallowing may be affected.

Each maxillary bone contains a **maxillary sinus (antrum of Highmore)** that empties into the nasal cavity (see Figure 6-8). The **alveolar process** (*alveolus* = hollow) contains the bony sockets into which the teeth are set. The **palatine process** is a horizontal projection of the maxilla that forms the anterior and larger part of the hard palate, or anterior portion of the roof of the oral cavity. The **infraorbital foramen,** which can be seen in the anterior view of the skull in Figure 6-2, is an opening in the maxilla inferior to the orbit. The infraorbital nerve and artery are transmitted through this opening. Another prominent fossa in the maxilla is the **incisive fossa** just posterior to the incisor teeth. Through it pass branches of the descending palatine vessels and the nasopalatine nerve. A final fossa associated with the maxilla and sphenoid bone is the **inferior orbital fissure.** It is located between the greater wing of the sphenoid and the maxilla (see Figure 6-4). It transmits the maxillary branch of the trigeminal nerve (V) and the infraorbital vessels.

PARANASAL SINUSES

Cavities, called **paranasal sinuses,** are located in certain bones near the nasal cavity (Figure 6-8). The paranasal sinuses are lined with mucous membranes that are continuous with the lining of the nasal cavity. Cranial bones containing paranasal sinuses are the frontal bone, the sphenoid, the ethmoid, and the maxillae. The ethmoid sinus consists of a series of small cavities called ethmoid ''cells,'' which range in number from 3 to 18.

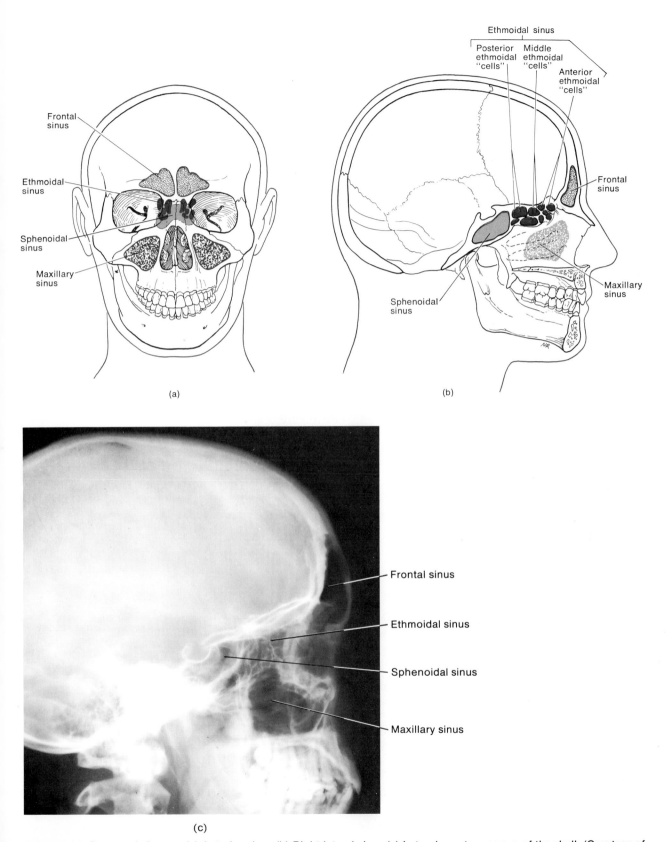

FIGURE 6-8 Paranasal sinuses. (a) Anterior view. (b) Right lateral view. (c) Lateral roentgenogram of the skull. (Courtesy of Eastman Kodak Company.)

ZYGOMATIC BONES

The two **zygomatic bones (malars),** commonly referred to as the cheekbones, form the prominences of the cheeks and part of the outer wall and floor of the orbits (Figure 6-2b).

The **temporal process** of the zygomatic bone projects posteriorly and articulates with the zygomatic process of the temporal bone. These two processes form the **zygomatic arch.** A foramen associated with the zygomatic bone is the **zygomaticofacial foramen** near the center of the bone (see Figure 6-2). It transmits the zygomaticofacial nerve and vessels.

MANDIBLE

The **mandible** or lower jawbone is the largest, strongest facial bone (Figure 6-9). It is the only movable bone in the skull.

In the lateral view you can see that the mandible consists of a curved, horizontal portion called the **body** and two perpendicular portions called the **rami.** The **angle** of the mandible is the area where each ramus meets the body. Each ramus has a **condylar process** that articulates with the mandibular fossa and articular tubercle of the temporal bone to form the temporo-

mandibular joint. It also has a **coronoid process** to which the temporalis muscle attaches. The depression between the coronoid and condylar processes is called the **mandibular notch.** The **mental foramen**(*mentum* = chin) is approximately below the first molar tooth. The mental nerve and vessels pass through this opening. Dentists inject anesthetics through this foramen. The **alveolar process,** like that of the maxillae, is an arch containing the sockets for the teeth. Another foramen associated with the mandible is the **mandibular foramen** on the medial surface of the ramus. It transmits the inferior alveolar nerve and vessels.

LACRIMAL BONES

The paired **lacrimal bones** (*lacrimal* = tear) are thin bones roughly resembling a fingernail in size and shape. They are the smallest bones of the face. These bones are posterior and lateral to the nasal bones in the medial wall of the orbit. They can be seen in the anterior and lateral views of the skull in Figure 6-2. The lacrimal bones form a part of the medial wall of the orbit. They also contain the **lacrimal foramina** through which the tear ducts pass into the nasal cavity (see Figure 6-2).

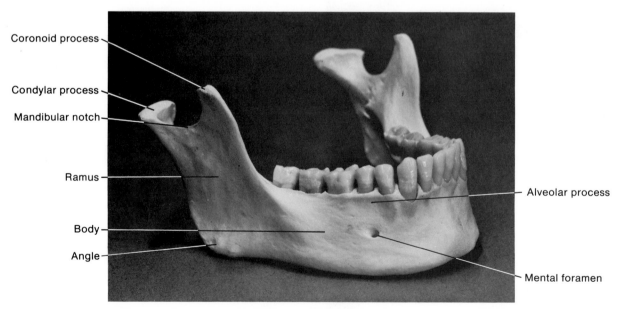

FIGURE 6-9 Right lateral photograph of the mandible. (Courtesy of Vincent P. Destro, Mayo Foundation.)

PALATINE BONES

The two **palatine bones** are L-shaped and form the posterior portion of the hard palate, part of the floor and lateral walls of the nasal cavities, and a small portion of the floor of the orbit. The posterior portion of the hard palate, which separates the nasal cavity from the oral cavity, is formed by the **horizontal plates** of the palatine bones. These can be seen in Figure 6-4. Two foramina associated with the palatine bones are the greater and lesser palatine foramina. The **greater palatine foramen,** at the posterior angle of the hard palate, transmits the greater palatine nerve and descending palatine vessels (see Figure 6-4). The **lesser palatine foramina,** usually two or more on each side, are posterior to the greater palatine foramina. They transmit the lesser palatine nerve (see Figure 6-4).

INFERIOR NASAL CONCHAE

Refer to the views of the skull in Figure 6-2a and 6-6a. The two **inferior nasal conchae** (*concha* = shell) are scroll-like bones that form a part of the lateral wall of the nasal cavity and project into the nasal cavity inferior to the superior and middle nasal conchae of the ethmoid bone. They serve the same function as the superior and middle nasal conchae; that is, they allow for the circulation and filtration of air before it passes into the lungs. The inferior nasal conchae are separate bones and not part of the ethmoid.

VOMER

The **vomer,** which means *plowshare,* is a roughly triangular bone that forms the inferior and posterior part of the nasal septum. It is clearly seen in the anterior view of the skull in Figure 6-2. The inferior border of the vomer articulates with the cartilage septum that divides the nose into a right and left nostril. Its superior border articulates with the perpendicular plate of the ethmoid bone. Thus the structures that form the *nasal septum,* or partition, are the perpendicular plate of the ethmoid, the septal cartilage, and the vomer (Figure 6-7a). If the vomer is pushed to one side—that is, deviated—the nasal chambers are of unequal size. See the skull viewed from below (Figure 6-4) for another view of the vomer.

Before continuing your study of the bones of the axial skeleton, refer to Exhibit 6-2 on page 128, which contains a summary of the foramina of the skull.

Hyoid Bone

The single **hyoid bone** (*hyoid* = U-shaped) is a unique component of the axial skeleton because it does not articulate with any other bone (Figure 6-10). Rather, it is suspended from the styloid process of the temporal bone by ligaments. The hyoid is located in the neck between the mandible and larynx. It supports the tongue and provides attachment for some of its muscles. Refer to the anterior and lateral views of the skull in Figure 6-2 to see the position of the hyoid bone.

The hyoid consists of a horizontal **body** and paired projections called the **lesser cornu** and the **greater cornu.** Muscles and ligaments attach to these paired projections.

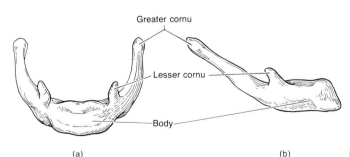

Greater cornu

Lesser cornu

Body

(a)　　　　　　(b)

FIGURE 6-10 Hyoid bone. (a) Anterior view. (b) Right lateral view.

EXHIBIT 6-2
SUMMARY OF FORAMINA OF THE SKULL

FORAMEN	LOCATION	STRUCTURES PASSING THROUGH
Carotid (Figure 6-4)	Petrous portion of temporal	Internal carotid artery
Greater palatine (Figure 6-4)	Posterior angle of hard palate	Greater palatine nerve and descending palatine vessels
Hypoglossal (Figure 6-5)	Superior to base of occipital condyles	Hypoglossal nerve and branch of ascending pharyngeal artery
Incisive (Figure 6-7)	Posterior to incisor teeth	Branches of descending palatine vessels and nasopalatine nerve
Inferior orbital (Figure 6-4)	Between greater wing of sphenoid and maxilla	Maxillary branch of trigeminal nerve (V), zygomatic nerve, and infraorbital vessels
Infraorbital (Figure 6-2a)	In maxilla inferior to orbit	Infraorbital nerve and artery
Jugular (Figure 6-4)	Posterior to carotid canal between petrous portion of temporal and occipital	Internal jugular vein, glossopharyngeal nerve (IX), vagus nerve (X), and accessory nerve (XI)
Lacerum (Figure 6-5)	Bounded anteriorly by sphenoid, posteriorly by petrous portion of temporal, and medially by the sphenoid and occipital	Internal carotid artery and branch of ascending pharyngeal artery
Lacrimal (Figure 6-2d)	Lacrimal bone	Lacrimal (tear) duct
Lesser palatine (Figure 6-4)	Posterior to greater palatine foramen	Lesser palatine nerves
Magnum (Figure 6-4)	Occipital bone	Medulla oblongata and its membranes, the accessory nerve (XI), and the vertebral and spinal arteries
Mandibular	Medial surface of ramus of mandible	Inferior alveolar nerve and vessels
Mastoid (Figure 6-4)	Posterior border of mastoid process of temporal bone	Vein to transverse sinus and branch of occipital artery to dura mater
Mental (Figure 6-9)	Inferior to second premolar tooth in mandible	Mental nerve and vessels
Olfactory (Figure 6-5)	Cribriform plate of ethmoid	Olfactory nerve (I)
Optic (Figure 6-5)	Between upper and lower portions of small wing of sphenoid	Optic nerve (II) and ophthalmic artery
Ovale (Figure 6-5)	Greater wing of sphenoid	Mandibular branch of trigeminal nerve (V)
Rotundum (Figure 6-5)	Junction of anterior and medial parts of sphenoid	Maxillary branch of trigeminal nerve (V)
Spinosum (Figure 6-5)	Posterior angle of sphenoid	Middle meningeal vessels
Stylomastoid (Figure 6-4)	Between styloid and mastoid processes of temporal	Facial nerve (VII) and stylomastoid artery
Superior orbital (Figure 6-5)	Between greater and lesser wings of sphenoid	Oculomotor nerve (III), trochlear nerve (IV), ophthalmic branch of trigeminal nerve (V), and abducens nerve (VI)
Supraorbital (Figure 6-2a)	Supraorbital margin of orbit	Supraorbital nerve and artery
Zygomaticofacial (Figure 6-2a)	Zygomatic bone	Zygomaticofacial nerve and vessels

Vertebral Column

The **vertebral column,** or **spine,** together with the sternum and ribs, constitutes the skeleton of the **trunk** of the body (Figure 6-11). The vertebral column is composed of a series of bones called **vertebrae.** In the average adult, the column measures about 71 cm (28 inches) in length. In effect, the vertebral column is a strong, flexible rod that moves anteriorly, posteriorly, and laterally. It encloses and protects the spinal cord, supports the head, and serves as a point of attachment for the ribs and the muscles of the back. Between the vertebrae are openings called **intervertebral foramina.** The nerves that connect the spinal cord to various parts of the body pass through these openings.

The adult vertebral column typically contains 26 vertebrae. These are distributed as follows: 7 **cervical vertebrae** in the neck region; 12 **thoracic vertebrae** posterior to the thoracic cavity; 5 **lumbar vertebrae** supporting the lower back; 5 **sacral vertebrae** fused into one bone called the **sacrum;** and usually four **coccygeal vertebrae** fused into one or two bones called the **coccyx.** Prior to the fusion of the sacral and coccygeal vertebrae, the total number of vertebrae is 33. Between the vertebrae are fibrocartilaginous **intervertebral discs.** These discs form strong joints and permit various movements of the column.

When viewed from the side, the vertebral column shows four curves. From the anterior view, these are alternately convex, meaning they curve out toward the viewer, and concave, meaning they curve away from the viewer. The curves of the column, like the curves in a long bone, are important because they increase its strength, help maintain balance in the upright position, absorb shocks from walking, and help protect the column from fracture.

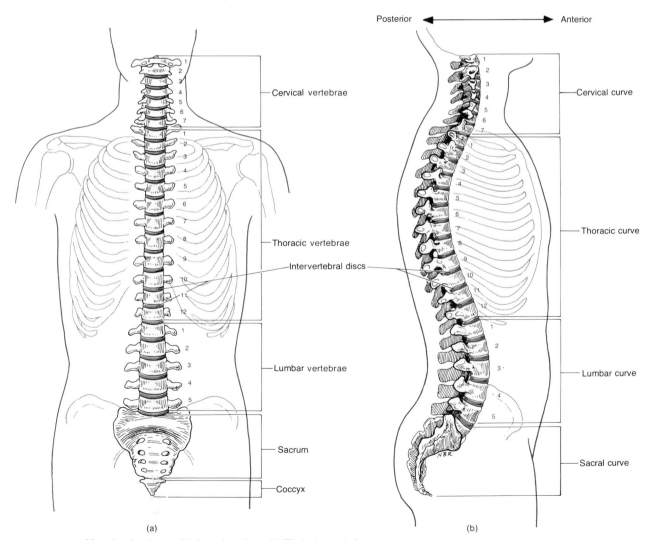

FIGURE 6-11 Vertebral column. (a) Anterior view. (b) Right lateral view.

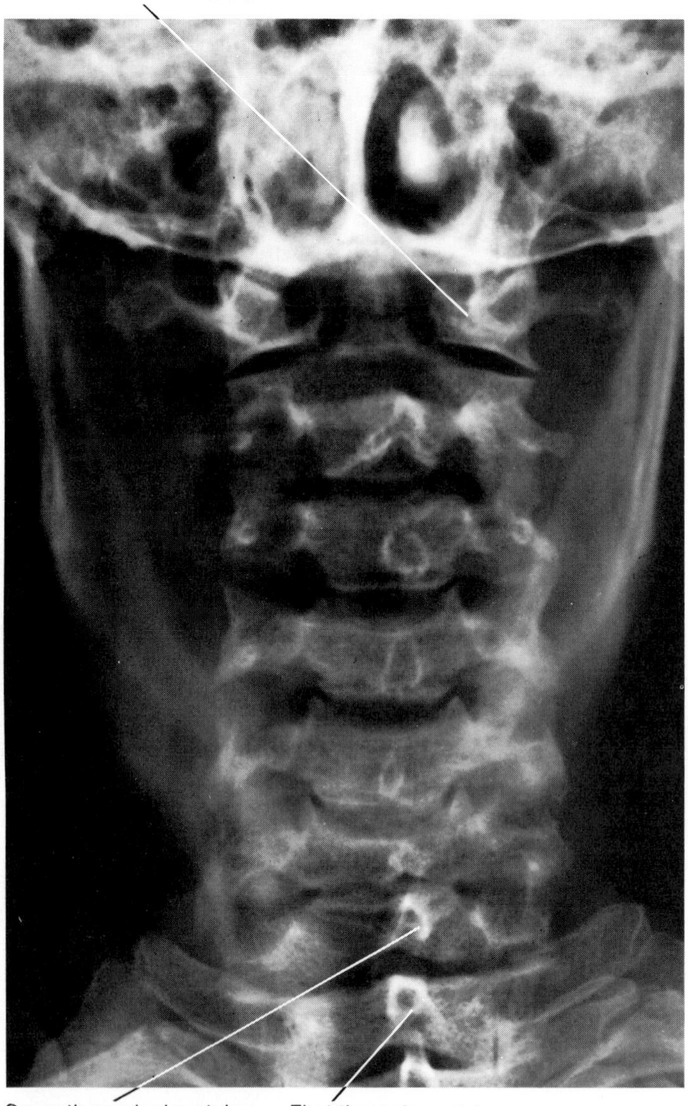

Seventh cervical vertebra

First thoracic vertebra

First cervical vertebra

Ribs

Seventh cervical vertebra   First thoracic vertebra
(c)

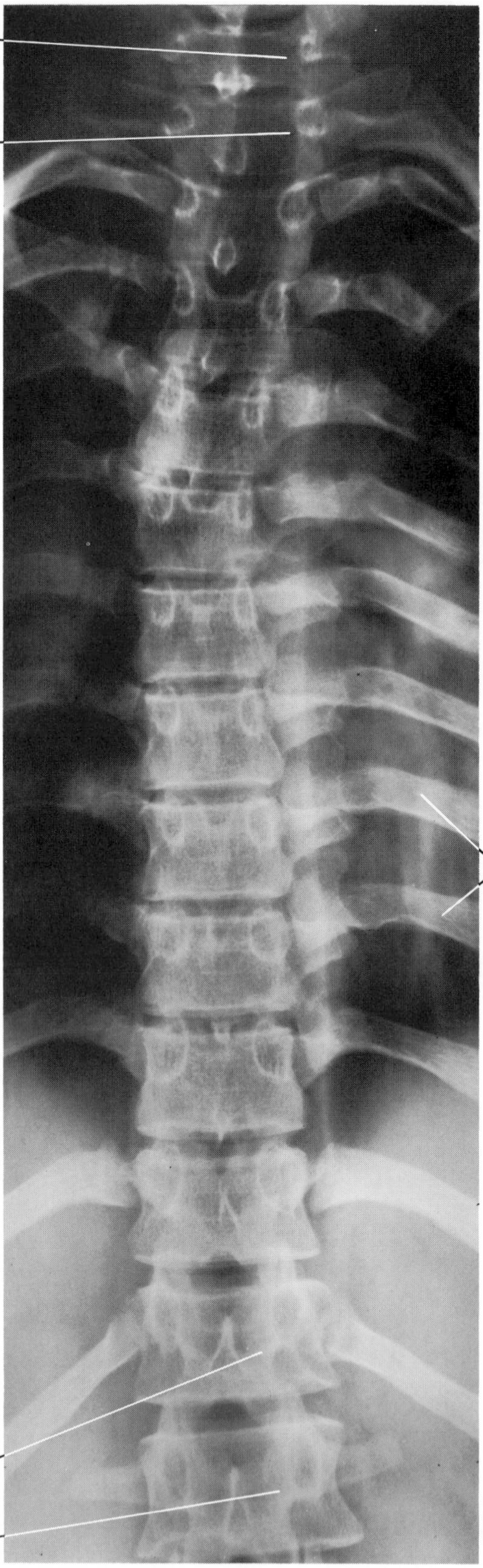

Twelfth thoracic vertebra

First lumbar vertebra

(d)

FIGURE 6-11 *(Continued)* Vertebral column. (c) Anteroposterior projection of the cervical vertebrae. (d) Anteroposterior projection of the thoracic vertebrae. (Courtesy of Eastman Kodak Company.) The term *projection* as used here and in subsequent legends of roentgenograms refers to which part of the body the x-ray beam enters and then exits. A *view,* by contrast, is determined by the side of the body closest to the film.

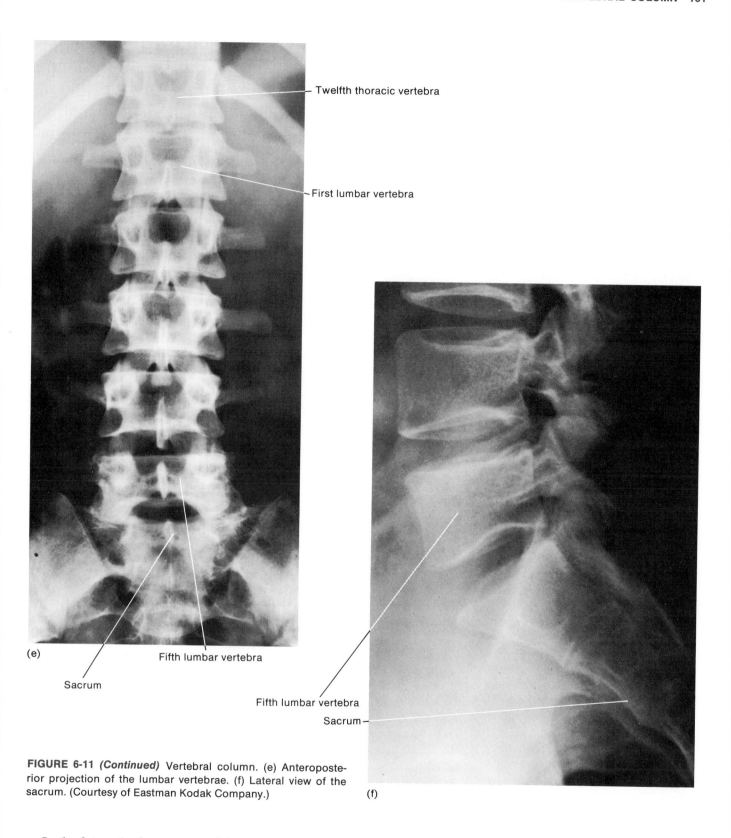

Twelfth thoracic vertebra

First lumbar vertebra

(e)

Fifth lumbar vertebra

Sacrum

Fifth lumbar vertebra

Sacrum

FIGURE 6-11 *(Continued)* Vertebral column. (e) Anteroposterior projection of the lumbar vertebrae. (f) Lateral view of the sacrum. (Courtesy of Eastman Kodak Company.)

(f)

In the fetus, the four curves of the vertebrae are not present. There is only a single curve that is anteriorly concave. At approximately the third postnatal month when an infant begins to hold its head erect, the **cervical curve** develops. Later, when the child stands and walks, the **lumbar curve** develops. The cervical and lumbar curves are convex anteriorly. Because they are modifications of the fetal positions, they are called **secondary curves.** The other two curves, the **thoracic curve** and the **sacral curve,** are anteriorly concave. Since they retain the anterior concavity of the fetus, they are referred to as **primary curves.**

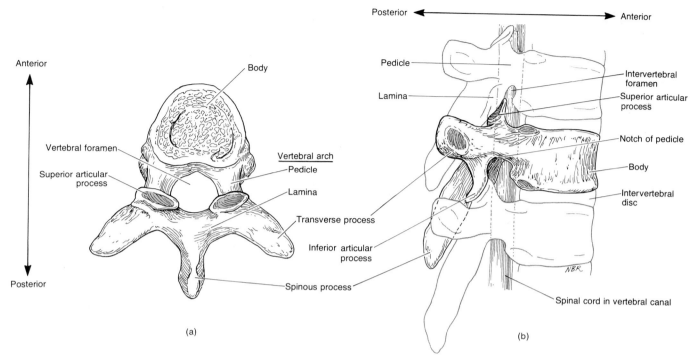

FIGURE 6-12 A typical vertebra. (a) Superior view. (b) Right lateral view.

TYPICAL VERTEBRA

All the vertebrae of the column are basically similar in structure (Figure 6-12). But there are differences in size, shape, and detail. A typical vertebra consists of the following components:

1. The **body** is the thick, disc-shaped anterior portion that is the weight-bearing part of a vertebra. Its superior and inferior surfaces are roughened for the attachment of intervertebral discs. The anterior and lateral surfaces contain nutrient foramina for blood vessels.

2. The **vertebral arch (neural arch)** extends posteriorly from the body of the vertebra. With the body of the vertebra, it surrounds the spinal cord. It is formed by two short, thick processes, the **pedicles,** which project posteriorly from the body to unite with the laminae. The **laminae** are the flat parts that join to form the posterior portion of the vertebral arch. The space that lies between the vertebral arch and body contains the spinal cord. This space is known as the **vertebral foramen.** The vertebral foramina of all vertebrae together form the **vertebral,** or spinal, **canal.** The pedicles are notched superiorly and inferiorly in such a way that, when they are arranged in the column, there is an opening between vertebrae on each side of the column. This opening, the **intervertebral foramen,** permits the passage of the spinal nerves.

3. Seven **processes** arise from the vertebral arch. At

the point where a lamina and pedicle join, a **transverse process** extends laterally on each side. A single **spinous process** or **spine** projects posteriorly and inferiorly from the junction of the laminae. These three processes serve as points of muscular attachment. The remaining four processes form joints with other vertebrae. The two **superior articular processes** of a vertebra articulate with the vertebra immediately superior to them. The two **inferior articular processes** of a vertebra articulate with the vertebra inferior to them.

CERVICAL REGION

When viewed from above, it can be seen that the bodies of **cervical vertebrae** are smaller than those of the thoracic vertebrae (Figure 6-13). The arches, however, are larger. The spinous processes of the second through sixth cervical vertebrae are often *bifid*—that is, with a cleft. Each cervical transverse process contains an opening, the **transverse foramen.** The vertebral artery and its accompanying vein and nerve fibers pass through it.

The first two cervical vertebrae differ considerably from the others. The first cervical vertebra, the **atlas,** is named for its support of the head. Essentially, the atlas is a ring of bone with **anterior** and **posterior arches** and large **lateral masses.** It lacks a body and a spinous process. The superior surfaces of the lateral masses, called **superior articular surfaces,** are concave and ar-

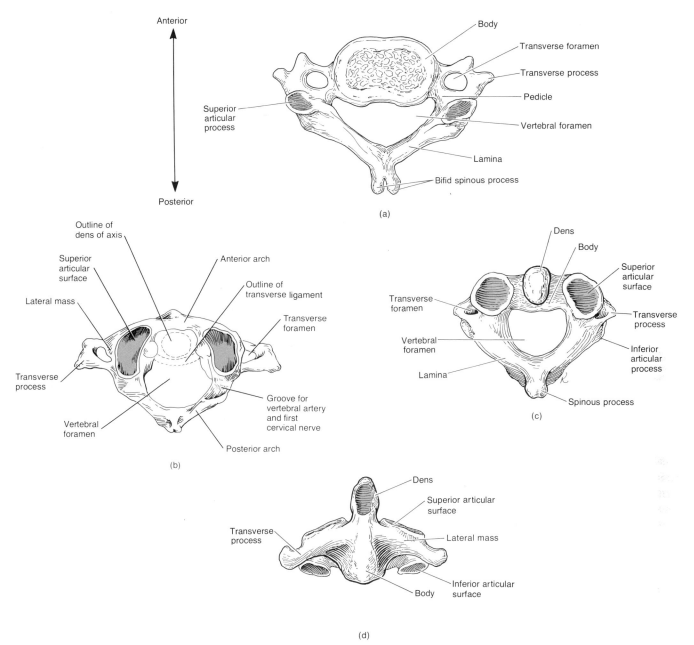

FIGURE 6-13 Cervical vertebrae. (a) Superior view of cervical vertebra. (b) Superior view of the atlas. (c) Superior view of the axis. (d) Anterior view of the axis.

ticulate with the occipital condyles of the occipital bone. This articulation permits the movement seen when nodding the head. The inferior surfaces of the lateral masses, the **inferior articular surfaces,** articulate with the second cervical vertebra. The transverse processes and **transverse foramina** of the atlas are quite large.

The second cervical vertebra, the **axis,** does have a **body.** A peglike process called the **dens,** or **odontoid process,** projects up through the ring of the atlas. The

dens makes a pivot on which the atlas and head rotate. This arrangement permits side-to-side rotation of the head.

The third through sixth cervical vertebrae correspond to the structural-pattern of the typical cervical vertebra shown. The seventh cervical vertebra, however, is somewhat different. It is called the **vertebra prominens** and is marked by a large, nonbifid spinous process that may be seen and felt at the base of the neck. (See Figure 6-11).

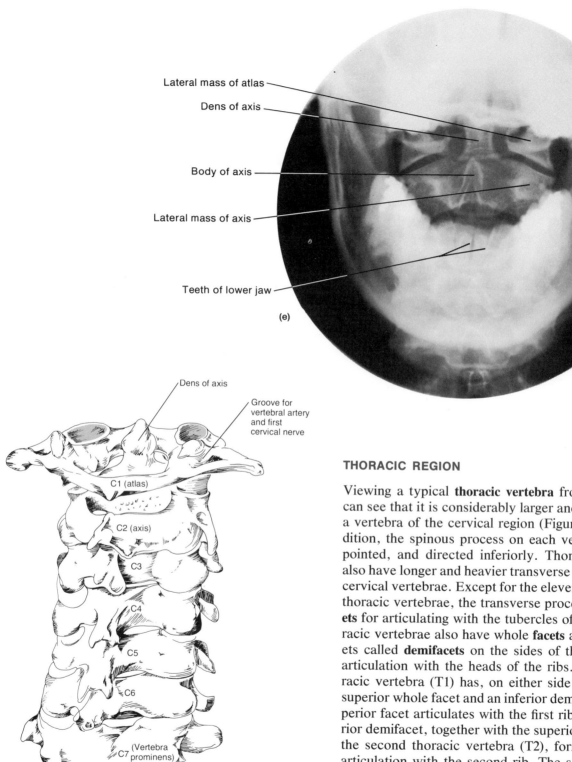

Lateral mass of atlas
Dens of axis
Body of axis
Lateral mass of axis
Teeth of lower jaw

(e)

Dens of axis
Groove for vertebral artery and first cervical nerve
C1 (atlas)
C2 (axis)
C3
C4
C5
C6
C7 (Vertebra prominens)

(f)

FIGURE 6-13 *(Continued)* Cervical vertebrae. (e) Anteroposterior projection of the atlas and axis taken through the open mouth. (Courtesy of John C. Bennett, St. Mary's Hospital, San Francisco.) (f) Cervical vertebrae articulated in posterior view.

THORACIC REGION

Viewing a typical **thoracic vertebra** from above, you can see that it is considerably larger and stronger than a vertebra of the cervical region (Figure 6-14). In addition, the spinous process on each vertebra is long, pointed, and directed inferiorly. Thoracic vertebrae also have longer and heavier transverse processes than cervical vertebrae. Except for the eleventh and twelfth thoracic vertebrae, the transverse processes have **facets** for articulating with the tubercles of the ribs. Thoracic vertebrae also have whole **facets** and/or half facets called **demifacets** on the sides of their bodies for articulation with the heads of the ribs. The first thoracic vertebra (T1) has, on either side of its body, a superior whole facet and an inferior demifacet. The superior facet articulates with the first rib, and the inferior demifacet, together with the superior demifacet of the second thoracic vertebra (T2), forms a facet for articulation with the second rib. The second through eighth thoracic vertebrae (T2–T8) have two demifacets on each side, a larger superior demifacet and a smaller inferior demifacet. When the vertebrae are articulated, they form whole facets for the heads of the ribs. The ninth thoracic vertebra (T9) has a single superior demifacet on either side of its body. Thoracic vertebrae ten through twelve (T10–T12) have whole facets on either side of their bodies (Figure 6-14b).

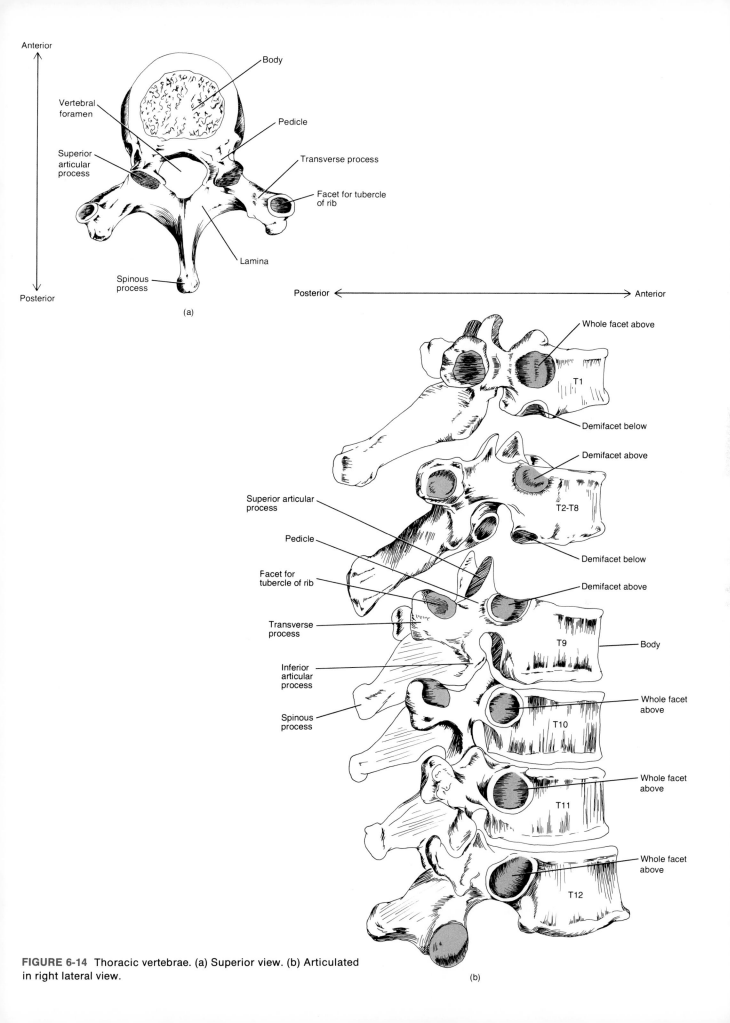

FIGURE 6-14 Thoracic vertebrae. (a) Superior view. (b) Articulated in right lateral view.

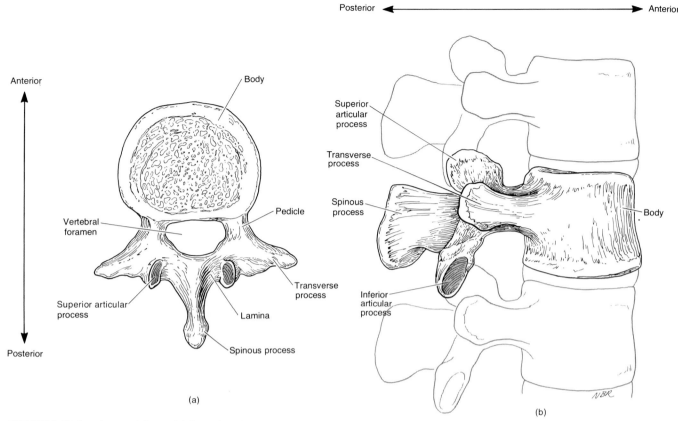

FIGURE 6-15 Lumbar vertebrae. (a) Superior view. (b) Articulated in right lateral view.

LUMBAR REGION

The **lumbar vertebrae** are the largest and strongest in the entire column (Figure 6-15). Their superior articular processes are directed medially instead of superiorly. And their inferior articular processes are directed laterally instead of inferiorly. The various projections of the lumbar vertebrae are short and thick, and the spinous process is quadrilateral in shape, thick and broad, and projects nearly straight posteriorly. The spinous processes are well adapted for the attachment of the large back muscles.

SACRUM AND COCCYX

The **sacrum** is a triangular bone formed by the union of five sacral vertebrae (Figure 6-16). These are indicated in the figure as S1 through S5. The sacrum serves as a strong foundation for the pelvic girdle. It is positioned at the posterior portion of the pelvic cavity between the two hipbones. Anterior and posterior views of the bone are shown here. The concave anterior side of the sacrum faces the pelvic cavity. It is

smooth and contains four **transverse lines** that mark the joining of the vertebral bodies. At the ends of these lines are four pairs of **pelvic foramina.** The convex, posterior surface of the sacrum is irregular. It contains a **median sacral crest,** a **lateral sacral crest,** and four pairs of **dorsal foramina.** These foramina communicate with the pelvic foramina through which nerves and blood vessels pass. The **sacral canal** is a continuation of the vertebral canal. The laminae of the fifth sacral vertebra, and sometimes the fourth, fail to meet. This leaves an inferior entrance to the vertebral canal called the **sacral hiatus.** On either side of the sacral hiatus are the **sacral cornua,** the inferior articular processes of the fifth lumbar vertebra. They are connected by ligaments to the coccygeal cornua of the coccyx. The superior border of the sacrum exhibits an anteriorly projecting border, the **sacral promontory.** It is an obstetrical landmark for measurements of the pelvis. An imaginary line running from the superior surface of the symphysis pubis to the sacral promontory separates the abdominal and pelvic cavities. Laterally, the sacrum has a large **auricular surface** for articulating with the ilium of the hipbone. Dorsal to the auricular

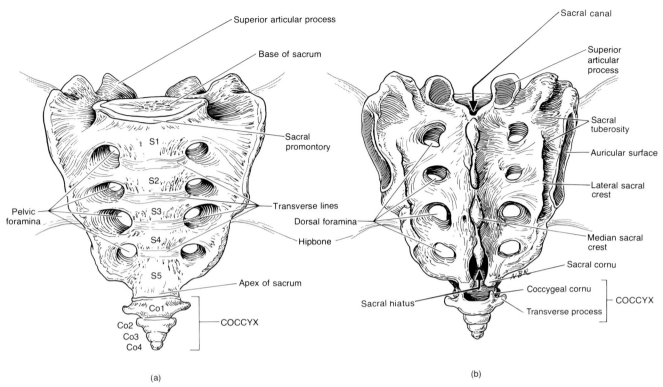

FIGURE 6-16 Sacrum and coccyx. (a) Anterior view. (b) Posterior view.

surface is a roughened surface, the **sacral tuberosity,** which contains depressions for the attachment of ligaments. Its **superior articular processes** articulate with the fifth lumbar vertebra.

The **coccyx** is also triangular in shape and is formed by the fusion of the coccygeal vertebrae, usually the last four. These are indicated in Figure 6-16 as Co1 through Co4. The dorsal surface of the body of the coccyx contains two long **coccygeal cornua** that are connected by ligaments to the sacral cornua. The coccygeal cornua are the pedicles and superior articular process of the first coccygeal vertebra. On the lateral surfaces of the body of the coccyx are a series of **transverse processes,** the first pair being the largest. The coccyx articulates superiorly with the sacrum. The coccyx is the most rudimentary part of the column, representing the vestige of a tail.

Thorax

Anatomically, the term **thorax** refers to the chest (Figure 6-17a). The skeletal portion of the thorax is a bony cage formed by the sternum, costal cartilage, ribs, and the bodies of the thoracic vertebrae. The thoracic cage is roughly cone-shaped, the narrow portion being superior and the broad portion inferior. It is flattened from front to back. The thoracic cage encloses and protects the organs in the thoracic cavity. It also provides support for the bones of the shoulder girdle and upper extremities.

STERNUM

The **sternum,** or breastbone, is a flat, narrow bone measuring about 15 cm (6 inches) in length. It is located in the median line of the anterior thoracic wall.

The sternum (see Figure 6-17a, b) consists of three basic portions: the **manubrium,** which is a triangular, superior portion; the **body,** which is the middle, largest portion; and the **xiphoid process,** which is the inferior, smallest portion. The manubrium has a depression on its superior surface called the **jugular (suprasternal) notch.** On each side of the jugular notch are **clavicular notches** that articulate with the medial ends of the clavicles. The manubrium also articulates with the first and second ribs. The body of the sternum articulates directly or indirectly with the second through tenth ribs. The xiphoid process has no ribs attached to it but provides attachment for some abdominal muscles.

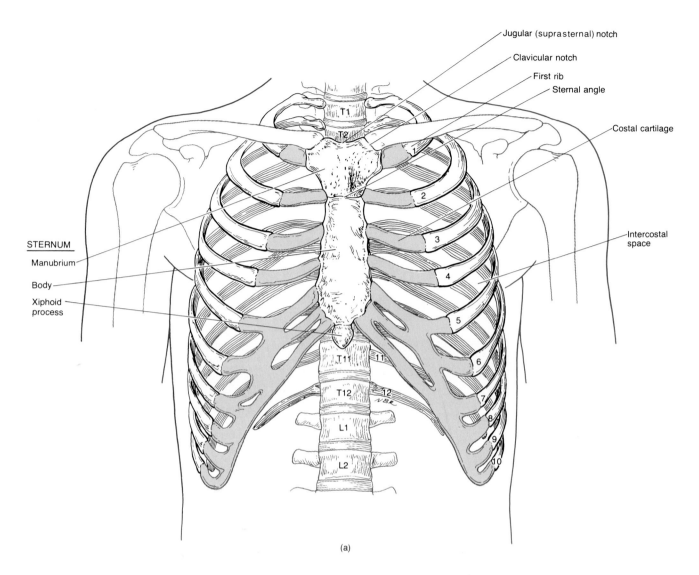

Jugular (suprasternal) notch

Clavicular notch

First rib

Sternal angle

Costal cartilage

Intercostal space

STERNUM

Manubrium

Body

Xiphoid process

(a)

FIGURE 6-17 The bony thorax. (a) Anterior view. (b) Right lateral view of the sternum.

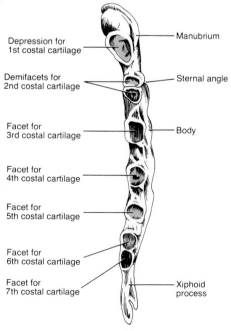

Manubrium

Depression for
1st costal cartilage

Demifacets for
2nd costal cartilage

Sternal angle

Facet for
3rd costal cartilage

Body

Facet for
4th costal cartilage

Facet for
5th costal cartilage

Facet for
6th costal cartilage

Facet for
7th costal cartilage

Xiphoid process

(b)

Diaphragm · Heart · Ribs · First rib · Clavicle

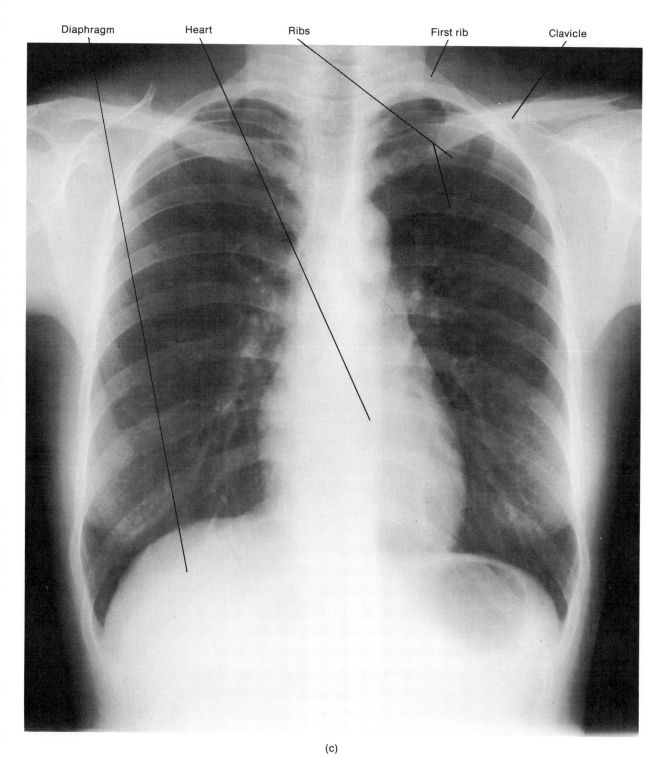

(c)

FIGURE 6-17 *(Continued)* The bony thorax. (c) Anteroposterior projection of the thorax. (Courtesy of John C. Bennett, St. Mary's Hospital, San Francisco.)

RIBS

Twelve pairs of **ribs** make up the sides of the thoracic cavity (see Figure 6-17a). The ribs increase in length from the first through seventh. Then they decrease in length to the twelfth rib. Each rib articulates posteriorly with its corresponding thoracic vertebra. The first through seventh ribs have a direct anterior attachment to the sternum by a strip of hyaline cartilage, called **costal cartilage** (*costa* = rib). These ribs are called **true ribs** or **vertebrosternal ribs.** The remaining five pairs of ribs are referred to as **false ribs** because their costal cartilages do not attach directly to the sternum. Of the false ribs, the cartilages of the eighth, ninth, and tenth ribs attach to each other and then to the cartilage of the seventh rib. These are thus called **vertebrochondral ribs.** The eleventh and twelfth ribs are designated as **floating ribs** because their anterior ends do not attach even indirectly to the sternum. They attach only to the muscles of the body wall.

Although there is some variation in rib structure, we will examine the parts of a typical rib when viewed from the right side and from behind (Figure 6-18). The **head** of a typical rib is projection at the posterior end of the rib. It is wedge shaped and consists of one or two **facets** that articulate with facets on the bodies of a single or two adjacent thoracic vertebrae. The facets on the head of a rib are separated by a horizontal **interarticular crest.** The inferior facet on the head of a rib is larger than the superior facet. The **neck** is a constricted portion just lateral to the head. A knoblike structure on the posterior surface where the neck joins the body is called a **tubercle.** It consists of a **nonarticular portion** which affords attachment to the ligament of the tubercle and an **articular portion** which articulates with the facet of a transverse process of a thoracic vertebra. The **body** or **shaft,** is the main part of the rib. The inner surface of the rib has a **costal groove** that protects blood vessels and a small nerve, artery, and vein. The posterior portion of the rib is therefore

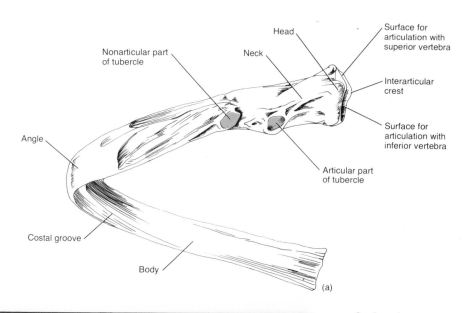

FIGURE 6-18 A typical rib. (a) Diagram of a left rib. (b) Photograph of the inner aspect of a portion of the fifth right rib. (Courtesy of Vincent P. Destro, Mayo Foundation.)

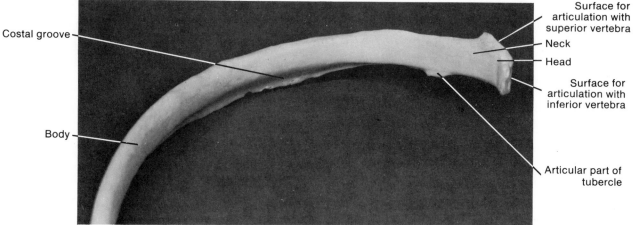

connected to a thoracic vertebra by its head and articular portion of a tubercle. The facet of the head fits into a facet on the body of a vertebra, and the articular portion of the tubercle articulates with the facet of the transverse process of the vertebra. Each of the second through ninth ribs articulates with the bodies of two adjacent vertebrae. The first, tenth, eleventh, and twelfth ribs articulate with only one vertebra each. On the eleventh and twelfth ribs, there is no articulation between the tubercles and the transverse processes of their corresponding vertebrae. Spaces between ribs called **intercostal spaces** are occupied by intercostal muscles.

STUDY OUTLINE

Types of Bones
1. On the basis of shape, bones are classified as long, short, flat, or irregular.
2. Wormian or sutural bones are found between the sutures of certain cranial bones. Sesamoid bones develop in tendons or ligaments.

Surface Markings
1. Markings are definitive areas on the surfaces of bones.
2. Each marking is structured for a specific function—joint formation, muscle attachment, or as a passageway for nerves and blood vessels.
3. Terms that describe markings include fissure, foramen, meatus, fossa, process, condyle, head, facet, tuberosity, crest, and spine.

Division of the Skeletal System
1. The axial skeleton consists of bones arranged along the longitudinal axis. The parts of the axial skeleton are the skull, hyoid bone, auditory ossicles, vertebral column, sternum, and ribs.
2. The appendicular skeleton consists of the bones of the girdles and the upper and lower extremities. The parts of the appendicular skeleton are the shoulder girdle, the bones of the upper extremities, the pelvic girdle, and the bones of the lower extremities.

Skull
1. Sutures are immovable joints between bones of the skull. Examples are coronal, sagittal, lambdoidal, and squamosal sutures.
2. Fontanels are membrane-filled spaces between the cranial bones of fetuses and infants. The major fontanels are the anterior, posterior, anterolateral, and posterolateral.
3. The skull consists of the cranium and the face. It is composed of 22 bones.
4. The eight cranial bones include the frontal, parietal (two), temporal (two), occipital, sphenoid, and ethmoid.
5. The 14 facial bones are the nasal (two), maxillae (two), zygomatic (two), mandible, lacrimal (two), palatine (two), inferior nasal conchae (two), and vomer.
6. Paranasal sinuses are cavities in bones of the skull that communicate with the nasal cavity. They are lined by mucous membranes.
7. The cranial bones containing the paranasal sinuses are the frontal, sphenoid, ethmoid, and maxilla.
8. The mastoid sinus is located in the temporal bone.

Vertebral Column
1. The vertebral column, the sternum, and the ribs constitute the skeleton of the trunk.
2. The bones of the adult vertebral column are the cervical vertebrae (7), thoracic vertebrae (12), lumbar vertebrae (5), the sacrum, and the coccyx.
3. The vertebral column contains primary curves (thoracic and sacral) and secondary curves (cervical and lumbar). These curves give strength, support, and balance.

Thorax
1. The thoracic skeleton consists of the sternum, the ribs and costal cartilages, and the thoracic vertebrae.
2. The thorax protects vital organs in the chest area.

REVIEW QUESTIONS

1. What are the four principal types of bones? Give an example of each.
2. What are surface markings? Describe and give an example of each.
3. Distinguish between the axial and appendicular skeletons. What subdivisions and bones are contained in each?
4. What bones compose the skull? The cranium? The face?
5. Define a suture. What are the four principal sutures of the skull? Where are they located?
6. What is a fontanel? Describe the location of the six major fontanels.
7. What is a paranasal sinus? Give examples of cranial bones that contain paranasal sinuses.
8. Identify each of the following: sinusitis, cleft palate, harelip, and mastoiditis.
9. What is the hyoid bone? In what respect is it unique? What is its function?
10. What bones form the skeleton of the trunk? Distinguish between the number of nonfused vertebrae found in the adult vertebral column and that of a child.
11. What is a curve in the vertebral column? How are primary and secondary curves differentiated from each other?
12. What bones form the skeleton of the thorax? What are the functions of the thoracic skeleton?

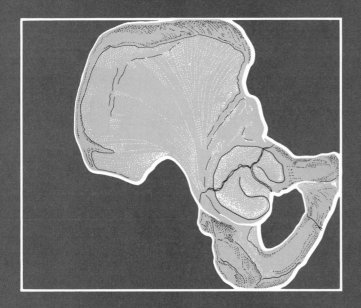

7

The Skeletal System: The Appendicular Skeleton

STUDENT OBJECTIVES

- Identify the bones of the shoulder girdle and their major markings.

- Identify the upper extremity, its component bones, and their markings.

- Identify the components of the pelvic girdle and their principal markings.

- Identify the lower extremity, its component bones, and their markings.

- Define the structural features and importance of the arches of the foot.

- Compare the principal structural differences between male and female skeletons.

This chapter discusses the bones of the appendicular skeleton, that is, the bones of the shoulder and pelvic girdles and extremities. The differences between male and female skeletons are also compared.

Shoulder Girdles

The **shoulder,** or **pectoral, girdles** attach the bones of the upper extremities to the axial skeleton (Figure 7-1). Structurally, each of the two shoulder girdles con-

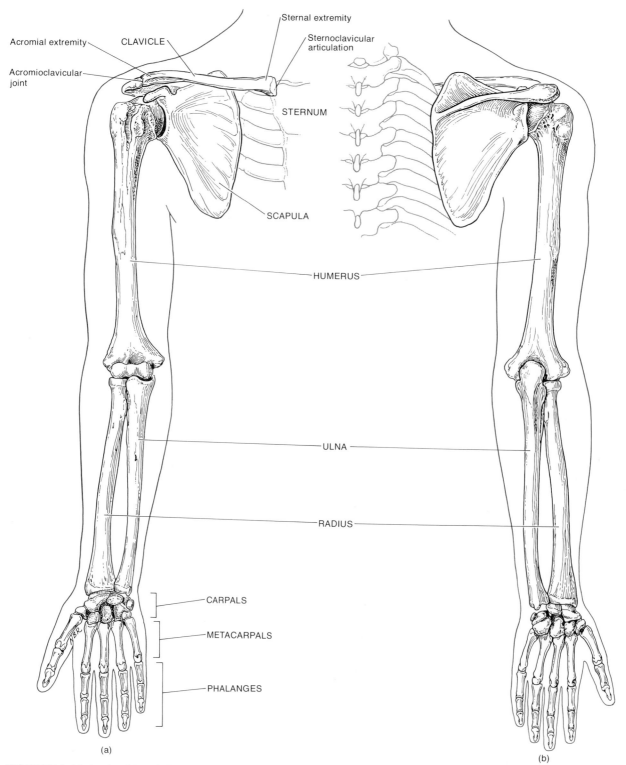

FIGURE 7-1 Right shoulder girdle and upper extremity. (a) Anterior view. (b) Posterior view.

FIGURE 7-2 Right clavicle viewed from below.

sists of two bones: a clavicle and a scapula. The shoulder girdles have no articulation with the vertebral column. The clavicle is the anterior component of the shoulder girdle and articulates with the sternum at the sternoclavicular joint. The posterior component, the scapula, which is positioned freely by complex muscle attachments, articulates with the clavicle and humerus. Although the shoulder joints are weak, they allow movement in many directions and are thus freely movable.

CLAVICLES

The **clavicles,** or collarbones, are long slender bones with a double curvature (Figure 7-2). The two bones lie horizontally in the superior and anterior part of the thorax superior to the first rib.

The medial end of the clavicle, the **sternal extremity,** is rounded and articulates with the sternum. The broad, flat, lateral end, the **acromial extremity,** articulates with the acromion process of the scapula. This joint is called the **acromioclavicular joint.** Refer to Figure 7-1 for a view of these articulations. The **conoid**

tubercle on the inferior surface of the lateral end of the bone serves as a point of attachment for a ligament. The **costal tuberosity** on the interior surface of the medial end also serves as a point of attachment for a ligament.

SCAPULAE

The **scapulae,** or shoulder blades, are large, triangular, flat bones situated in the dorsal part of the thorax between the levels of the second and seventh ribs (Figure 7-3). Their medial borders are located about 5 cm (2 inches) from the vertebral column.

A sharp ridge, the **spine,** runs diagonally across the posterior surface of the flattened, triangular **body.** The end of the spine projects as a flattened, expanded process called the **acromion.** This process articulates with the clavicle. Inferior to the acromion is a depression called the **glenoid cavity.** This cavity articulates with the head of the humerus to form the shoulder joint. The thin edge of the body near the vertebral column is the **medial** or **vertebral border.** The thick edge closer to the arm is the **lateral** or **axillary border.** The medial

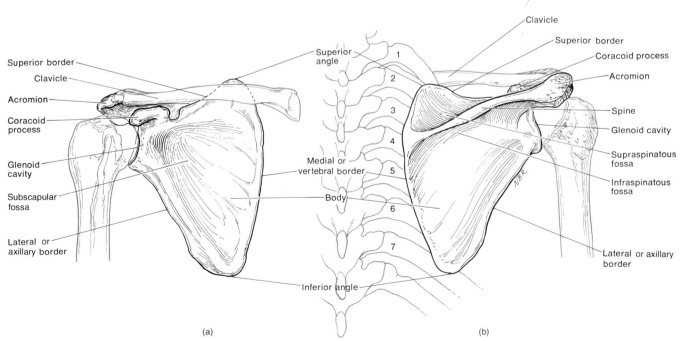

(a)   (b)

FIGURE 7-3 Right scapula. (a) Anterior view. (b) Posterior view.

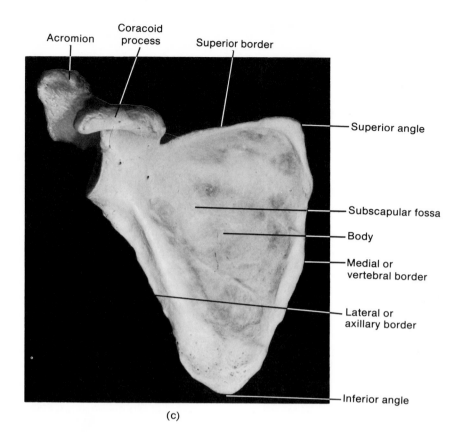

(c)

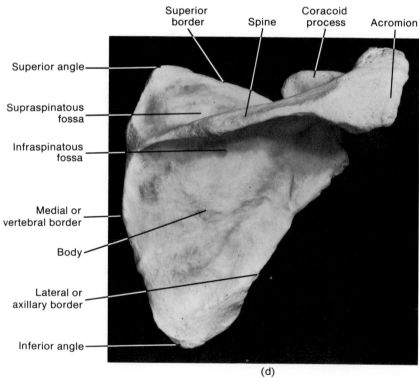

FIGURE 7-3 (Continued) Right scapula. (c) Photograph of anterior view. (d) Photograph of posterior view. (Courtesy of Lenny Patti.)

(d)

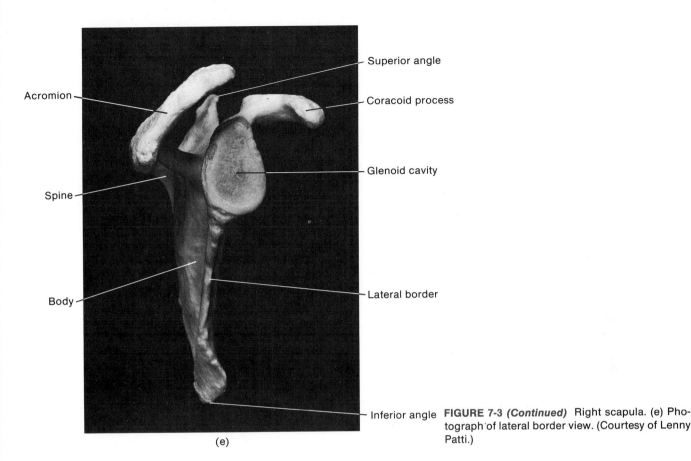

Acromion

Spine

Body

Superior angle

Coracoid process

Glenoid cavity

Lateral border

Inferior angle

(e)

FIGURE 7-3 (Continued) Right scapula. (e) Photograph of lateral border view. (Courtesy of Lenny Patti.)

and lateral borders joint at the **inferior angle.** The superior edge of the scapular body is called the **superior border.** At the lateral end of the superior border is a projection of the anterior surface called the **coracoid process** to which muscles attach. Above and below the spine are two fossae: the **supraspinatous fossa** and the **infraspinatous fossa,** respectively. Both serve as surfaces of attachment for shoulder muscles. On the anterior surface is a lightly hollowed-out area called the **subscapular fossa,** also a surface of attachment for shoulder muscles.

Upper Extremities

The **upper extremities** consist of 60 bones. The skeleton of the right upper extremity is shown in Figure 7-1. It includes a humerus in each arm, an ulna and radius in each forearm, carpals, or wrist bones, metacarpals, which are the palm bones, and phalanges in the fingers of each hand.

HUMERUS

The **humerus,** or arm bone, is the longest and largest bone of the upper extremity (Figure 7-4). It articulates proximally with the scapula and distally at the elbow with both ulna and radius.

The proximal end of the humerus consists of a **head** that articulates with the glenoid cavity of the scapula. It also has an **anatomical neck,** which is an oblique groove just distal to the head. The **greater tubercle** is a lateral projection distal to the neck. The **lesser tubercle** is an anterior projection. Between these tubercles runs an **intertubercular sulcus (bicipital groove). The surgical neck** is a constricted portion just distal to the tubercles and is named because of its liability to fracture. The **body** or shaft of the humerus is cylindrical at its proximal end. It gradually becomes triangular and is flattened and broad at its distal end. Along the middle portion of the shaft, there is a roughened, V-shaped area called the **deltoid tuberosity.** This area serves as a point of attachment for the deltoid muscle. The following parts are found at the distal end of the humerus. The **capitulum** is a rounded knob that articulates with the head of the radius. The **radial fossa** is a depression that receives the head of the radius when the forearm is flexed. The **trochlea** is a pulleylike surface that articulates with the ulna. The **coronoid fossa** is an anterior depression that receives part of the ulna when the forearm is flexed. The **olecranon fossa** is a posterior depression that receives the olecranon of the ulna when the forearm is extended. The **medial epicondyle** and **lateral epicondyle** are rough projections on either side of the distal end.

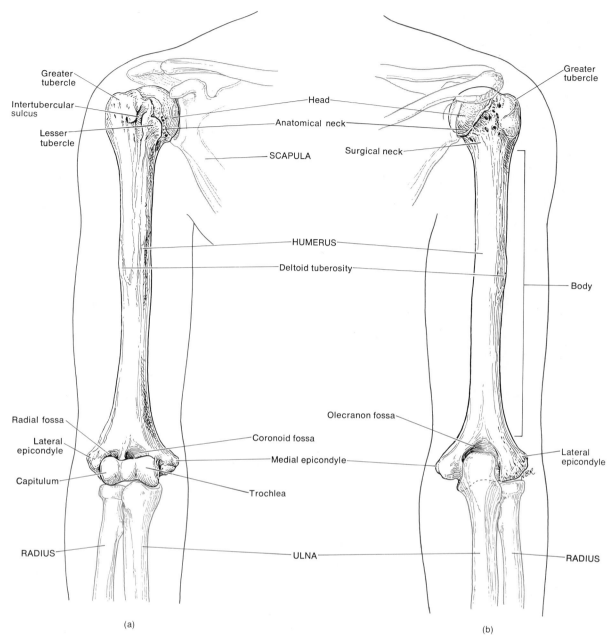

Greater tubercle

Intertubercular sulcus

Lesser tubercle

Head

Anatomical neck

SCAPULA

Surgical neck

Greater tubercle

HUMERUS

Deltoid tuberosity

Body

Radial fossa

Lateral epicondyle

Capitulum

Coronoid fossa

Medial epicondyle

Trochlea

Olecranon fossa

Lateral epicondyle

RADIUS

ULNA

RADIUS

(a)

(b)

FIGURE 7-4 Right humerus. (a) Anterior view. (b) Posterior view.

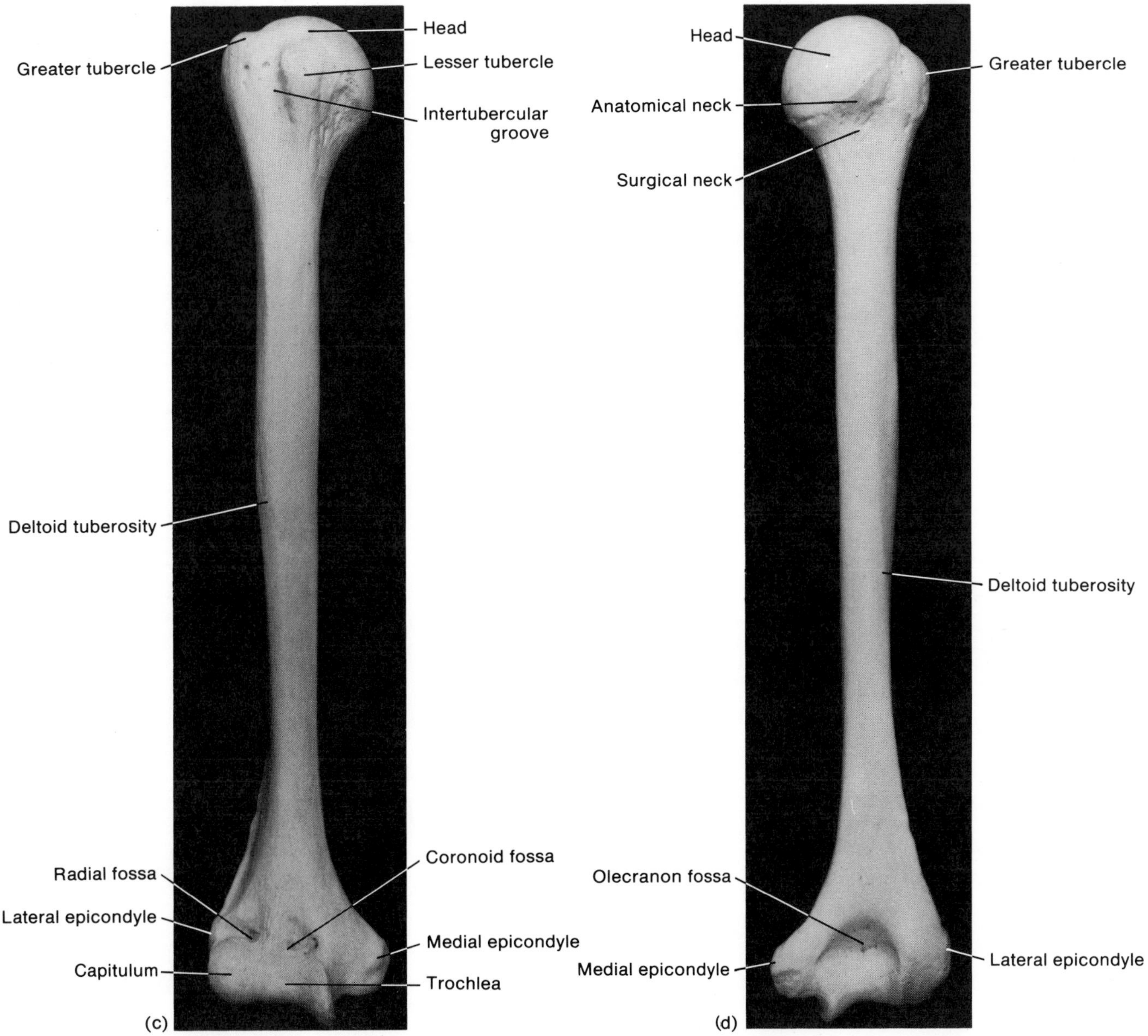

Greater tubercle — Head

— Lesser tubercle

— Intertubercular groove

Deltoid tuberosity —

Radial fossa — Coronoid fossa

Lateral epicondyle —

Capitulum — Medial epicondyle

Trochlea

(c)

Head —

Anatomical neck — Greater tubercle

Surgical neck —

Deltoid tuberosity

Olecranon fossa —

Medial epicondyle — Lateral epicondyle

(d)

FIGURE 7-4 (Continued) Right humerus. (c) Photograph in anterior view. (d) Photograph in posterior view. (Courtesy of Lenny Patti.)

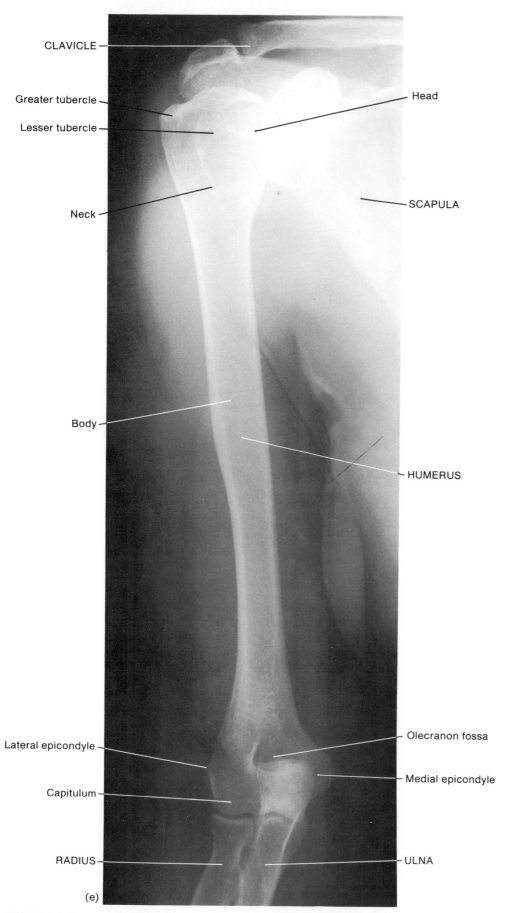

CLAVICLE —

Greater tubercle —

Lesser tubercle —

Neck —

Body —

Lateral epicondyle —

Capitulum —

RADIUS —

(e)

Head

SCAPULA

HUMERUS

Olecranon fossa

Medial epicondyle

ULNA

FIGURE 7-4 *(Continued)* Right humerus. (e) Anteroposterior projection. (Courtesy of John C. Bennett, St. Mary's Hospital, San Francisco.)

ULNA AND RADIUS

The **ulna** is the medial bone of the forearm (Figure 7-5). In other words, it is located on the small finger side.

The proximal end of the ulna presents an **olecranon (olecranon process),** which forms the prominence of the elbow. The **coronoid process** is an anterior projection that, together with the olecranon, receives the trochlea of the humerus. The **trochlear notch (semilunar notch)** is a curved area between the olecranon and the coronoid process. The trochlea of the humerus fits into this notch. The **radial notch** is a depression lo-

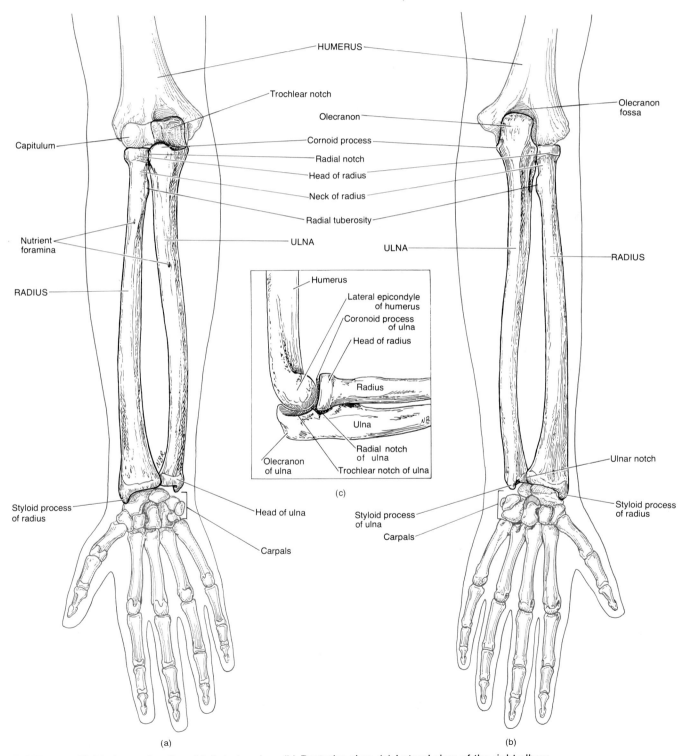

FIGURE 7-5 Right ulna and radius. (a) Anterior view. (b) Posterior view. (c) Lateral view of the right elbow.

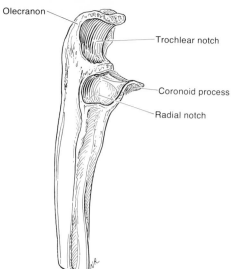

Olecranon

Trochlear notch

Coronoid process

Radial notch

(d)

FIGURE 7-5 *(Continued)* Right ulna and radius. (d) Details of proximal end of ulna. (e) Photograph of the radius in anterior view. (f) Photograph of the ulna in anterior view. (g) Photograph of the ulna in medial view. (Courtesy of Lenny Patti.)

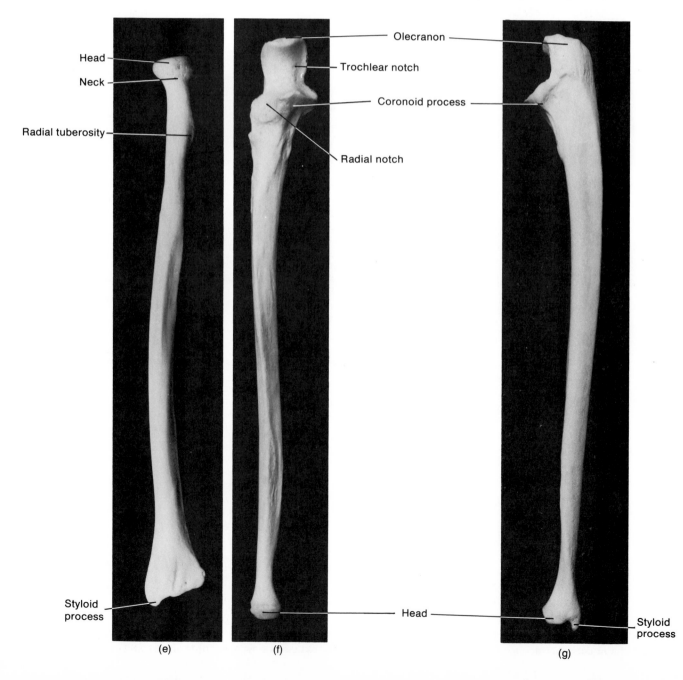

Head

Neck

Radial tuberosity

Styloid process

(e)

Olecranon

Trochlear notch

Coronoid process

Radial notch

Head

(f)

Olecranon

Coronoid process

Styloid process

(g)

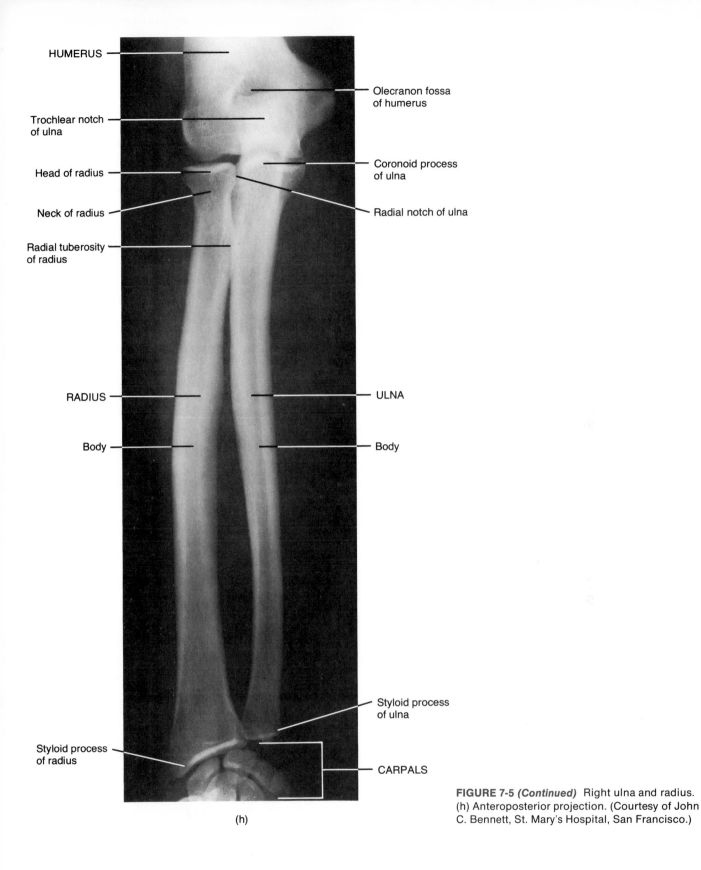

HUMERUS

Olecranon fossa
of humerus

Trochlear notch
of ulna

Head of radius

Coronoid process
of ulna

Neck of radius

Radial notch of ulna

Radial tuberosity
of radius

RADIUS

ULNA

Body

Body

Styloid process
of ulna

Styloid process
of radius

CARPALS

(h)

FIGURE 7-5 (Continued) Right ulna and radius.
(h) Anteroposterior projection. (Courtesy of John
C. Bennett, St. Mary's Hospital, San Francisco.)

cated laterally and inferiorly to the trochlear notch. It
receives the head of the radius. The distal end of the
ulna consists of a **head** that is separated from the wrist
by a fibrocartilage disc. A **styloid process** is on the posterior side of the distal end.

The **radius** is the lateral bone of the forearm; that is,
it is situated on the thumb side.

The proximal end of the radius has a disc-shaped
head that articulates with the capitulum of the humerus
and radial notch of the ulna. It also has a raised, rough-

ened area on the medial side called the **radial tuberosity.** This is a point of attachment for the biceps muscle. The shaft of the radius widens distally to form a concave inferior surface that articulates with two bones of the wrist called the lunate and navicular bones. Also at the distal end is a **styloid process** on the lateral side and a medial, concave **ulnar notch** for articulation with the distal end of the ulna. A common fracture of the radius called a **Colles' fracture** occurs along the shaft about 2.3 cm (1 inch) from the distal end of the bone.

CARPUS, METACARPUS, AND PHALANGES

The **carpus,** or wrist, consists of eight small bones united to each other by ligaments (Figure 7-6). The bones are arranged in two transverse rows, with four bones in each row. The proximal row, from the lateral to medial position, consists of the following bones: **navicular (scaphoid), lunate, triangular (triquetral),** and **pisiform.** In about 70 percent of cases involving carpal fractures, only the navicular is involved. The distal row of bones, from lateral to medial position, consists

of the following: **greater multangular (trapezium), lesser multangular (trapezoid), capitate,** and **hamate.**

The five bones of the **metacarpus** constitute the palm of the hand. Each metacarpal bone consists of a proximal **base,** a **shaft,** and a distal **head.** The metacarpal bones are numbered I to V, starting with the lateral bone. The bases articulate with the distal row of carpal bones and with one another. The heads articulate with the proximal phalanges of the fingers. The heads of the metacarpals are commonly called the ''knuckles'' and are readily visible when the fist is clenched.

The **phalanges,** or bones of the fingers, number 14 in each hand. Each consists of a proximal **base,** a **shaft,** and a distal **head.** There are two phalanges in the first digit, called the thumb or pollex, and three phalanges in each of the remaining four digits. The first row of phalanges, the **proximal row,** articulates with the metacarpal bones and second row of phalanges. The second row of phalanges, the **middle row,** articulates with the proximal row and the third row. The third row of phalanges, the **distal row,** articulates with the middle row. A single finger bone is referred to as a **phalanx.** The thumb has no middle phalanx.

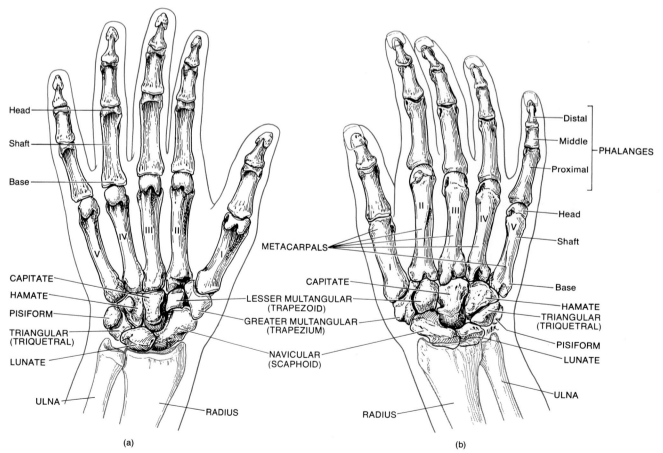

FIGURE 7-6 Right wrist and hand. (a) Anterior view. (b) Posterior view.

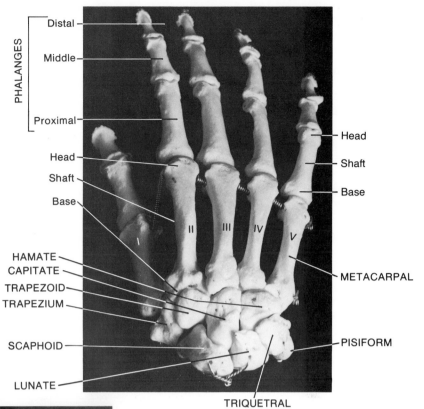

PHALANGES
- Distal
- Middle
- Proximal

Head
Shaft
Base

Head
Shaft
Base

HAMATE
CAPITATE
TRAPEZOID
TRAPEZIUM

I II III IV V

METACARPAL

SCAPHOID

PISIFORM

LUNATE

TRIQUETRAL

(c)

FIGURE 7-6 *(Continued)* Right wrist and hand. (c) Photograph in posterior view. (Photograph courtesy of Lenny Patti). (d) Anteroposterior projection. Note the sesamoid bone. (Courtesy of Daniel Sorrentino, R.T.)

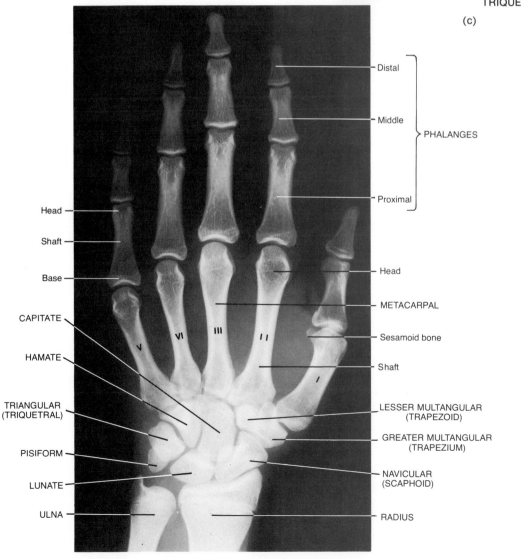

- Distal
- Middle
- Proximal
PHALANGES

Head
Shaft
Base

Head
METACARPAL
Sesamoid bone
Shaft

CAPITATE
HAMATE

TRIANGULAR
(TRIQUETRAL)

PISIFORM

LUNATE

ULNA

V VI III II I

LESSER MULTANGULAR
(TRAPEZOID)

GREATER MULTANGULAR
(TRAPEZIUM)

NAVICULAR
(SCAPHOID)

RADIUS

(d)

Pelvic Girdle

The **pelvic girdle** consists of the two **coxal bones,** commonly called the pelvic, innominate, or hipbones (Figure 7-7). The pelvic girdle provides a strong and stable support for the lower extremities on which the weight of the body is carried. The coxal bones are united to each other anteriorly at the symphysis pubis. They unite posteriorly to the sacrum.

Together with the sacrum and coccyx, the pelvic girdles form the basinlike structure called the **pelvis.** The pelvis is divided into a greater pelvis and a lesser pelvis. The **greater** or **false pelvis** represents the ex-

panded portion situated superior to the narrow bony ring called the **brim of the pelvis.** The greater pelvis consists laterally of the two ilia and posteriorly of the superior portion of the sacrum. There is no bony component in the anterior aspect of the greater pelvis. Rather, the front is formed by the walls of the abdomen. The **lesser** or **true pelvis** is inferior and posterior to the pelvic brim. It is formed by parts of the ilium, publis, sacrum, and coccyx. The lesser pelvis contains a superior opening called the **pelvic inlet** and an inferior opening called the **pelvic outlet.** *Pelvimetry* is the measurement of the size of the inlet and outlet of the birth canal.

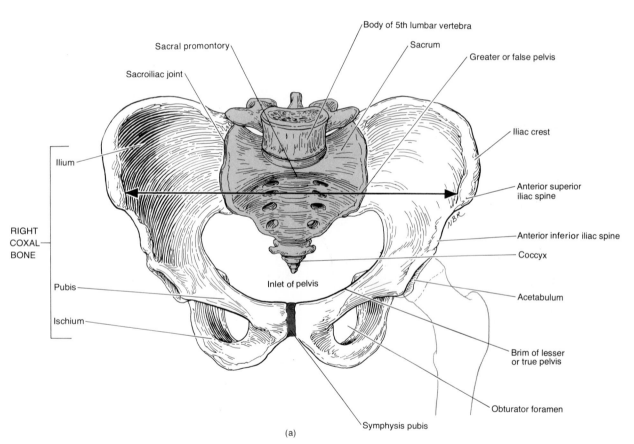

(a)

FIGURE 7-7 Pelvic girdle. (a) Anterior view.

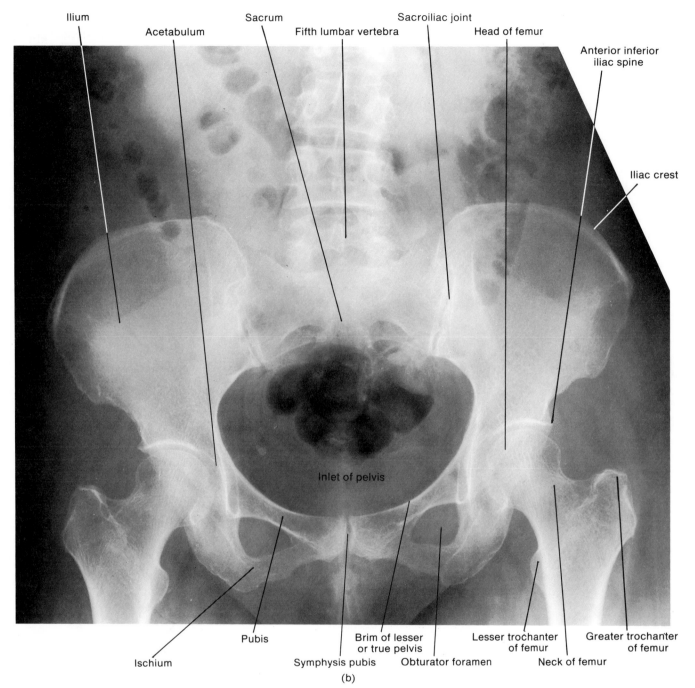

Ilium

Acetabulum

Sacrum

Fifth lumbar vertebra

Sacroiliac joint

Head of femur

Anterior inferior iliac spine

Iliac crest

Inlet of pelvis

Pubis

Ischium

Brim of lesser or true pelvis

Symphysis pubis

Obturator foramen

Lesser trochanter of femur

Neck of femur

Greater trochanter of femur

(b)

FIGURE 7-7 *(Continued)* Pelvic girdle. (b) Anteroposterior projection. (Courtesy of John C. Bennett, St. Mary's Hospital, San Francisco.)

COXAL BONES

The two **coxal bones (os coxae)** of a newborn consist of three components: a superior **ilium,** an inferior and anterior **pubis,** and an inferior and posterior **ischium** (Figure 7-8). Eventually, the three separate bones fuse into one. The area of fusion is a deep, lateral fossa called the **acetabulum.** This structure is the socket for the head of the femur. On the inferior portion of the acetabulum is a deep notch, the **acetabular notch.** Although the adult coxae are both single bones, it is common to discuss the bones as if they still consisted of three portions.

The ilium is the largest of the three subdivisions of the coxal bone. Its superior border, the **iliac crest,** ends anteriorly in the **anterior superior iliac spine.** The **anterior inferior iliac spine** is located inferior to the anterior superior spine. Posteriorly, the iliac crest ends in the **posterior superior iliac spine.** The **posterior inferior iliac spine** is just inferior. The spines serve as points of attachment for muscles of the abdominal wall. Slightly inferior to the posterior inferior iliac spine is the **greater sciatic notch.** The internal surface of the ilium seen from the medial side is the **iliac fossa.** It is a concavity where the iliacus muscle attaches. Posterior to this fossa are the **iliac tuberosity,** a point of attachment for the sacroiliac ligament, and the **auricular surface,** which articulates with the sacrum. The other conspicuous markings of the ilium are three arched lines on its gluteal (buttock) surface called the **posterior gluteal line,** the **anterior gluteal line,** and the **inferior gluteal line.** The gluteal muscles attach to the ilium between these lines.

The ischium is the inferior, posterior portion of the coxal bone. It contains a prominent **ischial spine,** a **lesser sciatic notch** below the spine, and an **ischial tuberosity.** The rest of the ischium, the **ramus,** joins with the pubis and together they surround the **obturator foramen.**

The pubis is the anterior and inferior part of the coxal bone. It consists of a **superior ramus,** an **inferior ramus,** and a **body** that contributes to the formation of the symphysis pubis. The **symphysis pubis** is the joint between the two coxal bones. It consists of fibrocartilage (Figure 7-7). The **acetabulum** is the socket formed by the ilium, ischium, and pubis. Two-fifths of the acetabulum is formed by the ilium, two-fifths by the ischium, and one-fifth by the pubis.

Lower Extremities

The **lower extremities** are composed of 60 bones (Figure 7-9). These include the femur of each thigh, each kneecap, the fibula and tibia in each leg, and ankle bones in each ankle, and the metatarsals and phalanges of each foot.

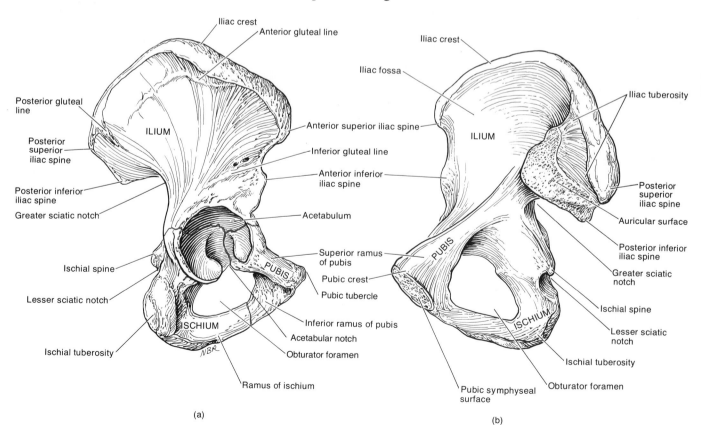

FIGURE 7-8 Right coxal bone. (a) Lateral view. (b) Medial view. The lines of fusion of the ilium, ischium, and pubis that are shown in color in (a) are not actually visible in an adult bone.

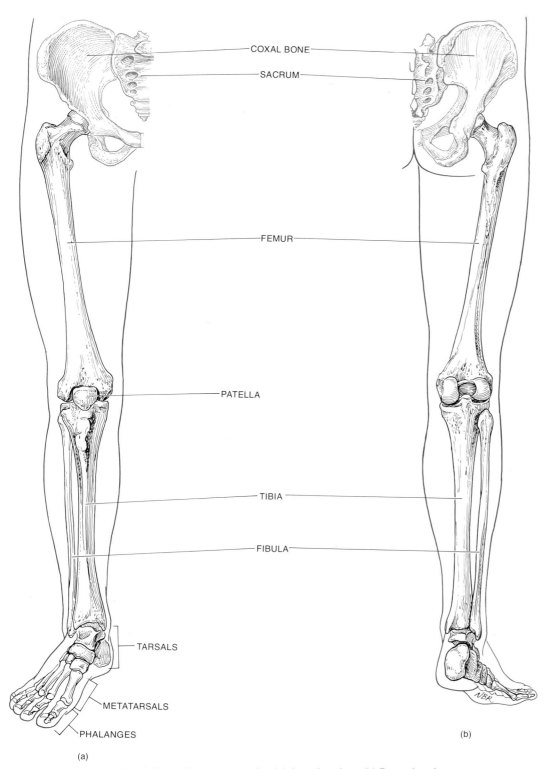

COXAL BONE

SACRUM

FEMUR

PATELLA

TIBIA

FIBULA

TARSALS

METATARSALS

PHALANGES

(a)

(b)

FIGURE 7-9 Right pelvic girdle and lower extremity. (a) Anterior view. (b) Posterior view.

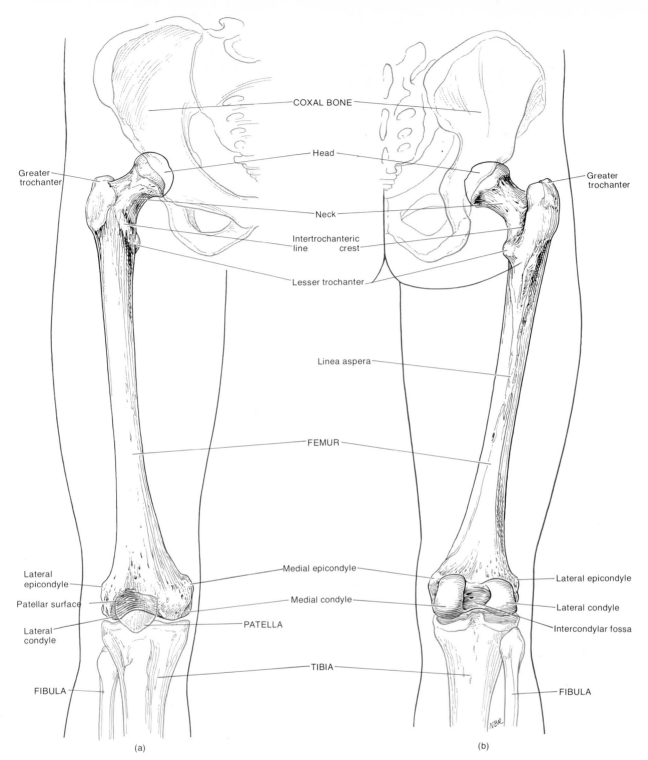

FIGURE 7-10 Right femur. (a) Anterior view. (b) Posterior view. A roentgenogram of the proximal end of the femur is shown in Figure 7-7b.

FEMUR

The **femur,** or thigh bone, is the longest and heaviest bone in the body (Figure 7-10). Its proximal end articulates with the coxal bone. Its distal end articulates with the tibia. The shaft of the femur bows medially so that it approaches the femur of the opposite thigh. As

a result of this convergence, the knee joints are brought nearer to the body's line of gravity. The degree of convergence is greater in the female because the female pelvis is broader.

The proximal end of the femur consists of a rounded **head** that articulates with the acetabulum of the coxal bone. The **neck** of the femur is a constricted region distal to the head. A fairly common fracture in the elderly occurs at the neck of the femur. Apparently the neck becomes so weak that it fails to support the body. The **greater trochanter** and **lesser trochanter** are projections that serve as points of attachment for some of the thigh and buttock muscles. Between the trochanters on the anterior surface is a narrow **intertrochanteric line.** Between the trochanters on the posterior surface is an **intertrochanteric crest.**

The shaft of the femur contains a rough vertical ridge on its posterior surface called the **linea aspera.** This ridge serves for the attachment of several thigh muscles.

The distal end of the femur is expanded and includes the **medial condyle** and **lateral condyle.** These articulate with the tibia. A depressed area between the condyles on the posterior surface is called the **intercondylar fossa.** The **patellar surface** is located between the condyles on the anterior surface. Lying superior to the condyles are the **medial epicondyle** and **lateral epicondyle.**

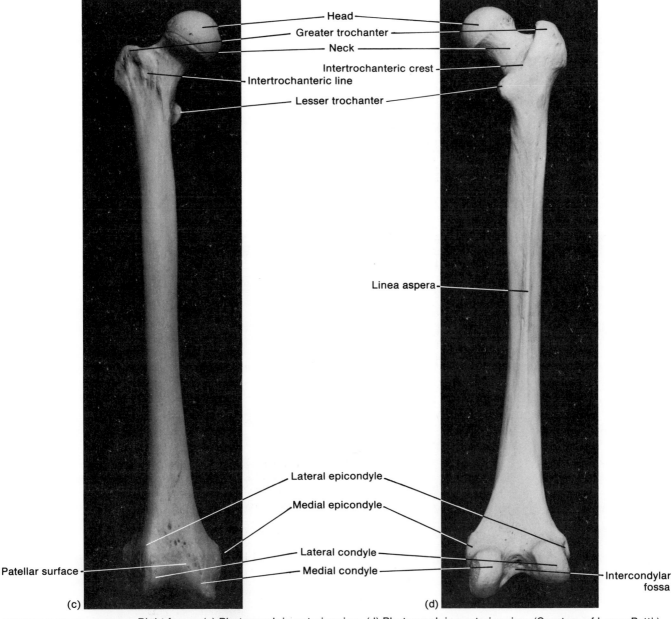

FIGURE 7-10 *(Continued)* Right femur. (c) Photograph in anterior view. (d) Photograph in posterior view. (Courtesy of Lenny Patti.)

PATELLA

The **patella,** or kneecap, is a small, triangular bone anterior to the knee joint (Figure 7-11). It develops in the tendon of the quadriceps femoris muscle. A bone that forms in a tendon, such as the patella, is called a *sesamoid bone*. The broad superior end of the patella is called the **base.** The pointed inferior end is the **apex.** The posterior surface contains two articular surfaces. These are the **articular facets.** for the medial and lateral condyles of the femur.

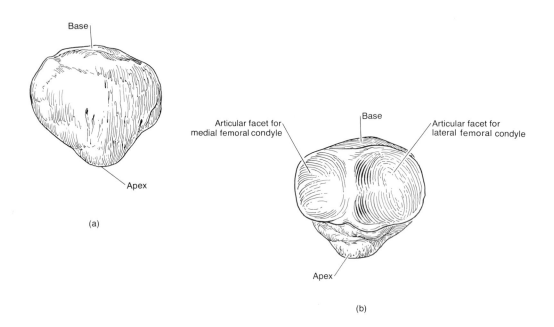

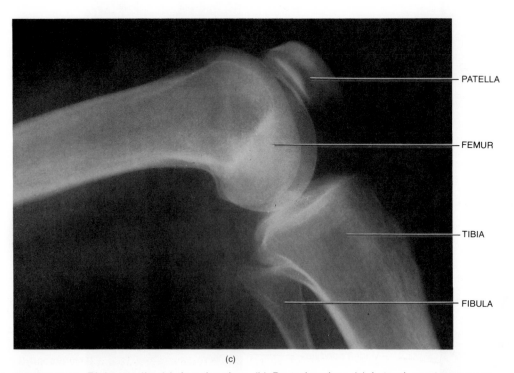

FIGURE 7-11 Right patella. (a) Anterior view. (b) Posterior view. (c) Lateral roentgenogram. (Courtesy of Daniel Sorrentino, R. T.)

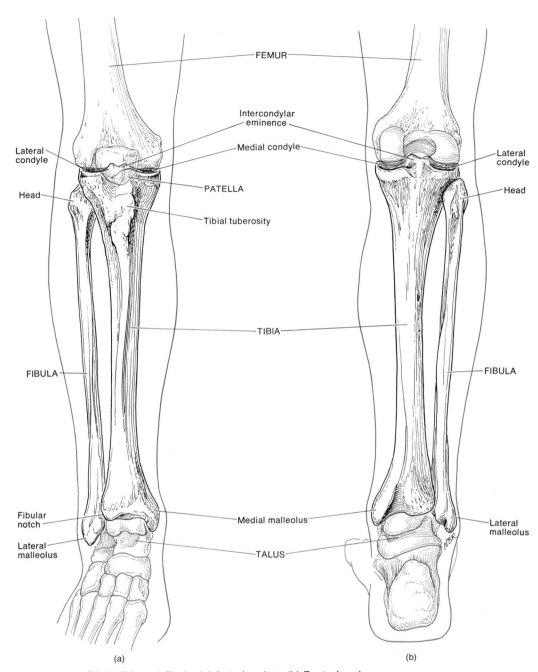

FIGURE 7-12 Right tibia and fibula. (a) Anterior view. (b) Posterior view.

TIBIA AND FIBULA

The **tibia,** or shinbone, is the larger, medial bone of the leg (Figure 7-12). It bears the major portion of the weight on the leg. The tibia articulates at its proximal end with the femur and at its distal end with the fibula of the leg and talus of the ankle.

The proximal end of the tibia is expanded into a **lateral condyle** and a **medial condyle.** These articulate with the condyles of the femur. The slightly concave condyles are separated by an upward projection called the **intercondylar eminence.** The **tibial tuberosity** on the anterior surface is a point of attachment for the patellar ligament.

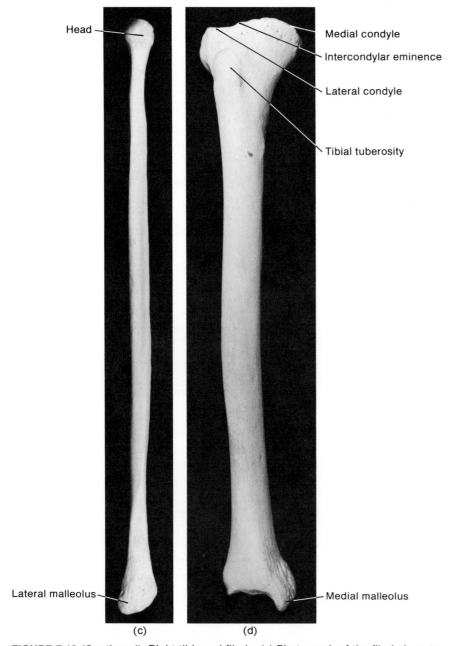

Head

Medial condyle

Intercondylar eminence

Lateral condyle

Tibial tuberosity

Lateral malleolus

Medial malleolus

(c)   (d)

FIGURE 7-12 (Continued) Right tibia and fibula. (c) Photograph of the fibula in anterior view. (d) Photograph of the tibia in anterior view. (Photographs courtesy of Lenny Patti.)

The medial surface of the distal end of the tibia forms the **medial malleolus.** This structure articulates with the talus bone of the ankle and forms the prominence that can be felt on the medial surface of your ankle. The **fibular notch** articulates with the fibula.

The **fibula** is parallel, and lateral, to the tibia. It is considerably smaller than the tibia.

The **head** of the fibula, the proximal end, articulates

with the lateral condyle of the tibia below the level of the knee joint. The distal end has a projection called the **lateral malleolus,** which articulates with the talus bone of the ankle. This forms the prominence on the lateral surface of the ankle. The inferior portion of the fibula also articulates with the tibia at the **fibular notch.** A fracture of the lower end of the fibula with injury to the tibial articulation is called a **Pott's fracture.**

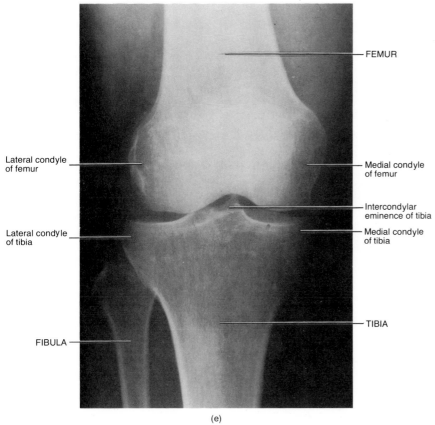

Lateral condyle of femur

FEMUR

Medial condyle of femur

Intercondylar eminence of tibia

Lateral condyle of tibia

Medial condyle of tibia

TIBIA

FIBULA

(e)

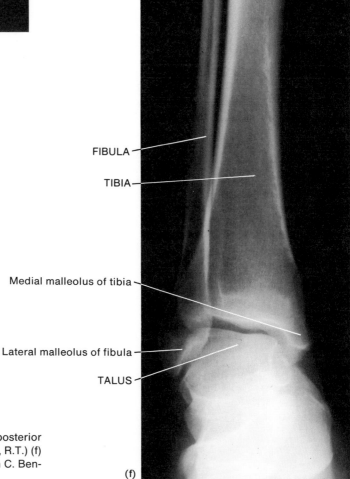

FIBULA

TIBIA

Medial malleolus of tibia

Lateral malleolus of fibula

TALUS

(f)

FIGURE 7-12 *(Continued)* Right tibia and fibula. (e) Anteroposterior projection of the proximal ends. (Courtesy of Daniel Sorrentino, R.T.) (f) Anteroposterior projection of the distal ends. (Courtesy of John C. Bennett, St. Mary's Hospital, San Francisco.)

TARSUS, METATARSUS, AND PHALANGES

The **tarsus** is a collective designation for the seven bones of the ankle (Figure 7-13). The term *tarsos* pertains to a broad, flat surface. The **talus** and **calcaneous** are located on the posterior part of the foot. The anterior part contains the **cuboid, navicular,** and three **cuneiform bones** called the first (medial), second (intermediate), and third (lateral) cuneiform. The talus, the uppermost tarsal bone, is the only bone of the foot that articulates with the fibula and tibia. It is surrounded on one side by the medial malleolus of the tibia and on the other side by the lateral malleolus of the fibula. During walking, the talus initially bears the entire weight of the extremity. About half the weight is then transmitted to the calcaneus. The remainder is transmitted to the other tarsal bones. The calcaneus, or heel bone, is the largest and strongest tarsal bone.

The **metatarsus** consists of five metatarsal bones numbered I to V from the medial to lateral position. Like the metacarpals of the palm of the hand, each metatarsal consists of a proximal **base,** a **shaft,** and a distal **head.** The metatarsals articulate proximally with the first, second, and third cuneiform bones and with the cuboid. Distally, they articulate with the proximal row of phalanges. The first metatarsal is thicker than the others because it bears more weight.

The **phalanges** of the foot resemble those of the hand both in number and arrangement. Each also consists of a proximal **base,** a **shaft,** and a distal **head.** The great (big) toe, or hallux, has two large, heavy phalanges called proximal and distal phalanges. The other four toes each have three phalanges. These are the proximal, middle, and distal phalanges.

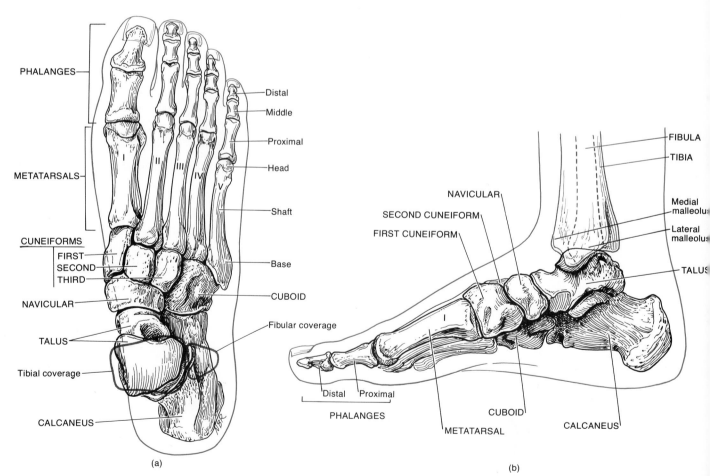

FIGURE 7-13 Right foot. (a) Superior view. (b) Medial view.

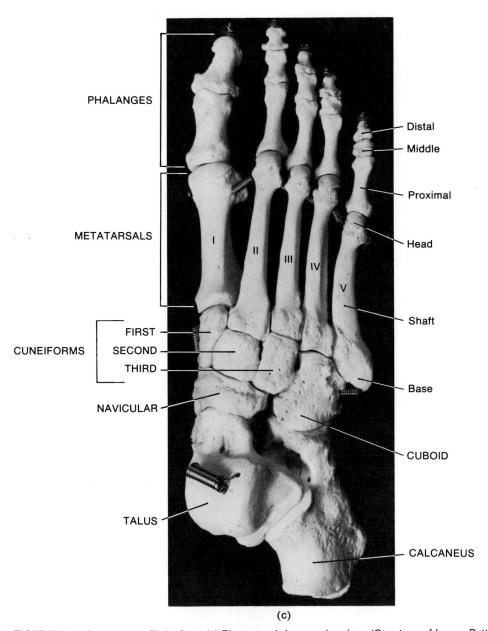

PHALANGES

Distal

Middle

Proximal

METATARSALS

Head

Shaft

I

II

III

IV

V

CUNEIFORMS

FIRST

SECOND

THIRD

Base

NAVICULAR

CUBOID

TALUS

CALCANEUS

(c)

FIGURE 7-13 *(Continued)* Right foot. (c) Photograph in superior view. (Courtesy of Lenny Patti.)

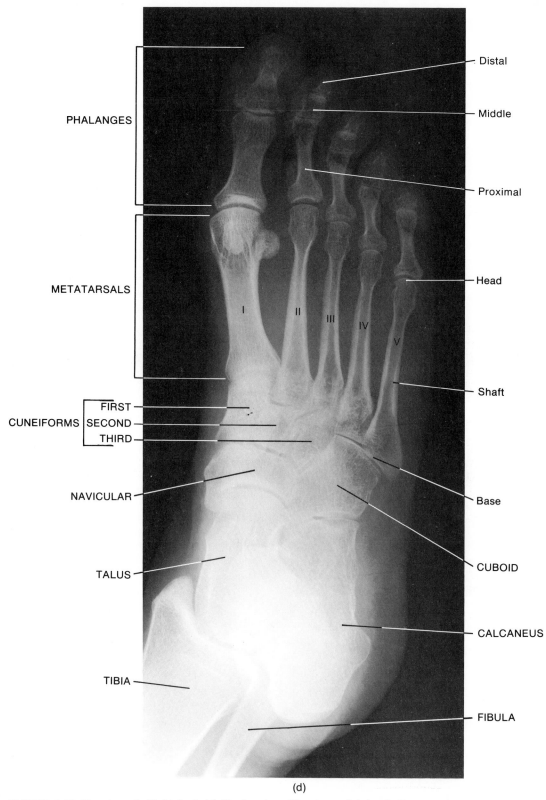

PHALANGES

METATARSALS

CUNEIFORMS
FIRST
SECOND
THIRD

NAVICULAR

TALUS

TIBIA

Distal

Middle

Proximal

Head

Shaft

Base

CUBOID

CALCANEUS

FIBULA

I   II   III   IV   V

(d)

FIGURE 7-13 *(Continued)* Right foot. (d) Plantar view. (Courtesy of John C. Bennett, St. Mary's Hospital, San Francisco.)

ARCHES OF THE FOOT

The bones of the foot are arranged in two arches (Figure 7-14). These arches enable the foot to support the weight of the body and provide leverage while walking. The arches are not rigid. They yield as weight is applied and spring back when the weight is lifted.

The **longitudinal arch** has two parts. Both consist of tarsal and metatarsal bones arranged to form an arch from the anterior to the posterior part of the foot. The **medial,** or inner, part of the longitudinal arch originates at the calcaneus. It rises to the talus and descends anteriorly through the navicular, the three cuneiforms, and the three medial metatarsals. The talus is the keystone of this arch. The **lateral,** or outer, part of the longitudinal arch also begins at the calcaneus. It rises at the cuboid and descends to the two lateral metatarsals. The cuboid is the keystone of the arch.

The **transverse arch** is formed by the calcaneus, navicular, cuboid, and the posterior parts of the five metatarsals.

The bones composing the arches are held in position by ligaments and tendons. If these ligaments and tendons are weakened, the height in the longitudinal arch may decrease or "fall." The result is *flatfoot*. A *bunion* is an abnormal lateral displacement of the big toe from its natural position. This condition produces an inflammatory reaction to the bursae that results in the formation of abnormal tissue.

Male and Female Skeletons

The bones of the male are generally larger and heavier than those of the female. The articular ends are also thicker when compared with the shafts. In addition, certain muscles of the male are larger than those of the female. Consequently, the male skeleton has larger tuberosities, lines, and ridges for the attachment of larger muscles.

One marked difference between male and female skeletons is the structure of the pelvis. The main differences between the male and female pelvis concern adaptations directly relating to childbearing (Figure 7-15). The female pelvis is wider, shallower, and lighter in structure than that of the male. The ilia of the female flare laterally to broaden the hips. The inlet of the true pelvis in the female is nearly oval, whereas that of the male is triangular or heart-shaped. The sacrum of the female is shorter, wider, and less curved than that of the male. The female coccyx is also more movable.

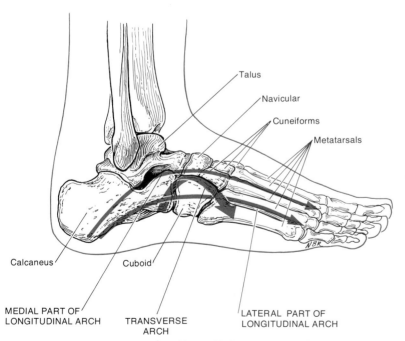

Talus

Navicular

Cuneiforms

Metatarsals

Calcaneus   Cuboid

MEDIAL PART OF
LONGITUDINAL ARCH   TRANSVERSE
ARCH   LATERAL PART OF
LONGITUDINAL ARCH

FIGURE 7-14 Arches of the right foot in lateral view.

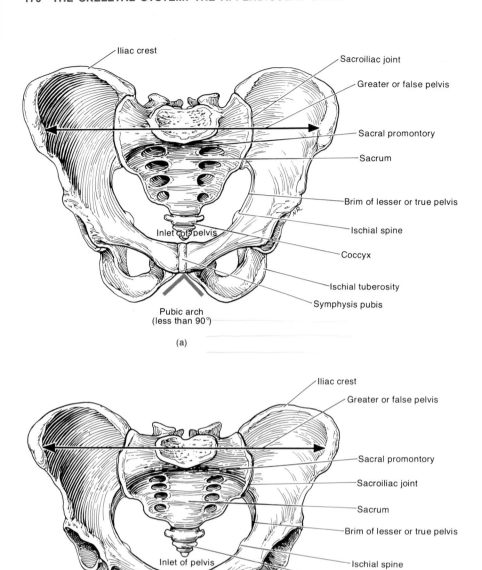

FIGURE 7-15 Pelvis. (a) Male pelvis in anterior view. (b) Female pelvis in anterior view.

The sciatic notches are wider and shallower in the female. The ischial spines and tuberosities of the female turn outward and are further apart than in the male. The pubic arch thus forms an obtuse angle rather than an acute angle as in the male. The pubic arch is the angle at which the right and left pubic portions of the coxal bones meet.

All these characteristics contribute to the wider outlet of the true pelvis in the female. This feature, of course, accommodates the birth of the child. In addition, the ligaments of the sacroiliac joint stretch during pregnancy and childbirth. Additional space is thus provided for the developing fetus, and delivery is facilitated.

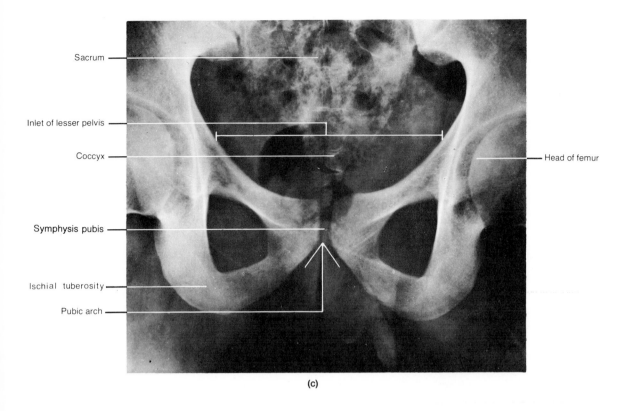

Sacrum

Inlet of lesser pelvis

Coccyx

Head of femur

Symphysis pubis

Ischial tuberosity

Pubic arch

(c)

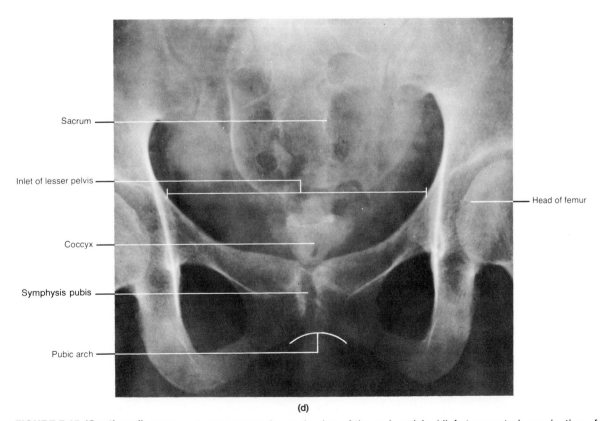

Sacrum

Inlet of lesser pelvis

Head of femur

Coccyx

Symphysis pubis

Pubic arch

(d)

FIGURE 7-15 (Continued)   Pelvis. (c) Anteroposterior projection of the male pelvis. (d) Anteroposterior projection of the female pelvis. (Courtesy of Eastman Kodak Company.)

STUDY OUTLINE

Shoulder Girdles

1. Each shoulder girdle or pectoral girdle consists of a clavicle and scapula.
2. Each attaches the upper extremity to the trunk.

Upper Extremities

1. The bones of each upper extremity include the humerus, ulna, radius, carpals, metacarpals, and phalanges.

Pelvic Girdle

1. The pelvic girdle consists of two coxal bones or hipbones.
2. It attaches the lower extremities to the trunk.

Lower Extremities

1. The bones of each lower extremity include the femur, tibia, fibula, tarsus, metatarsus, and phalanges.
2. The arches of the foot are bones arranged for support and leverage.
3. The two parts of the longitudinal arch are the higher medial and lower lateral arches. The other arch is the transverse arch.

Male and Female Skeletons

1. Male bones are generally larger than female bones.
2. The female pelvis is adapted for pregnancy and childbirth.

REVIEW QUESTIONS

1. What is a shoulder girdle? What are the bones of the upper extremity? What is a pelvic girdle? What are the bones of the lower extremity?
2. Define a Colles' fracture and a Pott's fracture.
3. Define an arch of the foot. What is its function? Distinguish between a longitudinal arch and a transverse arch.
4. How do bunions and flatfeet arise?
5. What are the principal structural differences between male and female skeletons?

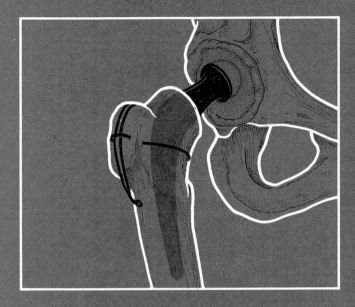

8

Articulations

STUDENT OBJECTIVES

- Define an articulation and identify the factors that determine the degree of movement at a joint.
- Contrast the structure, kind of movement, and location of fibrous, cartilaginous, and synovial joints.
- Discuss and compare the movements possible at various synovial joints.
- Describe the principal joints of the body with respect to the bones that enter into their formation, structural classification, and anatomical components.
- Describe the causes and symptoms of common joint disorders, including arthritis, rheumatism, rheumatoid arthritis, osteoarthritis, gouty arthritis, bursitis, and tendinitis.
- Define dislocation.
- Define a sprain.
- Describe the conditions that may cause a slipped disc.
- Define key medical terms associated with joints.

Bones are much too rigid to bend without damage. Fortunately, the skeletal system consists of many separate bones, which are held together at joints by flexible connective tissue. All movements that change the positions of the bony parts of the body, such as the extremities, occur at joints. You can understand the importance of joints if you imagine how a cast over the knee joint prevents flexing the leg or how a splint on a finger limits the ability to manipulate small objects.

The term **articulation** or **joint** refers to a point of contact between bones, cartilages, or between cartilage and bones. The joint's structure determines its function. Some joints permit no movement. Others permit slight movement. Still others afford considerable movement. In general, the more closely the bones fit together, the stronger the joint. At tightly fitted joints, however, movement is restricted. The greater the movement, the looser the fit. Unfortunately, loosely fitted joints are prone to dislocation. Movement at joints is also determined by the flexibility of the connective tissue that binds the bones together and by the position of ligaments, muscles, and tendons.

Classification

FUNCTIONAL

The functional classification of joints takes into account the degree of movement they permit. Functionally, joints are classified as **synarthroses,** which are immovable joints; **amphiarthroses,** which are slightly movable joints; and **diarthroses,** which are freely movable joints.

STRUCTURAL

The structural classification of joints is based on the presence or absence of a joint cavity (a space between the bones) and the kind of connective tissue that binds the bones together. Structurally, joints are classified as **fibrous,** in which there is no joint cavity and the bones are held together by fibrous connective tissue; **cartilaginous,** in which there is no joint cavity and the bones are held together by cartilage; and **synovial,** in which the joint contains a synovial cavity. We will discuss the joints of the body based upon their structural classification, but with continuous reference to their functional classification as well.

Fibrous Joints

Fibrous joints lack a joint cavity, and the articulating bones are held very closely together by fibrous connective tissue. They permit little or no movement. The three types of fibrous joints are (1) sutures, (2) syndesmoses, and (3) gomphoses.

Sutures are found between the bones of the skull and are united by a thin layer of dense fibrous connective tissue. Based upon the form of the margins of the bones, several types of sutures can be distinguished. In a *serrate suture,* the margins of the bones are serrated like the teeth of a saw. An example is the sagittal suture between the two parietal bones (see Figure 6-2a, b). In a *squamous suture,* the margin of one bone overlaps that of the adjacent bone. The squamosal suture between the parietal and temporal bones is an example (see Figure 6-2e, f). In a *plane suture,* the even, fairly regular margins of the adjacent bones are brought together. An example is the intermaxillary suture between maxillary bones (see Figure 6-2a, b). Since sutures are immovable, they are functionally classified as synarthroses. Some sutures, present during growth, are replaced by bone in the adult. In this case they are called *synostoses,* or bony joints, in which there is a complete fusion of bone across the suture line. An example is the joint between the left and right sides of the frontal bone. Synostoses are also functionally classified as synarthroses.

A **syndesmosis** is a fibrous joint in which there is much more of the uniting fibrous connective tissue than in a suture. The fibrous connective tissue forms an interosseous membrane or ligament. A syndesmosis is slightly movable because the bones are more separated than in a suture and some flexibility is permitted by the interosseous membrane or ligament. Thus, syndesmoses are functionally classified as amphiarthrotic. Examples of syndesmoses include the distal articulation of the tibia and fibula in which the bones are joined by an interosseous ligament and the articulation between the shafts of the ulna and radius in which the bones are united by an interosseous membrane (see Figure 8-23).

A **gomphosis** is a type of fibrous joint in which a cone-shaped peg fits into a socket. The intervening substance is the periodontal membrane. A gomphosis is functionally classified as synarthrotic; examples are the articulations of the roots of the teeth with the alveolar processes of the maxillae and mandible.

Cartilaginous Joints

Another joint that has no joint cavity is the **cartilaginous joint.** Here the articulating bones are tightly connected by cartilage. Like fibrous joints, they allow little or no movement (Figure 8-1). A **synchondrosis** is a cartilaginous joint in which the connecting material is hyaline cartilage, as in the epiphyseal plate. Such a

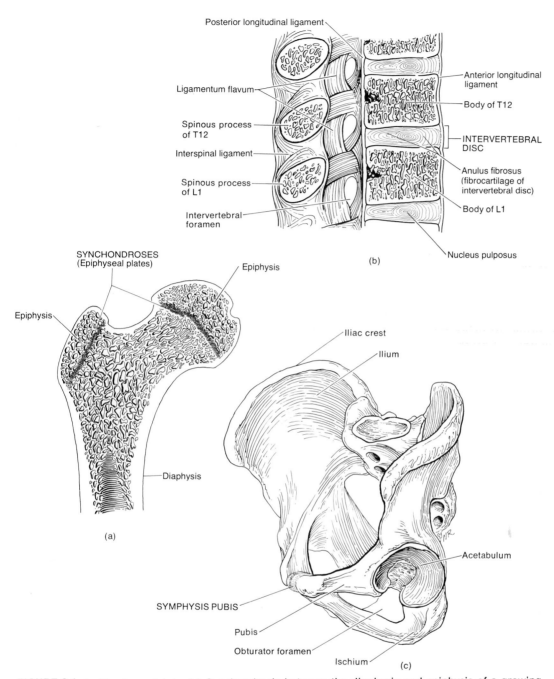

FIGURE 8-1 Cartilaginous joints. (a) Synchondrosis between the diaphysis and epiphysis of a growing femur. (b) Symphysis joint between the bodies of vertebrae seen in sagittal section. (c) Symphysis joint (symphysis pubis) between the coxal bones seen in an oblique view.

joint is found between the epiphysis and diaphysis of a growing bone and is immovable. Thus it is synarthrotic. Since the hyaline cartilage is eventually replaced by bone when growth ceases, the joint is temporary. It is replaced by a synostosis. A **symphysis** is a cartilaginous joint in which the connecting material is a broad, flat disc of fibrocartilage. This joint is found between bodies of vertebrae. A portion of the intervertebral disc is cartilaginous material. The symphysis pubis between the anterior surfaces of the coxal bones is another example of a symphysis-type joint. These joints are slightly movable, or amphiarthrotic.

Synovial Joints

When a joint cavity is present, the articulation is called a **synovial joint** (Figure 8-2). The cavity, called a **synovial** or **joint cavity,** is a space between the articulating bones. Because of this cavity and because no tissue exists between the articulating surfaces of the bones, synovial joints are freely movable. Thus, synovial joints are functionally classified as diarthrotic.

Synovial joints are also characterized by the presence of **articular cartilage,** which covers the surfaces of the articulating bones but does not bind the bones together. The articular cartilage of synovial joints is hyaline cartilage. In a few joints, such as the temporomandibular and sternoclavicular, a small articular disc (meniscus) of fibrocartilage is interposed between the articular cartilage coverings of the bones. Synovial joints are surrounded by a tubular **articular capsule** that surrounds the articular surfaces and encloses the joint cavity. The articular capsule is composed of two layers. The outer layer, the *fibrous capsule,* consists of dense connective (collagenous) tissue and is attached to the periosteum of the articulating bones at a variable distance from the edge of the articular cartilage. The flexibility of the fibrous capsule permits movement at a joint, whereas its great tensile strength resists dislocation. The fibers of some fibrous capsules are highly adapted to resist recurrent strain. In these instances, they are arranged in parallel bundles; they are then called ligaments and given special names. The fibrous capsule is one of the principal structures that holds bone to bone. The inner layer of the articular capsule is formed by a *synovial membrane* which is composed of loose connective tissue with elastic fibers and a variable amount of adipose tissue. The synovial membrane secretes synovial fluid which lubricates the joint and provides nourishment for the articular cartilage. Synovial fluid also contains phagocytic cells that remove microbes and debris resulting from wear and tear in the joint. Synovial fluid consists of hyaluronic acid and interstitial fluid formed from blood plasma and is similar in appearance and consistency to egg white. When there is no joint movement, the fluid is viscous; when there is increased joint movement, the fluid becomes less viscous. Although the amount of synovial fluid is variable in different joints of the body, even a large joint such as the knee contains only 0.5 ml. The amount present is sufficiently only to form a thin film over the surfaces within an articular capsule.

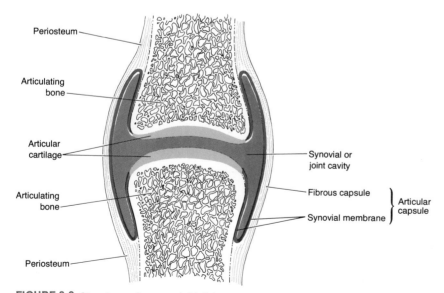

FIGURE 8-2 Structure of a synovial joint.

Many synovial joints also contain **accessory ligaments** known as extracapsular ligaments and intracapsular ligaments. *Extracapsular ligaments* are outside of the articular capsule. An example is the fibular collateral ligament of the knee joint (see Figure 8-22). *Intracapsular ligaments* occur within the articular capsule but are excluded from the joint cavity by reflections of the synovial membrane. Examples are the cruciate ligaments of the knee joint (see Figure 8-22).

Inside some synovial joints, there are pads of fibrocartilage that lie between the articular surfaces of the bones and are attached by their margins to the fibrous capsule. These pads are called **articular discs (menisci).** The discs usually subdivide the joint cavity into two separate spaces, thus providing a mechanism by which each bone can move independently. Articular discs help to maintain the stability of the joint and direct the flow of synovial fluid to areas of greatest friction.

The various movements of the body could create friction between moving parts. To reduce this friction, saclike structures called **bursae** are situated in the body tissues. These sacs resemble joints in that their walls consist of connective tissue lined by a synovial membrane. They are also filled with a fluid similar to synovial fluid. Bursae are located between the skin and bone in places where skin rubs over bone. They are also found between tendons and bones, muscles and bones, and ligaments and bones. As fluid-filled sacs, they cushion the movement of one part of the body over another.

The articular surfaces of synovial joints are kept in contact with each other by several factors. One factor is the fit of the articulating bones. This interlocking is very obvious at the hip joint where the head of the femur articulates with the acetabulum of the coxal bone. Another factor is the strength of the joint ligaments. This is especially important in the hip joint. A third factor is the tension of the muscles around the joint. For example, the fibrous capsule of the knee joint is formed principally from tendinous expansions by muscles acting on the joint.

MOVEMENTS AT SYNOVIAL JOINTS

The movement permitted at synovial joints is limited by several factors. One is the apposition of soft parts. For example, during bending of the elbow the anterior surface of the forearm is pressed against the anterior surface of the arm. This apposition limits movement. A second factor is the tension of ligaments. The different components of a fibrous capsule are tense only when the joint is in certain positions. Tense ligaments not only restrict movement, but also direct the movement of the articulating bones with respect to each

other. In the knee joint, for example, the major ligaments are lax when the knee is bent but tense when the knee is straightened. Also, when the knee is straightened the surfaces of the articulating bones are in fullest contact with each other. A third factor that restricts movement at a synovial joint is muscle tension. This reinforces the restraint placed on a joint by ligaments. A good example of the effect of muscle tension on a joint is seen at the hip joint. When the thigh is raised, the movement is restricted by the tension of the hamstring muscles on the posterior surface of the thigh if the knee is kept straight. But, if the knee is bent, the tension on the hamstring muscles is lessened and the thigh can be raised further.

Gliding

A **gliding movement** is the simplest kind that can occur at a joint. One surface moves back and forth and from side to side over another surface without angular or rotary motion. Some joints that glide are those between the carpals and between the tarsals. The heads and tubercles of ribs glide on the bodies and transverse process of vertebrae.

Angular

Angular movements increase or decrease the angle between bones. Among the angular movements are flexion, extension, abduction, and adduction (Figure 8-3). **Flexion** usually involves a decrease in the angle between the anterior surfaces of the articulating bones. An exception to this definition is flexion of the knee and the toe joints in which there is a decrease in the angle between the posterior surfaces of the articulating bones. Examples of flexion include bending the head forward, where the joint is between the occipital bone and the atlas, bending the elbow, and bending the knee. Flexion of the foot at the ankle joint is called **dorsiflexion. Extension** involves an increase in the angle between the anterior surface of the articulating bones, with the two exceptions noted above. Extension restores a body part to its anatomical position after it has been flexed. Examples of extension are returning the head to the anatomical position after flexion, straightening the arm after flexion, and straightening the leg after flexion. Continuation of extension beyond the anatomical position, as in bending the head backward, is called **hyperextension.** Extension of the foot at the ankle joint is **plantar flexion.**

Abduction usually means movement of a bone *away from* the midline of the body. An example of abduction is moving the arms upward and away from the body until they are held straight out at right angles to the

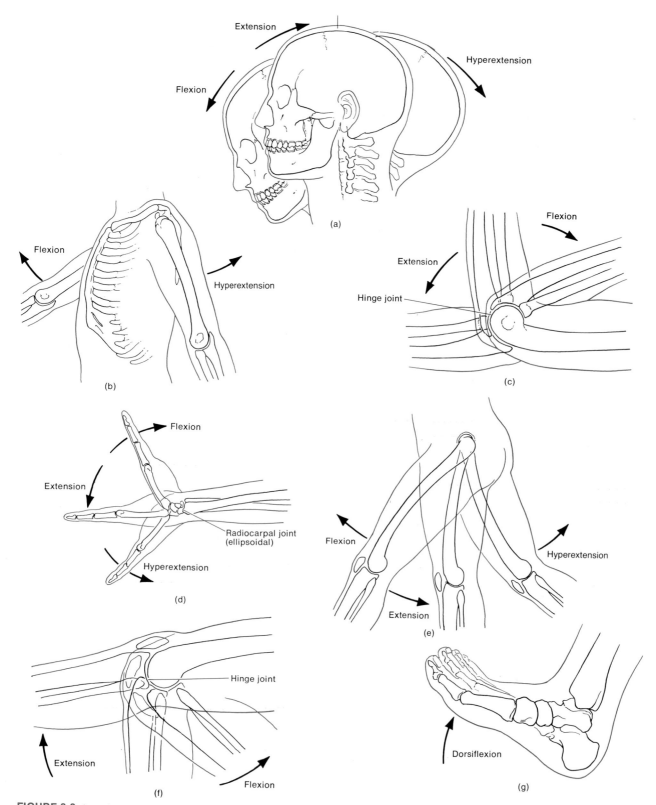

FIGURE 8-3 Angular movements at synovial joints.

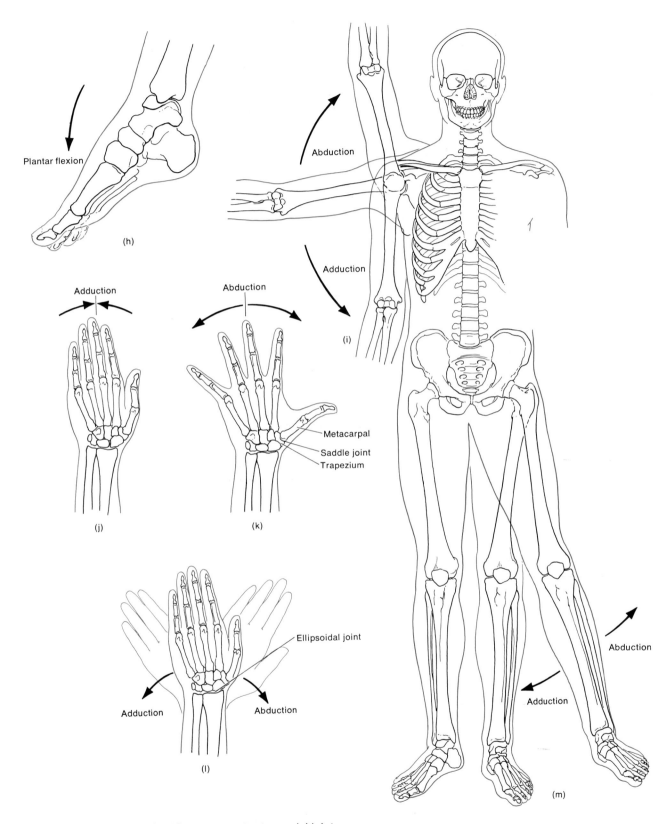

FIGURE 8-3 *(Continued)* Angular movements at synovial joints.

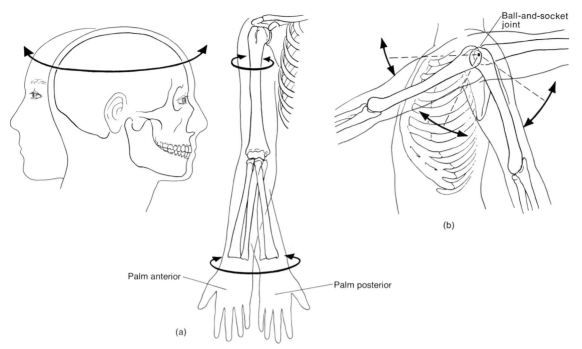

FIGURE 8-4 Rotation and circumduction. (a) Rotation at the atlantoaxial joint (left) and rotation of the humerus (right). (b) Circumduction of the humerus at the shoulder joint.

chest. With the fingers and toes, however, the midline of the body is not used as the line of reference. Abduction of the fingers is a movement away from an imaginary line drawn through the middle finger—in other words, spreading the fingers. Abduction of the toes is relative to an imaginary line drawn through the second toe. **Adduction** is usually movement of a part *toward* the midline of the body. An example of adduction is returning the arms to the sides after abduction. As in abduction, adduction of the fingers is relative to the middle finger. Adduction of the toes is relative to the second toe.

Rotation

Rotation is movement of a bone around its own axis. During rotation, no other motion is permitted. We rotate the atlas around the odontoid process of the axis when we shake the head from side to side. Moving from the shoulder and turning the forearm up, then palm down, and then palm up again is an example of slight rotation of the humerus (Figure 8-4a).

Circumduction

Circumduction is a movement in which the distal end of a bone moves in a circle while the proximal end remains stable. The bone describes a cone in the air.

Circumduction typically involves flexion, abduction, adduction, extension, and rotation. It involves a 360° rotation. An example is moving the outstretched arm in a circle to wind up to pitch a ball (Figure 8-4b).

Special

Special movements are those found only at the joints indicated in Figure 8-5. **Inversion** is the movement of the sole of the foot inward at the ankle joint. **Eversion** is the movement of the sole outward at the ankle joint. **Protraction** is the movement of the mandible or clavicle forward on a plane parallel to the ground. Thrusting the jaw outward is protraction of the mandible. Bringing your arms forward until the elbows touch requires protraction of the clavicle. **Retraction** is the movement of a protracted part of the body backward on a plane parallel to the ground. Pulling the lower jaw back in line with the upper jaw is retraction of the mandible. **Supination** is a movement of the forearm in which the palm of the hand is turned forward (anterior). To demonstrate supination, flex your arm at the elbow to prevent rotation of the humerus in the shoulder joint. **Pronation** is a movement of the flexed forearm in which the palm is turned backward (posterior). **Elevation** is a movement in which a part of the body moves upward. You elevate your mandible when you close your mouth. **Depression** is a movement in which

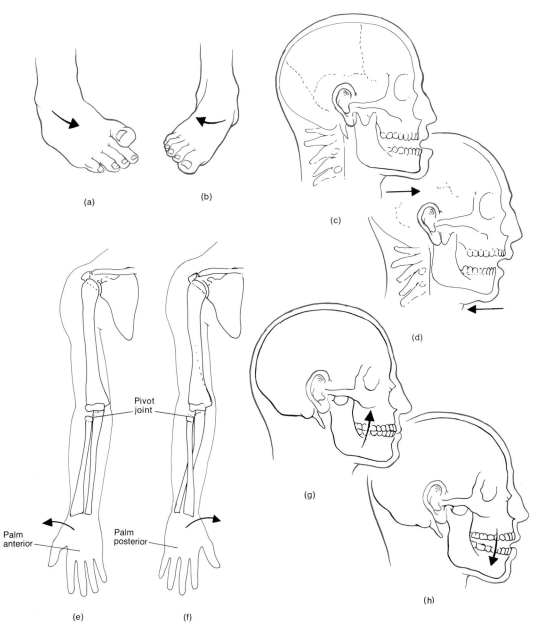

(a)

(b)

(c)

(d)

Pivot joint

Palm anterior

Palm posterior

(e)

(f)

(g)

(h)

FIGURE 8-5 Special movements. (a) Inversion. (b) Eversion. (c) Protraction. (d) Retraction. (e) Supination. (f) Pronation. (g) Elevation. (h) Depression.

a part of the body moves downward. You depress your mandible when you open your mouth. The shoulders can also be elevated and depressed.

TYPES OF SYNOVIAL JOINTS

Gliding

The articulating surfaces of bones in **gliding joints (arthrodia)** are usually flat. Only side-to-side and back-and-forth movements are permitted. Since this joint allows movement in two planes, it is called *biaxial*. Twisting and rotation are inhibited at gliding joints generally because ligaments or adjacent bones restrict the range of movement. Examples are the joints between carpal bones, tarsal bones, the sternum and clavicle, and the scapula and clavicle.

Hinge

A **hinge joint (ginglymus)** is characterized by the convex surface of one bone that fits into the concave surface of another bone. Movement is primarily in a single plane and is usually flexion and extension. The joint is therefore known as *monaxial*. The motion is similar to that of a hinged door. Examples of hinge joints include the elbow, knee, ankle, and interphalangeal joints. The movement allowed by a hinge joint is illustrated by flexion and extension at the elbow and knee. (See Figure 8-3c, f.)

Pivot

In a **pivot joint (trochoid),** a rounded, pointed, or conical surface of one bone articulates within a ring formed partly by bone and partly by a ligament. The primary movement permitted is rotation, and the joint is therefore monaxial. Examples include the joints between the atlas and axis (atlantoaxial) and between the proximal ends of the radius and ulna. Movement at a pivot joint is illustrated by supination and pronation of the palms and rotation of the head from side to side. (See Figure 8-4a.)

Ellipsoidal

In an **ellipsoidal joint (condyloid),** an oval-shaped condyle of one bone fits into an elliptical cavity of another bone. Since the joint permits side-to-side and back-and-forth movements, it is biaxial. The joint at the wrist between the radius and carpals is ellipsoidal. The movement permitted by such a joint is illustrated when you flex and extend and abduct and adduct the wrist. (See Figure 8-3d.)

Saddle

In a **saddle joint (sellaris),** the articular surfaces of both bones are saddle-shaped—in other words, concave in one direction and convex in the other. Essentially, the saddle joint is a modified ellipsoidal joint in which the movement is somewhat freer. Movements at a saddle joint are side to side and back and forth. Thus the joint is biaxial. The joint between the trapezium and metacarpal of the thumb is an example of a saddle joint. (see Figure 8-3k).

Ball-and-Socket

Ball-and-socket joints (spheroid) consist of a ball-like surface of one bone fitting into a cuplike depression of another bone. Such a joint permits *triaxial* movement. That is, there is movement in three planes of motion: flexion-extension, abduction-adduction, and rotation. Examples of ball-and-socket joints are the shoulder joint and hip joint. The range of movements at a ball-and-socket joint is illustrated by circumduction of the arm (Figure 8-4b).

The summary of joints presented in Exhibit 8-1 is based on the anatomy of the joints. Joints can also be classified according to movement. If we rearrange Exhibit 8-1 into a classification based on movement, we arrive at the following:

Synarthroses: immovable joints
   1. Suture
   2. Gomphosis
   3. Synchondrosis

Amphiarthroses: slightly movable joints
   1. Symphysis
   2. Syndesmosis

Diarthroses: freely movable joints
   1. Gliding
   2. Hinge
   3. Pivot
   4. Ellipsoidal
   5. Saddle
   6. Ball-and-socket

In the next section of the chapter, we will examine in some detail the principal articulations of the body. In order to simplify your learning efforts, a series of exhibits has been prepared. Each exhibit considers a specific articulation and contains (1) a definition; that is, a description of the bones that form the joint; (2) the type of joint (the structural classification of the joint); (3) the joint's anatomical components—a description of the major connecting ligaments, articular disc, articular capsule, and other distinguishing features of the joint; most exhibits also contain references that direct you to a labeled figure.

EXHIBIT 8-1
JOINTS

TYPE	DESCRIPTION	MOVEMENT	EXAMPLES
FIBROUS	No joint cavity; bones held together by a thin layer of fibrous tissue or dense fibrous tissue		
Suture	Found only between bones of the skull; articulating bones separated by a thin layer of fibrous tissue	None—synarthrotic	Lambdoidal suture between occipital and parietal bones
Syndesmosis	Articulating bones united by dense fibrous tissue	Slight—amphiarthrotic	Distal ends of tibia and fibula
Gomphosis	Cone-shaped peg fits into a socket	None—synarthrotic	Roots of teeth in alveolar processes
CARTILAGINOUS	No joint cavity; articulating bones united by cartilage		
Synchondrosis	Connecting material is hyaline cartilage.	None—synarthrotic	Temporary joint between the diaphysis and epiphyses of a long bone
Symphysis	Connecting material is a broad, flat disc of fibrocartilage.	Slight—amphiarthrotic	Intervertebral joints and symphysis pubis
SYNOVIAL	Joint cavity and articular cartilage present; articular capsule is composed of an outer fibrous capsule and an inner synovial membrane; may contain accessory ligaments, articular discs (menisci), and bursae.	Freely movable—diarthrotic	
Gliding	Articulating surfaces usually flat	Biaxial (flexion-extension, abduction-adduction)	Intercarpal and intertarsal joints
Hinge	Spool-like surface fits into a concave surface	Monaxial (flexion-extension)	Elbow, knee, ankle, and interphalangeal joints
Pivot	Rounded, pointed, or concave surface fits into a ring formed partly by bone and partly by a ligament.	Monaxial (rotation)	Atlantoaxial and radioulnar joints
Ellipsoidal	Oval-shaped condyle fits into an elliptical cavity	Biaxial (flexion-extension, abduction-adduction)	Radiocarpal joint
Saddle	Articular surfaces concave in one direction and convex in opposite direction	Biaxial (flexion-extension, abduction-adduction)	Carpometacarpal joint of thumb
Ball-and-socket	Ball-like surface fits into a cuplike depression.	Triaxial (flexion-extension, abduction-adduction, rotation).	Shoulder and hip joints

EXHIBIT 8-2
TEMPOROMANDIBULAR JOINT (Figure 8-6)

DEFINITION	Joint formed by mandibular condyle of the mandible and mandibular fossa and articular tubercle of temporal bone. The temporomandibular joint is the only movable joint between skull bones; all other skull joints are sutures and therefore immovable.
TYPE OF JOINT	Synovial: combined hinge (ginglymus) and gliding (arthrodial).
ANATOMICAL COMPONENTS	1. **Articular disc (meniscus).** Fibrocartilage disc that separates the joint cavity into a superior and inferior compartment, each of which has a synovial membrane. 2. **Articular capsule.*** Thin, fairly loose envelope around the circumference of the joint. 3. **Lateral ligament.** Two short bands on the lateral surface of the articular capsule that extend inferiorly and posteriorly from the lower border and tubercle of the zygomatic arch to the lateral and posterior aspect of the neck of the mandible; it is covered by the parotid gland and helps prevent displacement of the mandible. 4. **Sphenomandibular ligament.** Thin band that extends inferiorly and anteriorly from the spine of the sphenoid bone to the ramus of the mandible. 5. **Stylomandibular ligament.** Thickened band of deep cervical fascia that extends from the styloid process of the temporal bone to the inferior and posterior border of the ramus of the mandible; the ligament separates the parotid from the submandibular gland.

*Also referred to as the capsular ligament.

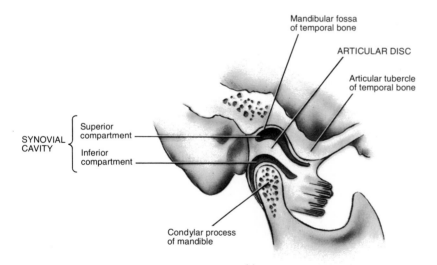

Mandibular fossa
of temporal bone

ARTICULAR DISC

Articular tubercle
of temporal bone

SYNOVIAL
CAVITY

Superior
compartment

Inferior
compartment

Condylar process
of mandible

(a)

FIGURE 8-6 Temporomandibular joint. (a) Sagittal section through the temporo-mandibular joint.

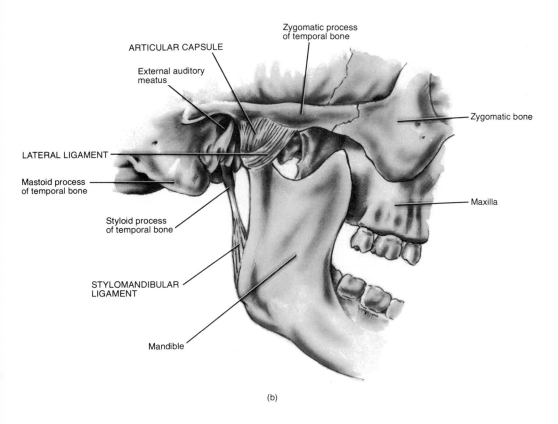

(b)

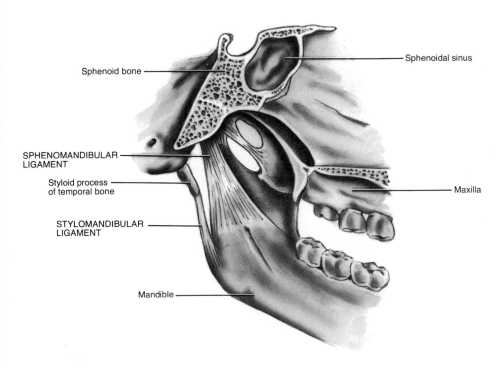

(c)

FIGURE 8-6 (Continued) Temporomandibular joint. (b) Right lateral view. (c) Median view.

EXHIBIT 8-3
ATLANTOOCCIPITAL JOINTS (Figure 8-7)

DEFINITION	Joints formed by the superior articular surfaces of the atlas and the occipital condyles of the occipital bone.
TYPE OF JOINT	Synovial of the ellipsoidal (condyloid) type.
ANATOMICAL COMPONENTS	1. **Articular capsules.** Thin, loose envelopes that surround the occipital condyles and attach them to the articular processes of the atlas.
	2. **Anterior atlantooccipital ligament.** Broad, thick ligament that extends from the anterior surface of the foramen magnum of the occipital bone to the anterior arch of the atlas.
	3. **Posterior atlantooccipital ligament.** Broad, thin ligament that extends from the posterior surface of the foramen magnum of the occipital bone to the posterior arch of the atlas; the inferior portion of the ligament is sometimes ossified.
	4. **Lateral atlantooccipital ligaments.** Thickened portions of the articular capsules that also contain fibrous tissue bundles and extend from the jugular process of the occipital bone to the transverse process of the atlas.

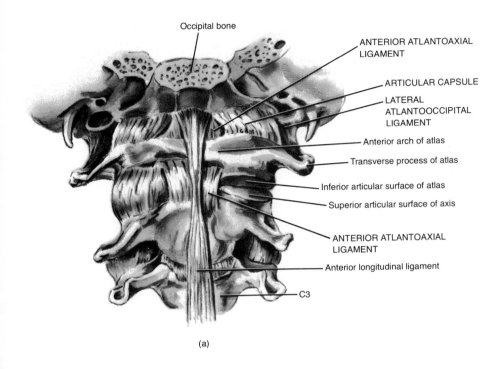

(a)

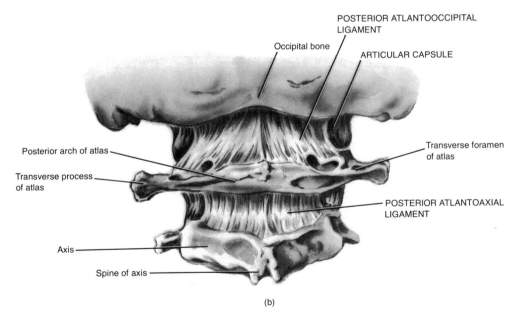

(b)

FIGURE 8-7 Atlantooccipital joints. (a) Anterior view. (b) Posterior view.

EXHIBIT 8-4
ATLANTOAXIAL JOINTS (Figure 8-8)

DEFINITION	Joints formed by the atlas and axis. There are two lateral atlantoaxial joints and two median atlantoaxial joints. The *lateral atlantoaxial joints* are formed by the inferior articular surfaces of the atlas and the superior articular surfaces of the axis. The *median atlantoaxial joints* are formed by the dens (odontoid process) of the axis and the ring formed by the anterior arch and transverse ligament of the atlas. The two synovial cavities are formed by (1) the posterior surface of the anterior arch and dens and (2) the anterior surface of the ligament and dens.
TYPE OF JOINT	The lateral atlantoaxial joints are synovial of the gliding (arthrodial) type and the median atlantoaxial joints are synovial of the pivot (trochoid) type.
ANATOMICAL COMPONENTS	1. **Articular capsules.** Thin, loose ligaments that extend from the lateral masses of the atlas to the posterior articular surfaces of the axis. Each articular capsule is provided with additional strength by an *accessory ligament* that extends from the base of the dens to the lateral mass of the atlas. 2. **Anterior atlantoaxial ligament.** Strong ligament that extends from the inferior border of the anterior arch of the atlas to the anterior surface of the body of the axis. (See Figure 8-7a.) 3. **Posterior atlantoaxial ligament.** Broad, thin ligament that extends from the inferior border of the posterior arch of the atlas to the superior edges of the laminae of the axis. (See Figure 8-7b.) 4. **Transverse ligament of the atlas.** A thick, strong ligament of the atlas that passes behind the dens between the tubercles on the medial sides of the lateral masses of the atlas. From the middle of the transverse ligament, longitudinal fibers extend superiorly to the anterior edge of the foramen magnum and inferiorly to the posterior surface of the body of the axis. These longitudinal fibers, together with the transverse ligament, are known as the *cruciform (cruciate) ligament of the atlas.*

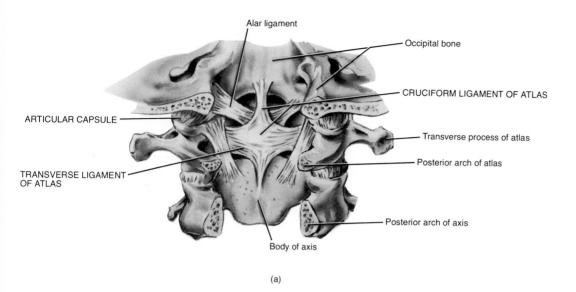

(a)

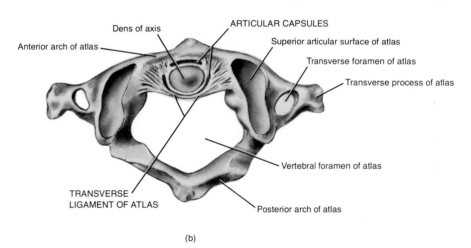

(b)

FIGURE 8-8 Atlantoaxial joints. (a) Posterior view. (b) Superior view.

EXHIBIT 8-5 ▰▰▰▰▰▰▰▰▰▰▰▰▰▰▰▰▰▰▰▰▰▰▰▰▰
INTERVERTEBRAL JOINTS (Figure 8-9)

DEFINITION	Joints formed between (1) vertebral bodies and (2) vertebral arches.
TYPE OF JOINT	Joints between vertebral bodies are cartilaginous joints of the symphysis type, whereas joints between vertebral arches are synovial joints of the gliding (arthrodial) type.

ANATOMICAL COMPONENTS OF JOINTS BETWEEN VERTEBRAL BODIES

1. **Anterior longitudinal ligament.** Broad, strong band that extends along the anterior surfaces of the vertebral bodies from the axis to the sacrum. It is firmly attached to the intervertebral discs.
2. **Posterior longitudinal ligament.** Extends along the posterior surfaces of the vertebral bodies within the vertebral canal from the axis to the sacrum. The free surface of the ligament is separated from the spinal dura mater by loose areolar tissue.
3. **Intervertebral discs.** The discs unite with adjacent surfaces of the vertebral bodies from the axis to the sacrum. Each disc is composed of a peripheral *anulus fibrosus* consisting of fibrous tissue and fibrocartilage and a central *nucleus pulposus* composed of a soft, pulpy, highly elastic substance.

ANATOMICAL COMPONENTS OF JOINTS BETWEEN VERTEBRAL ARCHES

1. **Articular capsules.** Thin, loose ligaments attached to the margins of the articular processes of adjacent vertebrae.
2. **Ligamenta flava.** Contain elastic tissue and connect the laminae of adjacent vertebrae from the axis to the first segment of the sacrum.
3. **Supraspinous ligament.** Strong fibrous cord that connects the spinous processes from the seventh cervical vertebra to the sacrum.
4. **Ligamentum nuchae.** Fibrous ligament that represents an enlargement of the supraspinous ligament in the neck. It extends from the external occipital protuberance of the occipital bone to the spinous process of the seventh cervical vertebra.
5. **Interspinous ligaments.** Relatively weak bands that run between adjacent spinous processes.
6. **Intertransverse ligaments.** Bands between transverse processes that are readily apparent in the lumbar region.

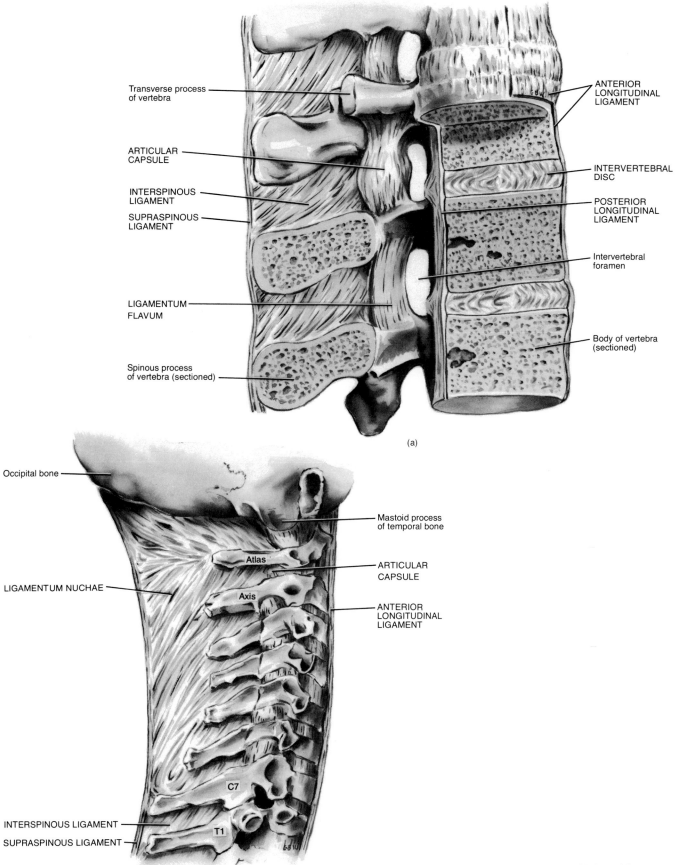

Transverse process of vertebra

ARTICULAR CAPSULE

INTERSPINOUS LIGAMENT

SUPRASPINOUS LIGAMENT

LIGAMENTUM FLAVUM

Spinous process of vertebra (sectioned)

ANTERIOR LONGITUDINAL LIGAMENT

INTERVERTEBRAL DISC

POSTERIOR LONGITUDINAL LIGAMENT

Intervertebral foramen

Body of vertebra (sectioned)

(a)

Occipital bone

LIGAMENTUM NUCHAE

INTERSPINOUS LIGAMENT

SUPRASPINOUS LIGAMENT

Mastoid process of temporal bone

Atlas

Axis

ARTICULAR CAPSULE

ANTERIOR LONGITUDINAL LIGAMENT

C7

T1

(b)

FIGURE 8-9 Intervertebral joints. (a) Median view. (b) Right lateral view.

EXHIBIT 8-6
LUMBOSACRAL JOINT (Figure 8-10)

DEFINITION	Joint formed by the body of the fifth lumbar vertebra and the superior surface of the first sacral vertebra of the sacrum.
TYPE OF JOINT	The joint between the bodies of the fifth lumbar vertebra and the first sacral vertebra is a cartilaginous joint of the symphysis type, whereas the joint between the articular processes is a synovial joint of the gliding (arthrodial) type.
ANATOMICAL COMPONENTS	1. The lumbosacral joint is similar to the joints between typical vertebrae, united by an intervertebral disc, anterior and posterior longitudinal ligaments, ligamenta flava, interspinous and supraspinous ligaments, and articular capsular between articular processes. The lumbosacral joint also contains an iliolumbar ligament. 2. **Iliolumbar ligament.** Strong ligament that connects the transverse process of the fifth lumbar vertebra with the base of the sacrum and the iliac crest.

EXHIBIT 8-7
SACROCOCCYGEAL JOINT

DEFINITION	Joint between the apex of the sacrum and the base of the coccyx.
TYPE OF JOINT	Cartilaginous joint of the symphysis type.
ANATOMICAL COMPONENTS	1. **Fibrocartilage disc.** Between the articulating surfaces of the sacrum and coccyx. 2. **Ventral sacrococcygeal ligament.** Thin ligament that connects the ventral surface of the sacrum to the coccyx. (See Figure 8-10a.) 3. **Dorsal sacrococcygeal ligament.** Flat ligament that connects the sacral canal with the dorsal surface of the coccyx. (See Figure 8-10b.) 4. **Lateral sacrococcygeal ligament.** Connects the transverse process of the coccyx to the lateral, inferior surface of the sacrum. (See Figure 8-10a-b.)

EXHIBIT 8-8
SACROILIAC JOINT

DEFINITION	Joint between the auricular surfaces of the sacrum and the ilium.
TYPE OF JOINT	Cartilaginous joint of the synchondrosis type.
ANATOMICAL COMPONENTS	1. **Ventral sacroiliac ligament.** Thin bands that unite the ventral surface of the lateral part of the sacrum with the ilium. (See Figure 8-10a.) 2. **Dorsal sacroiliac ligament.** Thick, extremely strong ligament that connects the dorsal surface of the sacrum to the tuberosity of the ilium. (See Figure 8-10b.) 3. **Interosseus sacroiliac ligament.** Short, strong fibers that connect the sacral and iliac tuberosities.

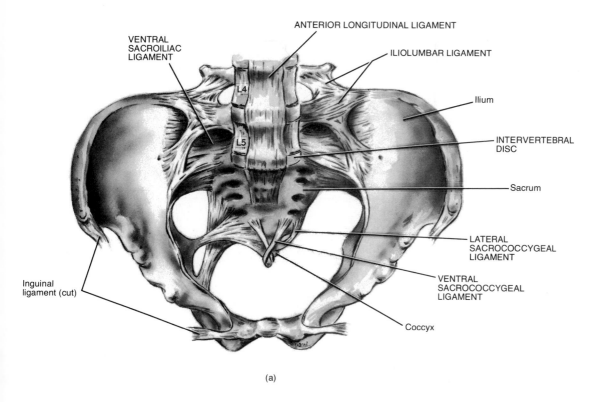

ANTERIOR LONGITUDINAL LIGAMENT

VENTRAL SACROILIAC LIGAMENT

ILIOLUMBAR LIGAMENT

Ilium

L4

L5

INTERVERTEBRAL DISC

Sacrum

LATERAL SACROCOCCYGEAL LIGAMENT

VENTRAL SACROCOCCYGEAL LIGAMENT

Inguinal ligament (cut)

Coccyx

(a)

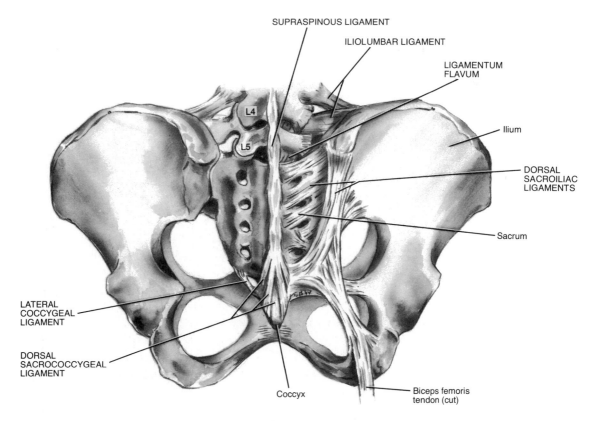

SUPRASPINOUS LIGAMENT

ILIOLUMBAR LIGAMENT

LIGAMENTUM FLAVUM

L4

L5

Ilium

DORSAL SACROILIAC LIGAMENTS

Sacrum

LATERAL COCCYGEAL LIGAMENT

DORSAL SACROCOCCYGEAL LIGAMENT

Coccyx

Biceps femoris tendon (cut)

(b)

FIGURE 8-10 Lumbosacral joint. (a) Anterior view. (b) Posterior view.

EXHIBIT 8-9
SYMPHYSIS PUBIS (Figure 8-11)

DEFINITION	The joint between the anterior articular surfaces of the two coxal bones.
TYPE OF JOINT	Cartilaginous joint of the symphysis type.

ANATOMICAL COMPONENTS

1. **Symphysis pubis (interpubic disc).** Disc of fibrocartilage between the anterior articulating surfaces of the coxal bones.
2. **Superior pubic ligament.** Connects the coxal bones superiorly and extends laterally to the pubic tubercles.
3. **Arcuate pubic ligament.** Thick band that connects the two coxal bones and forms the boundary of the pubic arch.

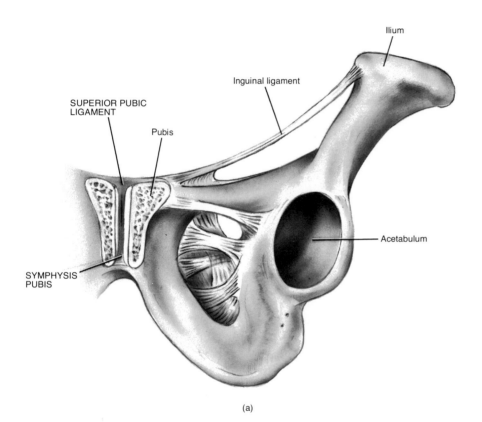

(a)

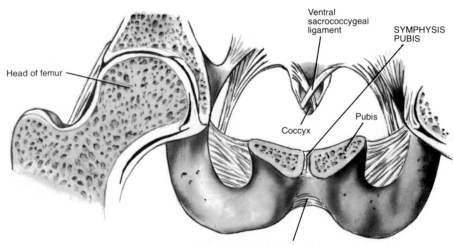

(b)

FIGURE 8-11 Symphysis pubis. (a) Frontal section. (b) Frontal section.

EXHIBIT 8-10
COSTOVERTEBRAL JOINTS (Figure 8-12)

DEFINITION	Joints formed between (1) heads of ribs and facets of bodies of thoracic vertebrae and the intervertebral discs between them and (2) necks and tubercles of ribs and the articular surfaces of the transverse processes of vertebrae.
TYPE OF JOINT	Both joints are synovial joints of the gliding (arthrodial) type.
ANATOMICAL COMPONENTS OF JOINTS BETWEEN HEADS OF RIBS AND BODIES OF VERTEBRAE	1. **Articular capsule.** Short, strong fibers around the joint that connect the head of the rib with the articular cavity formed by the adjacent vertebrae and intervertebral disc. 2. **Radiate ligament.** Extends from the anterior part of the head of each rib to the sides of the bodies of two vertebrae and the intervertebral disc between them. 3. **Intraarticular ligament.** Short, thick band within the joint that extends between the interarticular crest separating the two articular facets on the head of a rib and the intervertebral disc.
ANATOMICAL COMPONENTS OF JOINTS BETWEEN THE TUBERCLE OF A RIB AND THE ARTICULAR SURFACE OF THE TRANSVERSE PROCESS OF A VERTEBRA	1. **Articular capsule.** Thin membrane attached to the circumference of the articular surfaces. 2. **Superior costotransverse ligament.** Extends between the superior border of the neck of a rib and the inferior border of the transverse process of a vertebra immediately below. 3. **Posterior costotransverse ligament.** Extends from the neck of a rib to the transverse process and inferior articular process of the vertebra immediately above. 4. **Ligament of the neck of the rib.** Short, strong fibers that connect the back of the neck of a rib with the anterior surface of the adjacent transverse process of a vertebra. 5. **Ligament of the tubercle of the rib.** Short, thick band that extends from the transverse process of a vertebra to the nonarticular portion of the tubercle of a rib.

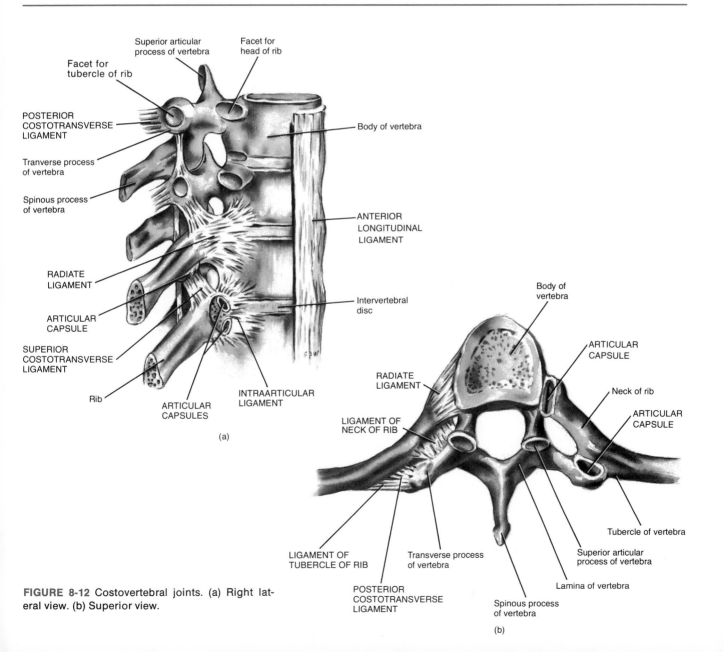

FIGURE 8-12 Costovertebral joints. (a) Right lateral view. (b) Superior view.

EXHIBIT 8-11
STERNOCOSTAL JOINTS (Figure 8-13)

DEFINITION

Joints formed between the cartilages of the true ribs, except the first pair, and the facets of the sternum. The costal cartilages of the first pair of ribs are directly united to the sternum and thus represent cartilaginous joints of the synchondrosis type. The cartilages of the false ribs, except for the eleventh and twelfth pairs, articulate with each other.

TYPE OF JOINT

The joints between the second through seventh costal cartilages and the facets on the sides of the sternum are synovial joints of the gliding (arthrodial) type.

ANATOMICAL COMPONENTS

1. **Articular capsules.** Surround the joints between the cartilages of the true ribs, except the first pair, and the facets of the sternum.
2. **Radiate sternocostal ligaments.** Thin, broad bands that connect the sternal ends of the costal cartilages to the anterior and posterior surfaces of the sternum.
3. **Intraarticular sternocostal ligament.** Typically found only between the second costal cartilages and the sternum.
4. **Costoxiphoid ligaments.** Connect the seventh, and sometimes the sixth, costal cartilage to the xiphoid process of the sternum.
5. The costal cartilages of the fifth, sixth, seventh, and eighth, and sometimes the ninth and tenth ribs, articulate with each other. These joints contain articular capsules lined by a synovial membrane. The joints are strengthened by *interchondral ligaments*.

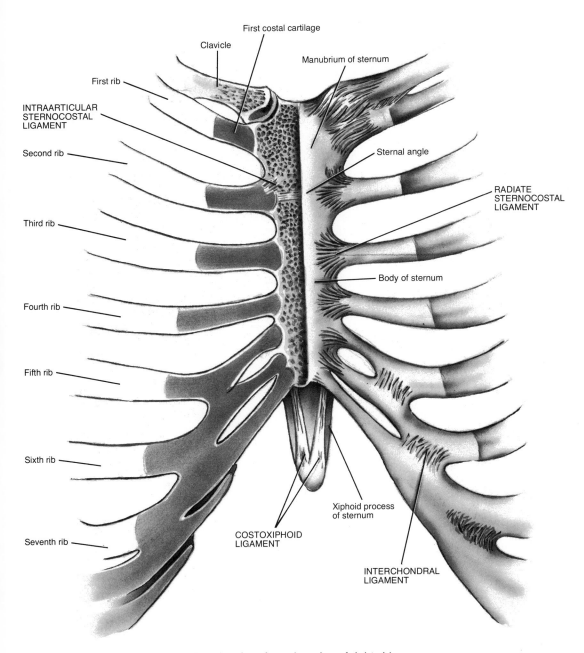

FIGURE 8-13 Sternocostal joints. Anterior view: frontal section of right side.

EXHIBIT 8-12
STERNOCLAVICULAR JOINT (Figure 8-14)

DEFINITION

Joint formed by the sternal extremity of the clavicle, the superior and lateral portions of the manubrium of the sternum, and the cartilage of the first rib.

TYPE OF JOINT

Synovial joint of the gliding (arthrodial) type.

ANATOMICAL COMPONENTS

1. **Articular disc (meniscus).** Flat, circular fibrocartilage disc between the articular surfaces of the clavicle and sternum. It divides the joint into two cavities, each of which has a synovial membrane.
2. **Articular capsule.** Surrounds the whole joint.
3. **Anterior sternoclavicular ligament.** Broad band covering the anterior surface of the joint extending from the sternal extremity of the clavicle to the manubrium of the sternum.
4. **Posterior sternoclavicular ligament.** Broad band covering the posterior surface of the joint extending from the sternal extremity of the clavicle to the manubrium of the sternum.
5. **Interclavicular ligament.** Flattened band that extends from the sternal extremity of one clavicle to the same portion of the other clavicle. It is also attached to the superior portion of the manubrium.
6. **Costoclavicular ligament.** Short, flat band that extends from the cartilage of the first rib to the costal tuberosity of the clavicle.

EXHIBIT 8-13
STERNAL JOINTS

DEFINITION

Joints formed between (1) the manubrium and body of the sternum called the *manubriosternal joint,* and (2) the body and xiphoid process of the sternum called the *xiphisternal joint.*

TYPE OF JOINT

The manubriosternal joint is a cartilaginous joint of the symphysis type. The manubrium and body join at an angle called the *sternal angle.* After maturity is reached, the joint usually ossifies. The xiphisternal joint is also a cartilaginous joint of the symphysis type.

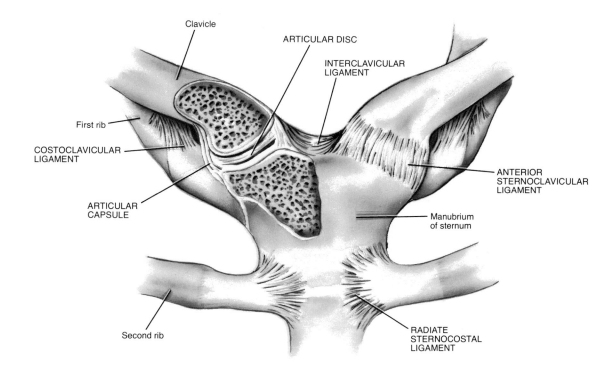

(a)

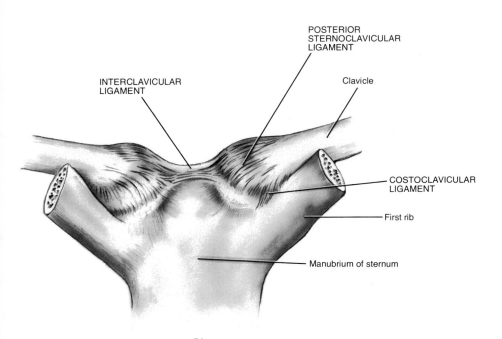

(b)

FIGURE 8-14 Sternoclavicular joint. (a) Anterior view. (b) Posterior view.

EXHIBIT 8-14
ACROMIOCLAVICULAR JOINT (Figure 8-15)

DEFINITION	Joint formed by the acromial extremity of the clavicle and the medial border of the acromion of the scapula.
TYPE OF JOINT	Synovial joint of the gliding (arthrodial) type.
ANATOMICAL COMPONENTS	1. **Articular capsule.** Completely envelops the joint.

1. **Articular capsule.** Completely envelops the joint.
2. **Acromioclavicular ligament.** The superior portion of the ligament covers the superior part of the joint and extends from the acromial extremity of the clavicle to the acromion of the scapula. The inferior portion of the ligament covers the inferior part of the joint and extends from the acromial extremity of the clavicle to the acromion of the scapula.
3. **Articular disc (meniscus).** Wedge-shaped piece of fibrocartilage attached to the superior part of the joint. The disc is frequently absent and, when present, rarely forms a complete partition.
4. **Coracoclavicular ligament.** Powerful ligament that connects the lateral end of the clavicle to the coracoid process of the scapula. It is divisible into two parts on the basis of shape. The *conoid ligament* is a dense band of fibers that attaches the angle between the two parts of the coracoid process to the conoid tubercle of the clavicle. The *trapezoid ligament* is broad and thin and extends from posterior part of the coracoid process to the inferior surface of the acromial extremity.

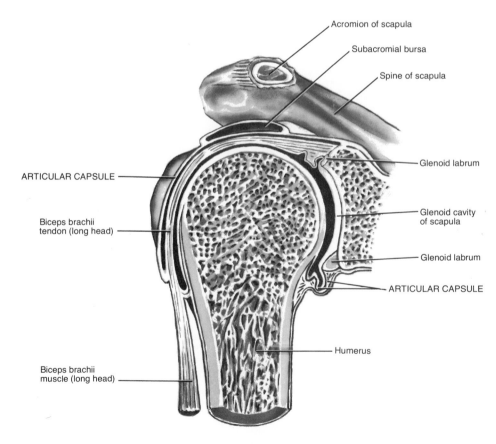

(a)

FIGURE 8-15 Acromioclavicular joint. (a) Frontal section.

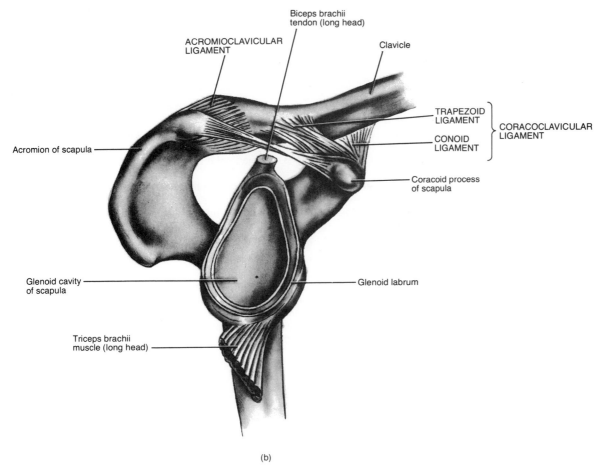

(b)

FIGURE 8-15 *(Continued)* Acromioclavicular joint. (b) Lateral view.

EXHIBIT 8-15
HUMEROSCAPULAR (SHOULDER) JOINT (Figure 8-16)

DEFINITION   Joint formed by the head of the humerus and the glenoid cavity of the scapula.

TYPE OF JOINT   Synovial joint of the ball-and-socket (spheroid) type.

ANATOMICAL COMPONENTS

1. **Articular capsule.** Loose sac that completely envelops the joint. It extends from the circumference of the glenoid cavity to the anatomical neck of the humerus.
2. **Coracohumeral ligament.** Strong, broad ligament that extends from the coracoid process of the scapula to the greater tubercle of the humerus.
3. **Glenohumeral ligaments.** Three thickenings of the articular capsule over the ventral surface of the joint.
4. **Transverse humeral ligament.** Narrow sheet extending from the greater tubercle to lesser tubercle of the humerus.
5. **Glenoid labrum.** Narrow rim of fibrocartilage around the edge of the glenoid cavity. (See Figure 8-15 a-b.)
6. **Bursae.** Among the bursae associated with the shoulder joint are:

   Subscapular bursa between the tendon of the subscapularis muscle and the underlying joint capsule
   Subdeltoid bursa between the deltoid muscle and joint capsule
   Subacromial bursa between the acromion and joint capsule (See Figure 8-15a also.)
   Subcoracoid bursa either lies between the coracoid process and joint capsule or appears as an extension from the subacromial bursa

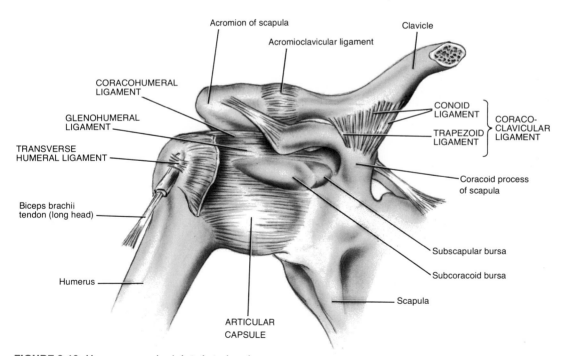

FIGURE 8-16 Humeroscapular joint. Anterior view.

EXHIBIT 8-16
ELBOW JOINT (Figure 8-17)

DEFINITION	Joint formed by the trochlea of the humerus, the trochlear notch of the ulna, the capitulum of the humerus and the head of the radius.
TYPE OF JOINT	Synovial joint of the hinge (ginglymus) type.
ANATOMICAL COMPONENTS	1. **Articular capsule.** The anterior part covers the anterior part of the joint from the radial and coronoid fossae of the humerus to the coronoid process of the ulna and the annular ligament of the radius. The posterior part extends from the capitulum, olecranon fossa, and lateral epicondyle of the humerus to the annular ligament of the radius, the olecranon of the ulna, and the ulna posterior to the radial notch.
	2. **Ulnar collateral ligament.** Thick, triangular ligament that extends from the medial epicondyle of the humerus to the coronoid process and olecranon of the ulna.
	3. **Radial collateral ligament.** Strong, triangular ligament that extends from the lateral epicondyle of the humerus to the annular ligament of the radius and the radial notch of the ulna.

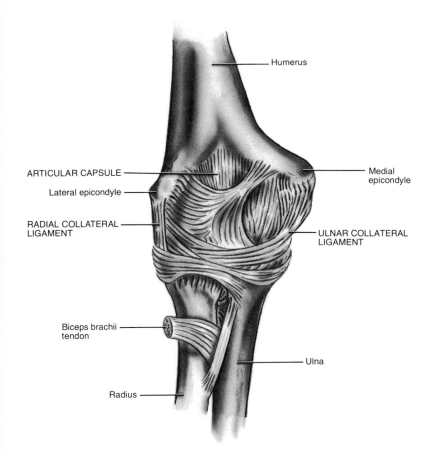

FIGURE 8-17 Elbow joint. Anterior view.

EXHIBIT 8-17
RADIOULNAR JOINTS (Figure 8-18)

DEFINITION	The *proximal radioulnar joint* is formed by the head of the radius and the ring formed by the radial notch of the ulna and the annular ligament of the radius. The *distal radioulnar joint* is formed by the head of the ulna and the ulna notch of the radius. The shafts of the radius and ulna are held together by (1) a small *oblique cord* that extends from the tubercle of the ulna to the radial tuberosity of the radius and (2) the *interosseous membrane* that connects the interosseous borders between the radius and ulna.
TYPE OF JOINT	Synovial joint of the pivot (trochoid) type.
ANATOMICAL COMPONENTS OF THE PROXIMAL RADIOULNAR JOINT	1. **Annular ligament of the radius.** Strong, curved ligament that encircles the head of the radius and keeps it in contact with the radial notch of the ulna. 2. **Quadrate ligament.** A thickened ligament extending from the inferior border of the annular ligament to the neck of the radius.
ANATOMICAL COMPONENTS OF THE DISTAL RADIOULNAR JOINT	1. **Articular capsule.** Attached to the margins of the ulnar notch and the head of the ulna. 2. **Articular disc.** Triangular piece of fibrocartilage that holds the distal ends of the ulna and radius together.

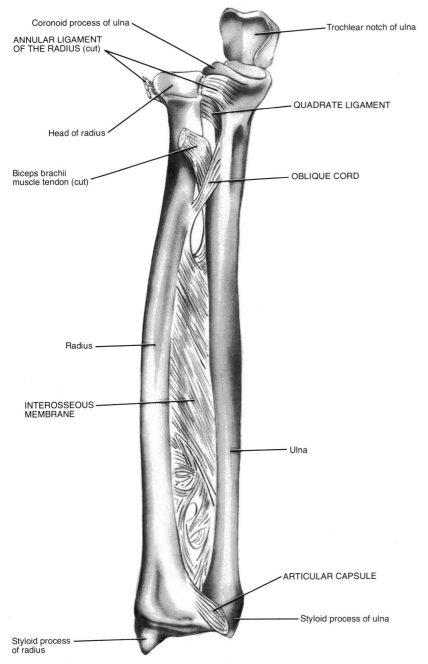

Coronoid process of ulna

ANNULAR LIGAMENT
OF THE RADIUS (cut)

Head of radius

Biceps brachii
muscle tendon (cut)

Radius

INTEROSSEOUS
MEMBRANE

Styloid process
of radius

Trochlear notch of ulna

QUADRATE LIGAMENT

OBLIQUE CORD

Ulna

ARTICULAR CAPSULE

Styloid process of ulna

FIGURE 8-18 Radioulnar joints. Anterior view.

EXHIBIT 8-18
RADIOCARPAL (WRIST) JOINT (Figure 8-19)

DEFINITION
Joint formed by the distal end of the radius and distal surface of the articular disc, and the navicular, lunate, and triangular carpal bones.

TYPE OF JOINT
Synovial joint of the ellipsoidal (condyloid) type.

ANATOMICAL COMPONENTS
1. **Articular capsule.** Extends from the styloid processes of the ulna and radius to the proximal row of carpal bones.
2. **Palmar radiocarpal ligament.** Thick, strong ligament that extends from the anterior border of the distal end of the radius to the proximal row of carpals. Some longer fibers extend to the capitate in the distal row.
3. **Dorsal radiocarpal ligament.** Extends from the posterior border of the distal end of the radius to the proximal row of carpals, especially the triangular.
4. **Ulnar collateral ligament.** Extends from the styloid process of the ulna to the triangular and pisiform bones and the transverse carpal ligament.
5. **Radial collateral ligament.** Extends from the styloid process of the radius to the navicular and pisiform bones.

EXHIBIT 8-19
INTERCARPAL JOINTS

DEFINITION
The intercarpal joints may be divided into three groups: (1) joints between carpals of the proximal row (navicular, lunate, and triangular); (2) joints between carpals of the distal row (greater multangular, lesser multangular, capitate, and hamate; and (3) joints between the proximal and distal rows of carpals, called *midcarpal joints.*

TYPE OF JOINT
Synovial joints of the gliding (arthrodial) type.

ANATOMICAL COMPONENTS OF JOINTS BETWEEN CARPALS OF THE PROXIMAL ROW
1. **Dorsal intercarpal ligaments.** Two ligaments that entend transversely and connect the navicular and lunate and the lunate and triangular. (See Figure 8-19a.)
2. **Palmar intercarpal ligaments.** Two ligaments that connect the navicular and lunate and the lunate and triangular.
3. **Interosseous intercarpal ligaments.** Two ligaments; one connects the navicular with the lunate, the other connects the lunate with the triangular.
4. The pisiform is joined to the triangular by the *articular capsule,* to the hamate by the *pisohamate ligament,* and to the base of the fifth metacarpal by the *pisometacarpal ligament.* (See Figure 8-19b.)

ANATOMICAL COMPONENTS OF JOINTS BETWEEN CARPALS OF THE DISTAL ROW
1. **Dorsal intercarpal ligaments.** Three ligaments that extend transversely on the dorsal surface and connect the greater multangular with the lesser multangular, the lesser multangular with the capitate, and the capitate with the hamate. (See Figure 8-19a.)
2. **Palmar intercarpal ligaments.** Three ligaments that extend transversely on the palmar surface and connect the same three carpals as the dorsal intercarpal ligaments.
3. **Interosseous intercarpal ligaments.** Three ligaments; one connects the capitate and hamate, one connects the capitate and lesser multangular, and one connects the greater multangular and lesser multangular.

ANATOMICAL COMPONENTS OF THE MIDCARPAL JOINTS
1. **Palmar intercarpal ligaments.** Extend from the palmar surfaces of the proximal carpals to the capitate.
2. **Dorsal intercarpal ligaments.** Connect the dorsal surfaces of the carpals of the proximal row with the carpals of the distal row. (See Figure 8-19a.)
3. **Ulnar collateral ligament.** Connects the triangular and hamate on the ulna side of the carpus. (See Figure 8-19a-b.)
4. **Radial collateral ligament.** Connects the navicular and greater multangular on the radial side of the carpus. The ulna and radial collateral ligaments are continuous with the collateral ligaments of the wrist joint. (See Figure 8-19a-b.)

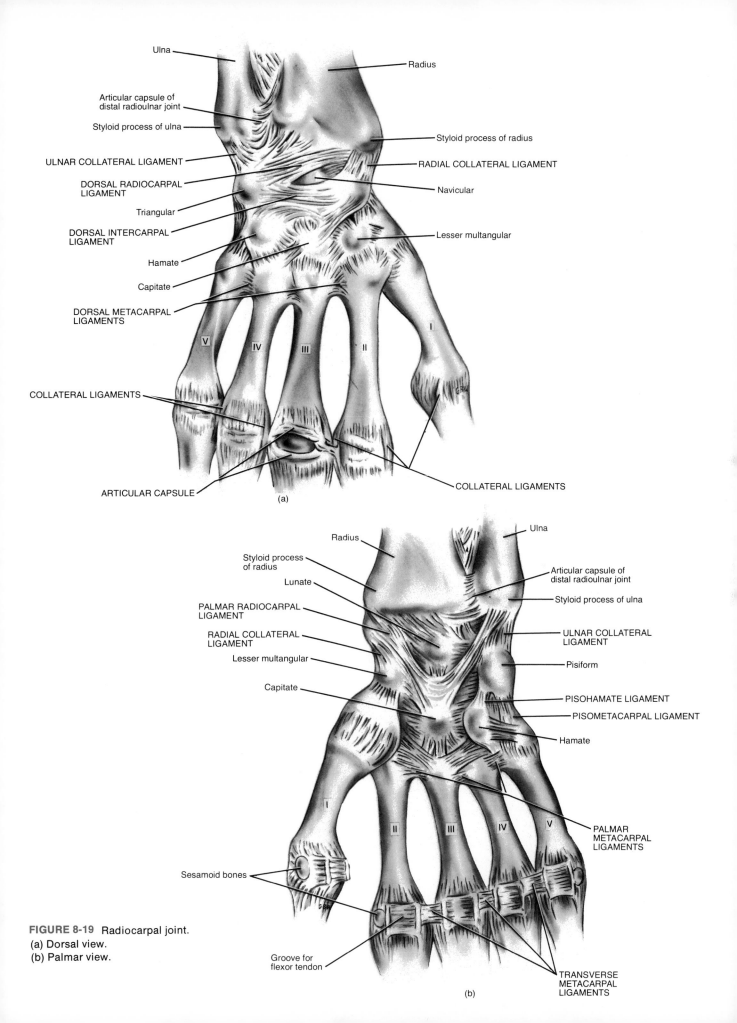

Ulna

Articular capsule of
distal radioulnar joint

Styloid process of ulna

ULNAR COLLATERAL LIGAMENT

DORSAL RADIOCARPAL
LIGAMENT

Triangular

DORSAL INTERCARPAL
LIGAMENT

Hamate

Capitate

DORSAL METACARPAL
LIGAMENTS

Radius

Styloid process of radius

RADIAL COLLATERAL LIGAMENT

Navicular

Lesser multangular

COLLATERAL LIGAMENTS

ARTICULAR CAPSULE

COLLATERAL LIGAMENTS

(a)

Radius

Styloid process
of radius

Lunate

PALMAR RADIOCARPAL
LIGAMENT

RADIAL COLLATERAL
LIGAMENT

Lesser multangular

Capitate

Sesamoid bones

Groove for
flexor tendon

Ulna

Articular capsule of
distal radioulnar joint

Styloid process of ulna

ULNAR COLLATERAL
LIGAMENT

Pisiform

PISOHAMATE LIGAMENT

PISOMETACARPAL LIGAMENT

Hamate

PALMAR
METACARPAL
LIGAMENTS

TRANSVERSE
METACARPAL
LIGAMENTS

(b)

FIGURE 8-19   Radiocarpal joint.
(a) Dorsal view.
(b) Palmar view.

EXHIBIT 8-20 ▰▰▰▰▰▰▰▰▰▰▰▰▰▰▰▰▰▰▰▰▰▰▰▰▰▰▰▰▰▰▰▰
CARPOMETACARPAL JOINTS

DEFINITION	Joints formed between (1) the base of metacarpal I and the greater multangular and (2) the base of metacarpals II–V and the distal surfaces of the greater multangular, lesser multangular, capitate, and hamate.
TYPE OF JOINT	The joint between metacarpal I and the greater multangular is a synovial joint of the saddle (sellaris) type and the joints between metacarpals II–V and the greater multangular, lesser multangular, capitate, and hamate are synovial joints of the gliding (arthrodial) type.
ANATOMICAL COMPONENTS OF THE JOINT BETWEEN METACARPAL I AND THE GREATER MULTANGULAR	1. **Articular capsule.** Surrounds the joint and extends from the base of metacarpal I to the edge of the greater multangular. 2. **Radial carpometacarpal ligament.** Attached to radial side of metacarpal I and the greater multangular. 3. **Oblique carpometacarpal ligaments.** Attached to the ulnar side of metacarpal I and the greater multangular.
ANATOMICAL COMPONENTS OF THE JOINT BETWEEN METACARPALS II–V AND THE DISTAL CARPALS	1. **Dorsal carpometacarpal ligaments.** Strongest carpal ligaments that connect the dorsal surfaces of the distal row of carpals to metacarpals. 2. **Palmar carpometacarpal ligaments.** Connect the palmar surfaces of the distal row of carpals to the metacarpals. 3. **Interosseous carpometacarpal ligments.** Short, thick ligaments that connect the hamate and capitate to metacarpals III and IV.

EXHIBIT 8-21 ▰▰▰▰▰▰▰▰▰▰▰▰▰▰▰▰▰▰▰▰▰▰▰▰▰▰▰▰▰▰▰▰
INTERMETACARPAL JOINTS

DEFINITION	Joints between the bases of metacarpals II–V with one another.
TYPE OF JOINT	Synovial joint of the gliding (arthrodial) type.
ANATOMICAL COMPONENTS	1. **Dorsal metacarpal ligaments.** Pass transversely from one bone to another on the dorsal surface. (See Figure 8-19a.) 2. **Palmar metacarpal ligaments.** Pass transversely from one bone to another on the palmar surface. (See Figure 8-19b.) 3. **Interosseous metacarpal ligaments.** Extend between the apposed surfaces of the bones, just distal to the articular facets. 4. **Transverse metacarpal ligament.** Extends across the palmar surfaces of the heads of metacarpals II–V. (See Figure 8-19b and 8-20.)

EXHIBIT 8-22
METACARPOPHALANGEAL JOINTS (Figure 8-20)

DEFINITION	The joints formed between the (1) head of metacarpal I and the base of the proximal phalanx (thumb) and (2) the heads of metacarpals II–V and the bases of the proximal phalanges of the four medial digits.
TYPE OF JOINT	The joint between the head of metacarpal I and the base of the proximal phalanx is a synovial joint of the hinge (ginglymus) type, whereas the joints between the heads of metacarpals II–V and the bases of the proximal phalanges of the four medial digits are synovial joints of the elliposidal (condyloid) type.
ANATOMICAL COMPONENTS	1. **Palmar ligaments.** Thick, fibrocartilage plates on the palmar surfaces of the joints that are connected to the metacarpal bones and the bases of the proximal phalanges. 2. **Collateral ligaments.** Two strong ligaments that attach the metacarpals and phalanges at the sides of the joints. (See Figure 8-19a also.)

EXHIBIT 8-23
INTERPHALANGEAL JOINTS

DEFINITION	Joints between the adjacent phalanges. There is one in the thumb and two in each of the other four digits.
TYPE OF JOINT	Synovial joints of the hinge (ginglymus) type.
ANATOMICAL COMPONENTS	Each joint has a palmar ligament and two collateral ligaments and they are constructed in exactly the same fashion as the metacarpophalangeal joints. (See Figure 8-20.)

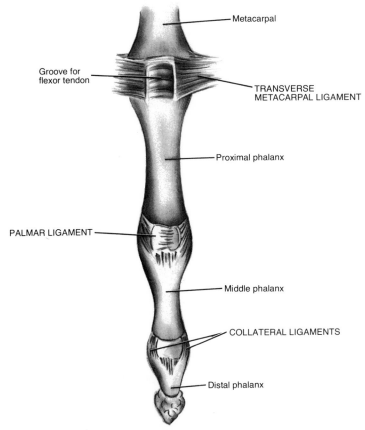

FIGURE 8-20 Metacarpophalangeal joints. Palmar view.

EXHIBIT 8-24
COXAL (HIP) JOINT (Figure 8-21)

DEFINITION	Joint formed between the head of the femur and acetabulum of the coxal bone.
TYPE OF JOINT	Synovial of the ball-and-socket (spheroid) type.

ANATOMICAL COMPONENTS

1. **Articular capsule.** One of the strongest ligaments of the body. It extends from the rim of the acetabulum to the neck of the femur. The capsule consists of circular and longitudinal fibers. The circular fibers, called the *zona orbicularis,* form a collar around the neck of the femur. The longitudinal fibers are reinforced by accessory ligaments known as the iliofemoral ligament, the pubofemoral ligament, and the ischiofemoral ligament.
2. **Iliofemoral ligament.** Thickened portion of the articular capsule that extends from the anterior inferior iliac spine of the coxal bone to the intertrochanteric line of the femur.
3. **Pubofemoral ligament.** Thickened portion of the articular capsule that extends from the pubic part of the rim of the acetabulum to the neck of the femur.
4. **Ischiofemoral ligament.** Thickened portion of the articular capsule that extends from the ischial wall of the acetabulum to the neck of the femur.
5. **Ligament of the head of the femur.** Flat, triangular band that extends from the fossa of the acetabulum to the head of the femur.
6. **Acetabular labrum.** Fibrocartilage rim attached to the margin of the acetabulum.
7. **Transverse ligament of the acetabulum.** Strong ligament which crosses over the acetabular notch, converting it to a foramen. It supports part of the acetabular labrum and is connected with the ligament of the head of the femur and the articular capsule.

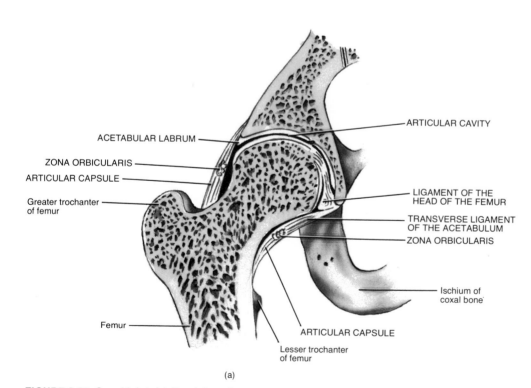

(a)

FIGURE 8-21 Coxal joint. (a) Frontal section.

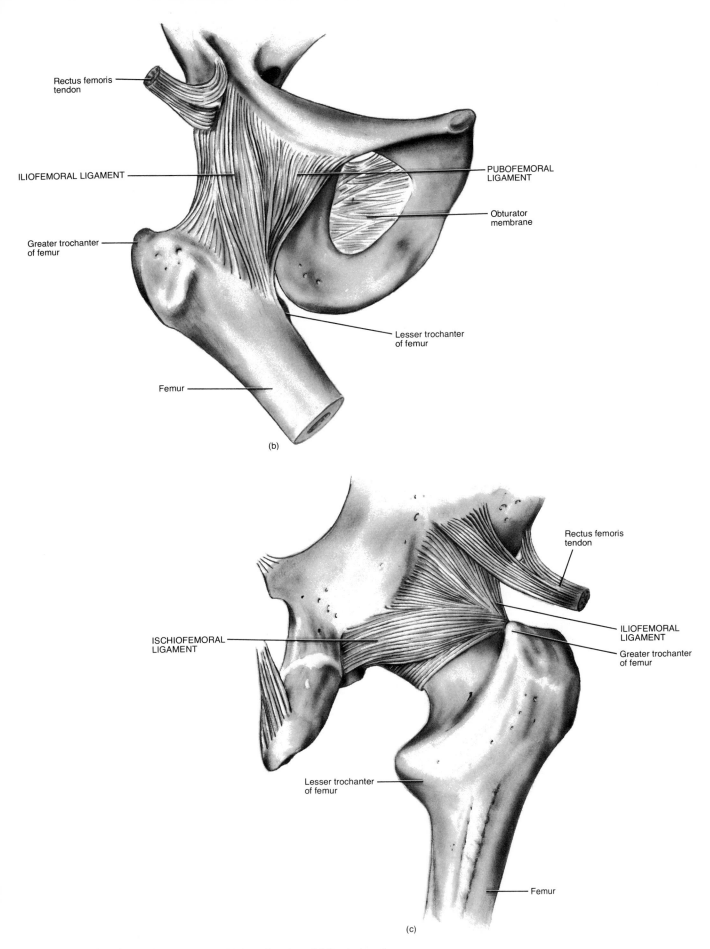

Rectus femoris
tendon

ILIOFEMORAL LIGAMENT

Greater trochanter
of femur

Femur

PUBOFEMORAL
LIGAMENT

Obturator
membrane

Lesser trochanter
of femur

(b)

ISCHIOFEMORAL
LIGAMENT

Rectus femoris
tendon

ILIOFEMORAL
LIGAMENT

Greater trochanter
of femur

Lesser trochanter
of femur

Femur

(c)

FIGURE 8-21 *(Continued)* Coxal joint. (b) Anterior view. (c) Posterior view.

EXHIBIT 8-25
TIBIOFEMORAL (KNEE) JOINT (Figure 8-22)

DEFINITION

The largest joint of the body, actually consisting of three joints: (1) an *intermediate patellofemoral joint* between the patella and the patellar surface of the femur; (2) a *lateral tibiofemoral joint* between the lateral condyle of the femur, lateral meniscus, and lateral condyle of the tibia; and (3) a *medial tibiofemoral joint* between the medial condyle of the femur, medial meniscus, and medial condyle of the tibia.

TYPE OF JOINT

The patellofemoral joint is a partly synovial joint of the gliding (arthrodial) type, and the lateral and medial tibiofemoral joints are synovial joints of the hinge (ginglymus) type.

ANATOMICAL COMPONENTS

1. **Articular capsule.** There is no complete, independent capsule uniting the bones. The ligamentous sheath surrounding the joint consists mostly of muscle tendons or expansions of them. There are, however, some capsular fibers connecting the articulating bones.
2. **Medial and lateral patellar retinacula.** Fused tendons of insertion of the quadriceps femoris muscle and the fascia lata that strengthen the anterior surface of the joint.
3. **Patellar ligament.** Central portion of the common tendon of insertion of the quadriceps femoris muscle that extends from the patella to the tibial tuberosity. This also strengthens the anterior surface of the joint. The posterior surface of the ligament is separated from the synovial membrane of the joint by an *infrapatellar fat pad*.
4. **Oblique popliteal ligament.** Broad, flat ligament that connects the intercondylar fossa of the femur to the head of the tibia. The tendon of the semimembranosus muscle is superficial to the ligament and passes from the medial condyle of the tibia to the lateral condyle of the femur. The ligament and tendon afford strength for the posterior surface of the joint.
5. **Arcuate popliteal ligament.** Extends from the lateral condyle of the femur to the styloid process of the head of the fibula. It strengthens the lower lateral part of the posterior surface of the joint.
6. **Tibial collateral ligament.** Broad, flat ligament on the medial surface of the joint that extends from the medial condyle of the femur to the medial condyle of the tibia. The ligament is crossed by tendons of the sartorius, gracilis, and semitendinosus muscles, all of which strengthen the medial aspect of the joint.
7. **Fibular collateral ligament.** Stong, rounded ligament on the lateral surface of the joint that extends from the lateral condyle of the femur to the lateral side of the head of the fibula. The ligament is covered by the tendon of the biceps femoris muscle. The tendon of the popliteus muscle is deep to the tendon.
8. **Intraarticular ligaments.** Ligaments within the capsule that connect the tibia and femur.
   a. **Anterior cruciate ligament.** Extends posteriorly and laterally from the area anterior to the intercondylar eminence of the tibia to the posterior part of the medial surface of the lateral condyle of the femur.
   b. **Posterior cruciate ligament.** Extends anteriorly and medially from the posterior intercondylar fossa of the tibia and lateral meniscus to the anterior part of the medial surface of the medial condyle of the femur.
9. **Menisci.** Fibrocartilage discs between the tibial and femoral condyles. They help to compensate for the incongruence of the articulating bones.
   a. **Medial meniscus.** Semicircular piece of fibrocartilage. Its anterior end is attached to the anterior intercondylar fossa of the tibia, in front of the anterior cruciate ligament. Its posterior end is attached to the posterior intercondylar fossa of the tibia between the attachments of the posterior cruciate ligament and lateral meniscus.
   b. **Lateral meniscus.** Circular piece of fibrocartilage. Its anterior end is attached anterior to the intercondylar eminence of the tibia, lateral and posterior to the anterior cruciate ligament. Its posterior end is attached posterior to the intercondylar eminence of the tibia and anterior to the posterior end of the medial meniscus. The medial and lateral menisci are connected to each other by the *transverse ligament* and to the margins of the head of the tibia by the *coronary ligaments*.
10. **Bursae.** The principal bursae of the knee include the following:
    a. **Anterior bursae:** (1) between the patella and skin *(prepatellar bursa)*, (2) between upper part of tibia and patellar ligament *(infrapatellar bursa)*, (3) between lower part of tibial tuberosity and skin, and (4) between lower part of femur and deep surface of quadriceps femoris muscle *(suprapatellar bursa)*.
    b. **Medial bursae:** (1) between medial head of gastrocnemius muscle and the articular capsule, (2) superficial to the tibial collateral ligament between the ligament and tendons of the sartorius, gracilis, and semitendinosus muscles, (3) deep to the tibial collateral ligament between the ligament and the tendon of the semimembranosus muscle, (4) between the tendon of the semimembranosus muscle and the head of the tibia, and (5) between the tendons of the semimembranosus and semitendinosus muscles.
    c. **Lateral bursae:** (1) between the lateral head of the gastrocnemius muscle and articular capsule, (2) between the tendon of the biceps femoris muscle and fibular collateral ligament, (3) between the tendon of the popliteus muscle and fibular collateral ligament, and (4) between the lateral condyle of the femur and the popliteus muscle.

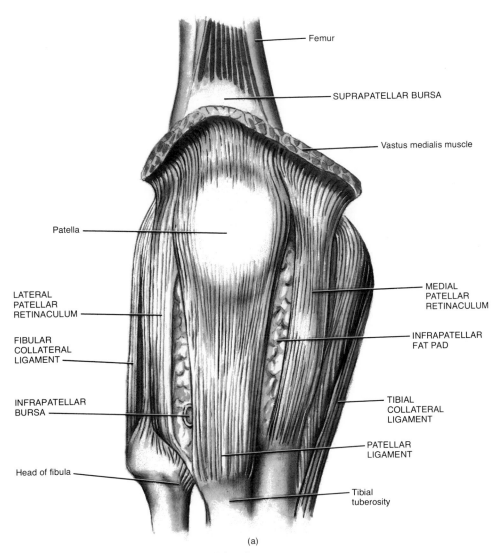

Femur

SUPRAPATELLAR BURSA

Vastus medialis muscle

Patella

LATERAL
PATELLAR
RETINACULUM

FIBULAR
COLLATERAL
LIGAMENT

INFRAPATELLAR
BURSA

Head of fibula

MEDIAL
PATELLAR
RETINACULUM

INFRAPATELLAR
FAT PAD

TIBIAL
COLLATERAL
LIGAMENT

PATELLAR
LIGAMENT

Tibial
tuberosity

(a)

FIGURE 8-22 Tibiofemoral joint. (a) Anterior view.

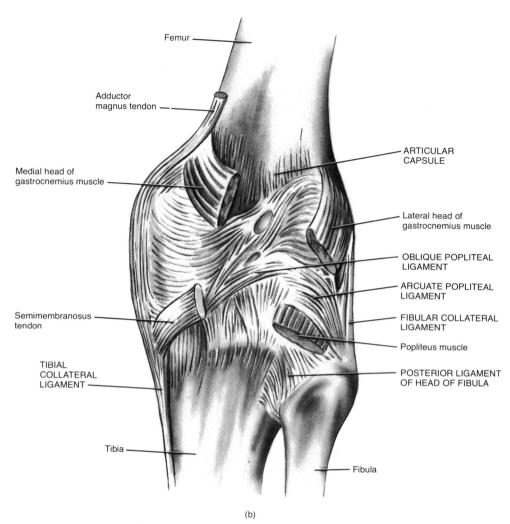

Femur

Adductor
magnus tendon

Medial head of
gastrocnemius muscle

Semimembranosus
tendon

TIBIAL
COLLATERAL
LIGAMENT

Tibia

ARTICULAR
CAPSULE

Lateral head of
gastrocnemius muscle

OBLIQUE POPLITEAL
LIGAMENT

ARCUATE POPLITEAL
LIGAMENT

FIBULAR COLLATERAL
LIGAMENT

Popliteus muscle

POSTERIOR LIGAMENT
OF HEAD OF FIBULA

Fibula

(b)

FIGURE 8-22 *(Continued)* Tibiofemoral joint. (b) Posterior view.

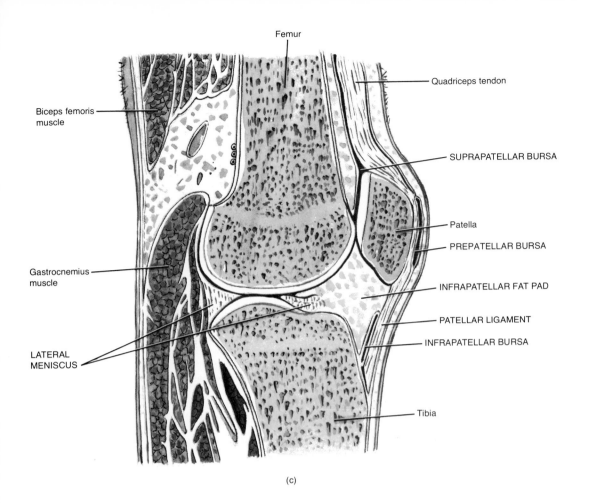

(c)

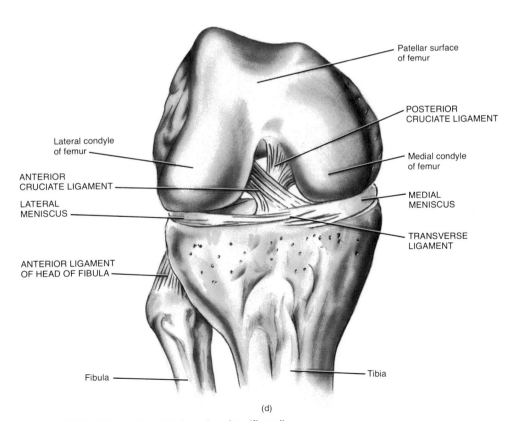

(d)

FIGURE 8-22 (Continued) Tibiofemoral joint. (c) Sagittal section. (d) Anterior view (flexed).

EXHIBIT 8-26
TIBIOFIBULAR JOINTS (Figure 8-23)

DEFINITION	The *proximal tibiofibular joint* is formed by the lateral condyle of the tibia and the head of the fibula. The *distal tibiofibular joint* is formed by the fibular notch of the tibia and the medial side of the distal end of the fibula. The shafts of the tibia and fibula are held together by the (1) *interosseous membrane,* which is attached to the interosseous borders of the tibia and fibula, and (2) the *interosseous ligament,* which also unites the interosseous borders of the bones and is continuous proximally with the interosseous membrane.
TYPE OF JOINT	The proximal tibiofibular joint is a synovial joint of the gliding (arthrodial) type, whereas the distal tibiofibular joint is a cartilaginous joint of the syndesmosis type.
ANATOMICAL COMPONENTS OF THE PROXIMAL TIBIOFIBULAR JOINT	1. **Articular capsule.** Surrounds the joint and is attached to the borders of the articular facets of the tibia and fibula. 2. **Anterior ligament of the head of the fibula.** Several broad, flat ligaments that extend from the head of the fibula to the anterior part of the lateral condyle of the tibia. 3. **Posterior ligament of the head of the fibula.** Single, thick ligament that extends from the head of the fibula to the posterior part of the lateral condyle of the tibia. (See Figure 8-22b.)
ANATOMICAL COMPONENTS OF THE DISTAL TIBIOFIBULAR JOINT	1. **Anterior tibiofibular ligament.** Flat ligament that extends between the adjacent margins of the tibia and fibula on the anterior surface of the joint. 2. **Posterior tibiofibular ligament.** Extends between the adjacent margins of the tibia and fibula on the posterior surface of the joint. (See Figure 8-24.) 3. **Transverse tibiofibular ligament.** Strong, thick ligament anterior to the posterior tibiofibular ligament that extends from the lateral malleolus of the fibula to nearly the medial malleolus of the tibia. (See Figure 8-24.)

EXHIBIT 8-27
TALOCRURAL (ANKLE) JOINTS (Figure 8-24)

DEFINITION	Joints formed between the (1) distal end of the tibia and its medial malleolus and the talus, and (2) the lateral malleolus of the fibula and the talus.
TYPE OF JOINT	Both are synovial joints of the hinge (ginglymus) type.
ANATOMICAL COMPONENTS	1. **Articular capsule.** Surrounds the joint and extends from the borders of the tibia and malleoli to the talus. 2. **Deltoid ligament.** Strong, triangular ligament that extends from the medial malleolus of the tibia to the navicular, calcaneus, and talus. 3. **Anterior talofibular ligament.** Extends from the anterior margin of the lateral malleolus of the fibula to the talus. 4. **Posterior talofibular ligament.** Extends from the posterior margin of the lateral malleolus of the fibula to the talus. 5. **Calcaneofibular ligament.** Extends from the apex of the lateral malleolus of the fibula to the talus.

FIGURE 8-23 Tibiofibular joints. Anterior view.

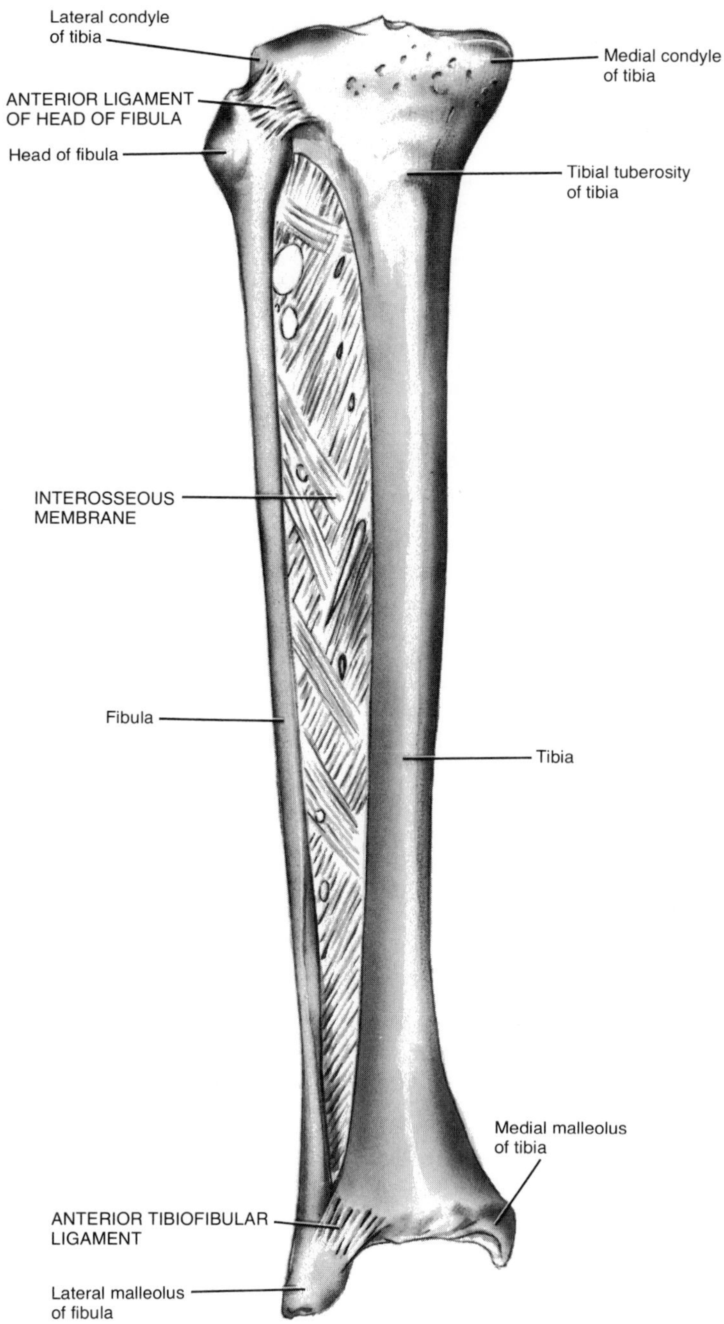

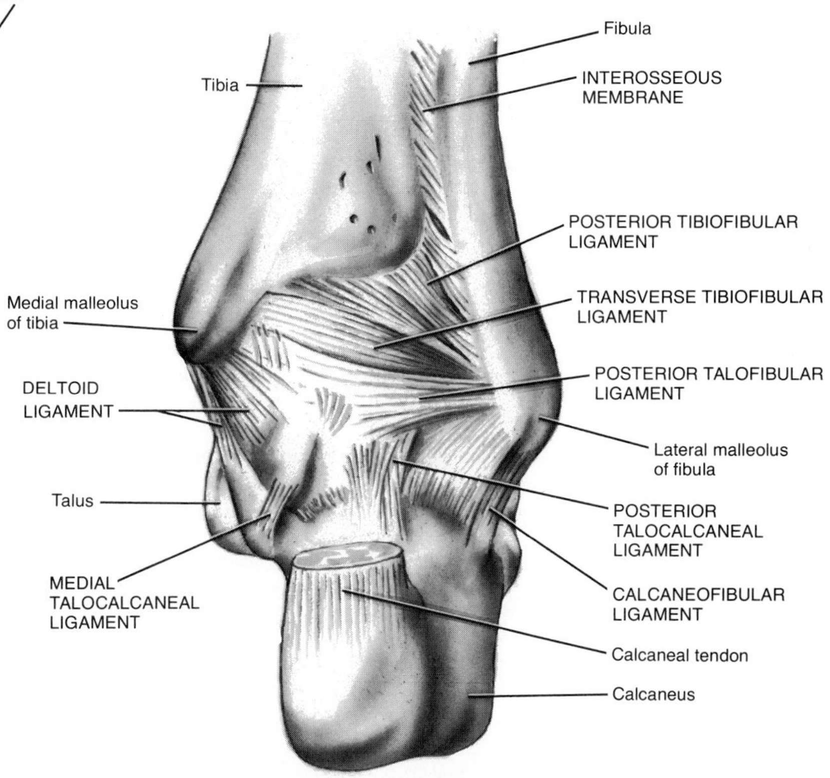

Lateral condyle of tibia

Medial condyle of tibia

ANTERIOR LIGAMENT OF HEAD OF FIBULA

Head of fibula

Tibial tuberosity of tibia

INTEROSSEOUS MEMBRANE

Fibula

Tibia

Medial malleolus of tibia

ANTERIOR TIBIOFIBULAR LIGAMENT

Lateral malleolus of fibula

Fibula

INTEROSSEOUS MEMBRANE

Tibia

POSTERIOR TIBIOFIBULAR LIGAMENT

TRANSVERSE TIBIOFIBULAR LIGAMENT

POSTERIOR TALOFIBULAR LIGAMENT

Medial malleolus of tibia

Lateral malleolus of fibula

DELTOID LIGAMENT

POSTERIOR TALOCALCANEAL LIGAMENT

Talus

CALCANEOFIBULAR LIGAMENT

MEDIAL TALOCALCANEAL LIGAMENT

Calcaneal tendon

Calcaneus

FIGURE 8-24 Talocrural joints. Posterior view.

EXHIBIT 8-28
INTERTARSAL JOINTS (Figure 8-25)

DEFINITION	Joints between tarsal bones of the ankle. The joints include (1) the *subtalar joint,* which consists of an anterior *talocalcaneonavicular joint* between the talus, calcaneus, and navicular bones, and a posterior *talocalcaneal joint* between the talus and calcaneus; (2) the *calcaneocuboid joint* between the calcaneus and cuboid; (3) the *cuneonavicular joint* between the three cuneiforms and navicular; (4) the *cuboideonavicular joint* between the cuboid and navicular; (5) the *intercuneiform joints* between the three cuneiforms; and (6) the *cuneocuboid joint* between the third (lateral) cuneiform and cuboid.
TYPE OF JOINT	All intertarsal joints are synovial joints of the gliding (arthrodial) type, except for the calcaneocuboid joint which is a synovial joint of the saddle (sellaris) type.
ANATOMICAL COMPONENTS OF THE TALOCALCANEONAVICULAR JOINT	1. **Articular capsule.** Binds the bones and encloses the common articular cavity. 2. **Dorsal talonavicular ligament.** Broad ligament that connects the talus to the navicular.
ANATOMICAL COMPONENTS OF THE TALOCALCANEAL JOINT	1. **Articular capsule.** Completely surrounds the joint. 2. **Anterior talocalcaneal ligament.** Connects the anterior and lateral surfaces of the talus to the superior surface of the calcaneus. 3. **Posterior talocalcaneal ligament.** Extends from the lateral surface of the talus to the proximal and medial parts of the calcaneus. (See Figure 8-24 also.) 4. **Lateral talocalcaneal ligament.** Connects the lateral surface of the talus to the lateral surface of the calcaneus. 5. **Medial talocalcaneal ligament.** Extends from the medial surface of the talus to the posterior surface of the calcaneus. 6. **Interosseous talocalcaneal ligament.** Principal ligament between the bones that connects the distal surface of the talus to the proximal surface of the calcaneus.
ANATOMICAL COMPONENTS OF THE CALCANEOCUBOID JOINT	1. **Articular capsule.** Completely surrounds the joint. 2. **Dorsal calcaneocuboid ligament.** Connects the dorsal surfaces of the calcaneus and cuboid. 3. **Bifurcated ligament.** Extends from the proximal surface of the calcaneus to the medial side of the cuboid and the lateral side of the navicular. 4. **Long plantar ligament.** Longest tarsal ligament that connects the plantar surface of the calcaneus to the plantar surface of the cuboid. (See Figure 8-26 also.) 5. **Plantar calcaneocuboid ligament.** Extends from the distal part of the plantar surface of the calcaneus to the plantar surface of the cuboid. (See Figure 8-26.)
ANATOMICAL COMPONENTS OF THE CUNEONAVICULAR JOINT	1. **Dorsal cuneonavicular ligaments.** Three ligaments, each attaching one of the cuneiforms to the navicular at the upper and medial aspects of the joints. 2. **Plantar cuneonavicular ligaments.** Three ligaments, each attaching one of the cuneiforms to the navicular at the plantar aspects of the joints.
ANATOMICAL COMPONENTS OF THE CUBOIDEONAVICULAR JOINT	1. **Dorsal cuboideonavicular ligament.** Extends distally and laterally from the navicular to the cuboid. 2. **Plantar cuboideonavicular ligament.** Connects the navicular and cuboid transversely. (See Figure 8-26.) 3. **Interosseous cuboideonavicular ligament.** Strong transverse fibers that connect the navicular and cuboid.
ANATOMICAL COMPONENTS OF THE INTERCUNEIFORM JOINTS	1. **Dorsal intercuneiform ligaments.** Weak transverse connections between the dorsal surfaces of the cuneiforms. 2. **Interosseous intercuneiform ligaments.** Strong, transverse ligaments that connect the cuneiforms. 3. **Plantar intercuneiform ligaments.** Strong, transverse ligaments that connect the plantar surfaces of the cuneiforms.
ANATOMICAL COMPONENTS OF THE CUNEOCUBOID JOINT	1. **Dorsal cuneocuboid ligaments.** Weak ligaments between the dorsal surfaces of the lateral cuneiform and cuboid. 2. **Interosseous cuneocuboid ligaments.** Strong, transverse ligaments that connect the lateral cuneiform and cuboid. 3. **Plantar cuneocuboid ligaments.** Strong transverse ligaments that connect the plantar surfaces of the lateral cuneiform and cuboid.

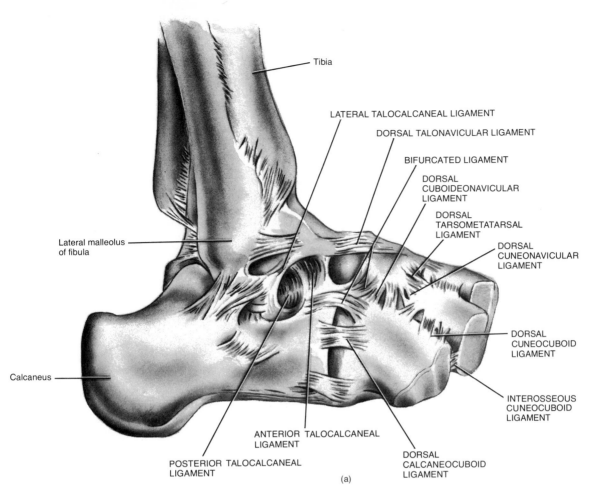

Tibia

LATERAL TALOCALCANEAL LIGAMENT

DORSAL TALONAVICULAR LIGAMENT

BIFURCATED LIGAMENT

DORSAL CUBOIDEONAVICULAR LIGAMENT

DORSAL TARSOMETATARSAL LIGAMENT

DORSAL CUNEONAVICULAR LIGAMENT

Lateral malleolus of fibula

DORSAL CUNEOCUBOID LIGAMENT

Calcaneus

INTEROSSEOUS CUNEOCUBOID LIGAMENT

ANTERIOR TALOCALCANEAL LIGAMENT

DORSAL CALCANEOCUBOID LIGAMENT

POSTERIOR TALOCALCANEAL LIGAMENT

(a)

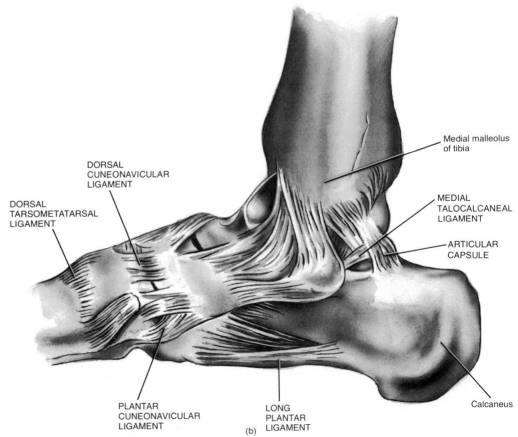

DORSAL CUNEONAVICULAR LIGAMENT

Medial malleolus of tibia

DORSAL TARSOMETATARSAL LIGAMENT

MEDIAL TALOCALCANEAL LIGAMENT

ARTICULAR CAPSULE

Calcaneus

PLANTAR CUNEONAVICULAR LIGAMENT

LONG PLANTAR LIGAMENT

(b)

FIGURE 8-25 Intertarsal joints. (a) Lateral view. (b) Medial view.

EXHIBIT 8-29
TARSOMETATARSAL JOINTS (Figure 8-26)

DEFINITION

The tarsometatarsal joints consist of the following articulations: (1) the joint between metatarsal I and the first (medial) cuneiform; (2) the joint between metatarsal II and the second (intermediate) cuneiform; (3) the joint between metatarsal III and the third (lateral) cuneiform, (4) the joint between metatarsal IV and the third (lateral) cuneiform and cuboid; and (5) the joint between metatarsal V and the cuboid.

TYPE OF JOINT

All tarsometatarsal joints are synovial joints of the gliding (arthrodial) type.

ANATOMICAL COMPONENTS

1. **Dorsal tarsometatarsal ligaments.** Strong, flat ligaments that connect the adjoining dorsal surfaces of the tarsal and metatarsal bones. (See Figure 8-25a-b.)
2. **Plantar tarsometatarsal ligaments.** Connect the adjoining plantar surfaces of the tarsal and metatarsal bones.
3. **Interosseous tarsometatarsal ligaments.** One, the strongest, connects the lateral surface of the first (medial) cuneiform to the lateral surface of metatarsal II. A second connects the third (lateral) cuneiform to the lateral aspect of metatarsal II. A third connects the lateral aspect of the third (lateral) cuneiform with the lateral surface of the base of metatarsal III.

EXHIBIT 8-30
INTERMETATARSAL JOINTS

DEFINITION

Joints between the bases of metatarsals II–V.

TYPE OF JOINT

Synovial joint of the gliding (arthrodial) type.

ANATOMICAL COMPONENTS

1. **Dorsal intermetatarsal ligaments.** Pass transversely from one metatarsal to another on the dorsal surface.
2. **Plantar intermetatarsal ligaments.** Pass transversely from one metatarsal to another on the palmar surface. (See Figure 8-26.)
3. **Interosseous intermetatarsal ligaments.** Strong transverse fibers that connect the nonarticular portions of the apposed surfaces.
4. **Transverse metatarsal ligament.** Narrow ligament that connects the heads of all metatarsals along the plantar surface. (See Figure 8-26.)

EXHIBIT 8-31
METATARSOPHALANGEAL JOINTS

DEFINITION

Joints formed by the heads of the metatarsals and the bases of the proximal phalanges.

TYPE OF JOINT

Synovial joints of the ellipsoidal (condyloid) type.

ANATOMICAL COMPONENTS

1. **Collateral ligaments.** Strong, rounded ligaments on either side of each joint that extend from the metatarsal to the proximal phalanx.
2. **Plantar ligaments.** Thick, dense ligaments on the plantar surfaces of the joints between and attached to the collateral ligaments. (See Figure 8-26.)
3. **Deep transverse metatarsal ligaments.** Transverse ligaments between the plantar ligaments of all the metatarsophalangeal joints. (See Figure 8-26.)

EXHIBIT 8-32
INTERPHALANGEAL JOINTS

DEFINITION

Joints between the adjacent phalanges. There is one in the great toe and two in each of the other four toes.

TYPE OF JOINT

Synovial joints of the hinge (ginglymus) type.

ANATOMICAL COMPONENTS

Each joint has a plantar ligament and two collateral ligaments and they are constructed in exactly the same fashion as the metatarsophalangeal joints. (See Figure 8-26.)

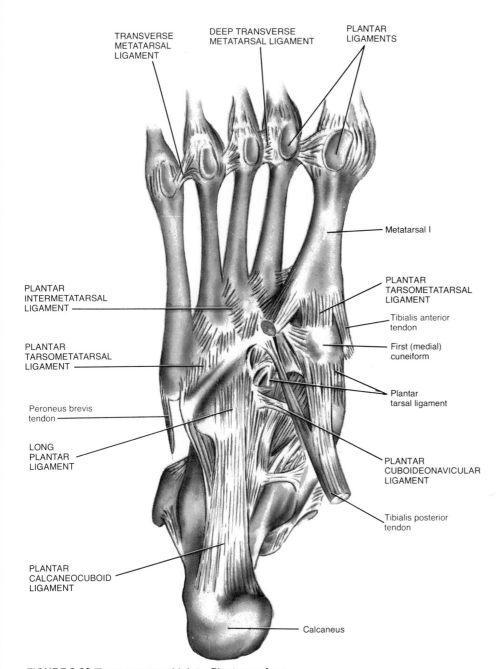

TRANSVERSE
METATARSAL
LIGAMENT

DEEP TRANSVERSE
METATARSAL LIGAMENT

PLANTAR
LIGAMENTS

Metatarsal I

PLANTAR
TARSOMETATARSAL
LIGAMENT

Tibialis anterior
tendon

PLANTAR
INTERMETATARSAL
LIGAMENT

First (medial)
cuneiform

PLANTAR
TARSOMETATARSAL
LIGAMENT

Plantar
tarsal ligament

Peroneus brevis
tendon

LONG
PLANTAR
LIGAMENT

PLANTAR
CUBOIDEONAVICULAR
LIGAMENT

Tibialis posterior
tendon

PLANTAR
CALCANEOCUBOID
LIGAMENT

Calcaneus

FIGURE 8-26 Tarsometatarsal joints. Plantar surface.

Applications to Health

ARTHRITIS

The term **arthritis** refers to at least 25 different diseases, the most common of which are rheumatoid arthritis, osteoarthritis, and gouty arthritis. All these ailments are characterized by inflammation in one or more joints. Inflammation, pain, and stiffness may also be present in adjacent parts of the body, such as the muscles near the joint.

The causes of arthritis are unknown. In some cases, it has followed the stress of sprains, infections, and joint injury. Some researchers think that the cause is a bacterium, a virus, or an allergy. Others believe the nervous system, hormones, or a metabolic disorder might be involved. Still others believe that certain types of prolonged psychological stress, such as inhibited hostility, can upset homeostatic balance and bring on arthritic attacks.

RHEUMATISM

Rheumatism refers to any painful state of the supporting structures of the body, its bones, ligaments, joints, tendons, or muscles. Arthritis is a form of rheumatism in which the joints have become inflamed.

Rheumatism is not necessarily related to rheumatoid arthritis. About 25 percent of all Americans—most of them over age 40—have rheumatism that is diagnosed as some form of osteoarthritis.

RHEUMATOID ARTHRITIS

Rheumatoid arthritis is the most common inflammatory form of arthritis. It involves inflammation of the joint, swelling, pain, and a loss of function. Usually this form occurs bilaterally—if your left knee is affected, your right knee may also be affected, although usually not to the same degree.

The disease may afflict at any age, but especially between 30 and 50 years old. The frequency of occurrence is approximately 57 percent in females and 43 percent in males. The symptoms in females are more severe than in males. One form of rheumatoid arthritis afflicts children shortly after birth—*juvenile rheumatoid arthritis*—and other forms can occur in later childhood.

The primary symptom of rheumatoid arthritis is inflammation of the synovial membrane. If it is completely untreated, the following sequential pathology may occur (Figure 8-27). The membrane thickens and synovial fluid accumulates. The resulting pressure causes pain and tenderness. The membrane then produces an abnormal tissue called pannus, which adheres to the surface of the articular cartilage. The pannus formation sometimes erodes the cartilage completely. When the cartilage is destroyed, fibrous tissue joins the exposed bone ends. The tissue ossifies and fuses the joint so that it is immovable—the ultimate crippling effect of rheumatoid arthritis. Most cases do not progress to this stage. But the range of motion of the joint is greatly inhibited by the severe inflammation and swelling.

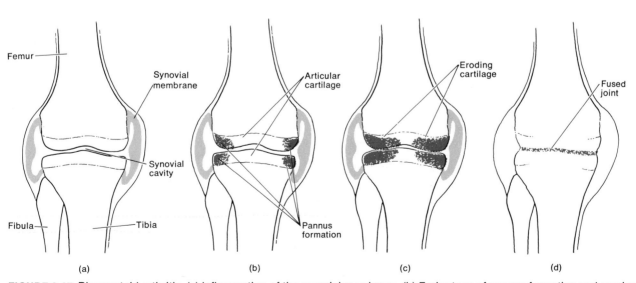

FIGURE 8-27 Rheumatoid arthritis. (a) Inflammation of the synovial membrane. (b) Early stage of pannus formation and erosion of articular cartilage. (c) Advanced stage of pannus formation and further erosion of articular cartilage. (d) Obliteration of joint cavity and fusion of articulating bones.

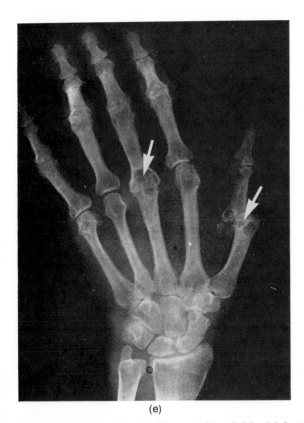

(e)

FIGURE 8-27 *(Continued)* Rheumatoid arthritis. (e) Anteroposterior projection of the hand and wrist showing changes characteristic of rheumatoid arthritis. The arrows indicate obliteration of the joint cavities and fusion of the bones. (Courtesy of Lester W. Paul and John H. Juhl, *The Essentials of Roentgen Interpretation*, 3d ed. Harper & Row, Publishers, Inc., New York, 1972.)

Damaged joints may be surgically replaced, either partly or entirely, with artificial joints (Figure 8-28). The artificial parts are inserted after removal of the diseased portion of the articulating bone and its cartilage. The new metal or plastic joint is fixed in place with a special acrylic cement. When freshly mixed in the operating room, it hardens as strong as bone in minutes. These new parts function nearly as well as a normal joint—and much better than a diseased joint.

Joint surgery, especially hip and knee replacement, holds much promise for the future. The technique is being extended to other joints—wrist, elbow, shoulder, fingers, ankle—with great optimism.

OSTEOARTHRITIS

A degenerative joint disease far more common than rheumatoid arthritis, and usually less damaging, is **osteoarthritis.** It apparently results from a combination of aging, irritation of the joints, and wear and abrasion.

Degenerative joint disease is a noninflammatory, progressive disorder of movable joints, particularly weight-bearing joints. It is characterized pathologically by the deterioration of articular cartilage and by formation of new bone in the subchondral areas and at the margins of the joint. The cartilage slowly degenerates, and as the bone ends become exposed, they deposit small bumps, or *spurs*, of new osseous tissue. These spurs decrease the space of the joint cavity and restrict joint movement. Unlike rheumatoid arthritis, osteoarthritis usually affects only the articular cartilage. The synovial membrane is rarely destroyed, and other tissues are unaffected.

GOUTY ARTHRITIS

Uric acid is a waste product produced during the metabolism of the nucleic acid purine. Normally, all the acid is quickly excreted in the urine. In fact, it gives urine its name. The person who suffers from *gout* either produces excessive amounts of uric acid or is not able to excrete normal amounts. The result is a buildup of uric acid in the blood. This excess acid then reacts with sodium to form a salt called sodium urate. Crystals of this salt are deposited in soft tissues. Typical sites are the kidneys and the cartilage of the ears and joints.

In **gouty arthritis,** sodium urate crystals are deposited in the soft tissues of the joints. The crystals irritate the cartilage, causing inflammation, swelling, and acute pain. Eventually, the crystals destroy all the joint tissues. If the disorder is not treated, the ends of the articulating bones fuse and the joint becomes immovable.

Gout occurs primarily in males of any age. It is believed to be the cause of two to five percent of all chronic joint diseases. Numerous studies indicate that gout is sometimes caused by an abnormal gene. As a result of this gene, the body manufactures purine by a mechanism that produces unusually large amounts of uric acid. Diet and environmental factors such as stress and climate are also suspected causes of gout.

BURSITIS

An acute or chronic inflammation of a bursa is called **bursitis.** The condition may be caused by trauma, by an acute or chronic infection (including syphilis and tuberculosis), or by rheumatoid arthritis. Repeated excessive friction often results in a bursitis with local inflammation and the accumulation of fluid. Bunions are frequently associated with a friction bursitis over the head of the first metatarsal bone. Symptoms include pain, swelling, tenderness, and the limitation of motion involving the inflamed bursa.

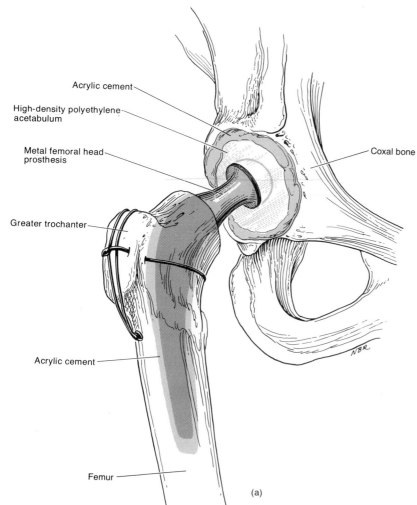

Acrylic cement

High-density polyethylene acetabulum

Metal femoral head prosthesis

Greater trochanter

Coxal bone

Acrylic cement

Femur

(a)

FIGURE 8-28 Total hip and total knee replacement. (a) Charnley's technique and prosthesis for total hip replacement. The arthritic portions of the acetabulum and head of the femur are replaced by a prefabricated joint. The acetabulum is made of a polyethylene substance of high density, and the metallic femoral head is cemented into the femur with acrylic cement. The greater trochanter of the femur is reattached after surgery.

TENDINITIS

Tendinitis or **tenosynovitis** frequently occurs as inflammation involving the tendon sheaths and synovial membrane surrounding certain joints. The wrists, shoulders, elbows (tennis elbow), finger joints (trigger finger), ankles, and associated tendons are most often affected. The affected sheaths may become visibly swollen because of fluid accumulation or they may remain dry. Local tenderness is variable and there may be disabling pain with movement of the body part. The condition often follows some form of trauma, strain, or excessive exercise. Treatment is with analgesic/anti-inflammatory drugs and, if warranted, cortisone injections.

DISLOCATION

A **dislocation** or **luxation** is the displacement of a bone from a joint. The most common dislocations are those involving a finger, thumb, or shoulder. Those of the mandible, elbow, knee, or hip are less common. Symptoms include loss of motion, temporary paralysis of the involved joint, pain, swelling, and occasional shock. A partial or incomplete dislocation is called a **subluxation.** A dislocation is usually caused by a blow or fall, although unusual physical effort may lead to this condition.

SPRAIN

A **sprain** is the forcible wrenching or twisting of a joint with partial rupture or other injury to its attachments without luxation. There may be damage to the associated blood vessels, muscles, tendons, ligaments, or nerves. A sprain is more serious than a *strain*, which is simply the overstretching of a muscle without swelling. Severe sprains may be so painful that the joint cannot be moved.

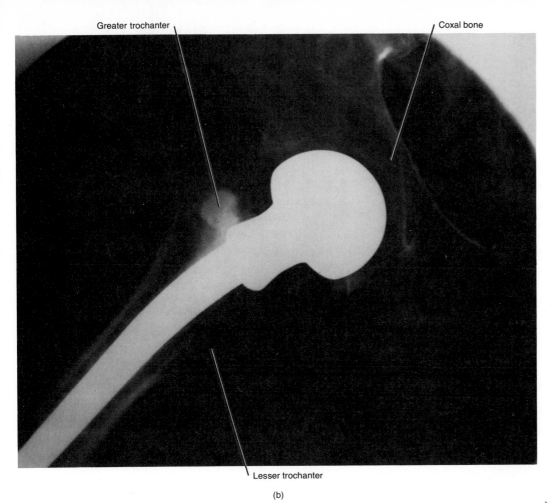

Greater trochanter

Coxal bone

Lesser trochanter

(b)

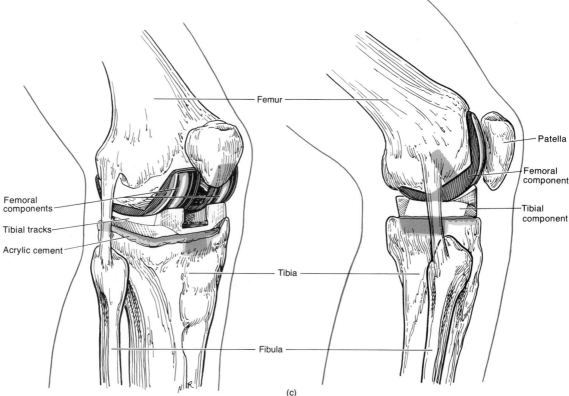

Femur

Patella

Femoral component

Tibial component

Femoral components

Tibial tracks

Acrylic cement

Tibia

Fibula

(c)

FIGURE 8-28 (Continued) Total hip and total knee replacement. (b) Anteroposterior projection of the hip replacement technique. (Courtesy of Arthur Provost, R. T.) (c) Illustrated are two total knee replacement prostheses. In the technique on the left, called the polycentric type, there are two femoral and two tibial components. In the technique on the right, the Waldius type, there is one femoral component and one tibial component.

SLIPPED DISC

Intervertebral discs are located between the bodies of adjacent vertebrae from the axis to the sacrum. Each disc is composed of an outer fibrous ring consisting of fibrocartilage called the *anulus fibrosus* and an inner soft, pulpy, highly elastic structure called the *nucleus pulposus*. (See Figure 8-1b.) The intervertebral discs absorb vertical shock. Under compression, they flatten, broaden, and bulge from their intervertebral spaces. The discs between the fourth and fifth lumbar vertebrae and between the fifth lumbar vertebra and sacrum are subject to great compressional forces. If the anterior and posterior ligaments of the discs become injured or weakened, the disc may become her-

niated—that is, the pressure developed in the nucleus pulposus is great enough to rupture the surrounding fibrocartilage. If this occurs, the nucleus pulposus may protrude posteriorly or into one of the adjacent vertebral bodies. This condition is called a **slipped disc.** Most often the nucleus pulposus slips posteriorly toward the spinal cord and spinal nerves. This movement exerts pressure on the spinal nerves causing considerable, sometimes very acute, pain. If the root of the sciatic nerve, which passes from the spinal cord to the foot, is pressured, the pain radiates down the back of the thigh, through the calf, and occasionally into the foot. If pressure is exerted on the spinal cord itself, nervous tissue may be destroyed.

KEY MEDICAL TERMS ASSOCIATED WITH JOINTS

Ankylosis (*agkyle* = stiff joint; *osis* = condition) Severe or complete loss of movement at a joint.

Arthralgia (*algia* = pain) Pain in a joint.

Arthrosis (*arth* = joint) Refers to an articulation; also a disease of a joint.

Bursectomy (*ectomy* = removal of) Removal of a bursa.

Chondritis (*chondro* = cartilage) Inflammation of a cartilage.

Deterioration The process or state of growing worse; disintegration or wearing away.

Detritus Particulate matter produced by or remaining after

the wearing away or disintegration of a substance or tissue; scales, crusts, and loosened skin.

Insidious Hidden, not apparent—as a disease that does not exhibit distinct symptoms of its arrival.

Pyogenic (*pyo* = pus) Producing suppuration (pus).

Reduce To replace in normal position—as to reduce a fracture.

Rheumatology The medical specialty devoted to arthritis.

Septic (*sept* = poison) Indicating the presence of microorganisms or their toxins.

Synovitis Inflammation of a synovial membrane in a joint.

STUDY OUTLINE

Classification
1. A joint or articulation is a point of contact between bones, cartilages, or cartilage and bone.
2. Closely fitting bones are strong but not freely movable. Loosely fitting joints are weaker but freely movable.

Fibrous Joints
1. Bones held by fibrous connective tissue, with no joint cavity, are fibrous joints.
2. These joints include immovable sutures (found in the skull), slightly movable syndesmoses (such as the tibiofibular articulation), and immovable gomphoses (roots of teeth in alveolar processes).

Cartilaginous Joints
1. Bones held together by cartilage, with no joint cavity, are cartilaginous joints.
2. These joints include immovable synchondroses united by hyaline cartilage (temporary cartilage between diaphysis and epiphyses) and partially movable symphyses united by fibrocartilage (the symphysis pubis).

Synovial Joints
1. These joints contain a joint cavity, articular cartilage,

and articular capsule; some also contain accessory ligaments, articular discs, and bursae.
2. All synovial joints are freely movable.
3. Types of synovial joints include gliding joints (wrist bones), hinge joints (elbow), pivot joints (radioulnar), ellipsoidal joints (radiocarpal), saddle joints (carpometacarpal), and ball-and-socket joints (shoulder and hip).
4. Planes of movement at synovial joints include monaxial, biaxial, and triaxial planes.
5. Types of movements at synovial joints include gliding movements, angular movements, rotation, circumduction, and special movements such as inversion, eversion, protraction, retraction, supination, pronation, elevation, and depression.

Principal Joints of the Body
1. Temporomandibular, between the mandible and temporal bone.
2. Atlantooccipital, between the atlas and occipital bones.
3. Atlantoaxial, between the atlas and axis
4. Intervertebral, between vertebral bodies and between vertebral arches
5. Lumbosacral, between the fifth lumbar vertebra and the first sacral vertebra of the sacrum

6. Sacrococcygeal, between the sacrum and coccyx
7. Sacroiliac, between the sacrum and ilium
8. Symphysis pubis, between the articular surfaces of the coxal bones
9. Costovertebral, between the ribs and vertebrae
10. Sternocostal, between the ribs and sternum
11. Sternoclavicular, between the sternum and clavicle
12. Sternal, between the manubrium and body and between the body and xiphoid process
13. Acromioclavicular, between the scapula and clavicle
14. Humeroscapular, between the humerus and scapula
15. Elbow, between the humerus and ulna and between the humerus and radius
16. Radioulnar, between the radius and ulna
17. Radiocarpal, between the radius and navicular, lunate, and triangular carpal bones
18. Intercarpal, between the proximal carpals, between the distal carpals, and between the proximal and distal carpals
19. Carpometacarpal, between the greater multangular and metacarpal I and between the distal carpals and metacarpals II–V
20. Intermetacarpal, between metacarpals II–V
21. Metacarpophalangeal, between metacarpal I and the thumb and between metacarpals II–V and the proximal phalanges of the medial four digits
22. Interphalangeal, between adjacent phalanges
23. Coxal, between the femur and coxal bone
24. Tibiofemoral, between the patella and femur and between the femur and tibia
25. Tibiofibular, between the tibia and fibula
26. Talocrural, between the tibia and talus and between the fibula and talus
27. Intertarsal, between the tarsal bones
28. Tarsometatarsal, between the tarsals and metatarsals
29. Intermetatarsal, between metatarsals II–V
30. Metatarsophalangeal, between the metatarsals and phalanges
31. Interphalangeal, between adjacent phalanges

Applications to Health

1. Arthritis refers to several disorders characterized by inflammation of joints, often accompanied by stiffness of adjacent structures.
2. Rheumatism is a painful state of supporting body structures such as bones, ligaments, tendons, joints, and muscles.
3. Rheumatoid arthritis refers to inflammation of a joint accompanied by pain, swelling, and loss of function.
4. Osteoarthritis is a degenerative joint disease characterized by deterioration of articular cartilage and spur formation.
5. In gouty arthritis, sodium urate crystals are deposited in the soft tissues of joints and eventually destroy the tissues.
6. Bursitis is an acute or chronic inflammation of bursae.
7. Tendinitis is an inflammation of tendon sheaths and synovial membranes.
8. A dislocation is a displacement of a bone from its joint; a partial dislocation is called subluxation.
9. A sprain is the forcible wrenching or twisting of a joint with partial rupture to its attachments without dislocation.
10. The protrusion of the nucleus pulposus from an intervertebral disc into one of the adjacent vertebral bodies is known as a slipped disc.

REVIEW QUESTIONS

1. Define an articulation. What factors determine the degree of movement at joints?
2. Distinguish among the three kinds of joints on the basis of structure and function. List the subtypes. Be sure to include degree of movement and specific examples.
3. Explain the components of a synovial joint. Indicate the relationship of ligaments and tendons to the strength of the joint and restrictions on movement.
4. Explain how the articulating bones in a synovial joint are held together.
5. What is an accessory ligament? Define the two principal types.
6. What is an articular disc? Why are they important?
7. What are bursae? What is their function?
8. Define the following principal movements: gliding, angular, rotation, circumduction, and special. Name a joint where each occurs.
9. Have another person assume the anatomical position and execute for you each of the movements at joints discussed in the text. Reverse roles, and see if you can execute the same movements.
10. Contrast monaxial, biaxial, and triaxial planes of movement. Give examples of each, and name a joint at which each occurs.
11. For each joint of the body discussed in Exhibits 8-2 through 8-32, be sure that you can define the bones that form the joint, identify the joint by type, and list the anatomical components of the joint.
12. Define arthritis. What are some suspected causes of arthritis?
13. What is rheumatism?
14. Distinguish between rheumatioid arthritis, osteoarthritis, and gouty arthritis with respect to causes and symptoms.
15. Define bursitis. How is it caused?
16. What is tendinitis? What are some of its causes?
17. Define dislocation. What are the symptoms of dislocation? How is a dislocation caused?
18. Distinguish between a sprain and a strain.
19. Describe the structure of an intervertebral disc. What is a slipped disc?
20. Refer to the glossary of key medical terms at the end of the chapter and be sure that you can define each term.

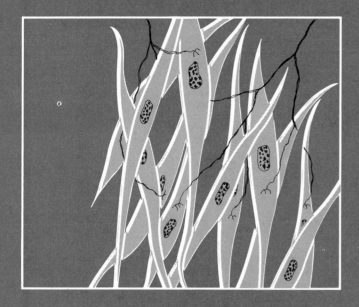

9
Muscle Tissue

STUDENT OBJECTIVES

- List the characteristics and functions of muscle tissue.

- Compare the location, microscopic appearance, nervous control, and functions of the three kinds of muscle tissue.

- Define fascia, epimysium, perimysium, endomysium, tendons, and aponeuroses, and list their modes of attachment to muscles.

- Describe the relationship of blood vessels and nerves to skeletal muscles.

- Identify the histological characteristics of skeletal muscle tissue.

- Describe the physiology of contraction by listing the events associated with the sliding-filament theory.

- Describe the importance of the motor unit.

- Describe the source of energy for muscular contraction.

- Define the all-or-none principle of muscular contraction.

- Contrast cardiac muscle tissue with smooth muscle tissue.

- Define such common muscular disorders as fibrosis, fibrositis, muscular dystrophy, and myasthenia gravis.

- Compare spasms, cramps, convulsions, tetany, and fibrillation as abnormal muscular contractions.

- Define key medical terms associated with the muscular system.

Although bones and joints provide leverage and form the framework of the body, they are not capable of moving the body by themselves. Motion is an essential body function that results from the contraction and relaxation of muscles. Muscle tissue constitutes about 40 to 50 percent of the total body weight and is composed of highly specialized cells with four striking characteristics.

Characteristics

Irritability is the ability of muscle tissue to receive and respond to stimuli. A stimulus is a change in the internal or external environment strong enough to initiate a nerve impulse. A second characteristic of muscle is **contractility,** the ability to shorten and thicken, or contract, when a sufficient stimulus is received. Muscle tissue also exhibits **extensibility**—it can be stretched. Many skeletal muscles are arranged in opposing pairs. While one is contracting, the other is undergoing extension. Another characteristic of muscle tissue is **elasticity,** the ability of muscle to return to its original shape after contraction or extension.

Functions

Through contraction, muscle performs three important functions: (1) motion, (2) maintenance of posture, and (3) heat production.

The most obvious body motions are walking, running, and locomotion. Other movements, such as grasping a pencil or nodding the head, may be localized to certain parts of the body. These movements rely on the integrated functioning of the bones, joints, and muscles attached to the bones. Less noticeable kinds of motion produced by muscles are the beating of the heart, the churning of food in the stomach, the pushing of food through the intestines, the contraction of the gallbladder to release bile, and the contraction of the urinary bladder to expel urine.

In addition to the movement function, muscle tissue also enables the body to maintain posture. The contraction of skeletal muscles holds the body in stationary positions, such as standing and sitting.

The third function of muscle tissue is heat production. Skeletal muscle contractions produce heat and are thereby important in maintaining normal body temperature.

Kinds

Three kinds of muscle tissue are recognized: **skeletal, visceral,** and **cardiac.** These three types are further categorized by location, microscopic structure, and nervous control. **Skeletal muscle tissue,** which is named for its location, is attached to bones. It is *striated* muscle tissue because striations, or bandlike structures,

are visible when the tissue is examined under a microscope. It is a *voluntary* muscle tissue because it can be made to contract by conscious control. **Visceral** or **smooth, muscle tissue** is located in the walls of hollow internal structures, such as blood vessels, the stomach, and the intestines. It is *involuntary* muscle tissue because its contraction is usually not under conscious control, and it is nonstriated. **Cardiac muscle tissue** forms the walls of the heart and is named for its location. Cardiac muscle tissue is also striated and is involuntary. Thus all muscle tissues are classified in the following ways: (1) skeletal, striated, voluntary muscle, (2) cardiac, striated, involuntary muscle, and (3) visceral, smooth, involuntary muscle tissue.

Skeletal Muscle Tissue

To understand the fundamental mechanisms of muscle movement, you will need some knowledge of its connective tissue components, its nerve and blood supply, and its histology, or microscopic structure.

FASCIA

The term **fascia** is applied to a sheet or broad band of fibrous connective tissue beneath the skin or around muscles and other organs of the body. Fasciae may be divided into three types: superficial, deep, and subserous. The *superficial fascia,* or *subcutaneous layer,* is immediately deep to the skin. It covers the entire body and varies in thickness in different regions. On the back or dorsum of the hand it is quite thin, whereas over the inferior abdominal wall it is thick. The superficial fascia is composed of adipose tissue and loose connective tissue. The outer layer usually contains fat and varies considerably in thickness. The inner layer is thin and elastic. Between the two layers of superficial fascia are found arteries, veins, lymphaics, nerves, the mammary glands, and the facial muscles. The *deep fascia* is by far the most extensive of the three types. It is a dense connective tissue. Unlike the superficial fascia, the deep fascia does not contain fat. The deep fascia lines the body wall and extremities and holds muscles together, separating them into functioning groups.

The *subserous fascia* is located between the internal investing layer of deep fascia and a serous membrane. It covers the external surfaces of viscera in the thoracic and abdominal cavities.

CONNECTIVE TISSUE COMPONENTS

Skeletal muscles are further protected, strengthened, and attached to other structures by several connective tissue components. The entire muscle is usually wrapped with a substantial quantity of fibrous connective tissue called the **epimysium** (Figure 9-1). The epi-

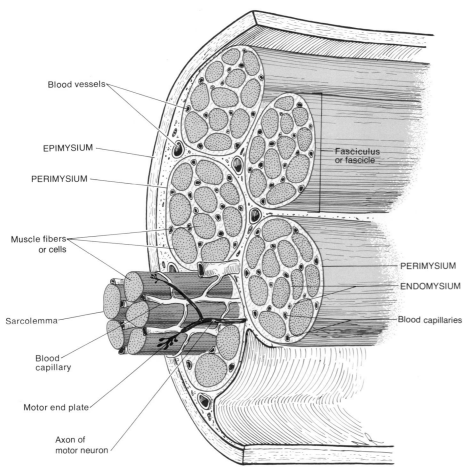

FIGURE 9-1 Relationships of connective tissue to skeletal muscle. Shown is a cross section of a skeletal muscle indicating the relative positions of the epimysium, perimysium, and endomysium. Compare this figure with the photomicrograph in Figure 9-3b.

mysium is an extension of deep fascia. When the muscle is cut in cross section, invaginations of the epimysium are seen to divide the muscle into bundles called *fasciculi* or *fascicles*. These invaginations of the epimysium are called the **perimysium.** Perimysium, like epimysium, is an extension of deep fascia. In turn, invaginations of the perimysium, called **endomysium,** penetrate into the interior of each fascicle and separate each muscle cell. Endomysium is also an extension of deep fascia. The epimysium, perimysium, and endomysium are all continuous with the connective tissue that attaches the muscle to another structure, such as bone or other muscle. All three elements can be extended beyond the muscle cells as a *tendon*—a cord of connective tissue that attaches a muscle to the periosteum of a bone. In other cases, the connective tissue elements may extend as a broad, flat band of tendons called an *aponeurosis*. This structure also attaches to the coverings of a bone or another muscle. When a muscle contracts, the tendon and its corresponding bone or muscle are pulled toward the contracting mus-

cle. In this way skeletal muscles produce movement. Certain tendons, especially those of the wrist and ankle, are enclosed by tubes of fibrous connective tissue called *tendon sheaths*. They are lined by a synovial membrane that permits the tendon to slide easily within the sheath. The sheaths also prevent the tendons from slipping out of place.

NERVE AND BLOOD SUPPLY

Skeletal muscles are well supplied with nerves and blood vessels. This heavy innervation and vascularization is directly related to contraction, the chief characteristic of muscle. For a skeletal muscle cell to contract, it must first be stimulated by an impulse from a nerve cell. Muscle contraction also requires a good deal of energy—meaning large amounts of nutrients and oxygen. Moreover, the waste products of these energy-producing reactions must be eliminated. Thus muscle action depends on the blood supply.

Generally, an artery and one or two veins accom-

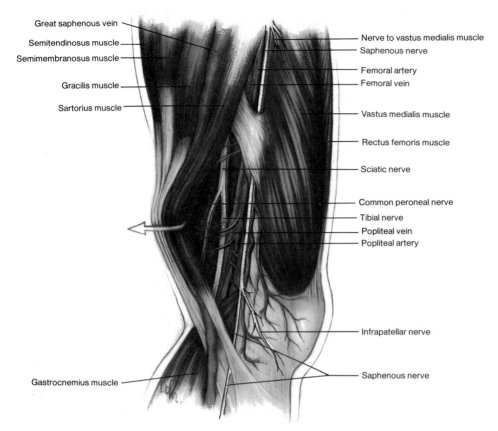

FIGURE 9-2 Relationship of blood vessels and nerves to skeletal muscles of the left thigh and knee in medial view.

pany each nerve that penetrates a skeletal muscle. The larger branches of the blood vessel accompany the nerve branches through the connective tissue of the muscle (Figure 9-2). Microscopic blood vessels called capillaries are arranged in the endomysium. Each muscle cell is thus in close contact with one or more capillaries. Each skeletal muscle cell usually makes contact with a portion of a nerve cell.

HISTOLOGY

Muscle tissue attached to bones is generally termed **skeletal muscle tissue.** When a typical skeletal muscle is teased apart and viewed microscopically, it can be seen to consist of many elongated, cylindrical cells called **muscle fibers** (Figure 9-3a, b). These fibers lie parallel to each other and range from 10 to 100 μm in diameter. Some fibers may reach lengths of 30 cm (12 inches) or more. Each muscle fiber is surrounded by a plasma membrane called the **sarcolemma** (*sarco* = flesh; *lemma* = sheath). The sarcolemma contains a quantity of cytoplasm called **sarcoplasm.** Within the sarcoplasm of a muscle fiber and lying close to the sarcolemma are many nuclei and a number of mitochon-

dria. Also within a muscle fiber is the **sarcoplasmic reticulum,** a network of membrane-enclosed tubules comparable to smooth endoplasmic reticulum (Figure 9-3c). Running transversely through the fiber and perpendicularly to the sarcoplasmic reticulum are **T tubules.** The tubules open to the outside of the fiber. A **triad** consists of a T tubule and the segments of sarcoplasmic reticulum on either side.

A highly magnified view of skeletal muscle fibers reveals threadlike structures, about 1 or 2 μm in diameter, called **myofibrils** (Figure 9-3c, d, e). The prefix *myo* means muscle. The myofibrils run longitudinally through the muscle fiber and consist of two kinds of even smaller structures called **myofilaments.** The **thin myofilaments** are about 6 nm in diameter. The **thick myofilaments** are about 16 nm in diameter.

The myofilaments of a myofibril do not extend the entire length of a muscle fiber—they are stacked in compartments called **sarcomeres.** Sarcomeres are partitioned by **Z lines,** which are narrow zones of dense material recognizable under the microscope. Each sarcomere is about 2.6 μm long. In a relaxed muscle fiber—that is, one that is not contracting—the thin and thick myofilaments overlap and form a dark, dense

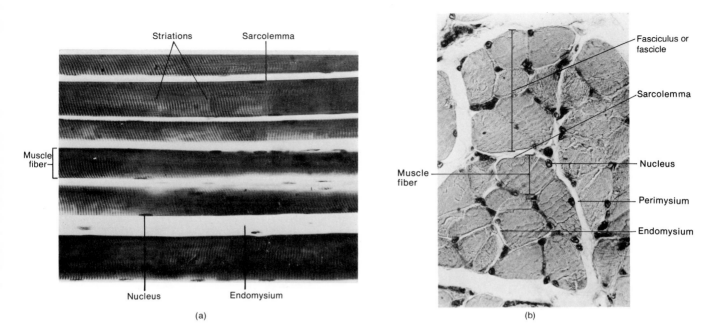

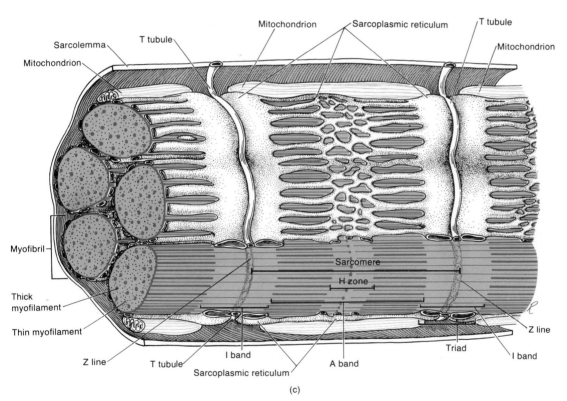

FIGURE 9-3 Histology of skeletal muscle tissue. (a) Photomicrograph of several muscle fibers in longitudinal section at a magnification of 640×. (b) Photomicrograph of several muscle fibers in cross section at a magnification of 640×. (Courtesy of Edward J. Reith, from *Atlas of Descriptive Histology,* by Edward J. Reith and Michael H. Ross, Harper & Row, Publishers, Inc., New York, 1970.) (c) Enlarged aspect of several muscle fibers based on an electron micrograph.

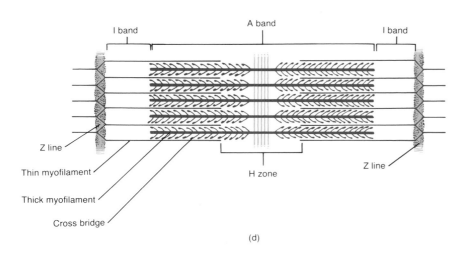

(d)

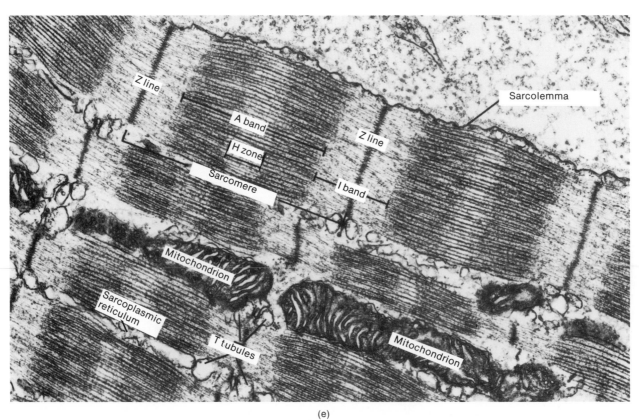

(e)

FIGURE 9-3 *(Continued)* Histology of skeletal muscle tissue. (d) Details of a sarcomere showing thin and thick myofilaments and various internal zones. (e) Electron micrograph of several sarcomeres at a magnification of 35,000×. (Courtesy of D. E. Kelly, from *Introduction to the Musculoskeletal System,* by Cornelius Rosse and D. Kay Clawson, Harper & Row, Publishers, Inc., New York, 1970.)

band called the *anisotropic* or *A band*. Each A band is about 1.6 μm long. A light-colored, less dense area called the *isotropic* or *I band* is composed of thin myofilaments only. The I bands are about 1 μm long. This combination of alternating dark and light bands gives the muscle fiber its striated or striped appearance. A narrow *H zone* contains thick myofilaments only. The H zone is about 0.5 μm long.

The thin myofilaments are composed mostly of a protein called *actin*. The actin molecules are arranged in a double-stranded coil that gives the thin myofilaments their characteristic shape (Figure 9-4a). Besides actin, the thin myofilaments contain two other molecules called *tropomyosin* and *troponin*. Together they are referred to as a *tropomyosin-troponin complex*.

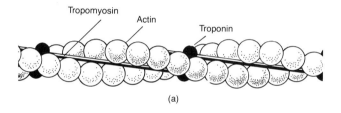

(a)

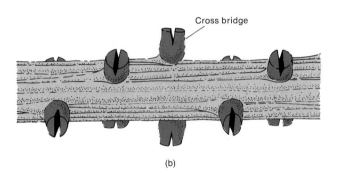

(b)

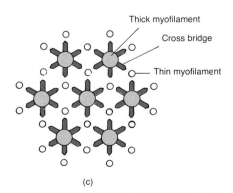

(c)

FIGURE 9-4 Detailed structure of myofilaments. (a) Thin myofilament. (b) Thick myofilament. (c) Cross section of several thin and thick myofilaments showing the arrangement of cross bridges. Note that each thick myofilament is surrounded by six thin myofilaments. (Parts (a) and (b) have been redrawn by permission from Arthur W. Ham, *Histology,* 7th ed., J. B. Lippincott Company, Philadelphia, 1974.)

The thick myofilaments are composed mostly of a protein called *myosin*. A myosin molecule is shaped like a rod with a round head. These rods form the long axis of the thick myofilaments, and the heads form projections called *cross bridges* (Figure 9-4b). The cross bridges are arranged in pairs and seem to spiral around the main axis of the thick myofilament. The relationship between thin and thick myofilaments is shown in Figure 9-4c.

Contraction

SLIDING-FILAMENT THEORY

During muscle contraction, the thin myofilaments slide inward toward the H zone, causing the sarcomere to shorten. However, the lengths of the thin and thick myofilaments do not change. The cross bridges of the thick myofilaments connect with portions of actin of the thin myofilaments. The myosin cross bridges move like the oars of a boat on the surface of the thin myofilaments—and the thin and thick myofilaments slide past each other (Figure 9-5). As the thin myofilaments move past the thick myofilaments, the width of the H zone between the ends of the thin myofilaments gets

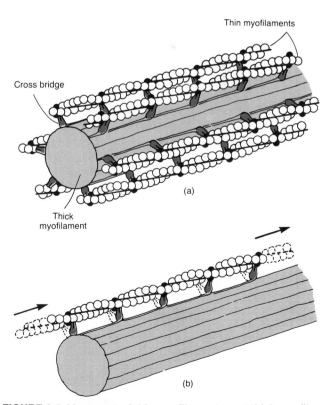

FIGURE 9-5 Movement of thin myofilaments past thick myofilaments. (a) Attachment of myosin cross bridges to actin of thin myofilaments. (b) Mechanism of movement of myosin cross bridges resulting in sliding of thin myofilaments toward H zone.

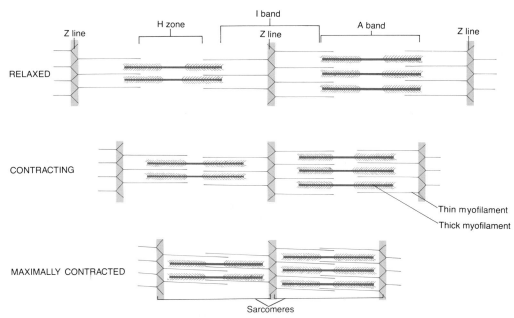

FIGURE 9-6 Sliding-filament theory of muscle contraction. Shown are the positions of the various parts of two sarcomeres in relaxed, contracting, and maximally contracted states. Note the movement of the thin myofilaments and the relative size of the H zone.

smaller and may even disappear when the thin myofilaments meet at the center of the sarcomere. In fact, the cross bridges may pull the thin myofilaments of each sarcomere so far inward that their ends overlap (Figure 9-6) As the thin myofilaments slide inward, the Z lines are drawn toward the A band and the sarcomere is shortened. The sliding of myofilaments and shortening of sarcomeres causes the shortening of the muscle fibers. All these events associated with the movement of myofilaments are known as the **sliding-filament theory** of muscle contraction. Let's examine the theory in more detail.

Physiology

When a muscle fiber is relaxed, the concentration of calcium ions (Ca^{2+}) in the sarcoplasm is low—these ions are stored in the sarcoplasmic reticulum. Moreover, molecules of ATP are attached to the myosin cross bridges. The cross bridges are prevented from combining with actin of the thin myofilaments by the tropomyosin-troponin complex while the complex is attached to actin and the ATP is bound to the myosin cross bridges. In other words, the muscle fiber remains relaxed as long as there are no calcium ions in the sarcoplasm, the tropomyosin-troponin complex is attached to actin, and ATP is bound to the myosin cross bridges.

When a nerve impulse reaches a muscle fiber, the neuron releases acetylcholine, which causes an elec-

trical change in the sarcolemma of the muscle fiber. This change travels over the surface of the sarcolemma and into the T tubules. When the impulse is conveyed from the T tubules to the sarcoplasmic reticulum, the reticulum releases the calcium ions from storage into the sarcoplasm surrounding the myofilaments. The calcium ions move to the myosin cross bridges and activate the myosin so that it can catalyze the breakdown of ATP into ADP + P. In the presence of calcium ions, myosin acts as an enzyme that breaks down ATP. Calcium ions also permit the tropomyosin-troponin complex to split from the thin myofilament so that the free receptor site of actin is now permitted to attach to the myosin cross bridge. The energy released from the breakdown of ATP is used for the attachment and movement of the myosin cross bridges and thus the sliding of the myofilaments. As the thin myofilaments slide past the thick myofilaments, the Z lines are drawn toward each other and the sarcomere shortens—the muscle contracts.

What happens when a muscle fiber goes from a contracted state back to a relaxed state? After the nerve impulse ends, the calcium ions return to the sarcoplasmic reticulum for storage. With their removal from the sarcoplasm, the enzymatic activity of myosin stops. The ADP is resynthesized into ATP, which again binds to the myosin cross bridges. The tropomyosin-troponin complex is reattached to the actin of the thin myofilaments. And the myosin cross bridges separate from the actin. Since the myosin cross bridges are broken, the thin myofilaments slip back to their resting position. The sarcomeres are thereby returned to their resting lengths and the muscle fiber resumes its resting state.

Energy

Contraction of a muscle requires energy. When a nerve impulse stimulates a muscle fiber, ATP, in the presence of ATPase (activated myosin), breaks down into ADP + phosphate (P) and energy is released. As far as we know, ATP is always the immediate source of energy for muscle contraction.

Like the other cells of the body, muscle cells synthesize ATP as follows:

$$ADP + P + energy \rightarrow ATP$$

The energy for replenishing ATP is derived from the breakdown of digested foods. However, unlike most other cells of the body, muscle fibers alternate between great activity and virtual inactivity. When a muscle is contracting, its energy requirements are high and the synthesis of ATP is accelerated. If the exercise is strenuous, ATP is used up even faster than it can be manufactured. Thus muscles must be able to build up a reserve supply of energy. They do this in two ways. A resting muscle needs little energy and produces much more ATP than it can use. At first, the muscle fiber stores the excess ATP on the thick myofilaments. When the fiber runs out of storage space for the ATP molecules, it combines the remainder of the ATP with a substance called *creatine*. Creatine, which is produced in the liver, can accept a high-energy phosphate from ATP to become the high-energy compound *creatine phosphate*. This reaction is anaerobic—it takes place without oxygen. It occurs as follows:

$$ATP + creatine \rightarrow creatine\ phosphate + ADP$$

Creatine phosphate is produced only when the muscle fibers are resting. During strenuous contraction, the reaction reverses itself. This reaction, also anaerobic, is shown in the following equation:

$$ADP + creatine\ phosphate \rightarrow ATP + creatine$$

Motor Unit

For a skeletal muscle fiber to contract, a stimulus must be applied to it. Such a stimulus is normally transmitted by nerve cells called neurons. A neuron has a threadlike process called a fiber, or axon, that may run 91 cm (3 ft) or more to a muscle. A bundle of such fibers from many different neurons composes a nerve. A neuron that transmits a stimulus to muscle tissue is called a motor neuron.

Upon entering a skeletal muscle, the axon of the motor neuron branches into fine endings that come into close approximation at grooves on the muscle membrane. The portion of the muscle membrane directly under the end of the axon is called a **motor end plate** (Figure 9-7). The area of contact between neuron and muscle fiber is called a **neuromuscular junction,** or **myoneural junction.** When a nerve impulse reaches a motor end plate, small vesicles in the terminal branches of the nerve fiber release a chemical called acetylcholine (ACh). The ACh transmits the nerve impulse from the neuron, across the myoneural junction, to the muscle fiber, thus stimulating it to contract.

A motor neuron, together with the muscle cells it stimulates, is referred to as a **motor unit.** A single motor neuron innervates about 150 muscle fibers. This means that stimulation of the one neuron will tend to cause the simultaneous contraction of about 150 muscle fibers. In addition, all the muscle fibers of a motor unit that are sufficiently stimulated will contract and relax together. Muscles that control precise movements, such as the eye muscles, have fewer than ten muscle fibers to each motor unit. Some muscles have as few as one muscle fiber per motor unit. Muscles of the body that are responsible for gross movements may have as many as 500 muscle fibers in each motor unit.

ALL-OR-NONE-PRINCIPLE

According to the **all-or-none principle,** individual muscle fibers of a motor unit will contract to their fullest extent or will not contract at all, depending on the extent of stimulation from the motor neuron. In other words, muscle fibers do not partly contract. The principle does not imply that the strength of the contraction is the same every time the fiber is stimulated. The strength of contraction may be decreased by fatigue, lack of nutrients, or lack of oxygen. The weakest stimulus from a neuron that can initiate a contraction is called a **threshold** or **liminal stimulus.** A stimulus of lesser intensity, or one that cannot initiate contraction, is referred to as a **subthreshold** or **subliminal stimulus.** Each muscle fiber in a motor unit has its own threshold level.

Cardiac Muscle Tissue

The principal constituent of the heart wall is **cardiac muscle tissue.** It has a striated appearance similar to skeletal muscle tissue but it is involuntary. The cells of cardiac muscle tissue are roughly quadrangular and have only a single nucleus (Figure 9-8). The individual fibers are covered by a thin, poorly defined sarcolemma, and the internal myofibrils produce the characteristic striations. Cardiac muscle cells have the

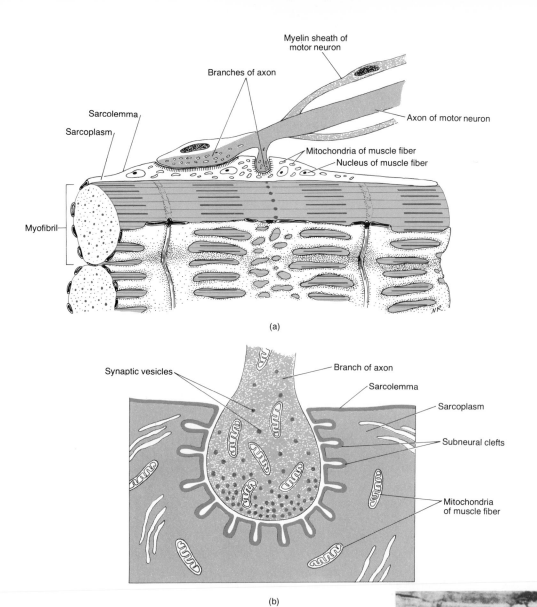

FIGURE 9-7 Motor end plate. (a) Diagram as seen by a light microscope. (b) Enlarged aspect as seen by an electron microscope. (c) Photomicrograph at a magnification of 640×. (Courtesy of Fisher Scientific Company and S.T.E.M. Laboratories, Inc., Copyright 1975.)

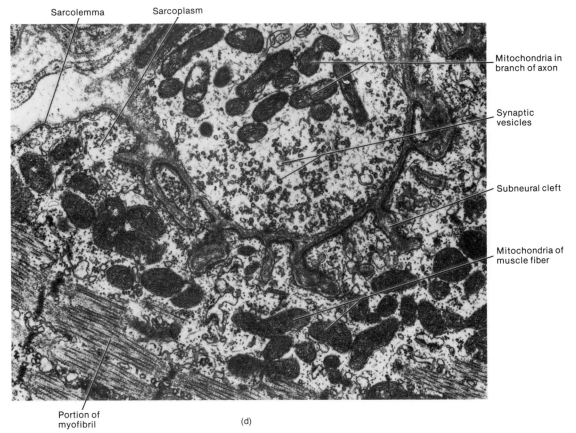

Sarcolemma

Sarcoplasm

Mitochondria in branch of axon

Synaptic vesicles

Subneural cleft

Mitochondria of muscle fiber

Portion of myofibril

(d)

FIGURE 9-7 (Continued) Motor end plate. (d) Electron micrograph at a magnification of 30,000×. (Courtesy of Cornelius Rosse and D. Kay Clawson, from *Introduction to the Musculoskeletal System,* Harper & Row, Publishers, Inc., New York, 1970.)

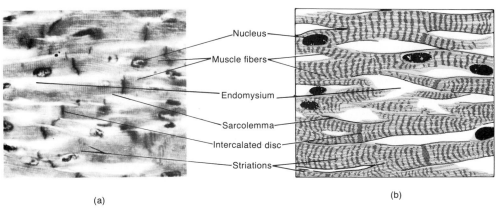

Nucleus

Muscle fibers

Endomysium

Sarcolemma

Intercalated disc

Striations

(a)   (b)

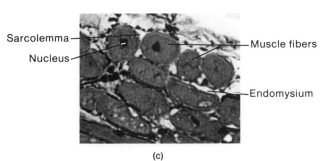

Sarcolemma

Nucleus

Muscle fibers

Endomysium

(c)

FIGURE 9-8 Histology of cardiac muscle tissue. (a) Photomicrograph of cardiac muscle tissue in longitudinal section at a magnification of 640×. (Courtesy of Edward J. Reith, from *Atlas of Descriptive Histology,* by Edward J. Reith and Michael H. Ross, Harper & Row, Publishers, Inc., New York, 1970.) (b) Diagram of the photomicrograph. (c) Photomicrograph of cardiac muscle tissue in cross section at a magnification of 100×. (Courtesy of Victor B. Eichler, Wichita State University.)

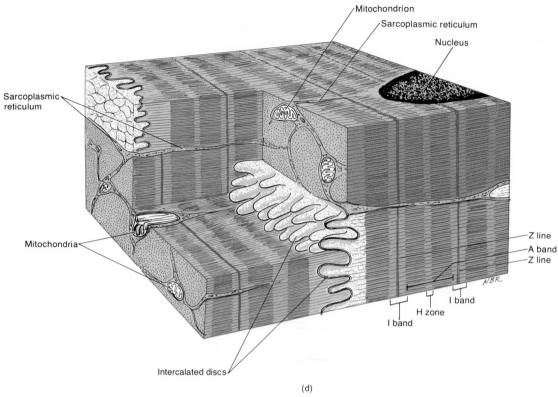

FIGURE 9-8 *(Continued)* Histology of cardiac muscle tissue. (d) Diagram based on an electron micrograph.

same basic arrangement of actin, myosin, and sarcoplasmic reticulum that is found in skeletal muscle cells. In addition, they contain a system of transverse tubules similar to the T tubules of skeletal muscle. The nuclei in cardiac cells, however, are centrally located compared to the peripheral location of nuclei in skeletal muscle.

While the groups of skeletal muscle fibers are arranged in a parallel fashion, those of cardiac muscle branch freely with other fibers to form two separate networks. The muscular walls and septum of the upper chambers of the heart (atria) compose one network. The muscular walls and septum of the lower chambers of the heart (ventricles) compose the other network. When a single fiber of either network is stimulated, all the fibers in the network become stimulated as well. Thus each network contracts as a functional unit. The fibers of each network were once thought to be fused together into a multinucleated mass called a syncytium. But it is now known that each fiber in a network is separated from the next fiber by an irregular transverse thickening of the sarcolemma called an **intercalated disc.** These discs strengthen the cardiac muscle tissue and aid in impulse conduction.

Under normal conditions, cardiac muscle tissue rhythmically contracts about 70–80 times a minute.

This is a major physiological difference between cardiac and skeletal muscle tissue. Another difference is the source of stimulation. Skeletal muscle tissue ordinarily contracts only when stimulated by a nerve impulse. In contrast, cardiac muscle tissue never requires nerve stimulation to induce contraction. Its source of stimulation is a conducting tissue of specialized muscle within the heart. About 70–80 times a minute, this tissue transmits electrical impulses that stimulate cardiac contraction. Nerve stimulation causes the conducting muscle tissue to increase or decrease its rate of discharge and strength of contraction.

Smooth Muscle Tissue

Like cardiac muscle tissue, **smooth muscle tissue** is usually involuntary. However, it is nonstriated. A single fiber of smooth muscle tissue is about 5 to 10 μm in diameter and 30 to 200 μm long. It is spindle-shaped and within the fiber is a single, oval, centrally located nucleus (Figure 9-9). Smooth muscle cells also contain actin and myosin filaments, but the filaments are not so orderly as in skeletal and cardiac muscle tissue. Thus well-differentiated striations do not occur.

Two kinds of smooth muscle tissue, visceral and multiunit, are recognized. The more common type is

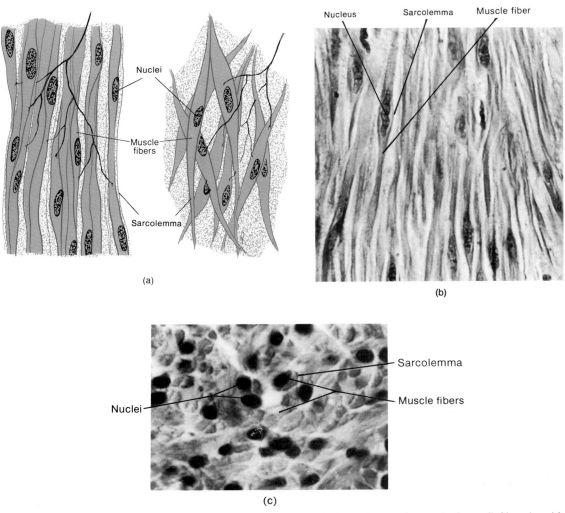

FIGURE 9-9 Histology of smooth muscle tissue. (a) Diagram of visceral smooth muscle tissue (left) and multi-unit smooth muscle tissue (right). (b) Photomicrograph of several smooth muscle fibers in longitudinal section at a magnification of 640×. (Courtesy of Edward J. Reith, from *Atlas of Descriptive Histology,* by Edward J. Reith and Michael H. Ross, Harper & Row, Publishers, Inc., New York, 1970.) (c) Photomicrograph of several smooth muscle fibers in cross section at a magnification of 200×. (Courtesy of Victor B. Eichler, Wichita State University.)

called *visceral muscle tissue.* It is found in wrap-around sheets that form part of the walls of the hollow viscera such as the stomach, intestines, uterus, and urinary bladder. The terms *smooth muscle tissue* and *visceral muscle tissue* are sometimes used interchangeably. The fibers in visceral muscle tissue are tightly bound together to form a continuous network. When a neuron stimulates one fiber, the impulse travels over the other fibers so that contraction occurs in a wave over many adjacent fibers.

The second kind of smooth muscle tissue, *multiunit smooth muscle tissue,* consists of individual fibers each with its own motor-nerve endings. Whereas stimulation of a single visceral muscle fiber causes contraction of many adjacent fibers, stimulation of a single multiunit muscle tissue is like skeletal muscle tissue. Multiunit smooth muscle tissue is found in the walls of

blood vessels, in the arrector pili muscles that attach to hair follicles, and in the intrinsic muscles of the eye, such as the iris.

Both kinds of smooth muscle tissue contract and relax more slowly than skeletal muscle tissue. This characteristic is probably due to the arrangement of the thin and thick myofilaments of smooth muscle. Whereas skeletal muscle cells contract as individual units, visceral muscle cells contract in sequence as the impulse spreads from one to another.

Applications to Health

FIBROSIS

The formation of fibrous tissue in locations where it normally does not exist is called **fibrosis.** Skeletal and cardiac muscle fibers cannot undergo mitosis, and dead muscle fibers are normally replaced with fibrous connective tissue. Fibrosis, then, is often a consequence of muscle injury or degeneration.

FIBROSITIS

Fibrositis is an inflammation of fibrous tissue. If it occurs in the lumbar region, it is termed *lumbago*. Fibrositis is a common condition characterized by pain, stiffness, or soreness of fibrous tissue, especially in the muscle coverings. It is not destructive or progressive. It may persist for years or spontaneously disappear. Attacks of fibrositis may follow an injury, repeated muscular strain, or prolonged muscular tension. **Myositis** is an inflammation of muscle cells.

MUSCULAR DYSTROPHY

The term **muscular dystrophy** applies to a number of inherited myopathies, or muscle-destroying diseases. The word *dystrophy* means degeneration. The disease is characterized by degeneration of the individual muscle cells, which leads to a progressive atrophy (reduction in size) of the skeletal muscle. Usually the voluntary skeletal muscles are weakened equally on both sides of the body, whereas the internal muscles, such as the diaphragm, are not affected. Histologically, the changes that occur include the variation in muscle fiber size, degeneration of fibers, and deposition of fat.

MYASTHENIA GRAVIS

Myasthenia gravis is a weakness of the skeletal muscles. It is caused by an abnormality at the neuromuscular junction that prevents the muscle fibers from contracting. Recall that motor neurons stimulate the skeletal muscle fibers to contract by releasing acetyl-choline. Myasthenia gravis is caused by failure of the neurons to release acetylcholine or by the release from the muscle fibers of an excess amount of cholinesterase, a chemical that destroys acetylcholine. As the disease progresses, more neuromuscular junctions become affected. The muscle becomes increasingly weaker and may eventually cease to function altogether.

The cause of myasthenia gravis is unknown. It is more common in females, occurring most frequently between the ages of 20 to 50. The muscles of the face and neck are most apt to be involved. Initial symptoms include a weakness of the eye muscles and difficulty in swallowing. Later, the individual has difficulty chewing and talking. Eventually, the muscles of the limbs may become involved. Death may result from paralysis of the respiratory muscles, but usually the disorder does not progress to this stage.

ABNORMAL CONTRACTIONS

One kind of abnormal contraction of a muscle is **spasm:** a sudden, involuntary contraction of short duration. A **cramp** is a painful spasmodic contraction of a muscle. It is an involuntary complete tetanic contraction. **Convulsions** are violent involuntary contractions of an entire group of muscles. Convulsions occur when motor neurons are stimulated by fever, poisons, hysteria, or changes in body chemistry due to withdrawl of certain drugs. The stimulated neurons send many bursts of seemingly disordered impulses to the muscle fibers. **Fibrillation** is the uncoordinated contraction of individual muscle fibers preventing the smooth contraction of the muscle.

A **tic** is a spasmodic twitching made involuntarily by muscles that are ordinarily under voluntary control. Twitching of the eyelid or face muscles are examples. In general, tics are of psychological origin.

KEY MEDICAL TERMS ASSOCIATED WITH THE MUSCULAR SYSTEM

Gangrene Death of tissue that results from interruption of its blood supply.

Myology (*myo* = muscle) Study of muscles.

Myomalacia (*malaco* = soft) Softening of a muscle.

Myopathy (*pathos* = disease) Any disease of muscle tissue.

Myosclerosis (*scler* = hard) Hardening of a muscle.

Myospasm Spasm of a muscle.

Myotonia Increased muscular irritability and contractility with decreased power of relaxation; tonic spasm of the muscle.

Trichinosis A myositis caused by the parasitic worm *Tri-chinella spiralis*, which may be found in the muscles of humans, rats, and pigs. People contract the disease by eating infected pork that is insufficiently cooked.

Volkmann's contracture (*contra* = against) Permanent contraction of a muscle due to replacement of destroyed muscle cells with fibrous tissue that lacks ability to stretch. Destruction of muscle cells may occur from interference with circulation caused by a tight bandage, a piece of elastic, or a cast.

Wryneck or torticollis Complete tetanus of one of the muscles of the neck; produces twisting of the neck and an unnatural position of the head.

STUDY OUTLINE

Characteristics
1. Irritability is the property of receiving and responding to stimuli.
2. Contractility is the ability to shorten and thicken, or contract.
3. Extensibility is the ability to be stretched or extended.
4. Elasticity is the ability to return to original shape after contraction or extension.

Functions
1. Through contraction, muscle tissue performs the three important functions of motion, maintenance of posture, and heat production.

Kinds
1. Skeletal muscle tissue is attached to bones. It is striated and voluntary.
2. Smooth muscle tissue is located in viscera. It is nonstriated and involuntary.
3. Cardiac muscle tissue forms the walls of the heart. It is striated and involuntary.

Skeletal Muscle Tissue
Fascia
1. The term *fascia* is applied to a sheet or broad band of fibrous connective tissue underneath the skin or around muscles and organs of the body.
2. Fascia is divided into three types; superficial, deep, and subserous.

Connective Tissue Components
1. The entire muscle is covered by the epimysium. Fasciculi are covered by perimysium. Fibers are covered by endomysium.
2. Tendons and aponeuroses attach muscle to bone.

Nerve and Blood Supply
1. Nerves convey impulses, and blood provides nutrients and oxygen for contraction.

Histology
1. The muscle consists of fibers covered by a sarcolemma. The fibers contain sarcoplasm, nuclei, sarcoplasmic reticulum, and T tubules.
2. Each fiber contains myofilaments (thin and thick). The myofilaments are compartmentalized into sarcomeres.

Contraction
Sliding-Filament Theory
1. A nerve impulse travels over the sarcolemma and enters the T tubules and sarcoplasmic reticulum.
2. The wave of depolarization leads to the release of calcium ions from the sarcoplasmic reticulum, triggering the contractile process.
3. Actual contraction is brought about when the thin myofilaments of one sarcomere slide toward each other.

Physiology
1. When a nerve impulse reaches the motor end plate, the neuron releases acetylocholine, which causes an electrical change.
2. This change releases calcium ions that activate the myosin, catalyzing the breakdown of ATP.
3. The energy released from the breakdown of ATP causes the sliding of the myofilaments.

Energy
1. The only direct source of energy is ATP.
2. When muscles are resting, ATP combines anaerobically with creatine to form creatine phosphate, which breaks down to produce ATP when muscles contract strenuously.

Motor Unit
1. A motor neuron transmits the stimulus to a skeletal muscle for contraction.
2. The region of the sarcolemma specialized to receive the stimulus is the motor end plate.
3. The area of contact between a motor neuron and muscle fiber is a neuromuscular or myoneural junction.
4. A motor neuron and the muscle fibers it stimulates form a motor unit.

All-or-None Principle
1. Muscle fibers of a motor unit contract to their fullest extent or not at all.
2. The weakest stimulus capable of causing contraction is a liminal or threshold stimulus.
3. A stimulus not capable of inducing contraction is a subliminal or subthreshold stimulus.

Cardiac Muscle Tissue
1. This muscle is found only in the heart. It is straited and involuntary.
2. The cells are quadrangular and contain centrally placed nuclei.
3. The fibers form a continuous, branching network that contracts as a functional unit.
4. Intercalated discs provide strength and aid impulse conduction.

Smooth Muscle Tissue
1. Smooth muscle is found in viscera. It is nonstriated and involuntary.
2. Visceral smooth muscle is found in the walls of viscera. The fibers are arranged in a network.
3. Multiunit smooth muscle is found in blood vessels and the eye. The fibers operate singly rather than as a unit.

Applications to Health
Fibrosis and Fibrositis
1. Fibrosis is the formation of fibrous tissue where it normally does not exist; it frequently occurs in damaged muscle tissue.

2. Fibrositis is an inflammation of fibrous tissue. If it occurs in the lumbar region, it is called lumbago. Myositis is muscle tissue inflammation.

Muscular Dystrophy and Myasthenia Gravis
1. Muscular dystrophy is a hereditary disease of muscles characterized by degeneration of individual muscle cells.

2. Myasthenia gravis is a disease exhibiting great muscular weakness and fatigability resulting from improper neuromuscular transmission.

Abnormal Contractions
1. Abnormal contractions include spasms, cramps, convulsions, fibrillation, and tics.

REVIEW QUESTIONS

1. How is the skeletal system related to the muscular system? What are the three kinds of motion that are accomplished by the muscular system?
2. What are the four characteristics of muscle tissue?
3. What criteria are employed for distinguishing the three kinds of muscle tissue?
4. Define epimysium, perimysium, endomysium, tendon, and aponeurosis. Describe the nerve and blood supply to a muscle.
5. Define fascia. Distinguish between superficial fascia, deep fascia, and subserous fascia on the basis of location.
6. Discuss the microscopic structure of skeletal muscle tissue.
7. In considering the contraction of skeletal muscle tissue, describe the following:
   a. Motor unit
   b. Sliding-filament theory
   c. Importance of calcium and troponin
   d. Sources of energy
8. What is the all-or-none principle? Relate it to a liminal and subliminal stimulus.
9. Compare cardiac and smooth muscle tissue with regard to microscopic structure, functions, and locations.
10. Define fibrosis. What is one of its causes? Define myositis.
11. What is muscular dystrophy?
12. What is myasthenia gravis? In this disease, why do the muscles not contract normally?
13. Define each of the following abnormal muscular contractions: spasm, cramp, convulsion, fibrillation, and tic.
14. Refer to the glossary of key medical terms associated with the muscular system. Be sure that you can define each term.

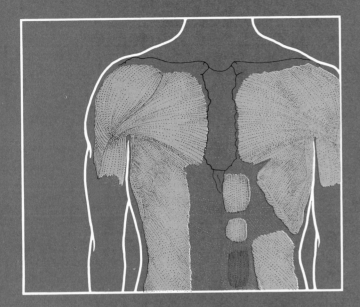

10

The Muscular System

The term *muscle tissue* refers to all the contractile tissues of the body: skeletal, cardiac, and smooth muscle. The **muscular system,** however, refers to the *skeletal* muscle system: the skeletal muscle tissue and connective tissues that make up individual muscle organs such as the biceps. Cardiac and smooth muscle tissues are classified with other organ systems. For instance, cardiac muscle tissue is located in the heart, an organ of the circulatory system. Smooth muscle tissue of the intestine is part of the digestive system. Smooth muscle tissue of the urinary bladder is part of the urinary system. In this chapter, we discuss only the muscular system. We will see how skeletal muscles produce movement and describe the principal skeletal muscles.

How Skeletal Muscles Produce Movement

ORIGIN AND INSERTION

Skeletal muscles produce movements by exerting force on tendons, which in turn pull on bones. Most muscles cross at least one joint and are attached to the articulating bones that form the joint (Figure 10-1a).

When such a muscle contracts, it draws one articulating bone toward the other. The two articulating bones usually do not move equally in response to the contraction. One is held nearly in its original position because other muscles contract to pull it in the opposite direction or because its structure makes it less movable. Ordinarily, the attachment of a muscle tendon to the stationary bone is called the **origin.** The attachment of the other muscle tendon to the movable bone is the **insertion.** A good analogy is a spring on a door. The part of the spring attached to the door represents the insertion; the part attached to the frame is the origin. The fleshy portion of the muscle between the tendons of the origin and insertion is called the **belly,** or **gaster.** The origin is usually proximal and the insertion distal, especially in the appendages. In addition, muscles that move a body part generally do not cover the moving part. Figure 10-1a shows that although contraction of the biceps muscle moves the forearm, the belly of the muscle lies over the humerus.

LEVER SYSTEMS

In producing a body movement, bones act as levers and joints function as fulcrums of these levers. A **lever**

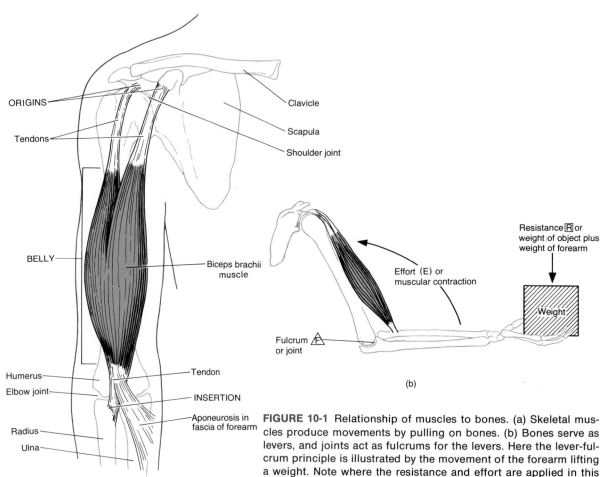

FIGURE 10-1 Relationship of muscles to bones. (a) Skeletal muscles produce movements by pulling on bones. (b) Bones serve as levers, and joints act as fulcrums for the levers. Here the lever-fulcrum principle is illustrated by the movement of the forearm lifting a weight. Note where the resistance and effort are applied in this example.

may be defined as a rigid rod that moves about on some fixed point called a **fulcrum.** A fulcrum may be symbolized as △. A lever is acted on at two different points by two different forces: the *resistance* □ and the *effort* (E). The resistance may be regarded as a force to be overcome, whereas the effort is exerted to overcome the resistance. The resistance may be the weight of the body part that is to be moved. The muscular effort (contraction) is applied to the bone at the insertion of the muscle and produces motion. Consider the biceps flexing the forearm at the elbow as a weight is lifted (Figure 10-1b). When the forearm is raised, the elbow is the fulcrum. The weight of the forearm plus the weight in the hand is the resistance. The shortening of the biceps pulling the forearm up is the effort.

Levers are categorized into three types according to the positions of the fulcrum, the effort, and the resistance.

1. In a **first-class lever,** the fulcrum is placed between the effort and resistance (Figure 10-2a). An example of a first-class lever is a seesaw. Examples of first-class levers in the body are not abundant, however. One is the head resting on the vertebral column. When the head is raised, the facial portion of the skull is the resistance. The joint between the atlas and occipital bone (atlantooccipital joint) is the fulcrum. The muscles of the back in contraction represent the effort.

2. Second-class levers operate like a wheelbarrow. The fulcrum is at one end, the effort is at the opposite end, and the resistance is in between (Figure 10-2b). Most authorities agree that there are no examples of second-class levers in the body. Some, however, consider that raising the body on the toes (resistance) and utilizing the ball of the foot as the fulcrum is an example. Here, the calf muscles pull the heel upward as it shortens (effort).

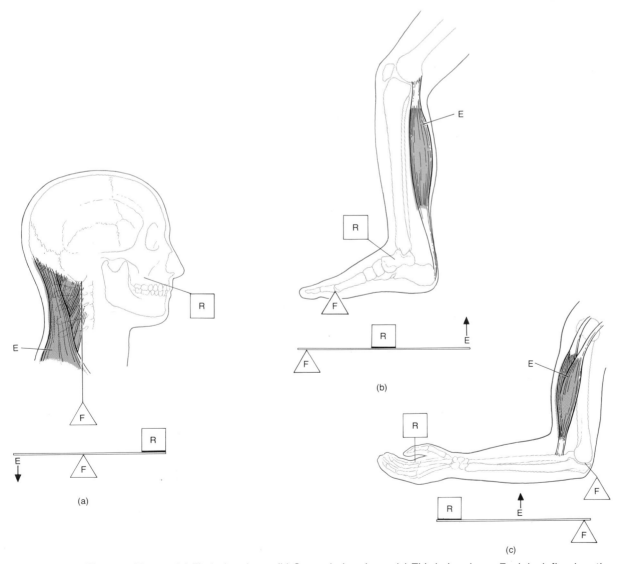

FIGURE 10-2 Classes of levers. (a) First-class lever. (b) Second-class lever. (c) Third-class lever. Each is defined on the basis of the placement of the fulcrum, effort, and resistance.

3. Third-class levers are the most common levers in the body. They consist of the fulcrum at one end, the resistance at the opposite end, and the effort between them (Figure 10-2c). A common example is flexing the forearm at the elbow. The weight of the forearm is the resistance, the contraction of the biceps is the effort, and the elbow joint is the fulcrum.

LEVERAGE

Leverage—the mechanical advantage gained by a lever—is largely responsible for a muscle's strength and range of movement. Consider strength first. Suppose we have two muscles of the same strength crossing and acting on a joint. Assume also that one is attached farther from the joint and one is closer. The muscle attached farther will produce the more powerful movement. Thus strength of movement depends on the placement of muscle attachments.

In considering the range of movement, again assume that we have two muscles of the same strength crossing and acting on a joint and that one is attached farther from the joint and one is closer. The muscle inserting closer to the joint will produce the greater range of movement. Thus range of movement depends on the placement of muscle attachments. Maximal strength and maximal range are therefore incompatible—strength and range vary inversely.

Arrangement of Fasciculi

Recall from Chapter 9 that skeletal muscle fibers are arranged within the muscle in bundles called fasciculi or fascicles. Although the muscle fibers are arranged in a parallel fashion, the arrangement of the parallel fibers in a fasciculus with respect to the tendons shows four characteristic patterns. The first is called **parallel.** In this pattern, the fasciculi are parallel with the longitudinal axis and terminate at either end in flat tendons. The muscle is typically quadrilateral in shape. An example is the stylohyoid muscle (see Figure 10-7). In a modification of the parallel arrangement, called *fusiform,* the fasciculi are nearly parallel with the longitudinal axis and terminate at either end in flat tendons, but the muscle tapers toward the tendons, where the diameter is less than that of the belly. An example is the biceps brachii muscle (see Figure 10-12). The second distinct pattern is called **convergent.** Here there is a broad origin of fasciculi that converge to a narrow, restricted insertion. Such a pattern gives the muscle a triangular shape. An example is the deltoid muscle (see Figure 10-11). The third distinct pattern is referred to as **pennate.** The fasciculi are short in relation to the entire length of the muscle and the fasciculi are directed obliquely toward the tendon, which extends nearly the entire length of the muscle. The fasciculi are

directed toward the tendon like the plumes of a feather. If the fasciculi are arranged on only one side of a tendon, as in the extensor digitorum muscle, the muscle is referred to as *unipennate* (see Figure 10-13). If, instead, the fasciculi are arranged on both sides of a centrally positioned tendon, as in the rectus femoris muscle, the muscle is referred to as *bipennate* (see Figure 10-16). The final distinct pattern is referred to as **circular.** The fasciculi are arranged in a circular pattern and enclose an orifice. An example is the orbicularis oris muscle (see Figure 10-4).

Fascicular arrangement is correlated with the power of a muscle and range of movement. When a muscle fiber contracts, it shortens to a length just slightly greater than half of its resting length. Thus, the longer the fibers in a muscle, the greater the range of movement it can produce. By contrast, the strength of a muscle depends on the total number of fibers it contains, since a short fiber can contract as forcefully as a long one. Since a given muscle can either contain a small number of long fibers or a large number of short fibers, fascicular arrangement represents a compromise between power and range of movement. Pennate muscles, for example, have a larger number of fasciculi distributed over their tendons, giving them greater power but a smaller range of movement. Parallel muscles, on the other hand, have comparatively few fasciculi that extend the length of the muscle. Thus, they have a greater range of movement, but less power.

GROUP ACTIONS

Most movements are coordinated by several skeletal muscles acting in groups rather than individually. Consider flexing the forearm at the elbow, for example. A muscle that causes a desired action is referred to as the **agonist** or **prime mover.** In this instance, the biceps brachii is the agonist (see Figure 10-12). Simultaneously with the contraction of the biceps brachii, another muscle, called the **antagonist,** is relaxing. In this movement, the triceps brachii serves as the antagonist (see Figure 10-12). The antagonist has an effect opposite to that of the agonist. That is, the antagonist relaxes and yields to the movement of the agonist. If we consider the extension of the forearm at the elbow, the triceps brachii would assume the role of the agonist and the biceps brachii would act as the antagonist. Most joints are operated by antagonistic groups of muscles. Still other muscles called **synergists,** or **fixators,** assist the agonist by reducing undesired action or unnecessary movements in the less mobile articulating bone. While flexing the forearm, the synergists, in this case the deltoid and pectoralis major muscles, hold the arm and shoulder in a suitable position for the flexing action (see Figure 10-10). Whereas the deltoid abducts the humerus, the pectoralis major adducts and medially rotates the humerus. Essentially, synergists contract at the same time as the prime mover and help the prime mover produce an effective movement. Many muscles are, at various times, prime movers, antagonists, or synergists, depending on the action.

Naming Skeletal Muscles

Most of the nearly 700 skeletal muscles are named on the basis of distinctive characteristics. If you understand them, you will remember the names of muscles. Some muscles are named on the basis of the *direction of the muscle fibers*. There are, for example, rectus (straight), transverse, and oblique muscles. Rectus fibers usually run parallel to the midline of the body. Transverse fibers run perpendicular to the midline. Oblique fibers are diagonal to the midline. Muscles named according to these three directions include the rectus abdominis, transversus abdominis, and external oblique.

Another characteristic is *location*. The temporalis is named from its proximity to the temporal bone. The tibialis anterior is near the tibia. *Size* is another criterion. The term *maximus* means largest, and *minimus* means smallest. *Longus* means long; *brevis* means short. Examples include the gluteus maximus, gluteus minimus, adductor longus, and peroneus brevis.

Some muscles such as the biceps, triceps, and quadriceps are named for their *number of origins*. The biceps has two origins. The triceps has three, and the quadriceps has four. Other muscles are named on the basis of *shape*. Common examples include the deltoid (meaning triangular) and trapezius (meaning trapezoid). Muscles may also be named after their *origin* and *insertion*. One example is the sternocleidomastoid, which originates on the sternum and clavicle and inserts at the mastoid process of the temporal bone. The stylohyoideus originates on the styloid process of the temporal bone and inserts at the hyoid bone.

Still another criterion used for naming muscles is *action*. Exhibit 10-1 lists the principal actions of muscles, their definitions, and examples of muscles that perform the actions. For convenience, the actions are grouped as antagonistic pairs where possible.

Principal Skeletal Muscles

Exhibits 10-2 through 10-16 list the muscles of the body in terms of their origins, insertions, actions, and innervations. Refer to Chapters 6 and 7 to review bone markings, since they serve as points of origin and insertion for muscles. By no means have all the muscles of the body been included. Only the major ones are discussed. As you study the skeletal muscles of the body in this chapter, you will also be introduced to their surface anatomy by means of labeled photographs.

Let us now investigate the principal skeletal muscles of the body by examining Exhibits 10-2 through 10-16. Figure 10-3 shows general anterior and posterior views of the muscular system. Do not try to mem-

EXHIBIT 10-1
MUSCLES NAMED ACCORDING TO ACTION

ACTION	DEFINITION	EXAMPLE
Flexor	Usually decreases the anterior angle at a joint; some decrease the posterior angle.	Flexor carpi radialis
Extensor	Usually increases the anterior angle at a joint; some increase the posterior angle.	Extensor carpi ulnaris
Abductor	Moves a bone away from the midline.	Abductor hallucis longus
Adductor	Moves a bone closer to the midline.	Adductor longus
Levator	Produces an upward movement.	Levator scapulae
Depressor	Produces a downward movement.	Depressor labii inferioris
Supinator	Turns the palm upward or anteriorly.	Supinator
Pronator	Turns the palm downward or posteriorly.	Pronator teres
Dorsiflexor	Flexes the foot at the ankle joint.	Tibialis anterior
Plantar flexor	Extends the foot at the ankle joint.	Plantaris
Invertor	Turns the sole of the foot inward.	Tibialis anterior
Evertor	Turns the sole of the foot outward.	Peroneus tertius
Sphincter	Decreases the size of an opening.	Orbicularis oculi
Tensor	Makes a body part more rigid.	Tensor fasciae latae
Rotator	Moves a bone around its longitudinal axis.	Obturator

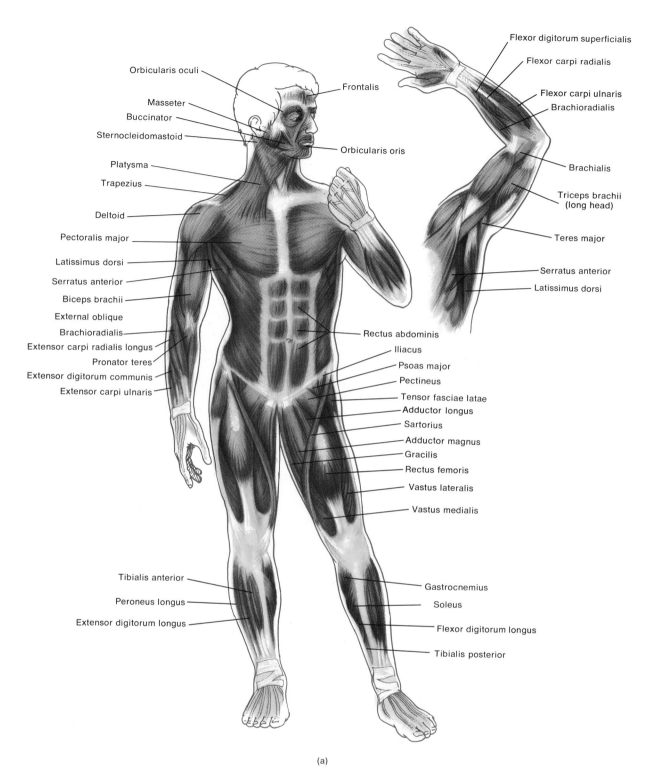

Orbicularis oculi

Masseter
Buccinator
Sternocleidomastoid

Platysma
Trapezius

Deltoid

Pectoralis major

Latissimus dorsi

Serratus anterior

Biceps brachii

External oblique
Brachioradialis
Extensor carpi radialis longus
Pronator teres
Extensor digitorum communis
Extensor carpi ulnaris

Frontalis

Orbicularis oris

Flexor digitorum superficialis
Flexor carpi radialis

Flexor carpi ulnaris
Brachioradialis

Brachialis

Triceps brachii
(long head)

Teres major

Serratus anterior
Latissimus dorsi

Rectus abdominis
Iliacus
Psoas major
Pectineus
Tensor fasciae latae
Adductor longus
Sartorius
Adductor magnus
Gracilis
Rectus femoris
Vastus lateralis
Vastus medialis

Tibialis anterior
Peroneus longus
Extensor digitorum longus

Gastrocnemius
Soleus
Flexor digitorum longus
Tibialis posterior

(a)

FIGURE 10-3 Principal superficial muscles. (a) Anterior view.

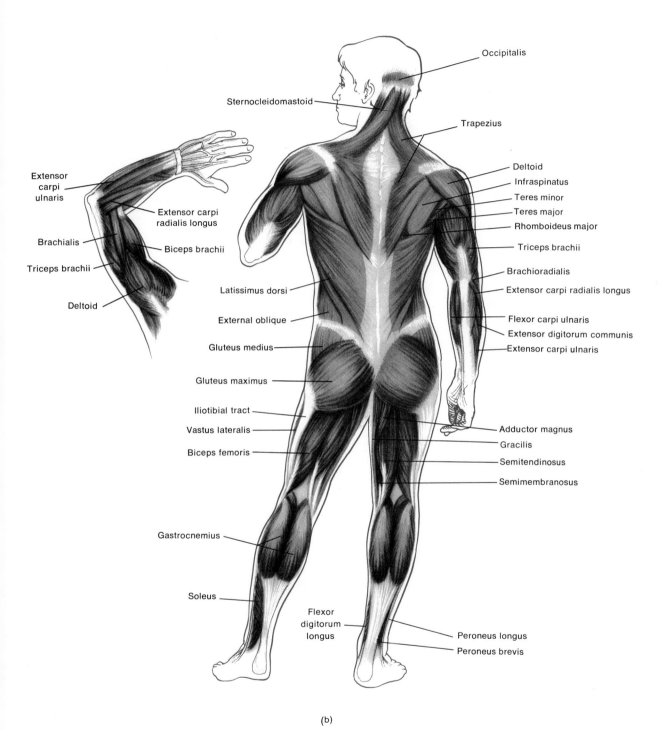

Occipitalis

Sternocleidomastoid

Trapezius

Deltoid

Infraspinatus

Teres minor

Teres major

Rhomboideus major

Triceps brachii

Brachioradialis

Extensor carpi radialis longus

Flexor carpi ulnaris

Extensor digitorum communis

Extensor carpi ulnaris

Extensor carpi ulnaris

Extensor carpi radialis longus

Brachialis

Biceps brachii

Triceps brachii

Deltoid

Latissimus dorsi

External oblique

Gluteus medius

Gluteus maximus

Iliotibial tract

Vastus lateralis

Biceps femoris

Adductor magnus

Gracilis

Semitendinosus

Semimembranosus

Gastrocnemius

Soleus

Flexor digitorum longus

Peroneus longus

Peroneus brevis

(b)

FIGURE 10-3 *(Continued)* Principal superficial muscles. (b) Posterior view.

orize all these muscles yet. As you study groups of muscles in subsequent exhibits, refer to Figure 10-3 to see how each group is related to all others.

An attempt has been made to indicate whether the muscles are superficial or deep, anterior or posterior,

and medial or lateral. An attempt has also been made to show the relationship of the muscles under consideration to other muscles in the area you are studying. If you have mastered the naming of muscles, their actions will have more meaning.

EXHIBIT 10-2
MUSCLES OF FACIAL EXPRESSION (Figure 10-4)

MUSCLE	ORIGIN	INSERTION	ACTION	INNERVATION
Epicranius (*epi* = over; *crani* = skull)	This muscle is divisible into two portions: the frontalis over the frontal bone and the occipitalis over the occipital bone. The two muscles are united by a strong aponeurosis, the galea aponeurotica, which covers the superior and lateral surfaces of the skull.			
Frontalis (*front* = forehead)	Galea aponeurotica	Skin superior to supraorbital line	Draws scalp forward, raises eyebrows, and wrinkles forehead horizontally	Facial nerve (VII)
Occipitalis (*occipito* = base of skull)	Occipital bone and mastoid process of temporal bone	Galea aponeurotica	Draws scalp backward	Facial nerve (VII)
Orbicularis oris (*orb* = circular; *or* = mouth)	Muscle fibers surrounding opening of mouth	Skin at corner of mouth	Closes lips, compresses lips against teeth, protrudes lips, and shapes lips during speech	Facial nerve (VII)
Zygomaticus major (*zygomatic* = cheek bone; *major* = greater)	Zygomatic bone	Skin at angle of mouth and orbicularis oris	Draws angle of mouth upward and outward as in smiling or laughing	Facial nerve (VII)
Levator labii superioris (*levator* = raises or elevates; *labii* = lip; *superioris* = upper)	Superior to infraorbital foramen of maxilla	Skin at angle of mouth and orbicularis oris	Elevates (raises) upper lip	Facial nerve (VII)
Depressor labii inferioris (*depressor* = depresses or lowers; *inferioris* = lower)	Mandible	Skin of lower lip	Depresses (lowers) lower lip	Facial nerve (VII)
Buccinator (*bucc* = cheek)	Alveolar processes of maxilla and mandible and pterygomandibular raphe (fibrous band extending from the pterygoid hamulus to the mandible)	Orbicularis oris	Major cheek muscle; compresses cheek as in blowing air out of mouth and causes the cheeks to cave in, producing the action of sucking	Facial nerve (VII)
Mentalis (*mentum* = chin)	Mandible	Skin of chin	Elevates and protrudes lower lip and pulls skin of chin up as in pouting	Facial nerve (VII)
Platysma (*platy* = flat, broad)	Fascia over deltoid and pectoralis major muscles	Mandible, muscles around angle of mouth, and skin of lower face	Draws outer part of lower lip downward and backward as in pouting; depresses mandible	Facial nerve (VII)
Risorius (*risor* = laughter)	Fascia over parotid (salivary) gland	Skin at angle of mouth	Draws angle of mouth laterally as in tenseness	Facial nerve (VII)
Orbicularis oculi (*ocul* = eye)	Medial wall of orbit	Circular path around orbit	Closes eye	Facial nerve (VII)

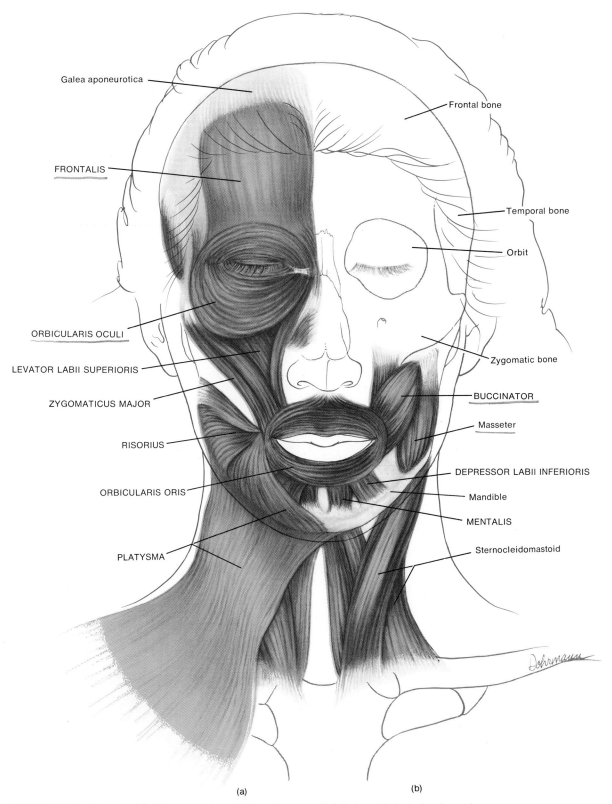

Galea aponeurotica

Frontal bone

FRONTALIS

Temporal bone

Orbit

ORBICULARIS OCULI

LEVATOR LABII SUPERIORIS

Zygomatic bone

ZYGOMATICUS MAJOR

BUCCINATOR

Masseter

RISORIUS

DEPRESSOR LABII INFERIORIS

ORBICULARIS ORIS

Mandible

MENTALIS

PLATYSMA

Sternocleidomastoid

(a)

(b)

FIGURE 10-4 Muscles of facial expression. (a) Anterior superficial view. (b) Anterior deep view.

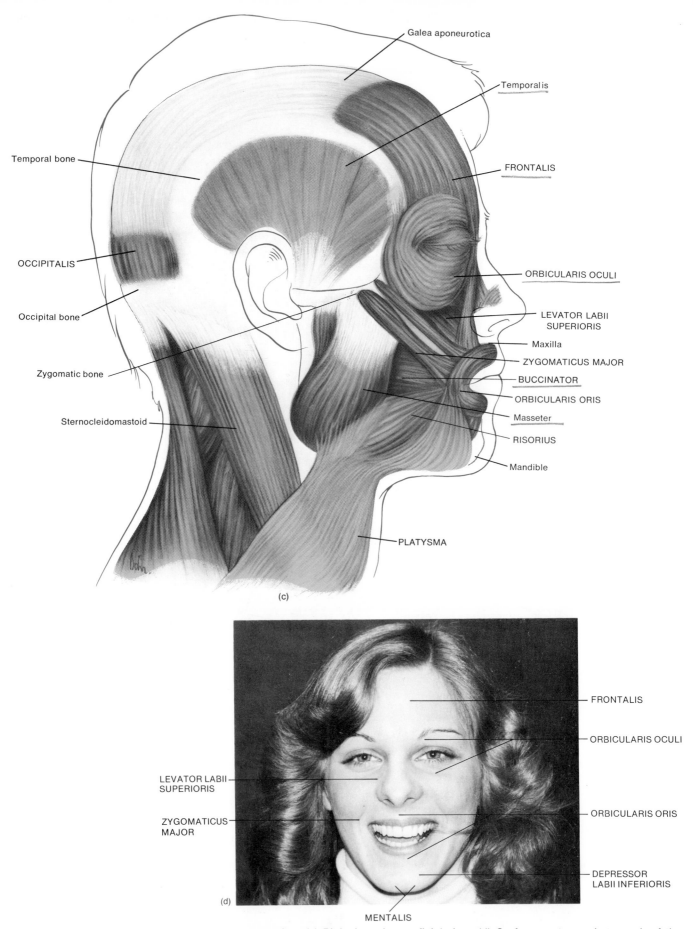

FIGURE 10-4 *(Continued)* Muscles of facial expression. (c) Right lateral superficial view. (d) Surface anatomy photograph of the anterior facial muscles. (Courtesy of Donald Castellaro and Deborah Massimi.)

EXHIBIT 10-3

MUSCLES THAT MOVE THE LOWER JAW (Figure 10-5)

MUSCLE	ORIGIN	INSERTION	ACTION	INNERVATION
Masseter (*maseter* = chewer)	Maxilla and zygomatic arch	Angle and ramus of mandible	Elevates mandible as in closing the mouth and protracts (protrudes) mandible	Mandibular branch of trigeminal nerve (V)
Temporalis (*tempora* = temples)	Temporal bone	Coronoid process of mandible	Elevates and retracts mandible	Temporal nerve from mandibular divison of trigeminal nerve (V)
Medial pterygoid (*medial* = closer to midline; *pterygoid* = like a wing; pterygoid plate of sphenoid)	Medial surface of lateral pterygoid plate of sphenoid; maxilla	Angle and ramus of mandible	Elevates and protracts mandible and moves mandible from side to side	Mandibular branch of trigeminal nerve (V)
Lateral pterygoid (*lateral* = farther from midline)	Greater wing and lateral surface of lateral pterygoid plate of sphenoid	Condyle of mandible; temporomandibular articulation	Protracts mandible, opens mouth, and moves mandible from side to side	Mandibular branch of trigeminal nerve (V)

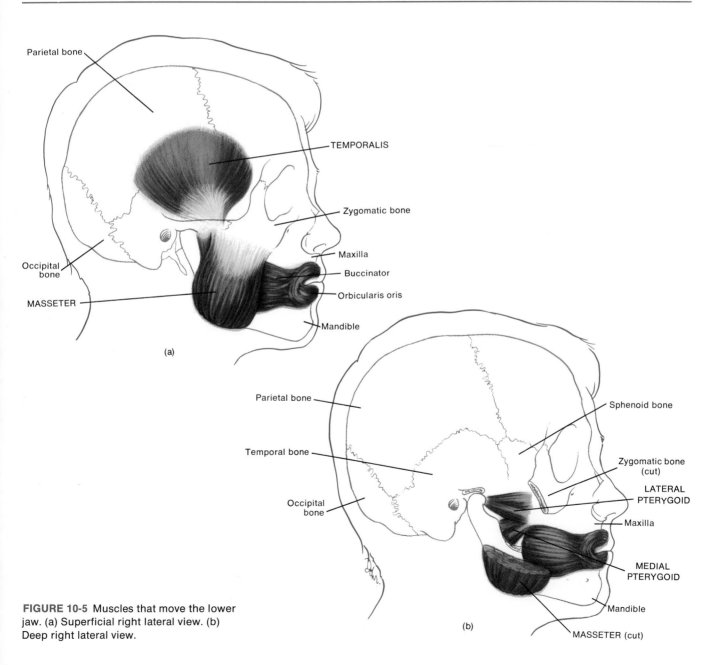

FIGURE 10-5 Muscles that move the lower jaw. (a) Superficial right lateral view. (b) Deep right lateral view.

EXHIBIT 10-4
MUSCLES THAT MOVE THE EYEBALLS—EXTRINSIC MUSCLES* (Figure 10-6)

MUSCLE	ORIGIN	INSERTION	ACTION	INNERVATION
Superior rectus (*superior* = above; *rectus* = in this case, muscle fibers running parallel to long axis of eyeball)	Tendinous ring attached to bony orbit around optic foramen	Superior and central part of eyeball	Rolls eyeball upward	Oculomotor nerve (III)
Inferior rectus (*inferior* = below)	Same as above	Inferior and central part of eyeball	Rolls eyeball downward	Oculomotor nerve (III)
Lateral rectus	Same as above	Lateral side of eyeball	Rolls eyeball laterally	Abducens nerve (VI)
Medial rectus	Same as above	Medial side of eyeball	Rolls eyeball medially	Oculomotor nerve (III)
Superior oblique (*oblique* = in this case, muscle fibers running diagonally to long axis of eyeball)	Same as above	Eyeball between superior and lateral recti	Rotates eyeball on its axis; directs cornea downward and laterally; note that it moves through a ring of fibrocartilaginous tissue called the trochlea (*trochlea* = pulley)	Trochlear nerve (IV)
Inferior oblique	Maxilla (front of orbital cavity)	Eyeball between superior and lateral recti	Rotates eyeball on its axis; directs cornea upward and laterally	Oculomotor nerve (III)

*Muscles situated on the outside of the eyeball

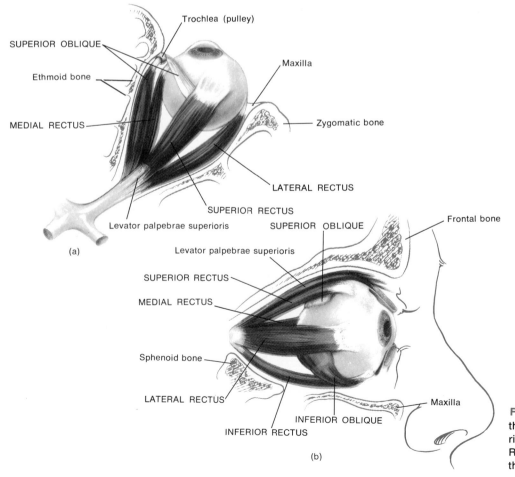

(a)

(b)

FIGURE 10-6 Extrinsic muscles of the eyeball. (a) Muscles of the right eyeball seen from above. (b) Right lateral view of muscles of the right eyeball.

EXHIBIT 10-5
MUSCLES THAT MOVE THE TONGUE (Figure 10-7)

MUSCLE	ORIGIN	INSERTION	ACTION	INNERVATION
Genioglossus (*geneion* = chin; *glossus* = tongue)	Mandible	Undersurface of tongue and hyoid bone	Depresses and thrusts tongue forward (protraction)	Hypoglossal nerve (XII)
Styloglossus (*stylo* = stake or pole; styloid process of temporal)	Styloid process of temporal bone	Side and undersurface of tongue	Elevates tongue and draws it backward (retraction)	Hypoglossal nerve (XII)
Stylohyoid (*hyoeides* = U-shaped; pertaining to hyoid bone)	Styloid process of temporal bone	Body of hyoid bone	Elevates and retracts tongue	Facial nerve (VII
Hyoglossus	Body of hyoid bone	Side of tongue	Depresses tongue and draws down its sides	Hypoglossal nerve (XII)

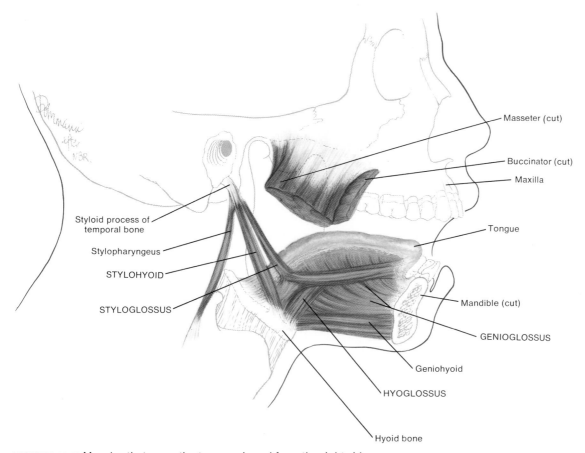

FIGURE 10-7 Muscles that move the tongue viewed from the right side.

EXHIBIT 10-6
MUSCLES THAT MOVE THE HEAD

MUSCLE	ORIGIN	INSERTION	ACTION	INNERVATION
Sternocleidomastoid (*sternum* = breastbone; *cleido* = clavicle; *mastoid* = mastoid process of temporal bone (see Figure 10-10)	Sternum and clavicle	Mastoid process of temporal bone	Contraction of both muscles flexes the neck on the chest; contraction of one muscle rotates face toward side opposite contracting muscle	Accessory nerve (XI); C2–C3
Semispinalis capitis (*semi* = half; *spine* = spinous process; *caput* = head) (see Figure 10-14)	Articular process of seventh cervical vertebra and transverse processes of first six thoracic vertebrae	Occipital bone	Both muscles extend head; contraction of one muscle rotates face toward same side as contracting muscle	Dorsal rami of spinal nerves
Splenius capitis (*splenion* = bandage) (see Figure 10-14)	Ligamentum nuchae and spines of seventh cervical vertebra and first four thoracic vertebrae	Occipital bone and mastoid process of temporal bone	Both muscles extend head; contraction of one rotates it to same side as contracting muscle	Dorsal rami of middle and lower cervical nerves
Longissimus capitis (*longissimus* = longest) (see Figure 10-14)	Transverse processes of last four cervical vertebrae	Mastoid process of temporal bone	Extends head and rotates face toward side opposite contracting muscle.	Dorsal rami of middle and lower cervical nerves

EXHIBIT 10-7
MUSCLES THAT ACT ON THE ANTERIOR ABDOMINAL WALL (Figure 10-8)

MUSCLE	ORIGIN	INSERTION	ACTION	INNERVATION
Rectus abdominis (*abdomino* = belly)	Pubic crest and symphysis pubis	Cartilage of fifth to seventh ribs and xiphoid process	Flexes vertebral column	Branches of 7–12 intercostal nerves
External oblique (*external* = closer to the surface)	Lower eight ribs	Iliac crest; linea alba (midline aponeurosis)	Both muscles compress abdomen; one side alone bends vertebral column laterally	Branches of 8–12 intercostal nerves, iliohypogastric and ilioinguinal nerves
Internal oblique (*internal* = farther from the surface)	Iliac crest, inguinal ligament, and thoracolumbar fascia	Cartilage of last three or four ribs	Compresses abdomen; one side alone bends vertebral column laterally	Branches of 8–12 intercostal nerves, iliohypogastric, and ilioinguinal nerves
Transversus abdominis (*transverse* = muscle fibers run transversely to midline)	Iliac crest, inguinal ligament, lumbar fascia, and cartilages of last six ribs	Xiphoid process, linea alba, and pubis	Compresses abdomen	Branches of 7–12 intercostal nerves, iliohypogastric, and ilioinguinal nerves

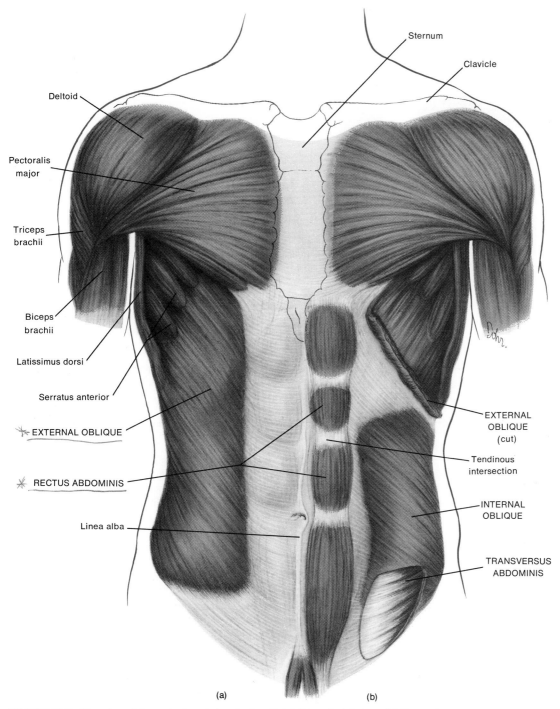

FIGURE 10-8 Muscles of the anterior abdominal wall. (a) Superficial view. (b) Deep view.

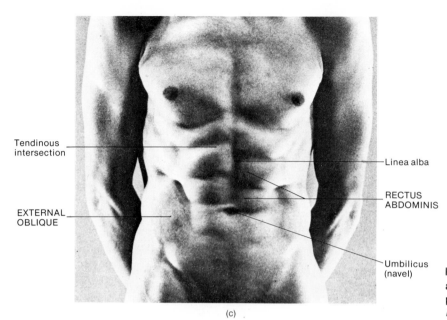

Tendinous intersection

Linea alba

RECTUS ABDOMINIS

EXTERNAL OBLIQUE

Umbilicus (navel)

(c)

FIGURE 10-8 *(Continued)* Muscles of the anterior abdominal wall. (c) Surface anatomy photograph. (Courtesy of R. D. Lockhart, *Living Anatomy,* Faber and Faber, 1974.)

EXHIBIT 10-8
MUSCLES USED IN BREATHING (Figure 10-9)

MUSCLE	ORIGIN	INSERTION	ACTION	INNERVATION
Diaphragm (*dia* = across; *phragma* = wall)	Xiphoid process, costal cartilages of last six ribs, and lumbar vertebrae	Central tendon	Forms floor of thoracic cavity; contraction pulls central tendon downward and increases vertical length of thorax during inspiration	Phrenic nerve
External intercostals (*inter* = between; *costa* = rib)	Inferior border of rib above	Superior border of rib below	Elevate ribs during inspiration and thus increase lateral and anteroposterior dimensions of thorax	Intercostal nerves
Internal intercostals	Superior border of rib below	Inferior border of rib above	Draw adjacent ribs together during forced expiration and thus decrease the lateral and anteroposterior dimensions of the thorax	Intercostal nerves

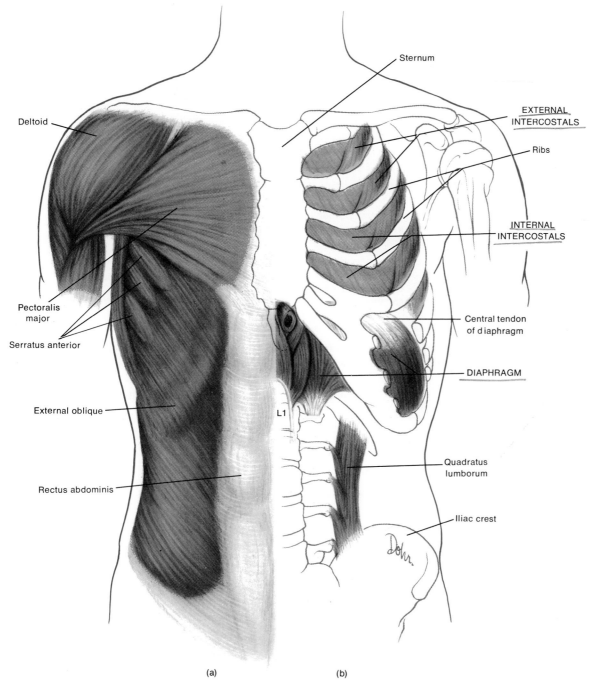

Deltoid

Sternum

EXTERNAL INTERCOSTALS

Ribs

INTERNAL INTERCOSTALS

Pectoralis major

Serratus anterior

Central tendon of diaphragm

DIAPHRAGM

External oblique

L1

Rectus abdominis

Quadratus lumborum

Iliac crest

(a)   (b)

FIGURE 10-9 Muscles used in breathing. (a) Superficial view. (b) Deep view.

EXHIBIT 10-9

MUSCLES THAT MOVE THE SHOULDER GIRDLE (Figure 10-10)

MUSCLE	ORIGIN	INSERTION	ACTION	INNERVATION
Subclavius (*sub* = under; *clavius* = clavicle)	First rib	Clavicle	Depresses clavicle	Nerve to subclavius
Pectoralis minor (*pectus* = breast, chest, thorax; *minor* = lesser)	Third through fifth ribs	Coracoid process of scapula	Depresses scapula, rotates shoulder joint anteriorly, and elevates third through fifth ribs during forced inspiration when scapula is fixed	Medial pectoral nerve
Serratus anterior (*serratus* = serrated; *anterior* = front)	Upper eight or nine ribs	Vertebral border and inferior angle of scapula	Rotates scapula laterally and elevates ribs when scapula is fixed	Long thoracic nerve
Trapezius (*trapezoides* = trapezoid-shaped)	Occipital bone, ligamentum nuchae, and spines of seventh cervical and all thoracic vertebrae	Acromion process of clavicle and spine of scapula	Elevates clavicle, adducts scapula, elevates or depresses scapula, and extends head	Acessory nerve (XI); C3–C4
Levator scapulae (*levator* = raises; *scapulae* = scapula)	Upper four or five cervical vertebrae	Vertebral border of scapula	Elevates scapula	Dorsal scapular nerve (C3–C5)
Rhomboideus major (*rhomboides* = rhomboid or diamond-shaped)	Spines of second to fifth thoracic vertebrae	Vertebral border of scapula	Adducts scapula and slightly rotates it upward	Dorsal scapular nerve
Rhomboideus minor	Spines of seventh cervical and first thoracic vertebrae	Superior angle of scapula	Adducts scapula	Dorsal scapular nerve

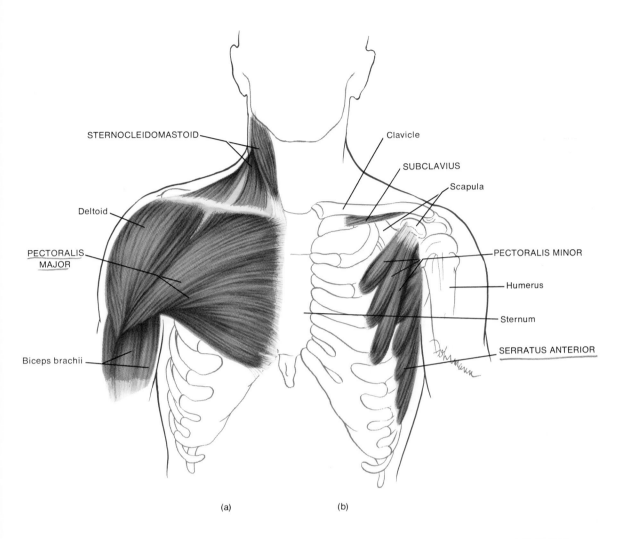

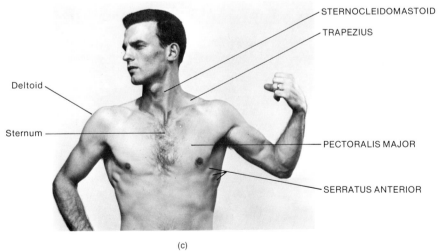

FIGURE 10-10 Muscles that move the shoulder girdle. (a) Anterior superficial view. (b) Anterior deep view. (c) Surface anatomy photograph of the neck and chest. (Courtesy of Vincent P. Destro, Mayo Foundation.)

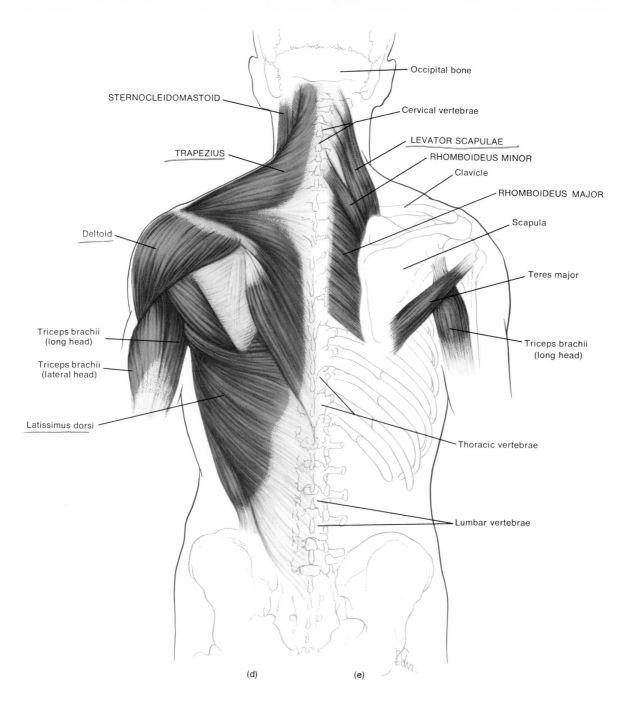

Occipital bone

STERNOCLEIDOMASTOID

Cervical vertebrae

LEVATOR SCAPULAE

TRAPEZIUS

RHOMBOIDEUS MINOR

Clavicle

RHOMBOIDEUS MAJOR

Scapula

Deltoid

Teres major

Triceps brachii
(long head)

Triceps brachii
(long head)

Triceps brachii
(lateral head)

Latissimus dorsi

Thoracic vertebrae

Lumbar vertebrae

(d)   (e)

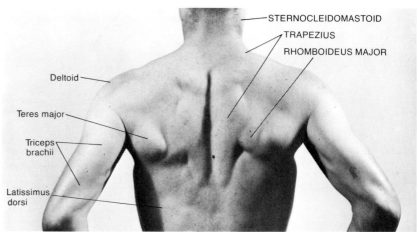

STERNOCLEIDOMASTOID

TRAPEZIUS

RHOMBOIDEUS MAJOR

Deltoid

Teres major

Triceps
brachii

Latissimus
dorsi

(f)

FIGURE 10-10 (Continued) Muscles that move the shoulder girdle. (d) Posterior superficial view. (e) Posterior deep view. (f) Surface anatomy photograph of the back. (Courtesy of Vincent P. Destro, Mayo Foundation.)

EXHIBIT 10-10
MUSCLES THAT MOVE THE ARM (Figure 10-11)

MUSCLE	ORIGIN	INSERTION	ACTION	INNERVATION
Pectoralis major (see Figure 10-10)	Clavicle, sternum, cartilages of second to sixth ribs	Greater tubercle of humerus	Flexes, adducts, and rotates arm medially	Medial and lateral pectoral nerve
Latissimus dorsi (*dorsum* = back)	Spines of lower six thoracic vertebrae, lumbar vertebrae, crests of sacrum and ilium, lower four ribs	Intertubercular groove of humerus	Extends, adducts, and rotates arm medially; draws shoulder downward and backward	Thoracodorsal nerve
Deltoid (*delta* = triangular)	Clavicle and acromion process and spine of scapula	Deltoid tuberosity of humerus	Abducts arm	Axillary nerve
Supraspinatus (*supra* = above; *spinatus* = spine of scapula)	Fossa superior to spine of scapula	Greater tubercle of humerus	Assists deltoid muscle in abducting arm	Suprascapular nerve
Infraspinatus (*infra* = below)	Fossa inferior to spine of scapula	Greater tubercle of humerus	Rotates arm laterally	Suprascapular nerve
Teres major (*teres* = long and round)	Inferior angle of scapula	Distal to lesser tubercle of humerus	Extends arm and draws it down; assists in adduction and medial rotation of arm	Lower subscapular nerve
Teres minor	Axillary border of scapula	Greater tubercle of humerus	Rotates arm laterally	Axillary nerve

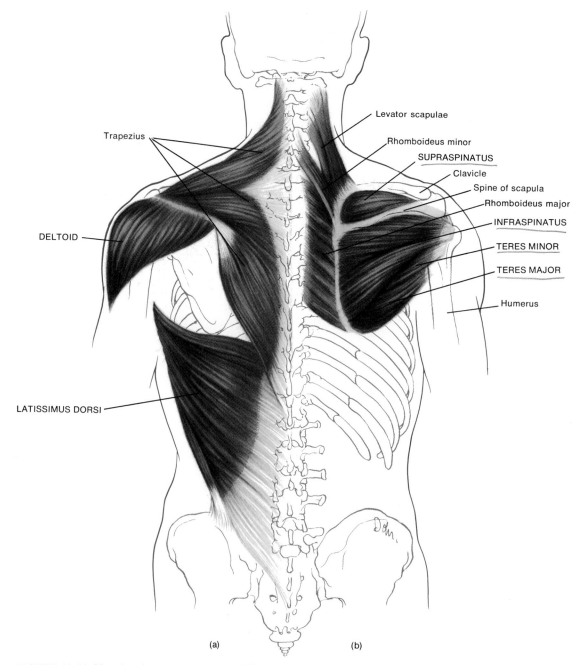

Trapezius

Levator scapulae

Rhomboideus minor

SUPRASPINATUS

Clavicle

Spine of scapula

Rhomboideus major

INFRASPINATUS

DELTOID

TERES MINOR

TERES MAJOR

Humerus

LATISSIMUS DORSI

(a)   (b)

FIGURE 10-11 Muscles that move the arm. (a) Posterior superficial view. (b) Posterior deep view.

(c)

FIGURE 10-11 *(Continued)* Muscles that move the arm. (c) Surface anatomy photograph of the back. (Courtesy of J. Royce, *Surface Anatomy,* Davis, 1965.)

Before you move on to Exhibit 10-11 (Muscles That Move the Forearm), refer to Figure 1-4b. This figure is a cross section of the trunk at the level of the heart and lungs. Several of the muscles you have studied up to this point are shown. Figure 1-4b will add to your understanding of how muscles are arranged with respect to each other from external to internal. It will also show you how muscles are oriented with regard to bones and viscera.

EXHIBIT 10-11
MUSCLES THAT MOVE THE FOREARM (Figure 10-12)

MUSCLE	ORIGIN	INSERTION	ACTION	INNERVATION
Biceps brachii (*biceps* = two heads of origin; *brachion* = arm)	Long head originates from tuberosity above glenoid cavity and short head originates from coracoid process of scapula	Radial tuberosity and bicipital aponeurosis	Flexes and supinates forearm.	Musculocutaneous nerve
Brachialis	Anterior surface of humerus	Tuberosity and coronoid process of ulna	Flexes forearm	Musculocutaneous, radial, and median nerves
Brachioradialis (*radialis* = radius) (see Figure 10-13 also)	Supracondyloid ridge of humerus	Superior to styloid process of radius	Flexes forearm	Radial nerve
Triceps brachii (*triceps* = three heads of origin)	Long head originates from infraglenoid tuberosity of scapula, lateral head originates from lateral and posterior surface of humerus superior to radial groove, medial head originates from posterior surface of humerus inferior to radial groove	Olecranon of ulna	Extends forearm	Radial nerve
Supinator (*supination* = turning palm upward or anteriorly)	Lateral epicondyle of humerus, ridge on ulna	Oblique line of radius	Supinates forearm	Deep radial nerve
Pronator teres (*pronation* = turning palm downward or posteriorly)	Medial epicondyle of humerus, coronoid process of ulna	Midlateral surface of radius	Pronates forearm	Median nerve

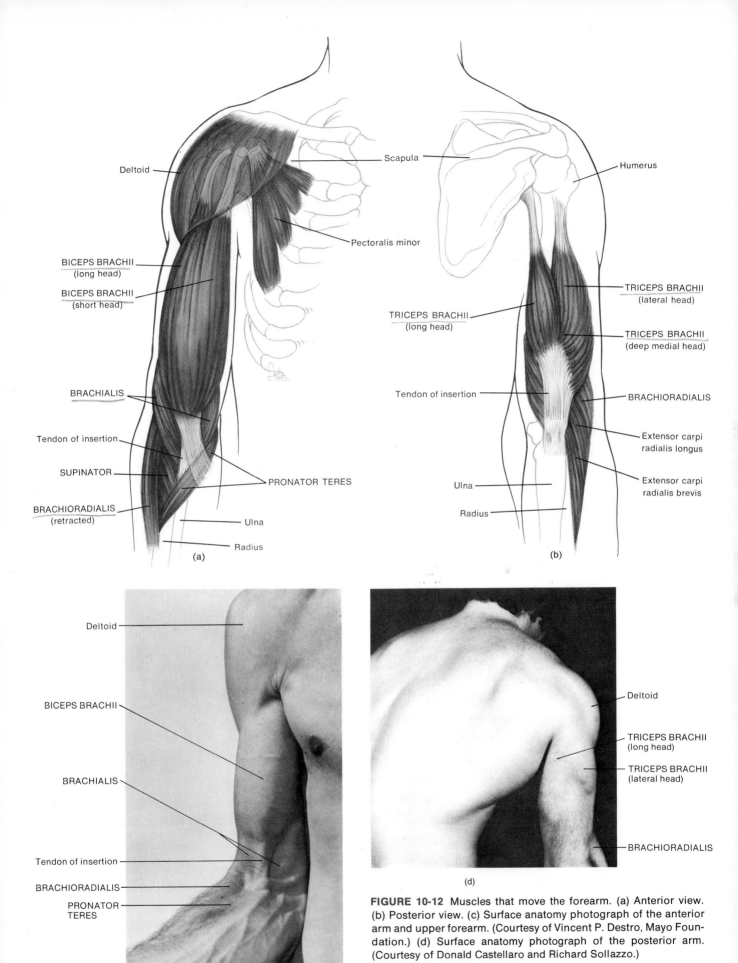

Deltoid

Scapula

Humerus

BICEPS BRACHII
(long head)

BICEPS BRACHII
(short head)

Pectoralis minor

TRICEPS BRACHII
(lateral head)

TRICEPS BRACHII
(long head)

TRICEPS BRACHII
(deep medial head)

BRACHIALIS

Tendon of insertion

Tendon of insertion

BRACHIORADIALIS

SUPINATOR

PRONATOR TERES

Extensor carpi
radialis longus

BRACHIORADIALIS
(retracted)

Ulna

Ulna

Radius

Radius

Extensor carpi
radialis brevis

(a)

(b)

Deltoid

BICEPS BRACHII

BRACHIALIS

Tendon of insertion

BRACHIORADIALIS

PRONATOR
TERES

Deltoid

TRICEPS BRACHII
(long head)

TRICEPS BRACHII
(lateral head)

BRACHIORADIALIS

(d)

FIGURE 10-12 Muscles that move the forearm. (a) Anterior view.
(b) Posterior view. (c) Surface anatomy photograph of the anterior
arm and upper forearm. (Courtesy of Vincent P. Destro, Mayo Foundation.) (d) Surface anatomy photograph of the posterior arm.
(Courtesy of Donald Castellaro and Richard Sollazzo.)

(c)

EXHIBIT 10-12

MUSCLES THAT MOVE THE WRIST AND FINGERS (Figure 10-13)

MUSCLE	ORIGIN	INSERTION	ACTION	INNERVATION
Flexor carpi radialis (*flexor* = decreases angle; *carpus* = wrist)	Medial epicondyle of humerus	Second and third metacarpals	Flexes and abducts wrist	Median nerve
Flexor carpi ulnaris (*ulnaris* = ulna)	Medial epicondyle of humerus and upper dorsal border of ulna	Pisiform, hamate, and fifth metacarpal	Flexes and adducts wrist	Ulnar nerve
Extensor carpi radialis longus (*extensor* = increases angle at a joint; *longus* = long)	Lateral epicondyle of humerus	Second metacarpal	Extends and abducts wrist	Radial nerve
Extensor carpi ulnaris	Lateral epicondyle of humerus and dorsal border of ulna	Fifth metacarpal	Extends and adducts wrist	Deep radial nerve
Flexor digitorum profundus (*digit* = finger or toe; *profundus* = deep)	Anterior medial surface of body of ulna	Bases of distal phalanges	Flexes distal phalanges of each finger	Median and ulnar nerves
Flexor digitorum superficialis (*superficialis* = superficial)	Medial epicondyle of humerus, coronoid process of ulna, oblique line of radius	Middle phalanges	Flexes middle phalanges of each finger	Median nerve
Extensor digitorum	Lateral epicondyle of humerus	Middle and distal phalanges of each finger	Extends phalanges	Deep radial nerve
Extensor indicis (*indicis* = index)	Dorsal surface of ulna	Tendon of extensor digitorum of index finger	Extends index finger	Deep radial nerve

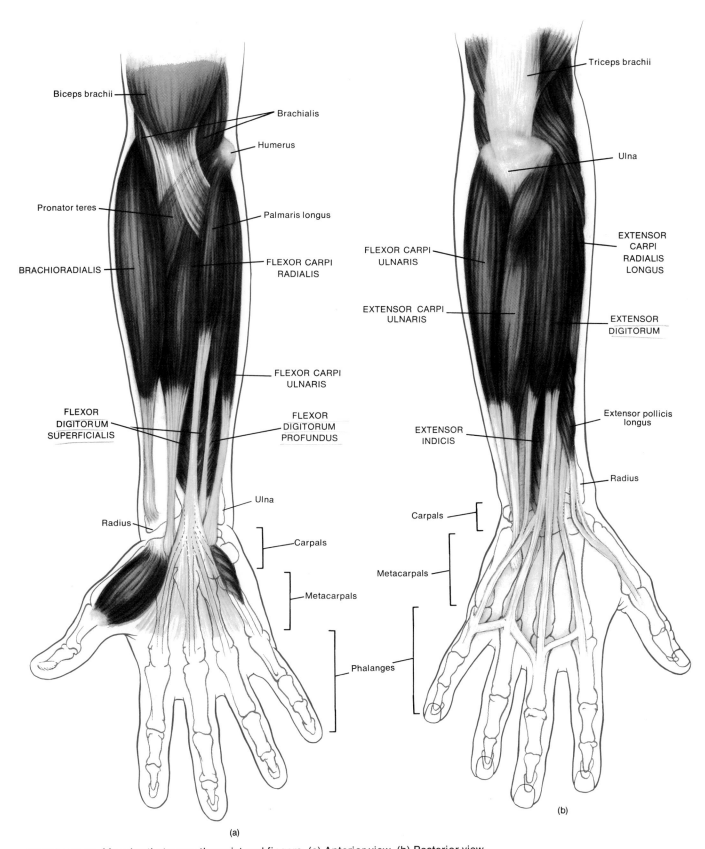

FIGURE 10-13 Muscles that move the wrist and fingers. (a) Anterior view. (b) Posterior view.

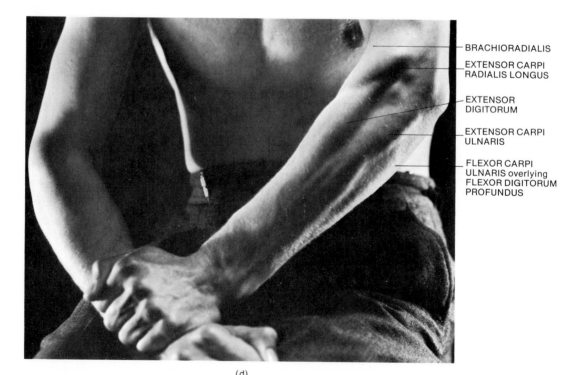

BRACHIORADIALIS

FLEXOR
CARPI
RADIALIS

Palmaris
longus

FLEXOR
CARPI
ULNARIS

(c)

BRACHIORADIALIS

EXTENSOR CARPI
RADIALIS LONGUS

EXTENSOR
DIGITORUM

EXTENSOR CARPI
ULNARIS

FLEXOR CARPI
ULNARIS overlying
FLEXOR DIGITORUM
PROFUNDUS

(d)

FIGURE 10-13 *(Continued)* Muscles that move the wrist and fingers. (c) Surface anatomy photograph of the anterior forearm. (Courtesy of Vincent P. Destro, Mayo Foundation.) (d) Surface anatomy photograph of the posterolateral forearm. (Courtesy of R. D. Lockhart, *Living Anatomy,* Faber and Faber, 1974.)

EXHIBIT 10-13

MUSCLES THAT MOVE THE VERTEBRAL COLUMN (Figure 10-14)

MUSCLE	ORIGIN	INSERTION	ACTION	INNERVATION
Rectus abdominis (see Figure 10-8)	Body of pubis of coxal bone	Cartilages of fifth through seventh ribs	Flexes vertebral column at lumbar spine and compresses abdomen	7–12 intercostal nerves
Quadratus lumborum (*quad* = four; *lumb* = lumbar region)	Iliac crest	Twelfth rib and upper four lumbar vertebrae	Flexes vertebral column laterally	T12–L1
Sacrospinalis (erector spinae)	This posterior muscle consists of three groupings: iliocostalis, longissimus, and spinalis. These groups, in turn, consist of a series of overlapping muscles. The iliocostalis group is laterally placed, the longissimus group is intermediate in placement, and the spinalis is medially placed.			
LATERAL				
Iliocostalis lumborum (*ilium* = flank; *lumbus* = loin)	Iliac crest	Lower six ribs	Extends lumbar region of vertebral column	Dorsal rami of lumbar nerves
Iliocostalis thoracis (*thorax* = chest)	Lower six ribs	Upper six ribs	Maintains erect position of spine	Dorsal rami of thoracic nerves
Iliocostalis cervicis (*cervix* = neck)	First six ribs	Transverse processes of fourth to sixth cervical vertebrae	Extends cervical region of vertebral column	Dorsal rami of cervical nerves
INTERMEDIATE				
Longissimus thoracis	Transverse processes of lumbar vertebrae	Transverse processes of all thoracic and upper lumbar vertebrae and ninth and tenth ribs	Extends thoracic region of vertebral column	Dorsal rami of spinal nerves
Longissimus cervicis	Transverse processes of fourth and fifth thoracic vertebrae	Transverse processes of second to sixth cervical vertebrae	Extends cervical region of vertebral column	Dorsal rami of spinal nerves
Longissimus capitis	Transverse processes of upper four thoracic vertebrae	Mastoid process of temporal bone	Extends head and rotates it to opposite side	Dorsal rami of middle and lower cervical nerves
MEDIAL				
Spinalis thoracis	Spines of upper lumbar and lower thoracic vertebrae	Spines of upper thoracic vertebrae	Extends vertebral column	Dorsal rami of spinal nerves

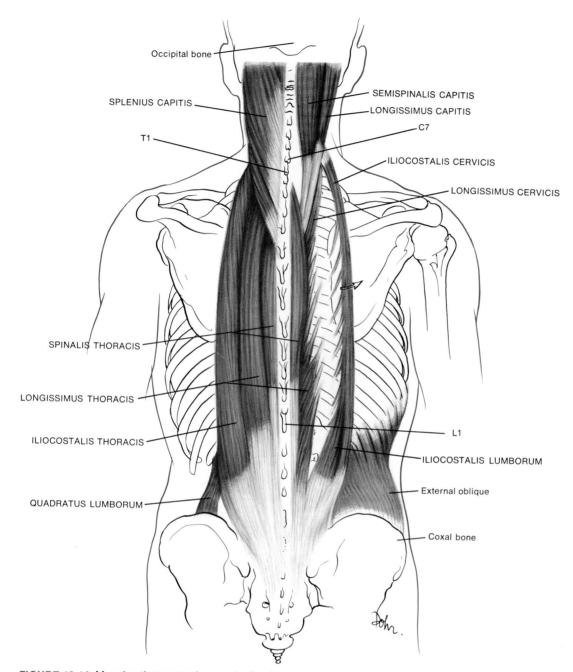

Occipital bone

SPLENIUS CAPITIS

SEMISPINALIS CAPITIS

LONGISSIMUS CAPITIS

T1

C7

ILIOCOSTALIS CERVICIS

LONGISSIMUS CERVICIS

SPINALIS THORACIS

LONGISSIMUS THORACIS

ILIOCOSTALIS THORACIS

L1

ILIOCOSTALIS LUMBORUM

External oblique

QUADRATUS LUMBORUM

Coxal bone

FIGURE 10-14 Muscles that move the vertebral column.

EXHIBIT 10-14
MUSCLES THAT MOVE THE THIGH (Figure 10-15)

MUSCLE	ORIGIN	INSERTION	ACTION	INNERVATION
Psoas major (*psoa* = muscle of loin)	Transverse processes and bodies of lumbar vertebrae	Lesser trochanter of femur	Flexes and rotates thigh laterally; flexes vertebral column	L2–L3
Iliacus (*iliac* = ilium) (Together the psoas major and iliacus are sometimes termed the iliopsoas muscle.)	Iliac fossa	Tendon of psoas major	Flexes and rotates thigh laterally; slight flexion of vertebral column	Femoral nerve
Gluteus maximus (*gloutos* = buttock; (*maximus* = largest)	Iliac crest, sacrum, coccyx, and aponeurosis of sacrospinalis	Iliotibial tract of fascia lata and gluteal tuberosity of femur	Extends and rotates thigh laterally.	Inferior gluteal nerve
Gluteus medius (*media* = middle)	Ilium	Greater trochanter of femur	Abducts and rotates thigh medially	Superior gluteal nerve
Gluteus minimus (*minimus* = small)	Ilium	Greater trochanter of femur	Abducts and rotates thigh laterally	Superior gluteal nerve
Tensor fasciae latae (*tensor* = makes tense; *fascia* = band; *latus* = wide)	Iliac crest	Tibia by way of the iliotibial tract	Flexes and abducts thigh	Superior gluteal nerve
Adductor longus (*adductor* = moves part closer to midline)	Pubic crest and symphysis pubis	Linea aspera of femur	Adducts, rotates, and flexes thigh	Obturator nerve
Adductor brevis (*brevis* = short)	Inferior ramus of pubis	Linea aspera of femur	Adducts, rotates, and flexes thigh	Obturator nerve
Adductor magnus (*magnus* = large)	Inferior ramus of pubis, ischium to ischial tuberosity	Linea aspera of femur	Adducts, flexes, and extends thigh (anterior part flexes, posterior part extends)	Obturator nerve
Piriformis (*pirum* = pear; *forma* = shape)	Sacrum	Greater trochanter of femur	Rotates thigh laterally and abducts it	S2 or S1–S2
Obturator internus (*obturator* = closed because it arises over obturator foramen, which is closed by heavy membrane; *internus* = inside)	Margin of obturator foramen, pubis, and ischium	Greater trochanter of femur	Rotates thigh laterally and abducts it	Obturator nerve
Pectineus (*pecten* = comb-shaped)	Fascia of pubis	Pectineal line of femur	Flexes, adducts, and rotates thigh laterally	Femoral nerve

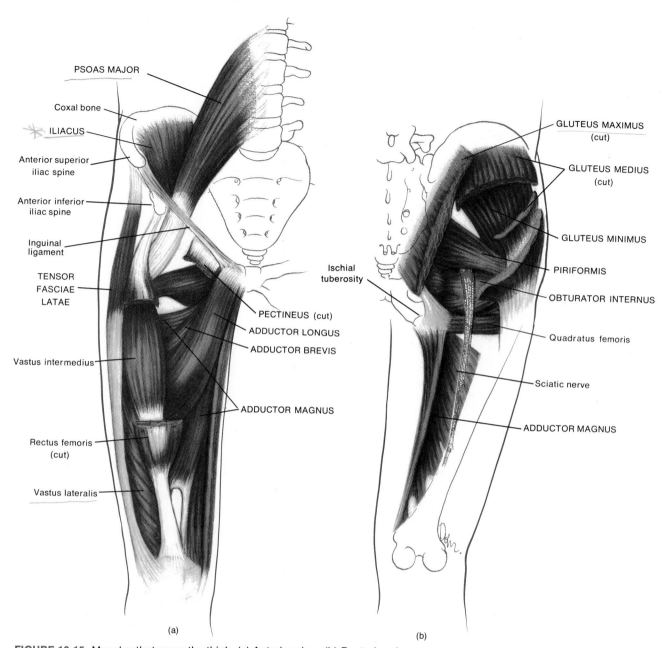

PSOAS MAJOR

Coxal bone

ILIACUS

Anterior superior
iliac spine

Anterior inferior
iliac spine

Inguinal
ligament

TENSOR
FASCIAE
LATAE

Vastus intermedius

Rectus femoris
(cut)

Vastus lateralis

PECTINEUS (cut)

ADDUCTOR LONGUS

ADDUCTOR BREVIS

ADDUCTOR MAGNUS

(a)

GLUTEUS MAXIMUS
(cut)

GLUTEUS MEDIUS
(cut)

GLUTEUS MINIMUS

PIRIFORMIS

OBTURATOR INTERNUS

Quadratus femoris

Sciatic nerve

ADDUCTOR MAGNUS

Ischial
tuberosity

(b)

FIGURE 10-15 Muscles that move the thigh. (a) Anterior view. (b) Posterior view.

EXHIBIT 10-15
MUSCLES THAT ACT ON THE LEG (see Figure 10-16)

MUSCLE	ORIGIN	INSERTION	ACTION	INNERVATION
Quadriceps femoris	A composite muscle that includes four distinct parts, usually described as four separate muscles. The common tendon that includes the patella and attaches to the tibial tuberosity is known as the patellar ligament.			
Rectus femoris (*rectus* = fibers parallel to midline; *femoris* = femur)	Anterior inferior iliac spine	Upper border of patella		Femoral nerve
Vastus lateralis (*vastus* = large; *lateralis* = lateral)	Greater trochanter and linea aspera of femur	Upper border and sides of patella; tibial tuberosity through patellar ligament (tendon of quadriceps)	All four heads extend leg; rectus portion alone also flexes thigh	Femoral nerve
Vastus medialis (*medialis* = medial)	Linea aspera of femur			Femoral nerve
Vastus intermedius (*intermedius* = middle)	Anterior and lateral surfaces of body of femur			Femoral nerve
Hamstrings	A collective designation for three separate muscles.			
Biceps femoris	Long head arises from ischial tuberosity; short head arises from linea aspera of femur	Head of fibula and lateral condyle of tibia	Flexes leg and extends thigh	Tibial nerve from sciatic nerve
Semitendinosus (*semi* = half; *tendo* = tendon)	Ischial tuberosity	Proximal part of medial surface of body of tibia	Flexes leg and extends thigh	Tibial nerve from sciatic nerve
Semimembranosus (*membran* = membrane)	Ischial tuberosity	Medial condyle of tibia	Flexes leg and extends thigh	Tibial nerve from sciatic nerve
Gracilis (*gracilis* = slender)	Symphysis pubis and pubic arch	Medial surface of body of tibia	Flexes leg and adducts thigh	Obturator nerve
Sartorius (*sartor* = tailor; refers to cross-legged position of tailors)	Anterior superior spine of ilium	Medial surface of body of tibia	Flexes leg; flexes thigh and rotates it laterally, thus crossing leg	Femoral nerve

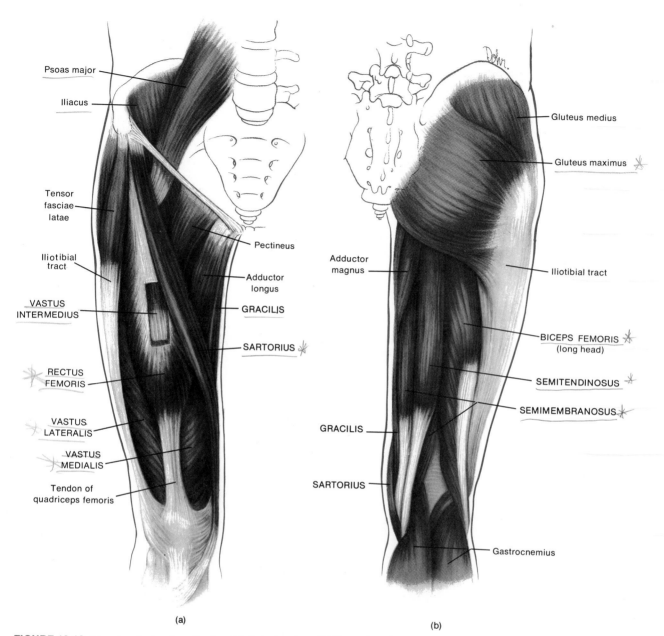

Psoas major

Iliacus

Tensor
fasciae
latae

Iliotibial
tract

VASTUS
INTERMEDIUS

RECTUS
FEMORIS

VASTUS
LATERALIS

VASTUS
MEDIALIS

Tendon of
quadriceps femoris

Pectineus

Adductor
longus

GRACILIS

SARTORIUS

Gluteus medius

Gluteus maximus

Iliotibial tract

Adductor
magnus

BICEPS FEMORIS
(long head)

SEMITENDINOSUS

SEMIMEMBRANOSUS

GRACILIS

SARTORIUS

Gastrocnemius

(a)   (b)

FIGURE 10-16 Muscles that act on the leg. (a) Anterior view. (b) Posterior view.

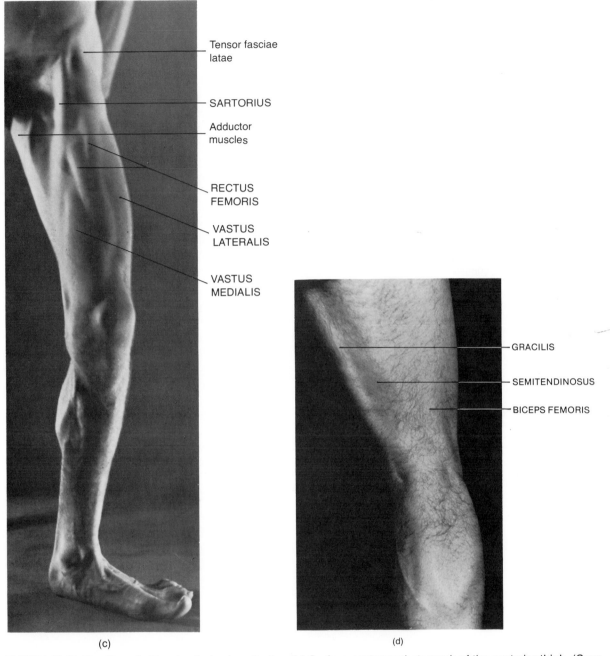

Tensor fasciae latae

SARTORIUS

Adductor muscles

RECTUS FEMORIS

VASTUS LATERALIS

VASTUS MEDIALIS

GRACILIS

SEMITENDINOSUS

BICEPS FEMORIS

(c)

(d)

FIGURE 10-16 *(Continued)* Muscles that act on the leg. (c) Surface anatomy photograph of the posterior thigh. (Courtesy of R. D. Lockhart, *Living Anatomy,* Faber and Faber, 1974.) (d) Surface anatomy photograph of the posterior thigh and leg. (Courtesy of Donald Castellaro and Richard Sollazzo.)

EXHIBIT 10-16

MUSCLES THAT MOVE THE FOOT AND TOES (Figure 10-17)

MUSCLE	ORIGIN	INSERTION	ACTION	INNERVATION
Gastrocnemius (*gaster* = belly; *kneme* = leg)	Lateral and medial condyles of femur and capsule of knee	Calcaneus by way of calcaneal ("Achilles") tendon	Plantar flexes foot	Tibial nerve
Soleus (*soleus* = sole of foot)	Head of fibula and medial border of tibia	Calcaneus by way of calcaneal ("Achilles") tendon	Plantar flexes foot	Tibial nerve
Peroneus longus (*perone* = fibula)	Head and body of fibula and lateral condyle of tibia	First metatarsal and first cuneiform bone	Plantar flexes and everts foot	Superficial peroneal nerve
Peroneus brevis	Body of fibula	Fifth metatarsal	Plantar flexes and everts foot	Superficial peroneal nerve
Peroneus tertius (*tertius* = third)	Distal third of fibula	Fifth metatarsal	Dorsiflexes and everts foot	Deep peroneal nerve
Tibialis anterior (*tibialis* = tibia)	Lateral condyle and body of tibia	First metatarsal and first cuneiform	Dorsiflexes and inverts foot	Deep peroneal nerve
Tibialis posterior (*posterior* = back)	Interosseus membrane between tibia and fibula	Second, third, and fourth metatarsals; navicular; third cuneiform, cuboid	Plantar flexes and inverts foot	Tibial nerve
Flexor digitorum longus (*digitorum* = digit, finger, or toe)	Tibia	Distal phalanges of four outer toes	Flexes toes and plantar flexes and inverts foot	Tibial nerve
Extensor digitorum longus	Lateral condyle of tibia and anterior surface of fibula	Middle and distal phalanges of four outer toes	Extends toes and dorsiflexes and everts foot	Deep peroneal nerve

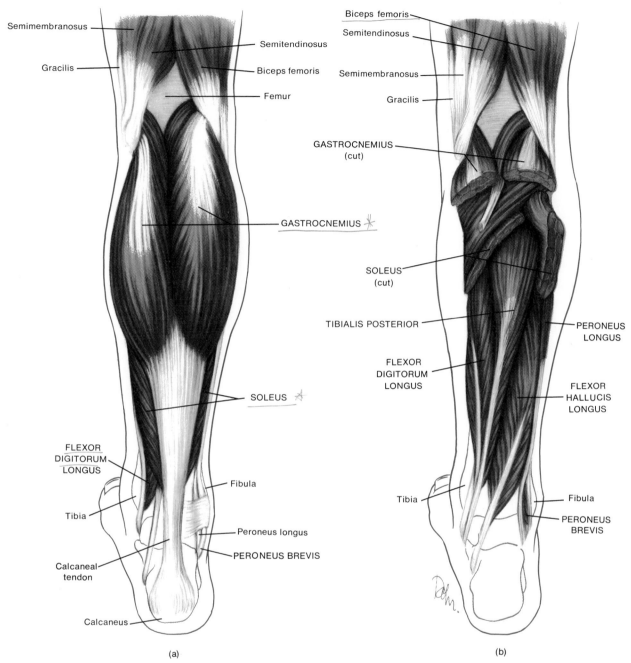

FIGURE 10-17 Muscles that move the foot and toes. (a) Superficial posterior view. (b) Deep posterior view.

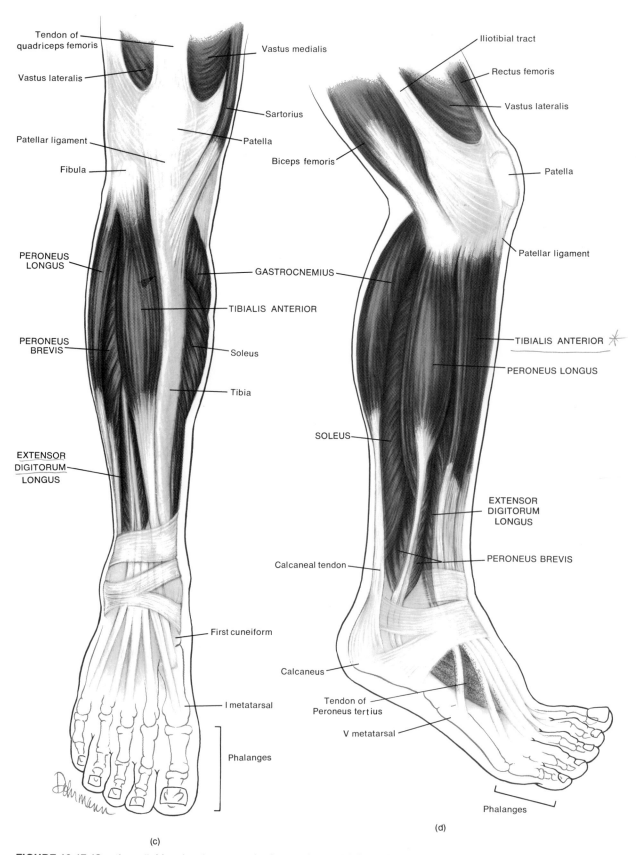

Tendon of quadriceps femoris

Vastus lateralis

Vastus medialis

Patellar ligament

Sartorius

Fibula

Patella

PERONEUS LONGUS

GASTROCNEMIUS

TIBIALIS ANTERIOR

PERONEUS BREVIS

Soleus

Tibia

EXTENSOR DIGITORUM LONGUS

First cuneiform

I metatarsal

Phalanges

(c)

Iliotibial tract

Rectus femoris

Vastus lateralis

Biceps femoris

Patella

Patellar ligament

TIBIALIS ANTERIOR

PERONEUS LONGUS

SOLEUS

EXTENSOR DIGITORUM LONGUS

Calcaneal tendon

PERONEUS BREVIS

Calcaneus

Tendon of Peroneus tertius

V metatarsal

Phalanges

(d)

FIGURE 10-17 *(Continued)* Muscles that move the foot and toes. (c) Superficial anterior view. (d) Superficial lateral view.

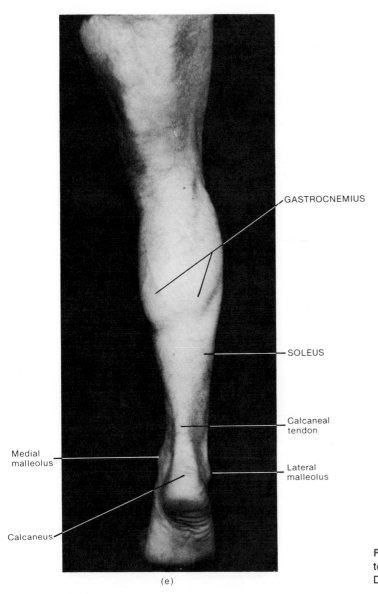

GASTROCNEMIUS

SOLEUS

Calcaneal
tendon

Medial
malleolus

Lateral
malleolus

Calcaneus

(e)

FIGURE 10-17 *(Continued)* Muscles that move the foot and toes. (e) Surface anatomy of the posterior leg. (Courtesy of Donald Castellaro and Richard Sollazzo.)

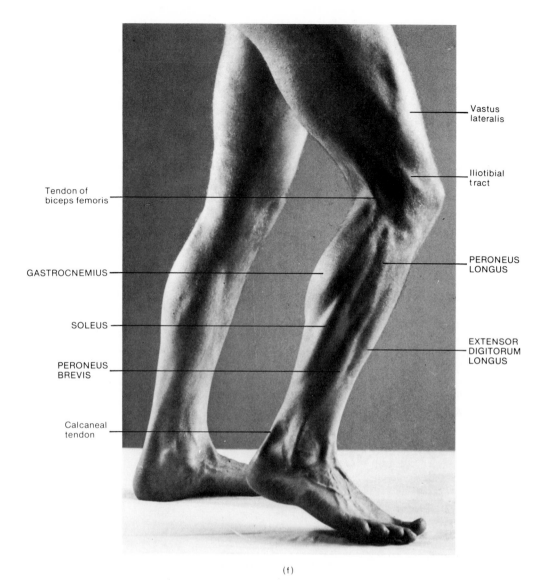

Vastus
lateralis

Iliotibial
tract

Tendon of
biceps femoris

PERONEUS
LONGUS

GASTROCNEMIUS

SOLEUS

EXTENSOR
DIGITORUM
LONGUS

PERONEUS
BREVIS

Calcaneal
tendon

(f)

FIGURE 10-17 *(Continued)* Muscles that move the foot and toes. (f) Surface anatomy photograph of the lateral leg. (Courtesy of R. D. Lockhart, *Living Anatomy,* Faber and Faber, 1974.)

Intramuscular Injections

Methods of administering drugs with a needle are called **parenteral.** Among the parenteral routes of administration are:

1. Intradermal or **intracutaneous.** The needle tip penetrates the epidermis and is inserted into the dermis.

2. Subcutaneous or **hypodermic.** The needle tip is inserted into the subcutaneous layer.

3. Intramuscular. The needle tip is inserted into muscle.

4. Intravenous. The needle tip is inserted directly into a vein.

5. Intraspinal. The needle tip is inserted into the vertebral canal.

At this point, we will discuss intramuscular injections only. Other parenteral routes are considered later.

An intramuscular injection penetrates the skin and subcutaneous tissue to enter the muscle itself. Intra-

muscular injections are preferred when prompt absorption is desired, when larger doses than can be given cutaneously are indicated, or when the drug is too irritating to give subcutaneously. The common sites for intramuscular injections include the buttock, lateral side of the thigh, and the deltoid region of the arm. Muscles in these areas, especially the gluteal muscles in the buttock, are fairly thick. Because of the large number of muscle fibers and extensive fascia, the drug has a large surface area for absorption. Absorption is further promoted by the extensive blood supply to muscles. Ideally, intramuscular injections should be given deep within the muscle and away from major nerves and blood vessels.

For many intramuscular injections, the preferred site is the gluteus medius muscle of the buttock (Figure 10-18a). The buttock should be divided into four quadrants and the upper outer quadrant used as the injection site. The iliac crest serves as a landmark for this quadrant. The spot for injection should be about 5 to 7.5 cm (2 to 3 inches) below the iliac crest. The upper outer quadrant is chosen because, in this area, the

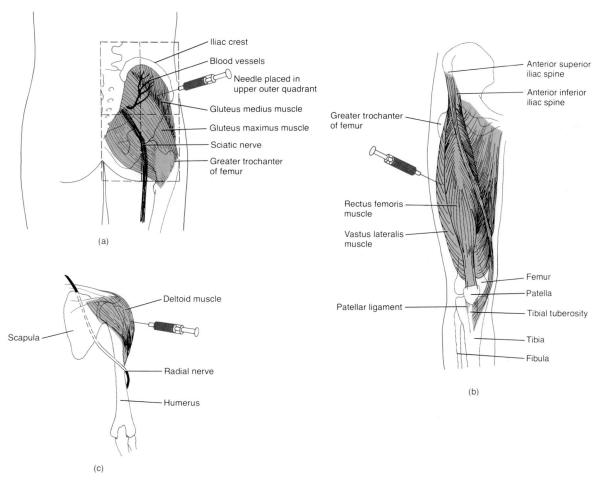

FIGURE 10-18 Intramuscular injections. Shown are the three common sites for intramuscular injections. (a) Buttock. (b) Lateral surface of the thigh. (c) Deltoid region of the arm.

muscle is quite thick with few nerves. Thus there is less chance of injuring the sciatic nerve. Injury to the nerve can cause paralysis of the lower extremity. The probability of injecting the drug into a blood vessel is also remote in this area. After the needle is inserted into the muscle, the plunger should be pulled up for a few seconds. If the syringe fills with blood, the needle is in a blood vessel and a different injection site on the opposite buttock should be chosen.

Injections given in the lateral side of the thigh are inserted into the midportion of the vastus lateralis muscle (Figure 10-18b). This site is determined by using the knee and greater trochanter of the femur as landmarks. The midportion of the muscle is located by measuring a handbreadth above the knee and a handbreadth below the greater trochanter.

The deltoid injection is given in the midportion of the muscle about two to three fingerbreadths below the acromion of the scapula and lateral to the axilla (Figure 10-18c).

STUDY OUTLINE

How Skeletal Muscles Produce Movement
1. Skeletal muscles produce movement by pulling on bones.
2. The stationary attachment is the origin. The movable attachment is the insertion.
3. Bones serve as levers and joints as fulcrums.
4. Levers are acted on by two different forces: resistance and effort. There are first-class, second-class, and third-class levers.
5. Fascicular arrangements include parallel, convergent, pennate, and circular. Fascicular arrangement is correlated with the power of a muscle and the range of movement.
6. The agonist or prime mover produces the desired action. The antagonist produces an opposite action. The synergist assists the agonist by reducing unnecessary movement.

Naming Skeletal Muscles
1. Skeletal muscles are named on the basis of distinctive criteria: direction of fibers, location, size, number of heads, shape, origin-insertion, and action.

Principal Skeletal Muscles
1. Surface anatomy is the study of the form and markings of the body surface.
2. A knowledge of surface anatomy will help you identify certain superficial structures by visual inspection or palpation through the skin.

Intramuscular Injections
1. Parenteral routes may be intradermal, subcutaneous, intramuscular, intravenous, or intraspinal.
2. Advantages of intramuscular injections are prompt absorption, use of larger doses than can be given cutaneously, and minimal irritation.
3. Common sites for intramuscular injections are the buttock, lateral side of the thigh, and deltoid region of the shoulder.

REVIEW QUESTIONS

1. What is meant by the muscular system? Explain fully.
2. Using the terms origin, insertion, and belly in your discussion, describe how skeletal muscles produce body movements by pulling on bones.
3. What is a lever? Fulcrum? Apply these terms to the body, and indicate the nature of the forces that act on levers.
4. Describe the three classes of levers, and provide one example for each in the body. Describe the various arrangements of fasciculi. How is fascicular arrangement correlated with the strength of a muscle and its range of movement?
5. Define the role of the agonist, antagonist, and synergist in producing body movements.
6. At the beginning of this chapter, several criteria for naming muscles were discussed. These were direction of fibers, location, size, number of heads, shape, origin, insertion, and action. Select at random the names of some muscles presented in Exhibits 10-2 through 10-16 and see if you can determine the criterion or criteria employed for each. In addition, refer to the prefixes, suffixes, roots, and definitions in each exhibit as a guide.

Select as many muscles as you wish, as long as you feel you understand the concept involved.
7. Discuss the muscles and their actions involved in facial expression.
8. What muscles would you use to do the following: (a) frown; (b) pout; (c) show surprise; (d) show you upper teeth; (e) pucker your lips; (f) squint; (g) blow up a balloon; (h) smile?
9. What are the principal muscles that move the mandible? Give the function of each.
10. What would happen if you lost tone in the masseter and temporalis muscles?
11. What muscles move the eyeball? In which direction does each muscle move the eyeball?
12. Describe the action of each of the muscles acting on the tongue.
13. What tongue, facial, and mandibular muscles would you use when chewing a piece of gum?
14. What muscles are responsible for moving the head, and how do they move the head?
15. Which of the muscles listed above would you use to signify "yes" and "no" by moving your head?

16. What muscles accomplish compression of the anterior abdominal wall?
17. What are the principal muscles involved in breathing? What are their actions?
18. In what directions is the shoulder girdle drawn? What muscles accomplish these movements?
19. What muscles are used to (a) raise your shoulders, (b) lower your shoulders, (c) join your hands behind your back, (d) join your hands in front of your chest?
20. What movements are possible at the shoulder joint? What muscles accomplish these movements?
21. What muscles move the arm? In which directions do these movements occur?
22. What muscles move the forearm and what actions are used when striking a match?
23. Discuss the various movements possible at the wrist and fingers. What muscles accomplish these movements?
24. How many muscles and actions of the wrist and fingers used when writing can you list?
25. Discuss the various muscles and movements of the vertebral column.
26. Can you perform an exercise that would involve the use of each of the muscles listed in Exhibit 10-13?
27. What muscles accomplish movements of the femur? What actions are produced by these muscles?
28. Review in your mind the various movements involved in your favorite kind of dancing. What muscles listed in Exhibit 10-14 would you be using and what actions would you be performing?
29. What muscles act at the knee joint? What kinds of movements do these muscles perform?
30. Determine the muscles and their actions listed in Exhibit 10-15 that you would use in climbing a ladder to a diving board, diving into the water, swimming the length of a pool, and then sitting at pool side.
31. Discuss the muscles that plantar flex, evert, pronate, dorsiflex, and supinate the foot.
32. In which directions are the toes moved? What muscles bring about these movements?
33. Define parenteral. What are the various routes of parenteral administration?
34. What are the advantages of intramuscular injections?
35. Describe how you would locate the sites for an intramuscular injection in the buttock, lateral side of the thigh, and deltoid region of the arm.

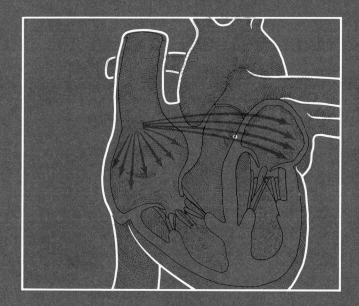

11

The Cardiovascular and Lymphatic Systems

STUDENT OBJECTIVES

- Contrast the general roles of blood, lymph, and interstitial fluid.
- Define the principal physical characteristics of blood and its functions in the body.
- Identify the plasma and formed element constituents of blood.
- Compare the origins of the formed elements in blood.
- Describe the structure of erythrocytes and their function in the transport of oxygen and carbon dioxide.
- Define erythropoiesis and describe erythrocyte production and destruction.
- Describe the importance of a reticulocyte count in the diagnosis of abnormal rates of erythrocyte production.
- List the structural features and types of leucocytes.
- Explain the significance of a differential count.
- Discuss the role of leucocytes in phagocytosis and antibody production.
- Discuss the structure of thrombocytes.
- List the components of plasma and explain their importance.
- Describe the location of the heart in the mediastinum.
- Distinguish between the structure and location of fibrous and serous pericardium.
- Contrast the structure of the epicardium, myocardium, and endocardium.
- Identify the blood vessels, chambers, and valves of the heart.
- Describe the initiation and conduction of nerve impulses through the electrical conduction system of the heart.
- Label and explain the deflection waves of a normal electrocardiogram.
- Describe the autonomic control of the heart.

- Describe the sounds of the heart and their clinical significance.
- Discuss the surface anatomy features of the heart.
- Contrast the structure and function of arteries, capillaries, and veins.
- Identify the principal arteries and veins of systemic circulation.
- Describe the route of blood in coronary circulation.
- Describe the importance and route of blood involved in hepatic portal circulation.
- Identify the major blood vessels of pulmonary circulation.
- Contrast fetal and adult circulation.
- Explain the fate of fetal circulation structures once postnatal circulation is established.
- Identify the components and functions of the lymphatic system.
- Compare the structure of veins and lymphatic vessels.
- Describe the structure and function of lymph nodes.
- Identify the principal groups of lymph nodes.
- Contrast the functions of the tonsils, spleen, and thymus gland as lymphatic organs.
- List the forces responsible for the circulation of lymph.
- Identify several common blood, cardiovascular, and lymphatic disorders.
- Define key medical terms associated with the cardiovascular and lymphatic systems.

The more specialized a cell becomes, the less capable it is of carrying on an independent existence. For instance, a specialized cell is less capable of protecting itself from extreme temperatures, toxic chemicals, and changes in pH. Often it cannot go looking for food or devour whole bits of food. And, if it is firmly implanted in a tissue, it cannot move away from its own wastes. The substance that bathes the cell and carries out these vital functions for it is called interstitial fluid (also known as intercellular or tissue fluid).

The interstitial fluid, in turn, must be serviced by blood and lymph. The blood picks up oxygen from the lungs, nutrients from the digestive tract, hormones from the endocrine glands, and enzymes from still other parts of the body. The blood then transports these substances to all the tissues where they diffuse from the capillaries into the interstitial fluid. In the interstitial fluid, the substances are passed on to the cells and exchanged for wastes.

Since the blood must service all the tissues of the body, it can be an important medium for the transport of disease-causing organisms. To protect itself from the spread of disease, the body has a lymphatic system—a collection of vessels containing a fluid called lymph. The lymph picks up materials, including wastes, from the interstitial fluid, cleanses them of bacteria, and returns them to the blood. The blood then carries the wastes to the lungs, kidneys, and sweat glands, where they are eliminated from the body. The blood also takes wastes to the liver, where they are detoxified and recycled.

The blood, heart, and blood vessels constitute the **cardiovascular system.** The lymph, lymph vessels, and lymph glands make up the **lymphatic system.** Let us first take a look at the substance known as blood.

Blood

The red body fluid that flows through all the vessels except the lymph vessels is called **blood.** Blood is a viscous fluid—it is thicker and more adhesive than water. Water is considered to have a viscosity of 1. The viscosity of blood, by comparison, ranges from 4.5 to 5.5. This means that it flows 4½ to 5½ times more slowly than water. The adhesive quality of blood, or its stickiness, may be felt by touching it. Blood is also slightly heavier than water. Other physical characteristics of blood include a temperature of about 38°C (100.4°F), a pH range of 7.35 to 7.45 (slightly alkaline), and a 0.85 to 0.90 percent concentration of salt (NaCl). Blood constitutes about 8 percent of the total body weight. The blood volume of an average-sized man is between 5 and 6 liters (5 to 6 qt). An average-sized woman has 4 to 5 liters.

Despite its simple appearance, blood is a complex liquid that performs a number of critical functions:

1. It transports oxygen from the lungs to all cells of the body.

2. It transports carbon dioxide from the cells to the lungs.

3. It transports nutrients from the digestive organs to the cells.

4. It transports waste products from the cells to the kidneys, lungs, and sweat glands.

5. It transports hormones from endocrine glands to the cells.

6. It transports enzymes to various cells.

7. It regulates body pH through buffers and amino acids.

8. It regulates normal body temperature because it contains a large volume of water (an excellent heat absorber and coolant).

9. It regulates the water content of cells, principally through dissolved sodium ions.

10. It prevents body fluid loss through the clotting mechanism.

11. It protects against toxins and foreign microbes through special combat-unit cells.

Microscopically, blood is composed of two portions: plasma, which is a liquid containing dissolved substances, and formed elements, which are cells and cell-like bodies suspended in the plasma.

FORMED ELEMENTS

In clinical practice, the most common classification of the **formed elements** of the blood is the following:

Erythrocytes (red blood cells)
Leucocytes (white blood cells)
   Granular leucocytes (granulocytes)
      Neutrophils
      Eosinophils
      Basophils
   Agranular leucocytes (agranulocytes)
      Lymphocytes
      Monocytes
Thrombocytes (platelets)

Origin

The process by which blood cells are formed is called **hemopoiesis** or **hematopoiesis.** During embryonic and fetal life, there are no clear-cut centers for blood cell production. The yolk sac, liver, spleen, thymus gland, lymph nodes, and bone marrow all participate at various times in producing the formed elements. In the adult, however, we can pinpoint the production process. Red bone marrow (myeloid tissue) is responsible for producing red blood cells, granular leucocytes, and

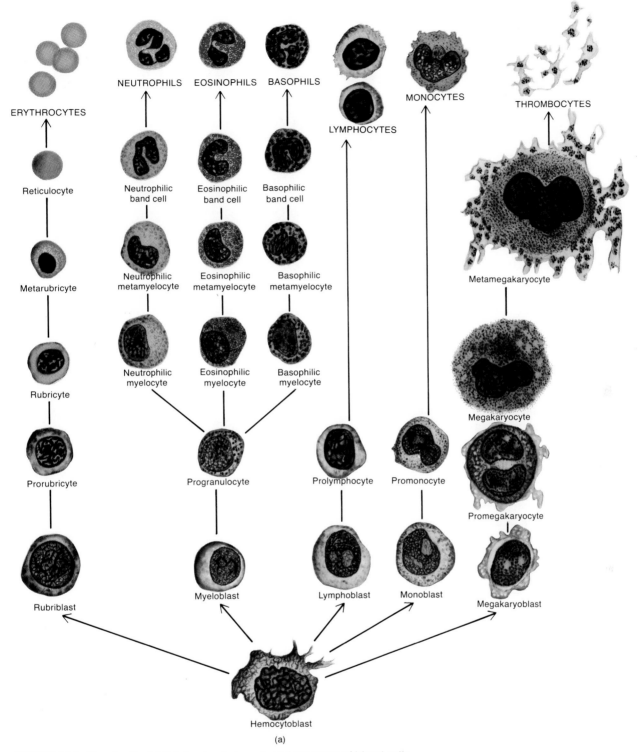

ERYTHROCYTES

Reticulocyte

Metarubricyte

Rubricyte

Prorubricyte

Rubriblast

NEUTROPHILS

Neutrophilic band cell

Neutrophilic metamyelocyte

Neutrophilic myelocyte

EOSINOPHILS

Eosinophilic band cell

Eosinophilic metamyelocyte

Eosinophilic myelocyte

BASOPHILS

Basophilic band cell

Basophilic metamyelocyte

Basophilic myelocyte

Progranulocyte

Myeloblast

LYMPHOCYTES

Prolymphocyte

Lymphoblast

MONOCYTES

Promonocyte

Monoblast

THROMBOCYTES

Metamegakaryocyte

Megakaryocyte

Promegakaryocyte

Megakaryoblast

Hemocytoblast

(a)

FIGURE 11-1 Blood cells. (a) Origin, development, and structure of blood cells.

platelets. Lymphoid tissue—spleen, tonsils, lymph nodes—shares the responsibility with myeloid tissue for producing agranular leucocytes. Undifferentiated mesenchymal cells in red bone marrow are transformed into **hemocytoblasts,** immature cells that are eventually capable of developing into mature blood cells (Figure 11-1). For example, the hemocytoblasts develop into:

1. Rubriblasts *(proerythroblasts),* that go on to form mature red blood cells

2. Myeloblasts, that go on to form mature neutrophils, eosinophils, and basophils

3. Megakaryoblasts, that go on to form mature platelets

4. Lymphoblasts, that eventually form lymphocytes

5. Monoblasts, that eventually form monocytes

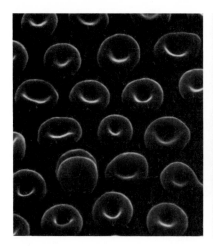
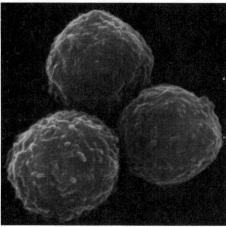
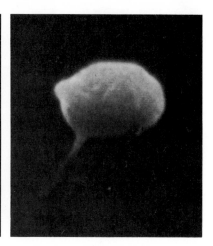

(b)

FIGURE 11-1 *(Continued)* Blood cells. (b) Scanning electron micrographs of erythrocytes (left) at a magnification of 5000×, leucocytes (center) at a magnification of 10,000×, and a platelet (right) at a magnification of 10,000×. (Courtesy of Fisher Scientific Company and S.T.E.M. Laboratories, Inc., Copyright 1975.)

Erythrocytes

Microscopically, **red blood cells,** or **erythrocytes,** appear as biconcave discs averaging about 7.7 μm in diameter (Figure 11-1). Mature red blood cells are quite simple in structure. They lack a nucleus and can neither reproduce nor carry on extensive metabolic activities. The cell contains a network of protein called the stroma, some cytoplasm, lipid substances, including cholesterol, and a red pigment called hemoglobin. *Hemoglobin,* which constitutes about 33 percent of the cell volume, is responsible for the red color of blood.

The erythrocytes combine with oxygen and carbon dioxide and transport them through the blood vessels. Red blood cells are highly specialized for this purpose. The hemoglobin molecule consists of a protein called globin and a pigment called heme, which contains iron. As the erythrocyte passes through the lungs, each of the four iron atoms in the hemoglobin molecule combines with a molecule of oxygen. The oxygen is transported in this state to other tissues of the body. In the tissues, the iron-oxygen reaction reverses, and the oxygen is released to diffuse into the interstitial fluid. On the return trip, the globin portion combines with a molecule of carbon dioxide from the interstitial fluid. This complex is transported to the lungs where the carbon dioxide is released and then exhaled. Red blood cells contain a great many hemoglobin molecules in order to increase their oxygen carrying capacity—280 million molecules of hemoglobin per erythrocyte, according to one estimate. Hemoglobin is contained in the stroma of a red blood cell because hemoglobin molecules are small and, if they were free in plasma, they would leak through the endothelial membranes of blood vessels and be lost in the urine. The biconcave shape of a red blood cell has a much greater surface area than, say, a sphere or cube. The erythrocyte thus presents the maximum surface area for the diffusion of gas molecules that must pass through the membrane to combine with hemoglobin.

The cell membrane of a red blood cell becomes fragile and the cell is nonfunctional in about 120 days. The main reason for its short life is its inability to replace enzymes, particularly carbonic anhydrase. This enzyme is involved in the transportation of carbon dioxide by red blood cells. A healthy male has about 5.4 million red blood cells per cubic millimeter of blood and a healthy female has about 4.8 million. The higher number in males is due to their higher rate of metabolism and the monthly blood loss during menstruation in females. To maintain normal quantities of erythrocytes, the body must produce new mature cells at the astonishing rate of 2 million per second. In the adult, production takes place in the red bone marrow in the spongy bone of the cranium, ribs, sternum, bodies of vertebrae, and proximal epiphyses of the humerus and femur. The process by which erythrocytes are formed is called **erythropoiesis.**

Erythropoiesis starts with the transformation of a hemocytoblast into a rubriblast. The *rubriblast* (proerythroblast) gives rise to a *prorubricyte* (early erythroblast), which then develops into a *rubricyte* (intermediate erythroblast), the first cell in the sequence that begins to synthesize hemoglobin. The rubricyte next develops into a *metarubricyte* (late erythroblast). In the metarubricyte, hemoglobin synthesis is at a maximum and the nucleus is lost by extrusion. In the next stage, the metarubricyte develops into a *reticulocyte,* which in turn becomes an *erythrocyte,* or mature red blood cell. Once the erythrocyte is formed, it is released from the marrow and circulates through the blood vessels.

When mature red blood cells die, their disintegrating bodies pose the danger of clogging small blood vessels. To prevent this, certain cells called *reticuloendothelial cells* clear away their bodies after they die. These cells are formed from primitive reticular cells. Reticuloendothelial cells enter the spleen, tonsils, lymph nodes, liver, lungs, and other organs and become highly specialized for phagocytosis. The reticuloendothelial cells in the lymph nodes are particularly active in destroying microbes and their toxins. The reticuloendothelial cells in the liver and spleen concentrate on ingesting dead blood cells. The hemoglobin molecules of red blood cells are split apart—the iron is reused and the rest of the molecule is converted into other substances for reuse or elimination.

Normally erythropoiesis and red cell destruction proceed at the same pace. But if the body suddenly needs more erythrocytes, or if erythropoiesis is not keeping up with red blood cell destruction, a homeostatic mechanism steps up erythrocyte production. The stimulus for this mechanism is oxygen deficiency in the kidney cells and other tissues, which is not surprising since one of the chief functions of the erythrocytes is to deliver oxygen. As soon as the kidney cells become oxygen-deficient, they release a hormone called **erythropoietin.** This hormone circulates through the blood to the red bone marrow, where it stimulates hemocytoblasts to develop into red blood cells.

A diagnostic test that informs the physician about the rate of erythropoiesis is the **reticulocyte count.** Some reticulocytes are normally released into the bloodstream before they become mature red blood cells. If the number of reticulocytes in a sample of blood is less than 0.5 percent of the number of mature red blood cells in the sample, erythropoiesis is occurring too slowly. A low reticulocyte count might confirm a diagnosis of nutritional or pernicious anemia. In this condition, a person is unable to absorb vitamin B_{12} because the stomach does not secrete a substance called the intrinsic factor. Lack of vitamin B_{12} inhibits the rate of red blood cell production. Or it might indicate a kidney disease that prevents the kidney cells from producing erythropoietin. If the reticulocytes number more than 1.5 percent of the mature red blood cells, erythropoiesis is abnormally rapid. Any number of problems may be responsible for a high reticulocyte count: anemia, oxygen deficiency, and uncontrolled red blood cell production caused by a cancer in the bone marrow. If the individual has been suffering from a nutritional or pernicious anemia, the high count may indicate that treatment has been effective and the bone marrow is making up for lost time.

Leucocytes

Unlike red blood cells, **leucocytes,** or **white blood cells,** have nuclei and do not contain hemoglobin (Figure 11-1). They are far less numerous, averaging from 5000 to 9000 cells per cubic millimeter of blood. Red blood cells, therefore, outnumber white blood cells about 700 to 1. Leucocytes fall into two major groups. The first group contains the *granular leucocytes.* These develop from red bone marrow. They have granules in the cytoplasm and possess lobed nuclei. Three kinds of granular leucocytes exist: *neutrophils (polymorphs), eosinophils,* and *basophils.* The second principal group of leucocytes is called the *agranular leucocytes.* They develop from lymphoid and myeloid tissue. No cytoplasmic granules can be seen under a light microscope. Their nuclei are usually spherical. The two kinds of agranular leucocytes are *lymphocytes* and *monocytes.*

The general function of the leucocytes is to combat inflammation and infection. Some leucocytes are actively **phagocytotic**—they can ingest bacteria and dispose of dead matter. Most leucocytes also possess, to some degree, the ability to crawl through the minute spaces between the cells that form the walls of capillaries, the smallest blood vessels, and through connective and epithelial tissue. This movement, like that of amoebas, is called **diapedesis.** First, part of the cell membrane stretches out like an arm. Then the cytoplasm and nucleus flow into the projection. Finally, the rest of the membrane snaps up into place. Another projection is made, and so on, until the cell has crawled to its destination.

Diagnosis of an injury or infection may involve a **differential count**—calculation of the number of each kind of white cell in 100 white blood cells. A normal differential count might appear as follows:

Neutrophils	60–70%
Eosinophils	2–4%
Basophils	0.5–1%
Lymphocytes	20–25%
Monocytes	3–8%
	100%

Particular attention is paid to the neutrophils in a differential count. The neutrophils are the most active white cells in response to tissue destruction by bacteria. Their major role is phagocytosis. They also release the enzyme lysozyme, which destroys certain bacteria. More often than not, a high neutrophil count indicates damage by invading bacteria. An increase in the number of monocytes generally indicates a chronic (of long duration) infection such as tuberculosis. Apparently monocytes take longer to reach the site of infection than do neutrophils, but once they arrive they do so in larger numbers and destroy more microbes. Monocytes, like neutrophils, are phagocytic. They clean up cellular debris following an infection. High eosinophil counts indicate allergic conditions, since eosinophils are believed to combat the allergens that cause allergies. Eosinophils leave the capillaries, enter the tissue fluid, and produce antihistamines that destroy antigen-antibody complexes. Basophils are also believed to be involved in allergic reactions. Basophils leave the capillaries, enter the tissues, and become the mast cells of the tissues, liberating heparin, histamine, and serotonin.

The term **leucocytosis** refers to an increase in the number of white blood cells. If the increase exceeds 10,000, a pathological condition is usually indicated. An abnormally low level of white blood cells (below 5000) is termed **leucopenia.**

Some leucocytes, called lymphocytes, are involved in the production of antibodies. **Antibodies** are special proteins that inactivate antigens. An **antigen** is a substance that will stimulate the production of antibodies that are capable of reacting specifically with the antigen. Most antigens are proteins, and most are not synthesized by the body. Many of the proteins that make up the cell structures and enzymes of bacteria are antigens. The toxins released by bacteria are also antigens. When antigens enter the body, they react chemically with substances in the lymphocytes and stimulate some lymphocytes, called B cells, to become *plasma cells* (Figure 11-2). The plasma cells then pro-

duce antibodies: globulin-type proteins that attach to antigens much as enzymes attach to substrates. Like enzymes, a specific antibody will generally attach only to a certain antigen. However, unlike enzymes, which enhance the reactivity of the substrate, antibodies "cover" their antigens so the antigens cannot come in contact with other chemicals in the body. In this way, bacterial poisons can be sealed up and rendered harmless. The bacteria themselves are destroyed by the antibodies. This process is called the **antigen-antibody response.** Eosinophils in tissues phagocytose the antigen-antibody complexes. The antigen-antibody response helps us to combat infection. It gives us immunity to some diseases. And it is responsible for blood types, allergies, and the body's rejection of organs transplanted from an individual with a different genetic makeup.

Foreign bacteria exist everywhere in the environment and have continuous access to the body through the mouth, nose, and pores of the skin. Furthermore, many cells, especially those of epithelial tissue, age and die, and their remains must be disposed of daily. Even when the body is healthy, the leucocytes actively ingest bacteria and debris. However, a leucocyte can phagocytose only a certain number of substances before they interfere with the leucocyte's

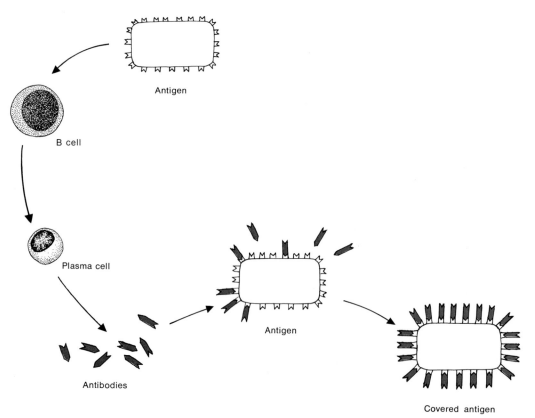

FIGURE 11-2 Antigen-antibody response. An antigen entering the body stimulates a B cell to develop into an antibody-producing plasma cell. The antibodies attach to the antigen, cover it, and render it harmless.

normal metabolic activities and bring on its death. Consequently, the life span of most leucocytes is very short. In a healthy body, some white blood cells will live a couple of days. During a period of infection they may live only a few hours.

Thrombocytes

If a hemocytoblast does not become an erythrocyte or granular leucocyte, it may develop into still another kind of cell called a megakaryoblast, that is transformed into a megakaryocyte. Megakaryocytes are large cells whose cytoplasm breaks up into fragments. Each fragment becomes enclosed by a piece of the cell membrane and is called a **thrombocyte** or **platelet** (Figure 11-1). Platelets are disc-shaped cells without a nucleus. They average from 2 to 4 μm in diameter. Between 250,000 and 500,000 platelets appear in each cubic millimeter of blood.

The platelets prevent fluid loss by initiating a chain of reactions that results in blood clotting. Like the other formed elements of the blood, platelets have a short life, probably only one week, because they are used up in clotting and are just too simple to carry on much metabolic activity.

PLASMA

When the formed elements are removed from blood, a straw-colored liquid called **plasma** is left. Exhibit 11-1 outlines the chemical composition of plasma. Note that 7 to 9 percent of the solutes are proteins. Some of these proteins are found elsewhere in the body, but in blood they are called *plasma proteins*. Albumins, which constitute the majority of plasma proteins, are responsible for blood's viscosity. The concentration of albumin is about four times higher in plasma than in interstitial fluid. Along with the electrolytes, albumins also help to regulate blood volume by preventing all the water in the blood from diffusing into the interstitial fluid. Recall that water moves by osmosis from an area of low solute (high water) concentration to an area of high solute (low water) concentration. Globulins, which are antibody proteins released by plasma cells, form a small component of the plasma proteins. Gamma globulin is especially well known because it is able to form an antigen-antibody complex with the proteins of the hepatitis and measles viruses and the tetanus bacterium. Fibrinogen, a third plasma protein, takes part in the blood-clotting mechanism along with the platelets. *Serum* is the term applied to plasma minus its fibrinogen.

EXHIBIT 11-1
CHEMICAL COMPOSITION AND DESCRIPTION OF SUBSTANCES IN PLASMA

CONSTITUENT	DESCRIPTION	CONSTITUENT	DESCRIPTION
WATER	Constitutes about 92 percent of plasma and is liquid portion of blood. Ninety percent of water is derived from absorption from digestive tract; 10 percent comes from cellular respiration. Water acts as solvent and suspending medium for solid components of blood and absorbs heat.	2. Nonprotein nitrogen (NPN) substances	Contain nitrogen but are not proteins. These substances include urea, uric acid, creatine, creatinine, ammonium salts. Represent breakdown products of protein metabolism and are carried by blood to organs of excretion.
SOLUTES		3. Food substances	Once foods are broken down in digestive tract, products of digestion are passed into blood for distribution to all body cells. These products include amino acids (from proteins), glucose (from carbohydrates) and fats (from lipids).
I. Proteins	Constitute 7 to 9 percent of solutes in plasma.		
Albumins	Constitute 55 to 64 percent of plasma proteins and are smallest plasma proteins. Produced by liver and provide blood with viscosity, a factor related to maintenance and regulation of blood pressure. Also exert considerable osmotic pressure to maintain water balance between blood and tissues and regulate blood volume.	4. Regulatory substances	Enzymes are produced by body cells and catalyze chemical reactions. Hormones, produced by endocrine glands, regulate growth and development in body.
		5. Respiratory gases	Oxygen and carbon dioxide are carried by blood. These gases are more closely associated with hemoglobin of red blood cells than plasma itself.
Globulins	Constitute about 15 percent of plasma proteins. Protein group to which antibodies produced by leucocytes belong. Gamma globulins attack measles and hepatitis viruses, tetanus bacteria, and possibly poliomyelitis virus.	6. Electrolytes	A number of ions constitute inorganic salts of plasma. Cations include Na^+, K^+, Ca^{2+}, Mg^{2+}. Anions include Cl^-, PO_4^{3-}, SO_4^{2-}, HCO_3^-. Salts help maintain osmotic pressure, normal pH, physiological balance between tissues and blood.
Fibrinogen	Represents small fraction of plasma proteins (4 percent). Produced by liver and plays essential role in clotting.		

INTERSTITIAL FLUID AND LYMPH

For all practical purposes interstitial fluid and lymph are the same. The major difference between the two is location. When the fluid bathes the cells, it is called **interstitial fluid, intercellular fluid,** or **tissue fluid.** When it flows through the lymphatic vessels, it is called **lymph** (Figure 11-3). Both fluids are similar in composition to plasma. The principal chemical difference is that they contain less protein than plasma because the larger protein molecules are not easily filtered through the cells that form the walls of the capillaries. Keep in mind that whole blood does not flow into the tissue spaces; it remains in closed vessels. Certain constituents of the plasma do move, however, and once they move out of the blood, they are called interstitial fluid. The transfer of materials between blood and interstitial fluid occurs by osmosis, diffusion, and filtration across the cells that make up the capillary walls. Both interstitial fluid and lymph contain variable numbers of leucocytes. Leucocytes can enter the tissue fluid by diapedesis, and the lymphoid tissue itself is the site of nongranular leucocyte production. However, interstitial fluid and lymph both lack erythrocytes and platelets.

Other substances, especially organic molecules, in interstitial fluid and lymph vary in relation to the location of the sample analyzed. The lymph vessels that drain the organs of the digestive tract, for example, contain a great deal of lipid absorbed from food.

Heart

The center of the body's cardiovascular system is the heart. The **heart** is a hollow, muscular organ that pumps blood through the blood vessels. It is situated obliquely between the lungs in the mediastinum, and about two-thirds of its mass lies to the left of the body's midline (Figure 11-4a). A roentgenogram of a normal heart is shown in Figure 11-4b. The heart is shaped like a blunt cone about the size of a closed fist—12 cm (5 inches) long, 9 cm (3.5 inches) wide at its broadest point, and 6 cm (2.5 inches) thick. Its pointed end, the *apex,* projects downward, foreward, and to the left, and lies superior to the central depression of the diaphragm. Its broad end, or *base,* projects upward, backward, and to the right and lies just inferior to the second rib. The major parts of the heart to be considered here are the parietal pericardium, the wall and chambers, and the valves.

PARIETAL PERICARDIUM (PERICARDIAL SAC)

The heart is enclosed in a loose-fitting serous membrane called the **parietal pericardium** or **pericardial sac** (Figure 11-4a, d). It consists of two layers: the fibrous layer and the serous layer. The *fibrous layer* or *fibrous pericardium* is the outer layer and consists of a tough, fibrous connective tissue. The fibrous pericardium is attached to the large blood vessels entering and leaving the heart, to the diaphragm, and to the inside of the sternal wall of the thorax. It also adheres to the parie-

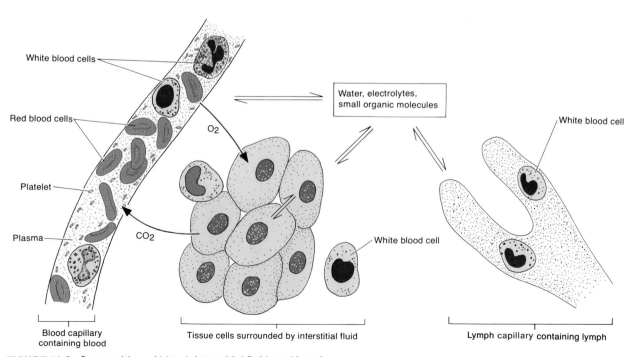

FIGURE 11-3 Composition of blood, interstitial fluid, and lymph.

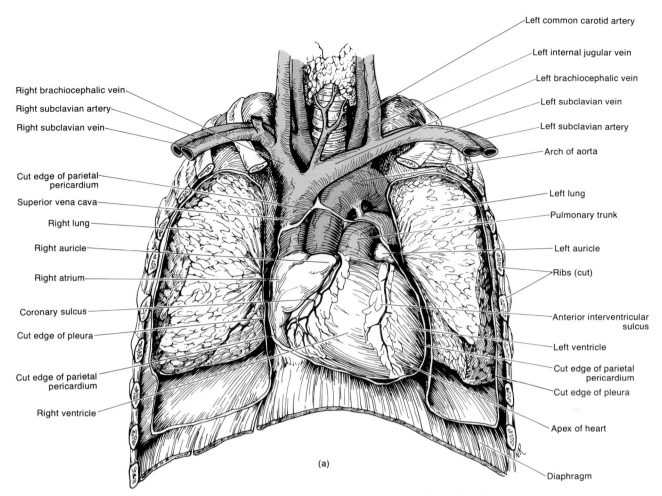

FIGURE 11-4 Heart. (a) Position of the heart and associated blood vessels in the thoracic cavity.

Labels, clockwise from top right:
Left common carotid artery
Left internal jugular vein
Left brachiocephalic vein
Left subclavian vein
Left subclavian artery
Arch of aorta
Left lung
Pulmonary trunk
Left auricle
Ribs (cut)
Anterior interventricular sulcus
Left ventricle
Cut edge of parietal pericardium
Cut edge of pleura
Apex of heart
Diaphragm
Right ventricle
Cut edge of parietal pericardium
Cut edge of pleura
Coronary sulcus
Right atrium
Right auricle
Right lung
Superior vena cava
Cut edge of parietal pericardium
Right subclavian vein
Right subclavian artery
Right brachiocephalic vein

(a)

tal pleurae. The fibrous pericardium prevents overdistension of the heart, provides a tough protective membrane around the heart, and anchors the heart in the mediastinum. The inner layer of the parietal pericardium is known as the *serous layer* or *serous pericardium*. This thin, more delicate membrane is continuous with the epicardium at the base of the heart and around the large blood vessels. The **epicardium** or **visceral pericardium** is the thin, transparent outer layer of the wall of the heart that is composed of serous tissue and mesothelium. Between the serous pericardium and the epicardium is a potential space called the *pericardial cavity*. The cavity contains a watery fluid, known as pericardial fluid, which prevents friction between the membranes as the heart moves.

WALL AND CHAMBERS

The wall of the heart (Figure 11-4d) is divided into three portions: the epicardium (external layer), the myocardium (middle layer), and the endocardium (inner layer). The **epicardium** is the same as the visceral pericardium. The **myocardium,** which is cardiac muscle tissue, comprises the bulk of the heart. Cardiac muscle fibers are involuntary, striated, and branched, and the tissue is arranged in interlacing bundles of fi-

bers. The myocardium is responsible for the contraction of the heart. The **endocardium** is a thin layer of endothelium overlying a thin layer of connective tissue pierced by tiny blood vessels and bundles of smooth muscle. It lines the inside of the myocardium and covers the valves of the heart and the tendons that hold them open. It is continuous with the endothelial lining of the large blood vessels of the heart. Inflammation of the endocardium is called *endocarditis*.

The interior of the heart is divided into four spaces or chambers, which receive the circulating blood (Figure 11-5). The two upper chambers are called the right and left **atria.** Each atrium has an appendage called an *auricle,* so named because of its resemblance to a dog's ear. The auricle increases the atrium's surface area. The lining of the atria is smooth, except for the anterior atrial walls and the lining of the auricles, which contain projecting muscle bundles that are parallel to each other and resemble the teeth of a comb: the *musculi pectinati*. These bundles give the lining of the auricles a ridged appearance. The atria are separated by a partition called the *interatrial septum*. On the posterior wall of the interatrial septum is an oval depression, the *fossa ovalis,* which corresponds to the site of the foramen ovale of the fetal heart. The two lower chambers, the right and left **ventricles,** are sep-

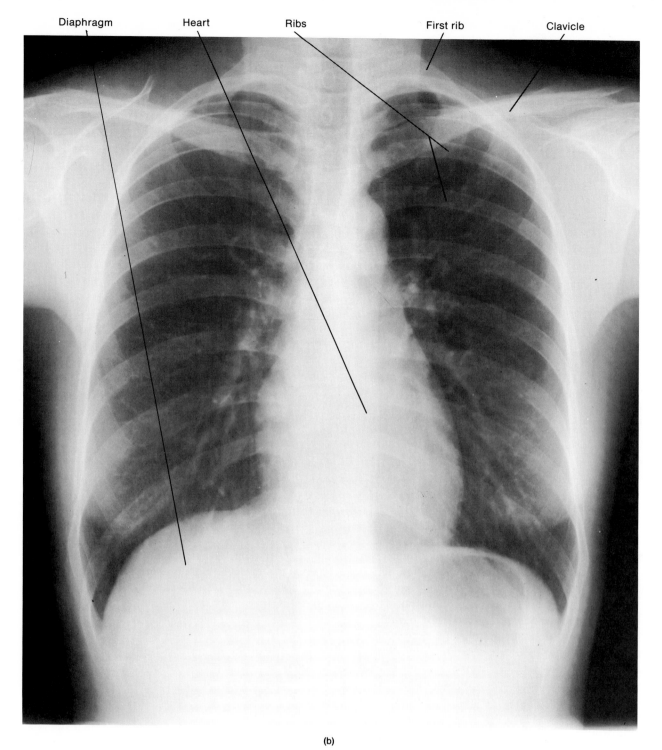

Diaphragm   Heart   Ribs   First rib   Clavicle

(b)

FIGURE 11-4 *(Continued)* Heart. (b) Anteroposterior projection of normal heart. (Courtesy of John C. Bennett, St. Mary's Hospital, San Francisco.)

arated by an *interventricular septum*. The muscle tissue of the atria and ventricles is separated by connective tissue that also forms the valves. This "cardiac skeleton" effectively divides the myocardium into two separate muscle masses. Externally, a groove known as the *coronary sulcus* separates the atria from the ventricles. It encircles the heart and houses the coronary sinus and circumflex branch of the left coronary artery. The *anterior interventricular sulcus* and *posterior interventricular sulcus* separate the right and left

ventricles externally. The sulci contain coronary blood vessels and a variable amount of fat (see Figure 11-4a).

The right atrium receives blood from all parts of the body except the lungs. It receives the blood through three veins. One of these veins is the *superior vena cava*, which brings blood from the upper portion of the body. Another is the *inferior vena cava*, which brings blood from the lower portions of the body. The third vein is the *coronary sinus*, which drains blood from most of the vessels supplying the walls of the heart.

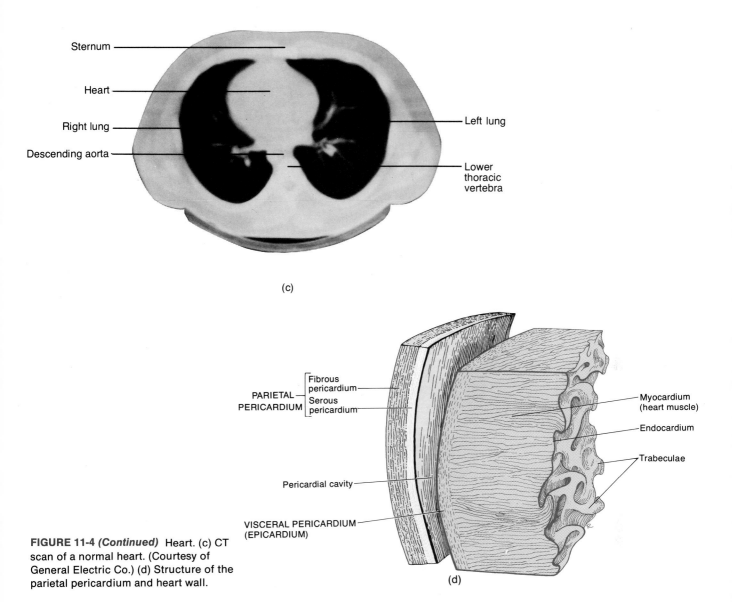

(c)

FIGURE 11-4 *(Continued)* Heart. (c) CT scan of a normal heart. (Courtesy of General Electric Co.) (d) Structure of the parietal pericardium and heart wall.

The right atrium then squeezes the blood into the right ventricle, which pumps it into the *pulmonary trunk*. The pulmonary trunk divides into a *right* and *left pulmonary artery,* each of which carries blood to the lungs. In the lungs, the blood releases its carbon dioxide and takes on oxygen. It returns to the heart via four *pulmonary veins* that empty into the left atrium. The blood is then squeezed into the left ventricle, which pumps the blood into the *ascending aorta.* From here aortic blood is passed into the *coronary arteries, arch of the aorta, descending thoracic aorta,* and *abdominal aorta.* These blood vessels transport the blood to all body parts except the lungs.

Note in Figure 11-5 that the right atrium is slightly larger than the left atrium. The thickness of the chamber walls varies too. The atria are thin-walled because they need only enough cardiac muscle tissue to squeeze the blood into the ventricles with the aid of a reduced pressure created by the expanding ventricles. The right ventricle has a much thicker layer of myocardium than the atria since it must send blood to the lungs and around back to the left atrium. The left ventricle has the thickest walls since it must pump blood at high pressure through literally thousands of miles of vessels in the head, trunk, and extremities.

VALVES

As each chamber of the heart contracts, it pushes a portion of blood into a ventricle or out of the heart through an artery. But as the walls of the chambers relax, some structure must prevent the blood from flowing back into the chamber. That structure is a **valve.**

Atrioventricular (AV) valves lie between the atria and their ventricles (Figure 11-5). The atrioventricular

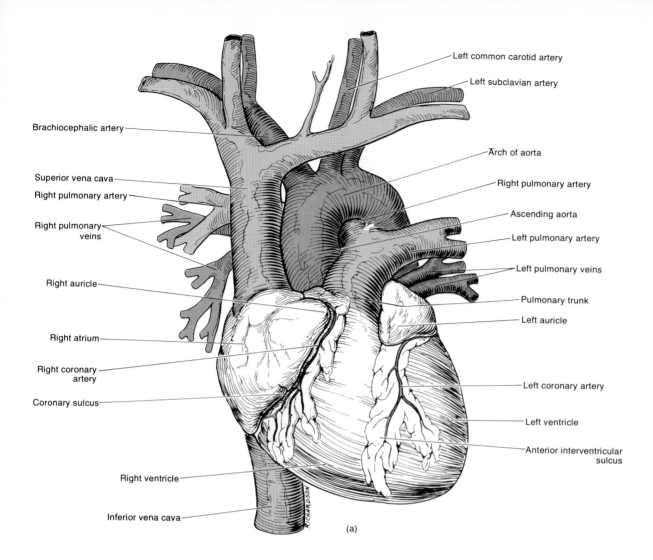

Left common carotid artery

Left subclavian artery

Brachiocephalic artery

Superior vena cava

Right pulmonary artery

Right pulmonary veins

Right auricle

Right atrium

Right coronary artery

Coronary sulcus

Right ventricle

Inferior vena cava

Arch of aorta

Right pulmonary artery

Ascending aorta

Left pulmonary artery

Left pulmonary veins

Pulmonary trunk

Left auricle

Left coronary artery

Left ventricle

Anterior interventricular sulcus

R. RICHARDSON

(a)

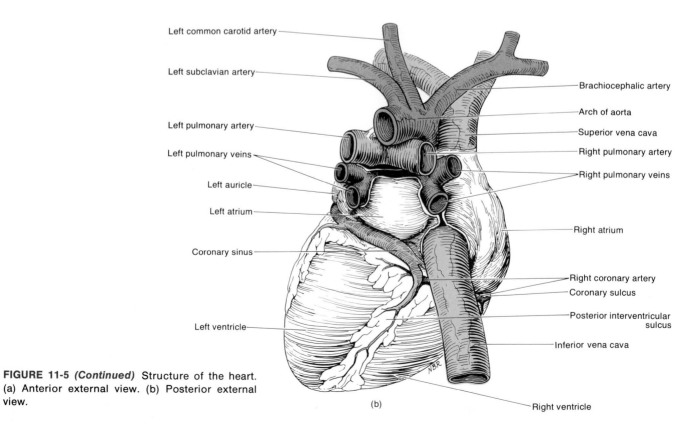

Left common carotid artery

Left subclavian artery

Left pulmonary artery

Left pulmonary veins

Left auricle

Left atrium

Coronary sinus

Left ventricle

Brachiocephalic artery

Arch of aorta

Superior vena cava

Right pulmonary artery

Right pulmonary veins

Right atrium

Right coronary artery

Coronary sulcus

Posterior interventricular sulcus

Inferior vena cava

Right ventricle

NBR

(b)

FIGURE 11-5 *(Continued)* Structure of the heart.
(a) Anterior external view. (b) Posterior external
view.

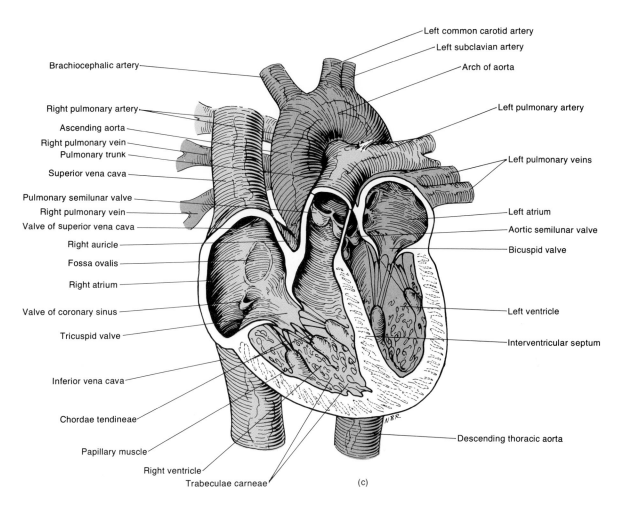

Left common carotid artery

Left subclavian artery

Arch of aorta

Brachiocephalic artery

Right pulmonary artery

Ascending aorta

Right pulmonary vein

Pulmonary trunk

Superior vena cava

Pulmonary semilunar valve

Right pulmonary vein

Valve of superior vena cava

Right auricle

Fossa ovalis

Right atrium

Valve of coronary sinus

Tricuspid valve

Inferior vena cava

Chordae tendineae

Papillary muscle

Right ventricle

Trabeculae carneae

Left pulmonary artery

Left pulmonary veins

Left atrium

Aortic semilunar valve

Bicuspid valve

Left ventricle

Interventricular septum

Descending thoracic aorta

(c)

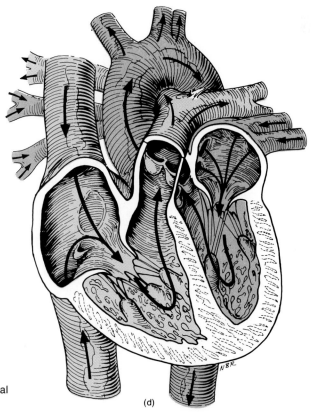

(d)

FIGURE 11-5 (Continued) Structure of the heart. (c) Anterior internal view. (d) Path of blood through the heart.

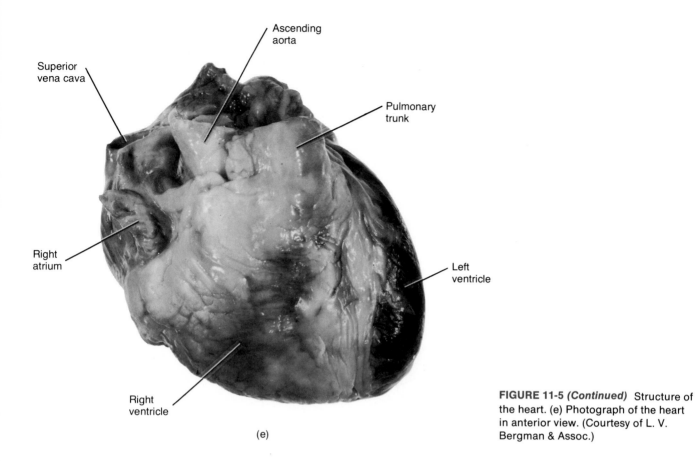

Ascending aorta

Superior vena cava

Pulmonary trunk

Right atrium

Left ventricle

Right ventricle

(e)

FIGURE 11-5 *(Continued)* Structure of the heart. (e) Photograph of the heart in anterior view. (Courtesy of L. V. Bergman & Assoc.)

valve between the right atrium and right ventricle is called the *tricuspid valve* because it consists of three flaps, or cusps. These flaps are fibrous tissues that grow out of the walls of the heart and are covered with endocardium. The pointed ends of the cusps project into the ventricle. Other names for this valve are the right atrioventricular or right AV valve. Cords called *chordae tendineae* connect the pointed ends to small conical projections—the *papillary muscles* (muscular columns)—located on the inner surface of the ventricles. The irregular surface of ridges and folds of the myocardium in the ventricles is known as the *trabeculae carneae*. The chordae tendineae and their muscles keep the flaps pointing in the direction of the blood flow. As the atrium relaxes and the ventricle squeezes the blood out of the heart, any blood driven back toward the atrium is pushed between the flaps and the ventricle walls (Figure 11-6a). This action drives the cusps upward until their edges meet and close the opening. At the same time, contraction of the papillary muscles prevents the valve from swinging upward into the atrium. The atrioventricular valve between the left atrium and left ventricle is called the *bicuspid* or *mitral valve*. It has two cusps that work in the same way as the cusps of the tricuspid valve. The bicuspid valve is also known as the left atrioventricu-

lar or left AV valve. Its cusps are also attached by way of the chordae tendineae to papillary muscles.

Each artery that leaves the heart has a valve that prevents blood from flowing back into the heart. These are the **semilunar valves**—*semilunar* meaning half-moon or crescent-shaped. The *pulmonary semilunar valve* lies in the opening where the pulmonary trunk leaves the right ventricle. The *aortic semilunar valve* is situated at the opening between the left ventricle and the aorta. Both valves consist of three semilunar cusps. Each cusp is attached by its convex margin to the artery wall. The free borders of the cusps curve outward and project into the opening inside the blood vessel (Figure 11-6b). Like the atrioventricular valves, the semilunar valves permit blood to flow in only one direction—in this case, from the ventricles into the arteries.

CONDUCTION SYSTEM

The heart is innervated by the autonomic nervous system (Chapter 15), but the autonomic neurons only increase or decrease the time it takes to complete a cardiac cycle; that is, they don't initiate contractions. The chamber walls can go on contracting and relaxing, contracting and relaxing, without any direct stimulus

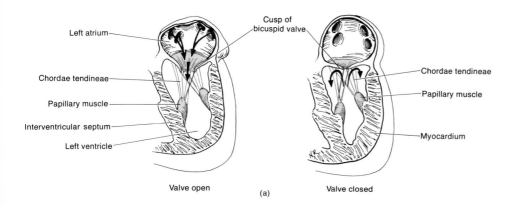

Left atrium

Chordae tendineae

Papillary muscle

Interventricular septum

Left ventricle

Valve open

Cusp of bicuspid valve

Chordae tendineae

Papillary muscle

Myocardium

Valve closed

(a)

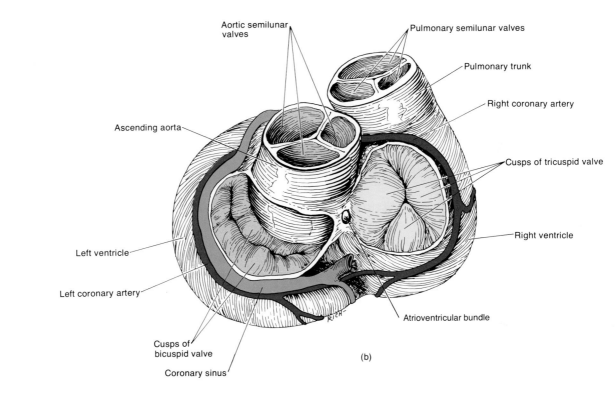

Aortic semilunar valves

Ascending aorta

Left ventricle

Left coronary artery

Cusps of bicuspid valve

Coronary sinus

Pulmonary semilunar valves

Pulmonary trunk

Right coronary artery

Cusps of tricuspid valve

Right ventricle

Atrioventricular bundle

(b)

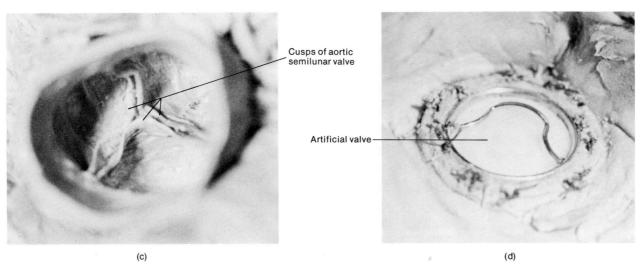

Cusps of aortic semilunar valve

(c)

Artificial valve

(d)

FIGURE 11-6 Valves of the heart. (a) Structure and function of the bicuspid valve. (b) Valves of the heart viewed from above. The atria have been removed to expose the tricuspid and bicuspid valves. (c) Photograph of a superior view of a semilunar valve. (d) Photograph of a superior view of an artificial valve. (Courtesy of John W. Eads, Univeristy of Arizona.)

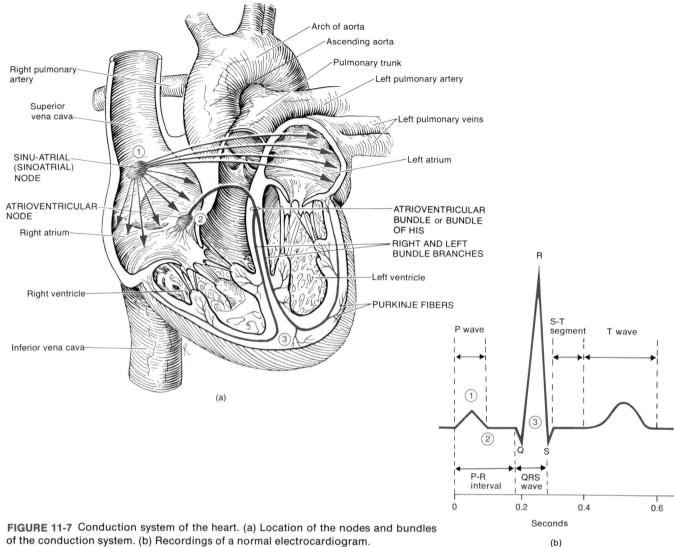

FIGURE 11-7 Conduction system of the heart. (a) Location of the nodes and bundles of the conduction system. (b) Recordings of a normal electrocardiogram.

from the nervous system. This action is possible because the heart has an intrinsic regulating system called the **conduction system.** The conduction system is composed of specialized muscle tissue that generates and distributes the electrical impulses which stimulate the cardiac muscle fibers to contract. These tissues are the sinu-atrial (sinoatrial) node, the atrioventricular node, the atrioventricular bundle, the bundle branches, and the Purkinje fibers. The cells of the conduction system develop during embryological life from certain cardiac muscle cells. These cells lose their ability to contract and become specialized for impulse transmission.

A **node** of the conducting system is a compact mass of conducting cells. The **sinu-atrial (sinoatrial) node,** known as the **SA node** or **pacemaker,** is located in the right atrial wall inferior to the opening of the superior vena cava (Figure 11-7a). The SA node initiates each cardiac cycle, and thereby sets the basic pace for the

heart rate—hence its common name, pacemaker. However, the rate set by the SA node may be altered by nervous impulses from the autonomic nervous system or by certain blood-borne chemicals such as thyroid hormone and epinephrine. Once an electrical impulse is initiated by the SA node, the impulse spreads out over both atria, causing them to contract and at the same time depolarizing the **atrioventricular (AV) node.** Because of its location near the inferior portion of the interatrial septum, the AV node is one of the last portions of the atria to be depolarized. From the AV node, a tract of conducting fibers called the **atrioventricular bundle** or **bundle of His** runs through the cardiac skeleton to the top of the interventricular septum. It then continues down both sides of the septum as the **right** and **left bundle branches.** The bundle of His distributes the charge over the medial surfaces of the ventricles. Actual contraction of the ventricles is stimulated by the **Purkinje fibers** that emerge from the bundle branches and pass into the cells of the myocardium.

ELECTROCARDIOGRAM

Impulse transmission through the conduction system generates electrical currents that may be detected on the body's surface. A recording of the electrical changes that accompany the cardiac cycle is called an **electrocardiogram (ECG).** The instrument used to record the changes is an *electrocardiograph.*

Each portion of the cardiac cycle produces a different electrical impulse. These impulses are transmitted from the electrodes to a recording needle that graphs the impulses as a series of up-and-down waves called *deflection waves.* In a typical record (Figure 11-7b), three clearly recognizable waves accompany each cardiac cycle. The first wave, called the **P wave,** is a small upward wave. It indicates atrial depolarization—the spread of an impulse from the SA node through the muscle of the two atria. A fraction of a second after the P wave begins, the atria contract. Then there is a deflection wave called the **QRS wave (complex).** It begins as a downward deflection, continues as a large, upright, triangular wave, and ends as a downward wave at its base. This deflection represents atrial repolarization and ventricular depolarization—that is, the spread of the electrical impulse through the ventricles. The third recognizable deflection is a dome-shaped **T wave.** This wave indicates ventricular repolarization. There is no deflection to show atrial repolarization because the stronger QRS wave masks this event.

In reading an electrocardiogram, it is important to note the size of the deflection waves and certain time intervals. Enlargement of the P wave, for example, indicates enlargement of the atrium, as in mitral stenosis. The *P-R interval* is measured from the beginning of the P wave to the beginning of the Q wave. It represents the conduction time from the beginning of atrial excitation to the beginning of ventricular excitation. The P-R interval is the time required for an impulse to travel through the atria and atrioventricular node to the remaining conducting tissues. The lengthening of this interval, as in arteriosclerotic heart disease and rheumatic fever, occurs because the heart tissue covered by the P-R interval, namely the atria and atrioventricular node, is scarred or inflamed. Thus the impulse must travel at a slower rate and the interval is lengthened. The normal P-R interval covers no more than 0.20 second.

An enlarged Q wave may indicate a myocardial infarction. An enlarged R wave generally indicates enlarged ventricles. The *S-T segment* begins at the end of the S wave and terminates at the beginning of the T wave. It represents the time between the end of the spread of the impulse through the ventricles and repolarization of the ventricles. The S-T segment is elevated in acute myocardial infarction and depressed when the heart muscle receives insufficient oxygen. The T wave represents ventricular repolarization. It is flat when the heart muscle is receiving insufficient oxygen, as in arteriosclerotic heart disease. It may be elevated when the body's potassium level is increased.

The ECG is invaluable in diagnosing abnormal cardiac rhythms and conduction patterns, detecting the presence of fetal life, determining the presence of several fetuses, and following the course of recovery from a heart attack.

AUTONOMIC CONTROL

The pacemaker receives nerves from the parasympathetic and sympathetic divisions of the autonomic nervous system. Parasympathetic neurons in the vagus nerve travel outside of the spine to the heart. When stimulated by the **cardioinhibitory center** of the medulla, the parasympathetic fibers inhibit the pacemaker and the rate of the heart beat is decreased. The sympathetic pathway originates in the **cardioacceleratory center** of the medulla. It travels in a tract down the spinal cord and then passes over sympathetic nerves to the heart. Sympathetic stimulation counteracts parasympathetic stimulation, quickens the heartbeat, and increases its strength of contractions. When neither cardiac center is stimulated by sensory neurons, the cardioacceleratory center tends to dominate. Sympathetic fibers then have a free rein to speed up heart rate until receptors intervene to stimulate the cardioinhibitory center.

SOUNDS

The first sound, which can be described as a **lubb** (ōō) sound, is a long, booming sound. The lubb is the sound created by the closure of the atrioventricular valves soon after ventricular systole (contraction) begins. The second sound, which is heard as a short, sharp sound, can be described as a **dupp** (ŭ) sound. Dupp is the sound created as the semilunar valves close toward the end of ventricular systole. A pause about two times longer comes between the second sound and the first sound of the next cycle. Thus the cardiac cycle can be heard as a lubb, dupp, pause; lubb, dupp, pause; lubb, dupp, pause. This is the sound of the heartbeat. But it comes primarily from turbulence in blood flow created by the closure of the valves and not from the contraction of the heart muscle.

Heart sounds provide valuable information about the valves. If the sounds are peculiar, they are called **murmurs.** Some murmurs are caused by the noise made by a little blood bubbling back up into an atrium because of improper closure of an atrioventricular

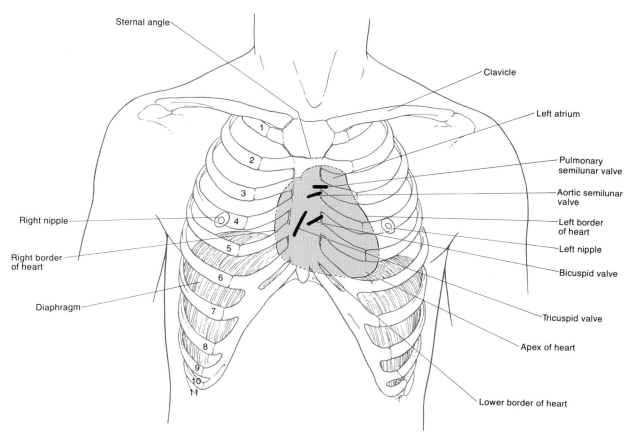

FIGURE 11-8 Surface projection of the heart.

valve. Murmurs do not always indicate a valve problem, however, and many have no clinical significance.

Although heart sounds are produced in part by the closure of valves, they are not necessarily heard best over these valves. Each sound tends to be clearest in a slightly different place closest to the body's surface.

SURFACE ANATOMY

The valves of the heart may be identified by surface projection (Figure 11-8). The pulmonary and aortic semilunar valves are represented on the surface by a line about 2.5 cm (1 inch) in length. The pulmonary semilunar valve lies horizontally behind the inner end of the left third costal cartilage and the adjoining part of the sternum. The aortic semilunar valve is placed obliquely behind the left side of the sternum, opposite the third intercostal space. The tricuspid valve lies behind the sternum, extending from the midline at the level of the fourth costal cartilage down toward the right sixth chondrosternal junction. The bicuspid lies behind the left side of the sternum obliquely at the level of the fourth costal cartilage. It is represented by a line about 3 cm in length.

Blood Vessels

The blood vessels form a network of tubes that carry blood away from the heart, transport it to the tissues of the body, and then return it to the heart. Blood vessels are called either arteries, arterioles, capillaries, venules, or veins. **Arteries** are the vessels that carry blood from the heart to the tissues. Two large arteries leave the heart and divide into medium-sized vessels that head toward the various regions of the body. The medium-sized arteries, in turn, divide into small arteries which, in turn, divide into vessels called **arterioles.** As the arterioles enter a tissue, they branch into countless microscopic vessels called **capillaries.** Through the walls of the capillaries, substances are exchanged between the blood and body tissues. Before leaving the tissue, groups of capillaries reunite to form small veins called **venules.** These, in turn, merge to form progressively larger tubes—the veins themselves. **Veins,** in other words, are blood vessels that convey blood from the tissues back to the heart. Since blood vessels require oxygen and nutrients just like other tissues of the body, they also have blood vessels in their own walls called **vasa vasorum.**

ARTERIES

Arteries and veins are fairly similar in construction (Figure 11-9a, d, e). Both have walls constructed of three coats or tunics and a hollow core, called a *lumen,* through which the blood flows. Arteries, however, are considerably thicker and stronger than veins. The pressure in an artery is always greater than in a vein. The inner coat of arterial wall is called the *tunica interna* or *intima.* It is composed of a lining of endothelium (simple squamous epithelium) that is in contact with the blood. It also has an overlying layer of areolar connective tissue and an outer layer of elastic tissue called the internal elastic membrane. The middle coat, or *tunica media,* is usually the thickest layer. It consists of elastic fibers and smooth muscle. The outer coat, the *tunica externa* or *adventitia,* is composed principally of loose connective tissue. The tunica externa contains elastic and collagenous fibers and a few smooth muscle fibers. An external elastic

membrane may separate the tunica media from the tunica externa.

As a result of the structure of the middle coat especially, arteries have two major properties: elasticity and contractility. When the ventricles of the heart contract and eject blood into the large arteries, the arteries expand to contain the extra blood. Then, as the ventricles relax, the elastic recoil of the arteries forces the blood onward. The contractility of an artery comes from its smooth muscle. The smooth muscle is arranged in rings around the lumen somewhat like a doughnut. As the muscle contracts, it squeezes the wall around the lumen and narrows the vessel. Such a decrease in the size of the lumen is called *vasoconstriction.* Conversely, if all the muscle fibers relax, the size of the arterial lumen increases. This increase is called *vasodilation* and is often due to the inhibition of vasoconstriction.

The contractility of arteries also serves a minor function in stopping bleeding. The blood flowing

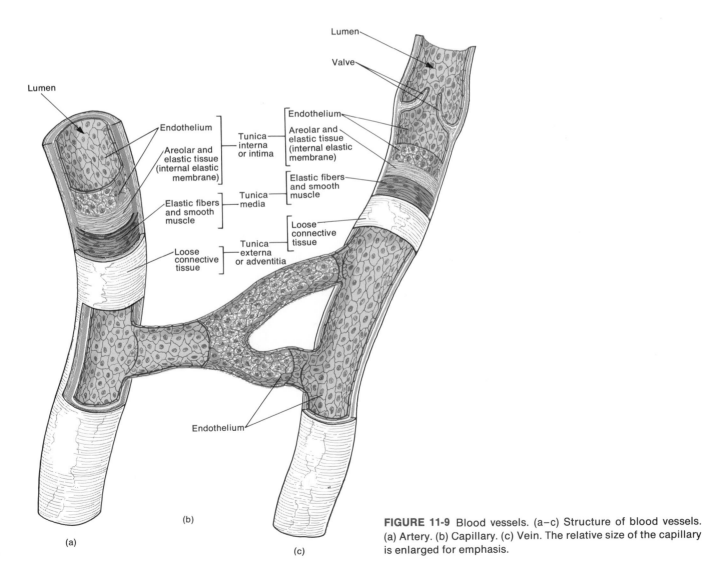

FIGURE 11-9 Blood vessels. (a–c) Structure of blood vessels. (a) Artery. (b) Capillary. (c) Vein. The relative size of the capillary is enlarged for emphasis.

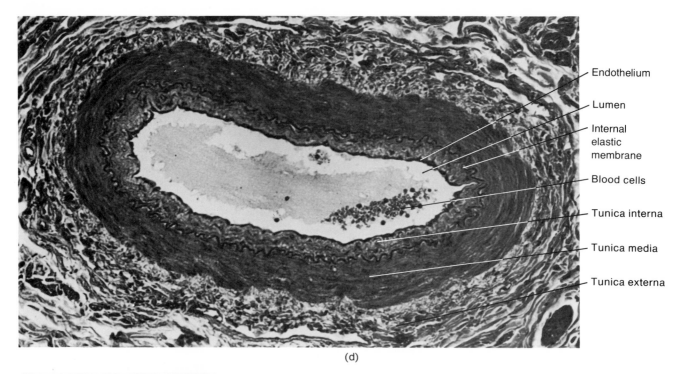

(d)

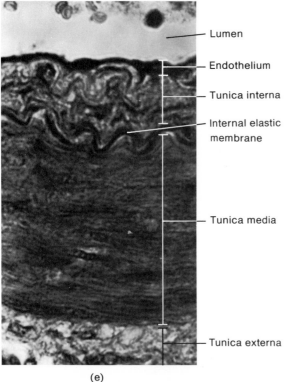

(e)

FIGURE 11-9 *(Continued)* Blood vessels. (d–f) Histology of arteries and veins. (d) Cross section of an artery at a magnification of 50×. (e) Enlarged aspect of an artery at a magnification of 200×. (Courtesy of Victor B. Eichler, Wichita State University.)

through an artery is under a great deal of pressure. Thus great quantities of blood can be quickly lost from a broken artery. When an artery is cut, its walls constrict so that blood does not escape quite so rapidly. However, there is a limit to how much vasoconstriction can help.

Most parts of the body receive branches from more than one artery. In such areas the distal ends of the vessels unite. The junction of two or more arteries supplying the same body region is called an **anastomosis.** Anastomoses may also occur between the origins of veins and between arterioles and venules. Anastomoses between arteries provide alternate routes by which blood can reach a tissue or organ. Thus if a vessel is occluded by disease, injury, or surgery, circulation to a part of the body is necessarily stopped. The alternate route of blood to a body part through an anastomosis is known as **collateral circulation.** An alternate blood route may also be from nonanastomosing vessels that supply the same region of the body.

CAPILLARIES

Capillaries are microscopic vessels measuring 4 to 12 μm in diameter and are found in close proximity to nearly every cell of the body. They usually connect arterioles with venules (Figure 11-9b) and permit the exchange of nutrients and gases between the blood and interstitial fluid. The structure of the capillaries is admirably suited for this purpose. First, the capillary

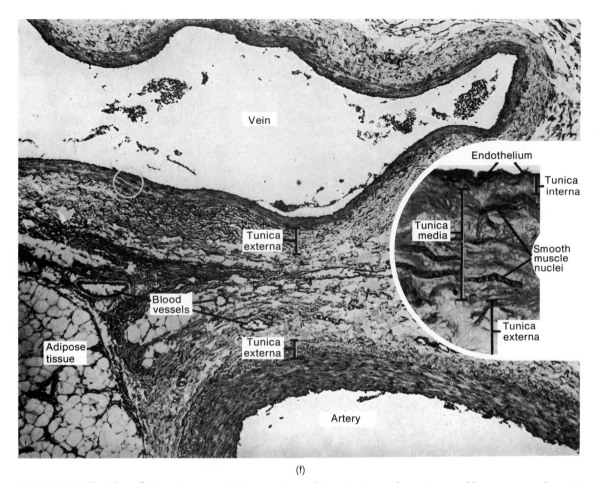

(f)

FIGURE 11-9 *(Continued)* Blood vessels. (f) Comparison of the structure of an artery and its accompanying vein at a magnification of 65×. The inset is an enlarged aspect of the layers of a vein at a magnification of 640×. (Courtesy of Edward J. Reith, from *Atlas of Descriptive Histology,* by Edward J. Reith and Michael H. Ross, Harper & Row, Publishers, Inc., New York, 1970.)

walls are composed of only a single layer of cells (endothelium). Thus a substance in the blood must pass through the plasma membrane of just one cell to reach the interstitial fluid. This vital exchange of materials occurs only through capillary walls—the thick walls of arteries and veins present too great a barrier. Capillaries are also well suited to their function since they form a *capillary network* throughout the tissue. The network increases the surface area for passage and thereby allows a rapid exchange of large quantities of materials. Though capillaries lack the elastic connective fibers of arteries, their walls are still capable of distension. Thus they can help adjust the amount and force of blood flowing through them.

Microscopic blood vessels in certain parts of the body, such as the liver, are termed *sinusoids.* They are wider than capillaries and more tortuous. Also, instead of the usual endothelial lining, sinusoids are lined largely by phagocytic cells. In the liver, such cells are called Kupffer cells. Like capillaries, sinu-

soids convey blood from arterioles to venules. Other organs containing sinusoids include the spleen, adenohypophysis, and parathyroid glands.

VEINS

Veins are composed of essentially the same three coats as arteries, but they have considerably less elastic tissue and smooth muscle (Figure 11-9c, f). However, veins do contain more white fibrous tissue. They are also distensible enough to adapt to variations in the volume and pressure of blood passing through them. Blood leaves a cut vein in an even flow rather than in the rapid spurts characteristic of arteries—by the time the blood leaves the capillaries and moves into the veins, it has lost a great deal of pressure. Most of the structural differences between arteries and veins reflect this pressure difference. For example, veins do not need walls as strong as those of arteries. The low pressure in veins, however, has its disadvantages.

When you stand, the pressure pushing blood up the veins in your lower extremities is barely enough to balance the force of gravity pushing it back down. For this reason, many veins, especially those in the limbs, contain valves that prevent backflow. Normal valves ensure the flow of blood toward the heart.

In people with weak valves, large quantities of blood are forced by gravity back down into distal parts of the vein. This pressure overloads the vein and pushes the walls outward. After repeated overloading, the walls lose their elasticity and become stretched and flabby. A vein damaged in this way is called a *varicose vein*. Because a varicosed wall is not able to exert a firm resistance against the blood, blood tends to accumulate in the pouched-out area of the vein, causing it to swell and forcing fluid into the surrounding tissue. Veins close to the surface of the legs are highly susceptible to varicosities. Veins that lie deeper are not so vulnerable because surrounding skeletal muscles prevent their walls from overstretching.

Varicosities are also common in the veins that lie in the walls of the anal canal. These varicosities are called *hemorrhoids*. Hemorrhoids may be caused in many people by constipation. Repeated straining during defecation forces blood down into the superior hemorrhoidal plexus, increasing pressure in these veins. Constipation is related to low-fiber diets, especially in North America. A recent hypothesis links increased intraabdominal pressure, caused by straining during evacuation of firm feces, directly to hemorrhoids or varicose veins and indirectly to thrombosis. A new development in the treatment of hemorrhoids is *cryosurgery* (*cryo* = cold)—external hemorrhoids can be destroyed by freezing them with a solution of nitrous oxide or liquid nitrogen.

Vascular (venous) sinuses, or simply *sinuses,* are veins with very thin walls—for example, the intracranial sinuses. These vessels consist of a wall of endothelium. They are supported by the dura mater, which replaces the tunica media and tunica externa. Intracranial vascular sinuses return deoxygenated blood from the brain to the heart. Another example of a vascular sinus is the coronary sinus of the heart.

Circulatory Routes

Figure 11-10 shows a number of basic **circulatory routes** through which the blood travels. The largest route by far is the **systemic circulation.** This route includes all the oxygenated blood that leaves the left ventricle through the aorta and returns to the right atrium after traveling to all the organs including the nutrient arteries to the lungs. Two of the many subdivisions of the systemic circulation are the **coronary circulation,** which supplies the myocardium of the heart,

and the **hepatic portal circulation,** which runs from the digestive tract to the liver. You may refer to Figure 11-21 for the details of coronary circulation. Blood leaving the aorta and traveling through the systemic arteries is a bright red color. As it moves through the capillaries, it loses its oxygen and takes on carbon dioxide, which gives the blood in the systemic veins its dark red color. When blood returns to the heart from the systemic route, it goes out of the right ventricle through the **pulmonary circulation** to the lungs. In the lungs, it loses its carbon dioxide and takes on oxygen. It is now bright red again. It returns to the left atrium of the heart and reenters the systemic circulation. Another major route—the **fetal circulation**—exists only in the fetus and contains special structures that allow the developing human to exchange materials with its mother. **Cerebral circulation** (circle of Willis) is discussed in Exhibit 11-4.

SYSTEMIC CIRCULATION

The flow of blood from the left ventricle to all parts of the body except the lungs and back to the right atrium is called the **systemic circulation.** The purpose of systemic circulation is to carry oxygen and nutrients to body tissues and to remove carbon dioxide and other wastes from the tissues. All systematic arteries branch from the *aorta,* which arises from the left ventricle of the heart.

As the aorta emerges from the left ventricle, it passes upward and deep to the pulmonary artery. At this point, it is called the *ascending aorta.* The ascending aorta gives off two coronary branches to the heart muscle. Then it turns to the left forming the *arch of the aorta* before descending to the level of the fourth lumbar vertebra as the *descending aorta.* The descending aorta lies close to the vertebral bodies, passes through the diaphragm, and terminates at the level of the fourth lumbar vertebra by dividing into two *common iliac arteries,* which carry blood to the lower extremities. The section of the descending aorta between the arch of aorta and the diaphragm is also referred to as the *thoracic aorta.* The section between the diaphragm and the common iliac arteries is termed the *abdominal aorta.* Each section of the aorta gives off arteries that continue to branch into distributing arteries leading to organs and finally into the capillaries that pierce the tissues.

Blood is returned to the heart through the systemic veins. All the veins of the systemic circulation flow into either the *superior* or *inferior venae cavae* or the *coronary sinus.* They in turn empty into the right atrium. The principal arteries and veins of systemic circulation are described and illustrated in Exhibits 11-2 to 11-13 and Figures 11-11 to 11-20.

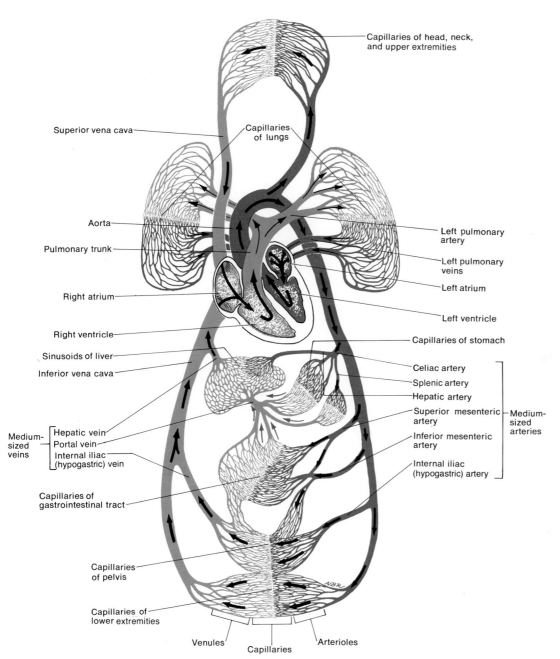

Capillaries of head, neck, and upper extremities

Superior vena cava

Capillaries of lungs

Aorta

Pulmonary trunk

Left pulmonary artery

Left pulmonary veins

Left atrium

Right atrium

Left ventricle

Right ventricle

Capillaries of stomach

Sinusoids of liver

Celiac artery

Inferior vena cava

Splenic artery

Hepatic artery

Superior mesenteric artery

Inferior mesenteric artery

Internal iliac (hypogastric) artery

Medium-sized veins

Hepatic vein
Portal vein
Internal iliac (hypogastric) vein

Medium-sized arteries

Capillaries of gastrointestinal tract

Capillaries of pelvis

Capillaries of lower extremities

Venules   Arterioles
          Capillaries

FIGURE 11-10 Circulatory routes. Systemic circulation is indicated by heavy black arrows; pulmonary circulation by thin black arrows; and hepatic portal circulation by thin colored arrows. Refer to Figure 11-21 for the details of coronary circulation and Figure 11-24 for the details of fetal circulation.

EXHIBIT 11-2
AORTA AND ITS BRANCHES (Figure 11-11)

DIVISION OF AORTA	ARTERIAL BRANCH	REGION SUPPLIED
Ascending aorta	Right and left coronary	Heart
Arch of aorta	Brachiocephalic ⟨ Right common carotid	Right side of head and neck
	Right subclavian	Right upper extremity
	Left common carotid	Left side of head and neck
	Left subclavian	Left upper extremity
Thoracic aorta	Intercostals	Intercostal and chest muscles, pleurae
	Superior phrenics	Posterior and superior surfaces of diaphragm
	Bronchials	Bronchi of lungs
	Esophageals	Esophagus
Abdominal aorta	Inferior phrenics	Inferior surface of diaphragm
	Celiac → Common hepatic	Liver
	Left gastric	Stomach and esophagus
	Splenic	Spleen, pancreas, stomach
	Superior mesenteric	Small intestine, cecum, ascending and transverse colons
	Suprarenals	Adrenal (suprarenal) glands
	Renals	Kidneys
	Gonadals ⟨ Testiculars	Testes
	Ovarians	Ovaries
	Inferior mesenteric	Transverse, descending, sigmoid colons; rectum
	Common iliacs ⟨ External iliacs	Lower extremities
	Internal iliacs (hypogastrics)	Uterus, prostate, muscles of buttocks, urinary bladder

EXHIBIT 11-3
ASCENDING AORTA (Figure 11-12)

BRANCH	DESCRIPTION AND REGION SUPPLIED
Coronary arteries	These two branches arise from ascending aorta just superior to aortic semilunar valve. They form crown around heart giving off branches to atrial and ventricular myocardium.

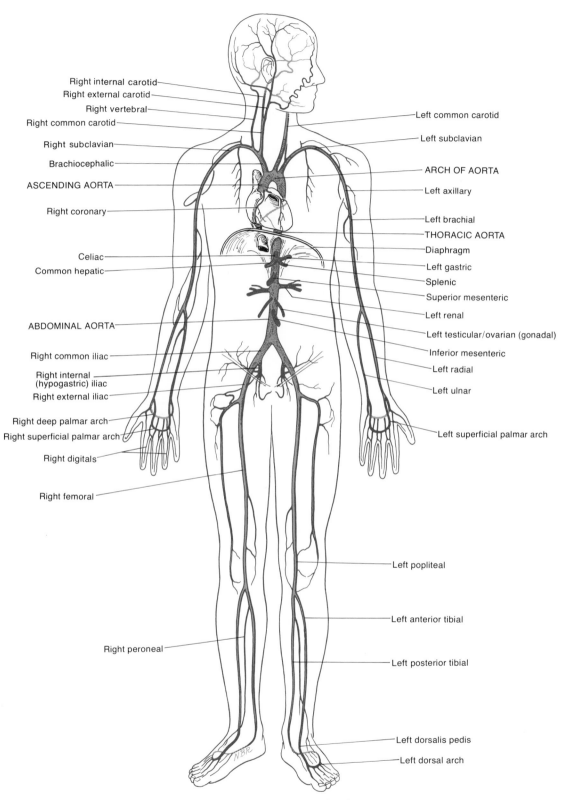

FIGURE 11-11 Aorta and its principal arterial branches in anterior view.

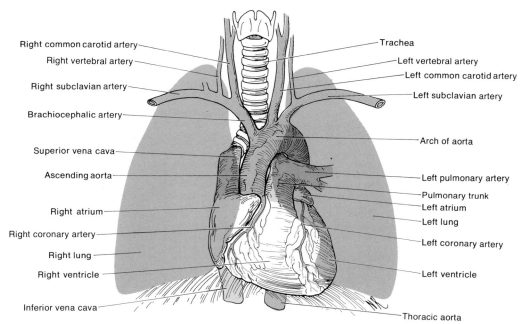

Right common carotid artery

Right vertebral artery

Right subclavian artery

Brachiocephalic artery

Superior vena cava

Ascending aorta

Right atrium

Right coronary artery

Right lung

Right ventricle

Inferior vena cava

Trachea

Left vertebral artery

Left common carotid artery

Left subclavian artery

Arch of aorta

Left pulmonary artery

Pulmonary trunk

Left atrium

Left lung

Left coronary artery

Left ventricle

Thoracic aorta

FIGURE 11-12 Ascending aorta and its arterial branches in anterior view.

EXHIBIT 11-4
ARCH OF AORTA (Figure 11-13)

BRANCH	DESCRIPTION AND REGION SUPPLIED
Brachiocephalic	**Brachiocephalic artery** is first branch off arch of aorta. Divides to form right subclavian artery and right common carotid artery. **Right subclavian artery** extends from brachiocephalic to first rib, passes into armpit, or axilla, and supplies arm, forearm, and hand. This artery is good example of giving same vessel different names as it passes through different regions. Continuation of right subclavian into axilla is called **axillary artery.** From here, it continues into arm as **brachial artery.** At bend of elbow, brachial artery divides into medial **ulnar** and lateral **radial arteries.** These vessels pass down to palm, one on each side of forearm. In palm, branches of two arteries anastomose to form two palmar arches—**superficial palmar arch** and **deep palmar arch.** From these arches arise **digital arteries,** which supply fingers and thumb.
	Before passing into axilla, right subclavian gives off major branch to brain called **vertebral artery.** Right vertebral artery passes through foramina of transverse processes of cervical vertebrae and enters skull through foramen magnum to reach undersurface of brain. Here it unites with left vertebral artery to form **basilar artery.**
	Anastomoses of left and right internal carotids along with basilar artery form arterial circle at base of brain called **circle of Willis.** From this anastomosis arise arteries supplying brain. Essentially, circle of Willis is formed by union of **anterior cerebral arteries** (branches of internal carotids) and **posterior cerebral arteries** (branches of basilar artery). Posterior cerebral arteries are connected with internal carotids by **posterior communicating arteries.** Anterior cerebral arteries are connected by **anterior communicating arteries.** Circle of Willis equalizes blood pressure to brain and provides alternate routes for blood to brain should arteries become damaged.
Right common carotid	**Right common carotid artery** passes upward in neck. At upper level of larynx, it divides into **right external** and **right internal carotid arteries.** External carotid supplies right side of thyroid gland, tongue, throat, face, ear, scalp, and dura mater. Internal carotid supplies brain, right eye, and right sides of forehead and nose.
Left common carotid	**Left common carotid** branches directly from arch of aorta. Corresponding to right common carotid, it divides into basically same branches with same names—except that arteries are now labeled "left" instead of "right."
Left subclavian	**Left subclavian artery** is third branch off arch of aorta. It distributes blood to left vertebral artery and vessels of left upper extremity. Arteries branching from left subclavian are named like those of right subclavian.

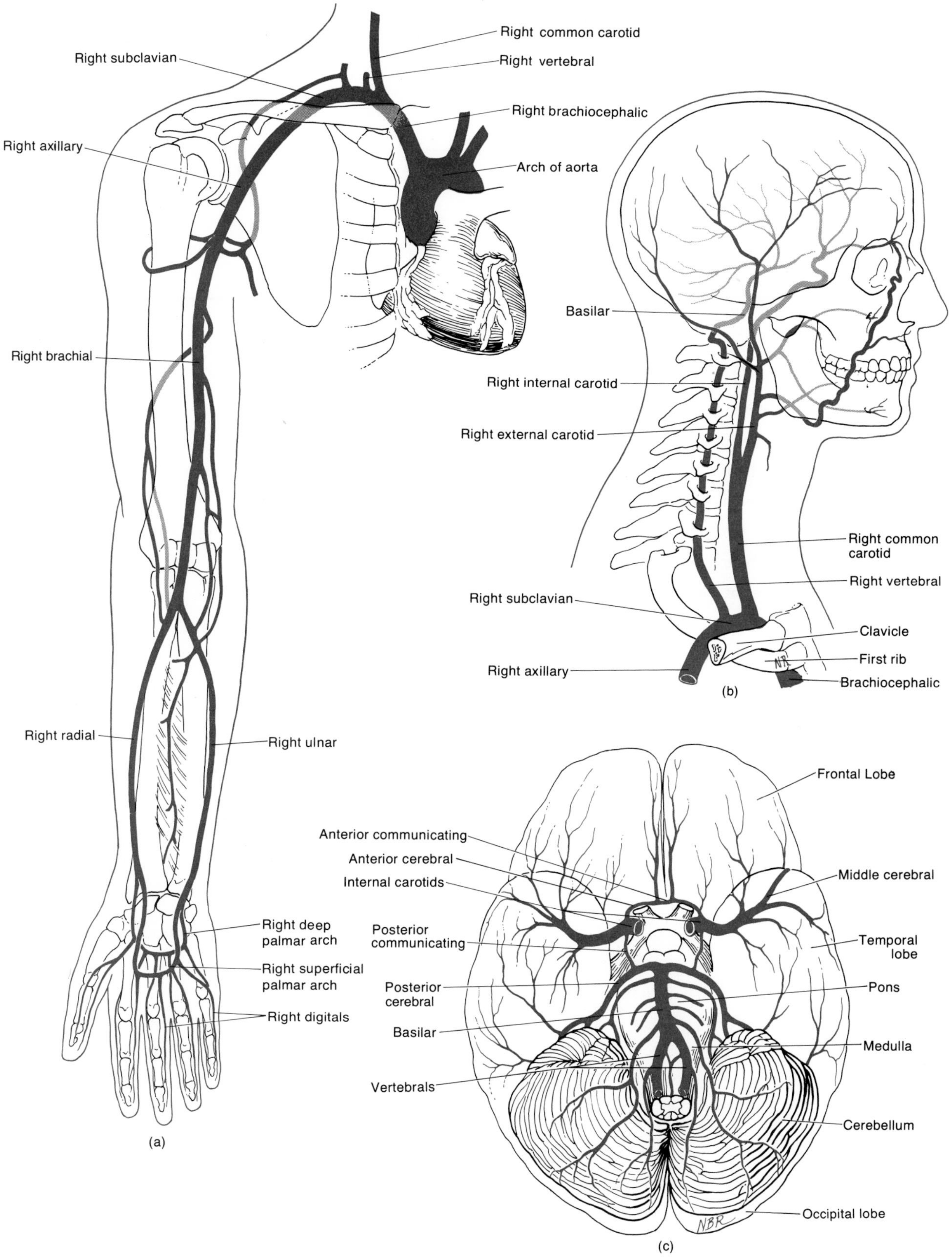

FIGURE 11-13 Arch of the aorta and its arterial branches. (a) Anterior view of the arteries of the right upper extremity. (b) Right lateral view of the arteries of the neck and head. (c) Arteries of the base of the brain.

EXHIBIT 11-5
THORACIC AORTA (Figure 11-14)

Thoracic aorta runs from fourth to twelfth thoracic vertebrae. Along its course, it sends off numerous small arteries to viscera and skeletal muscles of chest. *Visceral branches* supply pericardium around heart, bronchial tubes that lead from windpipe to lungs, cells of lungs (but not areas of lungs that oxygenate blood), esophagus, and tissue lining mediastinum. *Parietal branches* supply chest muscles, diaphragm, and mammary glands.

EXHIBIT 11-6
ABDOMINAL AORTA (Figure 11-14)

BRANCH	DESCRIPTION AND REGION SUPPLIED
VISCERAL	
Celiac	**Celiac artery** (trunk) is first visceral aortic branch below diaphragm. It has three branches: (1) **common hepatic artery,** which supplies tissues of liver; (2) **left gastric artery,** which supplies stomach; and (3) **splenic artery,** which supplies spleen, pancreas, and stomach.
Superior mesenteric	**Superior mesenteric artery** distributes blood to small intestine and part of large intestine.
Suprarenals	Right and left **suprarenal arteries** supply blood to adrenal (suprarenal) glands.
Renals	Right and left **renal arteries** carry blood to kidneys.
Testiculars	Right and left **testicular arteries** extend into scrotum and terminate in testes.
Ovarians	Right and left **ovarian arteries** are distributed to ovaries.
Inferior mesenteric	**Inferior mesenteric artery** supplies major part of large intestine and rectum.
PARIETAL	
Inferior phrenics	**Inferior phrenic arteries** are distributed to undersurface of diaphragm.
Lumbars	**Lumbar arteries** supply spinal cord and its meninges and muscles and skin of lumbar region of back.
Middle sacral	**Middle sacral artery** supplies sacrum, coccyx, gluteus maximus muscles, and rectum.

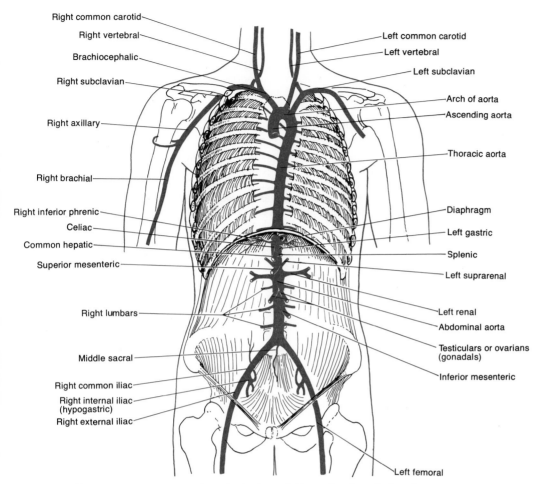

FIGURE 11-14 Abdominal aorta and its principal arterial branches in anterior view.

EXHIBIT 11-7
ARTERIES OF PELVIS AND LOWER EXTREMITIES (Figure 11-15)

BRANCH	DESCRIPTION AND REGION SUPPLIED
	At about level of fourth lumbar vertebra, abdominal aorta divides into right and left **common iliac arteries.** Each passes downward about 5 cm (2 inches) and gives rise to two branches: internal iliac and external iliac
Internal iliacs	**Internal iliac** or **hypogastric arteries** form branches that supply gluteal muscles, medial side of each thigh, urinary bladder, rectum, prostate gland, uterus, and vagina.
External iliacs	**External iliac arteries** diverge through pelvis, enter thighs, and here become right and left **femoral arteries.** Both femorals send branches back up to genitals and wall of abdomen. Other branches run to muscles of thigh. Femoral continues down medial and posterior side of thigh at back of knee joint, where it becomes **popliteal artery.** Between knee and ankle, popliteal runs down back of leg and is called **posterior tibial artery.** Below knee, **peroneal artery** branches off posterior tibial to supply structures on medial side of fibula and calcaneus. In calf, **anterior tibial artery** branches off popliteal and runs along front of leg. At ankle, it becomes **dorsalis pedis artery.** At ankle, posterior tibial divides into **medial** and **lateral plantar arteries.** These arteries anastomose with dorsalis pedis and supply blood to foot.

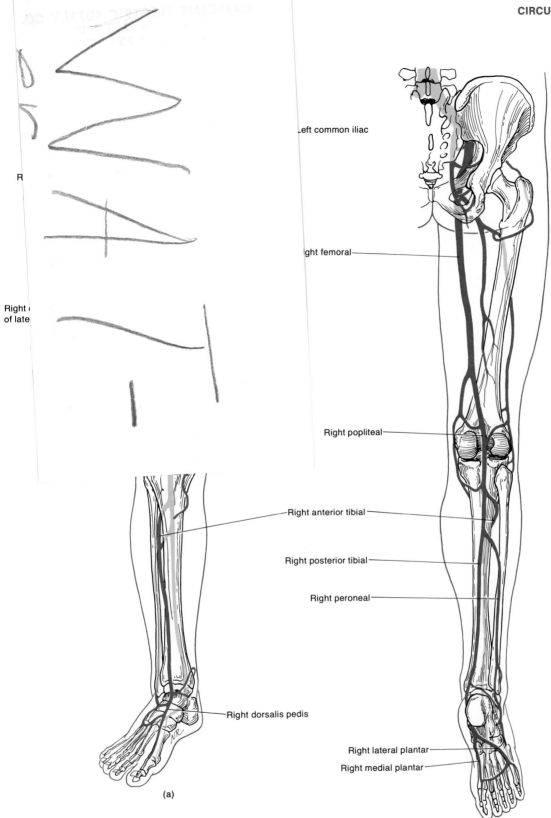

Left common iliac

ight femoral

Right popliteal

Right anterior tibial

Right posterior tibial

Right peroneal

Right dorsalis pedis

Right lateral plantar

Right medial plantar

(a)

(b)

FIGURE 11-15 Arteries of the pelvis and right lower extremity. (a) Anterior view. (b) Posterior view.

EXHIBIT 11-8
VEINS OF SYSTEMIC CIRCULATION (Figure 11-16)

Deep veins are located deep in body. They usually accompany arteries, and many have same names as corresponding arteries. **Superficial veins** are located just below skin and are visible.

All systemic veins return blood to right atrium of heart through one of three large vessels: coronary sinus, superior vena cava, and inferior vena cava. Return flow from coronary arteries is taken up by **cardiac veins,** which empty into large vein of heart called **coronary sinus.** From here, blood empties into right atrium of heart. Veins that empty into **superior vena cava** are veins of head and neck, upper extremities, thorax, and azygos veins. Veins that empty into **inferior vena cava** are veins of abdomen, pelvis, lower extremities, and azygos veins.

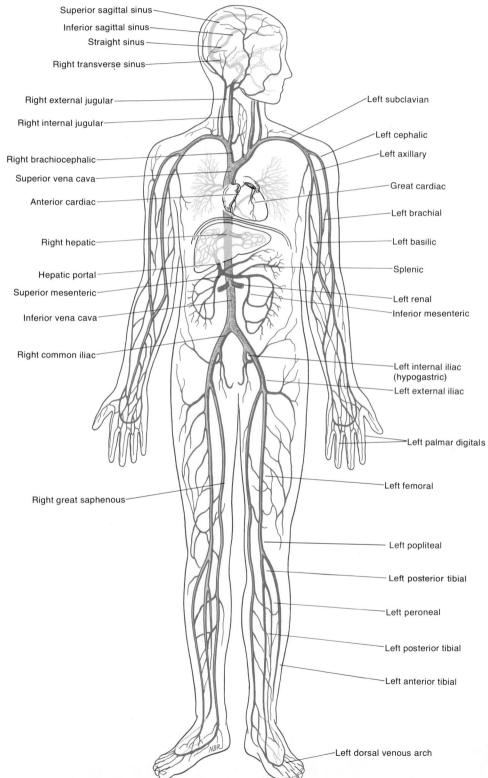

Superior sagittal sinus
Inferior sagittal sinus
Straight sinus
Right transverse sinus
Right external jugular
Right internal jugular
Right brachiocephalic
Superior vena cava
Anterior cardiac
Right hepatic
Hepatic portal
Superior mesenteric
Inferior vena cava
Right common iliac
Right great saphenous

Left subclavian
Left cephalic
Left axillary
Great cardiac
Left brachial
Left basilic
Splenic
Left renal
Inferior mesenteric
Left internal iliac (hypogastric)
Left external iliac
Left palmar digitals
Left femoral
Left popliteal
Left posterior tibial
Left peroneal
Left posterior tibial
Left anterior tibial
Left dorsal venous arch

FIGURE 11-16 Principal veins in anterior view.

EXHIBIT 11-9
VEINS OF HEAD AND NECK (Figure 11-17)

VEIN	DESCRIPTION AND REGION DRAINED
Internal jugulars	Right and left **internal jugular veins** arise as continuation of **sigmoid sinuses** at base of skull. Intracranial vascular sinuses are located between layers of dura mater and receive blood from brain. Other sinuses that drain into internal jugular include **superior sagittal sinus, inferior sagittal sinus, straight sinus,** and **transverse (lateral) sinuses.** Internal jugulars descend on either side of neck. They receive blood from superior part of face and neck and pass behind clavicles, where they join with right and left **subclavian veins.** Unions of internal jugulars and subclavians form right and left **brachiocephalic veins.** From here blood flows into **superior vena cava.**
External jugulars	Left and right **external jugular veins** run down neck along outside of internal jugulars. They drain blood from parotid (salivary) glands, facial muscles, scalp, and other superficial structures into **subclavian veins.**

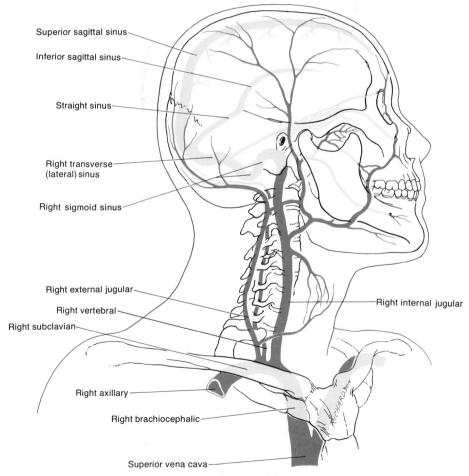

FIGURE 11-17 Veins of the neck and head in right lateral view.

EXHIBIT 11-10 ▰▰▰▰▰▰▰▰▰▰▰▰▰▰▰▰▰▰▰▰▰▰▰▰▰▰▰▰▰▰▰▰▰▰▰▰▰
VEINS OF UPPER EXTREMITIES (Figure 11-18)

VEIN	DESCRIPTION AND REGION DRAINED
	Blood from each upper extremity is returned to heart by deep and superficial veins. Both sets of veins contain valves.
DEEP VEINS	Deep veins run along side arteries and are called **brachial, axillary,** and **subclavian veins.**
SUPERFICIAL VEINS	Superficial veins anastomose extensively with each other and with deep veins.
Cephalics	**Cephalic vein** of each upper extremity begins in **dorsal arch** of hand and winds upward around radial border of forearm. Just below elbow, it unites with accessory cephalic vein to form cephalic vein of upper extremity. It eventually empties into axillary vein.
Basilics	**Basilic vein** of each upper extremity originates in ulnar part of dorsal arch. It extends along posterior surface of ulna to point below elbow where it joins **median cubital vein.** If a vein must be punctured for an injection, transfusion, or removal of a blood sample, median cubitals are preferred.
Axillaries	**Axillary vein** is continuation of basilic. It ends at about first rib, where it becomes subclavian.
Subclavians	Right and left **subclavian veins** unite with internal jugulars to form **brachiocephalic veins.** Thoracic duct of lymphatic system flows into left subclavian vein at junction with internal jugular. Right lymphatic duct enters right subclavian vein at corresponding junction.

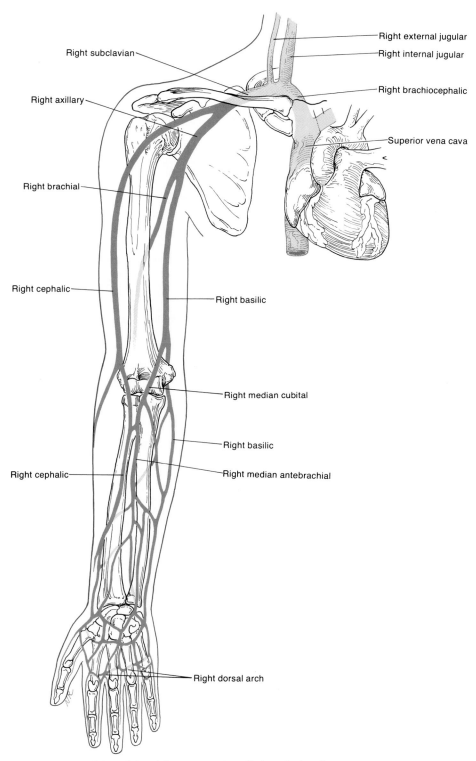

FIGURE 11-18 Veins of the right upper extremity in anterior view.

EXHIBIT 11-11
VEINS OF THORAX (Figure 11-19)

VEIN	DESCRIPTION AND REGION DRAINED
	Principal thoracic vessels that empty into superior vena cava are brachiocephalic and azygos veins.
BRACHIOCEPHALIC	Right and left **brachiocephalic veins,** formed by union of subclavians and internal jugulars, drain blood from head, neck, upper extremities, mammary glands, and upper thorax. Brachiocephalics unite to form **superior vena cava.**
AZYGOS	**Azygos veins,** besides collecting blood from thorax, may serve as bypass for inferior vena cava that drains blood from lower body. Several small veins directly link azygos veins with inferior vena cava. And large veins that drain lower extremities and abdomen dump blood into azygos. If inferior vena cava or hepatic portal vein becomes obstructed, azygos veins can return blood from lower body to superior vena cava.
Azygos vein	**Azygos vein** lies in front of vertebral column, slightly right of midline. It begins as continuation of right ascending lumbar vein. It connects with inferior vena cava, right common iliac, and lumbar veins. Azygos receives blood from right intercostal veins that drain chest muscles; from hemiazygos and accessory hemiazygos veins; from several esophageal, mediastinal, and pericardial veins; and from right bronchial vein. Vein ascends to fourth thoracic vertebra, arches over right lung, and empties into superior vena cava.
Hemiazygos vein	**Hemiazygos vein** is in front of vertebral column and slightly left of midline. It begins as continuation of left ascending lumbar vein. It receives blood from lower four or five intercostal veins and some esophageal and mediastinal veins. At level of ninth thoracic vertebra, it joins azygos vein.
Accessory hemiazygos vein	**Accessory hemiazygos vein** is also in front and to left of vertebral column. It receives blood from three or four intercostal veins and left bronchial vein. It joins azygos at level of eighth thoracic vertebra.

EXHIBIT 11-12
VEINS OF ABDOMEN AND PELVIS (Figure 11-19)

VEIN	DESCRIPTION AND REGION DRAINED
Inferior vena cava	**Inferior vena cava** is the largest vein of body. It is formed by union of two **common iliac veins** that drain lower extremities and abdomen. It extends upward through abdomen and thorax to right atrium. Numerous small veins enter inferior vena cava. Most carry return flow from branches of abdominal aorta, and names correspond to names of arteries: left and right **renal veins** from kidneys; right **testicular vein** from testes (left testicular vein empties into left renal vein); right **ovarian vein** from ovaries (left ovarian vein also empties into left renal vein); right **suprarenal veins** from adrenal glands (left inferior phrenic vein sends tributary to left renal vein) **inferior phrenic vein** from diaphragm (left inferior phrenic vein sends tributary to left renal vein); and **hepatic veins** from liver. In addition, a series of parallel **lumbar veins** drain blood from both sides of posterior abdominal wall. Lumbars connect at right angles with right and left ascending lumbar veins, which form origin of corresponding azygos or hemiazygos vein. Lumbars drain blood into ascending lumbars and then run to inferior vena cava, where they release remainder of flow.

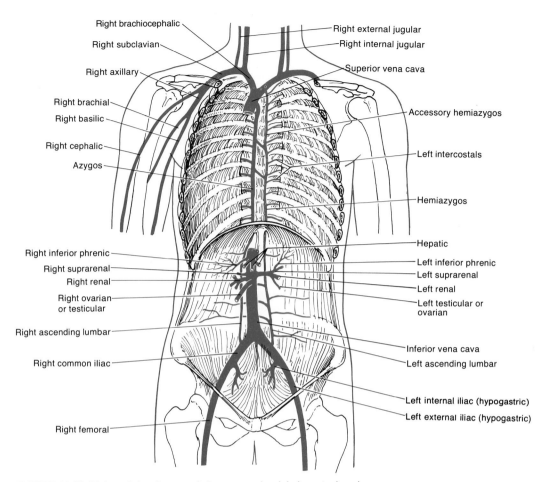

FIGURE 11-19 Veins of the thorax, abdomen, and pelvis in anterior view.

EXHIBIT 11-13
VEINS OF LOWER EXTREMITIES (Figure 11-20)

VEIN	DESCRIPTION AND REGION DRAINED
	Blood from each lower extremity is returned by superficial set and deep set of veins. Superficials are formed from extensive anastomoses close to surface. Deep veins follow large arterial trunks. Both sets have valves.
SUPERFICIAL VEINS	Main superficial veins are great saphenous and small saphenous. Both, especially great saphenous, frequently become varicosed.
Great saphenous	**Great saphenous vein,** longest vein in body, begins at medial end of **dorsal venous arch** of foot. It passes in front of medial malleolus and then upward along medial aspect of leg and thigh. It receives tributaries from superficial tissues and connects with deep veins as well. It empties into femoral vein in groin.
Small saphenous	**Small saphenous vein** begins at lateral end of dorsal venous arch of foot. It passes behind lateral malleolus and ascends under skin of back of leg. It receives blood from foot and posterior portion of leg. It empties into popliteal vein behind knee.
DEEP VEINS Posterior tibial	**Posterior tibial vein** is formed by union of **medial** and **lateral plantar veins** behind medial malleolus. It ascends deep in muscle at back of leg, receives blood from **peroneal vein,** and unites with anterior tibial vein just below knee.
Anterior tibial	**Anterior tibial vein** is upward continuation of **dorsalis pedis** veins in foot. It runs between tibia and fibula and unites with posterior tibial to form popliteal vein.
Popliteal	**Popliteal vein,** just behind knee, receives blood from anterior and posterior tibials and small saphenous vein.
Femoral	**Femoral vein** is upward continuation of popliteal just above knee. Femorals run up posterior of legs and drain deep structures of thighs. After receiving great saphenous veins in groin, they continue as right and left **external iliac veins.** Right and left **internal iliac veins** receive blood from pelvic wall and viscera, external genitals, buttocks, and medial aspect of thigh. Right and left **common iliac veins** are formed by union of internal and external iliacs. Common iliacs unite to form inferior vena cava.

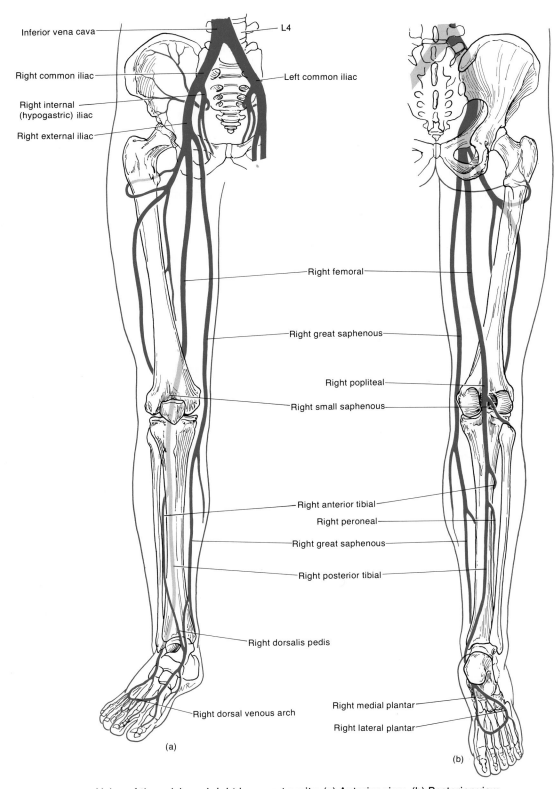

FIGURE 11-20 Veins of the pelvis and right lower extremity. (a) Anterior view. (b) Posterior view.

CORONARY CIRCULATION

The walls of the heart, like any other tissue, including large blood vessels, have their own blood vessels. Nutrients could not possibly diffuse through all the layers of cells that make up the heart tissue. And the blood in the left chambers of the heart would never supply enough oxygen. The flow of blood through the numerous vessels that pierce the myocardium is called the **coronary circulation** (Figure 11-21). The vessels that serve the myocardium include the *left coronary artery,* which originates as a branch of the ascending aorta. This artery runs under the left atrium and divides into the anterior interventricular and circumflex branches. The *anterior interventricular branch* follows the anterior interventricular sulcus and supplies oxygenated blood to the walls of both ventricles. The *circumflex branch* distributes oxygenated blood to the walls of the left ventricle and left atrium. The *right coronary artery* also originates as a branch of the ascending aorta. It runs under the right atrium and di-

vides into the posterior interventricular and marginal branches. The *posterior interventricular branch* follows the posterior interventricular sulcus and supplies the walls of the two ventricles with oxygenated blood. The *marginal branch* transports oxygenated blood to the myocardium of the right ventricle and right atrium. The left ventricle receives the most abundant blood supply because of the enormous work it must do.

As blood passes through the arterial system of the heart, it delivers oxygen and nutrients and collects carbon dioxide and wastes. Most of the deoxygenated blood, which carries the carbon dioxide and wastes, is collected by a large vein, the *coronary sinus,* which empties into the right atrium. The principal tributaries of the coronary sinus are the *great cardiac vein,* which drains the anterior aspect of the heart, and the *middle cardiac vein,* which drains the posterior aspect of the heart.

Most heart problems result from faulty coronary circulation. **Ischemia** is a reduced oxygen supply that weakens the cells but does not actually kill them. *An-*

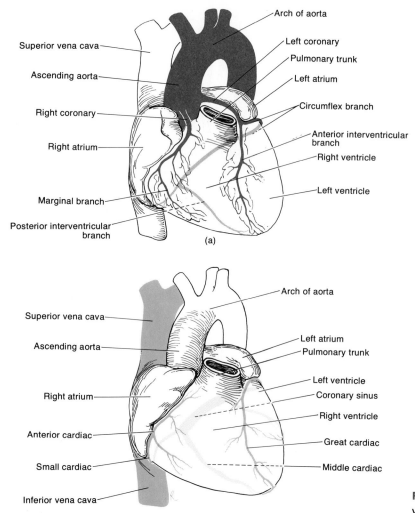

FIGURE 11-21 Coronary circulation. (a) Anterior view of arterial distribution. (b) Anterior view of venous drainage.

gina pectoris ("chest pain") is ischemia of the myocardium. (Remember that pain impulses originating from most visceral muscles are referred to an area on the surface of the body.) Angina pectoris occurs when coronary circulation is somewhat reduced for some reason. Stress, which produces constriction of vessel walls, is a common cause. Equally common is strenuous exercise after a heavy meal. When any quantity of food enters the stomach, the body increases blood flow to the digestive tract. The digestive glands can then receive enough oxygen for their increased activities, and the digested food can be quickly absorbed into the bloodstream. As a consequence, some blood is diverted away from other organs, including the heart. Exercise, however, increases heart muscle activity and thus increases its need for oxygen. Thus, doing heavy work while food is in the stomach can lead to oxygen deficiency in the myocardium. Angina pectoris weakens the heart muscle, but it does not produce a full-scale heart attack. The simple remedy of taking nitroglycerin, a drug that dilates coronary vessels and thereby increases the area of blood flow, brings coronary circulation back to normal and stops pain of angina. Because repeated attacks of angina can weaken the heart and lead to serious heart trouble, angina patients are told to avoid activities and stresses that bring on the attacks.

A much more serious problem is *myocardial infarction,* commonly called a "coronary" or "heart attack." **Infarction** means death of an area of tissue because of a drastically reduced or completely interrupted blood supply. Myocardial infarction results from a thrombus or embolus in one of the coronary arteries. The tissue on the downstream side of the obstruction dies and is replaced by noncontractile scar tissue. Thus, the heart muscle loses at least some of its strength. The aftereffects depend partly on the size and location of the infarcted, or dead, area.

HEPATIC PORTAL CIRCULATION

Blood enters the liver from two sources. The hepatic artery delivers oxygenated blood from the systemic circulation; the hepatic portal vein delivers deoxygenated blood from the digestive organs. The term **hepatic portal circulation** refers to this flow of venous blood from the digestive organs to the liver before returning to the heart (Figure 11-22). Hepatic portal blood is rich with substances absorbed from the digestive tract. The liver monitors these substances before they pass into the general circulation. For example, the liver stores nutrients such as glucose. It modifies other digested substances so they may be used by cells. And it detoxifies harmful substances that have been absorbed by

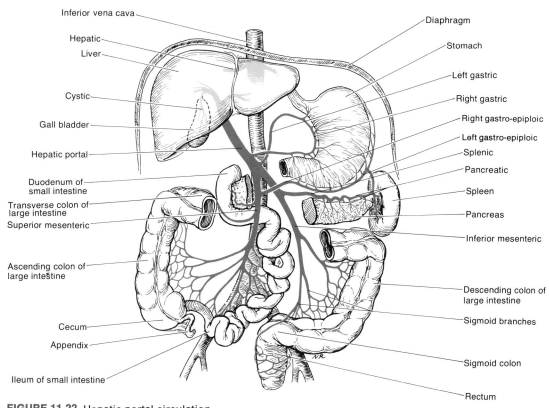

FIGURE 11-22 Hepatic portal circulation.

the digestive tract and destroys bacteria by phagocytosis.

The hepatic portal system includes veins that drain blood from the pancreas, spleen, stomach, intestines, and gallbladder and transport it to the hepatic portal vein of the liver. The *hepatic portal vein* is formed by the union of the superior mesenteric and splenic veins. The *superior mesenteric vein* drains blood from the small intestine and portions of the large intestine and stomach. The *splenic vein* drains the spleen and receives tributaries from the stomach, pancreas, and colon. The tributaries from the stomach are the *coronary, pyloric,* and *gastro-epiploic veins.* The *pancreatic veins* come from the pancreas, and the *inferior mesenteric veins* come from portions of the colon. Before the hepatic portal vein enters the liver, it receives the *cystic vein* from the gallbladder. Ultimately, blood leaves the liver through the *hepatic veins,* which enter the inferior vena cava.

PULMONARY CIRCULATION

The flow of deoxygenated blood from the right ventricle to the lungs and the return of oxygenated blood from the lungs to the left atrium is called the **pulmonary circulation** (Figure 11-23). Pulmonary circulation carries a flow of blood equal to all the systemic arteries. The *pulmonary trunk* emerges from the right ventricle and passes upward, backward, and to the left. It then divides into two branches. The *right pulmonary artery* runs to the right lung; the *left pulmonary artery* goes to the left lung. On entering the lungs, the

branches divide and subdivide. They get smaller and ultimately form capillaries around the alveoli in the lungs. Carbon dioxide is passed from the blood into the alveoli to be breathed out of the lungs. Oxygen breathed in by the lungs is passed from the alveoli into the blood. The capillaries then unite. They grow larger and become veins. Eventually, two *pulmonary veins* exit from each lung and transport the oxygenated blood to the left atrium. The pulmonary veins are the only postnatal veins that carry oxygenated blood. Contraction of the left ventricle then sends the blood into the systemic circulation.

FETAL CIRCULATION

The circulatory system of a fetus, called **fetal circulation,** differs from an adult's because the lungs, kidneys, and digestive tract of a fetus are nonfunctional. The fetus derives its oxygen and nutrients and eliminates its carbon dioxide and wastes through the maternal blood (Figure 11-24).

The exchange of materials between fetal and maternal circulation occurs through a structure called the *placenta.* It is attached to the navel of the fetus by the umbilical cord, and it communicates with the mother through countless small blood vessels that emerge from the uterine wall. The umbilical cord contains blood vessels that branch into capillaries in the placenta. Wastes from the fetal blood diffuse out of the capillaries, into spaces containing maternal blood (intervillous spaces) in the placenta, and finally into the mother's uterine blood vessels. Nutrients travel the

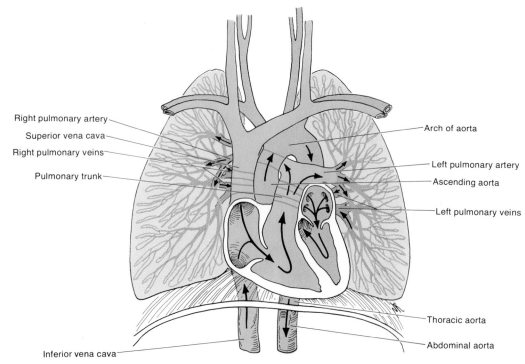

Right pulmonary artery
Superior vena cava
Right pulmonary veins
Pulmonary trunk

Arch of aorta
Left pulmonary artery
Ascending aorta
Left pulmonary veins

Thoracic aorta
Abdominal aorta
Inferior vena cava

FIGURE 11-23 Pulmonary circulation.

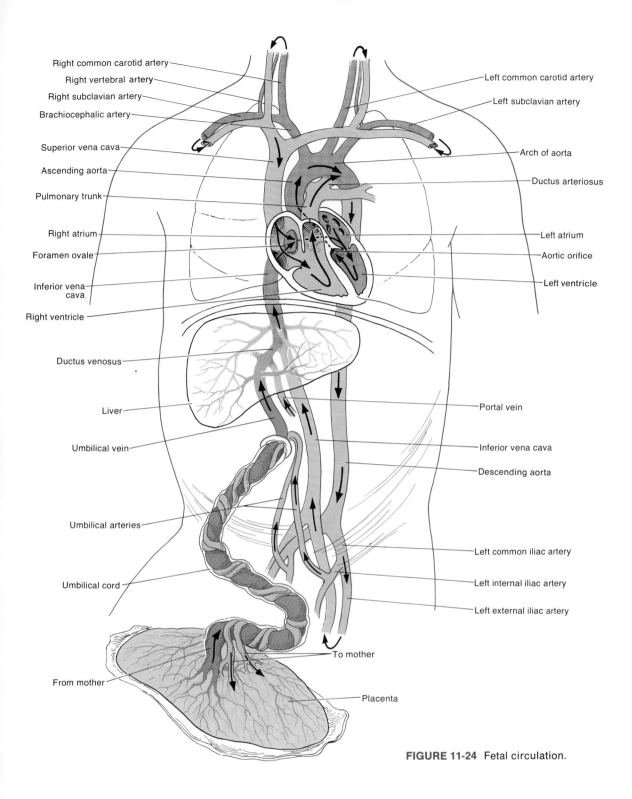

Right common carotid artery

Right vertebral artery

Right subclavian artery

Brachiocephalic artery

Superior vena cava

Ascending aorta

Pulmonary trunk

Right atrium

Foramen ovale

Inferior vena cava

Right ventricle

Ductus venosus

Liver

Umbilical vein

Umbilical arteries

Umbilical cord

From mother

Left common carotid artery

Left subclavian artery

Arch of aorta

Ductus arteriosus

Left atrium

Aortic orifice

Left ventricle

Portal vein

Inferior vena cava

Descending aorta

Left common iliac artery

Left internal iliac artery

Left external iliac artery

To mother

Placenta

FIGURE 11-24 Fetal circulation.

opposite route—from the maternal blood vessels to the intervillous spaces to the fetal capillaries. Normally there is no mixing of maternal and fetal blood since all exchanges occur through capillaries.

Blood passes from the fetus to the placenta via two *umbilical arteries*. These branches of the internal iliac arteries are included in the umbilical cord. At the placenta, the blood picks up oxygen and nutrients and eliminates carbon dioxide and wastes. The oxygenated blood returns from the placenta via a single *umbilical*

vein. This vein ascends to the liver of the fetus where it divides into two branches. Some blood flows through the branch that joins the hepatic portal vein and enters the liver. Although the fetal liver manufactures red blood cells, it does not function in digestion. Therefore, most of the blood flows into the second branch: the *ductus venosus*. The ductus venosus connects with the inferior vena cava.

In general, circulation through other portions of the fetus is not unlike the postnatal circulation. Deoxygenated blood returning from the lower regions is mingled with oxygenated blood from the ductus venosus in the inferior vena cava. This mixed blood then enters the right atrium. The circulation of blood through the upper portion of the fetus is also similar to the postnatal flow. Deoxygenated blood returning from the upper regions of the fetus is collected by the superior vena cava, and it also passes into the right atrium.

Unlike postnatal circulation, most of the blood does not pass through the right ventricle to the lungs since the fetal lungs do not operate. In the fetus, an opening called the *foramen ovale* exists in the septum between the right and left atria. A valve in the inferior vena cava directs most of the blood through the foramen ovale so that it may be sent directly into the systemic circulation. The blood that does descend into the right ventricle is pumped into the pulmonary trunk, but little of this blood actually reaches the lungs. Most blood in the pulmonary trunk is sent through the *ductus arteriosus.* This small vessel connecting the pulmonary trunk with the aorta enables blood in excess of nutrient requirements to bypass the fetal lungs. The blood in the aorta is carried to all parts of the fetus through its systemic branches. When the common iliac arteries branch into the external and internal iliacs, part of the blood flows into the internal iliacs. It then goes to the umbilical arteries and back to the placenta for another exchange of materials. The only vessel that carries fully oxygenated blood is the umbilical vein.

At birth, when lung, renal, digestive, and liver functions are established, the special structures of fetal circulation are no longer needed. Thus:

1. The umbilical arteries atrophy to become the lateral umbilical ligaments.
2. The umbilical vein becomes the round ligament of the liver.
3. The placenta is passed by the mother as the "afterbirth."
4. The ductus venosus becomes the ligamentum venosum, a fibrous cord in the liver.
5. The foramen ovale normally closes shortly after birth to become the fossa ovalis, a depression in the interatrial septum, (see Figure 11-5).
6. The ductus arteriosus closes, atrophies, and becomes the ligamentum arteriosum.

Usually the ductus arteriosus closes shortly after birth. When it fails to close or closes imperfectly, blood shuttles uselessly back and forth between heart and lungs. This condition, *patent ductus arteriosus,* is easily remedied by surgery.

Lymphatic System

Lymph, lymph vessels, a series of small masses of lymphoid tissue called lymph nodes, and three organs —tonsils, thymus, and spleen—make up the **lymphatic system.** The primary function of the lymphatic system is to drain, from the tissue spaces, protein-containing fluid that escapes from the blood capillaries. Such proteins cannot be directly reabsorbed. Other functions of the lymphatic system are to transport fats from the digestive tract to the blood, to produce lymphocytes, and to develop immunities.

LYMPHATIC VESSELS

Lympatic vessels originate as blind-end tubes that begin in spaces between cells. The tubes, which occur singly or in extensive plexuses, are called **lymph capillaries.** Lymph capillaries originate in most parts of the body. They are absent in avascular tissue and from the central nervous system. They are, however, slightly larger and more permeable than blood capillaries. Just as blood capillaries converge to form venules and veins, lymph capillaries unite to form larger and larger lymph vessels called **lymphatics** (Figure 11-25). Lymphatics resemble veins in structure but have thinner walls and more valves and contain lymph nodes at various intervals. Lymphatics of the skin travel in the loose subcutaneous tissue and generally follow veins. Lymphatics of the viscera generally follow arteries, forming plexuses around them. Ultimately, lymphatics converge into two main channels—the thoracic duct and the right lymphatic duct.

The **thoracic duct,** or **left lymphatic duct,** is about 38–45 cm (15-18 inches) long and begins as a dilation in front of the second lumbar vertebra. This dilation is called the *cisterna chyli.* The thoracic duct receives lymph from the left side of the head, neck, and chest, the left upper extremity, and the entire body below the ribs. It is the main collecting trunk of the lymphatic system. It then empties the lymph into the left subclavian vein. A pair of valves at this junction prevents the passage of venous blood into the thoracic duct. The **right lymphatic duct,** which is about 1.25 cm (0.5 inches) long, drains lymph from the upper right side of the body and empties it into the right subclavian vein at the junction of the right subclavian and jugular veins. Like the thoracic duct, the right lymphatic duct also has a pair of valves that prevent the passage of venous blood into it.

Lymphangiography is a procedure by which lymphatic vessels and lymph organs are filled with an opaque substance in order to be filmed. Such a film is called a *lymphangiogram.* Lymphangiograms are useful in detecting edema and carcinomas and in localizing lymph nodes for surgical or radiotherapeutic treatment. A normal lymphangiogram of lymphatic vessels and a few nodes in the upper thighs and pelvis is shown in Figure 11-26.

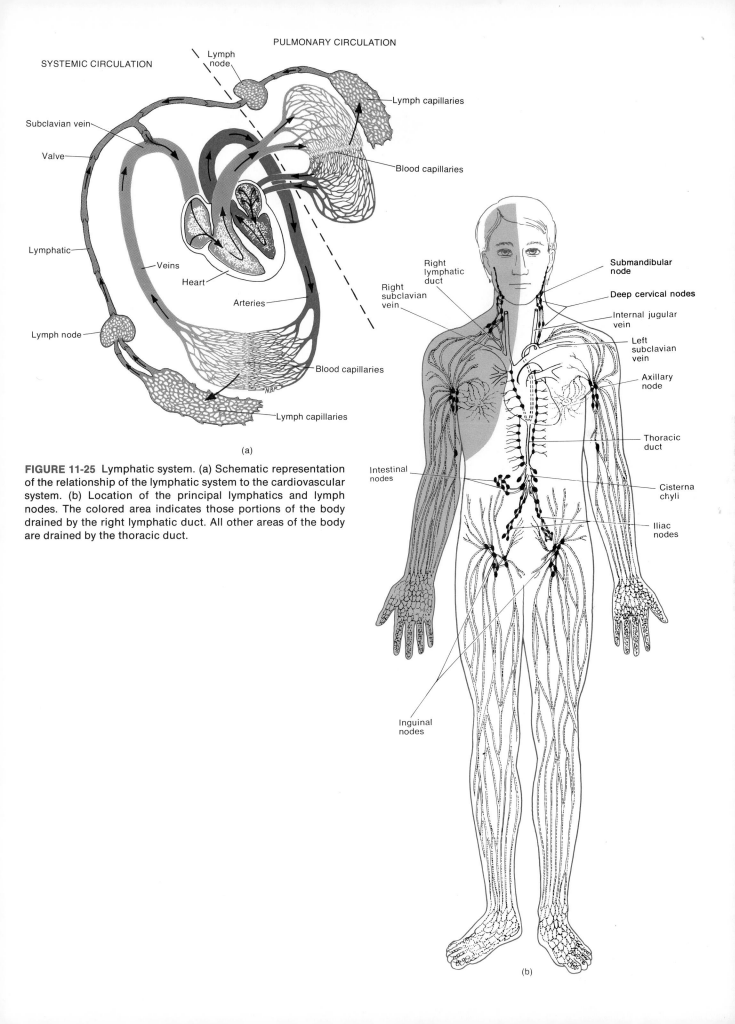

PULMONARY CIRCULATION

SYSTEMIC CIRCULATION

Lymph node

Subclavian vein

Valve

Lymphatic

Veins

Heart

Arteries

Lymph node

Blood capillaries

Lymph capillaries

Lymph capillaries

Blood capillaries

(a)

FIGURE 11-25 Lymphatic system. (a) Schematic representation of the relationship of the lymphatic system to the cardiovascular system. (b) Location of the principal lymphatics and lymph nodes. The colored area indicates those portions of the body drained by the right lymphatic duct. All other areas of the body are drained by the thoracic duct.

Right lymphatic duct

Right subclavian vein

Submandibular node

Deep cervical nodes

Internal jugular vein

Left subclavian vein

Axillary node

Thoracic duct

Intestinal nodes

Cisterna chyli

Iliac nodes

Inguinal nodes

(b)

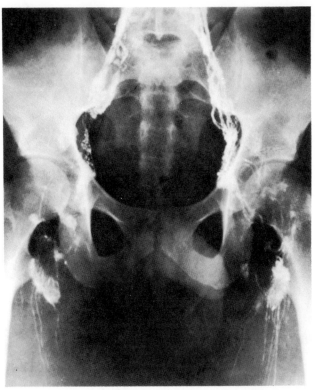

FIGURE 11-26 Normal lymphangiogram of the upper thighs and pelvis. Can you identify the lymphatics and lymph nodes? (Courtesy of Lester W. Paul and John H. Juhl, *The Essentials of Roentgen Interpretation,* 3d ed., Harper & Row, Publishers, Inc., New York, 1972.)

STRUCTURE OF LYMPH NODES

The oval or bean-shaped structures located along the length of lymphatics are called **lymph nodes** or **lymph glands.** They range from 1 to 25 mm (0.04 to 1 inch) in length. Structurally, a lymph node contains a slight depression on one side called a *hilus* where the blood vessels enter and leave the node (Figure 11-27). Each node is covered by a *capsule* of fibrous connective tissue that extends into the node. The capsular extensions are called *trabeculae.* The capsule, trabeculae, and hilus constitute the stroma (framework) of a lymph node. The parenchyma of a lymph node is specialized into two regions. The outer *cortex* contains densely packed lymphocytes arranged in masses called *lymph nodules.* The nodules often contain lighter-staining central areas, the *germinal centers,* where lymphocytes are produced. The inner region of a lymph node is called the *medulla.* In the medulla, the lymphocytes are arranged in strands called *medullary cords.*

The circulation of lymph through a node involves afferent lymphatic vessels, sinuses in the node, and efferent lymphatic vessels. *Afferent lymphatic vessels* enter the convex surface of the node at several points. They contain valves that open toward the node so that the circulation of lymph through the afferent lymphatic vessels is into the lymph node. Once inside the lymph node, the lymph enters the sinuses of the node, which

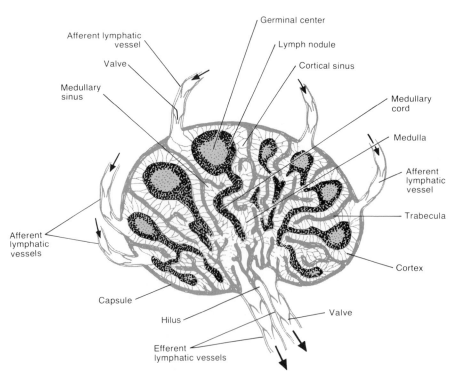

FIGURE 11-27 Structure of a lymph node. The path taken by circulating lymph is indicated by the arrows.

are a series of irregular channels. Lymph from the afferent lymphatic vessels enters the *cortical sinuses* under the capsule. From here the lymph circulates to the *medullary sinuses* between the medullary cords. From these sinuses, the lymph circulates into the *efferent lymphatic vessels*. These vessels are located at the hilus of the lymph node. The efferent vessels are wider and fewer in number than the afferent vessels and contain valves that open away from the lymph node. Thus they convey lymph out of the node.

As the lymph circulates through the nodes, it is processed by fixed phagocytic cells of the reticuloendothelial system called macrophages that line the sinuses. Macrophages filter lymph of bacteria, dirt, and cell debris. Lymph nodes also give rise to lymphocytes and/or plasma cells. The plasma cells produce antibodies. At times the number of entering microbes is so great that the node itself may become infected. It then becomes enlarged and tender. Histological features of lymphatics and lymph nodes are shown in Figure 11-28.

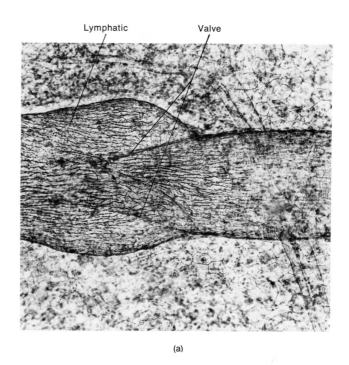

(a)

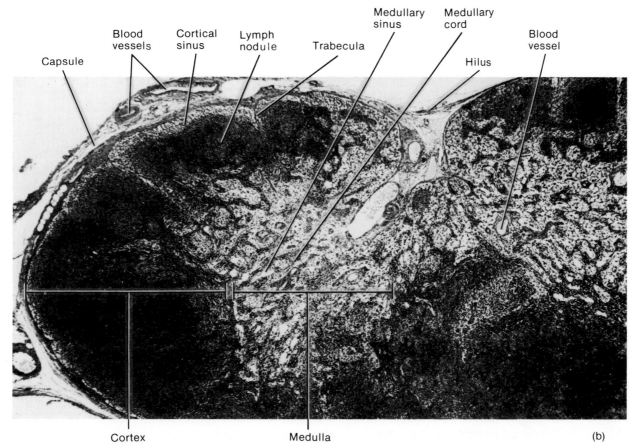

(b)

FIGURE 11-28 Histology of lymphatics and lymph nodes. (a) Longitudinal section of a lymphatic at a magnification of 50×. (Courtesy of Carolina Biological Supply Company.) (b) Section through a lymph node at a magnification of 40×. (Courtesy of Edward J. Reith, from *Atlas of Descriptive Histology,* by Edward J. Reith and Michael H. Ross, Harper & Row, Publishers, Inc., New York, 1970.)

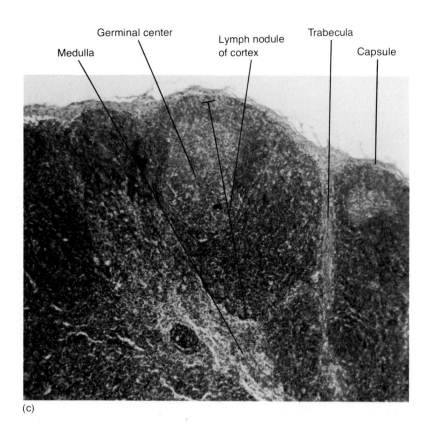

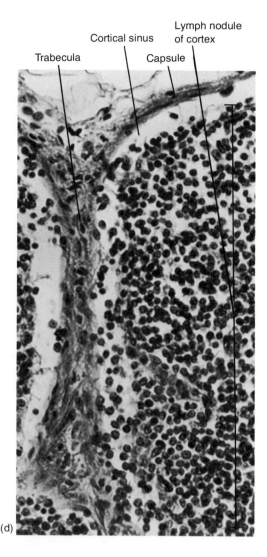

FIGURE 11-28 *(Continued)* Histology of lymphatics and lymph nodes. (c) Section through a lymph node at a magnification of 15×. (d) Enlarged aspect of a lymph node at a magnification of 90×. (Courtesy of Victor B. Eichler, Wichita State University.)

Plan of Lymph Circulation

When plasma is filtered by blood capillaries it passes into the interstitial spaces; it is then known as interstitial fluid. When this fluid passes from the interstitial spaces into lymph capillaries it is called lymph. Lymph from lymph capillary plexuses is then passed to lymphatics that run toward lymph nodes. At the nodes, afferent vessels penetrate the capsules at numerous points and the lymph passes through the sinuses of the nodes. Efferent vessels from the nodes either run with afferent vessels into another node of the same group or pass on to another group of nodes. From the most proximal group of each chain of nodes, the efferent vessels unite to form *lymph trunks*. The principal trunks, which will be described shortly, are named the *lumbar, intestinal, bronchomediastinal, subclavian,* and *jugular trunks*.

The principal trunks pass their lymph into the thoracic duct and right lymphatic trunk in the following way (Figure 11-29). The cisterna chyli receives lymph from the right and left lumbar trunks and from the intestinal trunk. The lumbar trunks drain lymph from the lower extremities, walls and viscera of the pelvis, kidneys, suprarenal nodes, and the deep lymphatics from most of the abdominal wall. The intestinal trunk drains lymph from the stomach, intestines, pancreas, spleen, and visceral surface of the liver. In the neck, the thoracic duct also receives lymph from the left jugular, left subclavian, and left bronchomediastinal trunks. The left jugular trunk drains lymph from the left side of the head and neck; the left subclavian trunk drains lymph from the left upper extremity; and the left bronchomediastinal trunk drains lymph from the left side of the deeper parts of the anterior thoracic wall, the upper part of the anterior abdominal wall, the anterior

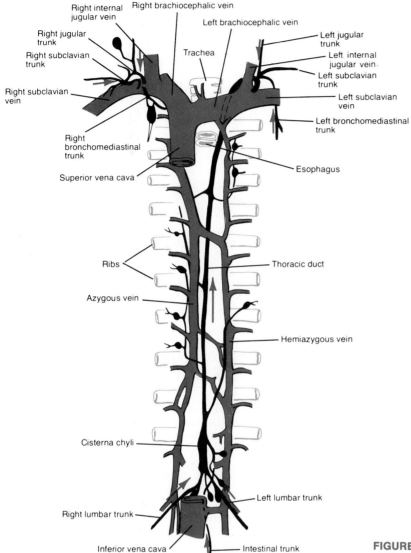

Right internal
jugular vein
Right brachiocephalic vein
Left brachiocephalic vein
Right jugular
trunk
Left jugular
trunk
Right subclavian
trunk
Trachea
Left internal
jugular vein
Left subclavian
trunk
Right subclavian
vein
Left subclavian
vein
Left bronchomediastinal
trunk
Right
bronchomediastinal
trunk
Superior vena cava
Esophagus
Ribs
Thoracic duct
Azygous vein
Hemiazygous vein
Cisterna chyli
Left lumbar trunk
Right lumbar trunk
Inferior vena cava
Intestinal trunk

FIGURE 11-29 Relationship of lymph trunks to the thoracic duct and right lymphatic duct.

part of the diaphragm, the left lung, and the left side of the heart.

The right lymphatic duct receives lymph from the right jugular trunk, which drains the right side of the head and neck, from the right subclavian trunk, which drains the right upper extremity, and from the right bronchomediastinal trunk, which drains the right side of the thorax, right lung, right side of the heart, and part of the convex surface of the liver.

Ultimately, the thoracic duct empties all of its lymph into the left subclavian vein and the right lymphatic duct empties all of its lymph into the right subclavian vein. Thus, lymph is drained back into the blood and the cycle repeats itself continuously.

Principal Groups of Lymph Nodes

Lymph nodes are scattered throughout the body, usually in groups. Typically, these groups are arranged in two sets: *superficial* and *deep*. Since lymph nodes may become enlarged and tender, an understanding of the regions drained by the nodes may be helpful in diagnosing the site of an infection.

Exhibits 11-14 through 11-18 list the principal groups of lymph nodes of the body by region, the general areas of the body they drain, and the destinations of the efferents—that is, where lymph is passed on its return to the thoracic duct and right lymphatic duct.

EXHIBIT 11-14
PRINCIPAL LYMPH NODES OF THE HEAD AND NECK (Figure 11-30)

LYMPH NODES	LOCATION, AREAS DRAINED, AND COURSE OF EFFERENTS
LYMPH NODES OF THE HEAD	
Occipital	Two to four nodes near the margin of the trapezius muscle and insertion of semispinalis capitis muscle; they drain the occipital portion of scalp and upper neck; efferents pass to superior deep cervical nodes.
Retroauricular	One or two nodes behind ear at insertion of sternocleidomastoid muscle; they drain skin of ear and posterior parietal region of scalp; efferents pass to superior deep cervical nodes.
Preauricular	One of three nodes anterior to tragus; they drain pinna and temporal region of the scalp; efferents pass to superior deep cervical nodes.
Parotid	Two sets of nodes, one embedded in parotid gland and one below gland; they drain root of nose, eyelids, anterior temporal region, external auditory meatus, tympanic cavity, nasopharynx, and posterior portions of nasal fossae; efferents pass to superior deep cervical nodes.
Facial	Three sets called **infraorbital** (below the orbit), **buccal** (at angle of mouth), and **mandibular** (over mandible); they drain eyelids, conjunctiva, and skin and mucous membrane of nose and cheek; efferents pass to submandibular nodes.
LYMPH NODES OF THE NECK	
Submandibular	Three to six nodes along inferior border of mandible; they drain chin, lips, nose, nasal fossae, cheeks, gums, lower surface of palate, and anterior portion of tongue; efferents pass to superior deep cervical nodes.
Submental	Two or three nodes between anterior bellies of digastric muscles; they drain chin, lower lip, cheeks, tip of tongue, and floor of mouth; efferents pass principally to submandibular nodes.
Superficial cervical	These nodes lie along external jugular vein; they drain lower part of ear and parotid region; efferents pass to superior deep cervical nodes.
Deep cervical	Largest group of nodes in neck, consisting of numerous large nodes forming a chain along carotid sheath extending from base of skull to root of neck; they are arbitrarily divided into superior deep cervical nodes and inferior deep cervical nodes. The **superior deep cervical nodes** are under sternocleidomastoid muscle along accessory nerve and internal jugular vein; they drain posterior head and neck, pinna, tongue, larynx, esophagus, thyroid gland, nasopharynx, nasal fossae, and palate; they receive efferents from all other nodes of head and neck, except for inferior deep cervical nodes; efferents pass to inferior deep cervical nodes and jugular trunk. **Inferior deep cervical nodes** are inferior to omohyoid muscle near subclavian vein; they drain posterior scalp and neck, superficial pectoral region, part of arm, and superior deep cervical nodes; efferents join efferents of superior deep cervical nodes to form jugular trunk; on right side jugular trunk terminates at junction of internal jugular and subclavian veins; on left side it terminates in thoracic duct.

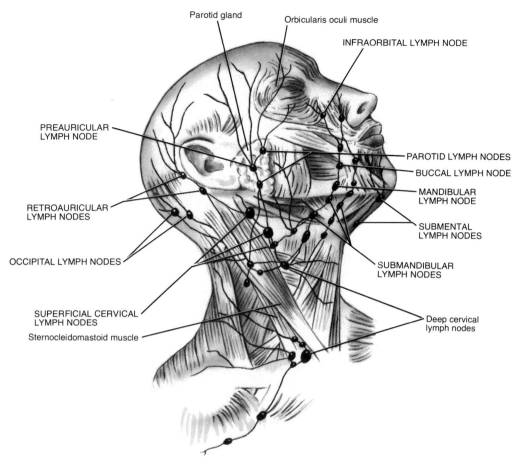

Parotid gland

Orbicularis oculi muscle

INFRAORBITAL LYMPH NODE

PREAURICULAR
LYMPH NODE

PAROTID LYMPH NODES

BUCCAL LYMPH NODE

MANDIBULAR
LYMPH NODE

RETROAURICULAR
LYMPH NODES

SUBMENTAL
LYMPH NODES

OCCIPITAL LYMPH NODES

SUBMANDIBULAR
LYMPH NODES

SUPERFICIAL CERVICAL
LYMPH NODES

Deep cervical
lymph nodes

Sternocleidomastoid muscle

FIGURE 11-30 Principal lymph nodes of the head and neck. Anterolateral view.

EXHIBIT 11-15
PRINCIPAL LYMPH NODES OF THE UPPER EXTREMITIES (Figure 11-31)

LYMPH NODES	LOCATION, AREAS DRAINED, AND COURSE OF EFFERENTS

SUPERFICIAL

Supratrochlear

One or two nodes above medial epicondyle of humerus; they drain medial fingers, palm, and forearm; efferents pass to deep nodes.

Deltopectoral

One or two nodes along cephalic vein below clavicle between pectoralis major and deltoid muscles; they drain lymphatic vessels on radial side of upper extremity; efferents pass to either subclavian or inferior deep cervical nodes.

DEEP

A few small deep lymph nodes are found in the forearm along the radial, ulnar, and interosseous arteries and in the arm alongside of the medial side of the brachial artery; efferents pass to the lateral axillary nodes.

Axillary

Most deep lymph nodes of the upper extremities are in the axilla and are called the axillary nodes. They are large in size, vary in number from 20 to 30 and may be grouped as follows:

Lateral. Four to six nodes on medial and posterior aspects of axillary artery; they drain whole upper extremity, except for those nodes around the cephalic vein; efferents pass to central (intermediate) and subclavicular (medial) nodes. Since infection or malignancy of upper extremity may cause tenderness and swelling in axilla, the axillary nodes, especially the lateral group, are clinically important since they filter lymph from much of upper extremity.

Pectoral (anterior). Four or five nodes located along inferior border of the pectoralis minor muscle; they drain skin and muscles of anterior and lateral thoracic walls and central and lateral portions of mammary gland; efferents pass to central (intermediate) and subclavicular (medial) nodes.

Subscapular (posterior). Six or seven nodes located along subscapular artery; they drain skin and muscles of posterior part of neck and thoracic wall; efferents pass to central (intermediate) nodes.

Central (intermediate). Three or four nodes near base of axilla embedded in adipose tissue; they drain lateral, pectoral (anterior), and subscapular (posterior) nodes; efferents pass to subclavicular (medial) nodes.

Subclavicular (medial). Six to twelve nodes located posterior and superior to pectoralis minor muscle; they drain deltopectoral nodes and have a direct connection with upper portion of mammary gland and all other axillary nodes; efferent vessels unite to form subclavian trunk which opens either into juction of internal jugular and subclavian veins, jugular lymphatic trunk, or thoracic duct; a few efferents pass into inferior deep cervical nodes.

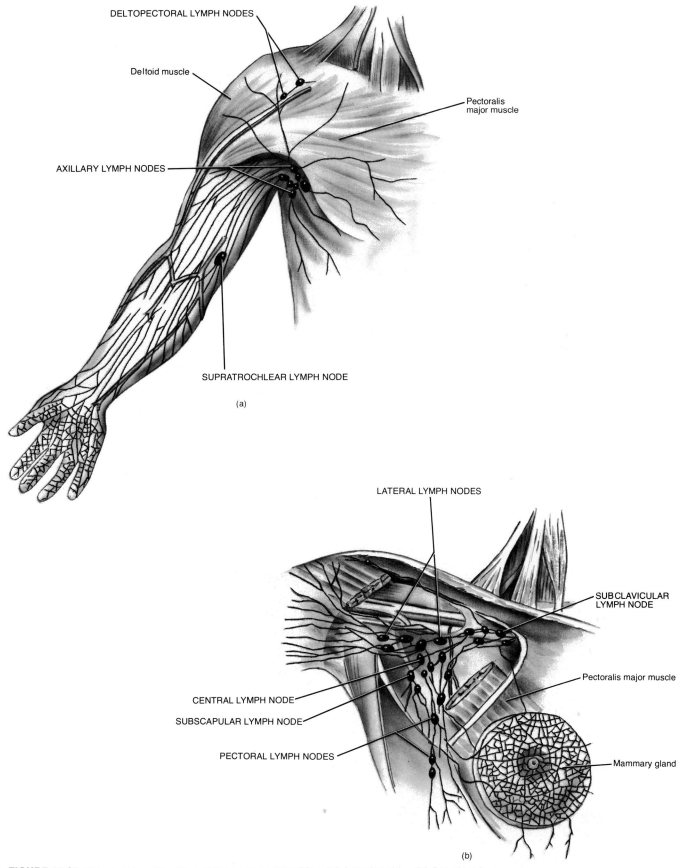

DELTOPECTORAL LYMPH NODES

Deltoid muscle

Pectoralis major muscle

AXILLARY LYMPH NODES

SUPRATROCHLEAR LYMPH NODE

(a)

LATERAL LYMPH NODES

SUBCLAVICULAR LYMPH NODE

Pectoralis major muscle

CENTRAL LYMPH NODE

SUBSCAPULAR LYMPH NODE

PECTORAL LYMPH NODES

Mammary gland

(b)

FIGURE 11-31 Principal lymph nodes of the upper extremities. (a) Anterior view. (b) Anterior view.

EXHIBIT 11-16
PRINCIPAL LYMPH NODES OF THE LOWER EXTREMITIES (Figure 11-32)

LYMPH NODES	LOCATION, AREAS DRAINED, AND COURSE OF EFFERENTS
Popliteal	Six or seven small nodes in adipose tissue in popliteal fossa; they drain anterior and posterior tibial lymphatics and knee; efferents pass to deep inguinal nodes.
Superficial inguinal	Ten to twenty nodes that parallel inguinal ligament and saphenous vein; they drain anterior and lateral abdominal wall to level of umbilicus, gluteal region, external genitals, perineal region, and entire superficial drainage of lower extremity; efferents pass to deep inguinal nodes and external iliac nodes.
Deep inguinal	One to three small nodes that lie medial to femoral vein; they drain deep lymphatics of lower extremity, penis, clitoris, and superficial inguinal nodes; efferents pass to external iliac nodes.

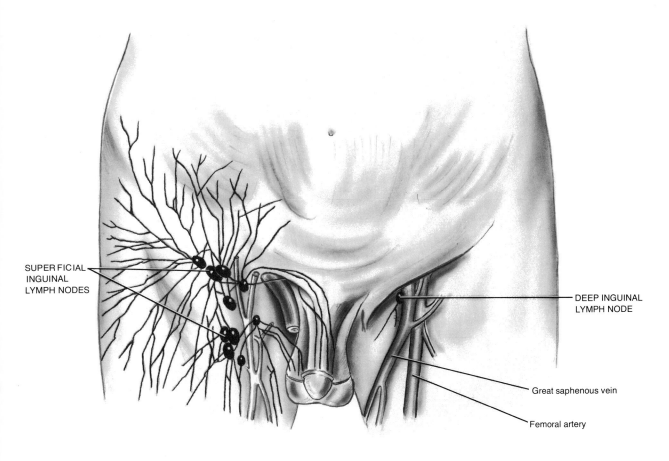

SUPERFICIAL
INGUINAL
LYMPH NODES

DEEP INGUINAL
LYMPH NODE

Great saphenous vein

Femoral artery

(a)

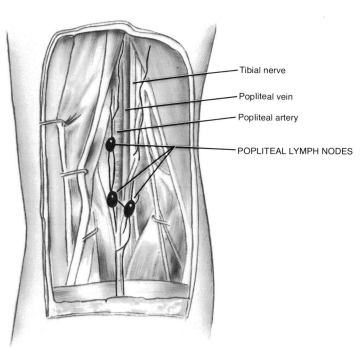

Tibial nerve

Popliteal vein

Popliteal artery

POPLITEAL LYMPH NODES

(b)

FIGURE 11-32 Principal lymph nodes of the lower extremities.
(a) Anterior view. (b) Posterior view.

EXHIBIT 11-17
PRINCIPAL LYMPH NODES OF THE ABDOMEN AND PELVIS (Figure 11-33)

LYMPH NODES	LOCATION, AREAS DRAINED, AND COURSE OF EFFERENTS
	Lymph nodes of the abdomen and pelvis are divided into parietal lymph nodes that are retroperitoneal (behind the parietal peritoneum) and in close association with larger blood vessels and visceral lymph nodes found in association with visceral arteries.
PARIETAL	
External iliac	Eight to ten nodes arranged about lateral, medial, and anterior aspects of external iliac vessels; they drain the inguinal and subinguinal nodes, deep lymphatics of abdominal wall below umbilicus, adductor region of thigh, urinary bladder, prostate gland, ductus deferens, seminal vesicles, prostatic and membranous urethra, uterine tube, uterus, and vagina; efferents pass to common iliac nodes and are interconnected by lymphatics to internal iliac nodes.
Common iliac	Four to six nodes arranged laterally, medially, and posteriorly along course of common iliac vessels; they drain external iliac and internal iliac nodes and pelvic viscera; efferents pass to lumbar nodes.
Internal iliac	These nodes are grouped near origin of branches of internal iliac artery; they drain pelvic viscera, perineum, gluteal region, and posterior surface of thigh; efferents pass to common iliac nodes.
Sacral	Five or six nodes located in hollow of sacrum; they drain rectum, prostate gland, and posterior pelvic wall; efferents pass to internal iliac and lumbar nodes.
Lumbar	These nodes extend in almost a complete chain from aortic bifurcation to aortic hiatus of diaphragm; they are related to four sides of aorta and are designated as **right lateral aortic nodes, left lateral aortic nodes, preaortic nodes,** and **retroaortic nodes;** they drain the efferents of common iliac nodes, lymphatics from testes, ovaries, uterine tubes, and body of uterus, lymphatics from kidneys, suprarenal glands, and abdominal surface of diaphragm, and efferent vessels of lateral abdominal wall; left lateral aortic nodes also receive some of efferents from inferior mesenteric nodes; most efferent vessels of lateral aortic nodes converge to form right and left lumbar trunks, which join cisterna chyli of thoracic duct; some efferents enter preaortic and retroaortic nodes, while others join caudal end of thoracic duct. The preaortic nodes are anterior to aorta and are divided into celiac, superior mesenteric, and inferior mesenteric groups located near origins of corresponding vessels; they drain viscera supplied by their respective arteries and lateral aortic nodes; some efferents pass to retroaortic nodes but most form intestinal trunk, which enters cisterna chyli. Retroaortic nodes are inferior to cisterna chyli near bodies of L3 and L4; they drain lateral and preaortic nodes; efferents pass to cisterna chyli.
VISCERAL	The visceral nodes are associated with branches of the celiac artery, superior mesenteric artery, and inferior mesenteric artery.
Celiac	The nodes related to the branches of the celiac artery form the gastric, hepatic, and pancreaticolienal sets of nodes.
Gastric	**Superior gastric nodes** accompany the left gastric artery and lie along cardiac half of lesser curvature of stomach between layers of lesser omentum; they drain lesser curvature of stomach; efferents pass to celiac nodes. **Inferior gastric nodes** lie within layers of greater omentum along pyloric half of greater curvature of stomach; they drain inferior, anterior, and posterior aspects of stomach; efferents pass to hepatic nodes.
Hepatic	These nodes lie along the hepatic artery; they drain stomach, duodenum, liver, gallbladder, and pancreas; efferents pass to celiac nodes.
Pancreaticolienal	These nodes lie along splenic artery; they drain stomach, spleen, and pancreas; efferents pass to celiac nodes.
Superior mesenteric	These nodes are divisible into mesenteric, ileocolic, and mesocolic groups.
Mesenteric	Numerous (more than 100) small nodes along trunk of superior mesenteric artery; they drain jejunum and all parts of ileum, except for terminal portion; efferents pass to preaortic nodes.
Ileocolic	Ten to twenty nodes forming a chain along ileocolic artery; they drain terminal portion of ileum, appendix, cecum, and ascending colon; efferents pass to preaortic nodes.
Mesocolic	Small nodes located between layers of transverse mesocolon; they drain descending iliac and sigmoid parts of colon; efferents pass to preaortic nodes.
Inferior mesenteric	These nodes are located near left colic and sigmoid arteries, superior rectal artery, and muscular tunic of rectum; they drain descending, iliac, and sigmoid parts of colon and superior part of rectum; efferents pass to preaortic nodes.

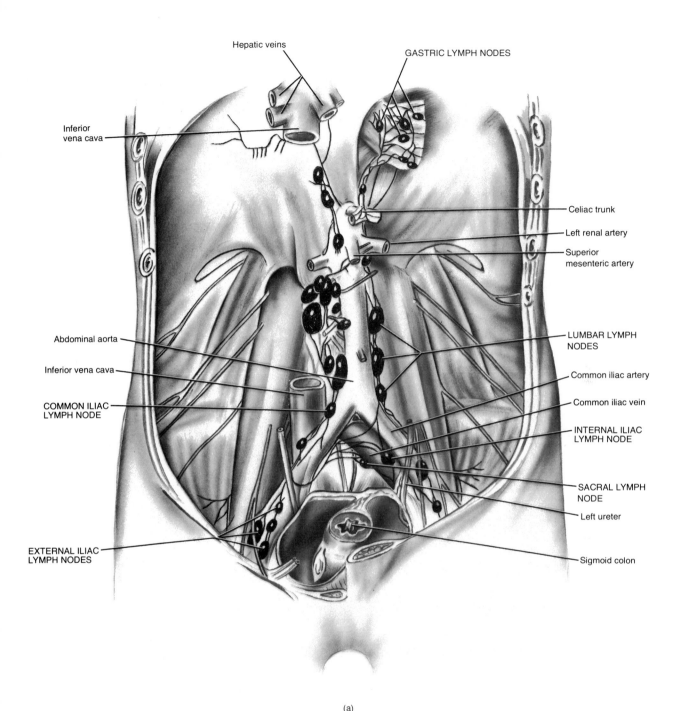

Hepatic veins

GASTRIC LYMPH NODES

Inferior
vena cava

Celiac trunk

Left renal artery

Superior
mesenteric artery

Abdominal aorta

LUMBAR LYMPH
NODES

Inferior vena cava

Common iliac artery

COMMON ILIAC
LYMPH NODE

Common iliac vein

INTERNAL ILIAC
LYMPH NODE

SACRAL LYMPH
NODE

Left ureter

EXTERNAL ILIAC
LYMPH NODES

Sigmoid colon

(a)

FIGURE 11-33   Principal lymph nodes of the abdomen and pelvis. (a) Anterior view.

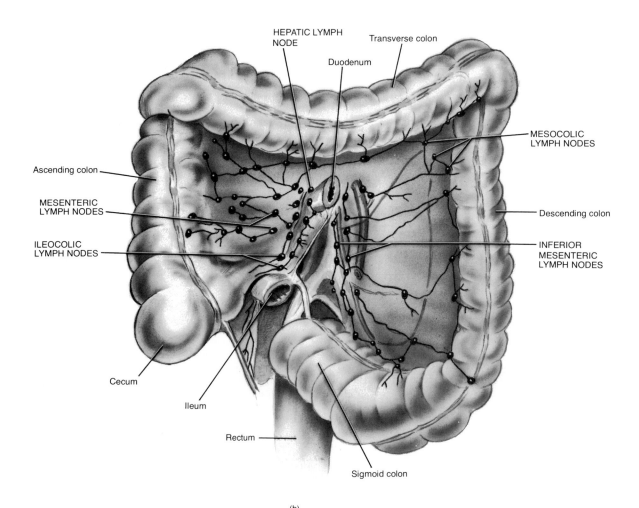

(b)

FIGURE 11-33 *(Continued)* Principal lymph nodes of the abdomen and pelvis. (b) Anterior view.

EXHIBIT 11-18
PRINCIPAL LYMPH NODES OF THE THORAX (Figure 11-34)

LYMPH NODES	LOCATION, AREAS DRAINED, AND COURSE OF EFFERENTS
	Lymph nodes of the thorax are divided into parietal lymph nodes that drain the wall of the thorax and visceral lymph nodes that drain the viscera.

PARIETAL

Sternal	Located alongside of internal thoracic artery at anterior ends of the three or four intercostal spaces; they drain central and lateral parts of mammary gland, deeper structures of anterior abdominal wall above umbilicus, diaphragmatic surface of liver, and deeper parts of anterior portion of thoracic wall; efferents typically unite to form single trunk which may contribute to bronchomediastinal trunk or empty directly into junction of internal jugular and subclavian veins; trunk may also empty into right lymphatic duct or thoracic duct on left side.
Intercostal	Located near heads of ribs at posterior parts of intercostal spaces; they drain posterolateral aspect of thoracic wall; some efferents (five lower intercostal spaces) unite to form a trunk that descends into cisterna chyli or beginning of thoracic duct; some efferents (intercostal spaces) enter thoracic duct, singly or in combination; other efferents (intercostal spaces 1 and 2) ascend and enter junction of internal jugular and subclavian veins.
Phrenic (diaphragmatic)	Located on thoracic aspect of the diaphragm and divisible into three sets called anterior phrenic, middle phrenic, and posterior phrenic.
	Anterior phrenic. Two or three nodes behind base of xiphoid process and one or two nodes on either side of xiphoid process posterior to costal cartilage; they drain convex surface of liver, diaphragm, anterior abdominal wall, and middle phrenic nodes; efferents pass to sternal nodes above.
	Middle phrenic. Two or three nodes on either side close to phrenic nerves where they pierce diaphragm; they drain middle part of diaphragm and convex surface of liver; efferents pass to posterior mediastinal nodes.
	Posterior phrenic. A few nodes located on back of diaphragm near aorta; they drain posterior part of diaphragm and a few efferent channels from middle phrenic nodes; efferents pass to lumbar nodes and posterior mediastinal nodes.

VISCERAL

Anterior mediastinal	Located in anterior part of superior mediastinum anterior to aortic arch; they drain thymus gland, pericardium, and sternal nodes; efferents unite with those of tracheobronchial nodes to form bronchomediastinal trunks.
Posterior mediastinal	Located posterior to pericardium with respect to esophagus and descending thoracic aorta; they drain esophagus, posterior aspect of the pericardium, diaphragm, and convex surface of liver; most efferents end in thoracic duct, a few join tracheobronchial nodes.
Tracheobronchial	These nodes are divisible into four groups:
	Tracheal, on either side of trachea
	Bronchial, at inferior part of trachea and in angle between two bronchi
	Bronchopulmonary, in the hilum of each lung
	Pulmonary, within lungs on larger bronchial branches
	The tracheobronchial nodes drain lungs, bronchi, thoracic part of trachea and heart, and some of posterior mediastinal nodes; efferents unite with those of sternal and anterior mediastinal nodes to form bronchomediastinal trunks; typically trunks empty into junction of internal jugular and subclavian veins on their respective sides.

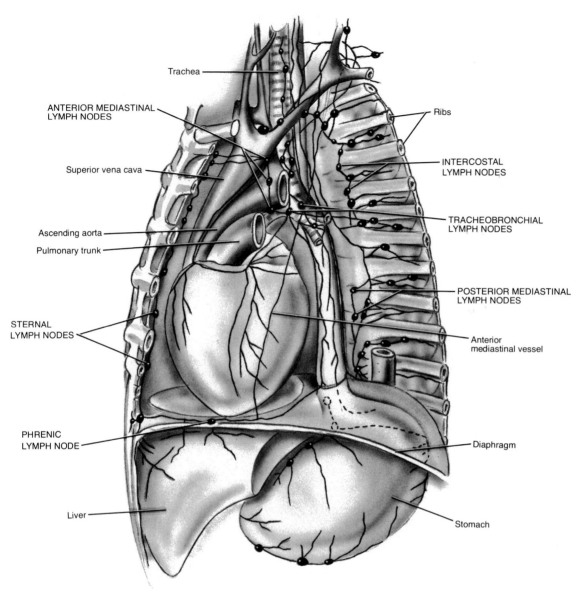

Trachea

ANTERIOR MEDIASTINAL
LYMPH NODES

Superior vena cava

Ascending aorta

Pulmonary trunk

STERNAL
LYMPH NODES

PHRENIC
LYMPH NODE

Liver

Ribs

INTERCOSTAL
LYMPH NODES

TRACHEOBRONCHIAL
LYMPH NODES

POSTERIOR MEDIASTINAL
LYMPH NODES

Anterior
mediastinal vessel

Diaphragm

Stomach

FIGURE 11-34 Principal lymph nodes of the thorax.

LYMPHATIC ORGANS

Tonsils

Tonsils are basically masses of lymphoid tissue embedded in mucous membrane. The *pharyngeal tonsils* are embedded in the posterior wall of the nasopharynx (see Figure 19-3). When they become enlarged, they are called adenoids. The *palatine tonsils* are situated in the tonsillar fossae between the pharyngopalatine and glossopalatine arches (see Figure 20-5). These are the ones commonly removed by a tonsillectomy. The *lingual tonsil* is located at the base of the tongue and may also have to be removed by a tonsillectomy (see Figure 20-5). The tonsils are supplied with reticuloendothelial cells and filter lymph.

Spleen

The oval **spleen** is the largest mass of lymphatic tissue in the body, measuring about 12 cm (5 inches) in length. It is situated in the left hypochondriac region inferior to the diaphragm and posterolateral to the stomach (see Figure 1-5d). Its visceral surface (Figure 11-35a) reflects the contours of the organs related to it, such as the gastric impression (stomach) and renal impression (left kidney). The left colic flexure, tail of the pancreas, and left suprarenal gland are also related to the visceral surface. The diaphragmatic surface (Figure 11-35b) is smooth and convex and conforms to the concave surface of the adjacent diaphragm.

The spleen is surrounded by a capsule of fibroelastic tissue and scattered smooth muscle. The capsule, in turn, is covered by a serous membrane, the peritoneum. Like lymph nodes, the spleen contains trabeculae and a hilus. The capsule, trabeculae, and hilus constitute the stroma of the spleen.

The parenchyma of the spleen consists of two different kinds of tissue called white pulp and red pulp. *White pulp* is essentially lymphoid tissue arranged around arteries. The clusters of lymphocytes surrounding the arteries at intervals and the expansions are referred to as *splenic nodules* or *Malpighian corpuscles (bodies)*. The *red pulp* consists of venous sinuses filled with blood and cords of splenic tissue called *splenic* or *Billroth's cords*. Veins are closely associated with the red pulp.

The splenic artery and vein and the efferent lymphatics pass through the hilus. The spleen phagocytoses bacteria and worn-out red blood cells and platelets. It also produces lymphocytes and plasma cells. In addition, the spleen stores and releases blood in case of hemorrhage. The release seems to be purely sympathetic. The sympathetic impulses cause the smooth muscle of the spleen to contract.

Thymus Gland

The **thymus gland** is a bilobed mass of lymphatic tissue in the upper thoracic cavity. It is found along the trachea behind the sternum (see Figure 17-15). The thymus is relatively large in children. It reaches its maximum size at puberty and then undergoes involution. Eventually, it is replaced by fat and connective tissue.

MAINTAINING LYMPH CIRCULATION

The flow of lymph is from tissue spaces to the large lymphatic ducts to the subclavian veins. The flow is

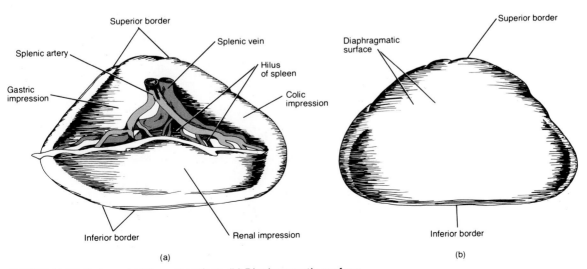

(a)   (b)

FIGURE 11-35 Spleen. (a) Visceral surface. (b) Diaphragmatic surface.

maintained primarily by the milking action of muscle tissue. Skeletal muscle contractions compress lymph vessels and force lymph toward the subclavian veins. Moreover, lymph vessels, like veins, contain valves, and the valves ensure the movement of lymph toward the subclavian veins. Another factor that maintains lymph flow is respiratory movements. These movements create a pressure gradient (difference) between the two ends of the lymphatic system. Lymph flows from the tissue spaces, where the pressure is higher, toward the thoracic region, where it is lower.

Edema, an excessive accumulation of lymph in tissue spaces, has several possible causes. One is an obstruction, such as an infected node, in the lymphatic channels between the lymphatic capillaries and the subclavian veins. Another is excessive lymph formation and increased permeability of blood capillary walls. A rise in capillary blood pressure, in which interstitial fluid is formed faster than it is passed into lymphatics, also may result in edema.

Applications to Health

BLOOD DISORDERS

The various blood disorders that can arise affect differing portions of the blood. With anemia, for example, the patient may have an abnormally low number of red blood cells, whereas with mononucleosis white blood cells are affected. There are numerous blood disorders, and all may have wide-ranging effects.

Anemia

Anemia is a sign, not a diagnosis. Many kinds of anemia exist, all characterized by insufficient erythrocytes or hemoglobin. These conditions lead to fatigue and intolerance to cold, both of which are related to lack of oxygen needed for energy and heat production, and paleness due to low hemoglobin content.

● *Hemorrhagic Anemia* An excessive loss of erythrocytes through bleeding is called **hemorrhagic anemia.** Common causes are large wounds, stomach ulcers, and heavy menstrual bleeding. If bleeding is extraordinarily heavy, the anemia is termed acute. Excessive blood loss can be fatal. Slow, prolonged bleeding is apt to produce a chronic anemia; its chief symptom is fatigue.

● *Hemolytic Anemia* If an erythrocyte cell membrane ruptures prematurely, the cell remains as a "ghost" and its hemoglobin pours out into the plasma. A characteristic sign of this condition is dis-

tortions in the shapes of erythrocytes that are progressing toward hemolysis. There may also be a sharp increase in the number of reticulocytes since the destruction of red blood cells stimulates erythropoiesis.

The premature destruction of red cells may result from inherent defects: hemoglobin defects, abnormal red cell enzymes, or defects of red cell membranes. Agents that may cause **hemolytic anemia** are parasites, toxins, and antibodies from incompatible blood (Rh⁻ mother and Rh⁺ fetus, for instance). Erythroblastosis fetalis of the newborn is an example of a hemolytic anemia.

● *Aplastic Anemia* Destruction or inhibition of the red bone marrow results in **aplastic anemia.** Typically, the marrow is replaced by fatty tissue, fibrous tissue, or tumor cells. Toxins, gamma ray radiation, and certain medications are causes. Many of the medications inhibit the enzymes involved in hemopoiesis.

● *Sickle Cell Anemia* The erythrocytes of a person with **sickle cell anemia** manufacture an abnormal kind of hemoglobin. When the erythrocyte gives up its oxygen to the interstitial fluid, its hemoglobin tends to lose its integrity in places of low oxygen tension and form long, stiff, rodlike structures that bend the erythrocyte into a sickle shape (Figure 11-36). The sickled cells rupture easily. Even though erythropoiesis is stimulated by the loss of the cells, it cannot keep pace with the hemolysis. The individual conse-

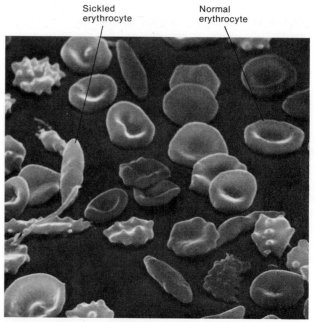

FIGURE 11-36 Scanning electron micrograph of erythrocytes in sickle cell anemia at a magnification of 5000×. (Courtesy of Fisher Scientific Company and S.T.E.M. Laboratories, Inc., Copyright 1975.)

quently suffers from a hemolytic anemia that reduces the amount of oxygen that can be supplied to the tissues. Prolonged oxygen reduction may eventually cause extensive tissue damage. Furthermore, because of the shape of the sickled cells, they tend to get stuck in blood vessels—and can cut off blood supply to an organ altogether.

Sickle cell anemia is inherited. The gene responsible for the tendency of the erythrocytes to sickle during hypoxia prevents erythrocytes from rupturing during a malarial crisis. Sickle cell genes are found primarily among populations, or descendents of populations, that live in the malaria belt around the world—including parts of Mediterranean Europe and subtropical Africa and Asia. A person with only one of the sickling genes is said to have sickle cell trait. Such an individual has a high resistance to malaria—a factor that may have tremendous survival value—but does not develop the anemia. Only people who inherit a sickling gene from both parents get sickle cell anemia.

Polycythemia

The term **polycythemia** refers to an abnormal increase in the number of red blood cells. Increases to 2 to 3 million cells per cubic millimeter above normal are considered to be polycythemic. The disorder is harmful because the blood's viscosity is greatly increased due to the extra red blood cells, and viscosity contributes to thrombosis and hemorrhage. It also causes a rise in blood pressure. The thrombosis results from too many red blood cells piling up as they try to enter smaller vessels. The hemorrhage is due to widespread hyperemia (unusually large amount of blood in an organ part).

A clinical test important in diagnosing polycythemia and anemia is the hematocrit. **Hematocrit** is the percentage of blood that is made up of red blood cells. It is determined by certrifuging blood and noting the ratio of red blood cells to whole blood. The average hematocrit for males is 47 percent. This means that in 100 ml of blood there are 47 ml of cells and 53 ml of plasma. The average hematocrit for females is 42 percent. Anemic blood may have a hematocrit of 15 percent; polycythemic blood may have a hematocrit of 65 percent. Athletes, however, may have elevated, hematocrits due to continual physical activity and not a pathological condition.

Infectious Mononucleosis

Infectious mononucleosis is a contagious disease thought to be of viral origin that occurs mainly in children and young adults. Its trademark is an elevated white blood cell count with an abnormally high percentage of lymphocytes and mononucleocytes. An increase in the number of monocytes usually indicates a chronic infection. Symptoms include slight fever, sore throat, brilliant red throat and soft palate, stiff neck, cough, and malaise. The spleen may enlarge. Secondary complications involving the liver, heart, kidneys, and nervous system may develop. There is no cure for mononucleosis, and treatment consists of watching for and treating complications. Usually the disease runs its course in a few weeks, and the individual generally suffers no permanent ill effects.

Leukemia

Also called "cancer of the blood," **leukemia** is an uncontrolled, greatly accelerated production of white cells. Many of the cells fail to reach maturity. As with most cancers, the symptoms result not so much from the cancer cells themselves as from their interference with normal body processes. The anemia and bleeding problems commonly seen in leukemia result from the crowding out of normal bone marrow cells, preventing normal production of red blood cells and platelets. The most common cause of death from leukemia is internal hemorrhaging, especially cerebral hemorrhage that destroys the vital centers in the brain. Another frequent cause of death is uncontrolled infection owing to lack of mature or normal white blood cells.

CARDIOVASCULAR DISORDERS

Diseases of the heart and blood vessels are the biggest single killers in the developed world. These diseases account for approximately 53 percent of all deaths. A recent comparison indicates that cardiovascular disease kills more people than cancer, accidents, pneumonia, influenza, and diabetes combined. Some of the cardiovascular problems involve aneurysms, atherosclerosis, hypertension, and various heart disorders.

Aneurysm

A blood-filled sac formed by an outpouching in an arterial or venous wall is called an **aneurysm.** Aneurysms may occur in any major blood vessel and include the following types:

1. Berry. A small aneurysm frequently in the cerebral artery. If it ruptures, it may cause a hemorrhage below the dura mater. Hemorrhaging is one cause of a stroke (Figure 11-37a).

2. Ventricular. A dilatation of a ventricle of the heart (Figure 11-37b)

3. Aortic. A dilatation of the aorta (Figure 11-37c).

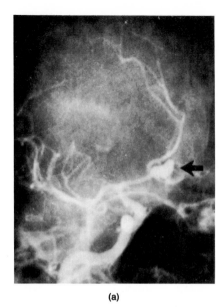

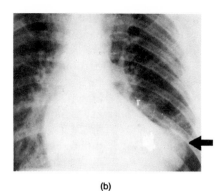

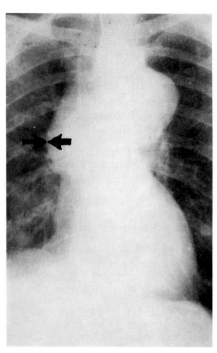

FIGURE 11-37 Aneurysms. (a) Aneurysm of the anterior cerebral artery. (b) Ventricular aneurysm. (c) Aortic aneurysm. (Courtesy of Lester W. Paul and John H. Juhl, *The Essentials of Roentgen Interpretation,* 3d ed., Harper & Row, Publishers; Inc., New York, 1972.)

Atherosclerosis

Atherosclerosis is a form of arteriosclerosis, which includes many diseases of the arterial wall. But atherosclerosis, the lipid-related arterial lesion, is the major disease responsible for the principal clinical complications. In this disorder, the tunica intima of an artery becomes thickened with soft fatty deposits called *atheromatous plaques* (Figure 11-38a).

An *atheroma* is an abnormal mass of fatty or lipid material deposited in an arterial wall. Atheromas involve the abdominal aorta and major leg arteries more extensively than the thoracic aorta. They may also be found in coronary, cerebral, and peripheral arteries. The atheroma looks like a pearly gray or yellow mound of tissue. As atheromas grow, they may impede blood flow in affected arteries and damage the tissues they supply. An additional danger is that the plaque may rupture and form a thrombus. If it breaks off and forms an embolus, it may obstruct small arteries, capillaries, and veins quite a distance from the site of formation. Moreover, the plaque may provide a roughened surface for clot formation.

The degree of arterial stenosis necessary to produce symptoms depends on the site of stenosis and the artery. The process causes a collateral circulation to develop around the stenotic block. These collaterals may prevent ischemic necrosis. When they do not, ischemia may progress to gangrene of the leg.

Atherosclerosis is generally a slow, progressive disease. It may start in childhood, and its development may produce absolutely no symptoms for 40 years or longer. Even if it reaches the advanced stages, the individual may feel no symptoms and the condition may be discovered only at postmortem examination. Diagnosis is possible by injecting radiopaque substances into the blood and then taking x-rays of the arteries. This technique is called *angiography* or *arteriography*. The film is called an *arteriogram* (Figure 11-38b).

Animal experiments have given us considerable scientific information about the plaques. It is possible to produce the streaks in many animals by feeding them a diet high in fat and cholesterol. This diet raises the blood lipid levels—a condition called *hyperlipidemia* (*hyper* = above; *lipo* = fat). Hyperlipidemia increases the risk of atherosclerosis. Patients with a high blood level of cholesterol should be treated with diet and drug therapy.

Hypertension

Hypertension, or high blood pressure, is the most common disease affecting the heart and blood vessels. Statistics from a recent National Health Survey indicate that hypertension afflicts at least 17 million American adults and perhaps as many as 22 million.

Primary hypertension, or essential hypertension, is a persistently elevated blood pressure that cannot be attributed to any particular organic cause. Specifically, the diastolic pressure continually exceeds 95 mm Hg. Approximately 85 percent of all hypertension cases fit this definition. The other 15 percent have *sec-*

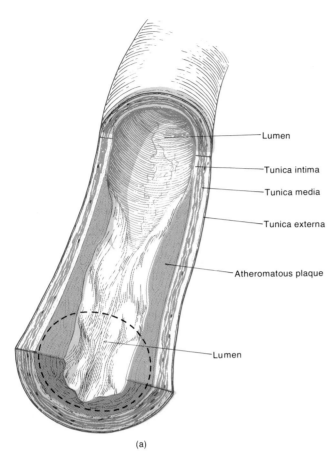

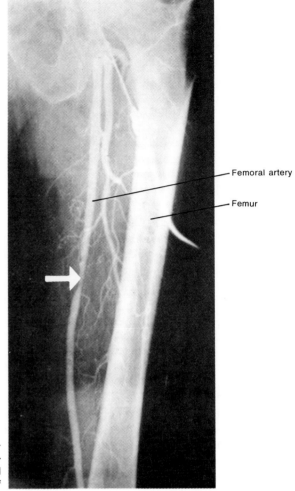

(a)

(b)

FIGURE 11-38 Atherosclerosis. (a) Atheromatous plaque formation. The broken circular line indicates the approximate size of a normal lumen. (b) Femoral arteriogram showing an atheromatous plaque (arrow) in the middle third of the thigh. (Courtesy of Lester W. Paul and John H. Juhl, *The Essentials of Roentgen Interpretation,* 3d ed., Harper & Row, Publishers, Inc., New York, 1972.)

ondary hypertension. Secondary hypertension is caused by disorders such as arteriosclerosis, kidney disease, and adrenal hypersecretion. Arteriosclerosis increases blood pressure by reducing the elasticity of the arterial walls and narrowing the space through which the blood can flow. Kidney disease and obstruction of blood flow may cause the kidney to release renin into the blood. This enzyme catalyzes the formation of angiotensin from a plasma protein. Angiotensin is a powerful blood-vessel constrictor—and the most potent agent known for raising blood pressure. Aldosteronism, the hypersecretion of aldosterone, may also cause an increase in blood pressure. Aldosterone is the adrenal cortex hormone that promotes the retention of salt and water by the kidneys. It thus tends to increase plasma volume. Pheochromocytoma is a tumor of the adrenal medulla. It produces and releases into the blood large quantities of norepinephrine and epinephrine. These hormones also raise blood pressure by stimulating the heart and constricting blood vessels.

The causes of primary hypertension are unknown. Medical science cannot yet cure it. However, almost all cases of hypertension, whether mild or severe, can be controlled by a variety of drugs that reduce blood pressure.

HEART DISORDERS

It is estimated that one in every five persons who reaches 60 will have a **heart attack.** And it is also estimated that one in every four persons between 30 and 60 has the potential to be stricken. Heart disease is epidemic in this country, despite the fact that some of the causes can be foreseen and prevented.

Risk Factors

The Framingham, Massachusetts, Heart Study, which began in 1950, is the longest and most famous study ever made of the susceptibility of a community to heart disease. Approximately 13,000 people in the town have participated by receiving examinations every two years since the study began. The results of this research indicate that people who develop combinations of certain risk factors eventually have heart attacks. These factors are high blood cholesterol level, high blood pressure, cigarette smoking, obesity, lack of exercise, and diabetes mellitus.

The first five risk factors all contribute to increasing the heart's workload. Cigarette smoking, through the effects of nicotine, stimulates the adrenal gland to oversecrete aldosterone, epinephrine, and norepinephrine—powerful vasocontrictors. Overweight people develop miles of extra capillaries to nourish fat tissue. The heart has to work harder to pump the blood through more vessels. Without exercise, venous return gets less help from contracting skeletal muscles. In addition, regular exercise strengthens the smooth muscle of blood vessels and enables them to assist general circulation. Exercise also increases cardiac efficiency and output. In diabetes mellitus, fat metabolism dominates glucose metabolism. As a result, cholesterol levels get progressively higher and result in plaque formation, a situation that may lead to high blood pressure. High blood pressure drives fat into the vessel wall, encouraging atherosclerosis.

Generally, the immediate cause of heart trouble is one of the following: inadequate coronary blood supply, anatomical disorders, or faulty electrical conduction in the heart.

● *Inadequate Coronory Blood Supply*   Angina pectoris and myocardial infarction result from insufficient oxygen supply to the myocardium. Coronary artery disease kills about one in twelve of all Americans who die between the ages of 25 and 34. It claims almost one in four of all those who die between 35 and 44. It has been reported that 50 to 65 percent of all sudden deaths are due to coronary heart disease.

At least half the deaths from myocardial infarction occur before the patient reaches the hospital. These early deaths could result from an irregular heart rhythm—an *arrhythmia.* Sometimes this condition progresses to the stage called *cardiac arrest* or ventricular fibrillation, in which the heart stops functioning. An arrhythmia is caused by disturbances in the conduction system. This abnormal rhythm of the heartbeat can result in cardiac arrest if the heart cannot supply its own oxygen demands, as well as those of the rest of the body. Serious arrhythmias can be controlled, and the normal heart rhythm can be reestablished, if they are detected and treated early enough. Coronary care units have reduced hospital mortality rates from acute myocardial infarctions to about 20 to 30 percent or less by preventing or controlling serious arrhythmias.

● *Anatomical Disorders*   Less than one percent of all new babies have a **congenital,** or **inborn, heart defect.** Even so, the total number of babies so born in this country each year is estimated to be 30,000 to 40,000. Some of these infants may live quite healthy and long lives without any need for repairing their hearts. But sometimes an inborn heart defect is so severe that an infant lives only a few hours. A common anatomical defect is *patent ductus arteriosus.* The connection between the aorta and the pulmonary artery remains open instead of closing completely after birth. This defect results in aortic blood flowing into the lower-pressure pulmonary trunk, thus increasing the pulmonary trunk blood pressure and overworking both ventricles and the heart.

A **septal defect** is an opening in the septum that divides the interior of the heart into left and right sides. *Atrial septal defect* is failure of the fetal foramen ovale to be closed off after birth, separating the two atria from one another. Because pressure in the right atrium is low, atrial septal defect generally allows a good deal of blood to flow from the left atrium to the right. This defect overloads the pulmonary circulation, produces fatigue and increases respiratory infections. If it occurs early in life, it inhibits growth because the systemic circulation may be deprived of a considerable portion of the blood destined for the organ and tissues of the body. *Ventricular septal defect* is caused by an abnormal development of the interventricular septum. Deoxygenated blood gets mixed with the oxygenated blood pumped into the systemic circulation. Consequently, the victim suffers *cyanosis,* a blue or dark purple discoloration of the skin. Cyanosis occurs whenever deoxygenated blood reaches the cells because of heart defect, lung defect, or suffocation. Septal openings can now be sewn shut or covered with synthetic patches.

Valvular stenosis is a narrowing, or *stenosis,* of one of the valves regulating blood flow in the heart. Narrowing may occur in the valve itself—most commonly in the mitral valve from rheumatic heart disease or in the aortic valve from sclerosis or rheumatic fever. Or it may occur near a valve. All stenoses are serious because they all place a severe workload on the heart by making it work harder to push the blood through the abnormally narrow valve openings. As a result of mitral stenosis, blood pressure is increased. Angina pectoris and heart failure may accompany this disorder. Most stenosed valves are totally replaced with artificial valves.

Tetralogy of Fallot is a combination of four defects causing a ''blue baby''—a ventricular septal opening, an aorta that emerges from both ventricles instead of from the left ventricle, a stenosed pulmonary semilunar valve, and an enlarged right ventricle (Figure 11-

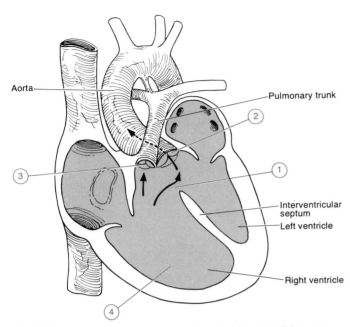

Aorta

Pulmonary trunk

②

③

①

Interventricular septum

Left ventricle

Right ventricle

④

FIGURE 11-39 Tetralogy of Fallot. The four abnormalities associated with this condition are indicated by numbers. (1) Opening in the interventricular septum. (2) Origin of the aorta in both ventricles. (3) Stenosed pulmonary semilunar valve. (4) Enlarged right ventricle.

39). Because of the ventricular septal defect, oxygenated and unoxygenated blood are mixed in the ventricles. However, the body tissues are much more starved for oxygen than are those of a child with simple ventricular septal defect. Because the aorta also emerges from the right ventricle and the pulmonary artery is stenosed, very little blood ever gets to the lungs and pulmonary circulation is bypassed almost completely. Today it is possible to correct cases of tetralogy of Fallot when the patient is of proper age and condition. Open-heart operations are performed in which the narrowed pulmonary valve is cut open and the septal defect is sealed with a Dacron patch.

● *Faulty Electrical Conduction* **Arrhythmia** arises when electrical impulses through the heart are blocked at critical points in the conduction system. One such cause of arrhythmia is called a **heart block.** Perhaps the most common blockage is in the atrioventricular node, which conducts impulses from the atria to the ventricles. This disturbance is called *atrioventricular (AV) block.* It usually indicates a myocardial infarction, arteriosclerosis, rheumatic heart disease, diphtheria, or syphilis. In a first-degree AV block, which can be detected only with an electrocardiograph, the transmission of impulses from the atria to the ventricles is delayed. Here the P-R interval is greater than it should be. In a second-degree AV block, impulses fail to reach the ventricles so the ventricular rate is about half that of the atrial rate. When ventricular contraction does not occur (dropped beat), oxygenated blood is not pumped to the body. The patient may feel faint or may collapse if there are many dropped ventricular beats. In a third-degree or complete AV block, practically no impulses reach the ventricles. Atrial and ventricular rates get out of synchronization (Figure 11-40a). The ventricles may go into systole at any time. This condition could occur when the atria are in systole or just before. Or the ventricles may rest for a few cardiac cycles.

With complete AV block, patients may have vertigo, unconsciousness, or convulsions. These symptoms result from a decreased cardiac output with diminished cerebral blood flow and cerebral hypoxia or lack of sufficient oxygen. Among the causes of AV block are excessive stimulation by the vagus nerves that depresses conductivity of the junctional fibers, destruction of the AV bundle as a result of coronary infarct, arteriosclerosis, myocarditis, or depression caused by various drugs. Other heart blocks include *intraatrial (IA) block, interventricular (IV) block,* and *bundle branch block (BBB).* In the latter condition, the ventricles do not contract together because of the delayed impulse in the blocked branch.

Flutter and Fibrillation

Two rhythms that indicate heart trouble are atrial flutter and fibrillation. In **atrial flutter** the atrial rhythm averages between 240 and 360 beats per minute. The condition is essentially rapid atrial contractions accompanied by a second-degree AV block. It generally indicates severe damage to heart muscle. Atrial flutter usually becomes fibrillation after a few days or weeks. **Atrial fibrillation** is asynchronous contraction of the atrial muscles that causes the atria to contract irregularly and still faster (Figure 11-40b). Atrial flutter and fibrillation occur in myocardial infarction, acute and chronic rheumatic heart disease, and hyperthyroidism. Atrial fibrillation results in complete uncoordination of atrial contraction so that atrial pumping ceases altogether. When the muscle fibrillates, the muscle fibers of the atrium quiver individually instead of contracting together. The quivering cancels out the pumping of the atrium. In a strong heart, atrial fibrillation reduces the pumping effectiveness of the heart by only 25 to 30 percent.

Ventricular fibrillation is another abnormality that indicates imminent cardiac arrest and death. It is characterized by asynchronous, haphazard ventricular muscle contractions. The rate may be rapid or slow.

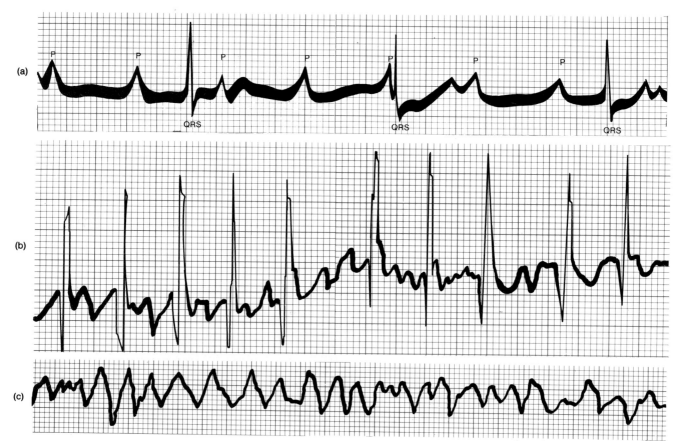

FIGURE 11-40 Abnormal electrocardiograms. (a) Complete heart block. There is no fixed ratio between atrial contractions (P waves) and ventricular contractions (QRS waves). (b) Atrial fibrillation. There is no regular atrial contraction and, therefore, no P wave. Since the ventricles contract irregularly and independently, the QRS wave appears at irregular intervals. (c) Ventricular fibrillation. In general, there is no rhythm of any kind.

The impulse travels to the different parts of the ventricles at different rates. Thus part of the ventricle may be contracting while other parts are still unstimulated. Ventricular contraction becomes ineffective and circulatory failure and death occur immediately unless the arrhythmia is reversed quickly (Figure 11-40c). Ventricular fibrillation may be caused by coronary occlusion. It sometimes occurs during surgical procedures on the heart or pericardium. It may be the cause of death in electrocution.

KEY MEDICAL TERMS ASSOCIATED WITH THE CARDIOVASCULAR AND LYMPHATIC SYSTEMS

Adenitis (*adeno* = gland) Enlarged, tender, and inflamed lymph nodes resulting from an infection.

Arteriography Recording of an image of arteries by x-ray revealed by the direct injection of dyes.

Blood plasma substitute This substance mimics the characteristics of plasma. It is used to maintain blood volume during emergency conditions (such as hemorrhage) until blood can be matched, or to prevent dehydration if a patient cannot swallow liquids. It is also used to replace fluid and electrolytes after the loss of blood that may occur during surgery.

Cardiac arrest Complete stoppage of the heartbeat.

Cardiomegaly (*megalo* = great) Heart enlargement.

Citrated whole blood Whole blood protected from coagulation by a citrate.

Cyanosis Slightly bluish, dark purple skin coloration due to oxygen deficiency in systemic blood.

Defibrillator A mechanical device for applying electrical shock to the heart to terminate abnormal cardiac rhythms.

Direct transfusion (immediate) Transfer of blood directly from one person to another without exposing the blood to air.

Elephantiasis Great enlargement of a limb (especially lower limbs) and scrotum resulting from obstruction of lymph glands or vessels caused by a tiny parasitic worm.

Epistaxis A nosebleed.

Exchange transfusion Removing blood from the recipient while simultaneously replacing it with donor blood. This method is used for erythroblastosis fetalis and poisoning.

Fractionated blood (*fract* = break) Blood that has been separated into its components. Only the part needed by the patient is given.

Hem, Hemo, Hema, Hemato (*heme* = iron) Various combining forms meaning blood.

Hematoma (hemangioma) (*hemo, hemato* = blood; *oma* = tumor) Leakage of blood from a vessel, which clots to form a solid mass or swelling in tissue.

Hemolysis (laking) A swelling and subsequent rupture of erythrocytes with the liberation of hemoglobin into the surrounding fluid.

Hemorrhage (*rrhage* = bursting forth) Bleeding, either internal (from blood vessels into tissues) or external (from blood vessels directly to the surface of the body).

Heparinized whole blood Whole blood in a heparin solution to prevent coagulation.

Hypersplenism (*hyper* = over, above) Abnormal splenic activity involving highly increased blood cell destruction.

Indirect transfusion (mediate) Transfer of blood from a donor to a container and then to the recipient, permitting blood to be stored for an emergency. The blood may be separated into its components so a patient receives only a needed part.

Lymphadenectomy (*ectomy* = removal) Removal of a lymph node.

Lymphadenopathy (*patho* = disease) Enlarged, sometimes tender lymph glands.

Lymphangitis Inflammation of the lymphatic vessels.

Lymphedema Accumulation of lymph fluid producing subcutaneous tissue swelling.

Lymphoma (*oma* = tumor) Any tumor composed of lymph tissue. Malignancy of reticuloendothelial cells of lymph nodes is called Hodgkin's disease.

Lymphangioma A benign tumor of the lymph vessels.

Lymphostasis (*stasis* = halt) A lymph flow stoppage.

Normal plasma Cell-free plasma containing normal concentrations of all solutes. It is used to bring blood volume up to normal when excessive numbers of blood cells have not been lost.

Occlusion The closure or obstruction of the lumen of a structure such as a blood vessel.

Phlebitis (*phleb* = vein) Inflammation of a vein.

Platelet concentrates Platelets obtained from freshly drawn whole blood and transfused for platelet-deficiency disorders such as hemophilia.

Reciprocal transfusion Blood is transferred from a person who has recovered from a contagious infection into the vessels of a patient suffering with the same infection. An equal amount of blood is returned from the patient to the well person. This method allows the patient to receive antibody-bearing lymphocytes from the recovered person.

Septicemia (*sep* = decay; *emia* = condition of blood) Toxins or disease-causing bacteria in the blood. Also called "blood poisoning."

Thrombophlebitis (*thrombo* = clot) Inflammation of a vein with clot formation.

Transfusion The transfer of whole blood, blood components (red blood cells only or plasma only), or bone marrow directly into the bloodstream.

Venesection Opening of a vein for withdrawal of blood.

Whole blood Blood containing all formed elements, plasma, and plasma solutes in natural concentration.

STUDY OUTLINE

Blood

1. The principal functions of blood are the transportation of O_2 and CO_2, nutrients and wastes, and hormones and enzymes. It regulates pH, normal body temperature, and water content of cells, and protects against disease. Blood consists of plasma and formed elements.

Formed Elements

1. Formed elements are erythrocytes, leucocytes, and thrombocytes.
2. Erthrocytes, or red blood cells, are biconcave discs without nuclei that contain hemoglobin. Erythrocyte formation is called erythropoiesis and occurs in adult red marrow of certain bones.
3. Leucocytes, or white blood cells, are nucleated cells. Two principal types are granular (neutrophils, eosinophils, and basophils) and agranular (lymphocytes and monocytes).
4. One function of leucocytes, especially neutrophils, is to combat inflammation and infection through phagocytosis.
5. In response to the presence of foreign proteins called antigens, lymphocytes are changed into plasma cells. Plasma cells produce antibodies, which cover antigens and render them harmless. This is called the antigen-antibody response and is important in combating infection and providing immunities.
6. Thrombocytes, or platelets, are disc-shaped structures without nuclei. They initiate clotting.

Plasma

1. The liquid portion of blood, called plasma, consists of 92 percent water and 8 percent solutes. Important solutes include proteins (albumins, globulins, and fibrinogen), foods, enzymes and hormones, gases, and electrolytes.

Interstitial Fluid and Lymph

1. Interstitial fluid bathes body cells, whereas lymph is found in lymphatic vessels.
2. These fluids are basically similar in chemical composition. They differ chemically from plasma in that both contain less protein, a variable number of leucocytes, and no platelets or erythrocytes.

Heart

1. The parietal pericardium, consisting of an outer fibrous layer and an inner serous layer, encloses the heart.
2. The wall of the heart has three layers called the epicardium, myocardium, and endocardium. The chambers include two upper atria and two lower ventricles.
3. All valves of the heart prevent the back flow of blood. The blood flows through the heart from the superior and inferior venae cavae, to the right atrium, through the tricuspid valve to the right ventricle, through the pulmonary artery to the lungs, through the pulmonary veins to the left atrium, through the bicuspid valve to the left ventricle, and out through the aorta.
4. The conduction system consists of tissue specialized for impulse conduction. Components are the sinu-atrial mode (pacemaker), atrioventricular node, bundle of His, bundle branches, and Purkinje fibers.
5. The record of electrical changes during cardiac cycle is referred to as an electrocardiogram (ECG). Normal ECG consists of a P wave (spread of impulse from SA node over atria), QRS wave (spread of impulse through ventricles), and T wave (ventricular repolarization).
6. The first heart sound (lubb) represents the closing of the atrioventricular valves. The second sound (dupp) represents the closing of the semilunar valves.
7. Surface landmarks are useful for identifying the position of the borders and valves of the heart.

Blood Vessels

1. Arteries carry blood away from the heart. Their walls are stronger and thicker than vein walls, consisting of a tunica interna, tunica media, and tunica externa.
2. Many arteries anastomose, which means that the distal ends of two or more vessels unite. An alternate blood route from an anastomosis is called collateral circulation.
3. Capillaries are microscopic blood vessels through which materials are exchanged between blood and interstitial fluid. They unite to form venules, which in turn form veins to carry blood back to the heart.
4. Veins have less elastic tissue and smooth muscle than arteries, and they contain valves to prevent back flow of blood.
5. Weak valves can lead to varicose veins or hemorrhoids.

Circulatory Routes

1. The systemic circulation takes oxygenated blood from the left ventricle through the aorta to all parts of the body except the lungs. It includes the coronary and hepatic portal circulations.
2. The coronary circulation takes oxygenated blood through the arterial system of the myocardium. Deoxygenated blood returns to the right atrium via the coronary sinus. Complications of this system are angina pectoris and myocardial infarction.
3. The hepatic portal circulation takes blood from the veins of the pancreas, spleen, stomach, intestines, and gallbladder to the hepatic portal vein of the liver. It enables the liver to utilize nutrients and detoxify harmful substances in the blood.
4. The pulmonary circulation takes deoxygenated blood from the right ventricle to the lungs and returns oxygenated blood from the lungs to the left atrium. It allows blood to be oxygenated for systemic circulation.
5. The fetal circulation involves the exchange of materials between the fetus and mother through the placenta.

Lymphatic System

1. This system consists of lymph vessels, lymph, lymph nodes, and lymph organs. Its primary function is to drain, from tissue spaces, protein-containing fluid which escapes from blood capillaries.
2. Lymphatic vessels are similar in structure to veins. All lymphatics deliver lymph to either the thoracic duct or right lymphatic duct.
3. Lymph nodes are oval-shaped structures located along lymphatics. Lymph passing through the nodes is filtered, and it picks up antibodies and agranular leucocytes.
4. Lymph organs that filter lymph and add white blood cells and antibodies are the tonsils, spleen, and thymus gland.
5. Lymph circulation is from tissue spaces to large lymphatic ducts to the subclavian veins.

Applications to Health

Blood Disorders

1. Anemia is indicated by a decreased erythrocyte count or hemoglobin deficiency. Kinds of anemia include hemorrhagic, hemolytic, aplastic, and sickle cell anemia.
2. Polycythemia is an abnormal increase in the number of erythrocytes.
3. Infectious mononucleosis is characterized by an elevated white cell count, especially the monocytes. The cause is unknown.
4. Leukemia is the uncontrolled production of white blood cells that interferes with normal clotting and vital body activities.

Cardiovascular Disorders

1. An aneurysm is a sac formed by an outpocketing of a portion of an arterial or venous wall.
2. Atherosclerosis is the hardening of the arteries caused by formation of plaques.
3. Hypertension is high blood pressure. Primary hypertension cannot be linked to a specific organic cause. Secondary hypertension may be caused by atherosclerosis, kidney disorders, excessive aldosterone secretion, and tumors.
4. Heart disorders related to inadequate blood supply are angina pectoris and myocardial infarction.
5. Congenital heart defects include patent ductus arteriosus, septal defects, valvular stenosis, and tetralogy of Fallot.
6. Heart conditions relative to conduction problems include heart blocks, atrial flutter and fibrillation, and ventricular fibrillation.

REVIEW QUESTIONS

1. Distinguish between the cardiovascular and lymphatic systems.
2. Define the principal physical characteristics of blood. List the functions of blood and their relationship to other systems of the body.
3. Distinguish between plasma and formed elements. Where are the formed elements produced?
4. Describe the microscopic appearance of erythrocytes. What is the essential function of erythrocytes?
5. What is a reticulocyte count? What is its diagnostic significance?
6. Describe the classification of leucocytes. What are their functions?
7. What is the importance of diapedesis and phagocytosis in fighting bacterial invasion?
8. What is a differential count? What is its significance?
9. Distinguish between leucocytosis and leucopenia.
10. Describe the antigen-antibody response. How is the response protective?
11. What are the major chemicals in plasma? What do they do?
12. What is the difference between plasma and serum?
13. Compare interstitial fluid and lymph with regard to location, chemical composition, and function.
14. Descibe the location of the heart in the mediastinum. Distinguish the subdivisions of the parietal pericardium. What is the purpose of this structure?
15. Compare the three portions of the wall of the heart. Define atria and ventricles. What vessels enter or exit the atria and ventricles?
16. Discuss the principal kinds of valves in the heart and how they operate.
17. Describe the components of the conduction system of the heart.
18. What is an electrocardiogram? Describe the importance of the deflection waves of an ECG.
19. Describe the various sounds of the heart and the importance of each.
20. Prepare a diagram to illustrate the surface landmarks used to locate the borders and valves of the heart.
21. Describe the structural and functional differences among arteries, capillaries, and veins.
22. Discuss the importance of the elasticity and contractility of arteries. What is an anastomosis?
23. Define varicose veins and hemorrhoids.
24. What is meant by a circulatory route? Define systemic circulation.
25. By means of a diagram, indicate the major divisions of the aorta, their principal arterial branches, and the regions supplied.
26. Trace a drop of blood from the arch of the aorta through its systemic circulatory route and back to the heart again. Remember that the major branches of the arch are the brachiocephalic artery, the left common carotid artery, and the left subclavian artery. In giving your answer, be sure to indicate which veins return the blood to the heart.
27. What is the circle of Willis? Why is it important?
28. What are visceral branches of an artery? Parietal branches?
29. What major organs are supplied by branches of the thoracic aorta? How is blood returned from these organs to the heart?
30. What organs are supplied by the celiac artery, superior mesenteric, renal, inferior mesenteric, inferior phrenic, and middle sacral? How is blood returned from these organs to the heart?
31. Trace a drop of blood from the common iliac arteries through their branches to the respective organs and back to the heart again.
32. What is a deep vein? A superficial vein? Define a venous sinus in relation to blood vessels. What are the three major groups of systemic veins?
33. Describe the route of blood in the coronary circulation. Distinguish between angina pectoris and myocardial infarction.
34. What is hepatic portal circulation? Describe the route by means of a diagram. Why is this route important?
35. Define pulmonary circulation. Prepare a diagram to indicate the route. What is the purpose of the route?
36. Discuss in detail the anatomy and physiology of fetal circulation. Be sure to indicate the function of the umbilical arteries, umbilical vein, ductus venosus, foramen ovale, and ductus arteriosus.
37. What is the fate of the special structures involved in fetal circulation once postnatal circulation is established?
38. Describe the cause and treatment of patent ductus arteriosus.
39. Identify the components of the lymph vascular system. What is its function?
40. How do lymphatic vessels originate? Compare veins and lymphatics with regard to structure.
41. Construct a diagram to indicate the role of the thoracic duct and right lymphatic duct in draining lymph from different regions of the body.
42. What is a lymphangiogram? What is its diagnostic value?
43. Describe the structure of a lymph node. What functions do lymph nodes serve?
44. Identify the principal lymph nodes in the head and neck, upper extremities, lower extremities, abdomen and pelvis, and thorax.
45. Compare the functions of the tonsils, spleen, and thymus gland as lymphatic organs.
46. Define anemia. Contrast the causes of hemorrhagic, hemolytic, aplastic, and sickle cell anemias.
47. What is leukemia, and what are the causes of some of its symptoms?
48. Define hematocrit. Compare the hematocrits of polycythemic and anemic blood.
49. What is infectious mononucleosis?
50. What is an aneurysm? Distinguish three types on the basis of location.
51. Discuss the causes, symptoms, and diagnosis of atherosclerosis.

52. Compare primary and secondary hypertension with regard to cause.
53. Describe the risk factors involved in the development of heart disease.
54. Distinguish among the following congenital heart disorders: patent ductus arteriosus, septal defects, valvular stenosis, and tetralogy of Fallot.
55. Define an arrhythmia. What is a heart block? Distinguish the various kinds of heart block.
56. Describe atrial flutter and atrial fibrillation. What is ventricular fibrillation?
57. Refer to the glossary of key medical terms associated with the cardiovascular and lymphatic systems. Be sure that you can define each term listed.

12
Nervous Tissue

STUDENT OBJECTIVES

- Describe the function of the nervous system in maintaining homeostasis.

- Classify the organs of the nervous system into central and peripheral divisions.

- Contrast the histological characteristics and functions of neuroglia and neurons.

- Categorize neurons by shape and function.

- Define a nerve impulse and describe its importance in the body.

The **nervous system** is the body's control center and communications network. In human beings it performs two broad functions. First, it stimulates movements. Second, it shares the maintenance of homeostasis with the endocrine system. Human life cannot exist without a functioning nervous system. Skeletal and most smooth muscle cells cannot contract until stimulated by a nerve impulse. If the intercostal muscles and diaphragm do not contract, we cannot breathe. If the digestive glands are not stimulated to release their secretions, we cannot digest our food. It is obvious, then, that our cells cannot receive nutrients unless the digestive system is connected to a functioning nervous system. But suppose our muscles and glands could stimulate themselves. Even then, we could not live very long without our nervous system. The nervous system *senses* changes within the body and in the outside environment. It then interprets these changes and may initiate action to maintain homeostasis.

Organization

The nervous system may be divided into two principal portions: the central nervous system and the peripheral nervous system (Figure 12-1).

The **central nervous system (CNS)** is the control center for the entire system and consists of the brain and spinal cord. All body sensations must be relayed from receptors to the central nervous system if they are to be interpreted and acted on. All the nerve impulses that stimulate muscles to contract and glands to secrete must also pass from the central nervous system.

The various nerve processes that connect the brain and spinal cord with receptors, muscles, and glands comprise the **peripheral nervous system (PNS).** The peripheral nervous system may be classified into an afferent system and an efferent system. The *afferent system* consists of nerve cells that convey information from receptors in the periphery of the body to the central nervous system. These nerve cells, called afferent (sensory) neurons, are the first cells to pick up incoming information. The *efferent system* consists of nerve cells that convey information from the central nervous system to muscles and glands. These nerve cells are called efferent (motor) neurons. The efferent system is subdivided into a somatic nervous system and an autonomic nervous system.

The *somatic nervous system* or *SNS* (*soma* = body) consists of efferent neurons that conduct impulses from the central nervous system to skeletal muscle tissue. Since the somatic nervous system produces movement only in skeletal muscle tissue, it is under conscious control and therefore voluntary.

By contrast, the *autonomic nervous system* or *ANS* contains efferent neurons that convey impulses from the central nervous system to smooth muscle tissue, cardiac muscle tissue, and glands. The autonomic system produces responses only in involuntary muscles and glands. Thus the autonomic system is usually considered to be involuntary. With few exceptions, the viscera receive nerve fibers from the two divisions of

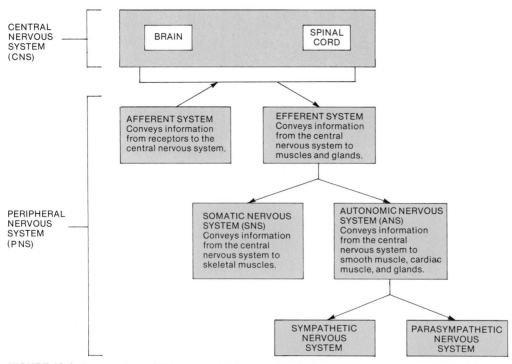

FIGURE 12-1 Organization of the nervous system.

the autonomic nervous system: the sympathetic division and the parasympathetic division.

Histology

Despite the organizational complexity of the nervous system, it consists of only two principal kinds of cells. The first of these, the neurons, make up the nervous tissue that forms the structural and functional portion of the system. Neurons are highly specialized for impulse conduction and for all special functions attributed to the nervous system: thinking, controlling muscle activity, regulating glands. The second type of cell, the neuroglia, serves as a special supporting and protective component of the nervous system.

NEUROGLIA

The cells of the nervous system that perform the functions of support and protection are called **neuroglia** or **glial cells** (*neuro* = nerve; *glia* = glue). About 90 percent of all brain cells are neuroglial cells. Many of the glial cells form a supporting network by twining around the nerve cells in the brain and spinal cord. Other glial cells bind nervous tissue to supporting structures and attach the neurons to their blood vessels. A few glial cells also serve specialized functions. For example, many nerve fibers are coated with a thick, fatty sheath produced by a particular type of neuroglia. Certain small glial cells are phagocytotic. They protect the central nervous system from disease by engulfing invading microbes and clearing away debris. These various functions performed by the neuroglia are divided among several different kinds of cells. Neuroglia are of clinical interest because they are a common source of tumors of the nervous system. Exhibit 12-1 lists the neuroglial cells and summarizes their functions.

NEURONS

Nerve cells, called **neurons,** are responsible for conducting impulses from one part of the body to another.

Structure

A neuron consists of three distinct portions: (1) the cell body, (2) dendrites, and (3) an axon (Figure 12-2a). The **cell body** or **perikaryon** contains a well-defined nucleus and nucleolus surrounded by a granular cytoplasm. Within the cytoplasm are typical organelles such as mitochondria and a Golgi apparatus. Also located in the cytoplasm are structures characteristic of neurons: Nissl bodies and neurofibrils. *Nissl bodies* are orderly arrangements of granular (rough) ER and free ribosomes whose function is protein synthesis. Newly

EXHIBIT 12-1
NEUROGLIA OF CENTRAL NERVOUS SYSTEM

TYPE	DESCRIPTION	MICROSCOPIC APPEARANCE	FUNCTION
Astrocytes (*astro* = star; *cyte* = cell)	Star-shaped cells with numerous processes		Twine around nerve cells to form supporting network in brain and spinal cord; attach neurons to their blood vessels
Oligodendrocytes (*oligo* = few; *dendro* = tree)	Resemble astrocytes in some ways, but processes are fewer and shorter		Give support by forming semirigid connective tissue rows between neurons in brain and spinal cord; produce a thick, fatty myelin sheath on neurons of central nervous system
Microglia (*micro* = small)	Small cells with few processes; normally stationary; if nervous tissue is damaged, they may migrate to injured area		Engulf and destroy microbes and cellular debris

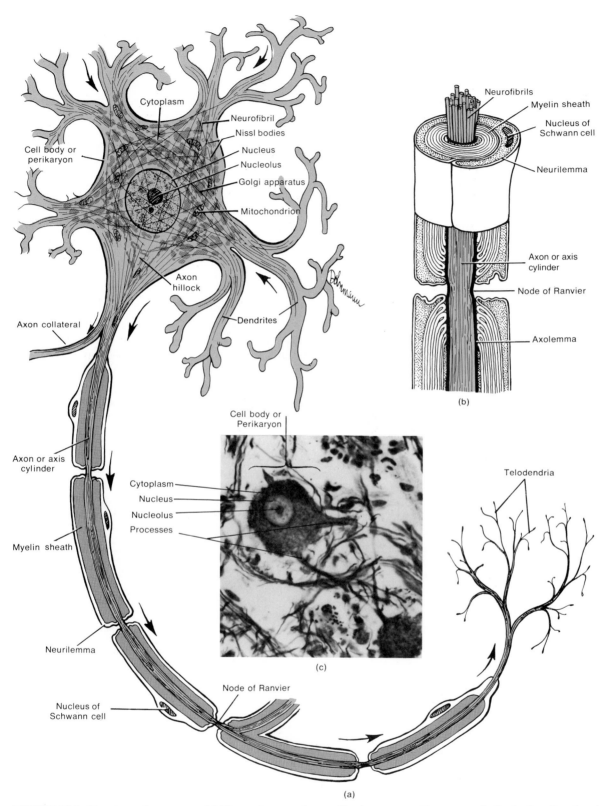

FIGURE 12-2 Structure of a neuron. (a) Shown in an entire multipolar neuron. The arrows indicate the direction in which a nerve impulse passes. (b) Cross section through a myelinated fiber. (c) Photomicrograph of a multipolar neuron from a sympathetic ganglion at a magnification of 640×. (Courtesy of Edward J. Reith, from *Atlas of Descriptive Histology,* by Edward J. Reith and Michael H. Ross, Harper & Row, Publishers, Inc., New York, 1970.)

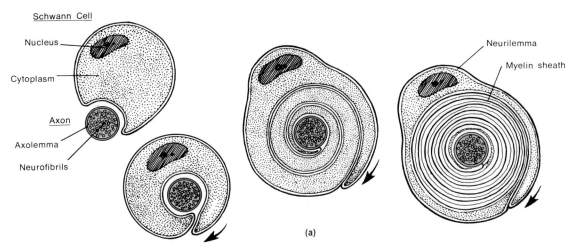

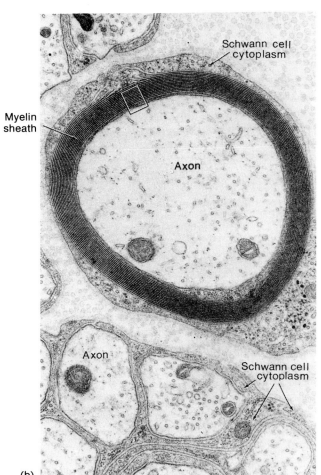

synthesized proteins pass from the perikaryon into the neuronal processes, mainly the axon, at the rate of about 1 mm (0.0394 inch)/day. These proteins replace those lost during metabolism and are used for growth of neurons and regeneration of peripheral nerve fibers. *Neurofibrils* are long, thin fibrils composed of microtubules. They may assume a function in support.

The cytoplasmic processes of neurons depend on the direction in which they conduct impulses. The processes are of two kinds: dendrites and axons. **Dendrites** are highly branched extensions of the cytoplasm of the cell body. A neuron usually has several main dendrites. Dendrites typically contain Nissl bodies and mitochondria. The function of dendrites is to conduct an impulse toward the cell body. The second type of cytoplasmic process, called an **axon** or **axis cylinder,** is a single, highly specialized, long process that conducts impulses away from the cell body to another neuron or tissue.

An axon usually originates from the cell body as a small conical elevation called the *axon hillock.* An axon contains mitochondria and neurofibrils but no Nissl bodies. Its cytoplasm, called *axoplasm,* is surrounded by a plasma membrane known as the *axolemma.* Axons vary in length from a few millimeters (1 mm = 0.0394 inch) in the brain to a meter (3.28 ft) or more between the spinal cord and toes. Along the course of an axon, there may be side branches called *axon collaterals.* The axon and its collaterals terminate by branching into many fine filaments called *telodendria.*

The term **nerve fiber** is applied to an axon and its sheaths. Figure 12-2b shows a cross section of a nerve fiber of the peripheral nervous system. Many axons, especially large, peripheral axons, are surrounded by a white, phospholipid, segmented covering called the *myelin sheath.* Axons containing such a covering are *myelinated,* while those without it are *unmyelinated.* Myelin is responsible for the color of the white matter in the nerves, brain, and spinal cord. The myelin sheath is produced by flattened cells, called *Schwann*

FIGURE 12-3 Myelin sheath. (a) Stages in the formation of a myelin sheath by a Schwann cell. (b) Electron micrograph of a nerve in cross section showing a myelinated nerve axon above and several unmyelinated nerve axons below at a magnification of 12,000×. (Courtesy of William Bloom and Don W. Fawcett, *A Textbook of Histology,* W. B. Saunders, Philadelphia, 1968.)

cells, located along the axon. These are neuroglial cells of the peripheral nervous system. In this process, a developing Schwann cell encircles the axon until its ends meet and overlap (Figure 12-3). The cell then winds around the axon several times and, in doing so, the cytoplasm and nucleus are pushed to the outside layer. The inner portion, consisting of several layers of Schwann cell membrane, is the myelin sheath. The function of the myelin sheath is to increase the speed of nerve impulse conduction and to insulate and maintain the axon. The *neurilemma* or *sheath of Schwann* is the peripheral nucleated cytoplasmic layer of the Schwann cell that encloses the myelin sheath (*lemma* = sheath). The neurilemma is found only around fibers of the peripheral nervous system. Its function is to assist in the regeneration of injured axons. Between the segments of the myelin sheath are unmyelinated gaps called *nodes of Ranvier*. Unmyelinated fibers are also enclosed by Schwann cells, but without multiple wrappings.

Nerve fibers of the central nervous system may be myelinated or unmyelinated. Myelination of central nervous system axons is accomplished by oligodendrocytes in somewhat the same manner that Schwann cells myelinate peripheral nervous system axons. Myelinated axons of the central nervous system also contain nodes of Ranvier, but they are not so numerous.

Classification

The different neurons in the body may be classified by structure and function.

The structural classification is based on the number of processes extending from the cell body. **Multipolar neurons** have several dendrites and one axon. Most neurons in the brain and spinal cord are of this type. **Bipolar neurons** have one dendrite and one axon and are found in the retina of the eye, the inner ear, and the olfactory area. The third structural type of neuron is the **unipolar neuron.** It has only one process extending from the cell body. The single process divides into a central branch, which functions as an axon, and a peripheral branch, which functions as a dendrite. Unipolar neurons originate in the embryo as bipolar neurons. During development, the axon and dendrite fuse into a single process.

The functional classification of neurons is based on the direction in which they transmit impulses. **Sensory neurons,** called **afferent neurons,** transmit impulses from receptors in the skin and sense organs to the brain and spinal cord. Sensory neurons also transmit impulses from receptors in the viscera. Sensory neurons are usually unipolar. **Motor neurons,** called **efferent neurons,** convey impulses from the brain and spinal cord to effectors, which may be either muscles or glands. Other neurons, called **association (connecting or internuncial) neurons,** carry impulses from sensory neurons to motor neurons and are located in the brain and spinal cord.

The processes of afferent and efferent neurons are arranged into bundles called *nerves.* Since nerves lie outside the central nervous system, they belong to the peripheral nervous system. The functional components of nerves are the nerve fibers, which may be grouped according to the following scheme:

1. General somatic afferent. These fibers conduct impulses from the skin, skeletal muscles, and joints to the central nervous system.

2. General somatic efferent. These fibers conduct impulses from the central nervous system to skeletal muscles. Impulses over these fibers cause the contraction of skeletal muscles.

3. General visceral afferent. These fibers convey impulses from the viscera and blood vessels to the central nervous system.

4. General visceral efferent. These fibers belong to the autonomic nervous system. Also called *autonomic fibers,* they convey impulses from the central nervous system to cause contractions of smooth muscle, cardiac muscle, and glands (causing secretion).

Structural Variation

Although all neurons conform to the general plan previously described, there are considerable differences in structure. For example, cell bodies range in diameter from 5 μm for the smallest cells to 135 μm for large motor neurons. The pattern of dendritic branching is also varied and distinctive for neurons in different parts of the body. Moreover, the axons of very small neurons are only a fraction of a millimeter in length and lack a myelin sheath, while some axons of large neurons are over a meter long and are usually enclosed in a myelin sheath.

A few patterns of diversity are shown in Figure 12-4. Note the structure of a typical afferent (sensory) neuron. Compare it to the typical efferent (motor) neuron. What structural differences do you observe? A few examples of association neurons are also shown: a *stellate cell, cell of Martinotti,* and a *horizontal cell of Cajal.* All are found in the cerebral cortex, the outer layer of the cerebrum. Note the *granule cell,* an association neuron in the cortex of the cerebellum.

Physiology

One of the most striking features of nervous tissue is its ability to send electrical messages called nerve impulses. Although it is beyond the scope of this text to describe the details of a nerve impulse, certain concepts must be understood in order to know how the

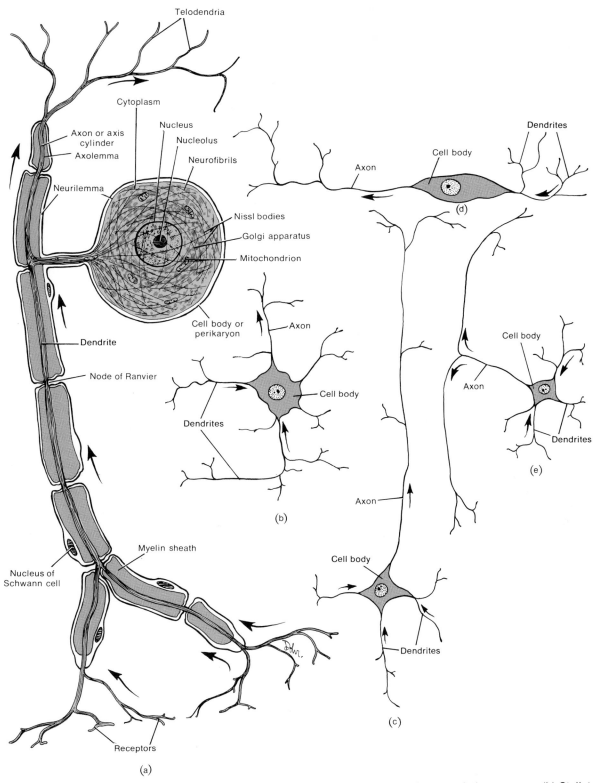

FIGURE 12-4 Varieties of neurons. (a) Typical afferent neuron. (b-e) Representative association neurons. (b) Stellate cell. (c) Cell of Martinotti. (d) Horizontal cell of Cajal. (e) Granule cell. Arrows indicate the direction of impulse conduction.

nervous system works. Very simply, a **nerve impulse** is a wave of electrical negativity that travels along the surface of the membrane of a neuron. Among other things, a nerve impulse depends upon the movement of sodium, potassium, and other ions between interstitial fluid and the inside of a neuron. For a nerve impulse to begin, a stimulus of adequate strength must be applied to the neuron. A stimulus is a change in the environment of sufficient strength to initiate an impulse. The ability of a neuron to respond to a stimulus and convert it into an impulse is known as **irritability.**

The nerve impulse is the most rapid way that the body can respond to environmental changes. It provides the quickest means for achieving homeostasis, the ability of the body to maintain normalcy. The speed of a nerve impulse is determined by the size, type, and physiological condition of the nerve fiber. For example, myelinated fibers with the largest diameters can transmit impulses at speeds up to about 100 m (328 ft)/second. Unmyelinated fibers with the smallest diameters can transmit impulses at the rate of about 0.5 m (1.5 ft)/second.

In addition to irritability, neurons are also capable of **conductivity,** the ability to transmit an impulse to another neuron or another tissue, such as a muscle or a gland. The junction between two neurons is called a **synapse,** and it is across the synapse that the impulses are conducted from one neuron to another. A neuron located before the synapse is known as a *presynaptic neuron,* and one located after the synapse is referred to as a *postsynaptic neuron.* Axons of presynaptic neurons release chemical transmitters into the synapse. Some transmitters permit the impulse to pass to the postsynaptic neuron or other tissue of the body. Other transmitters inhibit the impulse.

Impulse conduction at a synapse is one-way. The direction is from the presynaptic neuron to the postsynaptic neuron. In other words, impulses must move forward over their pathways. They cannot back up into another presynaptic neuron. Such a mechanism is very important in preventing impulse conduction along wrong pathways. Imagine the result if impulses transmitted along the motor neurons that move your hand could move back and stimulate the sensory neuron which relays information about heat. You would feel heat, cry in pain, and go through all the emotions of being burned every time you simply wanted to move your hand.

STUDY OUTLINE

Organization

1. The nervous system controls and integrates all body activities by sensing changes, interpreting them, and reacting to them.
2. The central nervous system consists of the brain and spinal cord.
3. The peripheral nervous system is classified into an afferent system and an efferent system.
4. The efferent system is subdivided into a somatic nervous system and an autonomic nervous system.
5. The somatic nervous system consists of efferent neurons that conduct impulses from the central nervous system to skeletal muscle tissue.
6. The autonomic nervous system contains efferent neurons that convey impulses from the central nervous system to smooth muscle tissue, cardiac muscle tissue, and glands.

Histology

Neuroglia

1. Neuroglia are specialized tissue cells that support neurons, attach neurons to blood vessels, produce the myelin sheath, and carry out phagocytosis.

Neurons

1. Neurons, or nerve cells, consist of a perikaryon or cell body, dendrites that pick up stimuli and convey impulses to the cell body, and usually a single axon. The axon transmits impulses from the neuron to the dendrites or cell body of another neuron or to an effector organ of the body.
2. On the basis of structure, neurons are multipolar, bipolar, and unipolar.
3. On the basis of function, sensory (afferent) neurons transmit impulses to the central nervous system; association neurons transmit impulses to other neurons including motor neurons; and motor (efferent) neurons transmit impulses to effectors.

Physiology

1. The ability of a neuron to respond to a stimulus and convert it into a nerve impulse is called irritability.
2. A nerve impulse is a wave of negativity that travels along the surface of the membrane of a neuron.
3. The speed of a nerve impulse is determined by the size, type, and physiological condition of the nerve fiber.
4. Conductivity is the ability of a neuron to transmit a nerve impulse to another neuron or another tissue.
5. The junction between neurons is called a synapse. Impulse conduction across a synapse requires the release of chemical transmitters by presynaptic neurons.
6. Impulse conduction at a synapse is one-way.

REVIEW QUESTIONS

1. What is the overall function of the nervous system?
2. Distinguish between the central and peripheral nervous system.
3. What are neuroglia? List their principal functions.
4. Define a neuron. Diagram a neuron, label it, and list the function next to each labeled part.
5. What is the myelin sheath? How is it formed?
6. Define the neurilemma. Why is it important?
7. Discuss the structural classification of neurons. Give an example of each.
8. Describe the functional classification of neurons.
9. Distinguish among the following kinds of fibers: general somatic afferent, general somatic efferent, general visceral afferent, and general visceral efferent.
10. What structural differences exist between a typical afferent and a typical efferent neuron?
11. Define a nerve impulse.
12. Compare irritability and conductivity.
13. What factors determine the rate of impulse conduction?
14. What is a synapse? How does an impulse cross a synapse?

13

The Spinal Cord and the Spinal Nerves

STUDENT OBJECTIVES

- Define white matter, gray matter, nerve, ganglion, tract, nucleus, and horn.

- Describe the gross anatomical features of the spinal cord.

- Explain how the spinal cord is protected by the meninges.

- Define a spinal puncture.

- Describe the structure of the spinal cord in cross section.

- Explain the functions of the spinal cord as a conduction pathway and a reflex center.

- List the location, origin, termination, and function of the principal ascending and descending tracts of the spinal cord.

- Describe the components of a reflex arc.

- Compare the mechanism of a stretch reflex, flexor reflex, and crossed extensor reflex.

- Define a spinal nerve.

- Describe the composition and coverings of a spinal nerve.

- Name the 31 pairs of spinal nerves.

- Explain how a spinal nerve branches upon leaving the intervertebral foramen.

- Define a plexus.

- Explain the composition and distribution of the cervical, brachial, lumbar, sacral, and coccygeal plexuses.

- Define an intercostal nerve.

- Define a dermatome and its clinical importance.

- Describe spinal cord injury and list the immediate and long-range effects.

- Describe the conditions necessary for peripheral nerve regeneration.

- Distinguish between sciatica and neuritis.

n this chapter, our main concern will be a study of the structure and function of the spinal cord and the nerves that originate from it.

Grouping of Neural Tissue

The term **white matter** refers to aggregations of myelinated axons from many neurons supported by neuroglia. The lipid substance, myelin, has a whitish color that gives white matter its name. The gray areas of the nervous system are called **gray matter.** They contain either nerve cell bodies and dendrites or bundles of unmyelinated axons and neuroglia.

A **nerve** is a bundle of fibers located outside the central nervous system. Since the dendrites of somatic afferent neurons and axons of somatic efferent neurons of the peripheral nervous system are myelinated, most nerves are white matter. Nerve cell bodies that lie outside the central nervous system are generally grouped with other nerve cell bodies to form **ganglia** (*ganglion* = knot). Ganglia, since they are made up principally of unmyelinated nerve cell bodies, are masses of gray matter.

A **tract** is a bundle of fibers in the central nervous system. Tracts may run long distances up and down the spinal cord. Tracts also exist in the brain and connect parts of the brain with each other and with the spinal cord. The chief spinal tracts that conduct impulses up the cord are concerned with sensory impulses and are called *ascending tracts.* By contrast, spinal tracts that carry impulses down the cord are motor tracts and are called *descending tracts.* The major tracts consist of myelinated fibers and are therefore white matter. A **nucleus** is a mass of nerve cell bodies and dendrites in the central nervous system. It consists of gray matter. **Horns** or **columns** are the chief areas of gray matter in the spinal cord. The term *horn* describes the two-dimensional appearance of the organization of gray matter in the spinal cord as seen in cross section. The term *column* describes the three-dimensional appearance of the gray matter in longitudinal columns. Since the white matter of the spinal cord is also arranged in columns, we will refer to the gray matter as being arranged in horns.

Spinal Cord

GENERAL FEATURES

The **spinal cord** is a cylindrical structure that is slightly flattened anteriorly and posteriorly. It begins as a continuation of the medulla oblongata, the inferior part of the brain stem, and extends from the foramen magnum of the occipital bone to the level of the second lumbar vertebra (Figure 13-1). The length of the adult spinal cord ranges from 42 to 45 cm (16 to 18 inches). The diameter of the cord varies at different levels.

When the cord is viewed externally, two conspicuous enlargements can be seen. The superior enlargement, the *cervical enlargement,* extends from the fourth cervical to the first thoracic vertebra. Nerves that supply the upper extremities arise from the cervical enlargement. The inferior enlargement, called the *lumbar enlargement,* extends from the ninth to twelfth thoracic vertebra. Nerves that supply the lower extremities arise from the lumbar enlargement.

Below the lumbar enlargement, the spinal cord tapers to a conical portion known as the *conus medullaris.* The conus medullaris lies at about the level of the first or second lumbar vertebra. Arising from the conus medullaris is the *filum terminale,* a nonnervous fibrous tissue of the spinal cord that extends inferiorly to the coccyx. The filum terminale consists mostly of pia mater, the innermost of three membranes that cover and protect the spinal cord and brain. Some nerves that arise from the lower portion of the cord do not leave the vertebral column immediately. They angle inferiorly in the vertebral canal like wisps of coarse hair flowing from the end of the cord. They are appropriately named the *cauda equina* (horse's tail).

The spinal cord is a series of 31 segments, each giving rise to a pair of spinal nerves. *Spinal segment* refers to a region of the spinal cord from which a pair of spinal nerves arises. Figure 13-3 shows that the cord is divided into right and left sides by two grooves. One of these, the *anterior median fissure,* is a deep, wide groove on the anterior (ventral) surface. The other is the *posterior median sulcus,* a shallower, narrow groove on the posterior (dorsal) surface.

PROTECTION AND COVERINGS

Vertebral Canal

The spinal cord is located in the vertebral canal of the vertebral column. The *vertebral canal* is formed by the vertebral foramina of all the vertebrae arranged on top of each other. Since the wall of the vertebral canal is essentially a ring of bone surrounding the spinal cord, the cord is well protected. A certain degree of protection is also provided by the meninges, the cerebrospinal fluid, and the vertebral ligaments.

Meninges

The *meninges* are coverings that run continuously around the spinal cord and brain. Those associated specifically with the cord are known as spinal me-

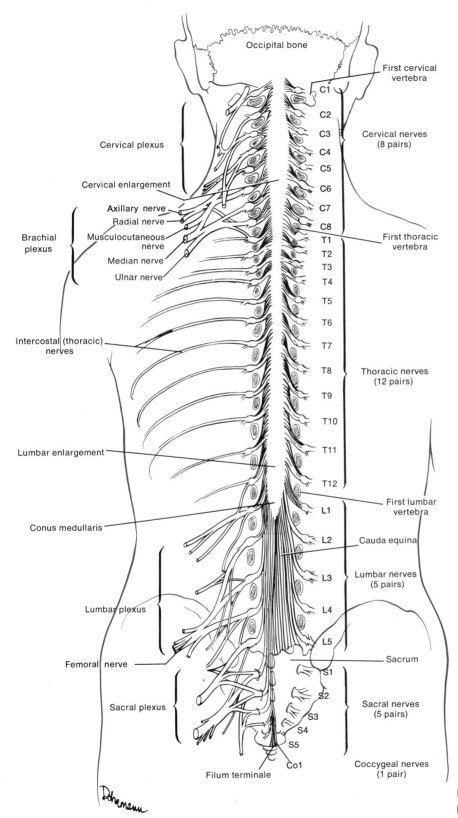

Occipital bone

First cervical vertebra

C1

Cervical plexus

C2
C3
C4
C5
C6
C7
C8

Cervical nerves (8 pairs)

Cervical enlargement

Axillary nerve
Radial nerve
Musculocutaneous nerve
Median nerve
Ulnar nerve

Brachial plexus

First thoracic vertebra

T1
T2
T3
T4
T5
T6
T7

Intercostal (thoracic) nerves

T8
T9
T10

Thoracic nerves (12 pairs)

T11

Lumbar enlargement

T12

Conus medullaris

First lumbar vertebra

L1
L2

Cauda equina

L3

Lumbar nerves (5 pairs)

L4

Lumbar plexus

L5

Femoral nerve

Sacrum

S1
S2

Sacral plexus

S3
S4
S5

Sacral nerves (5 pairs)

Co1

Filum terminale

Coccygeal nerves (1 pair)

FIGURE 13-1 Spinal cord and spinal nerves in posterior view.

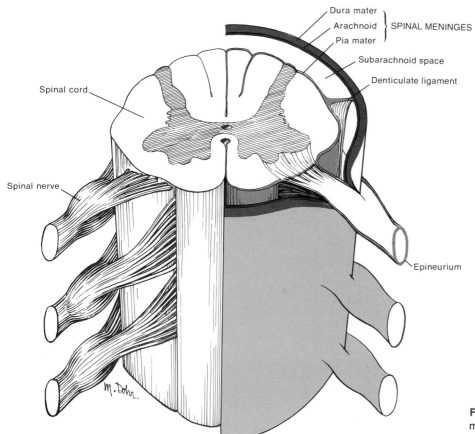

FIGURE 13-2 Location of the spinal meninges as seen on the left side in a cross section of the spinal cord.

ninges (Figure 13-2). The outer spinal meninx is called the *dura mater* (or tough mother) and forms a tube from the level of the second sacral vertebra, where it is fused with the filum terminale, to the foramen magnum, where it is continuous with the dura mater of the brain. The dura mater is composed of dense, fibrous connective tissue. Between the dura mater and the wall of the vertebral canal is the *epidural space,* which is filled with fat, connective tissue, and blood vessels. It serves as padding around the cord. The middle spinal meninx is called the *arachnoid* (or spider layer). It is a delicate connective tissue membrane that forms a tube inside the dura mater. It is also continuous with the arachnoid of the brain. Between the dura mater and the arachnoid is a space called the *subdural space,* which contains serous fluid. The inner meninx is known as the *pia mater* (or delicate mother). It is a transparent fibrous membrane that forms a tube around and adheres to the surface of the spinal cord and brain. It contains numerous blood vessels. *Meningitis,* or inflammation of the meninges, usually involves the arachnoid and pia mater. Between the arachnoid and the pia mater is the *subarachnoid space,* where the cerebrospinal fluid circulates. All three spinal meninges cover the spinal nerves as they exit the spinal column through the intervertebral fora-

mina. The spinal cord is suspended in the middle of its dural sheath by membranous extensions of the pia mater. These extensions, called the *denticulate ligaments,* are attached laterally along the length of the cord between the ventral and dorsal nerve roots on either side. The ligaments protect the spinal cord against shock and sudden displacement.

SPINAL PUNCTURE

The removal of cerebrospinal fluid from the subarachnoid space in the inferior lumbar region of the spinal cord is a *spinal (lumbar) puncture,* or *tap.* The procedure is normally performed between the third and fourth or fourth and fifth lumbar vertebrae. The spinous process of the fourth lumbar vertebra is easily located by drawing a line across the highest points of the iliac crests. This line will pass right through the spinous process of the fourth lumbar vertebra. A lumbar puncture is below the spinal cord and thus poses little danger to it. If the patient lies on one side, drawing the knees and chest together, the vertebrae separate slightly so that a needle can be conveniently inserted. Lumbar punctures are used to withdraw fluid for diagnostic purposes, to introduce antibiotics (as in the case of meningitis), and to administer anesthesia.

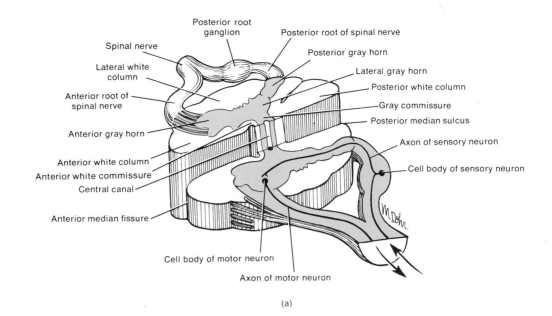

Posterior root
ganglion

Spinal nerve

Posterior root of spinal nerve

Lateral white
column

Posterior gray horn

Lateral gray horn

Posterior white column

Anterior root of
spinal nerve

Gray commissure

Posterior median sulcus

Anterior gray horn

Axon of sensory neuron

Anterior white column

Cell body of sensory neuron

Anterior white commissure

Central canal

Anterior median fissure

Cell body of motor neuron

Axon of motor neuron

(a)

Posterior
white
column

Posterior
median
sulcus

Central
canal

Posterior
gray horn

Lateral
white
column

Anterior
gray horn

Anterior
white
column

Anterior
median
fissure

Anterior
white
commissure

(b)

FIGURE 13-3 Spinal cord. (a) The organization of gray and white matter in the spinal cord as seen in a cross section of the spinal cord. The front of the figure has been sectioned at a lower level than the back so that you can see what is inside the posterior root ganglion, posterior root of the spinal nerve, anterior root of the spinal nerve, and the spinal nerve. (b) Photograph of the spinal cord at the seventh cervical segment, Weigert stain, at a magnification of 7×. (Courtesy of Murray L. Barr, *The Human Nervous System,* Harper & Row Publishers, Inc., New York, 1974.)

STRUCTURE IN CROSS SECTION

The spinal cord consists of both gray and white matter. Figure 13-3 shows that the gray matter lies in an area shaped like an H. The gray matter consists primarily of nerve cell bodies and unmyelinated axons and dendrites of association and motor neurons. The white matter surrounds the gray matter and consists of bundles of myelinated axons of motor and sensory neurons.

In the center of the gray matter is a cross bar of the H called the *gray commissure,* connecting the right and left portions of the H. In the center of the gray commissure is a small space called the *central canal.* This canal runs the length of the spinal cord and is continuous with the fourth ventricle of the medulla. It contains cerebrospinal fluid. Anterior to the gray commissure is the *anterior (ventral) white commissure,* which connects the white matter of the right and left sides of the spinal cord. The upright portions of the H are further subdivided into regions. Those closer to the front of the cord are called *anterior (ventral) gray horns*. They represent the motor part of the gray matter. The regions closer to the back of the cord are referred to as *posterior (dorsal) gray horns*. They represent the sensory part of the gray matter. The regions between the anterior and posterior gray horns are *lateral gray horns*. The lateral gray horns are most prominent in the thoracic and upper lumbar segments of the cord.

The gray matter of the cord also contains several nuclei that serve as relay stations for impulses and origins for certain nerves. Nuclei are clusters of nerve cell bodies and dendrites in the spinal cord and brain.

The white matter on each side of the cord, like the gray matter, is organized into regions. The anterior and posterior gray horns divide the white matter into three broad areas: *anterior (ventral) white column, posterior (dorsal) white column,* and *lateral white column*. Each column (or *funiculus*) in turn consists of distinct bundles of myelinated fibers that run the length of the cord. These bundles are called *fasciculi* or *tracts*. The longer *ascending tracts* consist of sensory axons that conduct impulses which enter the spinal cord and pass upward to the brain. The longer *descending tracts* consist of motor axons that conduct impulses from the brain downward through the spinal cord and out to muscles and glands. Thus the ascending tracts are sensory tracts and the descending tracts are motor tracts. Still other short tracts contain ascending or descending axons that convey impulses from one level of the cord to another.

FUNCTIONS

A major function of the spinal cord is to convey sensory impulses from the periphery to the brain and to conduct motor impulses from the brain to the periphery. A second principal function is to provide reflexes.

Conduction Pathway

The vital function of conveying sensory and motor information is carried out by the ascending and descending tracts of the cord. The names of the tracts indicate the white column (funiculus) in which the tract travels, where the cell bodies of the tract originate, where the axons of the tract terminate, and the direction of impulse conduction within the tract. For example, the anterior spinothalamic tract is located in the anterior white column, it originates in the spinal cord, it terminates in the thalamus (a region of the brain), and it is an ascending (sensory) tract since it conveys impulses from the cord upward to the brain.

Important ascending and descending tracts are shown in Figure 13-4. Exhibit 13-1 summarizes the principal tracts.

Reflex Center

The second principal function of the spinal cord is to provide reflexes. Spinal nerves are the paths of communication between the spinal cord tracts and the periphery. Figure 13-3 reveals that each pair of spinal nerves is connected to a segment of the cord by two points of attachment called roots. The **posterior** or **dorsal (sensory) root** contains sensory nerve fibers only and conducts impulses from the periphery to the spinal cord. These fibers extend into the posterior (dorsal)

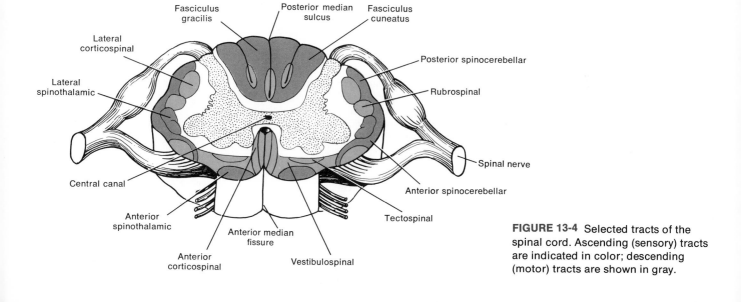

FIGURE 13-4 Selected tracts of the spinal cord. Ascending (sensory) tracts are indicated in color; descending (motor) tracts are shown in gray.

EXHIBIT 13-1 ▰▰▰▰▰▰▰▰▰▰▰▰▰▰▰▰▰▰▰▰▰▰
SELECTED ASCENDING AND DESCENDING TRACTS OF SPINAL CORD

TRACT	LOCATION (WHITE COLUMN)	ORIGIN	TERMINATION	FUNCTION
ASCENDING TRACTS				
Anterior (ventral) spinothalamic	Anterior (ventral) column	Posterior (dorsal) gray horn on one side of cord but crosses to opposite side	Thalamus; impulse eventually conveyed to cerebral cortex	Conveys sensations for crude touch and pressure from one side of body to opposite side of thalamus. Eventually sensations reach cerebral cortex.
Lateral spinothalamic	Lateral column	Posterior (dorsal) gray horn on one side of cord but crosses to opposite side	Thalamus; impulse eventually conveyed to cerebral cortex	Conveys sensations for pain and temperature from one side of body to opposite side of thalamus Eventually sensations reach cerebral cortex.
Fasciculus gracilis and fasciculus cuneatus	Posterior (dorsal) column	Axons of afferent neurons from periphery that enter posterior (dorsal) column and rise to same side of medulla	Nucleus gracilis and nucleus cuneatus of medulla; impulse eventually conveyed to cerebral cortex	Convey sensations from one side of body to same side of medulla for fine touch; two-point discrimination (ability to distinguish that two points on skin are touched even though close together); proprioception (awareness of precise position of body parts and their direction of movement); stereognosis (ability to recognize size, shape, and texture of object); weight discrimination (ability to assess weight of an object); and vibrations. Eventually sensations may reach cerebral cortex.
Posterior (dorsal) spinocerebellar	Posterior (dorsal) portion of lateral column	Posterior (dorsal) gray horn on same side of cord	Cerebellum	Conveys sensations from one side of body to same side of cerebellum for subconscious proprioception.
Anterior (ventral) spinocerebellar	Anterior (ventral) portion of column	Posterior (dorsal) gray horn on one side of cord; tract contains both crossed and uncrossed fibers	Cerebellum	Conveys sensations from both sides of body to cerebellum for subconscious proprioception.
DESCENDING TRACTS				
Lateral corticospinal	Lateral column	Cerebral cortex on one side of brain but crosses in base of medulla to opposite side of cord	Anterior (ventral) gray horn	Conveys motor impulses from one side of cortex to anterior gray horn of opposite side. Eventually impulses reach skeletal muscles on opposite side of body that coordinate precise, discrete movements.
Anterior (ventral) corticospinal	Anterior (ventral) column	Cerebral cortex on one side of brain, uncrossed in medulla, but crosses to opposite side of cord	Anterior (ventral) gray horn	Conveys motor impulses from one side of cortex to anterior gray horn of opposite side. Eventually impulses reach skeletal muscles on opposite side of body that coordinate precise, discrete movements.
Rubrospinal	Lateral column	Midbrain (red nucleus) but crosses to opposite side of cord	Anterior (ventral) gray horn	Conveys motor impulses from one side of midbrain to skeletal muscles on opposite side of body that are concerned with muscle tone and posture.
Tectospinal	Anterior (ventral) column	Midbrain but crosses to opposite side of cord	Anterior (ventral) gray horn	Conveys motor impulses from one side of midbrain to skeletal muscles on opposite side of body that control movements of head in response to auditory, visual, and cutaneous stimuli.
Vestibulospinal	Anterior (ventral) column	Medulla on one side of brain to same side of cord	Anterior (ventral) gray horn	Conveys motor impulses from one side of medulla to skeletal muscles on same side of body that regulate body tone in response to movements of head (equilibrium).

gray horn. Each dorsal root also has a swelling, the **posterior** or **dorsal (sensory) root ganglion,** which contains the cell bodies of the sensory neurons from the periphery. The other point of attachment of a spinal nerve to the cord is the **anterior** or **ventral (motor) root.** It contains motor nerve fibers only and conducts impulses from the spinal cord to the periphery. The cell bodies of the motor neurons are located in the gray matter of the cord. If the motor impulse supplies a skeletal muscle, the cell bodies are located in the anterior (ventral) gray horn. If, however, the impulse supplies smooth muscle, cardiac muscle, or a gland through the autonomic nervous system, the cell bodies are located in the lateral gray horn.

● *Reflex Arc* The path an impulse follows from its origin in the dendrites or cell body of a neuron in one part of the body to its termination elsewhere in the body is called a *conduction pathway*. All conduction pathways consist of circuits of neurons. One pathway is known as a **reflex arc,** the functional unit of the nervous system. A reflex arc contains two or more neurons over which impulses are conducted from a receptor to the brain or spinal cord and then to an effector. The basic components of a reflex arc are as follows:

1. Receptor. The distal end of a dendrite or a sensory structure associated with the distal end of a dendrite. Its role in the reflex arc is to respond to a change in the internal or external environment by initiating a nerve impulse in a sensory neuron.

2. Sensory neuron. Once stimulated, the sensory neuron passes the impulse from the receptor to its axonal termination in the central nervous system.

3. Center. A region, usually in the central nervous system, where an incoming sensory impulse generates an outgoing motor impulse. In the center, the impulse may be inhibited, transmitted, or rerouted. In the center of some reflex arcs, the sensory neuron directly generates the impulse in the motor neuron. The center may also contain an association neuron between the sensory neuron and the motor neuron leading to a muscle or a gland.

4. Motor neuron. Transmits the impulse generated by the sensory or association neuron in the center to the organ of the body that will respond.

5. Effector. The organ of the body, either muscle or gland, that responds to the motor impulse. This response is called a *reflex action* or *reflex*.

Reflexes are fast responses to changes in the internal or external environment to maintain homeostasis. Reflexes carried out by the spinal cord alone are called *spinal reflexes*. Reflexes that result in the contraction of skeletal muscles are known as *somatic reflexes*. Those that cause the contraction of smooth or cardiac muscle or secretion by glands are *visceral (autonomic) reflexes*. Our concern at this point is to examine a few somatic spinal reflexes: the stretch reflex, the flexor reflex, and the crossed extensor reflex.

● *Stretch Reflex* The **stretch reflex** is based on a *two-neuron, or monosynaptic, reflex arc*. Only two neurons are involved and there is only one synapse in the pathway (Figure 13-5). This reflex results in the contraction of a muscle when it is stretched. Slight stretching of a muscle stimulates receptors in the muscle called *neuromuscular spindles*. The spindles monitor changes in the length of the muscle. Once the spindle is stimulated, an impulse is sent along a sen-

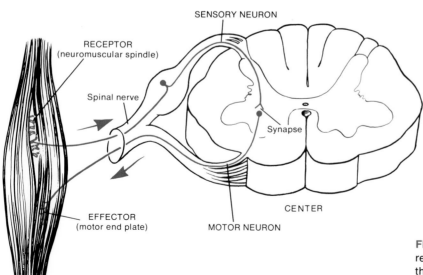

SENSORY NEURON

RECEPTOR
(neuromuscular spindle)

Spinal nerve

Synapse

EFFECTOR
(motor end plate)

MOTOR NEURON

CENTER

FIGURE 13-5 Stretch reflex. Notice that in a stretch reflex there are only two neurons involved and there is only one synapse in the pathway. Thus it is a monosynaptic reflex arc. Why is the reflex arc shown referred to as an ipsilateral reflex arc?

sory neuron to the spinal cord. The sensory neuron lies in the posterior root of a spinal nerve and synapses with a motor neuron in the anterior gray horn. The sensory neuron generates an impulse at the synapse that is transmitted along the motor neuron. The motor neuron lies in the anterior root of the spinal nerve and terminates in a skeletal muscle. Once the impulse reaches the stretched muscle, it contracts. Thus the stretch is counteracted by contraction. Since the sensory impulse enters the spinal cord on the same side that the motor impulse leaves the spinal cord, the reflex arc is called an *ipsilateral reflex arc*. All monosynaptic reflex arcs are ipsilateral. The stretch reflex is essential in maintaining muscle tone (firmness). Moreover, it is the basis for several tests used in neurological examinations. One such reflex is the *knee jerk*, or *patellar reflex*. This reflex is illustrated in Figure 13-5; it is tested by tapping the patellar ligament (stimulus). Neuromuscular spindles in the quadriceps femoris muscle attached to the ligament send the sensory impulse to the spinal cord and the returning motor impulse causes contraction of the muscle. The response is extension of the leg at the knee, or a knee jerk.

● *Flexor Reflex and Crossed Extensor Reflex*   Reflexes other than stretch reflexes involve association neurons in addition to the sensory and motor neuron— they are *polysynaptic reflex arcs*. One example of a reflex based on a polysynaptic reflex arc is the **flexor reflex,** or **withdrawal reflex** (Figure 13-6). Suppose you step on a tack. As a result of the painful stimulus, you immediately withdraw your foot. What has happened? A sensory neuron transmits an impulse from

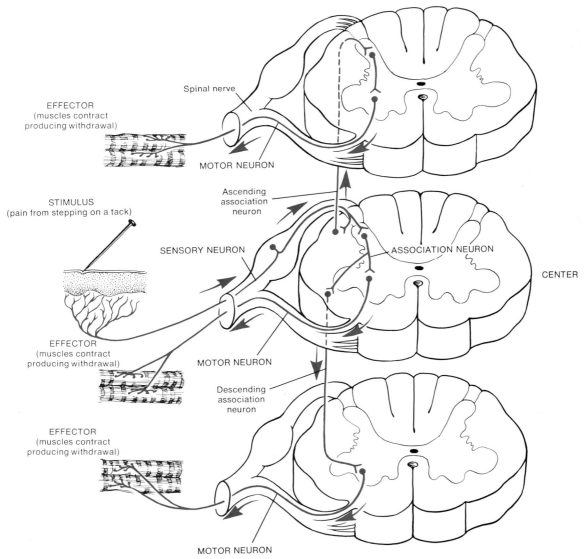

FIGURE 13-6 Flexor reflex. This reflex arc is a polysynaptic ipsilateral reflex arc because it involves more than one synapse. This is because it contains association neurons as well as sensory and motor neurons. Why is the reflex arc shown also an intersegmental reflex arc?

the receptor to the spinal cord. A second impulse is generated in an association neuron, which generates a third impulse in a motor neuron. The motor neuron stimulates the muscles of your foot and you withdraw it. Thus a flexor reflex is protective. It moves an extremity to avoid pain.

This stretch reflex is also ipsilateral. The incoming and outgoing impulses are on the same side of the spinal cord. The stretch reflex also illustrates another feature of reflex arcs. In the monosynaptic stretch reflex, the returning motor impulse affects only the quadriceps muscle of the thigh. When you withdraw your entire lower or upper extremity from a noxious stimulus, more than one muscle is involved. Therefore several motor neurons are simultaneously returning impulses to several upper and lower extremity muscles at the same time. Thus a single sensory impulse causes several motor responses. This kind of reflex arc, in which a single sensory neuron splits into ascending and descending branches, each forming a synapse with association neurons at different segments of the cord, is called an *intersegmental reflex arc*. Because of intersegmental reflex arcs, a single sensory neuron can activate several motor neurons and thereby cause stimulation of more than one effector.

Something else may happen when you step on a tack. You may lose your balance as your body weight shifts to the other foot. Then you do whatever you can to regain your balance so you do not fall. This means motor impulses are also sent to your unstimulated foot and both upper extremities. The motor impulses that travel to your unaffected foot cause extension at the knee so you can place your entire body weight on the foot. These impulses cross the spinal cord as shown in Figure 13-7. The incoming sensory impulse not only initiates the flexor reflex that causes you to withdraw, it also initiates an extensor reflex. The incoming sensory impulse crosses to the opposite side of the spinal cord through association neurons at that level and several levels above and below the point of sensory stimulation. From these levels, the motor neurons cause extension of the knee, thus maintaining balance. Unlike the flexor reflex, which passes over an ipsilateral reflex arc, the extensor reflex passes over a *contralateral reflex arc*—the impulse enters one side of the spinal cord and exits on the opposite side. The reflex just described in which extension of the muscles in one limb occurs as a result of flexion of the muscles of the opposite limb is simply called a **crossed extensor reflex.**

The flexor reflex and crossed extensor reflex also illustrate *reciprocal inhibition,* another feature of many reflexes. Reciprocal inhibition occurs when a reflex excites a muscle to cause its contraction and also inhibits another muscle to allow its extension. Thus, in this reflex, excitation and inhibition occur simultaneously.

In the flexor reflex, when the flexor muscles of your lower extremity are contracting, the extensor muscles of the same extremity are being extended. If both sets of muscles contracted at the same time, you would not be able to flex your limb because both sets of muscles would pull on the limb bones. But, because of reciprocal inhibition, one set of muscles contracts while the other is being extended.

In the crossed extensor reflex, reciprocal inhibition also occurs. While you are flexing the muscles of the limb that has been stimulated by the tack, the muscles of your other limb are producing extension to help maintain balance. Reciprocal inhibition is vital in coordinating body movements. In flexing the forearm at the elbow, there is a prime mover, an antagonist, and a synergist. The prime mover (biceps) contracts to cause flexion, the antagonist (triceps) extends to yield to the action of the prime mover, and the synergist (deltoid) helps the prime mover perform its role efficiently.

Spinal Nerves

COMPOSITION AND COVERINGS

A **spinal nerve** has two points of attachment to the cord: a posterior root and an anterior root. A short distance from the spinal cord the roots unite to form a spinal nerve. Since the posterior root contains sensory fibers and the anterior root contains motor fibers, a spinal nerve is a **mixed nerve.** The posterior (dorsal) root ganglion contains cell bodies of sensory neurons. The posterior and anterior roots unite to form the spinal nerve at the intervertebral foramen.

In Figure 13-8, you can see that the nerve contains many fibers surrounded by different coverings. The individual fibers, whether myelinated or unmyelinated, are wrapped in a connective tissue called the *endoneurium*. Groups of fibers with their endoneurium are arranged in bundles called fascicles, and each bundle is wrapped in connective tissue called the *perineurium*. The outermost covering around the entire nerve is the *epineurium*. The spinal meninges fuse with the epineurium as the nerve exits from the vertebral canal.

NAMES

The 31 pairs of spinal nerves are named and numbered according to the region and level of the spinal cord from which they emerge (see Figure 13-1). The first cervical pair emerges between the atlas and the occipital bone. All other spinal nerves leave the backbone from the intervertebral foramina between adjoining vertebrae. There are 8 pairs of cervical nerves, 12 pairs of thoracic nerves, 5 pairs of lumbar nerves, 5 pairs of sacral nerves, and 1 pair of coccygeal nerves.

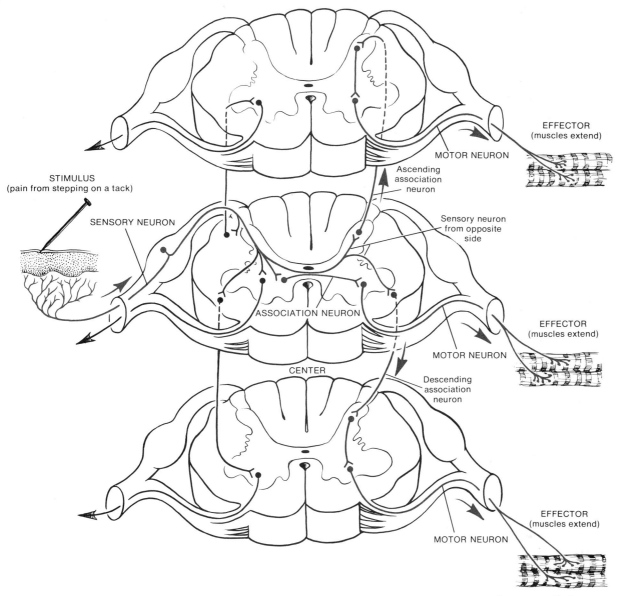

STIMULUS
(pain from stepping on a tack)

SENSORY NEURON

EFFECTOR
(muscles extend)

MOTOR NEURON

Ascending
association
neuron

Sensory neuron
from opposite
side

ASSOCIATION NEURON

EFFECTOR
(muscles extend)

MOTOR NEURON

CENTER

Descending
association
neuron

EFFECTOR
(muscles extend)

MOTOR NEURON

FIGURE 13-7 Crossed extensor reflex. Although the flexor reflex is shown on the left of the diagram so that you can correlate it with the crossed extensor reflex on the right, concentrate your attention on the crossed extensor reflex. Why is the crossed extensor reflex classified as a contralateral reflex arc?

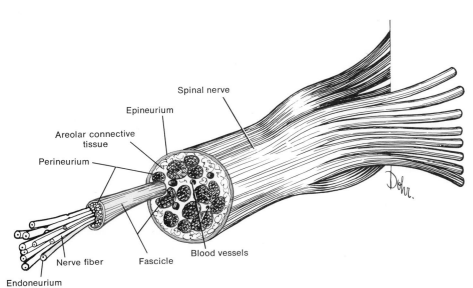

Spinal nerve

Epineurium

Areolar connective
tissue

Perineurium

Nerve fiber

Fascicle

Blood vessels

Endoneurium

FIGURE 13-8 Coverings of a spinal nerve.

During fetal life, the spinal cord and vertebral column grow at different rates, the cord growing more slowly. Thus not all the spinal cord segments are in line with their corresponding vertebrae. Remember that the spinal cord terminates near the level of the first or second lumbar vertebra. Thus the lower lumbar, sacral, and coccygeal nerves must descend more and more to reach their foramina before emerging from the vertebral column. This arrangement constitutes the cauda equina.

BRANCHES

Shortly after a spinal nerve leaves its intervertebral foramen, it divides into several branches (Figure 13-9). These branches are known as rami. The *dorsal ramus* innervates (supplies) the deep muscles and skin of the dorsal surface of the back. The *ventral ramus* of a spinal nerve innervates the superficial back muscles, all the structures of the extremities, and the lateral and ventral trunk. Except for thoracic nerves T2 to T12, the ventral rami of other spinal nerves enter into the formation of plexuses before supplying a part of the body. In addition to dorsal and ventral rami, spinal nerves also give off a *meningeal branch*. This branch supplies the vertebrae, vertebral ligaments, blood vessels of the spinal cord, and the meninges. Other branches of a spinal nerve are the *rami communicantes*.

PLEXUSES

The ventral rami of spinal nerves, except for T2 to T12, do not go directly to the structures of the body they supply. Instead, they form networks with adjacent nerves on either side of the body. Such networks are called **plexuses,** meaning braid. The principal plexuses are the cervical plexus, the brachial plexus, the lumbar plexus, and the sacral plexus. Emerging from the plexuses are nerves bearing names that are often descriptive of the general regions they supply or the course they take. Each of these nerves, in turn, may have several branches named for the specific structures they innervate.

Cervical Plexus

The **cervical plexus** is formed by the ventral rami of the first four cervical nerves (C1 to C4) with contributions from C5. There is one on each side of the neck alongside the first four cervical vertebrae (Figure 13-10 on page 384). The *roots* of the plexus indicated in the diagram are simply continuations of the ventral rami. The cervical plexus supplies the skin and muscles of the head, neck, and upper part of the shoulders. Branches of the cervical plexus also connect with cranial nerves XI (accessory) and XII (hypoglossal). A major pair of nerves arising from the cervical plexuses are the phrenic nerves supplying motor fibers to the diaphragm. Damage to the spinal cord above the origin of the phrenic nerves results in paralysis of the diaphragm since the phrenic nerves no longer send impulses to the diaphragm. Contractions of the diaphragm are essential for normal breathing.

Exhibit 13-2 on page 385 summarizes the nerves and distributions of the cervical plexus. The relationship of the cervical plexus to the other plexuses is shown in Figure 13-1.

Brachial Plexus

The brachial plexus is formed by the ventral rami of spinal nerves C5 to C8 and T1 with contributions from C4 and T2. On either side of the last four cervical and first thoracic vertebrae, the brachial plexus extends downward and laterally, passes over the first rib behind the clavicle, and then enters the axilla (Figure 13-11 on page 386). The brachial plexus constitutes the entire nerve supply for the upper extremities, as well as a number of neck and shoulder muscles.

The *roots* of the brachial plexus, like those of the cervical plexus, are continuations of the ventral rami of the spinal nerves. The roots of C5 and C6 unite to form the *upper trunk,* C7 becomes the *middle trunk,* and C8 to T1 form the *lower trunk.* Each trunk, in turn, divides into an *anterior division* and a *posterior division.* The divisions then unite to form cords. The *posterior cord* is formed by the union of the posterior divisions of the upper, middle, and lower trunks. The *medial cord* is formed as a continuation of the anterior division of the lower trunk. The *lateral cord* is formed by the union of the anterior divisions of the upper and middle trunk. The peripheral nerves arise from the cords. Thus the brachial plexus begins as roots that unite to form trunks. The trunks branch into divisions, the divisions form cords, and the cords give rise to the peripheral nerves.

A summary of the nerves and distributions of the brachial plexus is given in Exhibit 13-3 on page 387. The relationship of the brachial plexus to the other plexuses is shown in Figure 13-1.

Lumbar Plexus

The **lumbar plexus** is formed by the ventral rami of spinal nerves L1 to L4. It differs from the brachial plexus in that there is no intricate interlacing of fibers. It also consists of *roots* and an *anterior* and *posterior*

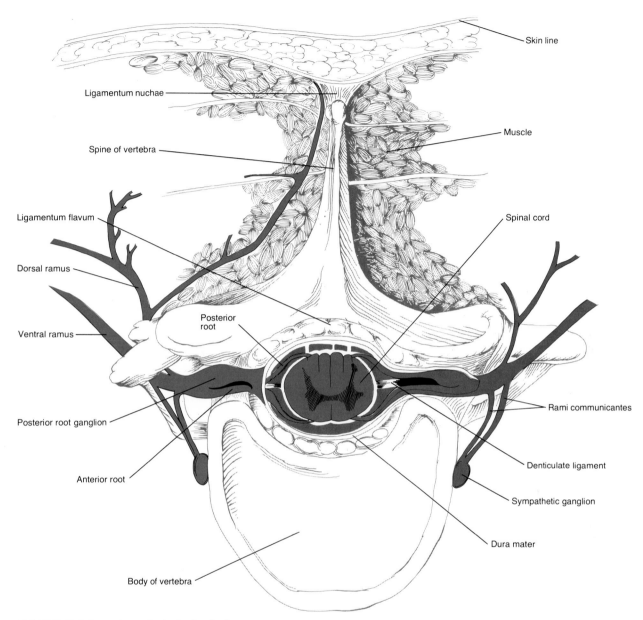

FIGURE 13-9 Branches of a typical spinal nerve.

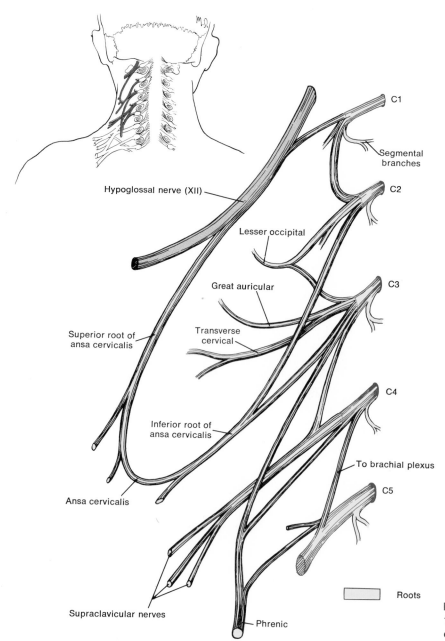

C1

Segmental
branches

C2

Hypoglossal nerve (XII)

Lesser occipital

Great auricular

C3

Superior root of
ansa cervicalis

Transverse
cervical

C4

Inferior root of
ansa cervicalis

To brachial plexus

C5

Ansa cervicalis

Supraclavicular nerves

Phrenic

Roots

FIGURE 13-10 Cervical plexus. Consult Exhibit 13-2 so that you can determine the distribution of each of the nerves of the plexus.

EXHIBIT 13-2
CERVICAL PLEXUS

NERVE	ORIGIN	DISTRIBUTION
SUPERFICIAL OR CUTANEOUS BRANCHES		
Lesser occipital	C2–C3	Skin of scalp behind and above ear
Greater auricular	C2–C3	Skin in front, below, and over ear
Transverse cervical	C2–C3	Skin over anterior aspect of neck
Supraclaviculars	C3–C4	Skin over upper portion of chest and shoulder
DEEP OR LARGELY MOTOR BRANCHES		
Ansa cervicalis	This nerve is divided into a superior root and an inferior root	
Superior root	C1–C2	Infrahyoid, thyrohyoid, and geniohyoid muscles of neck
Inferior root	C3–C4	Omohyoid, sternohyoid, and sternothyroid muscles of neck
Phrenic	C3–C5	Diaphragm between chest and abdomen
Segmental branches	C1–C5	Prevertebral (deep) muscles of neck, levator scapulae, sternocleidomastoid, and trapezius muscles.

division. On either side of the first four lumbar vertebrae, the lumbar plexus passes obliquely outward behind the psoas major muscle (posterior division) and anterior to the quadratus lumborum muscle (anterior division) and then gives rise to its peripheral nerves (Figure 13-12 on page 388). The lumbar plexus supplies the anterolateral abdominal wall, external genitals, and part of the lower extremity. The largest nerve arising from the lumbar plexus is the femoral nerve. Injury to the nerve is indicated by an inability to extend the leg and by loss of sensation in the skin over the anteromedial aspect of the thigh.

A summary of the nerves and distributions of the lumbar plexus is presented in Exhibit 13-4 on page 389. The relationship of the lumbar plexus to the other plexuses is shown in Figure 13-1.

Sacral Plexus

The **sacral plexus** is formed by the ventral rami of spinal nerves L4 to L5 and S1 to S4. It is situated largely in front of the sacrum (Figure 13-13 on page 390). Like the lumbar plexus, it contains *roots* and an *anterior* and *posterior division*. The sacral plexus supplies the buttocks, perineum, and lower extremities. The largest nerve arising from the sacral plexus—and, in fact, the largest nerve in the body—is the sciatic nerve. This nerve may be injured because of a slipped disc, dislocated hip, pressure from the uterus during pregnancy, or an improperly given gluteal intramuscular injection. The sciatic nerve supplies the entire musculature of the leg and foot.

A summary of the nerves and distributions of the sacral plexus is given in Exhibit 13-5 on page 391. The

relationship of the sacral plexus to the other plexuses is shown in Figure 13-1.

Intercostal (Thoracic) Nerves

Spinal nerves T2 to T12 do not enter into the formation of plexuses. Instead, these **intercostal** or **thoracic nerves** are distributed directly to the structures they supply (see Figure 13-1). After leaving the vertebral foramina, ventral rami of nerves T3 to T6 pass in the costal grooves of the ribs and are distributed to the intercostal muscles and skin of the anterior and lateral chest wall. Nerves T7 to T12 supply the intercostal muscles and the abdominal muscles and overlying skin. T2 supplies the intercostal muscles of the second interspace and the skin of the axilla and posteromedial aspect of the arm. The dorsal rami of the intercostal nerves supply the deep back muscles and skin of the dorsal aspect of the thorax.

DERMATOMES

The skin over the entire body is supplied segmentally by spinal nerves. This means that the spinal nerves innervate specific, constant segments of the skin. With the exception of spinal nerve C1, all other spinal nerves supply branches to the skin. The skin segment supplied by a spinal nerve is a **dermatome** (Figure 13-14 on page 392). In the neck and trunk, the dermatomes form consecutive bands of skin. In the trunk, there is an overlap of adjacent dermatome nerve supply. Thus there is little loss of sensation if only a single nerve supply to a dermatome is interrupted. Most of the skin of the face and scalp is supplied by cranial nerve V (trigeminal).

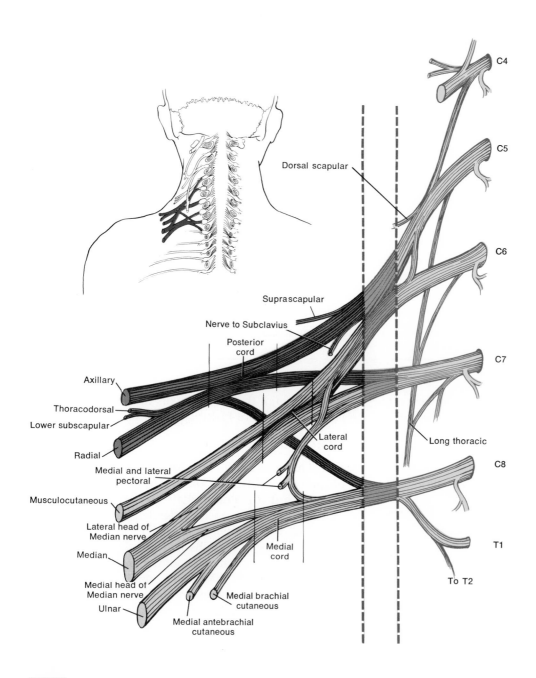

Dorsal scapular

Suprascapular

Nerve to Subclavius

Posterior cord

Axillary

Thoracodorsal

Lower subscapular

Radial

Medial and lateral pectoral

Musculocutaneous

Lateral head of Median nerve

Median

Medial head of Median nerve

Ulnar

Medial antebrachial cutaneous

Medial brachial cutaneous

Medial cord

Lateral cord

Long thoracic

C4

C5

C6

C7

C8

T1

To T2

 Roots

Trunks

Anterior divisions

Posterior divisions

FIGURE 13-11 Brachial plexus. Consult Exhibit 13-3 so that you can determine the distribution of each of the nerves of the plexus.

EXHIBIT 13-3 ▬▬▬▬▬▬▬▬▬▬▬▬▬▬▬▬▬▬▬▬▬▬▬▬▬▬▬▬
BRACHIAL PLEXUS

NERVE	ORIGIN	DISTRIBUTION
ROOT NERVES		
Dorsal scapular	C5	Levator scapulae, rhomboideus major, and rhomboideus minor muscles
Long thoracic	C5–C7	Serratus anterior muscle
TRUNK NERVES		
Nerve to subclavius	C5–C6	Subclavius muscle
Suprascapular	C5–C6	Supraspinatus and infraspinatus muscles
LATERAL CORD NERVES		
Musculocutaneous	C5–C7	Coracobrachialis, biceps brachii, and brachialis muscles
Median (lateral head)	C5–C7	See distribution for Median (medial head)
Lateral pectoral	C5–C7	Pectoralis major muscle
POSTERIOR CORD NERVES		
Upper subscapular	C5–C6	Subscapularis muscle
Thoracodorsal	C6–C8	Latissimus dorsi muscle
MEDIAL CORD NERVES		
Medial pectoral	C8—T1	Pectoralis major and pectoralis minor muscles
Medial brachial cutaneous	C8–T1	Skin of medial and posterior aspects of lower third of arm
Medial antebrachial cutaneous	C8–T1	Skin of medial and posterior aspects of forearm
Median (medial head)	C5–C8, T1	Medial and lateral heads of median nerve form median nerve. Distributed to flexors of forearm (pronator teres, flexor carpi radialis, flexor digitorum superficialis) except flexor carpi ulnaris and flexor digitorum profundus; skin of lateral two-thirds of palm of hand and fingers
Ulnar	C8–T1	Flexor carpi ulnaris and flexor digitorum profundus muscles; skin of medial side of hand, little finger, and medial half of ring finger
OTHER CUTANEOUS DISTRIBUTIONS		
Intercostobrachial	Second intercostal nerve	Skin over medial side of arm
Upper lateral brachial cutaneous	Axillary	Skin over deltoid muscle and down to elbow
Posterior brachial cutaneous	Radial	Skin over posterior aspect of arm
Lower lateral brachial cutaneous	Radial	Skin over lateral aspect of elbow
Lateral antebrachial cutaneous	Musculocutaneous	Skin over lateral aspect of forearm
Posterior antebrachial cutaneous	Musculocutaneous	Skin over posterior aspect of forearm

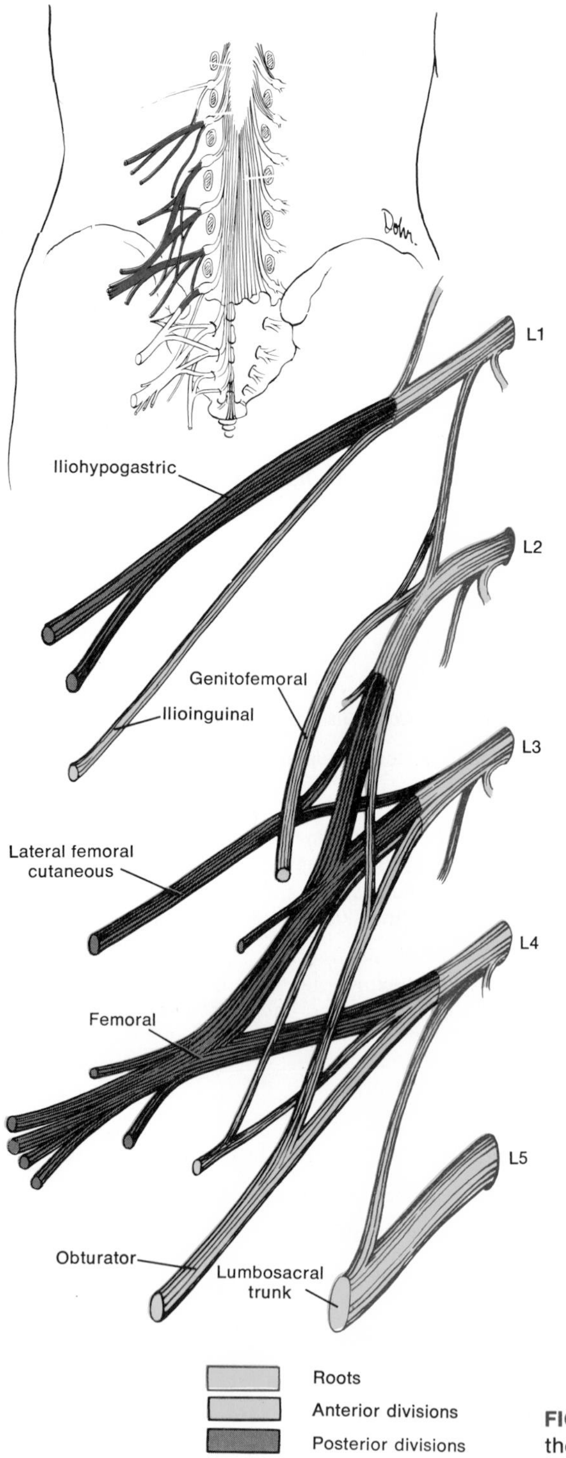

FIGURE 13-12 Lumbar plexus. Consult Exhibit 13-4 so that you can determine the distribution of each of the nerves of the plexus.

EXHIBIT 13-4 ▨
LUMBAR PLEXUS

NERVE	ORIGIN	DISTRIBUTION
Iliohypogastric	T12–L1	Muscles of anterolateral abdominal wall (external oblique, internal oblique, transversus abdominis); skin of lower abdomen and buttock
Ilioinguinal	L1	Muscles of anterolateral abdominal wall as indicated above; skin of upper median aspect of thigh, root of penis and scrotum in male, and labia majora and mons pubis in female
Genitofemoral	L1–L2	Cremaster muscle; skin over middle anterior surface of thigh, scrotum in male, and labia majora in female
Lateral femoral cutaneous	L2–L3	Skin over lateral, anterior, and posterior aspects of thigh
Femoral	L2–L4	Flexor muscles of thigh (iliacus, psoas major, pectineus, rectus femoris, sartorius); extensor muscles of leg (rectus femoris, vastus lateralis, vastus medialis; vastus intermedius); skin on front and over medial aspect of thigh and medial side of leg and foot
Obturator	L2–L4	Adductor muscles of leg (obturator externus, pectineus, adductor longus, adductor brevis, adductor magnus, gracilis); skin over medial aspect of thigh
Saphenous	L2–L4	Skin over medial aspect of leg

Since a physician knows that a particular dermatome is associated with a particular spinal nerve, it is possible to determine which segment of the spinal cord or spinal nerve is malfunctioning. If a dermatome is stimulated and the sensation is not perceived, it can be assumed that the nerve supplying the dermatome is involved.

Applications to Health

SPINAL CORD INJURY

The spinal cord may be damaged by fracture or dislocation of the vertebrae enclosing it or by wounds. All can result in **transection**—partial or complete severing of the spinal cord. Complete transection means that all ascending and descending pathways are cut. It results in loss of all sensation and voluntary muscular movement below the level of transection. In fact, individuals with complete cervical transections close to the base of the skull usually die of asphyxiation before treatment can be administered. This happens because impulses from the phrenic nerves to the breathing muscles are interrupted. If the upper cervical cord is partially transected, both the upper and lower extremities are paralyzed and the patient is classified as *quadriplegic*. Partial transection between the cervical and lumbar enlargements results in paralysis of the lower extremities only, and the patient is classified as *paraplegic*.

In the case of partial transection, **spinal shock** lasts from a few days to several weeks. During this period, all reflex activity is abolished, a condition called *areflexia*. In time, however, there is a return of reflex activity. The first reflex to return is the knee jerk. Its reappearance may take several days. Next the flexion reflexes return. This may take up to several months. Then the cross extensor reflexes return. Visceral reflexes such as erection and ejaculation are also affected by transection. Moreover, bladder and bowel function are no longer under voluntary control.

PERIPHERAL NERVE DAMAGE

If the cell body of a neuron is destroyed, the neuron cannot be replaced and its function is permanently lost. Axons that have a neurilemma can be repaired, however, as long as the cell body is intact and fibers are in association with Schwann cells. Most nerves that lie outside the brain and spinal cord consist of axons and dendrites that are covered with a neurilemma. A person who injures a nerve in the upper extremity, for example, has a good chance of regaining nerve function. Axons in the brain and spinal cord do not have a neurilemma. Injury there is permanent.

When there is damage to an axon (or to dendrites of somatic afferent neurons), there are usually changes in the cell body and always changes in the portions of the nerve processes distal to the site of damage. These changes associated with the cell body are referred to as the axon reaction or retrograde degeneration. Those associated with the distal portion of the cut fiber are called wallerian degeneration. The axon reaction occurs in essentially the same way, whether the damaged fiber is in the central or peripheral nervous system. The wallerian reaction, however, depends on whether the fiber is central or peripheral. We will consider the axon reaction first.

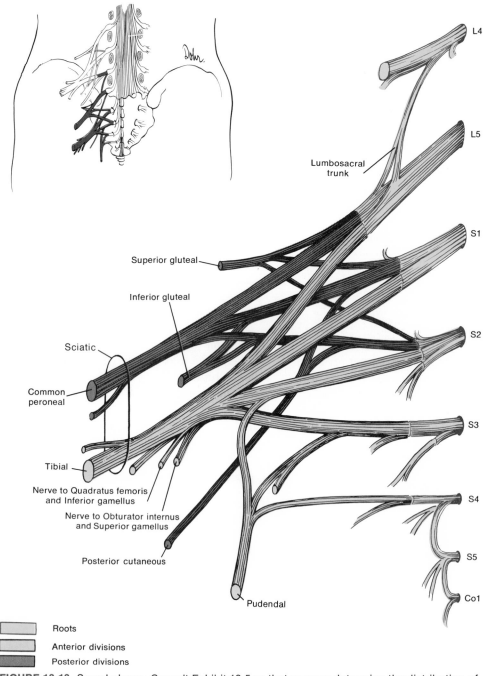

FIGURE 13-13 Sacral plexus. Consult Exhibit 13-5 so that you can determine the distribution of each of the nerves of the plexus.

EXHIBIT 13-5
SACRAL PLEXUS

NERVE	ORIGIN	DISTRIBUTION
Superior gluteal	L4–L5, S1	Gluteus minimus and gluteus medius muscles and tensor fasciae latae.
Inferior gluteal	L5–S2	Gluteus maximus muscle
Nerve to piriformis	S1–S2	Piriformis muscle
Nerve to quadratus femoris	L4–L5, S1	Inferior gemellus and quadratus femoris muscles
Nerve to obturator internus	L5–S2	Superior gemellus and obturator internus muscles
Perforating cutaneous	S2–S3	Skin over lower medial aspect of buttock
Posterior cutaneous	S1–S3	Skin over anal region, lower lateral aspect of buttock, upper posterior aspect of thigh, upper part of calf, scrotum in male, and labia majora in female
Sciatic	L4–S3	Actually two nerves: tibial and common peroneal, bound together by common sheath of connective tissue. It splits into its two divisions, usually at knee. (See below for distributions.) As sciatic nerve descends through thigh, it sends branches to hamstring muscles (biceps femoris, semitendinosus, semimembranosus) and adductor magnus.
Tibial (medial popliteal)	L4–S3	Gastrocnemius, plantaris, soleus, popliteus, tibialis posterior, flexor digitorum, and hallucis muscles. Branches of tibial nerve in foot are medial plantar nerve and lateral plantar nerve.
Medial plantar		Abductor hallucis, flexor digitorum brevis, and flexor hallucis muscles; skin over medial two-thirds of plantar surface of foot
Lateral plantar		Remaining muscles of foot not supplied by medial plantar nerve; skin over lateral third of plantar surface of foot
Common peroneal (lateral popliteal)	L4–S2	Divides into a superficial peroneal and a deep peroneal branch. (Distributions described below.)
Superficial peroneal		Peroneus longus and peroneus brevis muscles; skin over distal third of anterior aspect of leg and dorsum of foot
Deep peroneal		Tibialis anterior, extensor hallucis longus, peroneus tertius, and extensor digitorum brevis muscles; skin over great and second toes
Pudendal	S2–S4	Muscles of perineum; skin of penis and scrotum in male and clitoris, labia majora, labia minora, and lower vagina in female

Axon Reaction

When there is damage to an axon of a central or peripheral neuron, certain structural changes occur in the cell body—the **axon reaction.** One of the most significant features of the axon reaction occurs 24–48 hours after damage. The Nissl bodies, arranged in an orderly fashion in an uninjured cell body, break down into finely granular masses. This alteration is called *chromatolysis.* It begins between the axon hillock and nucleus but spreads throughout the cell body. As a result of chromatolysis, the cell body swells and the swelling reaches its maximum between 10 and 20 days after injury. Chromatolysis results in a loss of ribosomes by the rough endoplasmic reticulum and an increase in the number of free ribosomes. Another sign of the axon reaction is the off-center position of the nucleus in the cell body. This change makes it possible to identify the cell bodies of damaged fibers through a microscope. Following chromatolysis, there are signs of recovery in the cell body. There is an acceleration of RNA and protein synthesis, which favors regeneration of the axon. Recovery often takes several months and involves the restoration of normal levels of RNA and proteins and the Nissl bodies to their usual, uninjured patterns.

Wallerian Degeneration

The part of the axon distal to the damage becomes slightly swollen and then breaks up into fragments by the third to fifth day. The myelin sheath around the axon also undergoes degeneration. Degeneration of the distal portion of the axon and myelin sheath is called **wallerian degeneration.** Following degeneration, there is phagocytosis of the remains.

Even when there is degeneration of the axon and myelin sheath, the neurilemma of the Schwann cells remains. The Schwann cells on either side of the site of injury multiply by mitosis and grow toward each other. This growth results in the formation of a tube. This ''tunnel'' provides a means for new axons to grow from the proximal area, across the injured area, and into the distal area previously occupied by the

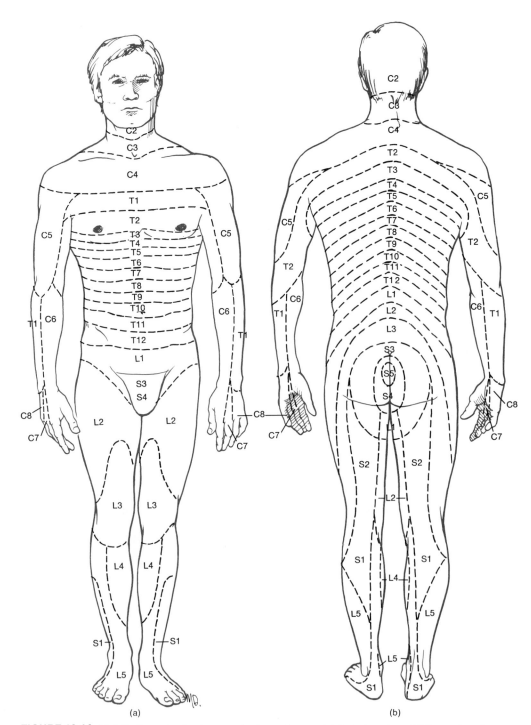

FIGURE 13-14 Distribution of spinal nerves to dermatomes. (a) Anterior view. (b) Posterior view.

original nerve fiber. The growth of the new axons will not occur if the gap at the site of injury is too large or if the gap becomes filled with dense collagenous fibers.

Regeneration

Accelerated protein synthesis is required for repair of the damaged axon. The proteins synthesized in the cell body pass into the axon at about the rate of 1 mm (0.039 inch)/day. The proteins assist in regenerating the damaged axon. During the first few days following damage, regenerating axons begin to invade the tube formed by the Schwann cells. Axons from the proximal area grow at the rate of about 1.5 mm (0.059 inch)/day across the area of damage, find their way into the distal neurilemmal tubes, and grow toward the distally located receptors and effectors. Thus sensory and motor connections are reestablished. In time, a new myelin sheath is also produced by the Schwann cells. However, function is never completely restored after a nerve is severed.

SCIATICA

Sciatica is a type of neuritis characterized by severe pain along the path of the sciatic nerve or its branches.

The term is commonly applied to a number of disorders affecting this nerve. Because of its length and size, the sciatic nerve is exposed to many kinds of injury. Inflammation of or injury to the nerve causes pain that passes from the back or thigh down its length into the leg, foot, and toes. Probably the most common cause of sciatica is a slipped or herniated disc. Other causes include irritation from osteoarthritis, back injuries, or pressure on the nerve from certain types of exertion. Other cases are idiopathic and sciatica may be associated with diabetes mellitus, gout, or vitamin deficiencies.

NEURITIS

Neuritis is inflammation of a single nerve, two or more nerves in separate areas, or many nerves simultaneously. It may result from irritation to the nerve produced by direct blows, bone fractures, contusions, or penetrating injuries. Additional causes include vitamin deficiency (usually thiamine) and poisons such as carbon monoxide, carbon tetrachloride, heavy metals, and some drugs. Neuritis exists in many forms and is usually considered a symptom rather than a disease. A thorough physical examination, along with laboratory studies, is necessary to discover its exact cause.

STUDY OUTLINE

The Grouping of Neural Tissue
1. White matter is an aggregation of myelinated axons.
2. Gray matter is a collection of nerve cell bodies and dendrites or unmyelinated axons.
3. A nerve is a bundle of fibers outside the central nervous system.
4. A ganglion is a collection of cell bodies outside the central nervous system.
5. A bundle of fibers of similar function inside the central nervous system forms a tract.
6. A mass of nerve cell bodies and dendrites inside the brain forms a nucleus.
7. A horn is an area of gray matter in the spinal cord.

The Spinal Cord
General Features
1. The spinal cord begins as a continuation of the medulla oblongata and terminates at about the second lumbar vertebra.
2. It contains a cervical and lumbar enlargement, which serve as points of origin for nerves to the extremities.
3. The tapered portion of the spinal cord is the conus medullaris, from which arises the filum terminale and cauda equina.
4. The spinal cord is partially divided into right and left sides by the anterior median fissure and posterior median sulcus.

Protection and Coverings
1. The spinal cord is protected by the wall of the vertebral canal, meninges, and cerebrospinal fluid.
2. The spinal meninges from outside to inside are the dura mater, arachnoid, and pia mater. Inflammation of the meninges is called meningitis.
3. The removal of cerebrospinal fluid from the subarachnoid space in the lumbar region of the cord is called a spinal puncture.

Structure
1. Parts of the spinal cord observed in cross section are the gray commissure; central canal; anterior, posterior, and lateral gray horns; anterior, posterior, and lateral white columns; and ascending and descending tracts.
2. The spinal cord conveys sensory and motor information by way of the ascending and descending tracts, respectively.
3. In serving as a reflex center, the posterior root, posterior root ganglion, and anterior root convey an impulse.

Reflex Arc
1. The reflex arc is the functional unit of the nervous system.
2. Components of a reflex arc include the receptor, sensory neuron, center, motor neuron, and effector.

3. Important somatic spinal reflexes include the stretch reflex, the flexor reflex, and the crossed extensor reflex.
4. Reflex arcs may be classified as ipsilateral, monosynaptic, polysynaptic, intersegmental, and contralateral.

Spinal Nerves
Composition and Coverings
1. Spinal nerves are attached to the spinal cord by means of a posterior root and an anterior root. All spinal nerves are mixed.
2. Spinal nerves are covered by endoneurium, perineurium, and epineurium.

Names and Branches
1. The 31 pairs of spinal nerves are named and numbered according to the region and level of the spinal cord from which they emerge.
2. Branches of a spinal nerve include the dorsal ramus, ventral ramus, meningeal branch, and rami communicantes.

Plexuses
1. The ventral rami of spinal nerves, except for T2 to T12, form networks of nerves called plexuses.
2. Emerging from the plexuses are nerves bearing names that are often descriptive of the general regions they supply or the course they take.

3. The principal plexuses are called the cervical, brachial, lumbar, sacral, and coccygeal plexuses.
4. Nerves that do not form plexuses are called intercostal nerves (T2 to T12). They are distributed directly to the structures they supply.

Dermatomes
1. With the exception of spinal nerve C1, all spinal nerves supply branches to the skin.
2. The skin segment supplied by a given spinal nerve is called a dermatome.
3. Knowledge of dermatomes helps a physician to determine which segment of the spinal cord or spinal nerve is malfunctioning.

Applications to Health
1. Complete or partial severing of the spinal cord is called transection. It may result in quadriplegia or paraplegia. Partial transection is followed by a period of loss of reflex activity called areflexia.
2. Axons with a neurilemma are capable of regeneration. The regeneration is preceded by the axon reaction and wallerian degeneration.
3. Sciatica is a neuritis characterized by severe pain along the sciatic nerve or its branches.
4. Neuritis is an inflammation of one or more nerves.

REVIEW QUESTIONS

1. Define the following terms: white matter, gray matter, nerve, ganglion, tract, nucleus, and horn.
2. Describe the location of the spinal cord. What are the cervical and lumbar enlargements?
3. Define conus medullaris, filum terminale, and cauda equina. What is a spinal segment? How is the spinal cord partially divided into a right and left side?
4. Describe the bony covering of the spinal cord.
5. Explain the location and composition of the spinal meninges. Describe the location of the epidural, subdural, and subarachnoid spaces. What are the denticulate ligaments?
6. Define meningitis.
7. What is a spinal puncture? Give several purposes served by a spinal puncture.
8. Based upon your knowledge of the structure of the spinal cord in cross section, define the following: gray commissure, central canal, anterior gray horn, lateral gray horn, posterior gray horn, anterior white column, lateral white column, posterior white column, ascending tract, and descending tract.
9. Describe the function of the spinal cord as a conduction pathway.
10. Using Exhibit 13-1 as a guide, be sure that you can list the location, origin, termination, and function of the principal ascending and descending tracts.
11. Describe how the spinal cord serves as a reflex center.
12. What is a reflex arc? List and define the components of a reflex arc.
13. Describe the mechanism of a stretch reflex, a flexor reflex, and a crossed extensor reflex.

14. Define the following terms: monosynaptic reflex arc, ipsilateral reflex arc, polysynaptic reflex arc, intersegmental reflex arc, contralateral reflex arc, and reciprocal inhibition.
15. Define a spinal nerve. Why are all spinal nerves classified as mixed nerves?
16. Describe how a spinal nerve is attached to the spinal cord.
17. Explain how a spinal nerve is enveloped by its several different coverings.
18. How are spinal nerves named and numbered?
19. Describe the branches and innervations of a typical spinal nerve.
20. What is a plexus?
21. Explain the location, origin, nerves, and distributions of the following plexuses: cervical, brachial, lumbar, and sacral.
22. What are intercostal nerves?
23. Define a dermatome. Why is a knowledge of dermatomes important?
24. What is transection? Distinguish between quadriplegia and paraplegia.
25. What is meant by spinal shock?
26. Explain the conditions necessary for peripheral nerve repair.
27. Distinguish between the axon reaction and wallerian degeneration.
28. Distinguish between sciatica and neuritis.

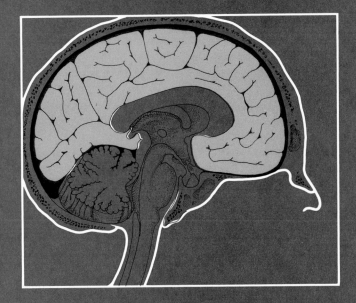

14

The Brain and the Cranial Nerves

STUDENT OBJECTIVES

- Identify the principal areas of the brain.

- Describe the location of the cranial meninges.

- Explain the formation and circulation of cerebrospinal fluid.

- Describe the blood supply to the brain and the concept of the blood-brain barrier.

- Compare the components of the brain stem with regard to structure and function.

- Identify the structure and functions of the diencephalon.

- Identify the structural features of the cerebrum.

- Describe the lobes, tracts, and basal ganglia of the cerebrum.

- Describe the structure and functions of the limbic system.

- Compare the motor, sensory, and association areas of the cerebrum.

- Describe the principle of the electroencephalograph and its significance in the diagnosis of certain disorders.

- Describe the anatomical characteristics and functions of the cerebellum.

- Define a cranial nerve.

- Identify the 12 pairs of cranial nerves by name, number, type, location, and function.

- Explain the effects of injury on cranial nerves.

- List the clinical symptoms of these disorders of the nervous system: poliomyelitis, syphilis, cerebral palsy, Parkinsonism, epilepsy, multiple sclerosis, cerebrovascular accidents, dyslexia, and Tay-Sachs disease.

- Define key medical terms associated with the central nervous system.

Now we will consider how the brain is protected, what its principal parts are, how it is related to the spinal cord, and how it is related to the 12 pairs of cranial nerves.

Brain

PROTECTION AND COVERINGS

The **brain** of an average adult is one of the largest organs of the body, weighing about 1300 g (3 lb). Figure 14-1 shows that the brain is mushroom-shaped. It is divided into four principal parts: brain stem, diencephalon, cerebrum, and cerebellum. The **brain stem,** the stalk of the mushroom, consists of the medulla oblongata, pons varolii, and midbrain. The lower end of the brain stem is a continuation of the spinal cord. Above the brain stem is the **diencephalon,** consisting primarily of the thalamus and hypothalamus. The **cerebrum** spreads over the diencephalon. The cerebrum constitutes about seven-eighths of the total weight of the brain and occupies most of the skull. Inferior to

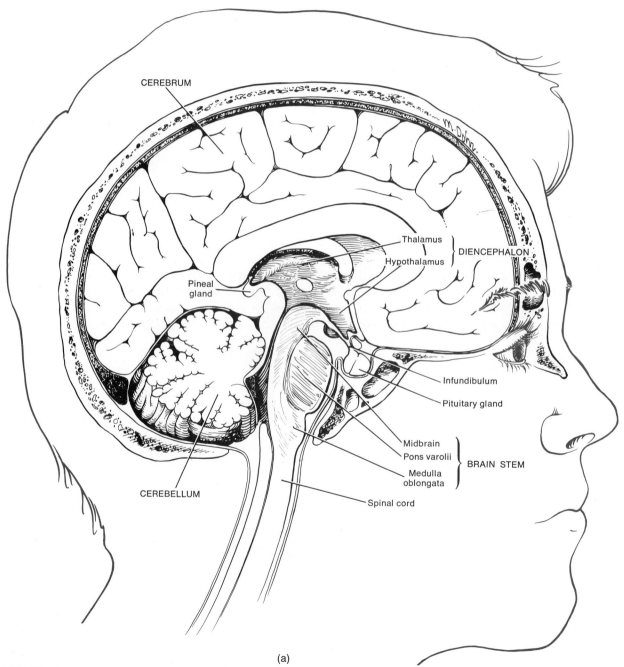

(a)

FIGURE 14-1 Brain. (a) Principal parts of the medial aspect of the brain seen in sagittal section. The infundibulum, pituitary gland, and pineal gland are discussed in conjunction with the endocrine system.

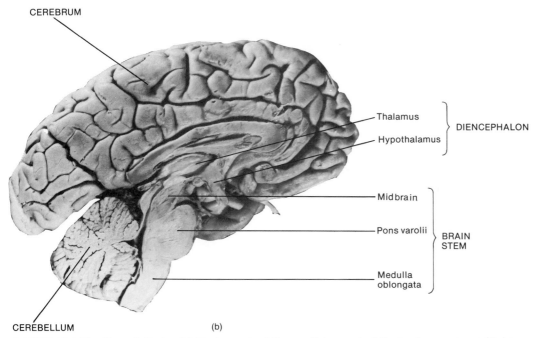

CEREBRUM

Thalamus } DIENCEPHALON
Hypothalamus

Midbrain
Pons varolii } BRAIN STEM
Medulla oblongata

CEREBELLUM   (b)

FIGURE 14-1 *(Continued)* Brain. (b) Photograph of the medial aspect of the brain seen in sagittal section. (Courtesy of Ernest Gardner *et al., Anatomy: A Regional Study of Human Structure,* 4th ed., W. B. Saunders, Philadelphia, 1975.)

the cerebrum and posterior to the brain stem is the **cerebellum.**

The brain is protected by the cranial bones (see Chapter 6). Like the spinal cord, the brain is also protected by meninges. The *cranial meninges* surround the brain and are continuous with the spinal meninges. The cranial meninges have the same basic structure and bear the same names as the spinal meninges: the outermost *dura mater,* middle *arachnoid,* and innermost *pia mater* (Figure 14-2). The cranial dura mater consists of two layers. The thicker, outer layer (periosteal layer) lightly adheres to the cranial bones and serves as a periosteum. The thinner, inner layer (meningeal layer) includes a mesothelial layer on its smooth surface. The spinal dura mater corresponds to the meningeal layer of the cranial dura mater.

CEREBROSPINAL FLUID

The brain, as well as the rest of the central nervous system, is further protected against injury by **cerebrospinal fluid.** This fluid circulates through the subarachnoid space around the brain and spinal cord and through the ventricles of the brain. The subarachnoid space is the area between the arachnoid and pia mater. The **ventricles** are cavities in the brain that communicate with each other, with the central canal of the spinal cord, and with the subarachnoid space. Each of the two *lateral ventricles* is located in a hemisphere (side) of the cerebrum under the corpus callosum (Figure 14-2). The *third ventricle* is a slit between and in-

ferior to the right and left halves of the thalamus and between the lateral ventricles. Each lateral ventricle communicates with the third ventricle by a narrow, oval opening: the *interventricular foramen,* or *foramen of Monro.* The *fourth ventricle* lies between the inferior brain stem and the cerebellum. It communicates with the third ventricle via the *cerebral aqueduct (aqueduct of Sylvius),* which passes through the midbrain. The roof of the fourth ventricle has three openings: a *median aperture (foramen of Magendie)* and two *lateral apertures (foramina of Luschka).* Through these openings, the fourth ventricle also communicates with the subarachnoid space of the brain and cord.

The entire central nervous system contains about 125 ml (4 oz) of cerebrospinal fluid. It is a clear, colorless fluid of watery consistency. Chemically, it contains proteins, glucose, urea, and salts. It also contains some white blood cells. The fluid serves as a shock absorber for the central nervous system. It also circulates nutritive substances filtered from the blood. Cerebrospinal fluid is formed primarily by filtration from networks of capillaries, called **choroid plexuses,** located in the ventricles (Figure 14-2). The fluid formed in the choroid plexuses of the lateral ventricles circulates through the interventricular foramen to the third ventricle, where more fluid is added by the choroid plexus in the third ventricle. It then flows through the cerebral aqueduct to the fourth ventricle. Here there are contributions from the choroid plexus in the fourth ventricle. The fluid then circulates through the apertures of the fourth ventricle into the subarachnoid

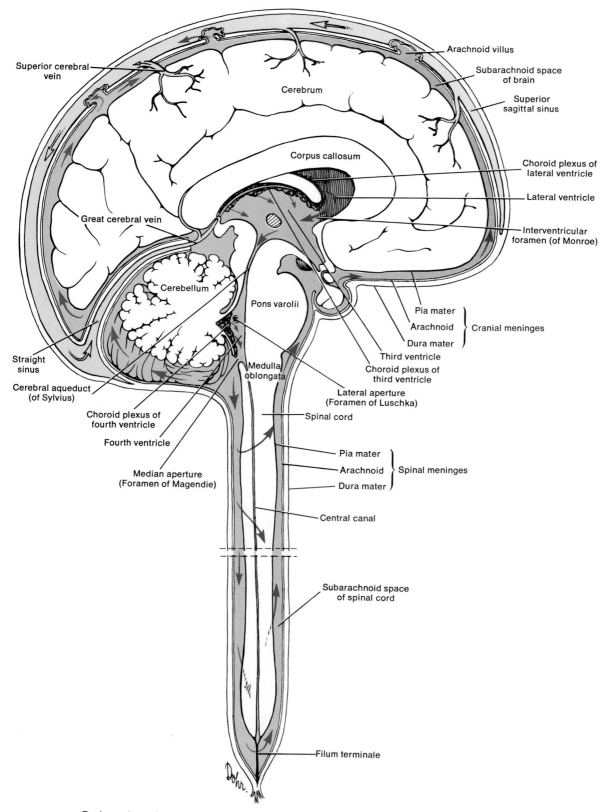

FIGURE 14-2 Brain and meninges seen in sagittal section. The direction of flow of cerebrospinal fuid is indicated by colored arrows.

space around the back of the brain. It also passes downward to the subarachnoid space around the posterior surface of the spinal cord, up the anterior surface of the spinal cord, and around the anterior part of the brain. From here it is gradually reabsorbed into veins. Some cerebrospinal fluid may be formed by ependymal (neuroglial) cells lining the central canal of the spinal cord. This small quantity of fluid ascends to reach the fourth ventricle. Most of the fluid is absorbed into the superior sagittal sinus (see Figure 11-17). The absorption actually occurs through **arachnoid villi**—fingerlike projections of the arachnoid that push into the superior sagittal sinus. Normally, cerebrospinal fluid is absorbed as rapidly as it is formed. Examine the three-dimensional view of the ventricles in Figure 14-3. It will give you a better understanding of the path taken by cerebrospinal fluid.

If an obstruction, such as a tumor, arises in the brain and interferes with the drainage of fluid from the ventricles into the subarachnoid space, large amounts of fluid accumulate in the ventricles. Fluid pressure inside the brain increases, and, if the fontanels have not yet closed, the head bulges to relieve the pressure. This condition is called **internal hydrocephalus** (*hydro* =water; *cephalo* = head). If an obstruction interferes

with drainage somewhere in the subarachnoid space and cerebrospinal fluid accumulates inside the space, the condition is termed **external hydrocephalus.**

BLOOD SUPPLY

The brain is well supplied with blood vessels of the circle of Willis, which supplies oxygen and nutrients (see Exhibit 11-4 and Figure 11-13c). Although the brain actually consumes less oxygen than most other organs of the body, it must receive a constant supply. If the blood flow to the brain is interrupted for even a few moments, unconsciousness may result. A one- or two-minute interruption may weaken the brain cells by starving them of oxygen. If the cells are totally deprived of oxygen for four minutes, many are permanently injured. Occasionally during childbirth the oxygen supply from the mother's blood is interrupted before the baby leaves the birth canal and can breathe. Often such babies are stillborn or suffer permanent brain damage that may result in mental retardation, epilepsy, and paralysis.

Blood supplying the brain also contains glucose, the principal source of energy for brain cells. Because carbohydrate storage in the brain is limited, the supply of

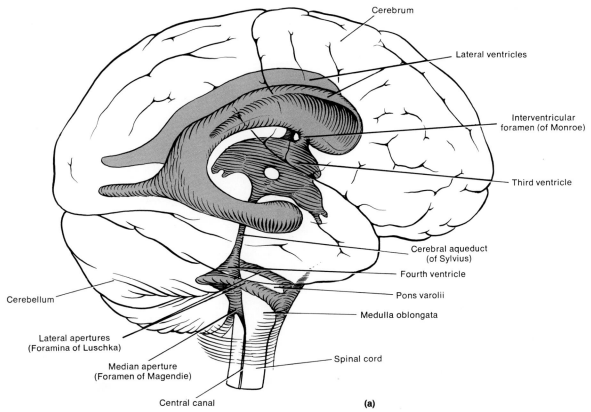

FIGURE 14-3 Ventricles of the brain and their interconnections. (a) Diagrammatic lateral projection.

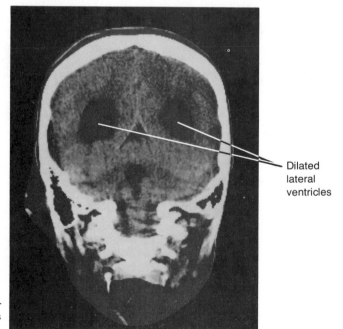

Posterior horns
lateral ventric

Lateral ventric

Third ventricle

(b)

Dilated
lateral
ventricles

(c)

FIGURE 14-3 *(Continued)* Ventricles of the brain and their interconnections. (b) Lateral view. (Courtesy of John C. Bennett, St. Mary's Hospital, San Francisco.) (c) CT scan of a frontal section of the brain. (Courtesy of General Electric Co.)

glucose must be continuous. If blood entering the brain has a low glucose level, mental confusion, dizziness, convulsions, and even loss of consciousness may occur.

Glucose, oxygen, and certain ions pass rapidly from the circulating blood into brain cells. Other substances, such as creatinine, urea, chloride, insulin, and sucrose, enter quite slowly. Still other substances—proteins and most antibiotics—do not pass at all from the blood into brain cells. The differential rates of passage of certain materials from the blood into the brain suggest a concept called the **blood-brain barrier.** Electron micrograph studies of the capillaries of the brain reveal that they differ structurally from other capillaries. Brain capillaries are constructed of more densely packed cells and are surrounded by large numbers of glial cells and a continuous basement membrane. These features form a barrier to the passage of certain materials. Thus substances that cross the barrier are either very small molecules or require the assistance of a carrier molecule to cross by active transport. The function of the blood-brain barrier is not known. It may protect brain cells from harmful substances.

BRAIN STEM

Medulla Oblongata

The **medulla oblongata,** or simply **medulla,** is a continuation of the upper portion of the spinal cord and forms the inferior part of the brain stem. Its position in relation to the other parts of the brain may be noted in Figure 14-1. It lies just superior to the level of the foramen magnum and extends upward to the inferior portion of the pons varolii. The medulla measures only 3 cm (about 1 inch) in length.

The medulla contains all ascending and descending tracts that communicate between the spinal cord and various parts of the brain. These tracts constitute the white matter of the medulla. Some tracts cross as they pass through the medulla. Let us see how this crossing occurs and what it means.

On the ventral side of the medulla are two roughly triangular structures called *pyramids* (Figure 14-4). The pyramids are composed of the largest motor tracts that pass from the outer region of the cerebrum (cerebral cortex) to the spinal cord. Just above the junction of the medulla with the spinal cord, most of the fibers in the left pyramid cross to the right side, and most of the fibers in the right pyramid cross to left. This crossing is called the **decussation of pyramids.** The adaptive value, if any, of this phenomenon is unknown. The principal motor fibers that undergo decussation belong to the lateral corticospinal tracts. These tracts originate in the cerebral cortex and pass inferiorly to the medulla. The fibers cross in the pyramids and descend

in the lateral columns of the spinal cord, terminating in the anterior gray horns. Here synapses occur with motor neurons that terminate in skeletal muscles. As a result of the crossing, fibers that originate in the left cerebral cortex activate muscles on the right side of the body, and fibers that originate in the right cerebral cortex activate muscles on the left side. Decussation explains why motor areas of one side of the cerebral cortex control muscular movements on the opposite side of the body.

The dorsal side of the medulla contains two pairs of prominent nuclei: the right and left *nucleus gracilis* and *nucleus cuneatus*. These nuclei receive sensory fibers from ascending tracts (right and left fasciculus gracilis and fasciculus cuneatus) of the spinal cord and relay the sensory information to the opposite side of the medulla. This information is conveyed to the thalamus and then to the sensory areas of the cerebral cortex. Nearly all sensory impulses received on one side of the body cross in the medulla or spinal cord and are perceived in the opposite side of the cerebral cortex.

In addition to its function as a conduction pathway for motor and sensory impulses between the brain and spinal cord, the medulla also contains an area of dispersed gray matter containing some white fibers. This region is called the **reticular formation.** Actually, portions of the reticular formation are located in the spinal cord, pons, midbrain, and diencephalon. The reticular formation functions in consciousness and arousal. Within the medulla are three vital reflex centers of the reticular system. The *cardiac center* regulates heart beat; the *medullary rhythmicity area* adjusts the basic rhythm of breathing; and the *vasoconstrictor* or *vasomotor center* regulates the diameter of blood vessels. Other centers in the medulla coordinate swallowing, vomiting, coughing, sneezing, and hiccuping.

The medulla also contains the nuclei of origin for several pairs of cranial nerves. These are the cochlear and vestibular branches of the vestibulocochlear nerves (VIII), which are concerned with hearing and equilibrium (there is also a nucleus for the vestibular branches in the pons); the glossopharyngeal nerves (IX), which relay impulses related to swallowing, salivation, and taste; the vagus nerves (X), which relay impulses to and from many thoracic and abdominal viscera; the accessory nerves (XI), which convey impulses related to head and shoulder movements (a part of this nerve also arises from the first five segments of the spinal cord); and the hypoglossal nerves (XII), which convey impulses that involve tongue movements.

On each lateral surface of the medulla is an oval projection called the *olive*. The olive contains an inferior olivary nucleus and two accessory olivary nuclei. The nuclei are connected to the cerebellum by fibers.

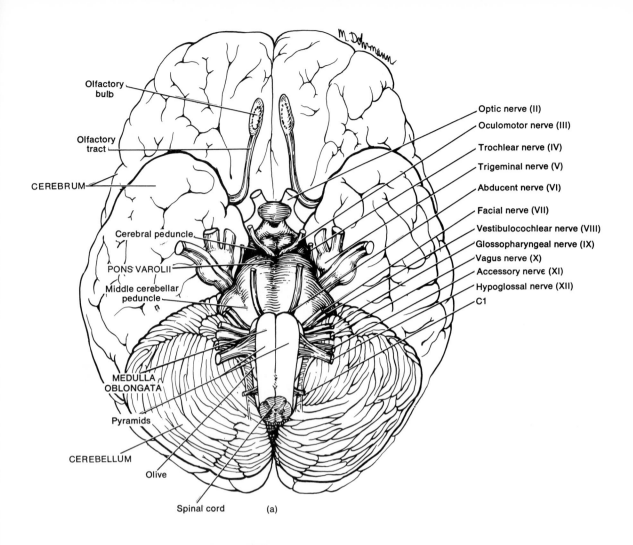

Olfactory bulb

Olfactory tract

CEREBRUM

Cerebral peduncle

PONS VAROLII

Middle cerebellar peduncle

MEDULLA OBLONGATA

Pyramids

CEREBELLUM

Olive

Spinal cord

Optic nerve (II)

Oculomotor nerve (III)

Trochlear nerve (IV)

Trigeminal nerve (V)

Abducent nerve (VI)

Facial nerve (VII)

Vestibulocochlear nerve (VIII)

Glossopharyngeal nerve (IX)

Vagus nerve (X)

Accessory nerve (XI)

Hypoglossal nerve (XII)

C1

(a)

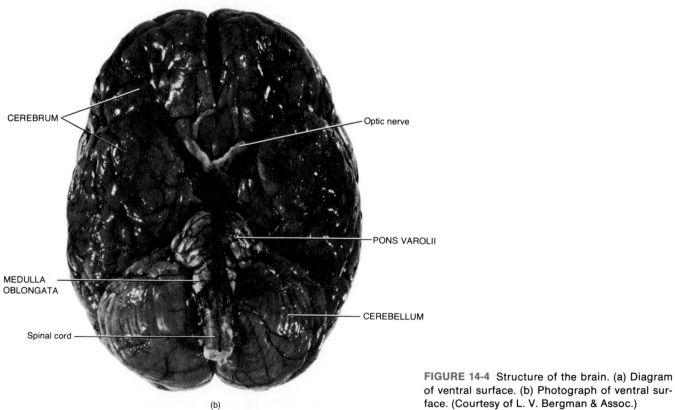

CEREBRUM

MEDULLA OBLONGATA

Spinal cord

Optic nerve

PONS VAROLII

CEREBELLUM

(b)

FIGURE 14-4 Structure of the brain. (a) Diagram of ventral surface. (b) Photograph of ventral surface. (Courtesy of L. V. Bergman & Assoc.)

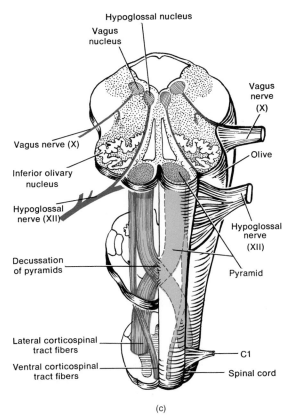

Hypoglossal nucleus

Vagus nucleus

Vagus nerve (X)

Inferior olivary nucleus

Hypoglossal nerve (XII)

Decussation of pyramids

Lateral corticospinal tract fibers

Ventral corticospinal tract fibers

Vagus nerve (X)

Olive

Hypoglossal nerve (XII)

Pyramid

C1

Spinal cord

(c)

FIGURE 14-4 *(Continued)* Structure of the brain. (c) Details of the medulla.

Pons Varolii

The relationship of the **pons varolii** or **pons** to other parts of the brain can be seen in Figures 14-1 and 14-4a. The pons, which means bridge, lies directly above the medulla and anterior to the cerebellum. It measures about 2.5 cm (1 inch) in length. Like the medulla, the pons consists of white fibers scattered throughout with nuclei. As the name implies, the pons is a bridge connecting the spinal cord with the brain and parts of the brain with each other. These connections are provided by fibers that run in two principal directions. The transverse fibers connect with the cerebellum through the middle cerebellar peduncles. The longitudinal fibers of the pons belong to the motor and sensory tracts that connect the spinal cord or medulla with the upper parts of the brain stem.

The nuclei for certain paired cranial nerves are also contained in the pons. These include the trigeminal nerves (V), which relay impulses for chewing and for sensations of the head and face; the abducens nerves (VI), which regulate certain eyeball movements; the facial nerves (VII), which conduct impulses related to taste, salivation, and facial expression; and the vestibular branches of the vestibulocochlear nerves (VIII), which are concerned with equilibrium.

Other important nuclei in the reticular formation of the pons are the *pneumotaxic area* and the *apneustic area*. Together with the medullary rhythmicity area in the medulla, they help control respiration.

Midbrain

The **midbrain** or **mesencephalon** extends from the pons to the lower portion of the diencephalon (Figure 14-1). It is about 2.5 cm (1 inch) in length. The cerebral aqueduct passes through the midbrain and connects the third ventricle above with the fourth ventricle below.

The ventral portion of the midbrain contains a pair of fiber bundles referred to as *cerebral peduncles*. The cerebral peduncles contain many motor fibers that convey impulses from the cerebral cortex to the pons and spinal cord. They also contain sensory fibers that pass from the spinal cord to the thalamus. The cerebral penduncles constitute the main connection for tracts between upper parts of the brain and lower parts of the brain and the spinal cord.

The dorsal portion of the midbrain is called the *tectum* and contains four rounded eminences: the *corpora quadrigemina*. Two of the eminences are known as the *superior colliculi*. These serve as a reflex center for movements of the eyeballs and head in response to visual and other stimuli. The other two eminences are the *inferior colliculi*. They serve as reflex centers for movements of the head and trunk in response to auditory stimuli. The midbrain also contains the *substantia nigra,* a large, heavily pigmented nucleus near the cerebral peduncles.

A major nucleus in the reticular formation of the midbrain is the *red nucleus*. Fibers from the cerebellum and cerebral cortex terminate in the red nucleus. The red nucleus is also the origin of cell bodies of the descending rubrospinal tract. Other nuclei in the midbrain are associated with cranial nerves. These include the oculomotor nerves (III), which mediate some movements of the eyeballs and changes in pupil size and lens shape, and the trochlear nerves (IV), which conduct impulses that move the eyeballs.

A structure called the *medial lemniscus* is common to the medulla, pons, and midbrain. The medial lemniscus is a band of white fibers containing axons that convey impulses for fine touch, proprioception, and vibrations from the medulla to the thalamus.

DIENCEPHALON

The **diencephalon** consists principally of the thalamus and hypothalamus. The relationship of these structures to the rest of the brain is shown in Figure 14-1.

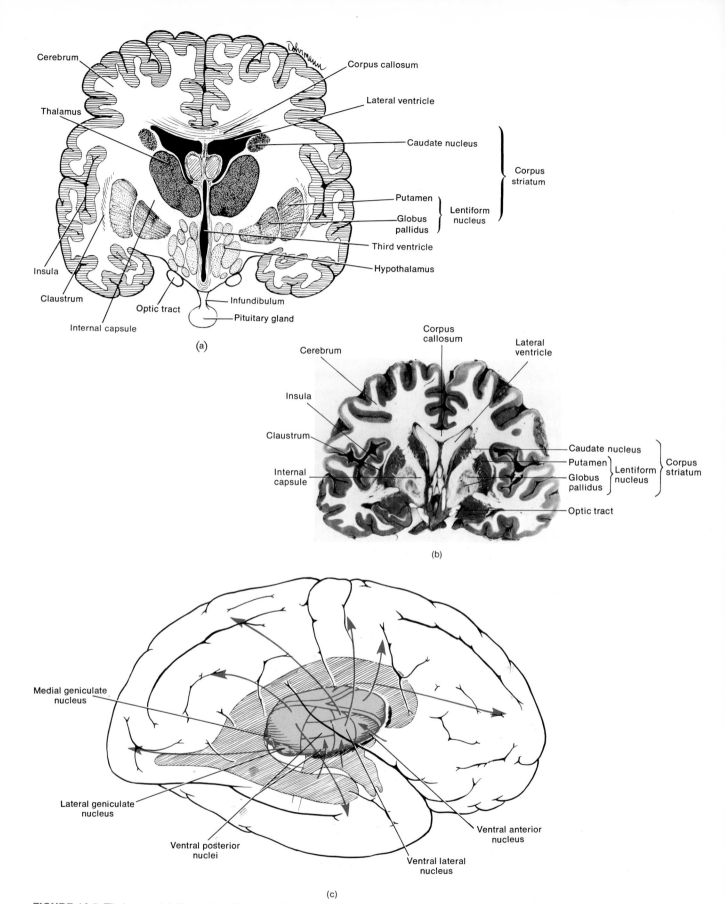

FIGURE 14-5 Thalamus. (a) Frontal section showing the thalamus and associated structures. (b) Photograph of a frontal section of the cerebrum anterior to the thalamus. (Courtesy of Murray L. Barr, *The Human Nervous System,* Harper & Row Publishers, Inc., 1974.) (c) Right lateral view of the thalamic nuclei.

Thalamus

The **thalamus** is a large oval structure above the midbrain (Figure 14-5a). It consists of two masses of gray matter covered by a thin layer of white matter. It measures about 3 cm (1 inch) in length and constitutes four-fifths of the diencephalon. The thalamus contains numerous nuclei organized into masses (Figure 14-5c). Some nuclei in the thalamus serve as relay stations for all sensory impulses, except smell, to the cerebral cortex. These include the *medial geniculate nuclei* (hearing), the *lateral geniculate nuclei* (vision), and the *ventral posterior nuclei* (general sensations and taste). Other nuclei are centers for synapses in the somatic motor system. These include the *ventral lateral nuclei* (voluntary motor actions) and *ventral anterior nuclei* (voluntary motor actions and arousal). The thalamus is the principal relay station for sensory impulses that reach the cerebral cortex from the spinal cord, brain stem, cerebellum, and parts of the cerebrum.

The thalamus also functions as an interpretation center. That is, some sensory impulses that enter the thalamus are interpreted there. At the thalamic level, one can have conscious recognition of pain and temperature and some awareness of crude touch and pressure. The thalamus also contains a *reticular nucleus* in its reticular formation and an *anterior nucleus* in the floor of the lateral ventricle.

Hypothalamus

The **hypothalamus** is a small portion of the diencephalon and its relationship to other parts of the brain is shown in Figures 14-1 and 14-5a. The hypothalamus forms the floor and part of the lateral walls of the third ventricle. The hypothalamus is protected by the sphenoid bone and indirectly by the sella turcica of the sphenoid bone. Despite its small size, nuclei in the hypothalamus control many body activities, most of them related to homeostasis. The chief functions of the hypothalamus are these:

1. It controls and integrates the autonomic nervous system, which stimulates smooth muscle, regulates the rate of contraction of cardiac muscle, and controls the secretions of many glands. Through the autonomic nervous system, the hypothalamus is the main regulator of visceral activities. It regulates heart rate, movement of food through the digestive tract, and contraction of the urinary bladder.

2. It is involved in the reception of sensory impulses from the viscera.

3. It is the principal intermediary between the nervous system and endocrine system—the two major control systems of the body. The hypothalamus lies just above the pituitary, the main endocrine gland.

When the hypothalamus detects certain changes in the body, it releases chemicals called regulating factors that stimulate or inhibit the anterior pituitary gland. The anterior pituitary then releases or holds back hormones that regulate carbohydrates, fats, proteins, certain ions, and sexual functions.

4. It is the center for the mind-over-body phenomenon. When the cerebral cortex interprets strong emotions, it often sends impulses along tracts that connect the cortex with the hypothalamus. The hypothalamus then directs impulses via the autonomic nervous system and also releases chemicals that stimulate the anterior pituitary gland. The result can be a wide range of changes in body activities. For instance, when you panic, impulses leave the hypothalamus to stimulate your heart to beat faster. Likewise, continued psychological stress can produce long-term abnormalities in body function that result in serious illness. These so-called psychosomatic disorders are definitely real.

5. It is associated with feelings of rage and aggression.

6. It controls normal body temperature. Certain cells of the hypothalamus serve as a thermostat—a mechanism sensitive to changes in temperature. If blood flowing through the hypothalamus is above normal temperature, the hypothalamus directs impulses along the autonomic nervous system to stimulate activities that promote heat loss. Heat can be lost through relaxation of the smooth muscle in the blood vessels causing vasodilation of cutaneous vessels and increased heat loss from the skin. Heat loss also occurs through sweating. Conversely, if the temperature of the blood is below normal, the hypothalamus generates impulses that promote heat retention. Heat can be retained through the contraction of cutaneous blood vessels, cessation of sweating, and shivering.

7. It regulates food intake through two centers. The *feeding center* is stimulated by hunger sensations from an empty stomach. When sufficient food has been ingested, the *satiety center* is stimulated and sends out impulses that inhibit the feeding center.

8. It contains a *thirst center*. Certain cells in the hypothalamus are stimulated when the extracellular fluid volume is reduced. The stimulated cells produce the sensation of thirst in the hypothalamus.

9. It is one of the centers that maintain the waking state and sleep patterns.

CEREBRUM

Supported on the brain stem and forming the bulk of the brain is the **cerebrum** (Figure 14-1). The surface of the cerebrum is composed of gray matter 2 to 4 mm (0.08 to 0.16 inch) thick and is referred to as the *cerebral cortex* (*cortex* = rind or bark). The cortex, con-

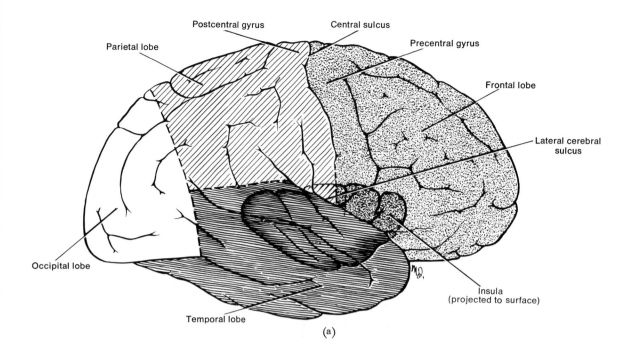

(a)

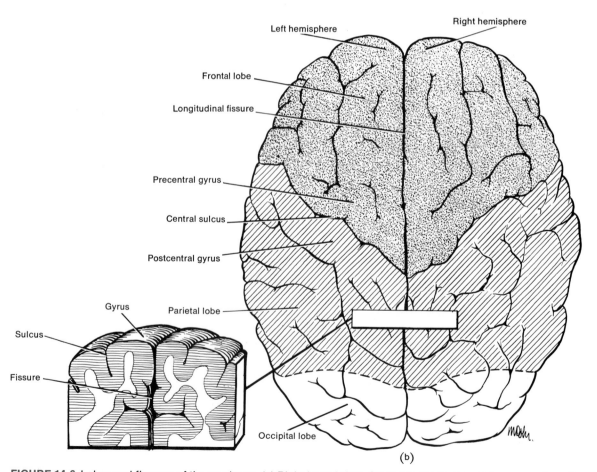

(b)

FIGURE 14-6 Lobes and fissures of the cerebrum. (a) Right lateral view. Since the insula cannot be seen externally, it has been projected to the surface. It can be seen in Figure 14-5a. (b) Superior view. The insert in (b) indicates the relative differences between a gyrus, sulcus, and fissure.

taining millions of cells, consists of six layers of nerve cell bodies. Beneath the cortex lies the cerebral white matter.

During embryonic development when there is a rapid increase in brain size, the gray matter of the cortex enlarges out of proportion to the underlying white matter. As a result, the cortical region rolls and folds upon itself. The upfolds are called *gyri* or *convolutions* (Figure 14-6b). The deep downfolds are referred to as *fissures;* the shallow downfolds are *sulci.* The most prominent fissure, the *longitudinal fissure,* nearly separates the cerebrum into right and left halves, or *hemispheres* (Figure 14-6b). The hemispheres, however, are connected internally by a large bundle of transverse fibers composed of white matter called the *corpus callosum.* Between the hemispheres is an extension of the cranial dura mater called the *falx cerebri.*

Lobes

Each cerebral hemisphere is further subdivided into four lobes by deep sulci or fissures (Figure 14-6). The *central sulcus,* or *fissure of Rolando,* separates the *frontal lobe* from the *parietal lobe.* A major gyrus, the *precentral gyrus,* is located immediately anterior to the central sulcus. The *lateral cerebral sulcus,* or *fissure of Sylvius,* separates the *frontal lobe* from the *temporal lobe.* The *parietooccipital sulcus* separates the *parietal lobe* from the *occipital lobe.* Another prominent fissure, the *transverse fissure,* separates cerebrum from the cerebellum. The frontal lobe, parietal lobe, temporal lobe, and occipital lobe are named after the bones that cover them. A fifth part of the cerebrum, the *insula (island of Reil),* lies deep within the lateral cerebral fissure, under the parietal, frontal, and temporal lobes. It cannot be seen in an external view of the brain (Figure 14-5a).

White Matter

The white matter underlying the cortex consists of myelinated axons running in three principal directions:

   1. Association fibers connect and transmit impulses between gyri in the same hemisphere.
   2. Commissural fibers transmit impulses from the gyri in one cerebral hemisphere to the corresponding gyri in the opposite cerebral hemisphere. Three important groups of commissural fibers are the *corpus callosum, anterior commissure,* and *posterior commissure.*
   3. Projection fibers form ascending and descending tracts that transmit impulses from the cerebrum to other parts of the brain and spinal cord.

Basal Ganglia

The **basal ganglia** or **cerebral nuclei** are paired masses of gray matter in each cerebral hemisphere (Figure 14-7). The largest of the basal ganglia of each hemisphere is the *corpus striatum.* It consists of the *caudate nucleus* and the *lentiform nucleus.* The lentiform nucleus, in turn, is subdivided into a lateral portion called the *putamen* and a medial portion called the *globus pallidus.* Figure 14-5a shows the two divisions of the lentiform nucleus and a structure called the *internal capsule.* It is made up of a group of sensory and motor

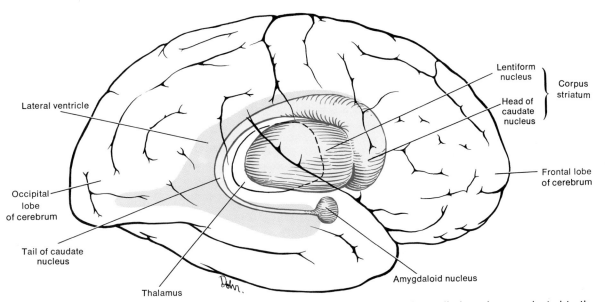

FIGURE 14-7 Basal ganglia. In this right lateral view of the cerebrum, the basal ganglia have been projected to the surface. Refer to Figure 14-5a to note the positions of the basal ganglia in the frontal section of the cerebrum.

white matter tracts that connect the cerebral cortex with the brain stem and spinal cord. The portion of the internal capsule passing between the lentiform nucleus and the caudate nucleus and between the lentiform nucleus and thalamus is sometimes considered part of the corpus striatum.

Other structures frequently considered part of the basal ganglia are the claustrum and amygdaloid nucleus. The *claustrum* is a thin sheet of gray matter lateral to the putamen. The *amygdaloid nucleus* is located at the tail end of the caudate nucleus. Some authorities also consider the *substantia nigra,* the *subthalamic nucleus,* and the *red nucleus* to be part of the basal ganglia. The substantia nigra is a large motor nucleus in the midbrain; its role in Parkinson's disease will be described later. The subthalamic nucleus lies against the internal capsule. Its major connection is with the globus pallidus. A lesion in the nucleus results in a motor disturbance on the opposite side of the body called *hemiballismus,* which is characterized by involuntary movements coming on suddenly with great force and rapidity. The movements are purposeless and generally of the withdrawal type, although they may be jerky. The spontaneous movements affect the proximal portions of the extremities most severely, especially the arms.

The basal ganglia are interconnected by many fibers. They are also connected to the cerebral cortex, thalamus, and hypothalamus. The caudate nucleus and the putamen control large unconscious movements of the skeletal muscles. An example of this is swinging the arms while walking. Such gross movements are also consciously controlled by the cerebral cortex. The globus pallidus is concerned with the regulation of muscle tone required for specific body movements. For example, if you wished to perform a very specific function with one of your hands, you might first position your body appropriately and then tense the muscles of the upper arm. Damage to the nuclei results in abnormal body movements, such as uncontrollable shaking, called tremor, and involuntary movements of skeletal muscle. Moreover, destruction of a substantial portion of the caudate nucleus almost totally paralyzes the opposite part of the body. The caudate nucleus is an area often affected by stroke.

Limbic System

Certain components of the cerebral hemispheres and diencephalon constitute the **limbic system.** It includes the following regions of gray matter:

1. Limbic lobe. Formed by two gyri of the cerebral hemisphere: the cingulate gyrus and the hippocampal gyrus.

2. Hippocampus. An extension of the hippocampal gyrus that extends into the floor of the lateral ventricle.

3. Amygdaloid nucleus. Located at the tail end of the caudate nucleus.

4. Hypothalamus. The regions of the hypothalamus that form part of the limbic system are the perifornical nuclei.

5. Anterior nucleus of the thalamus. Located in the floor of the lateral ventricle.

The limbic system functions in the emotional aspects of behavior related to survival. It also functions in memory. Although behavior is a function of the entire nervous system, the limbic system controls most of its involuntary aspects. Experiments on the limbic system of monkeys and other animals indicate that the amygdaloid nucleus assumes a major role in controlling the overall pattern of behavior.

Other experiments have shown that the limbic system is associated with pleasure and pain. When certain areas of the limbic system of the hypothalamus, thalamus, and midbrain are stimulated, experimental animals indicate they are experiencing intense punishment. When other areas are stimulated, the animals' reactions indicate they are experiencing extreme pleasure. In still other studies, stimulation of the perifornical nuclei of the hypothalamus result in a behavioral pattern called *rage.* The animal assumes a defensive posture—extending its claws, raising its tail, hissing, spitting, growling, and opening its eyes wide. Stimulating other areas of the limbic system results in an opposite behavioral pattern: docility, tameness, and affection.

Functional Areas of Cerebral Cortex

The functions of the cerebrum are numerous and complex. In a general way, the cerebral cortex is divided into motor, sensory, and association areas. The **motor areas** control muscular movement. The **sensory areas** interpret sensory impulses. And the **association areas** are concerned with emotional and intellectual processes.

● *Sensory Areas* The *general sensory area* or *somesthetic area* is located directly posterior to the central sulcus of the cerebrum on the postcentral gyrus. It extends from the longitudinal fissure on the top of the cerebrum to the lateral cerebral sulcus. In Figure 14-8a the general sensory area is designated by the areas numbered 1, 2, and 3.* The general sensory

---

*These numbers, as well as most of the others shown, are based on K. Brodmann's cytoarchitectural map of the cerebral cortex. His map, first published in 1909, is an attempt to correlate structure and function.

area receives sensations from cutaneous, muscular, and visceral receptors in various parts of the body. Each point of the general sensory area receives sensations from specific parts of the body. Essentially the entire body is spatially represented in the general sensory area. The portion of the sensory area receiving stimuli from body parts is not dependent on the size of the part but on the number of receptors. For example, a greater portion of the sensory area receives impulses from the lips than from the thorax. The major function of the general sensory area is to localize exactly the points of the body where the sensations originate. The thalamus is capable of localizing sensations in a general way. That is, the thalamus receives sensations from large areas of the body but cannot distinguish between specific areas of stimulation. This ability is reserved to the general sensory area of the cortex.

Posterior to the general sensory area is the *somesthetic association area*. It corresponds to the areas numbered 5 and 7 in Figure 14-8a. The somesthetic association area receives input from the thalamus, other lower portions of the brain, and the general sensory area. The somesthetic association area integrates and interprets sensations. This area permits you to determine the exact shape and texture of an object without looking at it, to determine the orientation of one object to another as they are felt, and to sense the relationship of one body part to another. Another role of the somesthetic association area is the storage of

memories of past sensory experiences. Thus you can compare sensations with previous experiences.

Other sensory areas of the cortex include:

1. Primary visual area (area 17). Located on the medial surface of the occipital lobe and occasionally extends around to the lateral surface. It receives sensory impulses from the eyes and interprets shape and color.

2. Visual association area (areas 18 and 19). Located in the occipital lobe and receives sensory signals from the primary visual area and the thalamus. It relates present to past visual experiences with recognition and evaluation of what is seen.

3. Primary auditory area (areas 41 and 42). Located in the superior part of the temporal lobe near the lateral cerebral sulcus. It interprets the basic characteristics of sound such as pitch and rhythm.

4. Auditory association area (area 22). Inferior to the primary auditory area in the temporal cortex. It determines if a sound is speech, music, or noise. It also interprets the meaning of speech by translating words into thoughts.

5. Primary gustatory area (area 43). Located at the base of the postcentral gyrus above the lateral cerebral sulcus in the parietal cortex. It interprets sensations related to taste.

6. Primary olfactory area. Located in the temporal lobe on the medial aspect and interprets sensations related to smell.

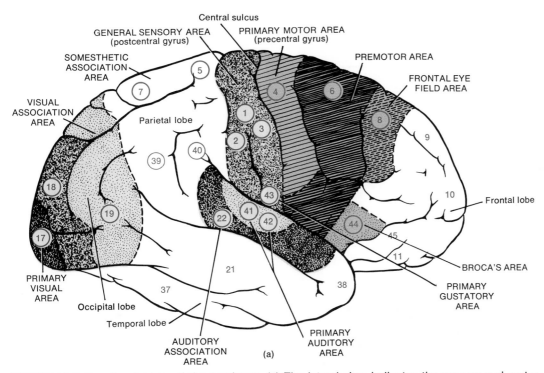

FIGURE 14-8 Functional areas of the cerebrum. (a) The lateral view indicates the sensory and motor areas. Although the right hemisphere is illustrated, Broca's area is in the left hemisphere of most people.

7. Gnostic area (areas 5, 7, 39, and 40). This *common integrative area* is located between the somesthetic, visual, and auditory association areas. The gnostic area receives impulses from these areas, as well as from the taste and smell areas, the thalamus, and lower portions of the brain stem. The gnostic area integrates all thoughts from the various sensory areas so that a common thought can be formed from the various sensory inputs. It then transmits signals to other parts of the brain to cause the appropriate response to the sensory signal.

● *Motor Areas* The *primary motor area* is located mainly in the precentral gyrus of the frontal lobe (Figure 14-8a). This region is also designated as area 4. Like the general sensory area, the primary motor area consists of points that control specific muscles or groups of muscles. Stimulation of a specific point of the primary motor area results in a muscular contraction, usually on the opposite side of the body.

The *premotor area* (area 6) is anterior to the primary motor area. It is concerned with learned motor activities of a complex and sequential nature. It generates impulses that cause a specific group of muscles to contract in a specific sequence. An example of this is writing. Thus the premotor area controls skilled movements.

The *frontal eye field area* (area 8) in the frontal cortex is sometimes included in the premotor area. This area controls voluntary scanning movements of the eyes—searching for a word in a dictionary, for instance.

The *language areas* are also significant parts of the motor cortex. When you listen to someone speaking, sounds are relayed to the primary auditory area of the cortex. The sounds are then interpreted as words in the auditory association area. The words are interpreted as thoughts in the gnostic area. Written words are interpreted by the visual association area and converted into thoughts by the gnostic area. Thus you can translate speech or written words into thoughts.

The translation of thoughts into speech involves *Broca's area* or the *motor speech area,* designated as area 44 and located in the frontal lobe just superior to the lateral cerebral sulcus. From this area, a sequence of signals is sent to the premotor regions that control the muscles of the larynx, throat, and mouth. The impulses from the premotor area to the muscles result in specific, coordinated contractions that enable you to speak. Simultaneously, impulses are sent from Broca's area to the primary motor area. From here, impulses reach your breathing muscles to regulate the proper flow of air past the vocal cords. The coordinated contractions of your speech and breathing muscles enable you to translate your thoughts into speech.

Broca's area is usually located in the left cerebral hemisphere of most individuals regardless of whether they are left-handed or right-handed. Injury to the sensory or motor speech areas results in *aphasia,* which is an inability to speak; *agraphia,* an inability to write; *word deafness,* an inability to understand spoken words; or *word blindness,* an inability to understand written words.

● *Association Areas* The *association areas* of the cerebrum are made up of association tracts that con-

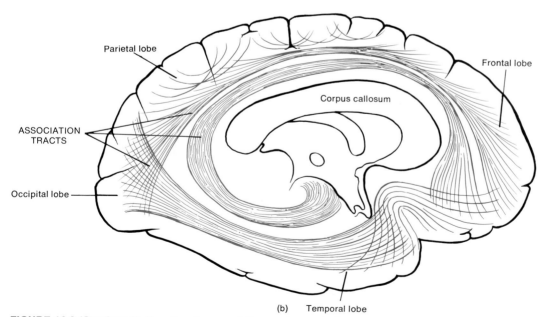

FIGURE 14-8 (Continued) Functional areas of the cerebrum. (b) The sagittal section shows the association tracts.

nect motor and sensory areas (see Figure 14-8b). The association region of the cortex occupies the greater portion of the lateral surfaces of the occipital, parietal, and temporal lobes and the frontal lobes anterior to the motor areas. The association areas are concerned with memory, emotions, reasoning, will, judgment, personality traits, and intelligence.

Brain Waves

Brain cells can generate electrical activity as a result of literally millions of action potentials of individual neurons. These electrical potentials are called **brain waves** and indicate activity of the cerebral cortex. Brain waves pass through the skull easily and can be detected by sensors called electrodes. A record of such waves is called an **electroencephalogram (EEG).** An EEG is obtained by placing electrodes on the head and amplifying the waves with an electroencephalograph. As indicated in Figure 14-9, four kinds of waves are produced by normal individuals:

1. **Alpha waves.** These rhythmic waves occur at a frequency of about 10 to 12 cycles/second. They are found in the EEGs of nearly all normal individuals when awake and in the resting state. These waves disappear entirely during sleep.

2. **Beta waves.** The frequency of these waves is between 15 and 60 cycles/second. Beta waves generally appear when the nervous system is active—that is, during periods of sensory input and mental activity.

3. **Theta waves.** These waves have frequencies of 5 to 8 cycles/second. Theta waves normally occur in children and in adults experiencing emotional stress.

4. **Delta waves.** The frequency of these waves is between 1 to 5 cycles/second. Delta waves occur during sleep. They are normal in an awake infant. When produced by an awake adult, they indicate brain damage.

Distinct EEG patterns appear in certain abnormalities. In fact, the EEG is used clinically in the diagnosis of epilepsy, infectious diseases, tumors, trauma, and hematomas. Electroencephalograms also furnish information regarding sleep and wakefulness.

CEREBELLUM

The **cerebellum** is the second largest portion of the brain and occupies the inferior and posterior aspects of the cranial cavity. Specifically, it is below the posterior portion of the cerebrum and is separated from it by the *transverse fissure* (see Figure 14-1). The cerebellum is also separated from the cerebrum by an extension of the cranial dura mater called the *tentorium cerebelli*. The cerebellum is shaped somewhat like a butterfly. The central constricted area is the *vermis,* which means worm-shaped, and the lateral "wings" are referred to as *hemispheres* (Figure 14-10). Between the hemispheres is another extension of the cranial dura mater: the *falx cerebelli.* It passes only a short distance between the cerebellar hemispheres.

The surface of the cerebellum, called the *cortex,* consists of gray matter in a series of slender, parallel ridges called *gyri.* These gyri are less prominent than those located on the cerebral cortex. Beneath the gray matter are white matter tracts *(arbor vitae)* that resemble branches of a tree. Deep within the white matter are masses of gray matter: the *cerebellar nuclei.*

The cerebellum is attached to the brain stem by three paired bundles of fibers called *cerebellar peduncles.* These are as follows:

1. **Inferior cerebellar peduncles,** which connect the cerebellum with the medulla at the base of the brain stem and with the spinal cord

2. **Middle cerebellar peduncles,** which connect the cerebellum with the pons

3. **Superior cerebellar penduncles,** which connect the cerebellum with the midbrain

The cerebellum is a motor area of the brain that produces certain subconscious movements in the skeletal muscles. These movements are required for coordination, for maintenance of posture, and for balance. The cerebellar peduncles are the fiber tracts that allow the cerebellum to perform its functions.

Let us now see how the cerebellum produces coordinated movement. Motor areas of the cerebral cortex

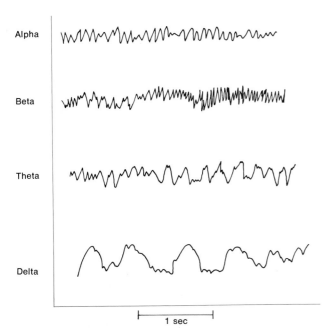

Alpha

Beta

Theta

Delta

|← 1 sec →|

FIGURE 14-9 Kinds of waves recorded in an electroencephalograph.

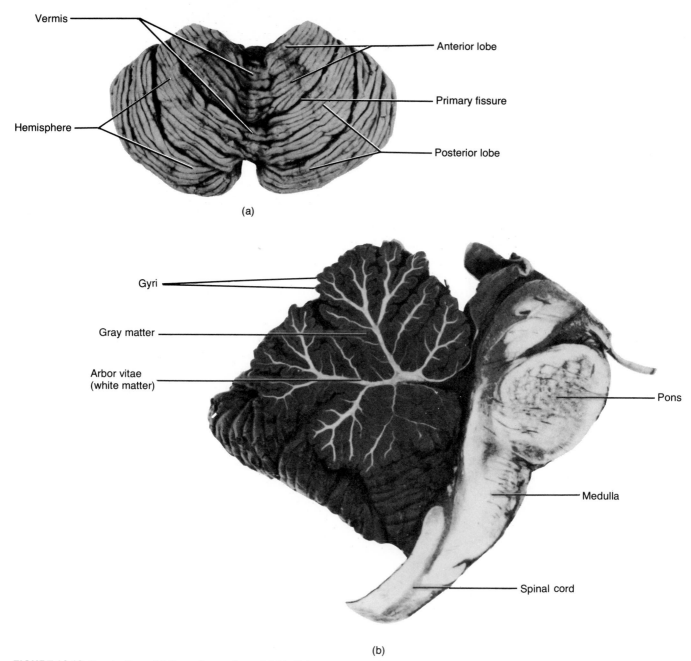

(a)

(b)

FIGURE 14-10 Cerebellum. (a) Superior surface. (b) Medial aspect seen in sagittal section. (Courtesy of M. L. Barr, from *The Human Nervous System,* Second Edition, 1974.)

voluntarily initiate muscle contraction. Once the movement has begun, the sensory areas of the cortex receive impulses from nerves in the joints. The impulses provide information about the extent of muscle contraction and the amount of joint movement. The term *proprioception* is applied to this sense of the position of one body part relative to another. The cerebral cortex uses the proprioceptive sensations to determine which muscles are required to contract next and with what strength they are to contract in order to

continue moving in the desired direction. Then a pattern of impulses is generated by the cerebral cortex along tracts to the pons and midbrain, which relay the impulses over the middle and superior cerebellar peduncles to the cerebellum. The cerebellum then generates subconscious motor impulses along the inferior cerebellar peduncles to the medulla and spinal cord. The impulses pass downward along the spinal cord and out the nerves that stimulate the prime movers and synergists to contract and that inhibit the contraction of the antagonists. The result is smooth, coordinated movement. A well-functioning cerebellum is essential for delicate movements such as playing the piano.

The cerebellum also transmits impulses that control postural muscles. That is, the cerebellum is required for maintaining normal muscle tone. The cerebellum also maintains body equilibrium. The inner ear contains structures that sense balance. Information such as whether the body is leaning to the left or right is transmitted from the inner ear to the cerebellum. The cerebellum then discharges impulses that cause the contraction of the muscles necessary for maintaining equilibrium.

Damage to the cerebellum through trauma or disease is characterized by certain symptoms involving skeletal muscles. There may be lack of muscle coordination, called *ataxia*. Blindfolded people with ataxia cannot touch the tip of their nose with a finger because they cannot coordinate movement with their sense of where a body part is located. Another sign of ataxia is a change in the speech pattern due to a lack of coordination of speech muscles. Cerebellar damage may also result in disturbances of gait in which the subject staggers or cannot coordinate normal walking movements.

Cranial Nerves

Of the 12 pairs of **cranial nerves,** 10 pairs originate from the brain stem, but all 12 pairs leave the skull through foramina in the base of the skull (see Figure 14-4a). The cranial nerves are designated in two ways—with Roman numerals and with names. The Roman numerals indicate the order in which the nerves arise from the brain (front to back). The names indicate the distribution or function of the nerves. Some cranial nerves are termed *mixed nerves.* They contain both sensory and motor fibers. Other cranial nerves contain sensory fibers only. The cell bodies of sensory fibers are located in ganglia outside the brain. The cell bodies of motor fibers lie in nuclei within the brain.

Some motor fibers control subconscious movements, yet the somatic nervous system has been defined as a *conscious* system. The reason for this apparent contradiction is that some fibers of the autonomic nervous system leave the brain bundled together with somatic fibers of the cranial nerves. Therefore subconscious functions transmitted by the autonomic fibers are described along with the conscious functions of the somatic fibers of the cranial nerves. Although the cranial nerves are mentioned singly in this discussion, remember that they are paired structures.

I. OLFACTORY

The **olfactory nerve** is entirely sensory and conveys impulses related to smell. It arises as bipolar neurons from the olfactory mucosa of the nasal cavity. The dendrites and cell bodies of these neurons are generally limited to the mucosa covering the superior nasal conchae and the adjacent nasal septum. Axons from the neurons pass through the cribriform plate of the ethmoid bone and synapse with other olfactory neurons in the *olfactory bulb,* an extension of the brain lying above the cribriform plate. The axons of these neurons make up the *olfactory tract.* The fibers from the tract terminate in the primary olfactory area in the cerebral cortex.

II. OPTIC

The **optic nerve** is also entirely sensory and conveys impulses related to vision. Impulses initiated by rods and cones of the retina are relayed by bipolar neurons to ganglion cells. Axons of the ganglion cells, the optic nerve fibers, enter the optic foramina where the two optic nerves unite to form the *optic chiasma.* Within the chiasma, fibers from the medial half of each retina cross to the opposite side; those from the lateral half remain on the same side. From the chiasma, the fibers pass posteriorly to the *optic tracts.* From the optic tracts, the majority of fibers terminate in a nucleus of the thalamus. They then synapse with neurons that pass to the visual areas of the cerebral cortex. Some fibers from the optic chiasma terminate in the superior colliculi of the midbrain. They synapse with neurons whose fibers terminate in the nuclei that convey impulses to the oculomotor (III), trochlear (IV), and abducens (VI) nerves—nerves that control the extrinsic (external) and intrinsic (internal) eye muscles. Through this relay, there are widespread motor responses to light stimuli.

III. OCULOMOTOR

The **oculomotor nerve** is a mixed cranial nerve. It originates from neurons in a nucleus in the ventral portion of the midbrain. It runs forward, divides into a superior and inferior branch, and passes through the superior orbital fissure in the orbit. The superior branch is distributed to the superior rectus (an extrinsic eyeball muscle) and the levator palpebrae superioris (the muscle of the upper eyelid). The inferior branch is distributed to the medial rectus, inferior rectus, and inferior oblique muscles—all extrinsic eyeball muscles. These distributions to the levator palpebrae superioris and extrinsic eyeball muscles constitute the motor portion of the oculomotor nerve. Through these distributions, impulses are sent that control movements of the eyeball and upper eyelid.

The inferior branch of the oculomotor nerve also sends a branch to the *ciliary ganglion,* a relay center of the autonomic nervous system that connects a nucleus in the midbrain with the intrinsic eyeball muscles. These intrinsic muscles include the ciliary muscle of the eyeball and the sphincter muscle of the iris. Through the ciliary ganglion, the oculomotor nerve controls the smooth muscle (ciliary muscle) responsible for accommodation of the lens for near vision and the smooth muscle (sphincter muscle of iris) responsible for constriction of the pupil.

The sensory portion of the oculomotor nerve consists of afferent fibers from proprioceptors in the eyeball muscles supplied by the nerve to the midbrain. These fibers convey impulses related to muscle sense (proprioception).

IV. TROCHLEAR

The **trochlear nerve** is a mixed cranial nerve. It is the smallest of the 12 cranial nerves. The motor portion originates in a nucleus in the midbrain, and axons from the nucleus pass through the superior orbital fissure of the orbit. The motor fibers innervate the superior oblique muscle of the eyeball, another extrinsic eyeball muscle. It controls movement of the eyeball.

The sensory portion of the trochlear nerve consists of afferent fibers that run from proprioceptors in the superior oblique muscle to the nucleus of the nerve in the midbrain. The sensory portion is responsible for muscle sense.

V. TRIGEMINAL

The **trigeminal nerve** is a mixed cranial nerve and the largest of the cranial nerves. As indicated by its name, the trigeminal nerve has three sensory branches: ophthalmic, maxillary, and mandibular. The trigeminal nerve contains two roots on the ventrolateral surface of the pons. The large sensory root has a swelling called the *semilunar (gasserian) ganglion* located in a fossa on the inner surface of the petrous portion of the temporal bone. From this ganglion, the *ophthalmic branch* enters the orbit via the superior orbital fissure, the *maxillary branch* enters the foramen rotundum, and the *mandibular branch* pierces the foramen ovale. The smaller motor root originates in a nucleus in the pons. The motor fibers join the mandibular branch and supply the muscles of mastication. These motor fibers, which control chewing movements, constitute the motor portion of the trigeminal nerve.

The sensory portion of the trigeminal nerve delivers impulses related to touch, pain, and temperature and consists of the ophthalmic, maxillary, and mandibular branches. The ophthalmic branch receives sensory fibers from the skin over the upper eyelid, eyeball, lacrimal glands, upper part of the nasal cavity, side of the nose, forehead, and anterior half of the scalp. The maxillary branch receives sensory fibers from the mucosa of the nose, palate, parts of the pharynx, upper teeth, upper lip, cheek, and lower eyelid. The mandibular branch transmits sensory fibers from the anterior two-thirds of the tongue (not taste), lower teeth, skin over the mandible and side of the head in front of the ear, and mucosa of the floor of the mouth. Sensory fibers from the three branches of the trigeminal nerve enter the semilunar ganglion and terminate in a nucleus in the pons. There are also sensory fibers from proprioceptors in the muscles of mastication.

VI. ABDUCENS

The **abducens nerve** is a mixed cranial nerve that originates from a nucleus in the pons. The motor fibers extend from the nucleus to the lateral rectus muscle of the eyeball, an extrinsic eyeball muscle. Impulses over the fibers bring about movement of the eyeball. The sensory fibers run from proprioceptors in the lateral rectus muscle to the pons and mediate muscle sense. The abducens nerve reaches the lateral rectus muscle through the superior orbital fissure of the orbit.

VII. FACIAL

The **facial nerve** is a mixed nerve. Its motor fibers originate from a nucleus in the pons and enter the petrous portion of the temporal bone. The motor fibers are distributed to facial, scalp, and neck muscles. Impulses along these fibers cause contraction of the muscles of facial expression. Some motor fibers are also distributed to the lacrimal, sublingual, and submandibular glands.

The sensory fibers extend from the taste buds of the anterior two-thirds of the tongue to the *geniculate ganglion,* a swelling of the facial nerve. From here, the fibers pass to a nucleus in the pons which sends fibers to the thalamus for relay to the gustatory area of the cerebral cortex. The sensory portion of the facial nerve conveys sensations related to taste. Proprioceptors are in the muscles of the face and scalp.

VIII. VESTIBULOCOCHLEAR

The **vestibulocochlear nerve** is a sensory cranial nerve. It consists of two branches: the cochlear (auditory) branch and the vestibular branch. The *cochlear branch,* which conveys impulses associated with hearing, arises in the spiral organ of Corti in the cochlear of the internal ear. The cell bodies of the cochlear branch are in the *spiral ganglion* of the cochlea. From here the axons pass through a nucleus in the medulla and terminate in the thalamus. Ultimately, the fibers synapse with neurons that relay the impulses to the auditory areas of the cerebral cortex.

The *vestibular branch* arises in the semicircular canals, the saccule, and the utricle of the inner ear. Fibers from the semicircular canals, saccule, and utricle extend to the *vestibular ganglion,* where the cell bodies are contained. The cell bodies of the fibers synapse in the ganglion with fibers that extend to a nucleus in the medulla and pons and terminate in the thalamus. Some fibers also enter the cerebellum. The vestibular branch transmits impulses related to equilibrium.

IX. GLOSSOPHARYNGEAL

The **glossopharyngeal nerve** is a mixed cranial nerve. Its motor fibers originate in a nucleus in the medulla. The nerve exits the skull through the jugular foramen. The motor fibers are distributed to the swallowing muscles of the pharynx and the parotid gland to mediate swallowing movements and the secretion of saliva. The sensory fibers of the glossopharyngeal nerve supply the pharynx and taste buds of the posterior third of the tongue. Some sensory fibers also originate from receptors in the carotid sinus, which assumes a major role in blood pressure regulation. The sensory fibers of the glossopharyngeal nerve terminate in a nucleus in the thalamus. There are also sensory fibers from proprioceptors in the muscles innervated by this nerve.

X. VAGUS

The **vagus nerve** is a mixed cranial nerve that is widely distributed from the head and neck into the thorax and abdomen. Its motor fibers originate in a nucleus of the medulla and terminate in the muscles of the pharynx, layrnx, respiratory passageways, lungs, heart, esophagus, stomach, small intestine, most of the large intestine, and gallbladder. Impulses along the motor fibers generate visceral, cardiac, and skeletal muscle movement. Sensory fibers of the vagus nerve supply essentially the same structures as the motor fibers. They convey impulses for various sensations from the larynx and viscera. The fibers terminate in the medulla and pons. There are also sensory fibers from proprioceptors in the muscles supplied by this nerve.

XI. ACCESSORY

The **accessory nerve** (formerly the spinal accessory nerve) is a mixed cranial nerve. It differs from all other cranial nerves in that it originates from both the brain stem and the spinal cord. The *bulbar (medullary) portion* originates from nuclei in the medulla, passes through the jugular foramen, and supplies the voluntary muscles of the pharynx, larynx, and soft palate that are used in swallowing. The *spinal portion* originates in the anterior gray horn of the first five segments of the cervical portion of the spinal cord. The fibers from the segments join, enter the foramen magnum, and exit through the jugular foramen along with the bulbar portion. The spinal portion conveys motor impulses to the sternocleidomastoid and trapezius muscles to coordinate head movements. The sensory fibers originate from proprioceptors in the muscles supplied by its motor neurons and terminate in upper cervical posterior root ganglia. The conduct impulses for proprioception.

XII. HYPOGLOSSAL

The **hypoglossal nerve** is a mixed nerve. The motor fibers originate in a nucleus in the medulla, pass through the hypoglossal canal, and supply the muscles of the tongue. These fibers conduct impulses related to speech and swallowing.

The sensory portion of the hypoglossal nerve consists of fibers originating from proprioceptors in the tongue muscles and terminating in the medulla. The sensory fibers conduct impulses for muscle sense.

Applications to Health

Many disorders can affect the central nervous system. Some are caused by viruses or bacteria. Others are caused by damage to the nervous system during birth. The origins of many conditions, however, are unknown. Here we discuss the origins and symptoms of some common central nervous system disorders.

POLIOMYELITIS

Poliomyelitis, also known as **infantile paralysis,** is a viral infection that is most common during childhood. Onset of the disease is marked by fever, severe headache, a stiff neck and back, deep muscle pain and weakness, and loss of certain somatic reflexes. The virus may spread via the respiratory passages and blood to the central nervous system where it destroys the motor nerve cell bodies, specifically those in the anterior horns of the spinal cord and in the nuclei of the cranial nerves. Injury to the spinal gray matter is the basis for the name of this disease (*polio* = gray matter; *myel* = spinal cord). Destruction of the anterior horns produces paralysis. The first sign of bulbar polio is difficulty in swallowing, breathing, and speaking. Poliomyelitis can cause death from respiratory or heart failure if the virus invades the brain cells of the vital medullary centers. In recent years, an immunization against the disease has been used.

SYPHILIS

Syphilis is a venereal disease caused by the *Treponema pallidum* bacterium. Venereal diseases are infectious disorders that can be spread through sexual contact. The disease progresses through several stages: primary, secondary, latent, and sometimes tertiary. During the *primary stage,* the chief symptom is an open sore, called a chancre, at the point of contact. The chancre eventually heals. About six weeks later, symptoms such as a skin rash, fever, and aches in the joints and muscles usher in the *secondary stage.* At this stage, syphilis can usually be treated with antibiotics. Even if individuals do not undergo treatment, their symptoms will eventually disappear. Within a few years, the disease will cease to be infectious. The symptoms of the disease disappear, but a blood test is generally positive. During this later "symptomless" period, called the *latent stage,* the bacteria may invade and slowly destroy body organs. Untreated syphilis is considered dangerous for this reason. When organ degeneration appears, the disease is said to be in the *tertiary stage.* If the syphilis bacteria attack the organs of the nervous system, the tertiary stage is called *neurosyphilis.* Neurosyphilis may take different forms, depending on the tissue involved. For instance, about two years after the onset of the disease, the bacteria may attack the meninges, producing meningitis. The blood vessels that supply the brain may also become infected. In this case, symptoms depend on the parts of the brain destroyed by oxygen and glucose starvation. Cerebellar damage is manifested by uncoordinated movements as in writing. As the motor areas become extensively damaged, victims may be unable to control urine and bowel movements. Eventually, they may become bedridden, unable even to feed themselves. Damage to the cerebral cortex produces memory loss and personality changes that range from irritability to hallucinations.

CEREBRAL PALSY

The term **cerebral palsy** refers to a group of motor disorders caused by damage to the motor areas of the brain during fetal life, birth, or infancy. One cause is infection of the mother with German measles during the first three months of pregnancy. During early pregnancy, certain cells in the fetus are dividing and differentiating in order to lay down the basic structures of the brain. These cells can be abnormally changed by toxin from the measles virus. Radiation during fetal life, temporary oxygen starvation during birth, and hydrocephalus during infancy may also damage brain cells.

Cases of cerebral palsy are categorized into three groups depending on whether the cortex, the basal ganglia of the cerebrum, or the cerebellum is affected most severely. Most cerebral palsy victims have at least some damage in all three areas. The location and extent of motor damage determine the symptoms. The victim may be deaf or partially blind. About 70 percent of cerebral palsy victims appear to be mentally retarded. The apparent mental slowness, however, is often due to the person's inability to speak or hear well. Such individuals are often more mentally acute than they appear.

Cerebral palsy is not a progressive disease. Thus it does not worsen as time elapses. Once the damage is done, however, it is irreversible.

PARKINSONISM

This disorder, also called **Parkinson's disease,** is a progressive malfunction of the basal ganglia of the cerebrum. The basal ganglia regulate subconscious contractions of skeletal muscles that aid activities desired by the motor areas of the cerebral cortex—swinging the arms when walking, for example. In Parkinsonism, the basal ganglia produce unnecessary skeletal movements that often interfere with voluntary movement. For instance, the muscles of the upper extremities may alternately contract and relax, causing the hands to shake. This shaking is called *tremor.* Other muscles may contract continuously, causing rigidity of the involved body part. *Rigidity* of the facial muscles gives the face a masklike appearance. The expression is characterized by a wide-eyed, unblinking stare and a slightly open mouth with uncontrolled drooling. Vision, hearing, and intelligence are unaffected by the disorder, indicating that Parkinsonism does not attack the cerebral cortex.

Parkinsonism seems to be caused by a malfunction at the neuron synapses. The motor neurons of the basal ganglia release the chemical transmitter acetylcholine. In normal people, the basal ganglia also produce a synaptic transmitter called dopamine, which quickly inactivates the acetylcholine and prevents continuous conduction across the synapse. People with Parkinsonism do not manufacture enough dopamine in their brains. As a result, stimulated basal ganglia neurons do not easily stop conducting impulses. Injections of dopamine are useless; the blood-brain barrier stops it. However, symptoms are somewhat relieved by a drug developed a few years ago, levodopa, and its successors carbidopa and bromocriptine—none without distressing side effects.

EPILEPSY

Epilepsy is a disorder characterized by short, recurrent, periodic attacks of motor, sensory, and/or psychological malfunction. The attacks, called **epileptic seizures,** are brought on by abnormal and irregular discharges of electricity by millions of neurons in the brain. The discharges stimulate many of the neurons to send impulses over their conduction pathways. As a result, a person undergoing an attack may contract skeletal muscles involuntarily. Lights, noise, or smells may be sensed when the eyes, ears, and nose actually have not been stimulated. The electrical discharges may also inhibit certain brain centers. For instance, the waking center in the brain may be depressed so that the person loses consciousness.

Many different types of epileptic seizures exist. The particular type of seizure depends on the area of the brain that is electrically stimulated and whether the stimulation is restricted to a small area or spreads throughout the brain. *Grand mal* seizures are brought on by a burst of electrical discharges that travel throughout the motor areas and spread to the areas of consciousness in the brain. The person loses consciousness, has spasms of voluntary muscles, and may also lose urinary and bowel control. Sensory and intellectual areas may also be involved. For instance, just as the attack begins, the person may sense a peculiar taste in the mouth or see flashes of light or have olfactory hallucinations. The unconsciousness and motor activity last a few minutes. Then the muscles relax, and the person awakens. Afterward, the individual may be mentally confused for a short period of time. Studies with EEGs show that grand mal attacks are characterized by rapid brain waves occurring at the rate of 25 to 30 per second (Figure 14-11a). The normal adult rate is 10 waves/second.

Many epileptics suffer from electrical discharges that are restricted to one or several relatively small areas of the brain. An example is the *petit mal* form, which apparently involves the thalamus and hypothalamus. Petit mal seizures are characterized by an abnormally slow brain wave pattern occurring at the rate of 3 waves/second (Figure 14-11b). The person may lose contact with the environment for anywhere from 5 to 30 seconds but does not undergo the loss of motor control that is typical of a grand mal seizure. The victim merely seems to be daydreaming. Some people ex-

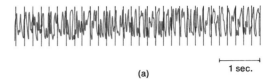

(a)   1 sec.

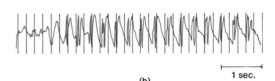

(b)   1 sec.

FIGURE 14-11 Electroencephalographs. (a) Grand mal seizure. (b) Petit mal seizure.

perience several hundred petit mal seizures each day. For them, the chief problems are a loss of productivity in school or work and periodic inattentiveness while driving a car.

Some epileptics experience motor seizures that are restricted to the precentral motor area of one cerebral hemisphere. These attacks consist of spasms that pass up or down one side of the body. People who suffer from sensory seizures may see lights or distorted objects if the discharge occurs in the occipital lobe. They may hear voices or a roaring in their ears if the discharge is located in the temporal lobe. Or they may taste something unpleasant if the discharge is in the parietal lobe. People undergoing attacks of localized motor or sensory disturbances may or may not lose consciousness. A form of epilepsy that is sometimes confused with mental illness is *psychomotor epilepsy*. The electrical outburst occurs in the temporal lobe, where it causes the person to lose contact with reality. It may spread to some of the motor areas and produce mild spasms in some of the voluntary muscles. These persons may stare into space and involuntarily smack their lips or clap their hands during an attack. If the motor areas are not involved, they may simply walk aimlessly. When they come back to reality, they are surprised to find themselves in a strange or different place.

The causes of epilepsy are varied. Many conditions can cause nerve cells to produce periodic bursts of impulses. Head injuries, tumors and abscesses of the brain, and childhood infections, such as mumps, whooping cough, and measles, are some of the causes. Epilepsy may also be idiopathic—without demonstrable cause.

It should be noted, however, that epilepsy almost never affects intelligence. If frequent severe seizures are allowed to occur over a long period of time, some cerebral damage may occasionally result. However, damage can be prevented by controlling the seizures with drug therapy. The seizures can be eliminated or alleviated by drugs that make neurons more difficult to stimulate. Many of these drugs change the permeability of the neuron cell membrane so that it does not depolarize as easily.

MULTIPLE SCLEROSIS

Multiple sclerosis causes progressive destruction of the myelin sheaths of neurons in the central nervous system. The sheaths deteriorate to *scleroses,* which are hardened scars or plaques, in multiple regions—hence the name. The destruction of myelin sheaths interferes with the transmission of impulses from one neuron to another, literally short-circuiting conduction pathways. Multiple sclerosis is one of the most common disorders of the central nervous system. Usually the first symptoms occur between the ages of 20 and 40. Early symptoms are generally produced by the formation of a few plaques and are, consequently, mild. Plaque formation in the cerebellum may produce lack of coordination in one hand. The patient's handwriting becomes strained and irregular. A short-circuiting of pathways in the corticospinal tract may partially paralyze the leg muscles so that the patient drags a foot when walking. Other early symptoms include double vision and urinary tract infections. Following a period of remission during which the symptoms temporarily disappear, a new series of plaques develop and the victim suffers a second attack. One attack follows another over the years. Each time the plaques form, certain neurons are damaged by the hardening of their sheaths. Other neurons are uninjured by their plaques. The result is a progressive loss of function interspersed with remission periods during which the undamaged neurons regain their ability to transmit impulses.

The symptoms of multiple sclerosis depend on the areas of the central nervous system most heavily laden with plaques. Sclerosis of the white matter of the spinal cord is common. As the sheaths of the neurons in the corticospinal tract deteriorate, the patient loses the ability to contract skeletal muscles. Damage to the ascending tracts produces numbness and short-circuits impulses related to position of body parts and flexion of joints. Damage to either set of tracts also destroys spinal cord reflexes.

As the disease progresses, most voluntary motor control is eventually lost and the patient becomes bedridden. Death occurs anywhere from 7 to 30 years after the first symptoms appear. The usual cause of death is a severe infection resulting from the loss of motor activity. Without the constricting action of the urinary bladder wall, for example, the bladder never totally empties and stagnant urine provides an environment for bacterial growth. Bladder infection may then spread to the kidney, damaging kidney cells.

Multiple sclerosis may be caused by a virus. The occasional appearance of more than one case in a family suggests such an infectious agent, but the same circumstances also suggest a genetic predisposition. Like other demyelinating diseases, multiple sclerosis is incurable. Electrical stimulation of the spinal cord can improve function in certain patients, however.

CEREBROVASCULAR ACCIDENTS

The most common brain disorder is a **cerebrovascular accident (CVA),** also called a **stroke** or **cerebral apoplexy.** A CVA is the destruction of brain tissue or infarction resulting from disorders in the vessels that supply the brain. Common causes of CVAs are intracerebral hemorrhages from aneurysms, emboli, and atherosclerosis of the cerebral arteries. An *intracerebral hemorrhage* is a rupture of a vessel in the pia mater or brain. Blood seeps into the brain and damages neurons by increasing intracranial fluid pressure. An *embolus* is a blood clot, air bubble, or bit of foreign material, most often debris from an inflammation, that becomes lodged in an artery and blocks circulation. *Atherosclerosis* is the formation of plaques in the artery walls. The plaques may slow down circulation by constricting the vessel. Both emboli and atherosclerosis cause brain damage by reducing the supply of oxygen and glucose needed by brain cells.

Many elderly people suffer mild CVAs as a result of short periods of reduced blood supply. Another cause is atherosclerosis. Another is *arteriosclerosis,* or hardening of the arteries, which occurs with aging. Damage is generally undetectable or very mild. During these mild CVAs the individual may have a short blackout, blurred vision, or dizziness and does not realize anything serious has occurred. A CVA can also cause sudden, massive damage, however. Severe CVAs cause about 21 percent of all deaths from cardiovascular disease. The person who recovers may suffer partial paralysis and mental disorders such as speech difficulty. The malfunction depends on the parts of the brain that were injured. Vascular disorders are more common after age 40.

DYSLEXIA

Dyslexia (*dys* = difficulty; *lexis* = words) is unrelated to basic intellectual capacity, but it causes a mysterious difficulty in handling words and symbols. Apparently some peculiarity in the brain's organizational pattern distorts the ability to read, write, and count. Letters in words seem transposed, reversed, or even topsy-turvy—*dog* becomes *god; b* changes identity with *d;* a sign saying "OIL" inverts into "710." Many dyslexics cannot orient themselves in the three dimensions of space and may show bodily awkwardness.

The cause of dyslexia is unknown, since it is unaccompanied by outward scars of detectable neurological damage and its symptoms vary from victim to victim. It occurs three times as often among boys as among girls. It has been variously attributed to defective vision, brain damage, lead in the air, physical trauma, or oxygen deprivation during birth, but remains an unsolved problem.

TAY-SACHS DISEASE

Tay-Sachs disease is a central nervous system affliction that brings death before age five. The Tay-Sachs gene is carried by normal-appearing individuals descended from the Ashkenazi Jews of Eastern Europe. Approximately one in 3600 of their offspring will be afflicted with Tay-Sachs disease. The disease is caused by the neuronal degeneration of the central nervous system because of excessive amounts of the sphingolipid known as ganglioside G_{m2} in the nerve cells of the brain. The afflicted child will develop normally until the age of four to eight months. Then symptoms follow a course of progressive degeneration: paralysis, blindness, inability to eat, decubitus ulcers, and death from infection. There is no known cure.

KEY MEDICAL TERMS ASSOCIATED WITH THE CENTRAL NERVOUS SYSTEM

Analgesia (*an* = without; *algia* = painful condition) Insensibility to pain.

Anesthesia (*esthesia* = feeling) Loss of feeling.

Aphasia (*a* = without; *phasis* = speech; *ia* = condition) Diminished or complete loss of ability to comprehend and/or express spoken or written words, due to injury or disease of the brain centers; most common cause is CVA.

Bacterial meningitis Acute inflammation of the meninges caused by bacteria.

Bradykinesia (*brady* = slow) Abnormal slowness of movement.

Coma Abnormally deep unconsciousness with an absence of voluntary response to stimuli; varying degrees of reflex activity remain. May be due to illness or to an injury.

Dyslexia Imparied ability to comprehend written language.

Epidural External to the dura mater.

Idiopathic Self-originated; occurring without known cause or due to some other condition already present.

Lethargy A condition of functional torpor or sluggishness.

Neuralgia (*neur* = nerve) Attacks of pain along the entire course or branch of a peripheral sensory nerve; one common type involves one or more branches of the trigeminal nerve and is called trigeminal neuralgia (tic douloureux).

Neuritis Inflammation of a nerve; can result from irritation to the nerve produced by trauma, bone fractures, nutritional deficiency (usually thiamine), poisons such as carbon monoxide and carbon tetrachloride, heavy metals such as lead, and some drugs. Neuritis of the facial nerve that results in paralysis of facial muscles is called Bell's palsy.

Paralysis Diminished or total loss of motor function resulting from damage to nervous tissue or a muscle.

Sciatica Severe pain along the sciatic nerve and its branches. Usually due to rupture of an intervertebral disc or to osteoarthritis of the lower spinal column: the disc or arthritic joint puts pressure on the nerve root supplying the sciatic nerve and thereby causes the pain.

Shingles Acute inflammation caused by a virus that attacks sensory cell bodies of dorsal root ganglia. Inflammation spreads peripherally along a spinal nerve and infiltrates dermis and epidermis over the nerve, producing a characteristic line of skin blisters.

Spastic (*spas* = draw or pull) Resembling spasms or convulsions.

Spina bifida (*bifid* = into two parts) An abnormality in one or many vertebral arches. The arches fail to fuse during embryonic development so that part of the spinal cord may be exposed.

Stupor Condition of unconsciousness, torpor, or lethargy with suppression of sense or feeling.

Torpor Abnormal inactivity or no response to normal stimuli.

Viral encephalitis An acute inflammation of the brain caused by a direct attack by various viruses or by an allergic reaction to any of the many viruses that are normally harmless to the central nervous system. If the virus affects the spinal cord as well, it is called *encephalomyelitis*.

STUDY OUTLINE

Brain

1. The brain consists of four principal parts: brain stem (medulla, pons, midbrain), diencephalon (thalamus and hypothalamus), cerebellum, and cerebrum.
2. The brain is protected by the cranial meninges and cerebrospinal fluid.
3. Cerebrospinal fluid is formed in the choroid plexuses and circulates through the subarachnoid space, ventricles, and central canal. Most of the fluid is absorbed by the arachnoid villi of the superior sagittal sinus.
4. The blood supply to the brain is via the circle of Willis. The blood continuously delivers oxygen and glucose to brain cells. The differential rates of passage of certain materials from the blood into the brain is based upon a blood-brain barrier.

Brain Stem

1. The medulla oblongata is continuous with the upper part of the spinal cord. Within it, the decussation of pyramids occurs. The medulla contains nuclei that are reflex cen-

ters for heartbeat, respiration, and vasoconstriction. The medulla also contains the nuclei of origin of cranial nerves VIII (cochlear and vertibular branches) to XII.

2. The pons varolii is superior to the medulla. It serves as a bridge between the spinal cord and other part of the brain and contains the nuclei of origin of cranial nerves V to VII and the vestibular branch of VIII.

3. The midbrain connects the pons and diencephalon. It conveys motor impulses from the cerebrum to the cerebellum and spinal cord and conveys sensory impulses from the spinal cord to the thalamus. It also coordinates auditory and visual stimuli and contains the nuclei of origin of cranial nerves III and IV.

Diencephalon

1. Consists of the thalamus and hypothalamus.

2. The thalamus is superior to the midbrain and contains nuclei that serve as relay stations for all sensory impulses, except smell, to the cerebral cortex. The thalamus also registers conscious recognition of pain and temperature and some awareness of crude touch and pressure.

3. The hypothalamus is inferior to the thalamus. It controls the autonomic nervous systems, connects the nervous and endocrine system, controls body temperature, food and fluid intake, the waking state, and sleep.

Cerebrum

1. The cerebrum is the largest part of the brain. Its cortex contains convolutions, fissures, and sulci.

2. The cerebral lobes are named the frontal, parietal, temporal, and occipital.

3. The white matter is under the cortex and consists of myelinated axons running in three principal directions.

4. The basal ganglia are paired masses of gray matter in the cerebral hemispheres. They help to control muscular movements.

5. The limbic system is found in the cerebral hemispheres and diencephalon. It functions in emotional aspects of behavior and memory.

6. The motor areas of the cerebral cortex are the regions that govern muscular movement. The sensory areas are concerned with the interpretation of sensory impulses. The association areas are concerned with emotional and intellectual processes.

7. Brain waves generated by the cerebral cortex are re-

corded by an EEG. They may be used to diagnose epilepsy, infections, and tumors.

Cerebellum

1. The cerebellum occupies the inferior and posterior aspects of the cranial cavity. It consists of two hemispheres and a central constricted vermis.

2. It is attached to the brain stem by three pairs of cerebellar peduncles.

3. The cerebellum functions in the coordination of skeletal muscles, the maintenance of posture, and keeping the body balanced.

Cranial Nerves

1. The 12 pairs of cranial nerves leave the skull through foramina in the base of the skull. Some are sensory and some are mixed.

2. They are named primarily on the basis of distribution and are numbered on the basis of order of origin.

Applications to Health

1. Poliomyelitis is a viral infection that results in paralysis.

2. Syphilis is caused by the bacterium *Treponema pallidum* and may result in blindness, memory defects, abnormal behavior, and loss of sensory functions in trunk and limbs.

3. Cerebral palsy includes a group of central nervous system disorders that primarily involve the cerebral cortex, cerebellum, and basal ganglia. The disorders damage motor centers.

4. Parkinsonism is a malfunction of the basal ganglia of the cerebrum caused by insufficient dopamine.

5. With epilepsy, the victim experiences convulsive seizures. It results from irregular electrical discharges of brain cells and may be diagnosed by an EEG.

6. Multiple sclerosis is the destruction of myelin sheaths of the neurons of the central nervous system. Impulse transmission is interrupted.

7. Cerebrovascular accidents are also called strokes. Brain tissue is destroyed due to hemorrhage, thrombosis, and arteriosclerosis.

8. Dyslexia involves an inability of an individual to comprehend written language.

9. Tay-Sachs disease is an inherited disorder that involves neurological degeneration of the central nervous system because of excessive amounts of ganglioside.

REVIEW QUESTIONS

1. Identify the four principal parts of the brain and the components of each, where applicable.

2. Describe the location of the cranial meninges.

3. Where is cerebrospinal fluid formed? Describe its circulation. Where is cerebrospinal fluid absorbed?

4. Distinguish between internal and external hydrocephalus.

5. Describe the blood supply to the brain. Explain the importance of oxygen and glucose to brain cells.

6. Explain what is meant by the blood-brain barrier? Is it of any advantage?

7. Describe the location and structure of the medulla. Define decussation of pyramids. Why is it important?

8. List the principal functions of the medulla.

9. Describe the location and structure of the pons. What are its functions?

10. Describe the location and structure of the midbrain. What are some of its functions?

11. Describe the location and structure of the thalamus. List some of the functions of the thalamus.

12. Where is the hypothalamus located? Explain some of the major functions of the hypothalamus.

13. Where is the cerebrum located? Describe the cortex, convolutions, fissures, and sulci of the cerebrum.
14. List and locate the lobes of the cerebrum. How are they separated from each other? What is the insula?
15. Describe the organization of cerebral white matter. Be sure to indicate the function of each group of fibers.
16. What are basal ganglia? Name the important basal ganglia and list the function of each.
17. Describe the effects of damage on the basal ganglia.
18. Define the limbic system. Explain several of its functions.
19. What is meant by a sensory area of the cerebral cortex? List, locate, and give the function of each sensory area.
20. What is meant by a motor area of the cerebral cortex? List, locate, and give the function of each motor area.
21. What is an association area of the cerebral cortex? What are its functions?
22. Define an electroencephalogram. List the types of waves recorded on an EEG and explain the importance of each. What is the diagnostic value of an EEG?
23. Describe the location of the cerebellum. List the principal parts of the cerebellum.
24. Describe the relationship of the dural extensions to the cerebellum.
25. What are cerebellar peduncles? List and explain the function of each.
26. Explain the functions of the cerebellum. What is ataxia?
27. Define a cranial nerve. How are cranial nerves named and numbered? Distinguish between a mixed and a sensory cranial nerve.
28. For each of the 12 pairs of cranial nerves, list (a) their name, number, and type; (b) their location; and (c) their function.
29. Define each of the following: poliomyelitis, syphilis, Parkinsonism, epilepsy, multiple sclerosis, cerebrovascular accident, dyslexia, and Tay-Sachs disease.

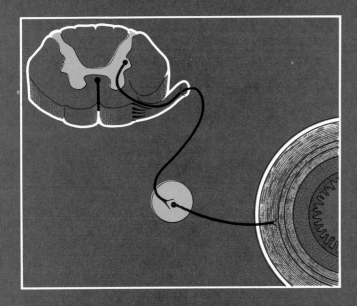

15

The Autonomic Nervous System

STUDENT OBJECTIVES

- Compare the structural and functional differences between the somatic efferent and autonomic portions of the nervous system.

- Identify the structural features of the autonomic nervous system.

- Compare the sympathetic and parasympathetic divisions of the autonomic nervous system in terms of structure, physiology, and chemical transmitters released.

- Describe a visceral autonomic reflex and its components.

- Explain the role of the hypothalamus and its relationship to the sympathetic and parasympathetic division.

- Explain the relationship between biofeedback and the autonomic nervous system.

- Describe the relationship between meditation and the autonomic nervous system.

The portion of the nervous system that regulates the activities of smooth muscle, cardiac muscle, and glands is the **autonomic nervous system.** Structurally, the system consists of visceral efferent neurons organized into nerves, ganglia, and plexuses. Functionally, it usually operates without conscious control. Physiologists originally thought the system functioned autonomously with no control from the central nervous system—hence its name, the autonomic nervous system. In truth, the autonomic system is neither structurally nor functionally independent of the central nervous system. It is regulated by centers in the brain, in particular by the cerebral cortex, hypothalamus, and medulla oblongata. However, the autonomic nervous system does differ from the somatic efferent in some ways. For convenience of study the two are separated.

Somatic Efferent and Autonomic Nervous Systems

Whereas the somatic efferent nervous system produces conscious movement in skeletal muscles, the autonomic nervous system (visceral efferent nervous system) regulates visceral activities. And it generally does so involuntarily and automatically. Examples of visceral activities regulated by the autonomic nervous system are changes in the size of the pupil, accommodation for near vision, dilatation and constriction of blood vessels, adjustment of the rate and force of the heartbeat, movements of the gastrointestinal tract, formation of gooseflesh, and secretion by most glands. These activities usually lie beyond conscious control. They are automatic.

The autonomic nervous system is entirely motor. All its axons are efferent fibers, which transmit impulses from the central nervous system to visceral effectors. Autonomic fibers are called **visceral efferent fibers. Visceral effectors** include cardiac muscle, smooth muscle, and glandular epithelium. This does not mean there are no afferent (sensory) impulses from visceral effectors, however. Impulses that give rise to visceral sensations pass over visceral afferent neurons that have cell bodies located outside but close to the central nervous system. Some functions of these afferent neurons were described with the cranial and spinal nerves. However, the hypothalamus, which largely controls the autonomic nervous system, also receives impulses from the visceral sensory fibers.

The autonomic nervous system consists of two principal divisions: the **sympathetic** and the **parasympathetic.** Many organs innervated by the autonomic nervous system receive visceral efferent neurons from both components of the autonomic system—one set

from the sympathetic division, another from the parasympathetic division. In general, impulses transmitted by the fibers of one division stimulate the organ to start or increase activity, whereas impulses from the other division decrease or halt the organ's activity. Organs that receive impulses from both sympathetic and parasympathetic fibers have *dual innervation.* In the somatic efferent nervous system, only one kind of motor neuron innervates an organ, which is always a skeletal muscle. When the somatic neurons stimulate the cells of the skeletal muscle, the muscle becomes active. When the neuron ceases to stimulate the muscle, contraction stops altogether. Skeletal muscle cells of each motor unit contract only when stimulated by their motor neuron. When the impulse stops, contraction stops.

Structure of the Autonomic Nervous System

The sympathetic and parasympathetic divisions of the autonomic nervous system are also referred to as the thoracolumbar and craniosacral divisions, respectively. Let us see what this means by discussing the general features applicable to both divisions.

VISCERAL EFFERENT NEURONS

Autonomic visceral efferent pathways always consist of two neurons. One runs from the central nervous system to a ganglion. The other runs directly from the ganglion to the effector.

The first of these visceral efferent neurons in an autonomic pathway is called a **preganglionic neuron** (Figure 15-1). Preganglionic neurons have their cell bodies in the brain or spinal cord. Their myelinated axons, called **preganglionic fibers,** pass out of the central nervous system as part of a cranial or spinal nerve. At some point, they leave these nerves and course to autonomic ganglia where they synapse with the dendrites or cell bodies of postganglionic neurons.

Postganglionic neurons, the second visceral efferent neurons in an autonomic pathway, lie entirely outside the central nervous system. Their cell bodies and dendrites (if they have dendrites) are located in the autonomic ganglia, where the synapse with the preganglionic fibers occurs. The axons of postganglionic neurons are called **postganglionic fibers.** Postganglionic fibers are nonmyelinated, and they terminate in visceral effectors.

Thus preganglionic neurons convey efferent impulses from the central nervous system to autonomic ganglia. Postganglionic neurons relay the impulses from the autonomic ganglia to visceral effectors.

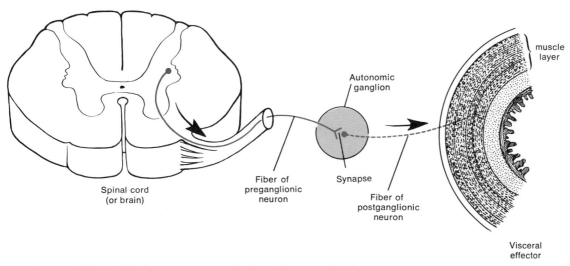

FIGURE 15-1 Relationship between preganglionic and postganglionic neurons.

PREGANGLIONIC NEURONS

In the sympathetic division, the preganglionic neurons have their cell bodies in the lateral gray horns of the thoracic segments and first two lumbar segments of the spinal cord (Figure 15-2). It is for this reason that the sympathetic division is also called the **thoracolumbar division** and the fibers of the sympathetic preganglionic neurons are known as the **thoracolumbar outflow.**

The cell bodies of the preganglionic neurons of the parasympathetic division are located in nuclei in the brain stem and in the lateral gray horns of the second through fourth sacral segments of the spinal cord (Figure 15-2)—hence the synonymous term **craniosacral division.** The fibers of the parasympathetic preganglionic neurons are referred to as the **craniosacral outflow.**

AUTONOMIC GANGLIA

Autonomic pathways also include **autonomic ganglia,** where synapses between visceral efferent neurons occur. Autonomic ganglia differ from posterior root ganglia. The latter contain cell bodies of sensory neurons and no synapses occur in them. The autonomic ganglia may be divided into three general groups (Figure 15-3). The *sympathetic trunk* or *vertebral chain ganglia* are a series of ganglia that lie in a vertical row on either side of the vertebral column, extending from the base of the skull to the coccyx. They are also known as *paravertebral* or *lateral ganglia.* They receive preganglionic fibers only from the thoracolumbar (sympathetic) division (Figure 15-2).

A second kind of ganglion of the sympathetic division of the autonomic nervous system is called a *prevertebral* or *collateral ganglion* (Figure 15-3). The

ganglia of this group lie anterior to the spinal column and close to the large abdominal arteries from which their names are derived. Examples of prevertebral ganglia so named are: the celiac ganglion, on either side of the celiac artery just below the diaphragm; the superior mesenteric ganglion, near the beginning of the superior mesenteric artery in the upper abdomen; and the inferior mesenteric ganglion, located near the beginning of the inferior mesenteric artery in the middle of the abdomen (Figure 15-2). Prevertebral ganglia receive preganglionic fibers from the thoracolumbar (sympathetic) division.

The third kind of autonomic ganglion belongs to the parasympathetic division and is called a *terminal* or *intramural ganglion.* The ganglia of this group are located at the end of a visceral efferent pathway very close to visceral effectors or within the walls of visceral effectors. Terminal ganglia receive preganglionic fibers from the craniosacral (parasympathetic) division. The preganglionic fibers do not pass through sympathetic trunk ganglia (Figure 15-2).

In addition to autonomic ganglia, the autonomic nervous system also contains **autonomic plexuses.** Slender nerve fibers from ganglia containing postganglionic nerve cell bodies arranged in a branching network constitute an autonomic plexus.

POSTGANGLIONIC NEURONS

Axons from preganglionic neurons of the sympathetic division pass to ganglia of the sympathetic trunk. They can either synapse in the sympathetic chain ganglia with postganglionic sympathetics or they can continue, without synapsing, through the chain ganglia to end at a prevertebral ganglion where synapses with the

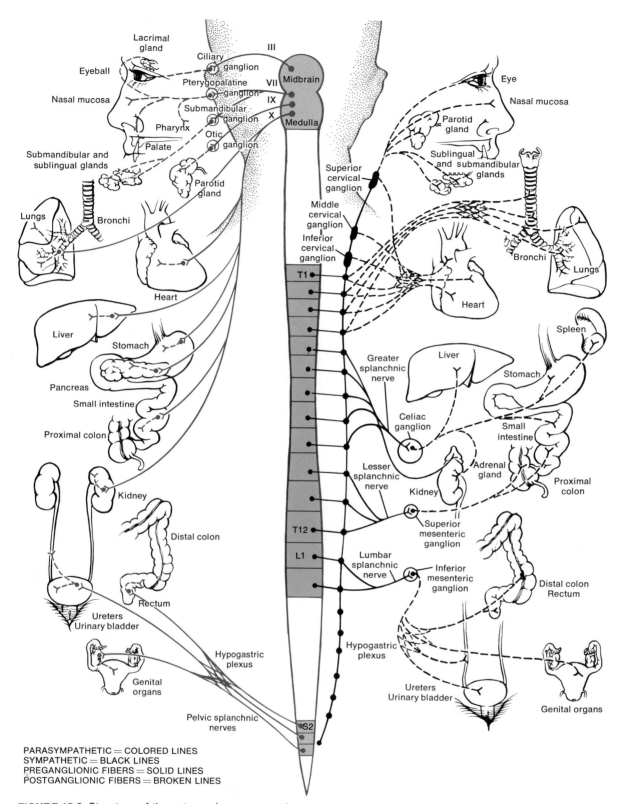

PARASYMPATHETIC = COLORED LINES
SYMPATHETIC = BLACK LINES
PREGANGLIONIC FIBERS = SOLID LINES
POSTGANGLIONIC FIBERS = BROKEN LINES

FIGURE 15-2 Structure of the autonomic nervous system.

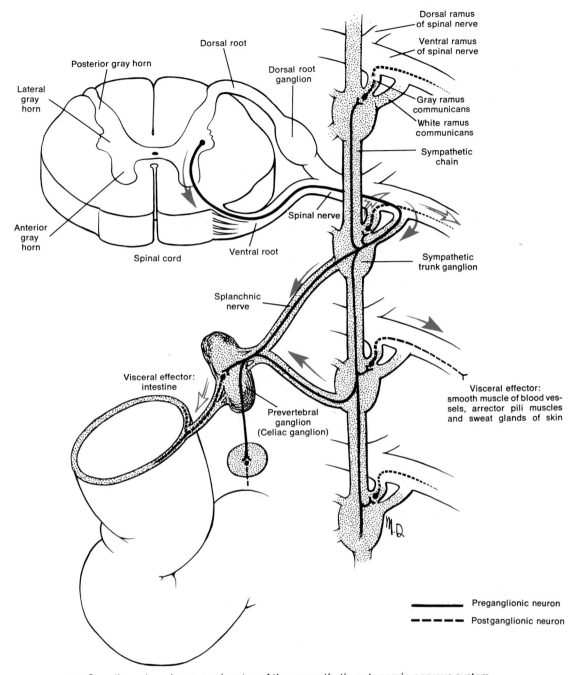

FIGURE 15-3 Ganglia and rami communicantes of the sympathetic autonomic nervous system.

postganglionic sympathetics can take place. Each sympathetic preganglionic fiber synapses with several postganglionic fibers in the ganglion, and the postganglionic fibers pass to several visceral effectors. Upon exiting their ganglia, the postsynaptic fibers innervate their visceral effectors.

Axons from preganglionic neurons of the parasympathetic division pass to terminal ganglia near or within a visceral effector. In the ganglion, the presynaptic neuron usually synapses with only four or five postsynaptic neurons to a single visceral effector.

Upon exiting their ganglia, the postsynaptic fibers supply their visceral effectors.

With this background in mind, we can now examine some specific structural features of the sympathetic and parasympathetic divisions of the autonomic nervous system.

SYMPATHETIC DIVISION

The preganglionic fibers of the sympathetic division have their cell bodies located in the lateral gray horn of the spinal cord in the thoracic and first two lumbar segments (Figure 15-2). The preganglionic fibers are myelinated and leave the spinal cord through the ventral root of a spinal nerve along with the somatic efferent fibers of the same segmental levels. After exiting through the intervertebral foramina, the preganglionic sympathetic fibers enter a white ramus to pass to the nearest sympathetic trunk ganglion on the same side. Collectively, the white rami are called the **white rami communicantes.** Their name indicates that they contain myelinated fibers. Only thoracic and upper lumbar nerves have white rami communicantes. The white rami communicantes connect the ventral ramus of the spinal nerve with the ganglia of the sympathetic trunk.

When a preganglionic fiber of a white ramus communicans enters the sympathetic trunk, it may terminate (synapse) in several ways. Some fibers synapse in the first ganglion at the level of entry. Others pass up or down the sympathetic trunk for a variable distance to form the fibers on which the ganglia are strung. These fibers, known as *sympathetic chains* (Figure 15-3), may not synapse until they reach a ganglion in the cervical or sacral area. Some postganglionic fibers leaving the sympathetic trunk ganglia pass directly to visceral effectors of the head, neck, chest, and abdomen. Most, however, rejoin the spinal nerves before supplying peripheral visceral effectors such as sweat glands and the smooth muscle in blood vessels and around hair follicles. The **gray ramus communicans** is the structure containing the postganglionic fibers that run from the ganglion of the sympathetic trunk to the spinal nerve (Figure 15-3). The term *gray* refers to the fact that the fiber is unmyelinated. All spinal nerves have gray rami communicantes. Gray rami communicantes outnumber the white rami since there is a gray ramus leading to each of the 31 pairs of spinal nerves.

In most cases, a sympathetic preganglionic fiber terminates by synapsing with a large number of postganglionic cell bodies in a ganglion, usually 20 or more. Often the postganglionic fibers then terminate in widely separated organs of the body. Thus an impulse that starts in a single preganglionic neuron may affect several visceral effectors. For this reason, most sympathetic responses have widespread effects on the body.

The sympathetic trunks are two in number, situated anterolaterally to the spinal cord, one on either side. Each consists of a series of ganglia arranged more or less segmentally. The divisions of the sympathetic trunk are named on the basis of location. Typically, there are 22 ganglia in each chain: 3 cervical, 11 thoracic, 4 lumbar, and 4 sacral. Although the trunk extends downward from the neck, thorax, and abdomen to the coccyx, it receives preganglionic fibers only from the thoracic and lumbar segments of the spinal cord (Figure 15-2).

The cervical portion of each sympathetic trunk is located in the neck anterior to the prevertebral muscles. It is subdivided into a superior, middle, and inferior ganglion (Figure 15-2). The *superior cervical ganglion* is behind the internal carotid artery on a level with the second or third cervical vertebra. Postganglionic fibers leaving the ganglion serve the head where they are distributed to the sweat glands, smooth muscle of the eye, the smooth muscle of the blood vessels of the face, nasal mucosa, and the submandibular, sublingual, and parotid salivary glands. The *middle cervical ganglion* is situated at the level of the sixth cervical vertebra. Postganglionic fibers from it innervate the heart. The *inferior cervical ganglion* is located near the first rib. Its postganglionic fibers also supply the heart.

The thoracic portion of each sympathetic trunk usually consists of 11 segmentally arranged ganglia, lying ventral to the necks of the corresponding ribs. This portion of the sympathetic trunk receives most of the sympathetic preganglionic fibers. Postganglionic fibers from the thoracic sympathetic trunk innervate heart, lungs, bronchi, and other thoracic viscera.

The lumbar portion of each sympathetic trunk is found on either side of the corresponding lumbar vertebrae. The sacral portion of the sympathetic trunk lies in the pelvic cavity on the medial side of the sacral foramina. Postganglionic fibers from the lumbar and sacral sympathetic chain ganglia are distributed with the respective spinal nerves via gray rami communicantes or they may join the hypogastric plexus via direct visceral branches.

Some preganglionic fibers pass through the sympathetic trunk without terminating in the trunk. Beyond the trunk, they form nerves known as *splanchnic nerves* (Figure 15-2). After passing through the trunk of ganglia, the splanchnic nerves terminate in the *celiac (solar) plexus.* In the plexus, the preganglionic fibers synapse in ganglia with postganglionic cell bodies. These ganglia are prevertebral ganglia. The greater splanchnic nerve passes to the celiac ganglion of the celiac plexus. From here, postganglionic fibers are distributed to the stomach, spleen, liver, kidney, and small intestine. The lesser splanchnic nerve passes through the celiac plexus to the superior mesenteric ganglion of the superior mesenteric plexus. Postganglionic fibers from this ganglion innervate the small intestine and colon. The lowest splanchnic nerve, not always present, enters the renal plexus. Postganglionics supply the renal artery and ureter. The lumbar splanchnic nerve enters the inferior mesenteric plexus. In the plexus, the preganglionic fibers synapse with postganglionic fibers in the inferior mesenteric ganglion. These fibers pass through the hypogastric plexus and supply the distal colon and rectum, urinary bladder, and genital organs. As noted earlier, the postganglionic fibers leaving the prevertebral ganglia follow the course of various arteries to abdominal and pelvic visceral effectors.

PARASYMPATHETIC DIVISION

The preganglionic cell bodies of the parasympathetic division are found in nuclei in the brain stem and the lateral gray horn of the second through fourth sacral segments of the spinal cord (Figure 15-2). Their fibers emerge as part of a cranial nerve or as part of the ventral root of a spinal nerve. The **cranial parasympathetic outflow** consists of preganglionic fibers that leave the brain stem by way of the oculomotor nerves (III), facial nerves (VII), glossopharyngeal nerves (IX), and vagus nerves (X). The **sacral parasympathetic outflow** consists of preganglionic fibers that leave the ventral roots of the second through fourth sacral nerves. The preganglionic fibers of both the cranial and sacral outflows end in terminal ganglia where they synapse with postganglionic neurons. We will first look at the cranial outflow.

Four pairs of cranial parasympathetic ganglia innervate structures in the head and are located close to the organs they innervate. The *ciliary ganglion* is near the back of an orbit lateral to each optic nerve. Preganglionic fibers pass with the oculomotor nerve (III) to the ciliary ganglion. Postganglionic fibers from the ganglion innervate smooth muscle cells in the eyeball. Each *pterygopalatine ganglion* is situated lateral to a sphenopalatine foramen. It receives preganglionic fibers from the facial nerve (VII) and transmits postganglionic fibers to the nasal mucosa, palate, pharynx, and lacrimal gland. Each *submandibular ganglion* is found near the duct of a submandibular salivary gland. It receives preganglionic fibers from the facial nerve (VII) and transmits postganglionic fibers that innervate the submandibular and sublingual salivary glands. The *otic ganglia* are situated just below each foramen ovale. The otic ganglion receives preganglionic fibers from the glossopharyngeal nerve (IX) and transmits postganglionic fibers that innervate the parotid salivary gland. Ganglia associated with the cranial outflow are classified as terminal ganglia. Since the terminal ganglia are close to their visceral effectors, postganglionic parasympathetic fibers are short. Postganglionic sympathetic fibers are relatively long.

The last component of the cranial outflow, the preganglionic fibers that leave the brain via the vagus nerves (X), has the most extensive distribution of the parasympathetic fibers. Each vagus nerve enters into the formation of several plexuses in the thorax and abdomen. As it passes through the thorax, it sends fibers to the *superficial cardiac plexus* in the arch of the aorta and the *deep cardiac plexus* anterior to the branching of the trachea. These plexuses contain ter-

EXHIBIT 15-1
STRUCTURAL FEATURES OF SYMPATHETIC AND PARASYMPATHETIC DIVISIONS

SYMPATHETIC	PARASYMPATHETIC
Forms thoracolumbar outflow	Forms craniosacral outflow
Contains sympathetic trunk and prevertebral ganglia	Contains terminal ganglia
Ganglia are close to the CNS and distant from visceral effectors.	Ganglia are near or within visceral effectors.
Each preganglionic fiber synapses with many postganglionic neurons that pass to many visceral effectors.	Each preganglionic fiber usually synapses with four or five postganglionic neurons that pass to a single visceral effector.
Distributed throughout the body, including the skin.	Distribution limited primarily to head and viscera of thorax, abdomen, and pelvis.

minal ganglia, and the postganglionic parasympathetic fibers emerging from them supply the heart. Also in the thorax is the *pulmonary plexus,* in front and behind the roots of the lungs and within the lungs themselves. It receives preganglionic fibers from the vagus and transmits postganglionic parasympathetic fibers to the lungs and bronchi. Other plexuses associated with the vagus nerve are described in later chapters in conjunction with the appropriate thoracic, abdominal, and pelvic viscera. Postganglionic fibers from these plexuses innervate viscera such as the liver, pancreas, stomach, kidney, small intestine, and part of the colon.

The sacral parasympathetic outflow consists of preganglionic fibers from the ventral roots of the second through fourth sacral nerves. Collectively, they form the *pelvic splanchnic nerves.* They pass into the hypogastric plexus. From ganglia in the plexus, parasympathetic postganglionic fibers are distributed to the colon, ureters, urinary bladder, and reproductive organs.

The salient structural features of the sympathetic and parasympathetic divisions are compared in Exhibit 15-1.

Physiology

CHEMICAL TRANSMITTERS

Autonomic fibers, like other axons of the nervous system, release chemical transmitters at synapses as well as at points of contact between autonomic fibers and visceral effectors. These latter points are called **neuroeffector junctions.** On the basis of the chemical transmitter produced, autonomic fibers may be classified as either cholinergic or adrenergic. **Cholinergic fibers** release *acetylcholine* and include the following: (1) all sympathetic and parasympathetic preganglionic axons, (2) all parasympathetic postganglionic axons, and (3) some sympathetic postganglionic axons. The cholinergic sympathetic postganglionic axons include those to sweat glands and those to blood vessels in skeletal muscles and the external genitalia. Since acetylcholine is quickly inactivated by the enzyme cholinesterase, the effects of cholinergic fibers are short-lived and local. **Adrenergic fibers** produce the chemical transmitter *norepinephrine* or *noradrenalin*. Most sympathetic postganglionic axons are adrenergic. Since norepinephrine is inactivated much more slowly than acetylcholine and norepinephrine may enter the bloodstream, the effects of sympathetic stimulation are longer lasting and more widespread than parasympathetic stimulation.

ACTIVITIES

Most visceral effectors have dual innervation. That is, they receive fibers from both the sympathetic and the parasympathetic divisions. In these cases, impulses from one division stimulate the organ's activities, whereas impulses from the other division inhibit the organ's activities. The stimulating division may be either the sympathetic or the parasympathetic, depending on the organ. For example, sympathetic impulses increase heart activity whereas parasympathetic impulses decrease heart activity. On the other hand, parasympathetic impulses increase digestive activities whereas sympathetic impulses inhibit them. A summary of the activities of the autonomic system is presented in Exhibit 15-2.

The parasympathetic division is primarily concerned with activities that restore and conserve body energy. It is a rest-repose system. Under normal body conditions, for instance, parasympathetic impulses to the digestive glands and the smooth muscle of the digestive system dominate over sympathetic impulses. Thus energy-supplying foods can be digested and absorbed by the body.

The sympathetic division, by contrast, is primarily concerned with processes involving the expenditure of energy. When the body is in homeostasis, the main function of the sympathetic division is to counteract the parasympathetic effects just enough to carry out normal processes requiring energy. During extreme stress, however, the sympathetic dominates the parasympathetic. When people are confronted with a dangerous situation, for example, their bodies become alert and they sometimes perform feats of unusual strength. Fear stimulates the sympathetic division. Activation of the sympathetic division sets into oper-

ation a series of physiological responses collectively called the *fight-or-flight response*. It produces the following effects: (1) The pupils of the eyes dilate. (2) The heart rate increases. (3) The blood vessels of the skin and viscera constrict. (4) The remainder of the blood vessels dilate. This reaction causes a rise in blood pressure and a faster flow of blood into the dilated blood vessels of skeletal muscles, cardiac muscle, lungs, and brain—organs involved in fighting off danger. Rapid breathing occurs as the bronchioles dilate to allow faster movement of air in and out of the lungs. Blood sugar level rises as liver glycogen is converted to glucose to supply the body's additional energy needs. The sympathetic division also stimulates the medulla of the adrenal gland to produce epinephrine and norepinephrine, hormones that intensify and prolong the sympathetic effects noted above. During this period of stress, the sympathetic effects inhibit other processes that are not essential for meeting the situation. Muscular movements of the gastrointestinal tract and digestive secretions are slowed down or even stopped.

Visceral Autonomic Reflexes

A **visceral autonomic reflex** adjusts the activity of a visceral effector. In other words, it results in the contraction of smooth or cardiac muscle or secretion by a gland. Such reflexes assume a key role in activities such as regulating heart action, blood pressure, respiration, digestion, defecation, and urinary bladder functions.

A visceral autonomic reflex arc consists of the following components:

1. Receptor. The receptor is the distal end of an afferent neuron in an exteroceptor or enteroceptor.

2. Afferent neuron. This neuron, either a somatic afferent or visceral afferent neuron, conducts the sensory impulse to the spinal cord or brain.

3. Association neurons. These neurons are found in the central nervous system.

4. Visceral efferent preganglionic neuron. In the thoracic and abdominal regions, this neuron is in the lateral gray horn of the spinal cord. The axon passes through the ventral root of the spinal nerve, the spinal nerve, and the white ramus communicans. It then enters a sympathetic trunk or prevertebral ganglion, where it synapses with a postganglionic neuron. In the cranial and sacral regions, the visceral efferent preganglionic axon leaves the central nervous system and passes to a terminal ganglion, where it synapses with a postganglionic neuron. The role of the visceral efferent preganglionic neuron is to convey a motor impulse from the brain or spinal cord to an autonomic ganglion.

EXHIBIT 15-2
ACTIVITIES OF AUTONOMIC NERVOUS SYSTEM

VISCERAL EFFECTOR	EFFECT OF SYMPATHETIC STIMULATION	EFFECT OF PARASYMPATHETIC STIMULATION
Eye		
Iris	Contracts dilator muscle of iris and brings about dilatation of pupil	Contracts sphincter muscle of iris and brings about constriction of pupil
Ciliary muscle	No innervation	Contracts ciliary muscle and accommodates lens for near vision
Glands		
Sweat	Stimulates secretion	No innervation
Lacrimal (tear)	No innervation	Normal or excessive secretion
Salivary	Vasoconstriction which decreases salivary secretion.	Stimulation of salivary secretion and vasodilation
Gastric	No known effect	Secretion stimulated
Intestinal	No known effect	Secretion stimulated
Adrenal medulla	Promotes epinephrine and norepinephrine secretion	No innervation
Lungs (bronchial tubes)	Dilatation	Constriction
Heart	Increases rate and strength of contraction; dilates coronary vessels that supply blood to heart muscle cells	Decreases rate and strength of contraction; constricts coronary vessels
Blood vessels		
Skin	Constriction	No innervation for most
Skeletal muscle	Dilatation	No innervation
Visceral organs (except heart and lungs)	Constriction	No innervation for most
Liver	Promotes glycogenolysis; decreases bile secretion	Promotes glycogenesis; increases bile secretion
Stomach	Decreases motility	Increases motility
Intestines	Decreases motility	Increases motility
Kidney	Constriction of blood vessels that results in decreased urine volume	No effect
Pancreas	Inhibits secretion	Promotes secretion
Spleen	Contraction and discharge of stored blood into general circulation	No innervation
Urinary bladder	Relaxes muscular wall	Contracts muscular wall
Arrector pili of hair follicles	Contraction results in erection of hairs ("goose pimples")	No innervation
Uterus	Inhibits contraction if nonpregnant; stimulates contraction if pregnant	Minimal effect
Sex organs	Vasoconstriction of ductus deferens, seminal vesicle, prostate; results in ejaculation	Vasodilation and erection

5. Visceral efferent postganglionic neuron. This neuron conducts a motor impulse from a visceral efferent preganglionic neuron to the visceral effector.

6. Visceral effector. A visceral effector is smooth muscle, cardiac muscle, or a gland.

The basic difference between a somatic reflex arc and a visceral autonomic reflex arc is that in a somatic reflex, only one efferent neuron is involved. In a visceral autonomic reflex arc, two efferent neurons are involved.

Visceral sensations do not always reach the cerebral cortex. Therefore they remain at subconscious levels. Under normal conditions, you are not aware of muscular contractions of the digestive organs, heartbeat, changes in the diameter of blood vessels, and pupil dilation and constriction. When your body is making adjustments in such visceral activities, they are handled by visceral reflex arcs whose centers are in the spinal cord or lower regions of the brain. Among such centers are the cardiac, respiratory, vasomotor, swallowing, and vomiting centers in the medulla and the temperature control center in the hypothalamus. Thus stimuli delivered by somatic or visceral afferent neurons synapse in these centers, and the returning motor impulses conducted by visceral efferent neurons bring about an adjustment in the visceral effector without conscious recognition. The impulses are interpreted and acted on subconsciously. Some visceral sensations do give rise to conscious recognition: hunger, nausea, and fullness of the urinary bladder and rectum.

Control by Higher Centers

The autonomic nervous system is not a separate nervous system. Axons from many parts of the central nervous system are connected to both the sympathetic and the parasympathetic divisions of the autonomic nervous system and thus exert considerable control over it. Autonomic centers in the cerebral cortex are connected to autonomic centers of the thalamus, for example. These, in turn, are connected to the hypothalamus. In this hierarchy of command, the thalamus sorts incoming impulses before they reach the cerebral cortex. The cerebral cortex then turns over control and integration of visceral activities to the hypothalamus. It is at the level of the hypothalamus that the major control and integration of the autonomic nervous system is exerted.

The hypothalamus is connected to both the sympathetic and the parasympathetic divisions of the autonomic nervous system. The posterior and lateral portions of the hypothalamus appear to control the sympathetic division. When these areas are stimulated, there is an increase in visceral activities—an increase in heart rate, a rise in blood pressure due to vasoconstriction of blood vessels, an increase in the rate and depth of respiration, dilatation of the pupils, and inhibition of the digestive tract. On the other hand, the anterior and medial portions of the hypothalamus seem to control the parasympathetic division. Stimulation of these areas results in a decrease in heart rate, lowering of blood pressure, constriction of the pupils, and increased motility of the digestive tract.

Control of the autonomic nervous system by the cerebral cortex occurs primarily during emotional stress. In extreme anxiety, the cerebral cortex can stimulate the hypothalamus. This stimulation, in turn, increases heart rate and blood pressure. You may have experienced this very reaction before taking an examination. If the cortex is stimulated by an extremely unpleasant sight, the stimulation causes vasodilation of blood vessels, a lowering of blood pressure, and fainting.

BIOFEEDBACK

In its simplest terms, **biofeedback** is a process in which people get constant signals, or feedback, on visceral body functions such as blood pressure, heart rate, and muscle tension. By using special monitoring devices, they can control these visceral functions consciously.

Suppose you are connected to a monitor that informs you of your heartbeat by means of lights. A red light indicates a fast heart beat, an amber light a normal rate, and a green light a slow rate. When you see the red light flash, you know your heart is beating too fast. You have been informed of a visceral response. This is the biofeedback. According to some researchers, you can be taught to slow down the heart rate by thinking of something pleasant and thus relaxing the body. The green light flashes and your reward is a slower heart rate. In similar experiments, some individuals have learned to control heart rhythm.

Researchers estimate that between 5 and 10 percent of the American population suffers from migraine headaches. Moreover, no effective treatment has been developed that does not have significant side effects and serious risks. One approach to alleviating migraine headaches is biofeedback.* The first clue that biofeedback could be used in this way came when a patient in the voluntary control laboratory of the Menninger Foundation demonstrated that with a 10°F rise in hand skin temperature (as a result of vasodilation and increased blood flow) she could spontaneously recover from a migraine headache.

In a study conducted at the Menninger Foundation, subjects suffering from migraine headaches received instructions in the use of a monitor that registers the skin temperature of the right index finger. Subjects were also given a typewritten sheet containing two sets of phrases. The first set was designed to help them relax the entire body. The second set was designed to bring about an increased flow of blood in the hands. The subjects practiced raising their skin temperature at home for 5 to 15 minutes a day. When skin termperature increased, the monitor emitted a high-pitched sound. In time, the monitor was abandoned.

Once the subjects learned how to vasodilate their blood vessels, the migraine headaches lessened. Since migraine headaches are believed to involve a distension of blood vessels in the head, the shunting of blood from head to hands relieved the distension and thus the pain.

Other experiments have shown that biofeedback can be applied to childbirth. Women were given mon-

*Much of the following discussion of the use of biofeedback for the treatment of migraine headaches is based upon the information provided by Dr. Joseph D. Sargent of the Menninger Foundation, Topeka, Kansas.

itors hooked up to their fingers and arms to measure electrical conductivity of the skin and skeletal muscle tension. Both conductivity and tension increase with nervousness and make labor more difficult. Muscle tension was recorded as a sirenlike sound that became louder with nervousness. Skin conductivity was recorded as a crackling noise that also increased with nervousness. The monitors kept the women informed of their nervousness. This was the biofeedback. Having pleasant thoughts reduced the sound levels. The reward was less nervousness. The results of the study indicate that the women needed less medication during labor and labor time itself was shortened.

There is no way to determine where biofeedback will lead. Perhaps the outstanding contribution of biofeedback research has been to demonstrate that the autonomic nervous system is not autonomous. Visceral responses can be controlled. Strong supporters of biofeedback point out that its possible applications in medicine are legion. They envision the use of biofeedback in lowering blood pressure in patients with hypertension, altering heart rates and rhythms, relieving pain from migraine headaches, making delivery easier, and controlling anxiety related to a host of illnesses that may be linked to stress. One researcher concludes that "an important trend is beginning to take place in the areas of psychosomatic disorders and medicine. This is the increasing involvement of the patient in his own treatment. The traditional doctor-patient relationship is giving way slowly to a shared responsibility."

MEDITATION

Yoga, which literally means union, is defined as a higher consciousness achieved through a fully rested and relaxed body and a fully awake and relaxed mind. One widely practiced technique for achieving higher consciousness is called **transcendental meditation.** One sits in a comfortable position with the eyes closed and concentrates on a suitable sound or thought.

Research indicates that transcendental meditation can alter physiological responses. Oxygen consumption decreases drastically along with carbon dioxide elimination. Subjects have experienced a reduction in metabolic rate and blood pressure. Researchers also observed a decrease in heart rate, an increase in the intensity of alpha brain waves, a sharp decrease in the amount of lactic acid in the blood, and an increase in the skin's electrical resistance. These last four responses are characteristic of a highly relaxed state of mind. Alpha waves are found in the EEGs of almost all individuals awake and in a resting state, but they disappear during sleep.

These responses have been called an **integrated response**—essentially, a hypometabolic state due to inactivation of the sympathetic division of the autonomic nervous system. This state is exactly the opposite of the fight-or-flight response described earlier—which is a hyperactive state of the sympathetic division. The integrated response suggests that the central nervous system does exert some control over the autonomic nervous system.

STUDY OUTLINE

Somatic Efferent and Autonomic Nervous Systems
1. The autonomic nervous system automatically regulates the activities of smooth muscle, cardiac muscle, and glands.
2. It usually operates without conscious control.
3. It is regulated by centers in the brain, in particular by the cerebral cortex, the hypothalamus, and the medulla oblongata.
4. The somatic efferent nervous system produces conscious movement in skeletal muscles. The autonomic nervous system (visceral efferent nervous system) regulates visceral activities.

Structure
1. The autonomic nervous system consists of visceral efferent neurons organized into nerves, ganglia, and plexuses.
2. It is entirely motor. All autonomic axons are efferent fibers.
3. Efferent neurons are preganglionic (with myelinated axons) and postganglionic (with unmyelinated axons).
4. The autonomic system consists of two principal divisions: the sympathetic and the parasympathetic (also called the thoracolumbar and craniosacral divisions).

5. Autonomic ganglia are classified as sympathetic trunk ganglia (on sides of spinal column), prevertebral ganglia (anterior to spinal column), and terminal ganglia (near or inside visceral effectors).
6. Sympathetic responses are widespread and, in general, concerned with energy expenditure. Parasympathetic responses are restricted and are typically concerned with energy restoration and conservation.

Physiology
1. Autonomic fibers release chemical transmitters at synapses. On the basis of the transmitter produced, these fibers may be classified as cholinergic or adrenergic.
2. Cholinergic fibers release acetylcholine. Adrenergic fibers produce norepinephrine (noradrenalin).

Visceral Autonomic Reflexes
1. A visceral autonomic reflex adjusts the activity of a visceral effector.
2. A visceral autonomic reflex arc consists of a receptor, afferent neuron, association neuron, visceral efferent preganglionic neuron, visceral efferent postganglionic neuron, and visceral effector.

Control by Higher Centers

1. The hypothalamus controls and integrates the autonomic nervous system.
2. The hypothalamus is connected to both the sympathetic and the parasympathetic divisions.
3. Biofeedback is a process by which people get constant signals on visceral body functions. It has been used to control heart rate and other functions.
4. Yoga is a higher consciousness achieved through a fully rested and relaxed body and a fully awake and relaxed mind.
5. Transcendental meditation produces the following physiological responses: decreased oxygen consumption and carbon dioxide elimination, reduced metabolic rate, decrease in heart rate, increase in the intensity of alpha brain waves, a sharp decrease in the amount of lactic acid in the blood, and an increase in the skin's electrical resistance.

REVIEW QUESTIONS

1. What are the principal components of the autonomic nervous system? What is its general function? Why is it called involuntary?
2. What is the principal anatomical difference between the voluntary nervous system and the autonomic nervous system?
3. Relate the role of visceral efferent fibers and visceral effectors to the autonomic nervous system.
4. Distinguish the following with respect to location and function: preganglionic neurons and postganglionic neurons.
5. What is an autonomic ganglion? Describe the location and function of the three types of autonomic ganglia. Define white and gray rami communicantes.
6. On what basis are the sympathetic and parasympathetic divisions of the autonomic nervous system differentiated anatomically and functionally?
7. Discuss the distinction between cholinergic and adrenergic fibers of the autonomic nervous system.
8. Give examples of the antagonistic effects of the sympathetic and parasympathetic divisions of the autonomic nervous system.
9. Summarize the principal functional differences between the voluntary nervous system and the autonomic nervous system.
10. Below is a diagram of the human body as it might appear in a fear situation. Specific parts of the body have been labeled. Write the *sympathetic response* for each labeled part.
11. Define a visceral autonomic reflex and give three examples.
12. Describe a complete visceral autonomic reflex in proper sequence.
13. Describe how the hypothalamus controls and integrates the autonomic nervous system.
14. Define biofeedback. Explain how it could be useful.
15. What is transcendental meditation? How is the integrated response related to the autonomic nervous system?

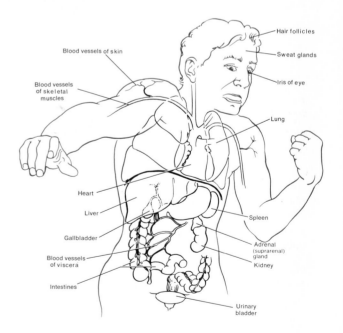

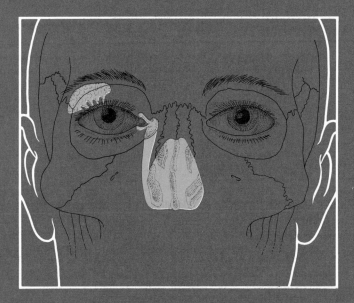

16

Sensory Structures

STUDENT OBJECTIVES

- Define a sensation and list the four prerequisities necessary for its transmission.

- Define projection, adaptation, afterimages, and modality as characteristics of sensations.

- Compare the location and function of exteroceptors, visceroceptors, and proprioceptors.

- Describe the distribution of cutaneous receptors by interpreting the results of the two-point discrimination test.

- List the location and function of the receptors for touch, pressure, cold, heat, pain, and proprioception.

- Distinguish among somatic, visceral, referred, and phantom pain.

- Identify the principal sensory and motor pathways.

- Locate the receptors for olfaction and describe the neural pathway for smell.

- Identify the gustatory receptors and describe the neural pathway for taste.

- Describe the structure and physiology of the accessory visual organs.

- List the structural divisions of the eye.

- Discuss retinal image formation by describing refraction, accommodation, constriction of the pupil, convergence, and inverted image formation.

- Define emmetropia, myopia, hypermetropia, and astigmatism.

- Describe the afferent pathway of light impulses to the brain.

- Define the anatomical subdivisions of the ear.

- List the principal events in the physiology of hearing.

- Identify the receptor organs for equilibrium.

- Discuss the receptor organs' roles in maintaining static and dynamic equilibrium.

- Contrast the causes and symptoms of cataracts, glaucoma, conjunctivitis, trachoma, Ménière's disease, and impacted cerumen.

- Define key medical terms associated with the sense organs.

Sensations

Your ability to sense stimuli is vital to your survival. If pain could not be sensed, burns would be common. An inflamed appendix or stomach ulcer would progress unnoticed. A lack of sight would increase and risk of injury from unseen obstacles, a loss of smell would allow a harmful gas to be inhaled, a loss of hearing would prevent recognition of automobile horns, and a lack of taste would allow toxic substances to be ingested. In short, if you could not "sense" your environment and make the necessary homeostatic adjustments, you would not survive on your own.

DEFINITION

In its broadest context, **sensation** refers to a state of awareness of external or internal conditions of the body. **Perception** refers to the conscious registration of a sensory stimulus, For a sensation to occur, four prerequisites must be fulfilled:

1. A **stimulus,** or change in the environment, capable of initiating a response by the nervous system must be present.

2. A **receptor** or **sense organ** must pick up the stimulus and convert it to a nerve impulse. A sense receptor or sense organ may be viewed as specialized nervous tissue that is extremely sensitive to internal or external conditions.

3. The impulse must be **conducted** along a nervous pathway from the receptor or sense organ to the brain.

4. A region of the brain must **translate** the impulse into a sensation.

Receptors are capable of converting a specific stimulus into a nerve impulse. The stimulus may be light, heat, pressure, mechanical energy, or chemical energy. Each stimulus is capable of causing the membrane of the receptor to depolarize. This depolarization is called a **generator potential (receptor potential).**

The generator potential is a graded response—within limits, the magnitude increases with stimulus strength and frequency. When the generator potential reaches the threshold level, it initiates an action potential (nerve impulse). Once initiated, the action potential is propagated along the nerve fiber. Whereas a generator potential is a local, graded response, an action potential is propagated and obeys the all-or-none principle. According to the all-or-none principle, if a stimulus is strong enough to generate an action potential, the impulse is transmitted along the length of the entire neuron at a constant and maximum strength for the existing conditions. The function of a generator potential is to initiate an action potential by transducing a stimulus into a nerve impulse.

A receptor may be quite simple. It may consist c the dendrites of a single neuron in the skin that ar sensitive to pain stimuli. Or it may be contained in complex organ such as the eye. Regardless of com plexity, all sense receptors contain the dendrites c sensory neurons. The dendrites occur either alone o in close association with specialized cells of other tis sues. Receptors are very excitable and their threshol stimulus is low. Except for pain receptors, each is spe cialized by having a low threshold to a specifi stimulus and a high threshold to all others.

Many sensory impulses are conducted to the sen sory areas of the cerebral cortex. Only in this regio can a stimulus produce conscious sensations. Sensor impulses that terminate in the spinal cord or brain sten can initiate motor activities, but they can never pro duce conscious sensations. Once a stimulus is re ceived by a receptor and converted into an impulse the impulse is conducted along an afferent pathwa that enters either the spinal cord or the brain.

CHARACTERISTICS

Conscious sensations or perceptions occur in the cor tical regions of the brain. In other words, you see hear, and feel pain in the brain. You seem to see witl your eyes, hear with your ears, and feel pain in an in jured part of your body only because the cortex inter prets the sensation as coming from the stimulated sense receptor. The term **projection** describes this pro cess by which the brain refers sensations to their poin of stimulation.

A second characteristic of many sensations is **ad aptation.** The perception of a sensation may disappea even though a stimulus is still being applied. When yo get into a tub of hot water, you might feel a burning sensation. But soon the sensation decreases to one o comfortable warmth, even though the stimulus (ho water) is still present. Adaptation is due to synaptic fatigue. When an impulse is transmitted from one neu ron to another, a chemical substance (excitatory trans mitter) must be released to get the impulse across the synapse, which is the point of contact between the two neurons. In synaptic fatigue, the chemical substance i exhausted and impulse transmission between neuron stops.

Sensations may also be characterized by **after images.** That is, some sensations persist even thougl the stimulus has been removed. This phenomenon i the reverse of adaptation. One common example of af terimage occurs when you look at a bright light and then look away or close your eyes. You still see the light for several seconds or minutes afterward.

Another characteristic of sensations is **modality:** the specific sensation felt. The sensation may be one o pain, pressure, touch, body position, equilibrium hearing, vision, smell, or taste. In other words, the distinct property by which one sensation may be distinguished from another is its modality.

CLASSIFICATION

One convenient method of classifying sensations is by location of the receptor. Receptors may be classified as exteroceptors, visceroceptors, and proprioceptors. **Exteroceptors** provide information about the external environment. They are sensitive to stimuli outside the body and transmit sensations of hearing, sight, smell, touch, pressure, temperature, and pain. Exteroceptors are located near the surface of the body.

Visceroceptors or **enteroceptors** provide information about the internal environment. These sensations arise from within the body and may be felt as pain, pressure, taste, fatigue, hunger, thirst, and nausea. Visceroceptors are located in blood vessels and viscera.

Proprioceptors allow us to feel position and movement. Such sensations give us information about muscle tension, the position and tension of our joints, and equilibrium. These receptors are located in muscles, tendons, joints, and the internal ear. They provide information about body position and movement.

Sensations may also be classified according to the simplicity or complexity of the receptor and the neural pathway involved. *General senses* involve a simple receptor and neural pathway. The receptors for general sensations are numerous and widespread. Examples include cutaneous sensations such as touch, pressure, heat, cold, and pain. *Special senses,* by contrast, involve complex receptors and neural pathways. The receptors for each special sense are found in only one or two specific areas of the body. Among the special senses are smell, taste, sight, and hearing.

General Senses

Skin contains the receptor organs for many general senses. Receptor organs are also located in muscles, tendons, joints, subcutaneous tissue, and viscera.

CUTANEOUS SENSATIONS

Cutaneous sensations include touch, pressure, cold, heat, and pain. The receptors for these sensations are in the skin, connective tissue, and the ends of the gastrointestinal tract. The cutaneous receptors are randomly distributed over the body's surface so that certain parts of the skin are densely populated with receptors and other parts contain only a few. Areas of the body that have few cutaneous receptors are insensitive; those containing many are very sensitive. This fact can be demonstrated by using the *two-point discrimination test* for touch. A compass is applied to the skin, and the distance in millimeters between the two points of the compass is varied. The subject then indicates when two points are felt and when only one is felt.

The compass may be placed on the tip of the tongue, an area where receptors are very densely packed. The distance between the two points can then be narrowed to 1.4 mm (0.06 inch). At this distance, the points are able to stimulate two different receptors and the subject feels touched by two objects. If the distance is less than 1.4 mm, the subject feels only one point, even though both points are touching the tongue. This is because the points are so close together that they reach only one receptor. The compass can then be placed on the back of the neck, where receptors are few and far apart. In this area of the skin, the subject feels two distinctly different points only if the distance between them is 36.2 mm (1.43 inch) or greater.

The results of this test indicate that the more sensitive the area, the closer the compass points may be placed and still be felt separately. The following order for these receptors, from greatest sensitivity to least, has been established: tip of tongue, tip of finger, side of nose, back of hand, back of neck.

Cutaneous receptors have simple structures. They consist of the dendrites of sensory neurons that may or may not be enclosed in a capsule of epithelial or connective tissue. Impulses generated by cutaneous touch receptors pass along somatic afferent neurons in spinal and cranial nerves, through the thalamus, to the general sensory area of the parietal lobe of the cortex.

Fine Touch

Cutaneous receptors for **fine** or **light touch** include Meissner's corpuscles, Merkel's discs, and root hair plexuses (Figure 16-1). *Meissner's corpuscles* are egg-shaped receptors containing a mass of dendrites enclosed by connective tissue. They are located in the papillae of the skin and enable us to detect touch in the skin. Meissner's corpuscles are most numerous in the fingertips, palms of the hand, and soles of the feet. They are also abundant in the eyelids, tip of the tongue, lips, nipples, clitoris, and tip of the penis. *Merkel's discs* are receptors for touch that consist of disclike formations of dendrites attached to deeper layers of epidermal cells. They are distributed in many of the same locations as Meissner's corpuscles. *Root hair plexuses* are dendrites arranged in networks around the roots of hairs. If a hair shaft is moved, the dendrites are stimulated. Since the root hair plexuses are not surrounded by supportive or protective structures, they are called *free,* or *naked, nerve endings.*

Deep Pressure

Sensations of **deep pressure** are longer lasting and have less variation in intensity than do sensations of touch.

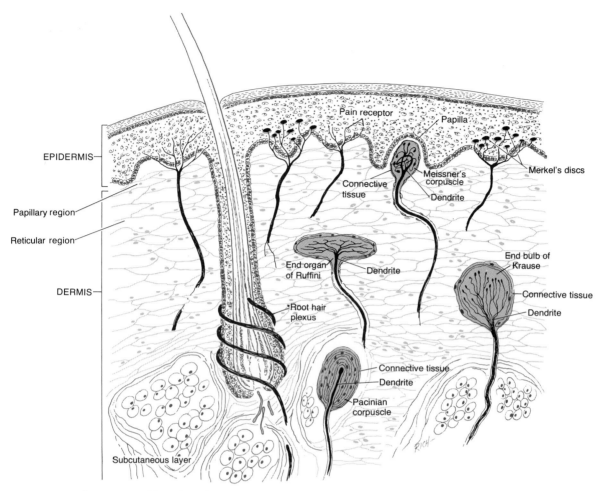

FIGURE 16-1 Structure and location of cutaneous receptors.

Moreover, light touch is felt in a small, "pinprick" area whereas deep pressure is felt over a much larger area. The deep pressure receptors are oval structures called *Pacinian corpuscles* (Figure 16-1). They are composed of a capsule resembling an onion and consisting of connective tissue layers enclosing dendrites. Pacinian corpuscles are located in the subcutaneous tissue under the skin, the deep subcutaneous tissues that lie under mucous membranes, in serous membranes, around joints and tendons, in the perimysium of muscles, in the mammary glands, in the external genitalia of both sexes, and in certain viscera.

Cold

The cutaneous receptors for **cold** are not known but were once thought to be *end bulbs of Krause* (Figure 16-1). The commonest form of these receptors is an oval connective tissue capsule containing dendrites. They are widely distributed in the dermis and subcutaneous connective tissue and are also located in the

conjunctiva of the eye, the tip of the tongue, and external genitals.

Heat

The cutaneous receptors for **heat** are also not known but were once thought to be *end organs of Ruffini* (Figure 16-1). These receptors are deeply embedded in the dermis and are less abundant than end bulbs of Krause.

PAIN SENSATIONS

The receptors for **pain** are simply the branching ends of the dendrites of certain sensory neurons. Pain receptors are found in practically every tissue of the body. They may be stimulated by any type of stimulus. Excessive stimulation of a sense organ causes pain. When stimuli for other sensations such as touch, pressure, heat, and cold reach a certain threshold, they

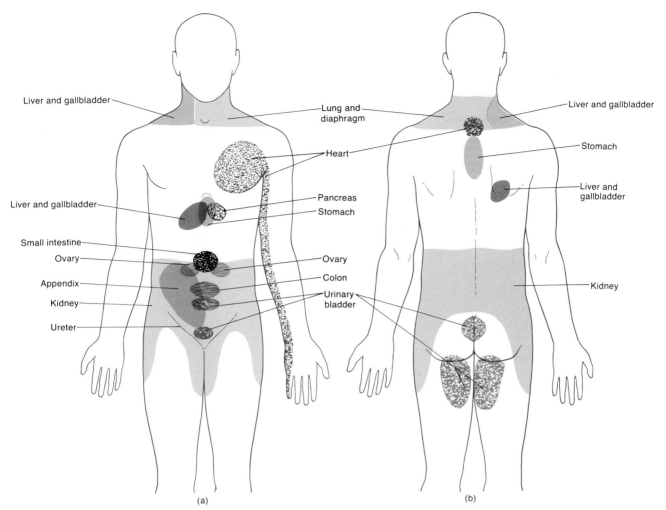

FIGURE 16-2 Referred pain. (a) Anterior view. (b) Posterior view. The shaded parts of the diagrams indicate cutaneous areas to which visceral pain is referred.

stimulate the sensation of pain as well. Additional stimuli for pain receptors include excessive distension or dilation of an organ, prolonged muscular contractions, muscle spasms, inadequate blood flow to an organ, or the presence of certain chemical substances. Pain receptors, because of their sensitivity to all stimuli, perform a protective function by identifying changes that may endanger the body. Pain receptors adapt only slightly or not at all. If there were adaptation to pain, it would be ignored and irreparable damage could result.

Sensory impulses for pain are conducted to the central nervous system along spinal and cranial nerves. The lateral spinothalamic tracts of the spinal cord relay impulses to the thalamus. From here the impulses may be relayed to the postcentral gyrus of the parietal lobe. Recognition of the kind and intensity of most pain is ultimately localized in the cerebral cortex. Some awareness of pain occurs at subcortical levels.

Pain may be divided into two types: somatic and visceral. **Somatic pain** arises from stimulation of the skin receptors. In this case, it is called *superficial somatic pain*. It may also arise from stimulation of receptors in skeletal muscles, joints, tendons, and fascia and is

thus called *deep somatic pain*. **Visceral pain** results from stimulation of receptors in the viscera.

The ability of the cerebral cortex to locate the origin of pain is related to past experience. In most instances of somatic pain and in some instances of visceral pain, the cortex accurately projects the pain back to the stimulated area. If you burn your finger, you feel the pain in your finger. If the lining of your pleural cavity is inflamed, you experience pain there. In most instances of visceral pain, however, the sensation is not projected back to the point of stimulation. Rather, the pain may be felt in or just under the skin that overlies the stimulated organ. The pain may also be felt in a surface area far from the stimulated organ. This phenomenon is called **referred pain.** In general, the area to which the pain is referred and the visceral organ involved receive their innervation from the same segment of the spinal cord. Consider the following example. Afferent fibers from the heart as well as from the skin over the heart and along the left arm enter spinal cord segments T1 to T4. Thus the pain of a heart attack is typically felt in the skin over the heart and along the left arm. Figure 16-2 illustrates cutaneous regions to which visceral pain may be referred.

A kind of pain frequently experienced by amputees is called **phantom pain**—pain in a limb that has been amputated. Suppose a foot has been amputated. A sensory nerve that originally terminated in the foot is severed during the operation but repairs itself and returns to function in the remaining leg. From past experience the brain has projected the stimulation of these neurons back to the distal end of the limb. Thus when the distal ends of the severed neurons are now stimulated, the brain continues to project the sensation back to the missing part. Even though a foot has been amputated, the patient still "feels" pain in the toes.

PROPRIOCEPTIVE SENSATIONS

An awareness of the activities of muscles, tendons, and joints is provided by the **proprioceptive** or **kinesthetic sense.** It informs us of the degree to which muscles are contracted and the amount of tension created in the tendons. The proprioceptive sense enables us to recognize the location and rate of movement of one body part in relation to others. It also allows us to estimate weight and determine the muscular work necessary to perform a task. With the proprioceptive sense, we can judge the position and movements of our limbs without using our eyes when we walk, type, or dress in the dark.

Proprioceptive receptors are located in muscles, tendons, joints, and the internal ear. The receptors for proprioception are of three types. The *joint kinesthetic receptors* are located in the capsules of joints. These receptors provide feedback information on the degree and rate of angulation (change of position) of a joint. The other two receptors for proprioception, neuromuscular spindles and Golgi tendon organs, provide feedback information from muscles. *Neuromuscular spindles* consist of the endings of sensory neurons that are wrapped around specialized muscle fibers. They are located in nearly all skeletal muscles and are numerous in the muscles of the extremities. Neuromuscular spindles provide feedback information on the degree of muscle stretch. This information is relayed to the central nervous system to assist in the coordination and efficiency of muscle contraction. Neuromuscular spindles are involved in the stretch and extensor reflexes. *Golgi tendon organs* are also proprioceptive receptors that provide information about skeletal muscles. These organs are located at the junction of muscle and tendon. They function by sensing the tension applied to a tendon. The degree of tension is related to the degree of muscle contraction and is translated by the central nervous system.

Proprioceptors adapt only slightly. This feature is advantageous since the brain must be apprised of the status of different parts of the body at all times so adjustments can be made to ensure coordination.

The afferent pathway for muscle sense consists of impulses generated by proprioceptors via cranial and spinal nerves to the central nervous system. Impulses for conscious proprioception pass along ascending tracts in the cord, where they are relayed to the thalamus and cerebral cortex. The sensation is registered in the general sensory area in the parietal lobe of the cortex posterior to the central fissure. Proprioceptive impulses that have resulted in reflex action pass to the cerebellum along spinocerebellar tracts.

LEVELS OF SENSATION

Sensory fibers terminating in the spinal cord can generate spinal reflexes without immediate action by the brain. Sensory fibers terminating in the lower brain stem bring about far more complex motor reactions than simple spinal reflexes. When sensory impulses reach the lower brain stem, they cause subconscious motor reactions. Sensory impulses that reach the thalamus can be localized crudely in the body. In fact, at the thalamic level sensations are sorted by modality—that is, when sensory information reaches the cerebral cortex, we experience precise localization. It is at this level that memories of previous sensory information are stored and the perception of sensation occurs on the basis of past experience. Let us now examine how sensory information is transmitted from receptors to the central nervous system.

Sensory Pathways

A review of Figure 13-4 and Exhibit 13-1 will help your understanding of this section.

Sensory information transmitted from the spinal cord to the brain is conducted along two general pathways: the posterior column pathway and the spinothalamic pathway.

In the **posterior column pathway** to the cerebral cortex there are three separate sensory neurons. The *first-order neuron* connects the receptor with the spinal cord and medulla on the same side of the body. The cell body of the first-order neuron is in the posterior root ganglion of a spinal or cranial nerve. The first-order neuron synapses with a *second-order neuron.* The second-order neuron passes from the medulla upward to the thalamus. The cell body of the second-order neuron is located in the nuclei cuneatus and gracilis of the medulla. Before passing into the thalamus, the second-order neuron crosses to the opposite side of the medulla and enters the medial lemniscus, a projection tract that terminates at the thalamus. In the thalamus, the second-order neuron synapses with a *third-order neuron.* The third-order neuron terminates in the somesthetic sensory area of the cerebral cortex. The posterior column pathway conducts impulses related to proprioception, fine touch, two-point discrimination, and vibrations.

The **spinothalamic pathway** is composed of three orders of sensory neurons also. The first-order neuron connects a receptor of the neck, trunk, and extremities with the spinal cord. The cell body of the first-order neuron is in the posterior root ganglion also. The first-order neuron synapses with the second-order neuron, which has its cell body in the posterior gray horn of the spinal cord. The fiber of the second-order neuron crosses to the opposite side of the spinal cord and passes upward to the brain stem in the lateral spinothalamic tract or ventral spinothalamic tract. The fibers from the second-order neuron terminate in the thalamus. There the second-order neuron synapses with a third-order neuron. The third-order neuron terminates in the somesthetic sensory area of the cerebral cortex. The spinothalamic pathway conveys sensory impulses for pain and temperature as well as crude touch and pressure.

The second-order neurons of the spinothalamic pathway enter the medulla, pons, and midbrain. Thus the spinothalamic pathway conducts sensory signals that result in subconscious motor reactions. By contrast, second-order neurons of the posterior column pathway have a direct connection with the thalamus and cerebral cortex. Thus the posterior column pathway conducts sensory information primarily into the conscious areas of the brain.

Now we can examine the anatomy of specific sensory pathways—for pain and temperature, for crude touch and pressure, and for fine touch, proprioception, and vibration.

PAIN AND TEMPERATURE

The sensory pathway for pain and temperature is called the **lateral spinothalamic pathway** (Figure 16-3). The first-order neuron conveys the impulse for pain or temperature from the appropriate receptor to the pos-

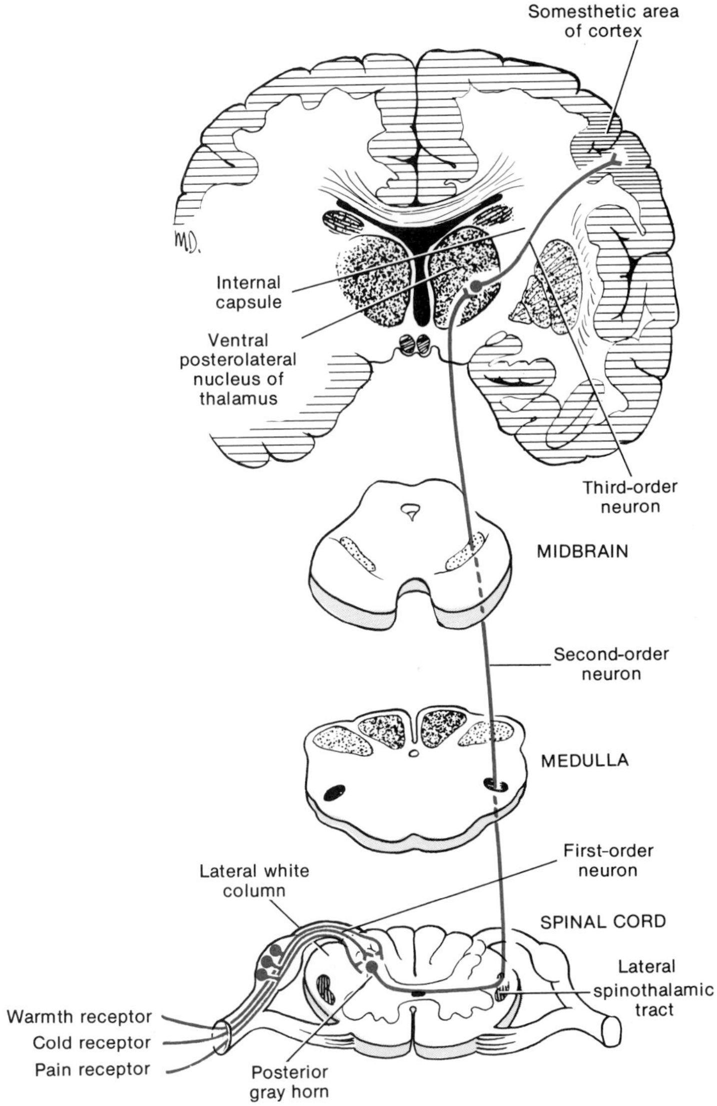

FIGURE 16-3 Sensory pathway for pain and temperature—the lateral spinothalamic pathway.

terior gray horn on the same side of the spinal cord. In the horn, the first-order neuron synapses with a second-order neuron. The axon of the second-order neuron crosses to the opposite side of the cord. Here it becomes a component of the *lateral spinothalamic tract* in the lateral white column. The second-order neuron passes upward in the tract through the brain stem to a nucleus in the thalamus called the ventral posterolateral nucleus. In the thalamus, conscious recognition of pain and temperature occurs. The sensory impulse is then conveyed from the thalamus through the internal capsule to the somesthetic area of the cortex by a third-order neuron. The cortex analyzes the sensory information for the precise source, severity, and quality of the pain and heat stimuli.

CRUDE TOUCH AND PRESSURE

The neural pathway that conducts impulses for crude touch and pressure is the **anterior (ventral) spinotha-lamic pathway** (Figure 16-4). By crude touch and pressure is meant the ability to perceive that something has touched the skin although its exact location, shape, size, or texture cannot be determined. The first-order neuron conveys the impulse from a crude touch or pressure receptor to the posterior gray horn on the same side of the spinal cord. In the horn, the first-order neuron synapses with a second-order neuron. The axon of the second-order neuron crosses to the opposite side of the cord and becomes a component of the *anterior spinothalamic tract* in the anterior white column. The second-order neuron passes upward in the tract through the brain stem to the ventral posterolateral nucleus of the thalamus. The sensory impulse is then relayed from the thalamus through the internal capsule to the somesthetic area of the cerebral cortex by a third-order neuron. Although there is some awareness of crude touch and pressure at the thalamic level, it is not fully perceived until the impulses reach the cortex.

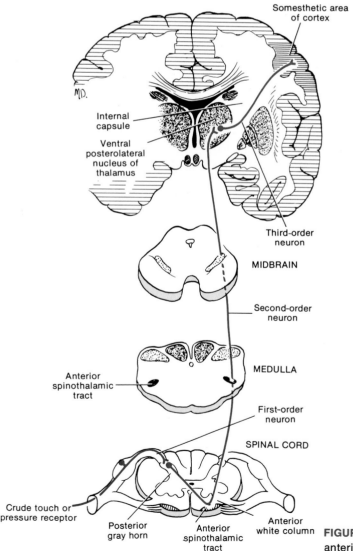

Somesthetic area
of cortex

Internal
capsule

Ventral
posterolateral
nucleus of
thalamus

Third-order
neuron

MIDBRAIN

Second-order
neuron

MEDULLA

Anterior
spinothalamic
tract

First-order
neuron

SPINAL CORD

Crude touch or
pressure receptor

Posterior
gray horn

Anterior
spinothalamic
tract

Anterior
white column

FIGURE 16-4 Sensory pathway for crude touch and pressure—the anterior spinothalamic pathway.

FINE TOUCH, PROPRIOCEPTION, VIBRATION

The neural pathway for fine touch, proprioception, and vibration is called the **posterior column pathway** (Figure 16-5). This pathway conducts impulses that give rise to several discriminating senses:

1. Fine touch: the ability to recognize the exact location of stimulation and to distinguish that two points are touched, even though they are close together (two-point discrimination).

2. Stereognosis: recognizing the size, shape, and texture of an object.

3. Weight discrimination: the ability to assess the weight of an object.

4. Proprioception: the awareness of the precise position of body parts and directions of movement.

5. The ability to sense **vibrations.**

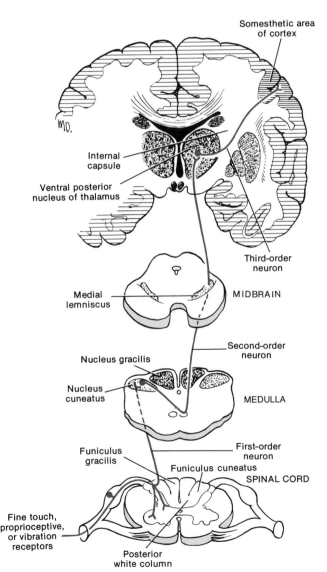

Somesthetic area of cortex

Internal capsule

Ventral posterior nucleus of thalamus

Third-order neuron

Medial lemniscus

MIDBRAIN

Nucleus gracilis

Second-order neuron

Nucleus cuneatus

MEDULLA

First-order neuron

Funiculus gracilis

Funiculus cuneatus

SPINAL CORD

Fine touch, proprioceptive, or vibration receptors

Posterior white column

FIGURE 16-5 Sensory pathway for fine touch, proprioception, and vibration—the posterior column pathway.

First-order neurons for the discriminating senses just noted follow a pathway different from those for pain and temperature and crude touch and pressure. Instead of terminating in the posterior gray horn, the first-order neurons from appropriate receptors pass upward in the fasciculus gracilis or fasciculus cuneatus in the posterior white column of the cord. From here the first-order neurons enter either the nucleus gracilis or nucleus cuneatus in the medulla where they synapse with second-order neurons. The axons of the second-order neurons cross to the opposite side of the medulla and ascend to the thalamus through the medial lemniscus, a projection tract of white fibers passing through the medulla, pons, and midbrain. The second-order neuron axons synapse with third-order neurons in the ventral posterior nucleus in the thalamus. In the thalamus, there is no conscious awareness of the discriminating senses, except for a possible crude awareness of vibrations. The third-order neurons convey the sensory impulses to the somesthetic area of the cerebral cortex. It is here that you perceive your sense of position and movement and fine touch.

CEREBELLAR TRACTS

The *posterior spinocerebellar tract* is an uncrossed tract that conveys impulses concerned with subconscious muscle sense. That is, the tract assumes a role in reflex adjustments for posture and muscle tone. The nerve impulses originate in neurons that run between proprioceptors in muscles, tendons, and joints and the posterior gray horn of the spinal cord. Here the neurons synapse with afferent neurons that pass to the ipsilateral lateral white column of the cord to enter the posterior cerebellar tract. The tract enters the inferior cerebellar peduncles from the medulla and ends at the cerebellar cortex. In the cerebellum, synapses are made that ultimately result in the transmission of impulses back to the spinal cord to the anterior gray horn to synapse with the lower motor neurons leading to skeletal muscles.

The *anterior spinocerebellar tract* also conveys impulses for subconscious muscle sense. It, however, is made up of both crossed and uncrossed nerve fibers. Sensory neurons deliver impulses from proprioceptors to the posterior gray horn of the spinal cord. Here a synapse occurs with neurons that make up the anterior spinocerebellar tracts. Some fibers cross to the opposite side of the spinal cord in the anterior white commissure. Others pass laterally to the ipsilateral anterior spinocerebellar tract and move upward, through the brain stem, to the pons to enter the cerebellum through the superior cerebellar peduncles. Here again the impulses concerned with subconscious muscle sense are registered.

Motor Pathways

The principal parts of the brain concerned with skeletal muscle control are the cerebral motor cortex, basal ganglia, reticular formation, and cerebellum. The motor cortex assumes the major role for controlling precise, discrete muscular movements. The basal ganglia largely integrate semivoluntary movements like walking, swimming, and laughing. The cerebellum, although not a control center, assists the motor cortex and basal ganglia by making body movements smooth and coordinated. Voluntary motor impulses are conveyed from the brain through the spinal cord by way of two major pathways: the pyramidal pathways and the extrapyramidal pathways.

PYRAMIDAL PATHWAYS

Voluntary motor impulses are conveyed from the motor areas of the brain to somatic efferent neurons leading to skeletal muscles via the **pyramidal pathways.** Most pyramidal fibers originate from cell bodies in the precentral gyrus. They descend through the internal capsule of the cerebrum and cross to the opposite side of the brain. They terminate in nuclei of cranial nerves that innervate voluntary muscles or in the anterior gray horn of the spinal cord. A short connecting neuron probably completes the connection of the pyramidal fibers with the motor neurons that activate voluntary muscles.

The pathways over which the impulses travel from the motor cortex to skeletal muscles have two components: *upper motor neurons (pyramidal fibers)* and *lower motor neurons (peripheral fibers)*. Here we consider three tracts of the pyramidal system:

1. Lateral corticospinal tract *(pyramidal tract proper)*. This tract begins in the motor cortex, descends through the internal capsule of the cerebrum, the cerebral peduncle of the midbrain, and then the pons on the same side as the point of origin (Figure 16-6). In the medulla, the fibers decussate to the opposite side to descend through the spinal cord in the lateral white column in the lateral corticospinal tract. Thus the motor cortex of the right side of the brain controls muscles on the left side of the body and vice versa. The upper motor neurons of the lateral corticospinal tract probably synapse with short association neurons in the anterior gray horn of the cord. These then synapse in the anterior gray horn with lower motor neurons that exit all levels of the cord via the ventral roots of spinal nerves. The lower motor neurons terminate in skeletal muscles.

2. Anterior corticospinal tract. About 15 percent of the upper motor neurons from the motor cortex do not cross in the medulla. These pass through the medulla

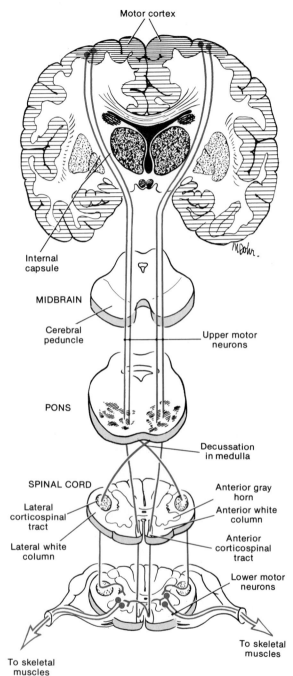

FIGURE 16-6 Pyramidal pathways.

and continue to descend on the same side to the anterior white column to become part of the *anterior (straight* or *uncrossed) corticospinal tract*. The fibers of these upper motor neurons decussate and probably synapse with association neurons in the anterior gray horn of the spinal cord of the side opposite the origin of the anterior corticospinal tract. The association neurons in the horn synapse with lower motor neurons that exit the cervical and upper thoracic segments of he cord via the ventral roots of spinal nerves. The lower motor neurons terminate in skeletal muscles that control muscles of the neck and part of the trunk.

3. Corticobulbar tract. The fibers of this tract begin in upper motor neurons in the motor cortex. They accompany the corticospinal tracts through the internal capsule to the brain stem, where they decussate and terminate in the nuclei of cranial nerves in the pons and medulla. These cranial nerves include the trigeminal (V), abducens (VI), facial (VII), glossopharyngeal (IX), vagus (X), accessory (XI), and hypoglossal (XII). The corticobulbar tract conveys impulses that largely control voluntary movements of the head and neck.

The various tracts of the pyramidal system convey impulses from the cortex that result in precise muscular movements.

EXTRAPYRAMIDAL PATHWAYS

The **extrapyramidal pathways** include all descending tracts other than the pyramidal tracts. Generally, these include tracts that begin in the basal ganglia and reticular formation. The main extrapyramidal tracts are as follows:

1. Rubrospinal tract. This tract originates in the red nucleus of the midbrain (after receiving fibers from the cerebellum), crosses over to descend in the lateral white column of the opposite side, and terminates in the anterior gray horns of the cervical and upper thoracic segments of the cord. The tract transmits impulses to skeletal muscles concerned with tone and posture.

2. Tectospinal tract. This tract originates in the superior colliculus of the midbrain, crosses to the opposite side, descends in the anterior white column, and enters the anterior gray horns in the cervical segments of the cord. Its function is to transmit impulses that control movements of the head in response to auditory, visual, and cutaneous stimuli.

3. Vestibulospinal tract. This tract originates in the vestibular nucleus of the medulla, descends on the same side of the cord in the anterior white column, and terminates in the anterior gray horns, mostly in the cervical and lumbosacral segments of the cord. It conveys impulses that regulate muscle tone in response to movements of the head. This tract, therefore, plays a major role in equilibrium.

Only one motor neuron carries the impulse from the cerebral cortex to the cranial nerve nuclei or spinal cord: an *upper motor neuron*. Only one motor neuron in the pathway actually terminates in a skeletal muscle:

the *lower motor neuron*. This neuron, a somatic efferent neuron, always extends from the central nervous system to the skeletal muscle. Since it is the final transmitting neuron in the pathway, it is also called the *final common pathway*. This neuron is important clinically. If it is damaged or diseased, there is neither voluntary nor reflex action of the muscle it innervated and the muscle remains in a relaxed state—a condition called *flaccid paralysis*. Injury or disease of upper motor neurons in a motor pathway is characterized by varying degrees of continued contraction of the muscle (*spasticity*) and exaggerated reflexes. Another characteristic is the sign of Babinski; stroking the plantar surface along the outer border of the foot produces a slow dorsiflexion of the great toe accompanied by fanning of the lateral toes. The normal reponse is plantar flexion of the toes.

Lower motor neurons are subjected to stimulation by many other presynaptic neurons. Some signals are excitatory; others are inhibitory. The algebraic sum of the opposing signals determines the final response of the lower motor neuron. It is not just a simple matter of the brain sending an impulse and the muscle always contracting.

Association neurons are of considerable importance in the motor pathways. Impulses from the brain are conveyed to association neurons before being received by lower motor neurons. These association neurons are integrators. They integrate the pattern of muscle contraction.

The basal ganglia have many connections with other parts of the brain. Through these connections, they help to control subconscious movements. The caudate nucleus controls gross intentional movements. The caudate nucleus and putamen, together with the cortex, control patterns of movement. The globus pallidus controls positioning of the body for performing a complex movement. The subthalamic nucleus is thought to control walking and possibly rhythmic movements. Many potential functions of the basal ganglia are held in check by the cerebrum. Thus if the cerebral cortex is damaged early in life, a person can still perform many gross muscular movements.

The role of the cerebellum is significant also. The cerebellum is connected to other parts of the brain that are concerned with movement. The vestibulocerebellar tract transmits impulses from the equilibrium apparatus in the ear to the cerebellum. The olivocerebellar tract transmits impulses from the basal ganglia to the cerebellum. The corticopontocerebellar tract conveys impulses from the cerebrum to the cerebellum. The spinocerebellar tracts relay proprioceptive information to the cerebellum. Thus the cerebellum receives considerable information regarding the overall physical status of the body. Using this information, the cerebellum generates impulses that integrate body responses.

Take tennis, for example. To make a good serve, you must bring your racket forward just far enough to make solid contact. How do you stop at the exact point without swinging too far? This is were the cerebellum comes in. It receives information about your body status while you are serving. Before you even hit the ball, the cerebellum has already sent information to the cerebral cortex and basal ganglia informing them that your swing must stop at an exact point. In response to cerebellar stimulation, the cortex and basal ganglia transmit motor impulses to your opposing body muscles to stop the swing. The cerebellar function of stopping overshoot when you want to zero in on a target is called its *damping function*. The cerebellum also helps you to coordinate different body parts while walking, running, and swimming. Finally, the cerebullum helps you maintain equilibrium.

Special Senses

In contrast to the general senses, the special senses of smell, taste, sight, hearing, and equilibrium have receptor organs that are structurally more complex. The sense of smell is the least specialized, as opposed to the sense of sight, which is the most specialized. Like the general senses, however, the special senses allow us to detect changes in our environment.

Olfactory Sensations

The receptors for the **olfactory sense** are located in the nasal epithelium in the superior portion of the nasal fossae (left and right chambers of the nasal cavity) on either side of the nasal septum (Figure 16-7a). The nasal epithelium consists of two principal kinds of cells: supporting and olfactory (Figure 16-7b,c). The *supporting cells* are columnar epithelial cells of the mucous membrane lining the nose. Olfactory glands in the mucosa keep the mucous membrane moist. The *olfactory cells* are bipolar neurons whose cell bodies lie between the supporting cells. The distal (free) end of each olfactory cell contains six to eight dendrites, called *olfactory hairs*. The unmyelinated axons of the olfactory cells unite to form the *olfactory nerves* (I) which pass through foramina in the cribriform plate of the ethmoid bone. The olfactory nerves terminate in paired masses of gray matter called the *olfactory bulbs*. The olfactory bulbs lie beneath the frontal lobes of the cerebrum on either side of the crista galli of the ethmoid bone. The first synapse of the olfactory neural pathway occurs in the olfactory bulbs between the axons of the olfactory nerves and the dendrites of neurons inside the olfactory bulbs. Axons of these neurons run posteriorly to form the *olfactory tract*. From here, impulses are conveyed to the olfactory portion of the cortex. In the cortex, the impulses are interpreted as odor and give rise to the sensation of smell.

A few of the surface anatomy features of the nose are indicated in Exhibit 1-3.

The mechanism by which the stimulus for smell is converted to a nerve impulse is explained by three widely accepted theories. One theory holds that substances capable of producing odors emit gaseous particles. On entering the nasal fossae, these particles become dissolved in the mucus of the nasal membrane. This fluid then acts chemically on the olfactory hairs to create a nerve impulse. According to the second theory, radiant energy given off by the molecules of the stimulating substance is the stimulus rather than the molecules themselves. The third theory purports that substances detected by smell are usually soluble in fat. Since the membrane of an olfactory hair is largely fat, it is assumed that molecules of substances to be smelled are dissolved in the membrane where they initiate a nerve impulse.

The sensation of smell happens quickly, but adaptation to odors also occurs rapidly. For this reason, we become accustomed to some odors and are also able to endure unpleasant ones. Rapid adaptation also accounts for the failure of a person to detect a gas that accumulates slowly in a room. The cortex stores memories of odors quite well. Once you have smelled a substance, you generally recognize its odor if you smell it again.

The supporting cells of the nasal epithelium and tear ducts are innervated by branches of the trigeminal nerve (V). The nerve receives stimuli of pain, cold, heat, tickling, and pressure. Olfactory stimuli, such as pepper, onions, ammonia, ether, and chloroform, are irritating and may cause tearing because they stimulate the receptors of the trigeminal nerve as well as the olfactory neurons.

Gustatory Sensations

The receptors for **gustatory sensations,** or sensations of taste, are located in the taste buds (Figure 16-8 b, c). Taste buds are most numerous on the tongue, but they are also found on the soft palate and in the throat. The *taste buds* are oval-shaped bodies consisting of two kinds of cells. The *supporting cells* are specialized epithelium that forms a capsule. Inside each capsule are 4 to 20 *gustatory cells*. Each gustatory cell contains a hairlike process *(gustatory hair)* that projects to the external surface through an opening in the taste bud called the *taste pore*. Gustatory cells make contact with taste stimuli through the taste pore.

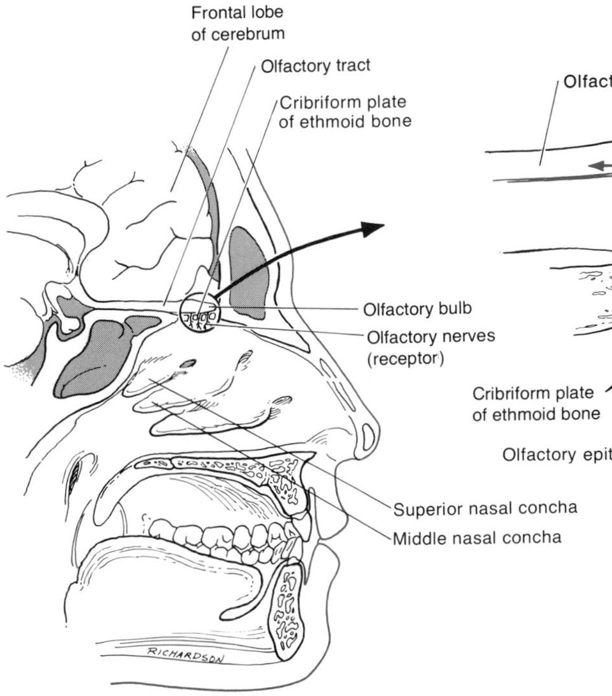

FIGURE 16-7 Olfactory receptors. (a) Location of receptors in left nasal fossa. (b) Enlarged aspect of olfactory receptors. (c) Photomicrograph of the olfactory mucosa at a magnification of 400×. (Courtesy of Donald I. Patt, from *Comparative Vertebrate Histology,* by Donald I. Patt and Gail R. Patt, Harper & Row, Publishers, Inc., New York, 1969.) (d) Scanning electron micrograph of the ciliated nasal mucosa at a magnification of 5000×. (Courtesy of Fisher Scientific Company and S.T.E.M. Laboratories, Inc., Copyright 1975.)

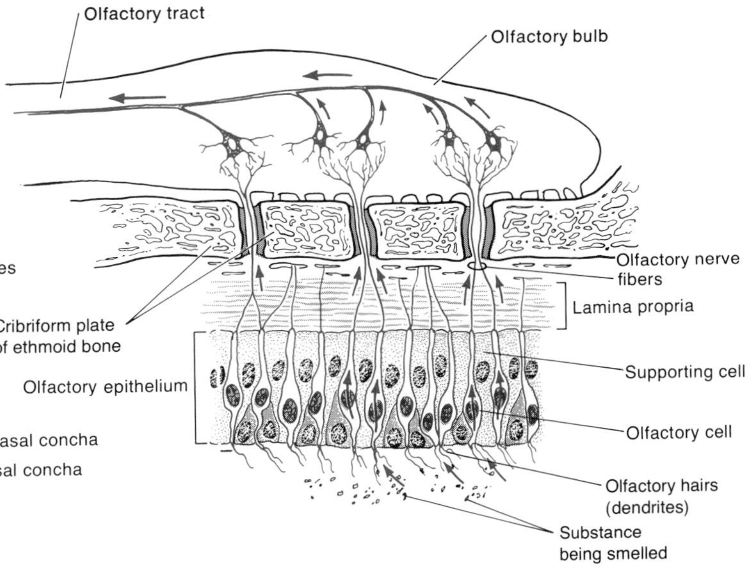

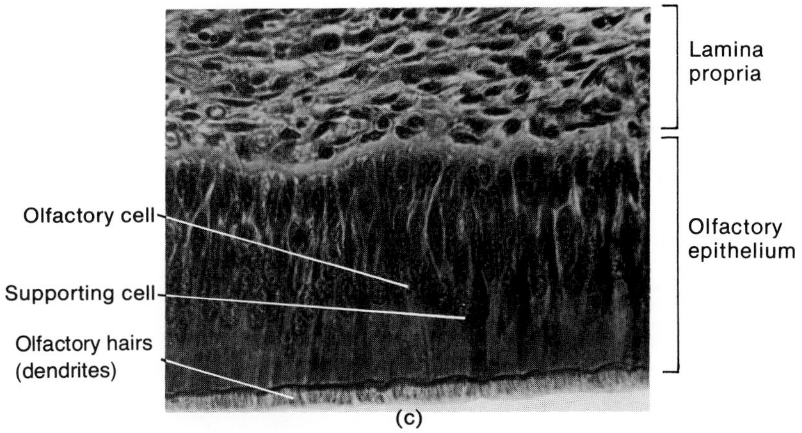

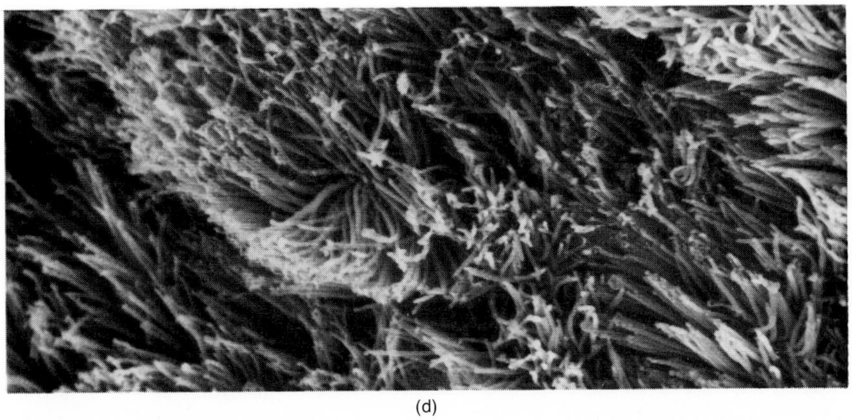

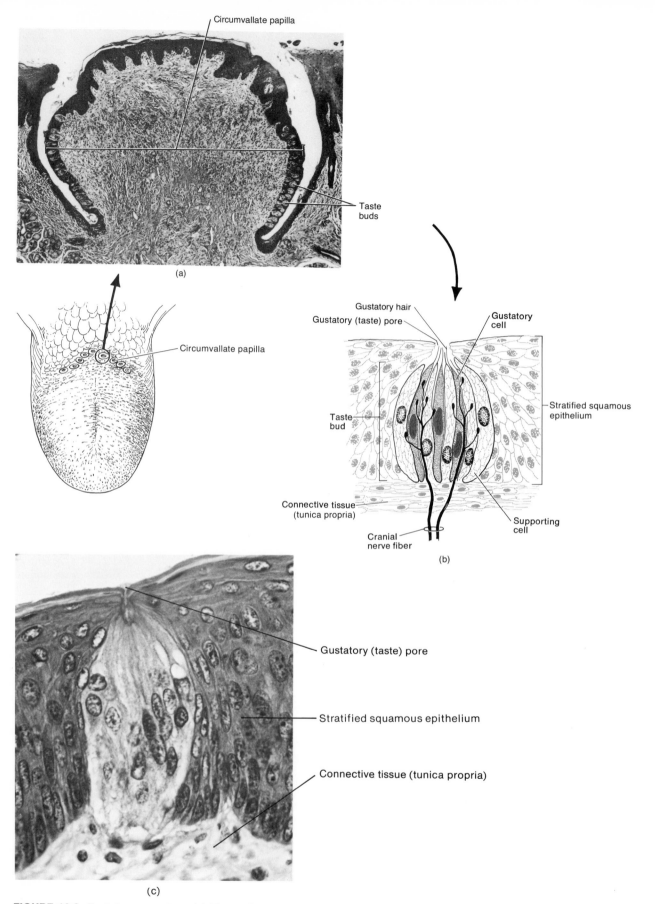

FIGURE 16-8 Gustatory receptors. (a) Photomicrograph of a circumvallate papilla showing the location of taste buds. (b) Structure of a taste bud. (c) Photomicrograph of a taste bud at a magnification of 575×. (Photomicrographs courtesy of Edward J. Reith, from *Atlas of Descriptive Histology,* by Edward J. Reith and Michael H. Ross, Harper & Row, Publishers, Inc., New York, 1970.)

Taste buds are found in some connective tissue elevations on the tongue called **papillae** (Figure 16-8a). The papillae give the upper surface of the tongue its characteristic rough appearance. *Circumvallate papillae,* the largest type, are circular and form an inverted V-shaped row at the posterior portion of the tongue. *Fungiform* (mushroom-shaped) *papillae* are knoblike elevations and are found primarily on the tip and sides of the tongue. All circumvallate and most fungiform papillae contain taste buds. *Filiform papillae* are threadlike structures that cover the anterior two-thirds of the tongue (see Figure 20-5).

In order for gustatory cells to be stimulated, the substances we taste must be in solution in the saliva so that they can enter the taste pores in the taste buds. Despite the many substances we taste, there are basically only four taste sensations: sour, salt, bitter, and sweet. Each taste is due to a different response to different chemicals. Certain regions of the tongue react more strongly than other regions to particular taste sensations. For example, although the tip of the tongue reacts to all four taste sensations, it is highly sensitive to sweet and salty substances. The posterior portion of the tongue is highly sensitive to bitter substances, and the lateral edges of the tongue are more sensitive to sour substances (see Figure 20-5).

The cranial nerves that supply afferent fibers to taste buds are the facial nerve (VII), which supplies the anterior two-thirds of the tongue; the glossopharyngeal (IX), which supplies the posterior one-third of the tongue; and the vagus (X), which supplies the area of the throat around the epiglottis. Taste impulses are conveyed from the gustatory cells in taste buds along the three nerves just cited to the medulla and then to the thalamus. They terminate in the parietal lobe of the cortex.

Visual Sensations

The structures related to vision are the eyeball, the optic nerve, the brain, and a number of accessory structures.

ACCESSORY STRUCTURES OF EYE

Among the accessory organs are the eyebrows, eyelids, eyelashes, and the lacrimal apparatus, which produces tears (Figure 16-9). The *eyebrows* form a transverse arch at the junction of the upper eyelid and forehead. Structurally, they resemble the hairy scalp. The skin of the eyebrows is richly supplied with sebaceous glands. The hairs are generally coarse and directed laterally. Deep to the skin of the eyebrows are the fibers of the orbicularis oculi muscles The eyebrows protect the eyeballs from falling objects, perspiration, and the direct rays of the sun.

The upper and lower *eyelids,* or *palpebrae,* consist of dense folds of skin. They shade the eyes during sleep, protect the eyes from excessive light and foreign objects, and spread lubricating secretions over the eyeballs. The upper eyelid is more movable than the lower and contains in its superior region a special levator muscle known as the *levator palpebrae superioris.* The space between the upper and lower eyelids that exposes the eyeball is called the *palpebral fissure.* Its angles are known as the *lateral canthus,* which is narrower and closer to the temporal bone, and the *medial canthus,* which is broader and nearer the nasal bone. In the medial canthus there is a small reddish elevation, the *lacrimal caruncle,* containing sebaceous and sudoriferous glands. A whitish material secreted by the caruncle collects in the medial canthus.

From superficial to deep, each eyelid consists of epidermis, dermis, subcutaneous areolar connective tissue, fibers of the orbicularis oculi muscle, a tarsal plate, tarsal glands, and a conjunctiva. The *tarsal plate* is a thick fold of connective tissue that gives form and support to the eyelid. Embedded in grooves on the deep surface of each tarsal plate is a row of elongated tarsal glands known as *Meibomian glands.* These are modified sebaceous glands, and their oily secretion helps keep the eyelids from adhering to each other. Infection of these glands produces a tumor or cyst on the eyelid called a *chalazion.* The *conjunctiva* is a thin mucous membrane called the *palpebral conjunctiva* when it lines the inner aspect of the eyelids. It is called the *bulbar* or *ocular conjunctiva* when it is reflected from the eyelids onto the eyeball to the periphery of the cornea.

Projecting from the border of each eyelid, anterior to the Meibomian glands, is a row of short, thick hairs: the *eyelashes.* In the upper lid, they are long and turn upward; in the lower lid, they are short and turn downward.

Sebaceous glands at the base of the hair follicles of the eyelashes called *glands of Zeis* pour a lubricating fluid into the follicles. Infection of these glands is called a *sty.*

The *lacrimal apparatus* is a term used for a group of structures that manufactures and drains away tears. These structures are the lacrimal glands, the excretory lacrimal ducts, the lacrimal canals, the lacrimal sacs, and the nasolacrimal ducts. A *lacrimal gland* is a compound tubuloacinar gland located at the superior lateral portion of each orbit. Each is about the size and shape of an almond. Leading from the lacrimal glands are 6 to 12 *excretory lacrimal ducts* that empty lacrimal fluid, or tears, onto the surface of the conjunctiva of the upper lid. From here the lacrimal fluid passes medially and enters two small openings (*puncta lacrimalia*) that appear as two small dots, one in each papilla of the eyelid, at the medial canthus of the eye.

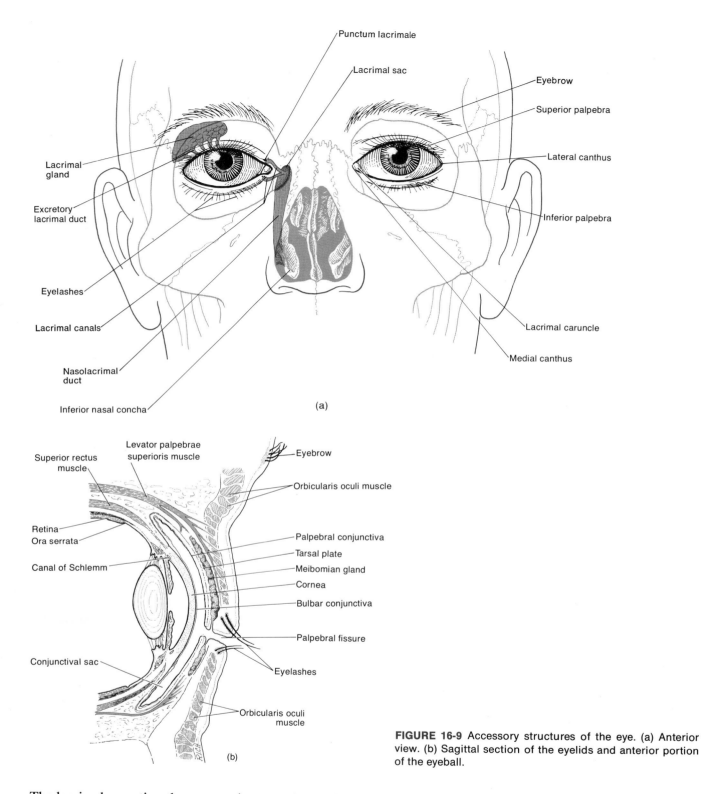

Punctum lacrimale

Lacrimal sac

Eyebrow

Superior palpebra

Lateral canthus

Inferior palpebra

Lacrimal gland

Excretory lacrimal duct

Eyelashes

Lacrimal canals

Nasolacrimal duct

Inferior nasal concha

Lacrimal caruncle

Medial canthus

(a)

Superior rectus muscle

Levator palpebrae superioris muscle

Eyebrow

Orbicularis oculi muscle

Retina

Ora serrata

Canal of Schlemm

Palpebral conjunctiva

Tarsal plate

Meibomian gland

Cornea

Bulbar conjunctiva

Palpebral fissure

Conjunctival sac

Eyelashes

Orbicularis oculi muscle

(b)

FIGURE 16-9 Accessory structures of the eye. (a) Anterior view. (b) Sagittal section of the eyelids and anterior portion of the eyeball.

The lacrimal secretion then passes into two ducts, the *lacrimal canals,* and is next conveyed into the lacrimal sac. The lacrimal canals are located in the lacrimal grooves of the lacrimal bones. The *lacrimal sac* is the superior expanded portion of the nasolacrimal duct, a canal that transports the lacrimal secretion into the inferior meatus of the nose.

The *lacrimal secretion* is a watery solution containing salts, some mucus, and a bactericidal enzyme called lysozyme. It cleans, lubricates, and moistens the eyeball. After being secreted by the lacrimal glands, it is spread over the surface of the eyeball by the blinking of the eyelids. Usually, 1 ml per day is produced. Normally, the secretion is carried away by evaporation or by passing into the lacrimal canals and then into the nasal cavities as fast as it is produced. If, however, an irritating substance makes contact with the conjunctiva, the lacrimal glands are stimulated to oversecrete. Tears then accumulate more rapidly than they can be carried away. This is a protective mechanism since the tears dilute and wash away the irritating substance. "Watery" eyes also occur when an inflammation of the nasal mucosa, such as a cold, obstructs the nasolacrimal ducts so that drainage of tears is blocked.

STRUCTURE OF EYEBALL

The adult **eyeball** measures about 2.5 cm (1 inch) in diameter. Of its total surface area, only the anterior one-sixth is exposed. The remainder is recessed and protected by the orbit into which it fits. Anatomically, the eyeball can be divided into three layers: fibrous tunic, vascular tunic, and retina (Figure 16-10).

The **fibrous tunic** is the outer coat of the eyeball. It can be divided into two regions: the posterior sclera and the anterior cornea. The *sclera,* the "white of the eye," is a white coat of fibrous tissue that covers all the eyeball except the anterior colored portion. The sclera gives shape to the eyeball and protects its inner

parts. Its posterior surface is pierced by the optic nerve. The anterior portion of the fibrous tunic is called the *cornea.* It is a nonvascular, nervous, transparent fibrous coat that covers the iris, the colored part of the eye. The outer surface of the cornea is covered by an epithelial layer that is continuous with the epithelium of the bulbar conjunctiva. At the junction of the sclera and cornea is a venous sinus known as the *canal of Schlemm.*

The **vascular tunic** is the middle layer of the eyeball and is composed of three portions: the posterior choroid, the anterior ciliary body, and the iris. Collectively, these three structures are called the *uvea.* The *choroid* is a thin, dark-brown membrane that lines most of the internal surface of the sclera. It contains numerous blood vessels and a large amount of pigment. The choroid absorbs light rays so they are not reflected back out of the eyeball. Through its blood supply, it nourishes the retina. The optic nerve also pierces the choroid at the back of the eyeball. The anterior portion of the choroid becomes the *ciliary body.* It is the thickest portion of the vascular tunic. It extends from the *ora serrata* of the retina (inner tunic) to a point just behind the sclerocorneal junction. The ora serrata is simply the jagged margin of the retina. This second division of the vascular tunic contains the *cili-*

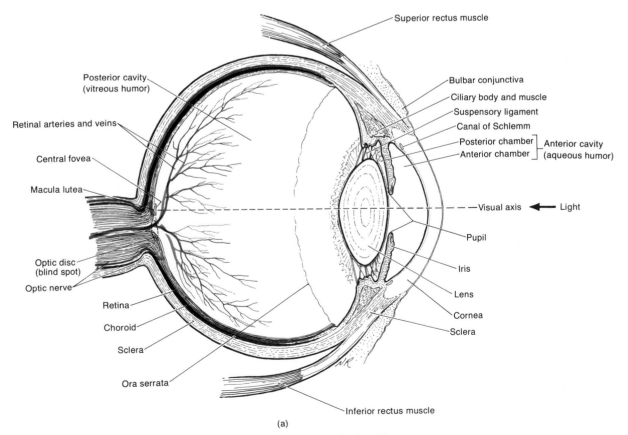

(a)

FIGURE 16-10 Structure of the eyeball. (a) Gross anatomy in sagittal section.

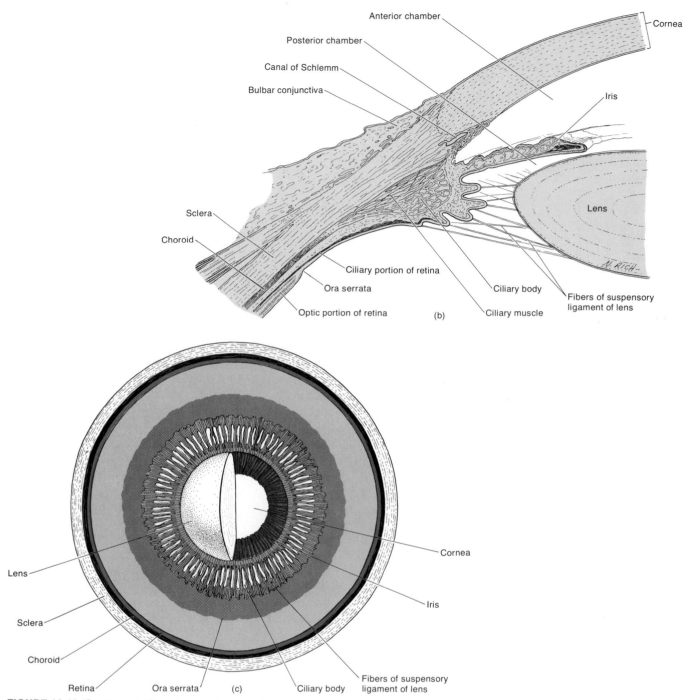

Anterior chamber

Posterior chamber

Canal of Schlemm

Bulbar conjunctiva

Cornea

Iris

Lens

Sclera

Choroid

Ciliary portion of retina

Ora serrata

Optic portion of retina

Ciliary body

Ciliary muscle

Fibers of suspensory
ligament of lens

(b)

Lens

Sclera

Choroid

Retina

Ora serrata

(c)

Ciliary body

Cornea

Iris

Fibers of suspensory
ligament of lens

FIGURE 16-10 (Continued) Structure of the eyeball. (b) Section through the anterior part of the eyeball at the sclerocorneal junction. (c) Anterior portion of the eyeball seen from behind.

ary muscle—a smooth muscle that alters the shape of the lens for near or far vision. The *iris* is the third portion of the vascular tunic. It consists of circular and radial smooth muscle fibers arranged to form a doughnut-shaped structure. The black hole in the center of the iris is the *pupil,* the area through which light enters the eyeball. This iris is suspended between the cornea and the lens and is attached at its outer margin to the cili-

ary body. A principal function of the iris is to regulate the amount of light entering the eyeball. When the eye is stimulated by bright light, the circular muscles of the iris contract and decrease the size of the pupil. When the eye must adjust to dim light, the radial muscles of the iris contract and increase the pupil's size.

The third and inner coat of the eye, the **retina,** lies only in the posterior portion of the eye. Its primary

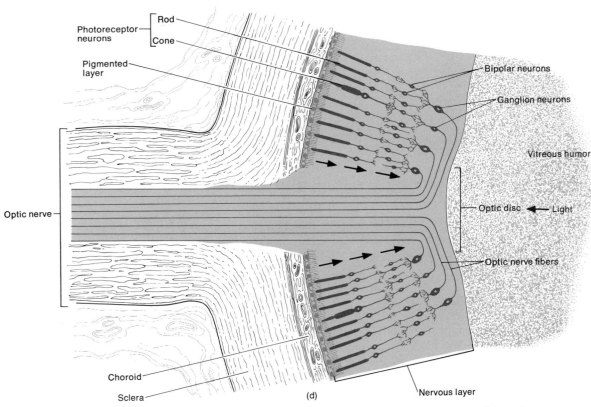

Photoreceptor neurons — Rod — Cone
Pigmented layer
Bipolar neurons
Ganglion neurons
Vitreous humor
Optic nerve
Optic disc — Light
Optic nerve fibers
Choroid
Sclera
(d)
Nervous layer

FIGURE 16-10 (Continued) Structure of the eyeball. (d) Diagram of the microscopic structure of the retina.

function is image formation. It consists of a nervous tissue layer and a pigmented layer. The retina covers the choroid. At the edge of the ciliary body, it appears to end in a scalloped border called the ora serrata. This is where the outer nervous layer or visual portion of the retina ends. The inner pigmented layer extends anteriorly over the back of the ciliary body and the iris as the nonvisual portion of the retina. The inner nervous layer contains three zones of neurons. These three zones, in the order in which they conduct impulses, are the photoreceptor neurons, bipolar neurons, and ganglion neurons.

The dendrites of the **photoreceptor neurons** are called rods and cones because of their shapes. They are visual receptors highly specialized for stimulation by light rays. **Rods** are specialized for vision in dim light. They also allow us to discriminate between different shades of dark and light and permit us to see shapes and movement. **Cones,** by contrast, are specialized for color vision and sharpness of vision *(visual acuity)*. Cones are stimulated only by bright light. This is why we cannot see color by moonlight. It is estimated that there are 7 million cones and somewhere between 10 and 20 times as many rods. Cones are most densely concentrated in the *central fovea,* a small depression in the center of the macula lutea. The *macula lutea,* or yellow spot, is in the exact center of the

retina. The fovea is the area of sharpest vision because of the high concentration of cones. Rods are absent from the fovea and macula, but they increase in density toward the periphery of the retina.

When impulses for sight have passed through the photoreceptor neurons, they are conducted across synapses to the **bipolar neurons** in the intermediate zone of the nervous layer of the retina. From here the impulses are passed to the ganglion neurons.

The axons of the **ganglion neurons** extend posteriorly to a small area of the retina called the *optic disc,* or *blind spot.* This region contains openings through which the fibers of the ganglion neurons exit as the optic nerve. Since this area contains neither rods nor cones, and only nerve fibers, no image falling on it is perceived. Thus it is called the blind spot.

In addition to the fibrous tunic, vascular tunic, and retina, the eyeball itself contains the lens, just behind the pupil and iris. The **lens** is constructed of numerous layers of protein fibers arranged like the layers of an onion. Normally, the lens is perfectly transparent. It is enclosed by a clear connective tissue capsule and held in position by the *suspensory ligament.*

The interior of the eyeball contains a large cavity divided into two smaller ones: the anterior cavity and the posterior cavity. They are separated from each other by the lens. The *anterior cavity,* in turn, has two

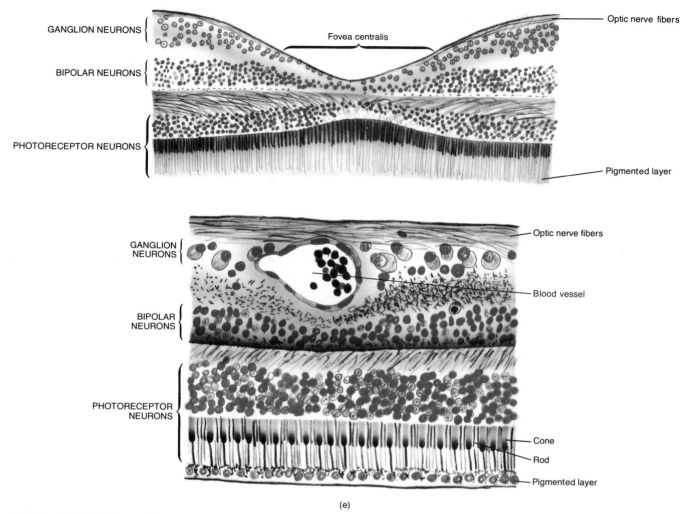

GANGLION NEURONS {

BIPOLAR NEURONS {

PHOTORECEPTOR NEURONS {

Fovea centralis

Optic nerve fibers

Pigmented layer

GANGLION NEURONS {

BIPOLAR NEURONS {

PHOTORECEPTOR NEURONS {

Optic nerve fibers

Blood vessel

Cone

Rod

Pigmented layer

(e)

FIGURE 16-10 (Continued) Structure of the eyeball. (e) Diagram of details of the retina through the center of the fovea centralis within the macula lutea (above) and details of the various regions of the retina at higher magnification (below).

subdivisions referred to as the anterior chamber and the posterior chamber. The *anterior chamber* lies behind the cornea and in front of the iris. The *posterior chamber* lies behind the iris and in front of the suspensory ligament and lens. The anterior cavity is filled with a watery fluid called the *aqueous humor*. The fluid is believed to be secreted into the posterior chamber by choroid plexuses of the ciliary processes of the ciliary bodies behind the iris. It is very similar to cerebrospinal fluid. From the posterior chamber, the fluid permeates the posterior cavity and then passes forward between the iris and the lens, through the pupil, into the anterior chamber. From the anterior chamber, the aqueous humor, which is continually produced, is drained off into the *canal of Schlemm* and then into the blood. The anterior chamber thus

serves a function similar to the subarachnoid space around the brain and spinal cord. The canal of Schlemm is analogous to a venous sinus of the dura mater. The pressure in the eye, called *intraocular pressure,* is produced mainly by the acqueous humor. The intraocular pressure keeps the retina smoothly applied to the choroid so the retina may form clear images. Besides maintaining normal intraocular pressure, the aqueous humor is also the principal link between the circulatory system and the lens and cornea. Neither the lens nor the cornea has blood vessels.

The second, and larger, cavity of the eyeball is the *posterior cavity*. It lies between the lens and the retina and contains a jellylike substance called the *vitreous humor*. This substance contributes to intraocular pressure, helps to prevent the eyeball from collapsing, and holds the retina flush against the internal portions of the eyeball. The vitreous humor, unlike the aqueous humor, does not undergo constant replacement. It is formed during embryonic life and is not replaced thereafter.

Exhibit 1-3 presents some surface features of the eye.

PHYSIOLOGY OF VISION

Before light can reach the rods and cones of the retina, it must pass through the cornea, aqueous humor, pupil, lens, and vitreous humor. Moreover, for vision to occur, light reaching the rods and cones must form an image on the retina. The resulting nerve impulses must then be conducted to the visual areas of the cerebral cortex. In discussing the physiology of vision, let us first consider retinal image formation.

Retinal Image Formation

The formation of an image on the retina requires four basic processes, all concerned with focusing light rays:

1. Refraction of light rays
2. Accommodation of the lens
3. Constriction of the pupil
4. Convergence of the eyes

Accommodation and pupil size are functions of the smooth muscle cells of the ciliary muscle and the dilator and sphincter muscles of the iris. They are termed *intrinsic eye muscles* since they are inside the eyeball. Convergence is a function of the voluntary muscles attached to the outside of the eyeball called the *extrinsic eye muscles*.

● *Refraction and Accommodation* When light rays traveling through a transparent medium (such as air) pass into a second transparent medium with a different density (such as water), they bend at the surface of the two media. This is *refraction* (Figure 16-11a). The eye has four such media of refraction: cornea, aqueous humor, lens, and vitreous humor (Figure 16-11b). Light rays entering the eye from the air are refracted at the following points: (1) the anterior surface of the cornea as they pass from the lighter air into the denser cornea; (2) the posterior surface of the cornea as they pass into the less dense aqueous humor; (3) the anterior surface of the lens as they pass from the aqueous humor into the denser lens; and (4) the posterior surface of the lens as they pass from the lens into the less dense vitreous humor.

When an object is 6 m (20 ft) or more away from the viewer, the light rays reflected from the object are nearly parallel to one another. The degree of refraction that takes place at each surface in the eye is very precise. Therefore the parallel rays are sufficiently bent to fall exactly on the central fovea, where vision is sharpest. However, light rays that are reflected from near objects are divergent rather than parallel. As a result, they must be refracted toward each other to a greater extent. This change in refraction is brought about by the lens of the eye.

If the surfaces of a lens curve outward, as in a convex lens, the lens will refract the rays toward each other so they eventually intersect (Figure 16-11d). The more the lens curves outward, the more acutely it bends the rays toward each other. Conversely, when the surfaces of a lens curve inward, as in a concave lens, the rays bend away from each other (Figure 16-11c). The lens of the eye is biconvex. Furthermore, it has the unique ability to change the focusing power of the eye by becoming moderately curved at one moment and greatly curved the next. When the eye is focusing on a close object, the lens curves greatly in order to bend the rays toward the central fovea. This increase in the curvature of the lens is called *accommodation* (Figure 16-12). The ciliary muscle contracts, pulling the ciliary body and choroid forward toward the lens. This action releases the tension on the lens and suspensory ligament. Due to its elasticity, the lens shortens, thickens, and bulges. In near vision, the ciliary muscle is contracted and the lens is bulging. In far vision, the ciliary muscle is relaxed and the lens is flatter. With aging, the lens loses elasticity and, therefore, its ability to accommodate.

The normal eye, known as an *emmetropic eye*, can sufficiently refract light rays from an object 6 m (20 ft) away to focus a clear object on the retina. Many individuals, however, do not have this ability because of abnormalities related to improper refraction. Among these abnormalities are *myopia* (nearsightedness), *hypermetropia* (farsightedness), and *astigmatism* (irregularities in the surface of the lens or cornea). The conditions are illustrated and explained in Figure 16-11c, d, e.

● *Constriction of Pupil* The muscles of the iris also assume a function in the formation of clear retinal images. Part of the accommodation mechanism consists of the contraction of the circular muscle fibers of the iris to constrict the pupil. Constricting the pupil means narrowing the diameter of the hole through which light enters the eye. This action occurs simultaneously with accommodation of the lens and prevents light rays from entering the eye through the periphery of the lens. Light rays entering at the periphery would not be brought to focus on the retina and would result in blurred vision. The pupil, as

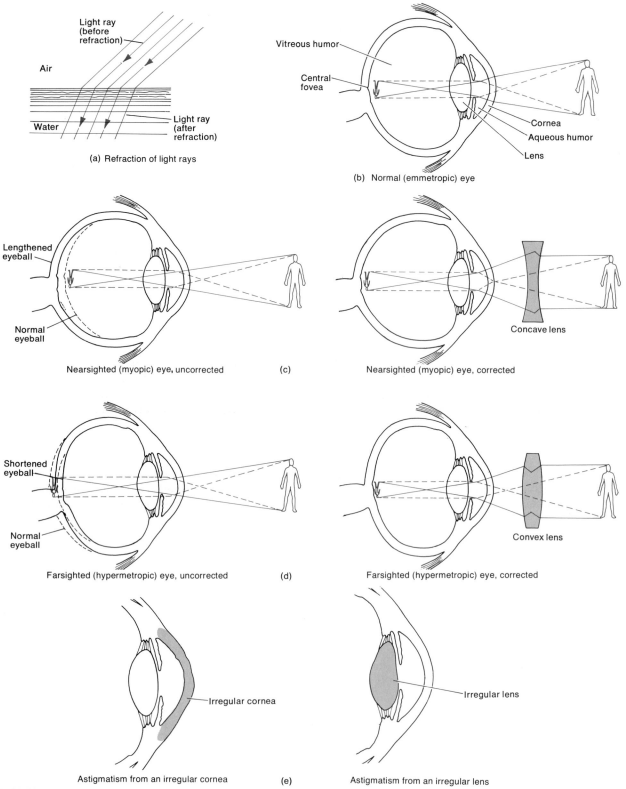

FIGURE 16-11 Normal and abnormal refraction in the eyeball. (a) Refraction of light rays passing from air into water. (b) In the normal or emmetropic eye, light rays from an object are bent sufficiently by the four refracting media and converged on the central fovea. A clear image is formed. (c) In the myopic eye, the image is focused in front of the retina. The condition may result from an elongated eyeball or a thickened lens. Correction is by use of a concave lens. (d) In the hypermetropic eye, the image is focused behind the retina. The condition may be the result of the eyeball being too short or the lens being too thin. Correction is by a convex lens. (e) Astigmatism. This is a condition in which the curvature of the cornea or lens is uneven. As a result, horizontal and vertical rays are focused at two different points on the retina. Suitable glasses correct the refraction of an astigmatic eye. On the left, astigmatism resulting from an irregular cornea. On the right, astigmatism resulting from an irregular lens. The image is not focused on the area of sharpest vision of the retina. This results in blurred or distorted vision.

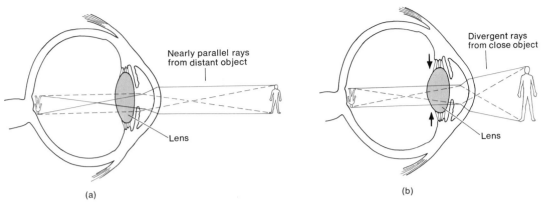

(a)   (b)

FIGURE 16-12 Accommodation. (a) For objects 6 m (20 ft) or more away. (b) For objects nearer than 6 m.

noted earlier, also constricts in bright light to protect the retina from sudden or intense stimulation.

● *Convergence* Because of the alignment of the orbits in the skull, many animals see a set of objects off to the left through one eye and an entirely different set off to the right through the other. This characteristic doubles their field of vision and allows them to detect predators behind them. In human beings, both eyes focus on only one set of objects—a characteristic called *single binocular vision*.

Single binocular vision occurs when light rays from an object are directed toward corresponding points on the two retinas. When we stare straight ahead at a distant object, the incoming light rays are aimed directly at both pupils and are refracted to identical spots on the retinas of both eyes. But as we move close to the object, our eyes must rotate medially—that is, become "crossed"—for the light rays from the object to hit the same points on both retinas. The term *convergence* refers to this medial movement of the two eyeballs to direct them both toward the object being viewed. The nearer the object, the greater the degree of convergence necessary to maintain single binocular vision. Convergence is brought about by the coordinated action of the extrinsic eye muscles.

● *Inverted Image* Images are focused upside down on the retina. They also undergo mirror reversal. That is, light reflected from the right side of an object hits the left side of the retina and vice versa. Note in Figure 16-11b that reflected light from the top of the object crosses light from the bottom of the object and strikes the retina below the central fovea. Reflected light from the bottom of the object crosses light from the top of the object and strikes the retina above the central fovea. The reason that we do not see an inverted world is that the brain learns early in life to coordinate visual images with the exact locations of objects. The brain stores memories of reaching and touching objects and automatically turns visual images right-side-up and right-side-around.

Stimulation of Photoreceptors

After an image is formed on the retina by refraction, accommodation, constriction of the pupil, and convergence, light impulses must be converted into nerve impulses by the rods and cones. Rods contain a reddish-purple pigmented compound called *rhodopsin,* or *visual purple*. This substance consists of the protein *scotopsin* plus *retinene (visual yellow),* a derivative of vitamin A. When light rays strike a rod, rhodopsin rapidly breaks down. This chemical breakdown stimulates impulse conduction by the rods (Figure 16-13).

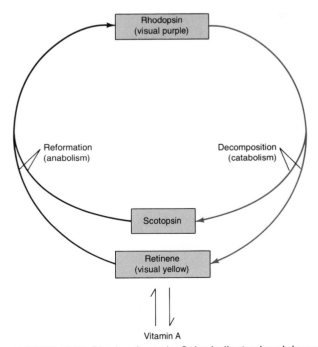

FIGURE 16-13 Rhodopsin cycle. Color indicates breakdown reactions in light. Black indicates reformation of rhodopsin in darkness.

Rhodopsin is highly light sensitive—even the light rays from the moon or a candle will break down some of it and thereby allow us to see. The rods, then, are uniquely specialized for night vision. However, they are of only limited help for daylight vision. In bright light, the rhodopsin is broken down faster than it can be manufactured. In dim light, production is able to keep pace with a slower rate of breakdown. These characteristics of rhodopsin are responsible for the experience of having to adjust to a dark room after walking in from the sunshine. The normal period of adjustment is the time it takes for the completely dissociated rhodopsin to reform. *Night blindness* is the lack of normal night vision following the adjustment period. It is most often caused by vitamin A deficiency.

Cones, which are the receptors for daylight and color, contain photosensitive chemicals that require bright light for their breakdown. Unlike rhodopsin, the photosensitive chemicals of the cones reform quickly. It is believed that there are three types of cones and that each contains a different visual pigment. Each pigment has a different maximum absorption of light of a different color so that each responds best to light of a given wavelength (color). One type of cone responds best to red light, the second to green light, and the third to blue light. Just as an artist can obtain almost any color by mixing primary colors, it is believed that cones can perceive any color by differential stimulation. Stimulation of a cone by two or more colors may produce any combination of colors.

Afferent Pathway to Brain

From the rods and cones, impulses are transmitted through bipolar neurons to ganglion cells. The cell bodies of the ganglion cells lie in the retina, and their axons leave the eyeball via the *optic nerve* (II) (Figure 16-14). The axons pass through the *optic chiasma,* a crossing point of the optic nerves. Some fibers cross to the opposite side. Others remain uncrossed. On passing through the optic chiasma, the fibers, now part of the *optic tract,* enter the brain and terminate in the thalamus. Here the fibers synapse with third-order neurons whose axons pass to the visual centers located in the occipital lobes of the cerebral cortex.

Analysis of the afferent pathway to the brain reveals that the visual field of each eye is divided into two regions: the *nasal* or *medial half* and the *temporal* or *lateral half.* For each eye, light rays from an object in the nasal half of the visual field fall on the temporal half of the retina. Light rays from an object in the temporal half of the vision field fall on the nasal half of the retina. Note that in the optic chiasma nerve fibers from the nasal halves of the retinas cross and continue on to the thalamus. Note also that nerve fibers from the temporal halves of the retinas do not cross but continue directly on to the thalamus. As a result, the visual center in the cortex of the right occipital lobe "sees" the left side of an object via impulses from the temporal half of the retina of the right eye and the nasal half of the retina of the left eye. The cortex of the left occipital lobe interprets visual sensations from the right side of an object via impulses from the nasal half of the right eye and the temporal half of the left eye.

Auditory Sensations and Equilibrium

In addition to containing receptors for sound waves, the **ear** also contains receptors for equilibrium. Anatomically, the ear is divided into three principal regions: the external or outer ear, the middle ear, and the internal or inner ear.

EXTERNAL OR OUTER EAR

The *external* or *outer ear* is structurally designed to collect sound waves and direct them inward (Figure 16-15a). It consists of the pinna, the external auditory canal, and the tympanic membrane, also called the eardrum. The *pinna,* or *auricle,* is a trumpet-shaped flap of elastic cartilage covered by thick skin. The rim of the pinna is called the helix; the inferior portion is the lobe (see Exhibit 1-3 for surface features of the external ear). The pinna is attached to the head by ligaments and muscles. The *external auditory canal* or *meatus* is a tube about 2.5 cm (1 inch) in length that lies in the external auditory meatus of the temporal bone. It leads from the pinna to the eardrum. The walls of the canal consist of bone lined with cartilage that is continuous with the cartilage of the pinna. The cartilage in the external auditory canal is covered with thin, highly sensitive skin. Near the exterior opening, the canal contains a few hairs and specialized sebaceous glands called *ceruminous glands,* which secrete *cerumen* (earwax). The combination of hairs and cerumen helps to prevent foreign objects from entering the ear. The *tympanic membrane* or *eardrum* is a thin, semi-transparent partition of fibrous connective tissue between the external auditory meatus and the middle ear. Its external surface is concave and covered with skin. Its internal surface is convex and covered with a mucous membrane.

MIDDLE EAR

Also called the *tympanic cavity,* the *middle ear* is a small, epithelial-lined, air-filled cavity hollowed out of the temporal bone (Figure 16-15a, b). The cavity is separated from the external ear by the eardrum and from the internal ear by a thin bony partition that con-

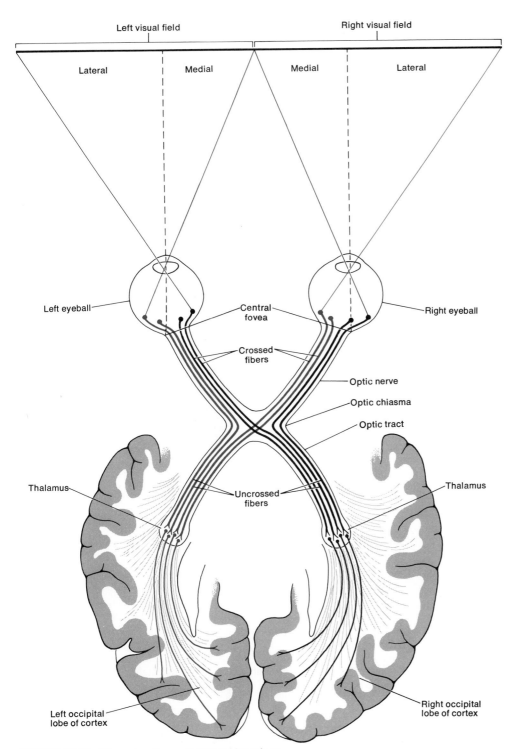

FIGURE 16-14 Afferent pathway for visual impulses.

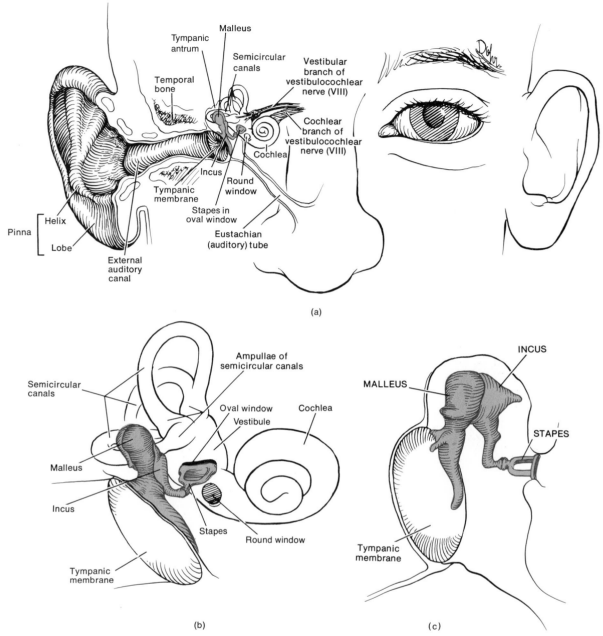

FIGURE 16-15 Structure of the auditory apparatus. (a) Divisions of the right ear into external, middle, and internal portions seen in a frontal section through the right side of the skull. The middle ear is shown in color. (b) Details of the middle ear and the bony labyrinth of the internal ear. (c) Ossicles of the middle ear.

tains two small openings: the oval window and the round window. The posterior wall of the cavity communicates with the mastoid cells of the temporal bone through a chamber called the *tympanic antrum*. This anatomical fact explains why a middle ear infection may spread to the temporal bone, causing mastoiditis, or even to the brain. The anterior wall of the cavity contains an opening that leads directly into the *eustachian tube*, also called the *auditory tube* or *meatus*. The eustachian tube connects the middle ear with the nose and nasopharynx of the throat. Through this pas-

sageway, infections may travel from the throat and nose to the ear. The function of the tube is to equalize air pressure on both sides of the tympanic membrane. Abrupt changes in external or internal air pressure might otherwise cause the eardrum to rupture. Since the tube opens during swallowing and yawning, these activities allow atmospheric air to enter or leave the middle ear until the internal pressure equals the external pressure. Any sudden pressure changes against the eardrum may be equalized by deliberately swallowing.

Extending across the middle ear are three exceedingly small bones called **auditory ossicles** (Figure 16-15c). These are called the malleus, incus, and stapes. According to their shapes, they are commonly named the hammer, anvil, and stirrup, respectively. The "handle" of the **malleus** is attached to the internal surface of the tympanic membrane. Its head articulates with the base of the incus. The **incus** is the intermediate bone in the series and articulates with the stapes. The base or footplate of the **stapes** fits into a small opening between the middle and inner ear called the *fenestra vestibuli,* or *oval window.* Directly below the oval window is another opening, the *fenestra cochlea,* or *round window.* This opening, which also separates the middle and inner ears, is enclosed by a membrane called the secondary tympanic membrane. The auditory ossicles are attached to the tympanic membrane,

to each other, and to the oval window by means of ligaments and muscles.

INTERNAL OR INNER EAR

The *internal* or *inner ear* is also called the *labyrinth* (Figure 16-16) because of its complicated series of canals. Structurally, it consists of two main divisions: (1) a bony labyrinth and (2) a membranous labyrinth that fits in the bony labyrinth. The *bony labyrinth* is a series of cavities in the petrous portion of the temporal bone. It can be divided into three areas named on the basis of shape: the vestibule, cochlea, and semicircular canals. The bony labyrinth is lined with periosteum and contains a fluid called *perilymph.* This fluid surrounds the *membranous labyrinth,* a series of sacs and

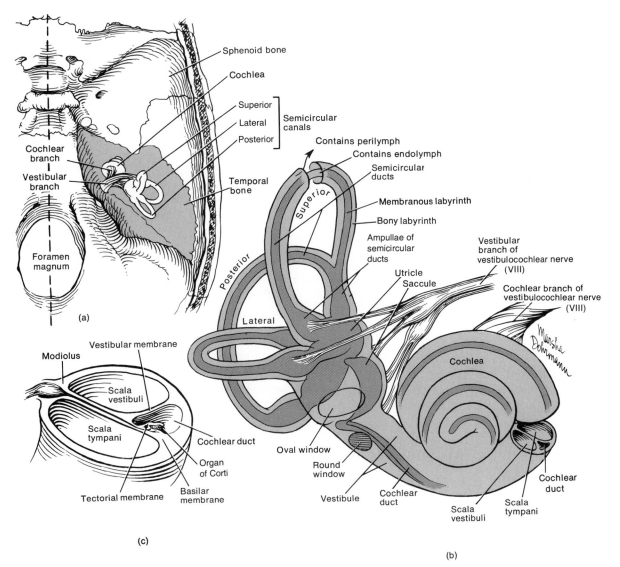

FIGURE 16-16 Details of the internal ear. (a) Relative position of the bony labyrinth projected to the inner surface of the floor of the skull. (b) The outer, gray area belongs to the bony labyrinth. The inner, colored area belongs to the membranous labyrinth. (c) Cross section through the cochlea.

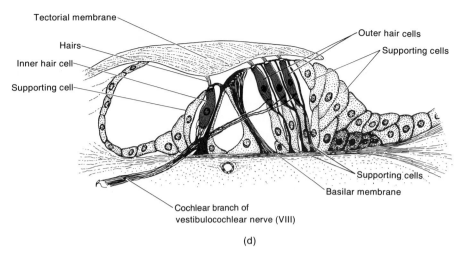

Tectorial membrane

Hairs

Inner hair cell

Supporting cell

Outer hair cells

Supporting cells

Supporting cells

Basilar membrane

Cochlear branch of
vestibulocochlear nerve (VIII)

(d)

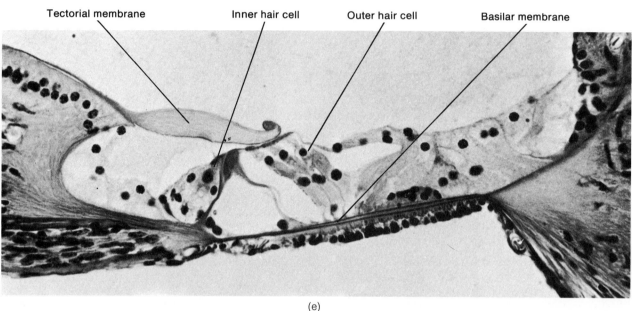

Tectorial membrane   Inner hair cell   Outer hair cell   Basilar membrane

(e)

FIGURES 16-16 *(Continued)* Details of the internal ear. (d) Enlargement of the organ of Corti. (e) Photomicrograph of the organ of Corti at a magnification of 380×. (Courtesy of Edward J. Reith, from *Atlas of Descriptive Histology,* by Edward J. Reith and Michael H. Ross, Harper & Row, Publishers, Inc., New York, 1970.)

tubes lying inside and having the same general form as the bony labyrinth. Epithelium lines the membranous labyrinth and contains a fluid called *endolymph*.

The *vestibule* constitutes the oval central portion of the bony labyrinth. The membranous labyrinth in the vestibule consists of two sacs called the *utricle* and *saccule*. These sacs are connected to each other by a small duct.

Projecting upward and posteriorly from the vestibule are the three bony *semicircular canals*. Each is arranged at approximately right angles to the other two. On the basis of their positions, they are called the superior, posterior, and lateral canals. One end of each canal enlarges into a swelling called the *ampulla*. Inside the bony semicircular canals lie portions of the

membranous labyrinth: the *semicircular ducts* or *membranous semicircular canals*. These structures are almost identical in shape to the bony semicircular canals and communicate with the utricle of the vestibule.

Lying in front of the vestibule is the *cochlea*, so designated because of its resemblance to a snail's shell. The cochlea consists of a bony spiral canal that makes about 2¾ turns around a central bony core called the *modiolus*. A cross section through the cochlea shows the canal is divided by partitions into three separate channels resembling the letter Y on its side (Figure 16-16b, c). The stem of the Y is a bony shelf that protrudes into the canal. The wings of the Y are composed of the bony labyrinth. The channel above the bony partition is the *scala vestibuli;* the channel below is the *scala tympani.* The cochlea adjoins the wall of the vestibule, into which the scala vestibuli opens. The scala tympani terminates at the round window. The perilymph of the vestibule is continuous with that of the scala vestibuli. The third channel (between the wings of the Y) is the membranous labyrinth: the *cochlear duct.* The cochlear duct is separated from the scala vestibuli by the *vestibular membrane,* also called *Reissner's membrane.* It is separated from the scala tympani by the *basilar membrane.* Resting on the basilar membrane is the *organ of Corti (spiral organ),* the organ of hearing. The organ of Corti is a series of epithelial cells on the inner surface of the basilar membrane. It consists of a number of supporting cells and hair cells, which are the receptors for auditory sensations. The inner hair cells are medially placed in a single row and extend the entire length of the cochlea. The outer hair cells are arranged in several rows throughout the cochlea. The hair cells have long hairlike processes at their free ends that extend into the endolymph of the cochlear duct. The basal ends of the hair cells are in contact with fibers of the cochlear branch of the vestibulocochlear nerve (VIII). Projecting over and in contact with the hair cells of the organ of Corti is the *tectorial membrane,* a delicate and flexible gelatinous membrane.

PHYSIOLOGY OF HEARING

Sound waves result from the alternate compression and decompression of air. They originate from a vibrating object and travel through air much as waves travel on water. The events involved in the physiology of hearing sound waves are illustrated in Figure 16-17 and listed below:

1. Sound waves that reach the ear are directed by the pinna into the external auditory canal.

2. When the waves strike the tympanic membrane, the alternate compression and decompression of the air cause the membrane to vibrate.

3. The central area of the tympanic membrane is connected to the malleus, which also starts to vibrate. The vibration is then picked up by the incus, which transmits the vibration to the stapes.

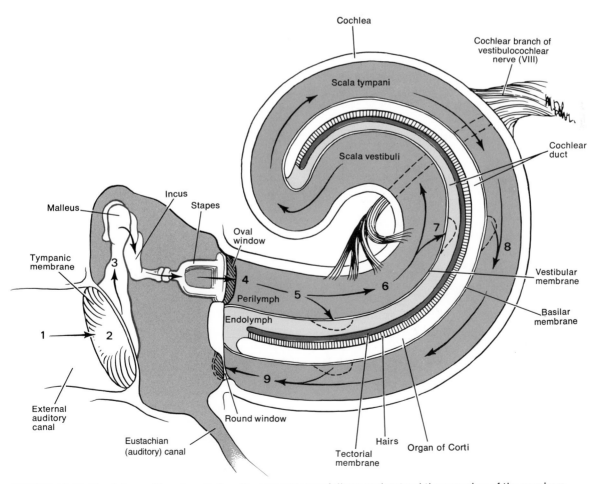

FIGURE 16-17 Physiology of hearing. Follow the text very carefully to understand the meaning of the numbers.

4. As the stapes moves back and forth, it pushes the oval window in and out.

5. The movement of the oval window sets up waves in the perilymph.

6. As the window bulges inward, it pushes the perilymph of the scala vestibuli up into the cochlea.

7. This pressure pushes the vestibular membrane inward and increases the pressure of the endolymph inside the cochlear duct.

8. The basilar membrane gives under the pressure and bulges out into the scala tympani.

9. The sudden pressure in the scala tympani pushes the perilymph toward the round window, causing it to bulge back into the middle ear. Conversely, as the sound wave subsides, the stapes moves backward and the procedure is reversed. That is, the fluid moves in the opposite direction along the same pathway and the basilar membrane bulges into the cochlear duct.

10. When the basilar membrane vibrates, the hair cells of the organ of Corti are moved against the tectorial membrane. In some unknown manner, the movement of the hairs stimulates the dendrites of neurons at their base and sound waves are converted into nerve impulses.

11. The impulses are then passed on to the cochlear branch of the vestibulocochlear nerve (Figure 16-18) and the medulla. Here some impulses cross to the opposite side and finally travel to the auditory area of the temporal lobe of the cerebral cortex.

If sound waves passed directly to the oval window without passing through the tympanic membrane and auditory bones, hearing would be inadequate. A minimal amount of sound energy is required to transmit sound waves through the perilymph of the cochlea. Since the tympanic membrane has a surface area about 22 times larger than that of the oval window, it can collect about 22 times more sound energy. This energy is sufficient to transmit sound waves through the perilymph.

PHYSIOLOGY OF EQUILIBRIUM

The term *equilibrium* has two meanings. One kind of equilibrium, called *static equilibrium,* refers to the orientation of the body (mainly the head) relative to the ground. The second kind, *dynamic equilibrium,* is the maintenance of body position (mainly the head) in response to sudden movements. The receptor organs for equilibrium are the saccule, utricle, and semicircular ducts.

The utricle and saccule each contain sensory hair cells that project into the cavity of the membranous labyrinth. The hairs are coated with a gelatinous layer which contains free particles of calcium carbonate, called *otoliths.* When the head tips downward, the otoliths slide with gravity in the direction of the ground. As the particles move, they exert a downward

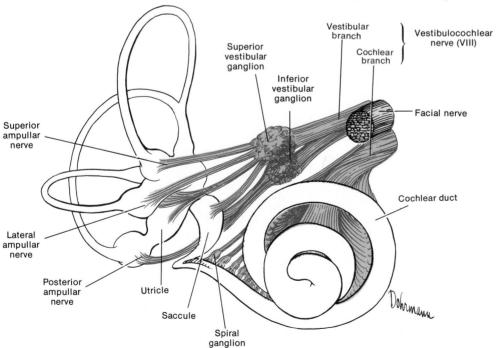

FIGURE 16-18 Principal nerves of the internal ear. Shown here is the right membranous labyrinth with the bony labyrinth removed. Nerve branches from each of the ampullae of the semicircular ducts and from the utricle and saccule form the vestibular branch of the vestibulocochlear nerve. The vestibular ganglia contain the cell bodies of the vestibular branch. The cochlear branch of the vestibulocochlear nerve arises in the organ of Corti. The spiral ganglion contains the cell bodies of the cochlear branch.

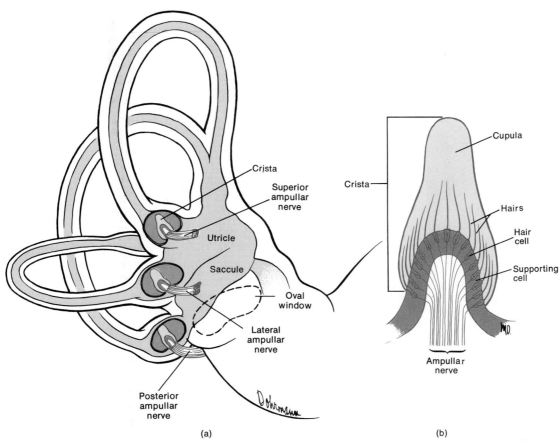

(a)   (b)

FIGURE 16-19 Semicircular ducts and dynamic equilibrium. (a) Position of the cristae relative to the membranous ampullae. (b) Enlarged aspect of a crista.

pull on the gelatinous mass, which in turn exerts a downward pull on the hairs and makes them bend. The movement of the hairs stimulates the dendrites at the bases of the hair cells. The impulse is then transmitted to the temporal lobe of the brain through the vestibular branch of the vestibulocochlear nerve (see Figure 16-18). The utricle and saccule are considered to be sense organs of static equilibrium. They provide information regarding the orientation of the head in space and are essential for maintaining posture.

The three semicircular ducts maintain dynamic equilibrium. They are positioned at right angles to each other in three planes: frontal (the superior duct), sagittal (the posterior duct), and lateral (the lateral duct). This positioning permits detection of an imbalance in three planes. In the ampulla, the dilated portion of each duct, there is a small elevation called the *crista* (Figure 16-19a). Each crista is composed of a group of hair cells covered by a mass of gelatinous material called the *cupula* (Figure 16-19b). When the head moves, the endolymph in the semicircular ducts flows over the hairs and bends them. The movement of the hairs stimulates sensory neurons, and the impulses pass over the vestibular branch of the vestibu-

locochlear nerve. The impulses then reach the temporal lobe of the cerebrum and are sent to the muscles that must contract in order to maintain body balance in the new position.

Applications to Health

The sense organs can be altered or damaged by numerous disorders. The causes of disorder can range from congenital origins to bacterial infections to the effects of old age. Here, we shall discuss only a few of the more common disorders of the eyes and ears.

CATARACT

The most prevalent disorder resulting in blindness is **cataract** formation. This disorder causes the lens or its capsule to lose its transparency (Figure 16-20). Cataracts can occur at any age, but we will discuss the type that develops with old age. As a person gets older the cells in the lenses may degenerate and be replaced with nontransparent fibrous protein. Or the lenses may start to manufacture nontransparent protein. The main symptom of cataract is a progressive, painless loss of

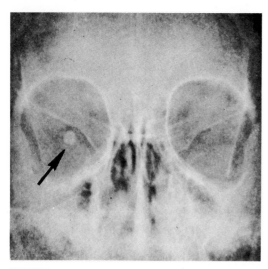

FIGURE 16-20 Anteroposterior projection of a cataract (arrow) in the lens of the right eye. (Courtesy of Lester W. Paul and John H. Juhl, *The Essentials of Roentgen Interpretation,* Harper & Row, Publishers, Inc., New York, 1972.)

vision. The degree of loss depends on the location and extent of the opacity. If vision loss is gradual, frequent changes in glasses may help maintain useful vision for a while. Eventually, though, the changes may be so extensive that light rays are blocked out altogether. At this point, surgery is indicated. Essentially, the surgical procedure consists of removing the opaque lens and substituting an artificial lens by means of eyeglasses or by implanting a plastic lens inside the eyeball to replace the natural one.

GLAUCOMA

The second most common cause of blindness, especially in the elderly, is **glaucoma.** This disorder is characterized by an abnormally high pressure of fluid inside the eyeball. The aqueous humor does not return into the bloodstream through the canal of Schlemm as quickly as it is formed. The fluid accumulates and, by compressing the lens into the vitreous humor, puts pressure on the neurons of the retina. If the pressure continues over a long period of time, it destroys the neurons and brings about blindness. It can affect a person of any age, but 95 percent of the victims are over 40. Glaucoma affects the eyesight of more than 1 million people in the United States.

CONJUNCTIVITIS

Many different eye inflammations exist, but the most common type is **conjunctivitis (pinkeye)**—an inflammation of the membrane that lines the insides of the eyelids and covers the cornea. Conjunctivitis can be caused by microorganisms—most often the pneumococci or staphylococci bacteria. In such cases, the in-

flammation is very contagious. It can also be caused by a number of irritants, in which case the inflammation is not contagious. Irritants include dust, smoke, wind, air pollution, and excessive glare. The condition may be acute or chronic. The epidemic type in children is extremely contagious, but normally it is not serious.

TRACHOMA

This chronic contagious conjunctivitis is caused by an organism called the TRIC agent. The organism has characteristics of both viruses and bacteria. **Trachoma** is characterized by many granulations or fleshy projections on the eyelids. If untreated, these projections can irritate and inflame the cornea and reduce vision. The disease produces an excessive growth of subconjunctival tissue and the invasion of blood vessels into the upper half of the front of the cornea. The disease progresses until it covers the entire cornea, bringing about a loss of vision because of corneal opacity.

MÉNIÈRE'S DISEASE

An important cause of deafness and loss of equilibrium in adults is **Ménière's disease** of the inner ear. It is a disturbance or malfunction of any part of the inner ear. It can be the result of many causes:

  1. Infection of the middle ear
  2. Trauma from brain concussion producing hemorrhage or splitting of the labyrinth
  3. Cardiovascular diseases, such as arteriosclerosis and blood vessel disturbances
  4. Congenital malformation of the labyrinth
  5. Excessive formation of endolymph
  6. Allergy

The last two causes can produce an increase in pressure in the cochlear duct and vestibular system. This pressure, in turn, causes a progressive atrophy of the hair cells of the cochlear or semicircular ducts.

If the cochlear duct is injured, typical symptoms are hissing, roaring, or ringing in the ears and deafness. If the semicircular ducts are involved, the person feels dizzy and nauseous. The dizzy spells can last from a few minutes to several days. Ménière's disease is a chronic disorder. It affects both sexes equally and usually begins in late middle life.

IMPACTED CERUMEN

Some people produce an abnormal amount of cerumen, or earwax, in the external auditory canal. Here it becomes impacted and prevents sound waves from reaching the tympanic membrane. The treatment for **impacted cerumen** is usually periodic ear irrigation or removal of wax with a blunt instrument.

KEY MEDICAL TERMS ASSOCIATED WITH SENSE ORGANS

Achromatopsia (*a* = without; *chrom* = color) Complete color blindness.

Ametropia (*ametro* = disproportionate; *ops* = eye) Refractive defect of the eye resulting in an inability to focus images properly on the retina.

Blepharitis (*blepharo* = eyelid) An eyelid inflammation.

Eustachitis Eustachian tube infection or inflammation.

Keratitis (*keralo* = cornea) An inflammation or infection of the cornea.

Keratoplasty (*plasty* = form) Corneal graft or transplant: an opaque cornea is removed and replaced with a normal transparent cornea to restore vision.

Labyrinthitis Inner ear or labyrinth inflammation.

Myringitis (*myringa* = eardrum) Inflammation of the eardrum; also called tympanitis.

Nystagmus A constant, rapid involuntary eyeball movement, possibly caused by a disease of the central nervous system.

Otalgia (*oto* = ear; *algia* = pain) Earache.

Otitis Inflammation of the ear.

Presbyopia (*presby* = old) Inability to focus on nearby objects due to loss of elasticity of the crystalline lens. The loss is usually caused by aging.

Ptosis (*ptosis* = fall) Falling or drooping of the eyelid. (This expression is also used for the slipping of any organ below its normal position.)

Retinoblastoma (*blast* = bud; *oma* = tumor) A common tumor arising from immature retinal cells and accounting for two percent of childhood malignancies.

Strabismus An eye muscle disorder, commonly called "crossed eyes." The eyeballs do not move in unison. It may be caused by lack of coordination of the extrinsic eye muscles.

Tinnitus A ringing in the ears.

STUDY OUTLINE

Sensations

Definition

1. Sensation is awareness of conditions and changes in these conditions inside and outside the body.
2. The prerequisites for sensation are receiving a stimulus, converting it into an impulse, conducting the impulse to the brain, and translating the impulse into a sensation.
3. A receptor generates a stimulus for a sensation.

Characteristics

1. Projection occurs when the brain refers a sensation to the point of stimulation.
2. Adaptation is the loss of sensation even though the stimulus is still applied.
3. An afterimage is the persistence of a sensation even though the stimulus is removed.
4. The modality is the property by which one sensation is distinguished from another.

Classification

1. Exteroceptors receive stimuli from the external environment.
2. Visceroceptors receive stimuli from blood vessels and viscera.
3. Proprioceptors receive stimuli from muscles, tendons, and joints for body position and movement.

General Senses

Cutaneous Sensations

1. Cutaneous sensations include touch, pressure, cold, heat, and pain. Receptors for these sensations are located in the skin, connective tissue, and the ends of the gastrointestinal tract.
2. Receptors for fine touch are Meissner's corpuscles, Merkel's discs, and root hair plexuses; Pacinian corpuscles are receptors for deep pressure; and Krause's end bulbs and Ruffini's end organs were once thought to be receptors for cold and heat respectively.

Pain Sensations

1. Receptors are located in nearly every body tissue.
2. Two kinds of pain recognized in the parietal lobe of the cortex are somatic and visceral.
3. Referred pain is felt in the skin near or away from the organ sending pain impulses.
4. With phantom pain, a person "feels" pain in a limb that has been amputated.

Proprioceptive Sensations

1. Receptors, found in muscles, tendons, and joints, inform us of muscle tone, movement of body parts, and body position.
2. This sense is called proprioception.
3. The receptors involved are joint kinesthetic receptors, neuromuscular spindles, and Golgi tendon organs.

Levels of Sensation

1. Sensory fibers teminating in the lower brain stem bring about far more complex motor reactions than simple spinal reflexes.
2. When sensory impulses reach the lower brain stem, they cause subconscious motor reactions.
3. Sensory impulses that reach the thalamus can be localized crudely in the body.
4. When sensory impulses reach the cerebral cortex, we experience precise localization.

Sensory Pathways

1. In the posterior column pathway and the spinothalamic pathway there are first-order, second-order, and third-order neurons.

2. The sensory pathway for pain and temperature is the lateral spinothalamic pathway.
3. The neural pathway that conducts impulses for crude touch and pressure is the anterior spinothalamic pathway.
4. The neural pathway for fine touch, proprioception, and vibration is the posterior column pathway.
5. The pathways to the cerebellum are the anterior and posterior spinocerebellar tracts.

Motor Pathways

1. Voluntary motor impulses are conveyed from the brain through the spinal cord along the corticospinal pathway and the extracorticospinal pathway.
2. Major extracorticospinal tracts are the rubrospinal, tectospinal, and vestibulospinal tracts.

Special Senses

Olfactory Sensations

1. Receptor cells in the nasal epithelium send impulses to the olfactory bulbs, olfactory tracts, and cortex.

Gustatory Sensations

1. Receptors in the taste buds send impulses to the cranial nerves, thalamus, and cortex.

Visual Sensations

1. Accessory structures of the eyes include the eyebrows, eyelids, eyelashes, and the lacrimal apparatus.
2. The eye is constructed of three coats: (a) fibrous tunic (sclera and cornea); (b) vascular tunic (choroid, ciliary body, and iris); and (c) retina, which contains rods and cones.
3. The anterior cavity contains aqueous humor, and the posterior cavity contains viterous humor.
4. Retinal image formation involves refraction of light, accommodation of lens, constriction of pupil, convergence, and inverted image formation.
5. Improper refraction may result from myopia (nearsightness), hypermetropia (farsightedness), and astigmatism (corneal or lens abnormalities).

6. Rods and cones convert light rays into visual nerve impulses; rhodopsin is necessary for the conversion.
7. Impulses from rods and cones are conveyed through retina to optic nerve, optic chiasma, optic tract, thalamus, and cortex.

Auditory Sensations and Equilibrium

1. The ear consists of three anatomical subdivisions: (a) the external or outer ear (pinna, external auditory canal, and tympanic membrane); (b) the middle ear (eustachian tube, ossicles, oval window, and round window); and (c) the internal or inner ear (bony labyrinth and membranous labyrinth). The internal ear contains the organ of Corti, the organ of hearing.
2. Sound waves are caused by the alternate compression and decompression of air.
3. Waves enter the external auditory canal, strike the tympanic membrane, pass through the ossicles, strike the oval window, set up waves in the perilymph, strike the vestibular membrane and scala tympani, increase pressure in the endolymph, strike the basilar membrane, and stimulate hairs on the organ of Corti. A sound impulse is then initiated.
4. Static equilibrium is the relationship of the body relative to the pull of gravity. The utricle and saccule are the sense organs of static equilibrium.
5. Dynamic equilibrium is equilibrium in response to movement of the body. The semicircular ducts are the sense organs of dynamic equilibrium.

Applications to Health

1. Cataract is the loss of transparency of the lens or capsule.
2. Glaucoma is abnormally high intraocular pressure, which destroys neurons of the retina.
3. Conjunctivitis is an inflammation of the conjunctiva.
4. Trachoma is a chronic, contagious inflammation of the conjunctiva.
5. Ménière's disease is the malfunction of the inner ear that may cause deafness and loss of equilibrium.
6. Impacted cerumen is an abnormal amount of earwax in the external auditory canal.

REVIEW QUESTIONS

1. Define a sensation and a sense receptor. What prerequisites are necessary for the perception of a sensation?
2. Describe the following characteristics of a sensation; projection, adaptation, afterimage, and modality.
3. Can you think of any examples of adaptation not discussed in the text?
4. Compare the location and function of exteroceptors, visceroceptors, and proprioceptors.
5. Distinguish between a general sense and a special sense.
6. What is a cutaneous sensation? How are cutaneous receptors distributed over the body? Relate your response to the two-point discrimination test.
7. For each of the following cutaneous sensations, describe the receptor involved in terms of structure, function, and location: touch, pressure, cold, and heat.

8. How do cutaneous sensations help you to survive?
9. Why are pain receptors important? Differentiate somatic pain, visceral pain, referred pain, and phantom pain.
10. What is proprioception?
11. Describe the structure, location, and function of the proprioceptive receptors.
12. What is the sensory pathway for pain and temperature?
13. Which pathway controls crude touch and pressure?
14. Which pathway is responsible for fine touch, proprioception, and vibrations?
15. Which pathways control voluntary motor impulses from the brain through the spinal cord?
16. Discuss the origin and path of an impulse that results in smelling.
17. How are papillae related to taste buds? Describe the

structure and location of the papillae. Discuss how an impulse for taste travels from a taste bud to the brain.

18. Describe the structure and importance of the following accessory structures of the eye: eyelids, eyelashes, eyebrows, and lacrimal apparatus.

19. By means of a labeled diagram, indicate the principal anatomical structures of the eye. How is the retina adapted to its function?

20. What is a sty?

21. How do extrinsic eye muscles differ from intrinsic eye muscles?

22. Describe the location and contents of the chambers of the eye. What is intraocular pressure? How is the canal of Schlemm related to this pressure?

23. Explain how each of the following events is related to the physiology of vision: (a) refraction of light, (b) accommodation, (c) constriction of the pupil, (d) convergence, and (e) inverted image formation.

24. Distinguish emmetropia, myopia, hypermetropia, and astigmatism by means of a diagram.

25. How is a light stimulus converted into an impulse?

26. What is night blindness? What causes it?

27. Describe the path of a visual impulse from the optic nerve to the brain.

28. Define visual field. Relate the visual field to image formation on the retina.

29. Diagram the principal parts of the outer, middle, and inner ear. Describe the function of each part labeled.

30. Explain the events involved in the transmission of sound from the pinna to the organ of Corti.

31. What is the afferent pathway for sound impulses from the vestibulocochlear nerve to the brain?

32. Compare the function of the semicircular ducts in maintaining dynamic equilibrium with the role of the saccule and utricle in maintaining static equilibrium.

33. Define each of the following: cataract, glaucoma, conjunctivitis, trachoma, Ménière's disease, and impacted cerumen.

34. Refer to the key medical terms associated with the sense organs. Be sure that you can define each term listed.

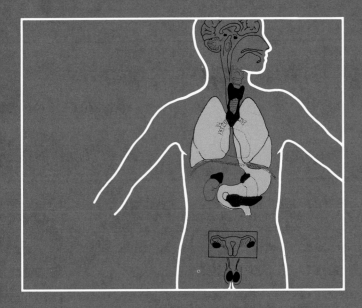

17

The Endocrine System

STUDENT OBJECTIVES

- Define an endocrine gland.
- Identify the relationship between an endocrine gland and a target organ.
- Explain the role of cyclic AMP in bringing about hormonal responses.
- Describe the location and blood supply of the anterior pituitary.
- List the hormones of the anterior pituitary, their target organs, and their functions.
- Describe the location and nerve supply of the posterior pituitary.
- Describe the source of hormones stored by the posterior pituitary, their target organs, and their functions.
- Describe the location and histology of the thyroid gland.
- Explain the physiological effects of the thyroid secretions.
- Describe the location and histology of the parathyroid glands.
- Explain the physiological effects of the parathyroid hormone.
- Identify the location and histology of the adrenal glands.
- Contrast the adrenal cortex and adrenal medulla with respect to hormones secreted and the effects of the hormones.
- Explain the correlation between the adrenal medulla and the sympathetic nervous system.
- Describe the location and histology of the pancreas.
- Explain the physiological effects of the pancreatic hormones.
- Identify the location, structure, and functions of the pineal gland.
- Identify the location, structure, and functions of the thymus gland.

- Describe pituitary dwarfism, Simmond's disease, giantism, and acromegaly as disorders of the anterior pituitary.
- Describe diabetes insipidus as a malfunctioning of the posterior pituitary.
- Explain cretinism, myxedema, exophthalmic goiter, and simple goiter as dysfunctions of the thyroid gland.
- Compare the effects of hypo- and hypersecretions of adrenocortical hormones.
- Describe diabetes mellitus and hyperinsulinism and dysfunctions of the pancreas.
- Define key medical terms associated with the endocrine system.

The nervous system controls the body through electrical impulses delivered over neurons. The body's other control system, the endocrine system, affects bodily activities by releasing chemical messengers, called hormones, into the bloodstream. Obviously, the body could not function if the two great control systems were to pull in opposite directions. The nervous and endocrine systems therefore coordinate their activities like an interlocking supersystem. Certain parts of the nervous system routinely stimulate or inhibit the release of hormones. The hormones, in turn, are quite capable of stimulating or inhibiting the flow of nerve impulses. In this chapter, you will study the endocrine glands—the body's means of chemical control.

Endocrine Glands

The body contains two kinds of glands: exocrine and endocrine. **Exocrine glands,** discussed in Chapter 3, secrete their products into ducts. The ducts then carry the secretions into body cavities or to the body's sur-face. They include sweat, sebaceous, mucous, and digestive glands. **Endocrine glands,** by contrast, secrete their products into the extracellular space around the secretory cells. Since they secrete internally, the term *endo,* meaning within, is used. The secretion passes into the capillaries to be transported in the blood. Since they have no ducts, endocrine glands are also called *ductless glands*. The endocrine glands of the body are the pituitary (hypophysis), thyroid, parathyroids, adrenals (suprarenals), pancreas, ovaries, testes, pineal (epiphysis cerebri), and thymus. The placenta or "afterbirth" is, in some ways, a temporary endocrine gland. The kidneys, stomach, and small intestine also produce hormones. The endocrine glands make up the **endocrine system.** The location of many organs of the endocrine system is illustrated in Figure 17-1.

The secretions of endocrine glands are called **hormones** (*hormone* = set in motion). A hormone may be a protein, an amine, or a steroid. Amines, like proteins, contain carbon, hydrogen, and nitrogen. Unlike proteins, they lack oxygen and contain no peptide

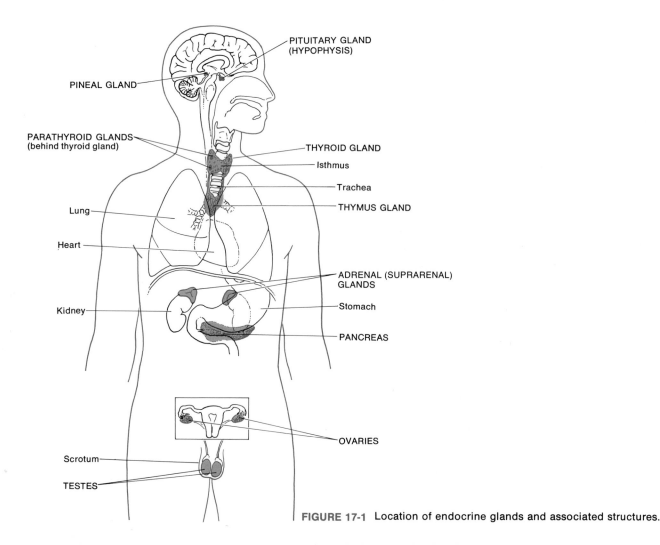

FIGURE 17-1 Location of endocrine glands and associated structures.

bonds. A steroid is a lipid with a cholesterol-type nucleus. The one thing all hormones have in common—whether protein, amine, or steroid—is the function of maintaining homeostasis by changing the physiological activities of cells. A hormone may stimulate changes in the cells of an organ or in groups of organs: the *target organs*. Or the hormone may directly affect the activities of all the cells in the body.

Hormones and Cyclic AMP

The general effects of most hormones are fairly well known. Insulin and glucagon regulate blood sugar level. Prolactin stimulates milk secretion by the mammary glands. The thyroid hormones, thyroxin (T_4) and triiodothyronine (T_3), help control metabolic rate. Progesterone helps prepare the uterus for implantation. However, the manner in which specific hormones affect different cells of the body is not entirely clear.

We have learned that one chemical substance seems to be involved in many reactions involving hormones: cyclic AMP. Cyclic AMP is synthesized from ATP, the principal energy-storing chemical in cells. This synthesis requires the enzyme adenyl cyclase. Current speculation is that a hormone circulating in the blood reaches a target cell and brings a specific message for that cell. The hormone is called the first messenger. To give a cell its message, the hormone must attach to a specific receptor site on the plasma membrane (Figure 17-2). This attachment increases the activity of adenyl cyclase in the plasma membrane. In the presence of the enzyme, ATP is converted into cyclic AMP in the cell. The cyclic AMP then diffuses throughout the cell and acts as a second messenger and performs a specific function according to the message indicated by the hormone.

In the cell, cyclic AMP activates the appropriate enzymes to get a specific job done. In liver or skeletal muscle cells, cyclic AMP activates certain enzymes that change glucose into glycogen to lower blood sugar

level according to a message from insulin but raises blood sugar level by activating other enzymes that change glycogen into glucose according to a message from glucagon. In cells of the thyroid gland, cyclic AMP stimulates the secretion of thyroid hormones in response to a message from another hormone called thyroid-stimulating hormone (TSH). In other words, cyclic AMP induces a specific target cell to perform a specific function based on the message it receives from the hormone that attaches to the plasma membrane. Thus target cells attract only specific hormones to the receptor sites of their membranes as the hormones circulate through the blood. In this way, different target cells accomplish different functions—a very efficient way for all your body cells to cooperate in maintaining homeostasis. It is also a very efficient way of ensuring that target cells function properly. Keep in mind that hormones are not enzymes; they may alter enzymatic activity, however.

High levels of cyclic AMP persist only briefly because it is degraded by an enzyme called cAMP phosphodiesterase. Prostaglandins may influence the formation of cyclic AMP in the cell membrane. Thus prostaglandins may help regulate hormonal action.

Hormones known to use cyclic AMP as a second messenger include human growth hormone, epinephrine, norepinephrine, glucagon, adrenocorticotropic hormone, luteinizing hormone, angiotensin, antidiuretic hormone, thyroid-stimulating hormone, melanocyte-stimulating hormone, thyroxin, gastrin, and serotonin.

Pituitary (Hypophysis)

The hormones of the **pituitary glands,** also called the **hypophysis,** regulate so many body activities that the pituitary has been nicknamed the "master gland." Surprisingly, the hypophysis is a small round structure measuring about 1.3 cm (0.5 inch) in diameter. It lies in the sella turcica of the sphenoid bone and is at-

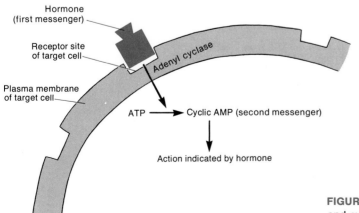

FIGURE 17-2 Proposed relationship between a hormone, target cell, and cyclic AMP.

tached to the hypothalamus of the brain via a stalklike structure. This structure is called the *infundibulum* (see Figure 14-1).

The pituitary is divided structurally and functionally into an anterior lobe and a posterior lobe. Both are connected to the hypothalamus. The *anterior lobe* contains many glandular epithelium cells and forms the glandular part of the pituitary. A system of blood vessels connects the anterior lobe with the hypothalamus. The *posterior lobe* contains axonic ends of neurons whose cell bodies are located in the hypothalamus. The nerve fibers that terminate in the posterior lobe are supported by neuroglial cells called pituicytes. Other nerve fibers connect the neurohypophysis directly with the hypothalamus.

The hypophysis is innervated by axons of the hypothalamic neurons via the infundibulum and by unmyelinated postganglionic axons from the superior cervical sympathetic ganglion. The glossopharyngeal nerve may contribute some parasympathetic fibers.

ADENOHYPOPHYSIS

The anterior lobe of the pituitary is also called the **adenohypophysis.** It releases hormones that regulate a whole range of bodily activities—from growth to re-

production. However, the release of these hormones is either stimulated or inhibited by chemical secretions from the hypothalamus of the brain. Such chemicals are called **regulating factors.** They constitute an important link between the nervous system and the endocrine system. The hypothalamic regulating factors are delivered to the adenohypophysis in the following way. The blood supply to the adenohypophysis and infundibulum is derived from several *superior hypophyseal arteries*. These arteries are branches of the internal carotid and posterior communicating arteries (Figure 17-3). The superior hypophyseal arteries form a network or plexus of capillaries, the *primary plexus*, in the infundibulum near the inferior portion of the hypothalamus. Regulating factors from the hypothalamus diffuse into this plexus. This plexus drains into veins, known as the *hypophyseal portal veins*, that pass down the infundibulum. At the inferior portion of the infundibulum, the veins form a *secondary plexus* in this adenohypophysis. From this plexus, hormones of the adenohypophysis pass into the anterior hypophyseal veins for distribution to tissue cells.

When the anterior lobe receives the proper chemical stimulation from the hypothalamus, its glandular cells secrete any one of seven hormones. The glandular cells themselves are called *acidophils, basophils,*

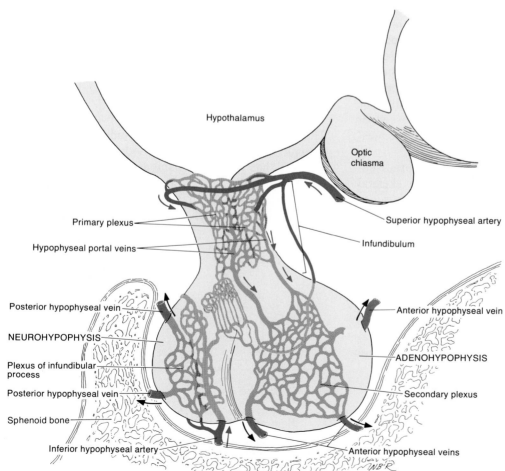

FIGURE 17-3 Blood supply of the pituitary gland.

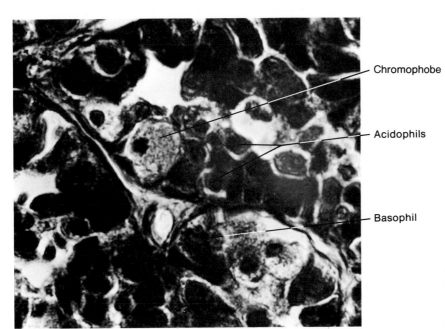

Chromophobe

Acidophils

Basophil

FIGURE 17-4 Histology of the adenohypophysis. The cytoplasm of the acidophils stains with acid dyes such as eosin. The cytoplasm of the basophils stains with basic dyes or hemtoxylin. The chromophobes stain poorly. The cells are shown at a magnification of 160X. (Courtesy of Victor B. Eichler, Wichita State University.)

or *chromophobes,* depending on the way their cytoplasm reacts to laboratory stains (Figure 17-4). The acidophils, which stain pink, secret two hormones: human growth hormone, which controls general body growth; and prolactin, which initiates milk secretion from the breasts. The basophils stain darkly and release the other five hormones. These are thyroid-stimulating hormone, which controls the thyroid gland; adrenocorticotropic hormone, which regulates the cortical regions of the adrenal glands; follicle-stimulating hormone, which stimulates the production of eggs and sperm in the reproductive organs; luteinizing hormone, which stimulates other sexual and reproductive activities; and melanocyte-stimulating hormone, which is related to skin pigmentation. The chromophobes may also be involved in the secretion of the adrenocorticotropic hormone.

Except for the growth hormone and melanocyte-stimulating hormone, all the secretions are referred to as *tropic hormones,* which means that their target organs are other endocrine glands. Prolactin, follicle-stimulating hormone, and luteinizing hormone are also called *gonadotropic hormones* because they regulate the functions of the gonads. The gonads (ovaries and testes) are the endocrine glands that produce sex hormones.

Human Growth Hormone (HGH)

The *human growth hormone (HGH)* is also known as *somatotropin* and the *somatotropic hormone (STH).* The word root *soma* means body, whereas *trop* means nourishment. Its principal function is to act on the hard and soft tissues to increase their rate of growth

and maintain their size once growth is attained. HGH causes cells to grow and multiply by directly increasing the rate at which amino acids enter cells and are built up into proteins. This process is accomplished through cyclic AMP. The building processes are called *anabolism.* Thus HGH is considered to be a hormone of protein anabolism since it increases the rate of protein synthesis. HGH also causes cells to switch from burning carbohydrates to burning fats for energy. For example, it stimulates adipose tissue to release fat. And it stimulates other cells to break down the released fat molecules. When chemical bonds are broken, energy is released. Since energy-releasing processes are referred to as *catabolism,* we can say that HGH promotes fat catabolism. At the same time, HGH accelerates the rate at which glycogen stored in the liver is converted into glucose and released into the blood. Since the cells are using fats for energy, however, they do not consume much glucose. The result is an increase in blood sugar level, a condition called *hyperglycemia.* Excessive amounts of the hormone may lead to diabetes. This process is called the *diabetogenic effect.*

Thyroid-Stimulating Hormone (TSH)

This hormone is also called *thyrotropin* and *TSH.* It stimulates the synthesis and secretion of the hormones produced by the thyroid gland.

Adrenocorticotropic Hormone (ACTH)

This hormone is also called *adrenocorticotropin* and *ACTH.* Its tropic function is to control the production and secretion of certain adrenal cortex hormones.

Follicle-Stimulating Hormone (FSH)

In the female, *follicle-stimulating hormone (FSH)* is transported from the adenohypophysis by the blood to the ovaries, where it stimulates the development of ova each month. FSH also stimulates cells in the ovaries to secrete estrogens, or female sex hormones. In the male, FSH stimulates the testes to produce sperm and secrete testosterone, a male sex hormone.

Luteinizing Hormone (LH or ICSH)

The *luteinizing hormone* is called *luteotropin (LH)* in the female and *interstitial cell stimulating hormone (ICSH)* in the male. In the female, together with estrogens, LH stimulates the ovary to release the developed ovum and prepares the uterus for implantation of a fertilized ovum. It also stimulates the secretion of progesterone (another female sex hormone) and readies the mammary glands for milk secretion. In the male, ICSH stimulates the interstitial cells in the testes to develop and secrete testosterone.

Prolactin (PR)

Prolactin (PR) or the *lactogenic hormone,* together with other hormones, initiates and maintains milk se-

cretion by the mammary glands, a process called lactation. Prolactin acts directly on tissues. By itself, it has little effect—it requires preparation by estrogens, progesterone, corticosteroids, and insulin. When the mammary glands have been primed by these hormones, prolactin brings about milk secretion.

Melanocyte-Stimulating Hormone (MSH)

The *melanocyte-stimulating hormone (MSH)* increases skin pigmentation by stimulating the dispersion of melanin granules in melanocytes. In the absence of the hormone, the skin may be pallid. An excess of MSH may cause darkening of the skin.

NEUROHYPOPHYSIS

In a strict sense, the posterior lobe, or **neurohypophysis,** is not an endocrine gland since it does not synthesize hormones. The posterior lobe consists of supporting cells called *pituicytes,* which are similar in appearance to the neuroglia of the nervous system. It also contains neuron fibers that establish an important connection with the hypothalamus (Figure 17-5). The cell bodies of the neurons originate in nuclei in the hypothalamus. The fibers project from the hypothala-

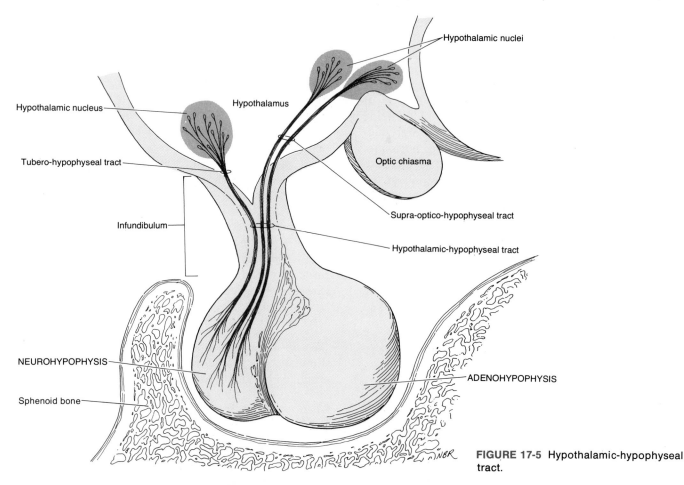

Hypothalamic nuclei

Hypothalamic nucleus

Hypothalamus

Tubero-hypophyseal tract

Optic chiasma

Infundibulum

Supra-optico-hypophyseal tract

Hypothalamic-hypophyseal tract

NEUROHYPOPHYSIS

ADENOHYPOPHYSIS

Sphenoid bone

FIGURE 17-5 Hypothalamic-hypophyseal tract.

mus, form the *hypothalamic-hypophyseal tract,* and terminate on blood capillaries in the neurohypophysis. The cell bodies of the neurons produce two hormones: oxytocin and the antidiuretic hormone. Following their production, the hormones are transported in the neuron fibers into the neurohypophysis and stored in the axon terminals. Later, when the hypothalamus is properly stimulated, it sends impulses over the neurons. The impulses cause the release of the hormones from the axon terminals into the blood.

The blood supply to the neurohypophysis is from the *inferior hypophyseal arteries,* derived from the internal carotid arteries. In the neurohypophysis, the inferior hypophyseal arteries form a plexus of capillaries called the *plexus of the infundibular process.* From this plexus hormones stored in the neurohypophysis pass into the *posterior hypophyseal veins* for distribution to tissue cells.

Oxytocin

Oxytocin stimulates the contraction of the smooth muscle cells in the pregnant uterus and the contractile cells around the ducts of the mammary glands. It is released in large quantities just prior to giving birth. When labor begins, the uterus and vagina are distended. This distension initiates afferent impulses to the hypothalamus that stimulate the secretion of more oxytocin by the hypothalamus. The oxytocin migrates along the nerve fibers of the hypothalamus to the neurohypophysis. The impulses also cause the neurohypophysis to release oxytocin into the blood. It is then carried by the blood to the uterus to reinforce uterine contractions. Oxytocin also affects milk ejection. Milk formed by the glandular cells of the breasts is stored until the baby begins active sucking. From about 30 seconds to one minute after nursing begins, the baby receives no milk. During this latent period, nerve impulses from the nipple are transmitted to the hypothalamus. The hypothalamus sends impulses down the neurosecretory neurons that release oxytocin from their axonic ends in the neurohypophysis. Oxytocin then flows from the neurohypophysis via the blood to the breasts, where it stimulates smooth muscle cells to contract and eject milk out of the mammary glands.

Antidiuretic Hormone (ADH)

An **antidiuretic** is any chemical substance that prevents excessive urine production. The principal physiological activity of ADH is its effect on urine volume. ADH causes the kidneys to remove water from newly forming urine and return it to the bloodstream—thus decreasing urine volume. In the absence of ADH, urine output may be increased tenfold.

ADH can raise blood pressure by bringing about constriction of arterioles. This effect is noted if large quantities of the purified hormone are injected. Only rarely, however, does the body secrete enough hormone to affect blood pressure significantly.

Thyroid

The endocrine organ located just below the larynx is called the **thyroid gland.** The right and left lateral lobes lie one on either side of the trachea and are connected by a mass of tissue called an *isthmus* that lies in front of the trachea just below the cricoid cartilage (Figure 17-6). The *pyramidal lobe,* when present, extends upward. The gland weighs about 25 g (almost 1 oz) and has a rich blood supply, receiving about 80 to 120 ml of blood per minute. The main blood supply is from the superior thyroid artery, a branch of the external carotid artery, and the inferior thyroid artery, a branch of the subclavian artery. The veins draining the thyroid are the superior and middle thyroid veins that pass into the internal jugular veins and the inferior thyroid veins that join the brachiocephalic veins. Histologically, the thyroid gland is composed of spherical-shaped sacs called *thyroid follicles* (Figure 17-7). The walls of each follicle consist of cells that reach the surface of the lumen of the follicle (*principal cells*) and cells that do not reach the lumen (*parafollicular cells*). The principal cells manufacture the hormones thyroxin (T_4 or tetraiodothyronine) and triiodothyronine (T_3). Together these hormones are referred to as the thyroid hormones. Thyroxin is considered to be the major hormone produced by the principal cells. The parafollicular cells produce the hormone thyrocalcitonin. The interior of each thyroid follicle is filled with *thyroid colloid,* a stored form of the thyroid hormones.

The nerve supply of the thyroid consists of postganglionic fibers from the superior and middle cervical sympathetic ganglia. Preganglionic fibers from the ganglia are derived from the second through seventh thoracic segments of the spinal cord.

One of the thyroid gland's unique features is its ability to store hormones and release them in a steady flow over a long period of time. The principal hormone, *thyroxin,* is synthesized from iodine and an amino acid called tyrosine. Synthesis usually occurs on a continuous basis. Thyroxin combines with a protein in the gland called thyroglobulin, the main constituent of the thyroid colloid. Thyroxin-thyroglobulin is stored in the thyroid follicles until thyroxin is needed by the body (Figure 17-8). Then thyroxin splits apart from thyroglobulin and is released into the blood. In the blood, thyroxin combines with plasma proteins for transportation to target tissue cells. Most thyroxin combines with a plasma protein called *thyroxin-binding globulin (TBG).* The thyroxin is released from the TBG as it

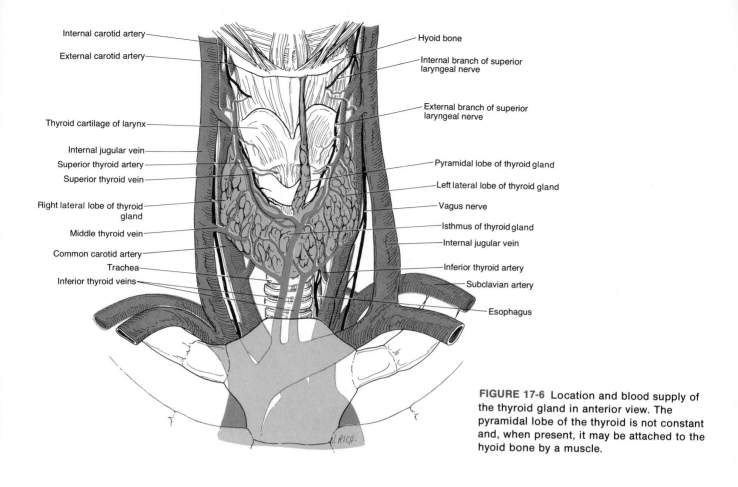

Internal carotid artery

External carotid artery

Thyroid cartilage of larynx

Internal jugular vein

Superior thyroid artery

Superior thyroid vein

Right lateral lobe of thyroid gland

Middle thyroid vein

Common carotid artery

Trachea

Inferior thyroid veins

Hyoid bone

Internal branch of superior laryngeal nerve

External branch of superior laryngeal nerve

Pyramidal lobe of thyroid gland

Left lateral lobe of thyroid gland

Vagus nerve

Isthmus of thyroid gland

Internal jugular vein

Inferior thyroid artery

Subclavian artery

Esophagus

FIGURE 17-6 Location and blood supply of the thyroid gland in anterior view. The pyramidal lobe of the thyroid is not constant and, when present, it may be attached to the hyoid bone by a muscle.

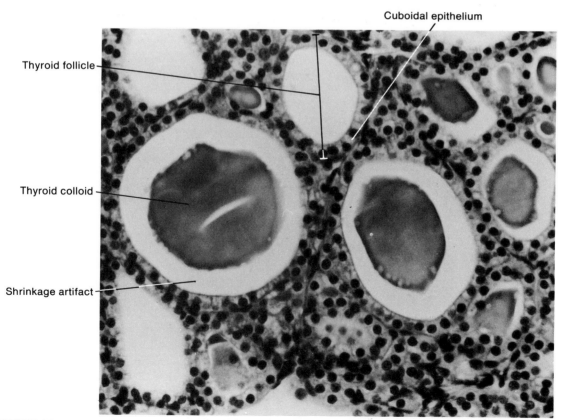

Cuboidal epithelium

Thyroid follicle

Thyroid colloid

Shrinkage artifact

FIGURE 17-7 Histology of the thyroid gland at a magnification of 90×. (Courtesy of Victor B. Eichler, Wichita State University.)

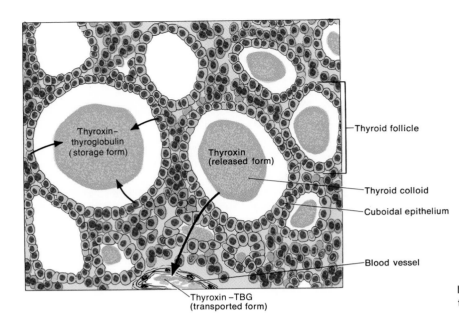

Thyroxin–thyroglobulin (storage form)

Thyroxin (released form)

Thyroid follicle

Thyroid colloid

Cuboidal epithelium

Blood vessel

Thyroxin –TBG (transported form)

FIGURE 17-8 Storage, release, and transportation of thyroxin.

enters tissue cells. Triiodothyronine is also synthesized from iodine and throsine and is stored in combination with thyroglobulin. When released from the follicles into the blood, triiodothyronine is also transported by TBG to tissue cells.

THYROXIN

The major function of thyroxin is to control metabolism by regulating the catabolic or energy-releasing processes and the anabolic or building-up processes. Thyroxin increases the rate at which carbohydrates are burned. And it stimulates cells to break down proteins for energy instead of using them for building processes. At the same time, thyroxin decreases the breakdown of fats. The overall effect, though, is to increase catabolism. It thus produces energy and raises body temperature as heat energy is given off. This process is called the *calorigenic effect*. Thyroxin also helps regulate tissue growth and development. It works with HGH to accelerate body growth, especially the growth of nervous tissue. Thyroxin deficiency during fetal development can result in fewer and smaller neurons, defective myelination of nerve fibers, and mental retardation. During the early years of life it can cause small stature and prevent certain organs from developing. Finally, thyroxin increases the reactivity of the nervous system, which causes an increase in heart rate and motility of the gastrointestinal tract.

THYROCALCITONIN (TCT)

The thyroid hormone produced by the parafollicular cells of the thyroid gland is *thyrocalcitonin (TCT)* or *calcitonin*. It is involved in the homeostasis of blood calcium level. Thyrocalcitonin lowers the amount of calcium in the blood by inhibiting bone breakdown and accelerating the absorption of calcium by the bones. It appears to exert its influence by antagonizing a number of bone resorptive agents such as vitamin D, vitamin A, and the parathyroid hormone (PTH). If thyrocalcitonin is administered to a person with a normal level of blood calcium, it causes *hypocalcemia* (low blood calcium level). Hypocalcemia is also a complication of magnesium deficiency. If thyrocalcitonin is given to a person with *hypercalcemia* (high blood calcium level), the level returns to normal.

Parathyroids

Typically embedded on the posterior surfaces of the lateral lobes of the thyroid are small, round masses of tissue called the **parathyroid glands.** Usually, two parathyroids, superior and inferior, are attached to each lateral thyroid lobe. They measure about 3 to 8 mm (0.1 to 0.3 inches) in length, 2 to 5 mm (0.07 to 0.2 inches) in width, and 0.5 to 2 mm (0.02 to 0.07 inches) in thickness. The combined weight of the four glands is between 0.05 and 3 g (Figure 17-9).

The parathyroids are abundantly supplied with blood from branches of the superior and inferior thyroid arteries. Blood is drained by the superior and middle thyroid veins which drain into the internal jugular vein and by the inferior thyroid veins which empty into the brachiocephalic or internal jugular veins. The nerve supply of the parathyroids is derived from the cervical sympathetic ganglia and the pharyngeal branch of the vagus. Histologically, the parathyroids

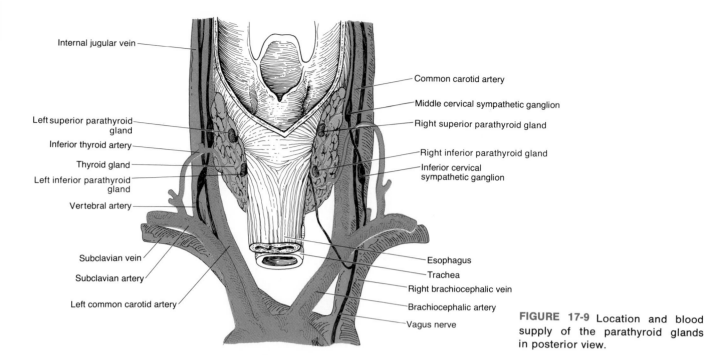

FIGURE 17-9 Location and blood supply of the parathyroid glands in posterior view.

contain two kinds of epithelial cells (Figure 17-10). The first kind, called *principal* or *chief cells,* is believed to be the major synthesizer of the parathyroid hormone. Some researchers believe that the other kind of cell, called an *oxyphil cell,* synthesizes a reserve capacity of hormone.

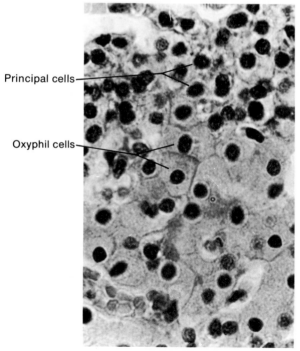

FIGURE 17-10 Histology of the parathyroid glands at a magnification of 50×. (Courtesy of Victor B. Eichler, Wichita State University.)

PARATHYROID HORMONE (PTH)

Parathyroid hormone (PTH) or *parathormone* controls the homeostasis of ions in the blood, especially the homeostasis of calcium and phosphate ions. First, if adequate amounts of vitamin D are present, PTH increases the rate at which calcium and some magnesium and phosphate ions are absorbed from the intestine into the blood. Second, PTH increases the number of osteoclasts, or bone-destroying cells. As a result, bone tissue is broken down, and calcium and phosphate ions are released into the blood. Recall that thyrocalcitonin secreted by the thyroid has the opposite effect. Finally, PTH produces two changes in the kidneys. (1) It increases the rate at which the kidneys remove calcium ions from the urine and return them to the blood. (2) It accelerates the transportation of phosphate ions from the blood into the urine for elimination. More phosphate is lost through the urine than is gained from the bones. The overall effect of PTH, then, is to decrease blood phosphate level and increase blood calcium level. As far as blood calcium level is concerned, PTH and thyrocalcitonin are antagonists.

Adrenals (Suprarenals)

The body has two **adrenal (suprarenal) glands,** one of which is located superior to each kidney (Figure 17-11). Each adrenal gland is structurally and functionally differentiated into two sections: the outer *adrenal cortex,* which makes up the bulk of the gland, and the inner *adrenal medulla* (Figure 17-12a). Covering the

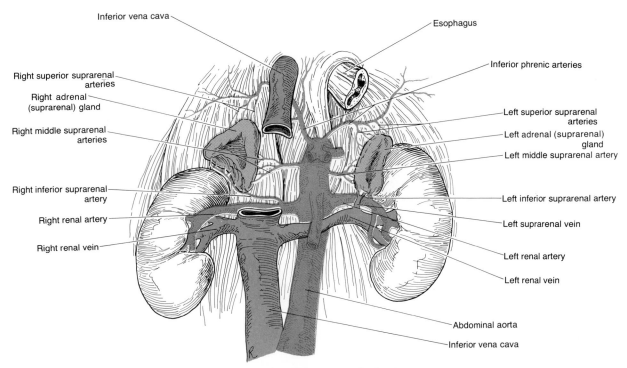

Inferior vena cava

Esophagus

Right superior suprarenal arteries

Inferior phrenic arteries

Right adrenal (suprarenal) gland

Left superior suprarenal arteries

Right middle suprarenal arteries

Left adrenal (suprarenal) gland

Left middle suprarenal artery

Right inferior suprarenal artery

Left inferior suprarenal artery

Right renal artery

Left suprarenal vein

Right renal vein

Left renal artery

Left renal vein

Abdominal aorta

Inferior vena cava

FIGURE 17-11 Location and blood supply of the adrenal (suprarenal) glands.

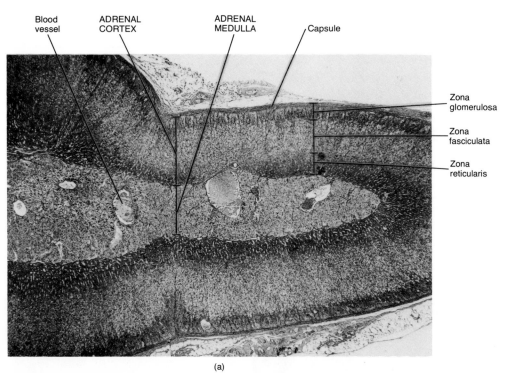

Blood vessel

ADRENAL CORTEX

ADRENAL MEDULLA

Capsule

Zona glomerulosa

Zona fasciculata

Zona reticularis

(a)

FIGURE 17-12 Histology of the adrenal (suprarenal) gland. (a) Section through the entire gland showing the cortex and medulla at a magnification of 30×.

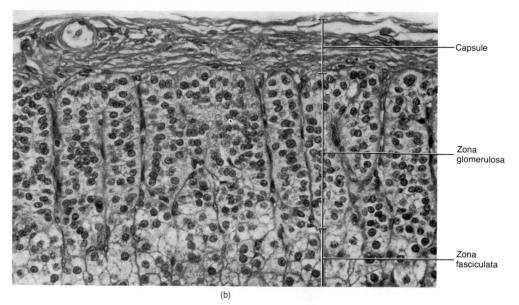

Capsule

Zona
glomerulosa

Zona
fasciculata

(b)

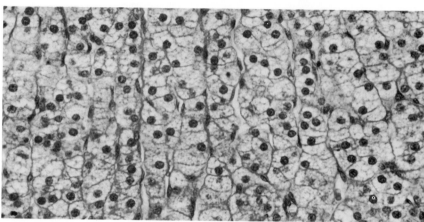

(c)

FIGURE 17-12 *(Continued)* Histology of the adrenal (suprarenal) gland. (b) Capsule and zona glomerulosa at a magnification of 400×. (c) Zona fasciculata at a magnification of 400×. (d) Zona reticularis at a magnification of 400×. (e) Medulla at a magnification of 400×. (Courtesy of Bioscience Media.)

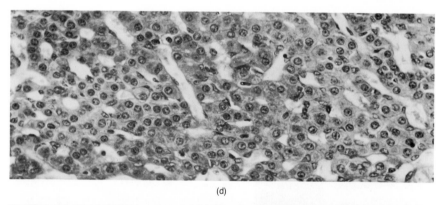

(d)

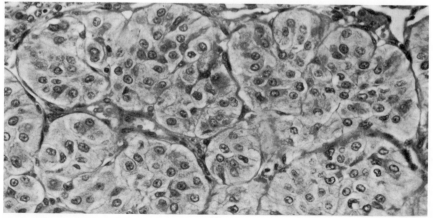

(e)

gland are an inner, thick layer of fatty connective tissue and an outer, thin fibrous capsule.

The average dimensions of the adult adrenal gland are about 50 mm (2 inches) in length, 30 mm (1.1 inches) in width, and 10 mm (0.4 inches) in thickness. The combined weight of the glands varies between 10 and 20 g (0.35 to 0.71 oz). The adrenals, like the thyroid, are among the most vascular organs of the body. The main arteries of supply are the three suprarenal arteries arising from the inferior phrenic artery, the aorta, and the renal arteries. The suprarenal vein of the right adrenal gland drains into the inferior vena cava, whereas the suprarenal vein of the left adrenal gland empties into the left renal vein.

The principal nerve supply to the adrenal glands is from preganglionic fibers from the splanchnic nerves and from the celiac and associated sympathetic plexuses. These myelinated fibers end on the secretory cells of the gland found in a region of the medulla.

ADRENAL CORTEX

Histologically, the cortex is subdivided into three zones (Figure 17-12b–d). Each zone has a different cellular arrangement and secretes different hormones. The outer zone, directly underneath the connective tissue covering, is referred to as the *zona glomerulosa.* Its cells are arranged in arched loops or round balls, and they secrete a group of hormones called mineralocorticoids. The middle zone, or *zona fasciculata,* is the widest of the three zones and consists of cells arranged in long, straight cords. The zona fasciculata secretes glucocorticoid hormones. The inner zone, the *zona reticularis,* contains cords of cells that branch freely. This zone synthesizes sex hormones, chiefly male hormones called androgens.

Mineralocorticoids

These hormones help control electrolyte homeostasis, particularly the concentrations of sodium and potassium. Although the adrenal cortex secretes at least three different hormones classified as mineralocorticoids, one of these hormones is responsible for about 95 percent of the mineralocorticoid activity. The name of this hormone is *aldosterone.* Aldosterone acts on the tubule cells in the kidneys and causes them to increase their reabsorption of sodium, with the result that sodium ions are removed from the urine and returned to the blood. In this manner, aldosterone prevents rapid depletion of sodium from the body. On the other hand, aldosterone decreases reabsorption of potassium, so large amounts of this ion are lost into the urine.

Glucocorticoids

The *glucocorticoids* are a group of hormones that are largely concerned with normal metabolism and the ability of the body to resist stress. Of the several known glucocorticoids, the most abundant and physiologically important one is *hydrocortisone (cortisol).* The glucocorticoids have the following effects on the body:

1. Glucocorticoids work with other hormones in promoting normal metabolism. Their role is to make sure that enough energy is provided. They increase the rate at which amino acids are removed from cells and transported to the liver. The amino acids may be synthesized into new proteins, such as the enzymes that are needed for the metabolic reactions. Or, if the body's reserves of glycogen and fat are low, the liver may convert the amino acids to glucose. This conversion of another substance into glucose is called gluconeogenesis. Glucocorticoids help the body to store fat.

2. Glucocorticoids work in many ways to provide resistance to stress. A sudden increase in available glucose by way of gluconeogenesis from amino acids makes the body more alert. Additional glucose gives the body energy for combating a range of stresses: fright, temperature extremes, high altitude, bleeding, infection. Glucocorticoids also make the blood vessels more sensitive to vessel-constricting chemicals. They thereby raise blood pressure. This effect is advantageous if the stress happens to be blood loss, which causes a drop in blood pressure.

3. Glucocorticoids decrease the blood vessel dilatation and edema associated with inflammations. They are thus anti-inflammatory compounds. Unfortunately, they also decrease connective-tissue regeneration and are thereby responsible for slow wound healing.

Gonadocorticoids

The adrenal cortices secrete both male and female *gonadocorticoids,* or *sex hormones.* But the concentration of sex hormones secreted by the adrenals is usually so low that it is insignificant. The exception is hypersecretion—an abnormality that will be described shortly.

ADRENAL MEDULLA

The adrenal medulla consists of hormone-producing cells, called *chromaffin cells,* which surround large blood-containing sinuses (Figure 17-12a, e). Chromaffin cells develop from the same source as the postganglionic cells of the sympathetic division of the nervous system. They are directly innervated by preganglionic cells of the sympathetic division of the autonomic nervous system and may be regarded as postganglionic cells that are specialized to secrete. In all other visceral effectors, preganglionic sympathetic fibers first synapse with postganglionic neurons before innervating the effector. In the adrenal medulla, however, the preganglionic fibers pass directly into the chromaffin cells of the gland. The secretion of hormones from the chromaffin cells is directly controlled by the autonomic nervous system, and innervation by the preganglionic fibers allows the gland to respond very rapidly to a stimulus.

Epinephrine and Norepinephrine

The two principal hormones synthesized by the adrenal medulla are *epinephrine* and *norepinephrine*. Epinephrine constitutes about 80 percent of the total secretion of the gland and is more potent in its action than norepinephrine. Both hormones are *sympathomimetic*—that is, they produce effects similar to those brought about by the sympathetic division of the autonomic nervous system. And, to a large extent, they are responsible for the fight-or-flight response. Like the glucocorticoids of the adrenal cortices, these hormones help the body resist stress situations. However, unlike the cortical hormones, the medullary hormones are not essential for life. Under stress conditions, impulses received by the hypothalamus are conveyed to sympathetic preganglionic neurons, which cause the chromaffin cells to increase their output of epinephrine and norepinephrine. Epinephrine increases blood pressure by increasing the heart rate and force of contraction and by constricting the blood vessels. It accelerates the rate of respiration, dilates respiratory passageways, decreases the rate of digestion, increases the efficiency of muscular contractions, increases blood sugar level, and stimulates cellular metabolism.

Pancreas

Because of its functions, the **pancreas** can be classified as both an endocrine and an exocrine gland. Since the exocrine functions of the gland will be discussed in the chapter on the digestive system, we shall treat only its endocrine functions at this point. The pancreas is a flattened organ located posterior and slightly inferior to the stomach (Figure 17-13). The adult pancreas consists of a head, body, and tail. Its average length is 12 to 15 cm (5 to 6 inches), and its average weight is about 85 g (3 oz). The arterial supply of the pancreas is from the superior and inferior pancreatoduodenal arteries and from the splenic and superior mesenteric arteries.

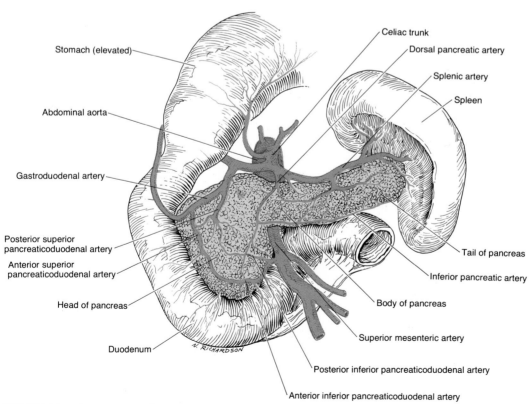

FIGURE 17-13 Location and blood supply of the pancreas.

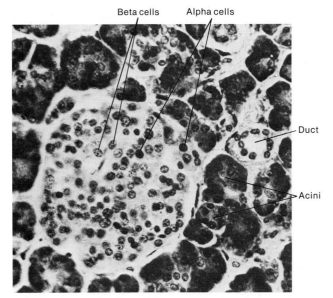

Beta cells   Alpha cells

Duct

Acini

FIGURE 17-14 Histology of the pancreas. Englarged aspect of a single islet of Langerhans and surrounding acini at a magnification of 100×. (Courtesy of Victor B. Eichler, Wichita State University.)

The nerves to the pancreas are branches of the celiac plexus. The glandular portion of the pancreas is innervated by the craniosacral division of the autonomic nervous system, whereas the blood vessels of the pancreas are innervated by the thoracolumbar division of the autonomic nervous system. The endocrine portion of the pancreas consists of clusters of cells called *islets of Langerhans* (Figure 17-14). Two kinds of cells are found in these clusters: (1) alpha cells, which constitute about 25 percent of the islet cells and secrete the hormone glucagon; and (2) *beta cells,* which constitute about 75 percent of the islet cells and secrete the hormone insulin. The islets are surrounded by blood capillaries and by the cells that form the exocrine part of the gland.

GLUCAGON AND INSULIN

The endocrine secretions of the pancreas—glucagon and insulin—are concerned with regulation of the blood sugar level.

Glucagon

The product of the alpha cells is *glucagon,* a hormone whose principal physiological activity is to increase the blood glucose level. Glucagon does this by accelerating the conversion of glycogen into glucose in the liver. The liver then releases the glucose into the blood, and the blood sugar level rises. If for some reason the alpha cells secrete glucagon continuously, hyperglycemia may result.

Insulin

The beta cells of the islets produce a hormone called *insulin.* Its chief physiological action is opposite that of glucagon. Insulin decreases blood sugar level. This is accomplished in two ways. First, insulin accelerates the transport of glucose from the blood into body cells, especially into the cells of the skeletal muscles. Second, insulin accelerates the conversion of glucose into glycogen. Insulin also increases the build-up of proteins in cells.

Ovaries and Testes

The female gonads, called the **ovaries,** are paired oval-shaped bodies located in the pelvic cavity (see Figure 17-1). The ovaries produce female sex hormones that are responsible for the development and maintenance of the female sexual characteristics. Along with the gonadotropic hormones of the pituitary, the sex hormones also regulate the menstrual cycle, maintain pregnancy, and ready the mammary glands for lactation. The male has two oval-shaped glands, called **testes,** that lie in the scrotum. The testes produce the male sex hormones that stimulate the development and maintenance of the male sexual characteristics. In Chapter 18, more will be said about the sex hormones and the anatomy of the testes and ovaries.

Pineal (Epiphysis Cerebri)

The cone-shaped gland located in the roof of the third ventricle is known as the **pineal gland,** or **epiphysis cerebri** (see Figure 17-1). The gland is about 5 to 8 mm (0.2 to 0.3 inches) long and 9 mm wide. It weighs about 0.2 g. It is covered by a capsule formed by the pia mater. It consists of masses of parenchymal and glial cells. Around the cells are scattered preganglionic sympathetic fibers. The pineal gland starts to degenerate at about age 7, and in the adult it is largely fibrous tissue. The posterior cerebral artery supplies the pineal with blood, and the great cerebral vein drains it.

Although many anatomical facts concerning the pineal gland have been known for years, its physiology is still somewhat obscure. One hormone secreted by the pineal gland is *melatonin,* which appears to affect the secretion of hormones by the ovaries. It has been known for years that light stimulates the sexual endocrine glands. Researchers have also discovered that blood levels of melatonin are low during the day and high at night. Putting these observations together, some investigators now believe that melatonin inhibits the activities of the ovaries. During daylight hours, light entering the eye stimulates neurons to transmit

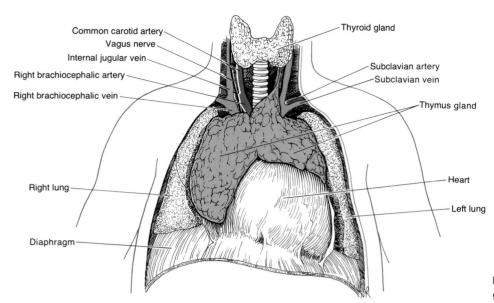

Common carotid artery
Vagus nerve
Internal jugular vein
Right brachiocephalic artery
Right brachiocephalic vein
Right lung
Diaphragm

Thyroid gland
Subclavian artery
Subclavian vein
Thymus gland
Heart
Left lung

FIGURE 17-15 Location of the thymus gland in a young child.

impulses to the pineal that inhibit melatonin secretion. Without melatonin interference, the ovaries are free to step up their hormone production. But at night, the pineal gland is able to release melatonin, and ovarian function is slowed down. One of the functions of the pineal gland might very well be regulation of the activities of the sexual endocrine glands, particularly the menstrual cycle.

Some evidence also exists that the pineal secretes a second hormone called *adrenoglomerulotropin*. This hormone may stimulate the adrenal cortex to secrete aldosterone. Still other functions attributed to the pineal gland are the secretion of growth-inhibiting factor and the secretion of a hormone called *serotonin* that is involved in normal brain physiology.

Thymus

Usually a bilobed organ, the **thymus gland** is located in the upper mediastinum posterior to the sternum and between the lungs (Figure 17-15). The gland is conspicuous in the infant, and during puberty it reaches its absolute maximum size, when it weighs between 30 and 40 g (1 to 1.3 oz). After puberty, the thymic tissue, which consists primarily of lymphocytes, is replaced by fat. By the time the person reaches maturity, the gland has atrophied. The thymus plays an essential role in the maturation of a population of lymphocytes called T cells (thymus-dependent cells) that are involved in one type of immunity.

The arterial supply of the thymus is derived mainly from the internal mammary and inferior thyroid vessels. The veins that drain the thymus are the internal mammary, brachiocephalic, and thyroid veins. Postganglionic sympathetic and parasympathetic fibers supply the gland.

Applications to Health

DISORDERS OF THE PITUITARY

Disorders of the endocrine system, in general, are based upon under- or overproduction of hormones. The term **hyposecretion** describes an underproduction, whereas the term **hypersecretion** means an oversecretion. The anterior pituitary gland produces many hormones. All these hormones, with the exception of the human growth hormone, directly control the activities of other endocrine glands. It is hardly surprising, then, that hypo- or hypersecretion of an anterior pituitary hormone produces widespread and complicated abnormalities.

Among the clinically interesting disorders related to the adenohypophysis are those involving the human growth hormone. Human growth hormone builds up cells, particularly those of bone tissue. If the hormone is hyposecreted during the growth years, bone growth is slow, and the epiphyseal plates close before normal height is reached. This is the condition called **pituitary dwarfism.** Other organs of the body also fail to grow, and the pituitary dwarf is childlike in many physical respects. Treating the condition requires administration of human growth hormone during childhood, before the epiphyseal plates close.

If secretion of human growth hormone is normal during childhood, but lower than normal during adult life, a rare condition called **pituitary cachexia (Simmond's disease)** occurs. The tissues of a person with Simmond's disease waste away, or **atrophy.** The victim becomes quite thin and shows signs of premature aging. For instance, as the connective tissue degenerates, it loses its elasticity, and the skin hangs and becomes wrinkled. The atrophy occurs because the person is not receiving enough human growth hormone to stimulate the protein-building activities that are required for replacing cells and cell parts.

Hypersecretion of the human growth hormone produces completely different disorders. For example, hyperactivity during childhood years results in **giantism,** which is an abnormal increase in the length of long bones. Hypersecretion during adulthood is called **acromegaly.** Acromegaly cannot produce further lengthening of the long bones because the epiphyseal plates are already closed. Instead, the bones of the hands, feet, cheeks, and jaws thicken. Other tissues also grow. For instance, the eyelids, lips, tongue, and nose enlarge, the skin thickens and furrows, especially on the forehead and soles of the feet.

The principal abnormality associated with dysfunction of the neurohypophysis is **diabetes** (overflow) **insipidus** (tasteless). This disorder should not be confused with diabetes mellitus (sugar), which is a disorder of the pancreas and is characterized by sugar in the urine. Diabetes insipidus is the result of a hyposecretion of ADH, usually caused by damage to the neurohypophysis or to the hypothalamus. Symptoms of the disorder include excretion of large amounts of urine and subsequent thirst. Diabetes insipidus is treated by administering ADH.

DISORDERS OF THE THYROID

Hyposecretion of thyroxin during the growth years results in a condition called **cretinism.** Two outstanding clinical symptoms of the cretin are dwarfism and mental retardation. The first is caused by failure of the skeleton to grow and mature. The second is caused by failure of the brain to develop fully. Recall that one of the functions of thyroid hormone is to control tissue growth and development. Cretins also exhibit retarded sexual development and a yellowish skin color. Flat pads of fat develop, giving the cretin the characteristic round face and thick nose, large, thick, protruding tongue, and protruding abdomen. Because the energy-producing metabolic reactions are so slow, the cretin has a low body temperature and general lethargy. Carbohydrates are stored rather than utilized. Heart rate is also slow. If the condition is diagnosed early, the symptoms are reversible following administration of the thyroid hormone.

Hypothyroidism during the adult years produces a disorder called **myxedema.** The name refers to the fact that thyroxin is a diuretic. Lack of thyroxin causes the body to retain water. One of the hallmarks of myxedema is an edema that causes the facial tissues to swell and look puffy. Another symptom caused by the retention of water is an increase in blood volume that frequently causes high blood pressure. Like the cretin, the person with myxedema also suffers from slow heart rate, low body temperature, muscular weakness, general lethargy, and a tendency to gain weight easily. The long-term combination of a slow heart rate and high blood pressure may overwork the heart muscles, causing the heart to enlarge. Because the brain has already reached maturity, the person with myxedema does not experience mental retardation. However, in moderately severe cases, nerve reactivity may be dulled so that the person lacks mental alertness. Myxedema occurs eight times more frequently in females than in males. Its symptoms are abolished by the administration of thyroxin.

Hypersecretion of thyroxin gives rise to a condition called **exophthalmic goiter.** This disease, like myxedema, is also more frequent in females, affecting eight females to every one male. One of its primary symptoms is an enlarged thyroid, called a *goiter,* which may be two to three times its original size. The two other characteristic symptoms are an edema behind the eyeball, which causes the eye to protrude (**exophthalmos),** and an abnormally high metabolic rate. The high metabolic rate produces a range of effects that are generally opposite to those of myxedema. The person has an increased pulse. The body temperature is high, and the skin is warm, moist, and flushed. Weight loss occurs, and the person is usually full of "nervous" energy. The thyroxin also increases the responsiveness of the nervous system. Thus a person with exophthalmic goiter may become irritable and may exhibit tremors of the fingers when they are extended The usual methods for treating hyperthyroidism are administering drugs that suppress thyroxin synthesis or surgically removing a part of the gland.

The term **goiter** simply means an enlargement of the thyroid gland. It is a symptom of many thyroid disorders. It may also occur if the gland does not receive enough iodine to produce sufficient thyroxin for the body's needs. The follicular cells then enlarge in a futile attempt to produce more thyroxin, and they secrete large quantities of colloid. This is called **simple goiter.** Simple goiter is most often caused by a lower-than-average amount of iodine in the diet. However, it may also develop if iodine intake is not increased during certain conditions that put a high demand on the body for thyroxin. Such conditions are frequent exposure to cold and high fat and protein diets.

DISORDERS OF THE PARATHYROIDS

A normal amount of calcium in the extracellular fluid is necessary to maintain the resting state of neurons. A deficiency of calcium caused by **hypoparathyroidism** causes neurons to depolarize without the usual stimulus. As a result, nervous impulses increase and result in muscle twitches, spasms, and convulsions. This condition is called **tetany.** The effects of hypocalcemic tetany are observed in the **Trousseau** and **Chvostek**

signs. Trousseau sign is observed when the binding of a blood pressure cuff around the upper arm produces contraction of the fingers and inability to open the hand. The Chvostek sign is a contracture of the facial muscles elicited by tapping the facial nerves at the angle of the jaw. Hypoparathyroidism results from surgical removal of the parathyroids or from parathyroid damage caused by parathyroid disease, infection, hemorrhage, or mechanical injury.

Hyperparathyroidism causes demineralization of bone. This condition is called **osteitis fibrosa cystica** because the areas of destroyed bone tissue are replaced by cavities that fill with fibrous tissue. The bones thus become deformed and are highly susceptible to fracture. Hyperparathyroidism is usually caused by a tumor in the parathyroids.

DISORDERS OF THE ADRENALS

Hypersecretion of the mineralocorticoid aldosterone results in a decrease in the body's potassium concentration. Potassium movement is involved in the transmission of nerve impulses. Consequently, if potassium depletion is great enough, neurons cannot depolarize and muscular paralysis results. Hypersecretion also brings about excessive retention of sodium and water. The water increases the volume of the blood and causes high blood pressure. It also increases the volume of the interstitial fluid, producing edema.

Disorders associated with glucocorticoids include Addison's disease and Cushing's syndrome. Hyposecretion of glucocorticoids results in the condition called **Addison's disease.** Clinical symptoms include hypoglycemia, which leads to muscular weakness, mental lethargy, and weight loss. In addition, increased potassium blood levels and decreased sodium blood levels lead to low blood pressure and dehydration. **Cushing's syndrome** is a hypersecretion of glucocorticoids, especially hydrocortisone and cortisone. The condition is characterized by the redistribution of fat. This results in spindly legs accompanied by a characteristic "moon face," "buffalo hump" on the back, and pendulous abdomen. The facial skin is flushed, and the skin covering the abdomen develops stretch marks. The individual also bruises easily, and wound healing is poor.

The **adrenogenital syndrome** results from overproduction of sex hormones, particularly the male androgens, by the adrenal cortex. Hypersecretion in male infants and young male children results in an enlarged penis. In young boys, it also causes premature development of male sexual characteristics. Hypersecretion in adult males is characterized by overgrowth of body hair, enlargement of the penis, and increased sexual drive. Hypersecretion in young girls results in prema-

ture sexual development. Hypersecretion in both girls and women usually produces a receding hairline, baldness, an increase in body hair, deepening of the voice, muscular arms and legs, small breasts, and an enlarged clitoris.

Tumors of the chromaffin cells of the adrenal medulla, called **pheochromocytomas,** cause hypersecretion of the medullary hormones. The oversecretion causes high blood pressure, high levels of sugar in the blood and urine, an elevated basal metabolic rate, nervousness, and sweating. Since the medullary hormones create the same effects as does sympathetic nervous stimulation, hypersecretion puts the individual into a prolonged version of the fight-or-flight response. Needless to say, this eventually wears out the body, and the individual eventually suffers from general weakness.

ENDOCRINE DISORDERS OF THE PANCREAS

Hyposecretion of insulin results in a number of clinical symptoms referred to as **diabetes mellitus.** Typically an inherited disease, diabetes mellitus is caused by the destruction or malfunction of the beta cells. Among the symptoms are hyperglycemia and excretion of glucose in the urine as hyperglycemia increases. There is also an inability to reabsorb water, resulting in increased urine production, dehydration, loss of sodium, and thirst. Although the cells need glucose for energy-releasing reactions, glucose cannot enter the cells without the help of insulin. The cells start breaking down large quantities of fats and proteins into glucose. When the fats are decomposed, organic acids called ketone bodies are formed as side products. Excessive decomposition of fats produces more ketone bodies than the body can neutralize through its buffer systems. As a result, the blood pH falls. This form of acidosis is called **ketosis.** The catabolism of stored fats and proteins also causes weight loss. As lipids are transported by the blood from storage depots to hungry cells, lipid particles are deposited on the walls of blood vessels. The deposition leads to atherosclerosis and a multitude of circulatory problems.

Hyperinsulinism is much rarer than hyposecretion and is generally the result of a malignant tumor in an islet. The principal symptom is a decreased blood glucose level, which stimulates the secretion of epinephrine, glucagon, and the growth hormone. As a consequence, anxiety, sweating, tremor, increased heart rate, and weakness occur. Moreover, brain cells do not have enough glucose to function efficiently. This leads to mental disorientation, convulsions, unconsciousness, shock, and eventual death as the vital centers in the medulla are affected.

KEY MEDICAL TERMS ASSOCIATED WITH THE ENDOCRINE SYSTEM

Aldosteronism A disorder caused by hypersecretion of adrenal mineralocorticoids; potassium depletion occurs, sometimes causing paralysis; sodium and water are retained, causing high blood pressure and edema.

Antidiuretic (*anti* = against; *diuresis* = increased production of urine) Any chemical substance that prevents excessive urine production.

Feminizing adenoma (*aden* = gland; *oma* = tumor) Malignant tumors of the adrenal gland that secrete abnormally high amounts of female sex hormones and produce female secondary sexual characteristics in the male.

Hyperplasia (*hyper* = over; *plas* = growth) Excessive development of tissue.

Hypoplasia (*hypo* = under) Defective development of tissue.

Neuroblastoma Malignant tumor arising from the adrenal medulla associated with metastases to bones.

Thyroid storm An aggravation of all symptoms of hyperthyroidism resulting from trauma, surgery, unusual emotional stress, or labor.

Virilism Masculinization.

Virilizing adenoma Malignant tumors of the adrenal gland that secrete high amounts of male sex hormones and produce male secondary sexual characteristics in the female.

STUDY OUTLINE

Endocrine Glands
1. Exocrine glands (sweat, sebaceous, digestive) secrete their products through ducts into body cavities or onto body surfaces.
2. Endocrine glands are ductless and secrete hormones into the blood.
3. Hormones are proteins, amines, or steroids that change the physiological activities of cells in order to maintain homeostasis. Many hormones exert their effect through cyclic AMP.
4. Organs that exhibit changes in response to hormones are called target organs.

Pituitary (Hypophysis)
1. This gland is differentiated into the adenohypophysis (the anterior lobe and glandular portion) and the neurohypophysis (the posterior lobe and nervous portion).
2. The adenohypophysis secretes tropic hormones and gonadotropic hormones. These hormones are controlled by regulating factors.
3. The blood supply to the adenohypophysis is from the superior hypophyseal arteries.
4. Hormones of the adenohypophysis are: (a) human growth hormone (regulates growth); (b) thyroid-stimulating hormone (regulates activities of thyroid); (c) adrenocorticotropic hormone (regulates adrenal cortex); (d) follicle stimulating hormone (regulates ovaries and testes); (e) luteinizing hormone (regulates female and male reproductive activities); (f) prolactin (initiates milk secretion); and (g) melanocyte-stimulating hormone (increases skin pigmentation).
5. The neural connection between the hypothalamus and neurohypophysis is via the hypothalamic-hypophyseal tract.
6. Hormones of the neurohypophysis are oxytocin (stimulates contraction of uterus and ejection of milk) and ADH (stimulates arteriole constriction and water reabsorption by the kidneys).

Thyroid
1. The gland synthesizes thyroxin which controls the rate of metabolism by increasing the catabolism of carbohydrates and proteins. It also produces triiodothyronine.
2. Throcalcitonin regulates the homeostasis of blood calcium.

Parathyroids
1. Parathyroid hormone regulates the homeostasis of calcium and phosphate by stimulating osteoclasts in response to hypercalcemia.

Adrenals (Suprarenals)
1. These glands consist of an outer cortex and inner medulla.
2. Cortical secretions are mineralocorticoids (regulate sodium reabsorption and potassium excretion); glucocorticoids (normal metabolism and resistance to stress); and gonadocorticoids (male and female sex hormones).
3. Medullary secretions are epinephrine and norepinephrine, which produce effects similar to sympathetic responses.

Pancreas
1. Alpha cells of the pancreas secrete glucagon (increases blood glucose level), and beta cells secrete insulin (decreases blood glucose level).

Ovaries and Testes
1. Ovaries are located in the pelvic cavity and produce sex hormones related to development and maintenance of female sexual characteristics.
2. Testes lie inside the scrotum and produce sex hormones related to the development and maintenance of male sexual characteristics.

Pineal (Epiphysis Cerebri)
1. This gland secretes melatonin (possibly regulates menstrual cycle), adrenoglomerulotropin (may stimulate adrenal cortex), and serotonin (involved in normal brain physiology).

Thymus
1. This gland is necessary for the maturation of T cells of the immune system.

Applications to Health
1. Pituitary disorders include pituitary dwarfism, pituitary cachexia, giantism, acromegaly, and diabetes insipidus.
2. Disorders associated with the thyroid include cretinism, myxedema, exophthalmic goiter, and simple goiter.

3. Parathyroid disorders include hypoparathyroidism and osteitis fibrosa cystica.
4. Addison's disease, adrenogenital syndrome, Cushing's syndrome, and pheochromocytomas are disorders associated with the adrenal gland.
5. Endocrine disorders of the pancreas include diabetes mellitus and hyperinsulinism.

REVIEW QUESTIONS

1. Distinguish between an endocrine gland and an exocrine gland. What is the relationship between an endocrine gland and a target organ?
2. What is a hormone? Distinguish between tropic and gonadotropic hormones. How do hormones exert their effects through cyclic AMP?
3. In what respect is the pituitary gland actually two separate glands? Describe the histology of the adenohypophysis.
4. Why does the anterior lobe of the pituitary gland have such an abundant blood supply?
5. What hormones are produced by the adenohypophysis, and what are their functions?
6. Discuss the histology of the neurohypophysis and the function and regulation of the hormones produced by the neurohypophysis.
7. Describe the structure and importance of the hypothalamic-hypophyseal tract.
8. Describe the location and histology of the thyroid gland.
9. List the functions of the thyroid hormones.
10. How is thyroxin synthesized, stored, and secreted?
11. Where are the parathyroids located? What is their histology? What are the functions of the parathyroid hormone?
12. Compare the adrenal cortex and adrenal medulla with regard to location and histology.
13. Describe the function of the hormones produced by the adrenal cortex.
14. What relationship does the adrenal medulla have to the

autonomic nervous system? What is the action of adrenal medullary hormones?
15. Describe the location of the pancreas and the histology of the islets of Langerhans. What are the actions of glucagon and insulin?
16. Where is the pineal gland located? What are its assumed functions?
17. Describe the location of the thymus gland. What is its proposed function?
18. Distinguish between hyposecretion and hypersecretion. What are the principal clinical symptoms of pituitary dwarfism, Simmond's disease, giantism, and acromegaly?
19. In diabetes insipidus, why does the patient exhibit high urine production and thirst?
20. What clinical symptoms are present in cretinism, myxedema, exophthalmic goiter, and simple goiter? Relate these symptoms to the normal activity of thyroxin.
21. Distinguish between the cause and symptoms of tetany and the cause and symptoms of osteitis fibrosa cystica.
22. What are the effects of hypersecretion of aldosterone? Describe Addison's disease, Cushing's syndrome, and the adrenogenital syndrome. What is a pheochromocytoma?
23. What are the principal effects of hypoinsulinism and hyperinsulinism?
24. Refer to the glossary of key medical terms associated with the endocrine system. Be sure that you can define each term.

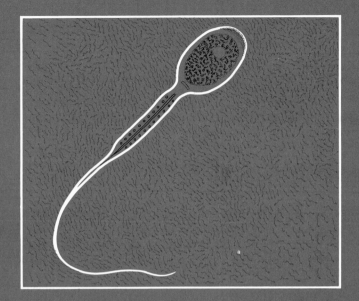

18

The Reproductive Systems

STUDENT OBJECTIVES

- Define reproduction and classify the organs of reproduction by function.

- Describe the structure and function of the scrotum.

- Explain the structure, histology, and functions of the testes.

- Describe the sequence of events involved in spermatogenesis.

- Describe the structure of a spermatozoan.

- Explain the physiological effects of testosterone.

- Describe the straight tubules and rete testis as components of the duct system of the testes.

- Describe the location, structure, histology, and functions of the ductus epididymis.

- Explain the structure and functions of the ductus deferens.

- Describe the procedure employed in a vasectomy.

- Describe the structure and function of the ejaculatory duct.

- List and describe the three subdivisions of the male urethra.

- Explain the location and functions of the seminal vesicles, prostate gland, and bulbourethral glands, the accessory reproductive glands.

- Describe the composition of semen.

- Explain the structure and functions of the penis.

- Describe the location, histology, and functions of the ovaries.

- Describe the sequence of events involved in oogenesis.

- Explain the location, structure, histology, and functions of the uterine tubes.

- Describe the location and ligamentous attachments of the uterus.

- Explain the histology and blood supply of the uterus.

- List and describe the physiological effects of estrogens and progesterone.

- Describe the location, structure, and functions of the vagina.

- List and explain the components of the vulva and note the function of each component.

- Describe the anatomical landmarks of the perineum.

- Explain the structure and histology of the mammary glands.

- Describe the symptoms of venereal diseases, prostate disorders, impotence, sterility, menstrual disorders, ovarian cysts, and breast and cervical cancer.

- Define key medical terms associated with the reproductive systems.

Reproduction is the mechanism by which the thread of life is sustained. It is the process by which a single cell duplicates its genetic material, allowing an organism to grow and repair itself. In this sense, reproduction enables the individual organism to maintain its own life. But reproduction is also the process by which genetic material is passed from generation to generation. In this regard, reproduction maintains the life of the species.

The male and female reproductive systems are organized into organs that may be grouped by function. The testes and ovaries, also called *gonads,* function in the production of gametes—sperm cells and ova, respectively. The gonads also secrete hormones. The *ducts* transport, receive, and store gametes. Still other reproductive organs, called *accessory glands,* produce materials that support gametes.

Male Reproductive System

The organs of the male reproductive system (Figure 18-1) are the testes, or male gonads, which produce sperm, a number of ducts that either store or transport sperm to the exterior, accessory glands that add secre-

tions comprising the semen, and several supporting structures, including the penis.

SCROTUM

The **scrotum** is a pouching of the abdominal wall consisting of loose skin and superficial fascia. It is the supporting structure for the testes. Externally, it looks like a single pouch of skin separated into lateral portions by a medial ridge called the *raphe.* Internally, it is divided by a septum into two sacs, each containing a single testis. The septum consists of superficial fascia and bundles of smooth muscle fibers: the *dartos.* The dartos is also found in the subcutaneous tissue of the scrotum. The testes are the organs that produce sperm. Sperm production and survival require a temperature that is lower than body temperature. Because the scrotum is outside of the body cavities, it supplies an environment about 3°F below body temperature. Exposure to cold causes contraction of the smooth muscle fibers, moving the testes closer to the pelvic cavity where they can absorb body heat. This contraction causes the skin of the scrotum to wrinkle. Exposure to warmth reverses the process. A muscle in the

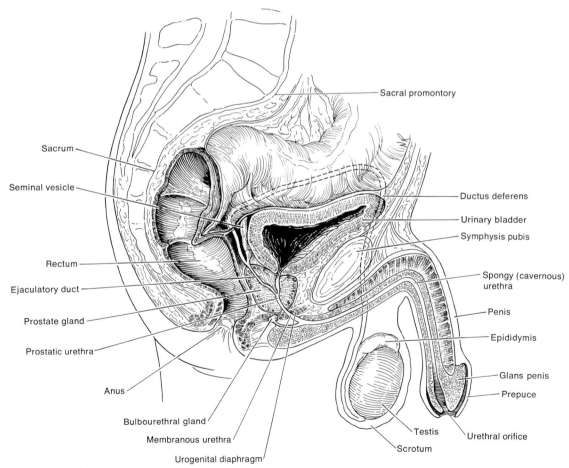

FIGURE 18-1 Male organs of reproduction seen in sagittal section.

spermatic cord, the cremaster muscle also elevates the testes upon exposure to cold.

The blood supply of the scrotum is derived from the internal pudendal branch of the internal iliac, the cremasteric branch of the inferior epigastric artery, and the external pudendal artery from the femoral artery. The scrotal veins follow the arteries. The scrotal nerves are derived from the pudendal, posterior cutaneous of the thigh, and ilioinguinal nerves.

TESTES

The **testes** are paired oval glands measuring about 5 cm (2 inches) in length and 2.5 cm (1 inch) in diameter. They weight between 10 and 15 g (Figure 18-2a). During most of fetal life they lie in the pelvic cavity, but about two months prior to birth they descend into the scrotum. When the testes do not descend, the condition is referred to as *cryptorchidism.* Cryptorchidism results in sterility because the sperm cells are destroyed by the higher body temperature of the pelvic cavity. Undescended testes can be placed in the scrotum by administering hormones or by surgical means prior to puberty without ill effects.

The testes are covered by a dense layer of white

fibrous tissue, the *tunica albuginea,* that extends inward and divides each testis into a series of internal compartments called *lobules.* Each lobule contains one to three tightly coiled tubules, the convoluted *seminiferous tubules,* that produce sperm by a process called *spermatogenesis.* A cross section through a seminiferous tubule reveals that it is packed with sperm cells in various stages of development (Figure 18-2b–d). The most immature cells, the spermatogonia, are located against the basement membrane. Toward the lumen in the center of the tube, one can see layers of progressively more mature cells. In order of advancing maturity, these are primary spermatocytes, secondary spermatocytes, and spermatids. By the time a **sperm cell,** or **spermatozoan,** has reached full maturity, it is in the lumen of the tubule and begins to be moved through a series of ducts. Embedded between the developing sperm cells in the tubules are *Sertoli cells* that produce secretions for supplying nutrients to the spermatozoa. Between the seminiferous tubules are clusters of *interstitial cells of Leydig.* These cells secrete the male hormone testosterone. Since the testes produce sperm and testosterone, they are both exocrine and endocrine glands.

The testes are supplied by the testicular arteries,

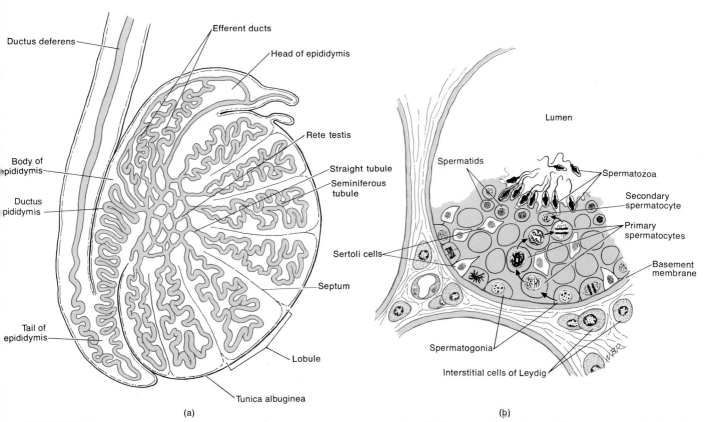

(a)

(b)

FIGURE 18-2 Testis. (a) Sectional view of a testis showing its system of tubes. (b) Cross section of a seminiferous tubule showing the stages of spermatogenesis.

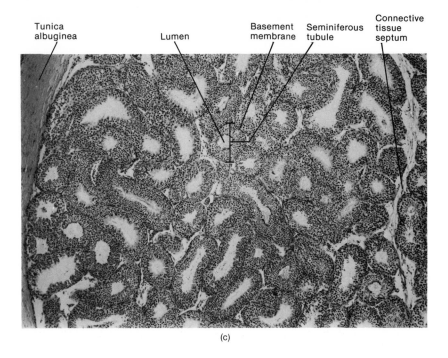

Tunica albuginea   Lumen   Basement membrane   Seminiferous tubule   Connective tissue septum

(c)

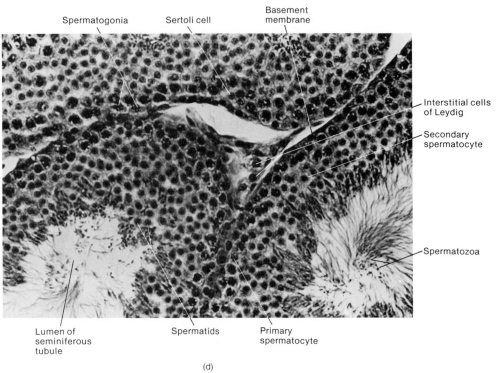

Spermatogonia   Sertoli cell   Basement membrane

Interstitial cells of Leydig

Secondary spermatocyte

Spermatozoa

Lumen of seminiferous tubule   Spermatids   Primary spermatocyte

(d)

FIGURE 18-2 (Continued) Testis. (c) Histology of several seminiferous tubules at a magnification of 65×. (Courtesy of Edward J. Reith, from *Atlas of Descriptive Histology,* by Edward J. Reith and Michael H. Ross, Harper & Row, Publishers, Inc., New York, 1970.) (d) Details of a seminiferous tubule at a magnification of 275×. (Courtesy of Victor B. Eichler, Wichita State University.)

which arise from the aorta immediately below the origin of the renal arteries. Blood from the testes is drained on the right side by the testicular vein which enters the inferior vena cava and on the left side by the testicular vein which enters the left renal vein. The nerve supply to the testes is from the testicular plexus which contains vagal parasympathetic fibers and sympathetic fibers from the tenth thoracic segment of the spinal cord.

SPERMATOGENESIS

Each human being develops from the union of an ovum and a sperm. Ova and sperm are collectively called **gametes,** and they differ radically from all the other body cells in that they have only half the normal number of chromosomes in their nuclei. **Chromosome number** is the number of chromosomes in each nucleated cell that is not a gamete. Chromosome numbers vary from species to species. The human chromosome number is 46—each brain cell, stomach cell, heart cell, and practically every other cell contains 46 chromosomes in its nucleus. In other words, there are 23 pairs of chromosomes in each cell other than a gamete. Two chromosomes that belong to a pair are called homologous chromosomes. The ovum or sperm has only one member of each pair. Of these 46 chromosomes, 23 contain the genes that are necessary for programming all the activities of the body. In a sense, the other 23 are a duplicate set. Another word for chromosome number is **diploid number** (*di-* = two), symbolized as *2n*.

A sperm containing 46 chromosomes that fertilizes an egg containing 46 chromosomes might be thought to create offspring with 92 chromosomes. In reality, the chromosome number does not double with each generation because of a special nuclear division called **meiosis.** Meiosis occurs in the process of producing sex cells. It causes a developing sperm or ovum to re-

linquish its duplicate set of chromosomes so that the mature gamete has only 23—this is the **haploid number,** meaning "one-half" and symbolized *n*.

In the testes, the formation of haploid spermatozoa by meiosis is called **spermatogenesis.** In the ovary, the formation of a haploid ovum by meiosis is referred to as **oogenesis.**

In humans, spermatogenesis takes about 2.5 months. The seminiferous tubules are lined with immature cells called *spermatogonia* or sperm mother cells (Figure 18-3). Spermatogonia contain the diploid chromosome number and are the precursor cells for all the spermatozoa the male will produce. At puberty, the spermatogonia embark on a lifetime of active division. As a result of their active mitosis, daughter cells are pushed inward toward the lumen of the seminiferous tubule. These cells lose contact with the basement membrane of the seminiferous tubule. And, as a result of certain developmental changes, these cells become known as *primary spermatocytes*. Primary spermato-

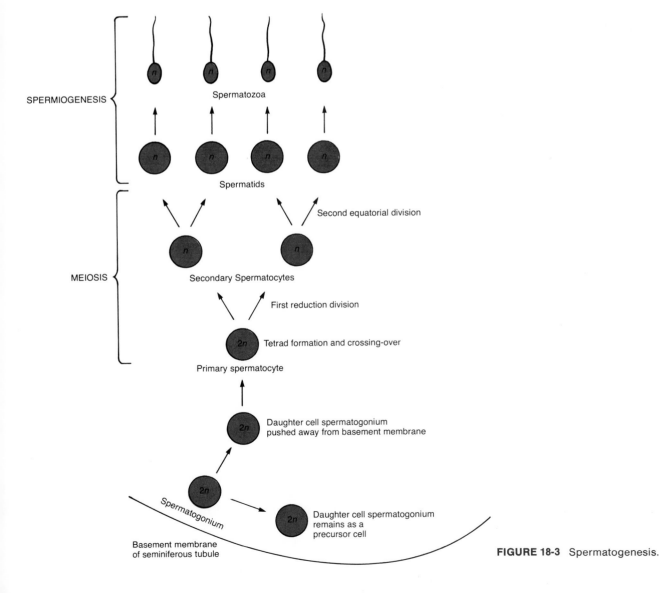

SPERMIOGENESIS

Spermatozoa

Spermatids

MEIOSIS

Second equatorial division

Secondary Spermatocytes

First reduction division

Tetrad formation and crossing-over

Primary spermatocyte

Daughter cell spermatogonium pushed away from basement membrane

Spermatogonium

Daughter cell spermatogonium remains as a precursor cell

Basement membrane of seminiferous tubule

FIGURE 18-3 Spermatogenesis.

cytes, like spermatogonia, are diploid, that is, they have 46 chromosomes. The other daughter cells formed by mitosis of the spermatogonia remain near the basement membrane and form a reservoir of precursor cells.

Each primary spermatocyte enlarges before dividing. This nuclear division is the first of two that take place as part of meiosis. In the first, DNA is replicated and 46 chromosomes form and move toward the equatorial plane of the nucleus. There they line up by homologous pairs so that there are 23 pairs of chromosomes (each of two chromatids) in the center of the nucleus. The four chromatids of each homologous pair then twist around each other to form a *tetrad*. In a tetrad, portions of one chromatid are exchanged with portions of another. This process, called *crossing-over,* permits an exchange of genes among chromatids (Figure 18-4) that results in the recombination of genes. Thus the spermatozoa eventually produced may be genetically unlike each other and unlike the cell that produced them—hence the great variation among humans. Following crossing-over the spindle forms and the threads attach to the centromeres of the paired chromosomes. As the pairs separate, one member of each pair migrates to opposite poles of the dividing nucleus. The cells thus formed by the first nuclear division (reduction division) are called *secondary spermatocytes*. Each cell has 23 chromosomes—the haploid number. Each chromosome, however, is made up of two chromatids. Moreover, the genes of the chromosomes of secondary spermatocytes are rearranged as a result of crossing-over.

The second nuclear division of meiosis is equatorial division. There is no replication of DNA. The chromosomes (each of two chromatids) line up in single file around the equatorial plane, and the chromatids of each chromosome separate from each other. The cells thus formed from the equatorial division are called *spermatids*. Each contains half the original chromosome number, or 23 chromosomes, and is haploid. Each primary spermatocyte therefore produces four spermatids by meiosis (reduction division and equatorial division). Spermatids lie close to the lumen of the seminiferous tubule.

The final stage of spermatogenesis—*spermiogenesis*—involves the maturation of spermatids into spermatozoa. Each spermatid embeds in a Sertoli cell and develops a head with an acrosome and a flagellum (tail). Sertoli cells nourish the developing spermatids. Since there is no cell division in spermiogenesis, each spermatid develops into a single spermatozoan.

In summary, then, a single primary spermatocyte develops into four spermatozoa by meiosis and spermiogenesis. Spermatozoa enter the lumen of the seminiferous tubule and migrate to the ductus epididymis. After about ten days, they complete their maturation and become capable of fertilizing an ovum. Many spermatozoa are probably stored in the ductus deferens.

Spermatozoa

Spermatozoa, once ejaculated, have a life expectancy of about 48 hours. A spermatozoan is highly adapted for reaching and penetrating a female ovum. It is composed of a head, a middle piece, and a tail (Figure 18-5). The head contains the nuclear material and the acrosome, which contains chemicals that effect penetration of the sperm cell into the ovum. Numerous mitochondria in the middle piece carry on the metabolism that provides energy for locomotion. The tail, a typical flagellum, propels the sperm along its way.

Testosterone

Secretions of the anterior pituitary gland assume a major role in the developmental changes associated with puberty. At the onset of puberty the anterior pituitary starts to secrete gonadotropic hormones called follicle-stimulating hormone (FSH) and interstitial cell stimulating hormone (ICSH). Their release is controlled from the hypothalamus by follicle-stimulating hormone releasing factor (FSHRF) and interstitial cell

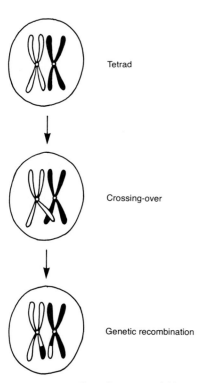

Tetrad

Crossing-over

Genetic recombination

FIGURE 18-4 Crossing-over within a tetrad resulting in genetic recombination.

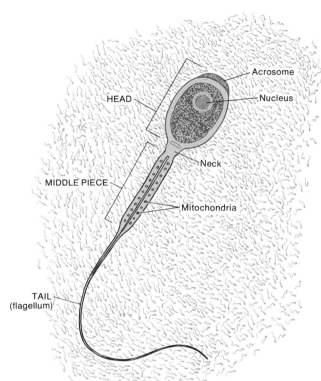

(a)

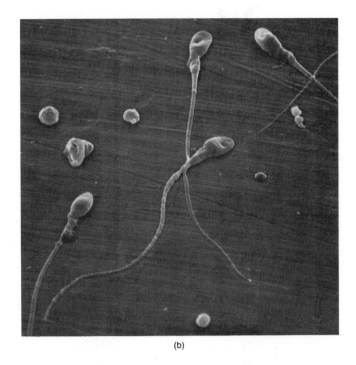

(b)

FIGURE 18-5 Spermatozoa. (a) Parts of a spermatozoan. (b) Scanning electron micrograph of several spermatozoa at a magnification of 2000×. (Courtesy of Fisher Scientific Company and S.T.E.M. Laboratories, Inc., Copyright 1975.)

stimulating hormone releasing factor (ICSHRF). Once secreted, the gonadotropic hormones have profound effects on male reproductive organs. FSH acts on the seminiferous tubules to initiate spermatogenesis. ICSH acts on the seminiferous tubules and further assists the tubules to develop mature sperm, but its chief function is to stimulate the interstitial cells of Leydig to secrete testosterone.

Testosterone has a number of effects on the body. It controls the development, growth, and maintenance of the male sex organs. It also regulates bone growth, protein anabolism, sexual behavior, sperm production, and the development of male secondary sex characteristics. These characteristics, which appear at puberty, include muscular development resulting in wide shoulders and narrow hips; body hair patterns that include pubic hair, axillary and chest hair (within hereditary limits), facial hair, and temporal hairline recession; and enlargement of the thyroid cartilage producing deepening of the voice.

DUCTS

When the sperm mature, they are moved through the convoluted seminiferous tubules to the **straight tubules.** The straight tubules lead to a network of ducts in the testis called the **rete testis.** Some of the cells lining the rete testis possess cilia that probably push the sperm along. The sperm are next transported out of the testis through a series of coiled **efferent ducts** that empty into a single tube called the **ductus epididymis.** (see Figure 18-2a). At this point, the sperm are morphologically mature.

Epididymis

The two **epididymides** are comma-shaped organs. Each lies along the posterior border of the testis (see Figures 18-1 and 18-2a) and consists mostly of a tightly coiled tube: the *ductus epididymis.* The larger, superior portion of the epididymis is known as the *head.* It consists of the efferent ducts that empty into the ductus epididymis. The *body* of the epididymis contains the ductus epididymis. The *tail* is the smaller, inferior portion. Within the tail, the ductus epididymis becomes the ductus deferens.

The ductus epididymis measures about 6 mm (20 ft) in length and 1 mm in diameter. It is tightly packed within the epididymis, which measures only about 3.8 cm (1.5 inches). The ductus epididymis is lined with pseudostratified columnar epithelium, and its wall contains smooth muscle. The free surfaces of the columnar cells contain long, branching microvilli called *stereocilia* (Figure 18-6). Functionally, the epididymis

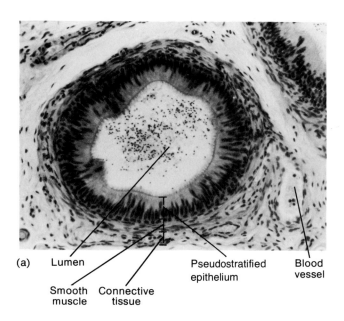

(a)

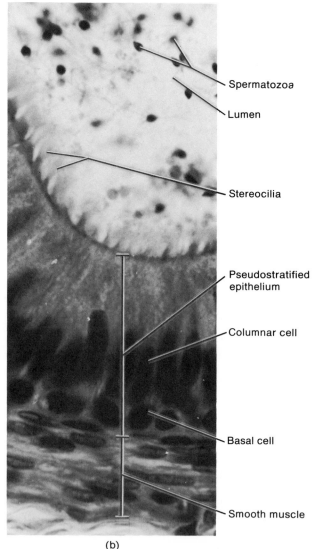

(b)

FIGURE 18-6 Histology of the ductus epididymis. (a) Cross section through the ductus epididymis at a magnification of 30×. (b) Enlarged aspect of the epithelium at a magnification of 200×. (Courtesy of Victor B. Eichler, Wichita State University.)

is the site of sperm maturation. It stores spermatozoa and propels them toward the urethra during ejaculation by peristaltic contractions of the smooth muscle.

Ductus Deferens

Within the tail of the epididymis, the ductus epididymis becomes less convoluted, its diameter increases, and at this point it is referred to as the **ductus (vas) deferens** or **seminal duct** (see Figure 18-2a). The ductus deferens, about 45 cm (18 inches) long, ascends along the posterior border of the testis, penetrates the inguinal canal, and enters the pelvic cavity, where it loops over the side and down the posterior surface of the urinary bladder. The dilated terminal portion of the ductus deferens is known as the *ampulla*. Histologically, the ductus deferens is lined with pseudostratified epithelium and contains a heavy coat of three layers of muscle (Figure 18-7). Peristaltic contractions of

the muscular coat propel the spermatozoa toward the urethra during ejaculation.

Traveling with the ductus deferens as it ascends in the scrotum are the testicular artery, autonomic nerves, veins that drain the testes, lymphatics, and a small circular band of skeletal muscle called the *cremaster muscle*. These structures constitute the **spermatic cord,** a supporting structure of the male reproductive system. The cremaster muscle elevates the testes during sexual stimulation and exposure to cold. The spermatic cord passes through the *inguinal canal,* a slitlike passageway in the anterior abdominal wall just superior to the medial half of the inguinal ligament. The area of the inguinal canal and spermatic cord represents a weak spot in the abdominal wall, and it is frequently the site of a *hernia*—a rupture or separation of a portion of the abdominal wall resulting in the protrusion of a part of a viscus.

One method of sterilization of males is called *vasec-*

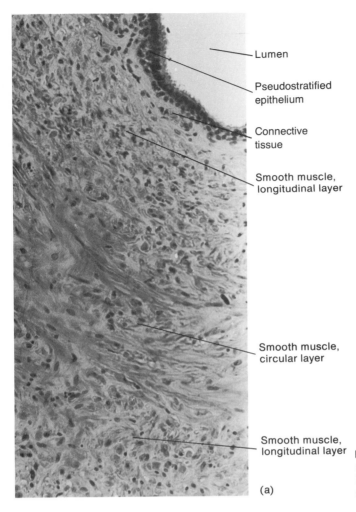

Lumen

Pseudostratified epithelium

Connective tissue

Smooth muscle, longitudinal layer

Smooth muscle, circular layer

Smooth muscle, longitudinal layer

(a)

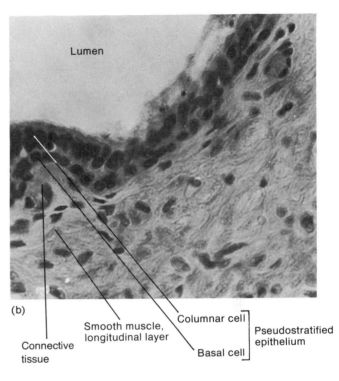

Lumen

(b)

Connective tissue

Smooth muscle, longitudinal layer

Columnar cell

Basal cell

Pseudostratified epithelium

FIGURE 18-7 Histology of the ductus deferens. (a) Cross section of a portion of the ductus deferens at a magnification of 40×. (b) Enlarged aspect of the epithelium and muscularis at a magnification of 100×. (Courtesy of Victor B. Eichler, Wichita State University.)

tomy, in which a portion of each ductus deferens is removed. In the procedure, an incision is made in the scrotum, the ducts are located, and each is tied in two places. Then the portion between the ties is excised. Although sperm production continues in the testes, the sperm cannot reach the exterior since the ducts are cut.

Ejaculatory Duct

Posterior to the urinary bladder, each ductus deferens joins its **ejaculatory duct** (Figure 18-8). Each duct is about 2 cm (1 inch) long. Both ejaculatory ducts eject spermatozoa into the prostatic urethra. The urethra is the terminal duct of the system, serving as a common passageway for both spermatozoa and urine.

Urethra

In the male, the **urethra** passes through the prostate gland, the urogenital diaphragm, and the penis. It mea-

sures about 20 cm (8 inches) in length and is subdivided into three parts (see Figures 18-1 and 18-9). The *prostatic portion* is 2 to 3 cm (1 inch) long and passes through the prostate gland. It continues inferiorly as the membranous portion as it passes through the urogenital diaphragm, a muscular partition between the two ischiopubic rami. The *membranous portion* is about 1 cm (0.5 inch) in length. After passing through the urogenital diaphragm, it is known as the *spongy (cavernous) portion* of the urethra. This portion is about 15 cm (6 inches) long. The spongy urethra enters the bulb of the penis and terminates at the external urethral orifice.

ACCESSORY GLANDS

Whereas the ducts of the male reproductive system store and transport sperm cells, the **accessory glands** secrete the liquid portion of semen. The paired **seminal vesicles** (Figure 18-8) are convoluted pouchlike struc-

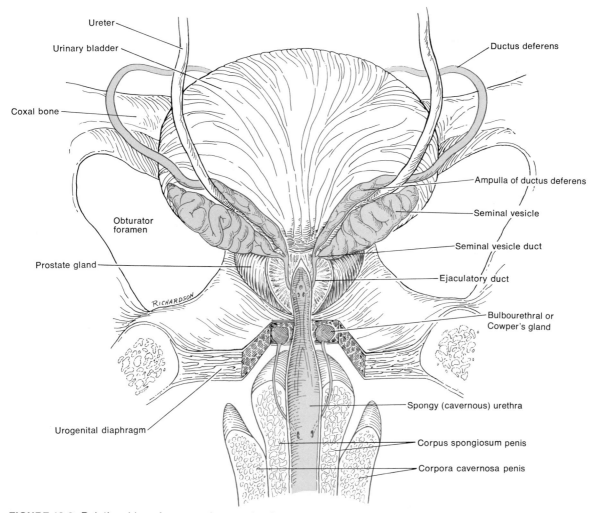

FIGURE 18-8 Relationships of some male reproductive organs: posterior view of urinary bladder.

tures, about 5 cm (2 inches) in length, lying posterior to and at the base of the urinary bladder in front of the rectum. They secrete the alkaline viscous component of semen rich in the sugar fructose and pass it into the ejaculatory duct. The seminal vesicles contribute about 60 percent of the volume of semen.

The **prostate gland** is a single, doughnut-shaped gland about the size of a chestnut (see Figure 18-8). It is inferior to the urinary bladder and surrounds the upper portion of the urethra. The prostate secretes an alkaline fluid that constitutes 13 to 33 percent of the semen into the prostatic urethra. The prostate may become enlarged or develop tumors in older men, obstructing urine flow and requiring surgical intervention.

The paired **bulbourethral** or **Cowper's glands** are about the size of peas. They are located beneath the prostate on either side of the membranous urethra (Figure 18-8). Like the prostate, the Cowper's glands secrete an alkaline fluid; their ducts open into the spongy urethra.

SEMEN

Semen, or **seminal fluid,** is a mixture of sperm and the secretions of the seminal vesicles, the prostate gland, and the bulbourethral glands. The average volume of semen for each ejaculation is 2.5–6 ml, and the average range of spermatozoa ejaculated is 50–100 million per milliliter. When the number of spermatozoa falls below approximately 20 million per milliliter, the male is likely to be sterile. Though only a single spermatozoan fertilizes an ovum, fertilization seems to require the combined action of a large number of spermatozoa. The ovum is enclosed by cells that present a barrier to the sperm. An enzyme called hyaluronidase is secreted by the acrosomes of sperm. Hyaluronidase is believed to dissolve intercellular materials of the cells covering the ovum, giving the sperm a passageway into the ovum.

Semen has a pH range of 7.35 to 7.50—slightly alkaline. The prostatic secretion gives semen a milky appearance, and fluids from the seminal vesicles and bulbourethral glands give it a mucoid consistency. Semen provides spermatozoa with a transportation medium and nutrients. It also neutralizes the acid environment of the male urethra and the female vagina. Semen also activates sperm and enzymes after ejaculation.

Once ejaculated into the vagina, liquid semen coagulates rapidly due to a clotting enzyme produced by the prostate that acts on a substance produced by the seminal vesicle. This clot liquifies in a few minutes because of another enzyme produced by the prostate gland. It is not clear why semen coagulates and liquifies in this manner.

PENIS

The **penis** is used to introduce spermatozoa into the vagina (Figure 18-9). The distal end of the penis is a slightly enlarged region called the *glans,* which means shaped like an acorn. Covering the glans is the loosely fitting *prepuce* or *foreskin*. Internally, the penis is composed of three cylindrical masses of tissue bound together by fibrous tissue. The two dorsolateral masses are called the *corpora cavernosa penis*. The smaller midventral mass, the *corpus spongiosum penis,* contains the spongy urethra. All three masses of tissue are spongelike and contain blood sinuses. Under the influence of sexual stimulation, the arteries supplying the penis dilate, and large quantities of blood enter the blood sinuses. Expansion of these spaces compresses the veins draining the penis so most entering blood is retained. These vascular changes result in an *erection*. The penis returns to its flaccid state when the arteries constrict and pressure on the veins is relieved. During ejaculation, the

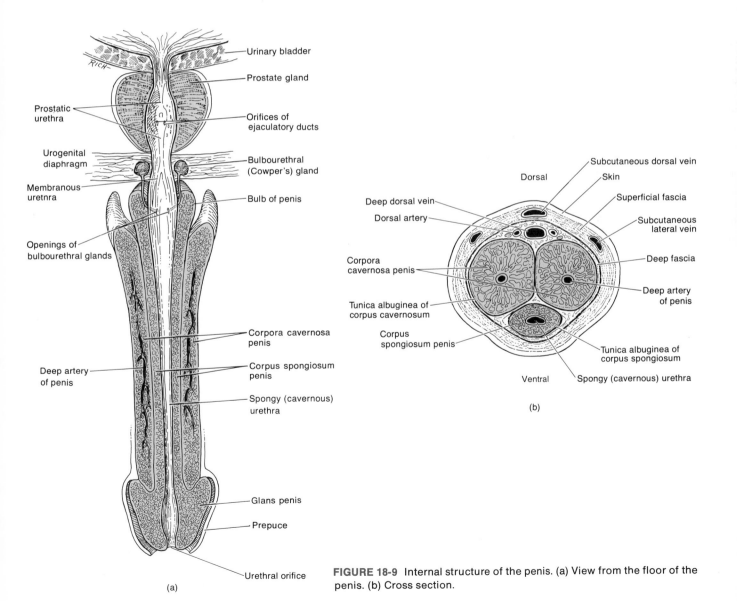

FIGURE 18-9 Internal structure of the penis. (a) View from the floor of the penis. (b) Cross section.

smooth muscle sphincter at the base of the urinary bladder is closed due to the higher pressure in the urethra caused by expansion of the corpus spongiosum penis. Thus urine is not expelled during ejaculation and semen does not enter the urinary bladder.

The penis has a very rich blood supply from the internal pudendal artery and the femoral artery. The veins drain into corresponding vessels. The sensory nerves to the penis are branches from the pudendal and ilioinguinal nerves. The corpora have a sympathetic and parasympathetic supply. As a result of parasympathetic stimulation, the blood vessels dilate and the muscles contract. The result is that blood is trapped within the penis and erection is maintained. The musculature of the penis is supplied by the pudendal nerve. The muscles include the bulbocavernosus muscle, which overlies the bulb of the penis; the ischiocavernosus muscles on either side of the penis; and the superficial transverse perineus muscles, on either side of the bulb of the penis (see Figure 18-20a).

Female Reproductive System

The female organs of reproduction (Figure 18-10) include the ovaries, which produce ova; the uterine or fallopian tubes, which transport the ova to the uterus (or womb); the vagina; and external organs that constitute the vulva or pudendum. The mammary glands, or breasts, also are considered part of the female reproductive system.

OVARIES

The **ovaries,** or female gonads, are paired glands resembling unshelled almonds in size and shape. They are positioned in the upper pelvic cavity, one on each side of the uterus. The ovaries are maintained in position by a series of ligaments (Figure 18-11). They are attached to the broad ligament of the uterus, which is itself part of the parietal peritoneum, by a fold of peritoneum called the *mesovarium;* anchored to the uterus by the *ovarian ligament;* and attached to the pelvic wall by the *suspensory ligament.* These ligaments can be seen in Figure 18-12, which shows the pelvic organs viewed from above. Each ovary also contains a *hilum,* the point of entrance for blood vessels and nerves.

The microscope reveals that each ovary consists of the following parts (Figure 18-13):

1. Germinal epithelium. This layer of simple cuboidal epithelium covers the free surface of the ovary.

2. Tunica albuginea. This is a capsule of collage-

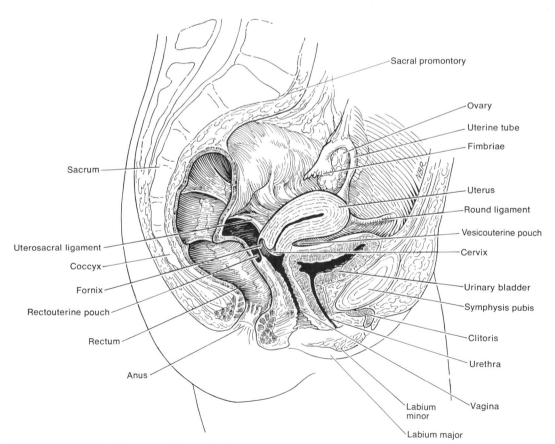

FIGURE 18-10 Female organs of reproduction seen in sagittal section.

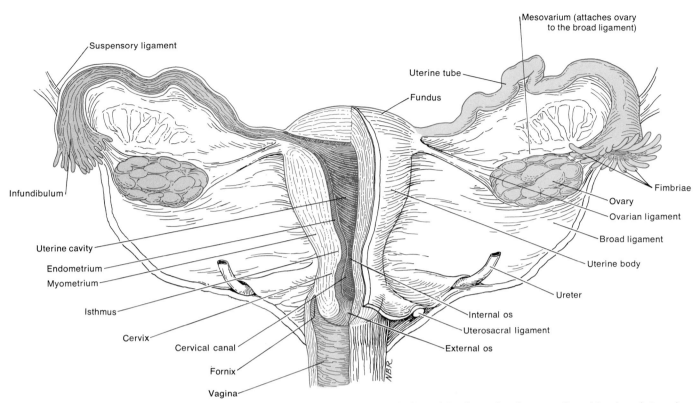

FIGURE 18-11 Uterus and associated female reproductive structures. The left side of the figure has been sectioned to show internal structures.

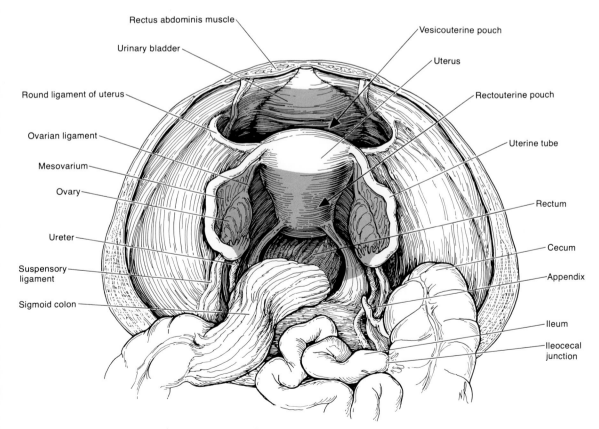

FIGURE 18-12 The female pelvis viewed from above.

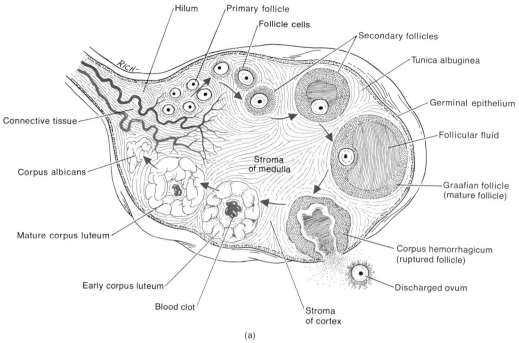

Hilum / Primary follicle

Follicle cells

Secondary follicles

Tunica albuginea

RICH-

Germinal epithelium

Connective tissue

Follicular fluid

Corpus albicans

Stroma
of medulla

Graafian follicle
(mature follicle)

Mature corpus luteum

Corpus hemorrhagicum
(ruptured follicle)

Early corpus luteum

Discharged ovum

Blood clot

Stroma
of cortex

(a)

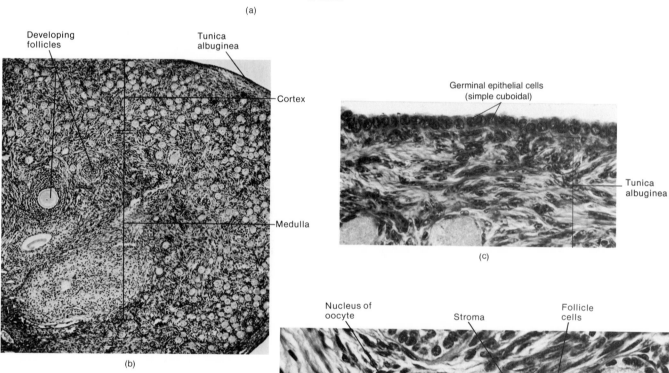

Developing
follicles

Tunica
albuginea

Germinal epithelial cells
(simple cuboidal)

Cortex

Tunica
albuginea

Medulla

(c)

(b)

Nucleus of
oocyte

Stroma

Follicle
cells

FIGURE 18-13 Ovary. (a) Parts of the ovary seen in
sectional view. The arrows indicate the sequence of
developmental stages that occur as part of the ovarian
cycle. (b) Sectional view of a portion of the ovary at a
magnification of 65×. (c) Enlarged aspect of the surface of the ovary at a magnification of 640×. (d) Primary follicles at a magnification of 640×. (Courtesy of
Edward J. Reith, from *Atlas of Descriptive Histology,*
by Edward J. Reith and Michael H. Ross, Harper &
Row, Publishers, Inc., New York, 1970.)

(d)

nous connective tissue immediately deep to the germinal epithelium.

3. Stroma. This is a region of connective tissue deep to the tunica albuginea. The stroma is composed of an outer, dense layer called the *cortex* and an inner, loose layer known as the *medulla*. The cortex contains ovarian follicles.

4. Ovarian follicles. These are ova and their surrounding tissues in various stages of development.

5. Graafian follicle This endocrine gland is made up of a mature ovum and its surrounding tissues. The Graafian follicle secretes hormones called estrogens.

6. Corpus luteum. This glandular body develops from a Graafian follicle after extrusion of an ovum (ovulation). The corpus luteum produces the hormones progesterone and estrogens.

The ovarian blood supply is furnished by the ovarian arteries and branches of the uterine arteries. The ovaries are drained by the ovarian veins. On the right side, they drain into the inferior vena cava, and on the left side, they drain into the renal vein. Sympathetic and parasympathetic nerve fibers to the ovaries are said to terminate on the blood vessels and not enter the substance of the ovaries.

The ovaries produce mature ova, discharge mature ova (ovulation), and secrete the female sexual hormones. The ovaries are analogous to the testes of the male.

OOGENESIS

The formation of a haploid ovum by meiosis in the ovary is referred to as **oogenesis.** With some exceptions, oogenesis occurs in essentially the same manner as spermatogenesis: It involves meiosis and maturation.

The precursor cell in oogenesis is a diploid cell called the *oogonium,* or egg mother cell (Figure 18-14). By the time the female fetus is ready for birth, the oogonia in the primary follicles have lost their ability to carry on mitosis. Such immature cells containing the diploid number of chromosomes are called *primary oocytes.* They remain in this stage until their follicular cells respond to FSH from the anterior pituitary, which in turn has responded to FSHRF from the hypothalamus. Starting with puberty, several follicles respond each month to the rising level of FSH. As the cycle proceeds and LH is secreted from the anterior pituitary, one of the follicles reaches a stage in which the diploid ovum, now a mature primary oocyte, undergoes its reduction division. Tetrad formation and crossing-over occur, and two cells of unequal size, both with 23 chromosomes of two chromatids each, are produced. The smaller cell is called the *first polar body* and is essentially a packet of discarded nuclear material. The larger cell, which receives most of the cytoplasm, is known as the *secondary oocyte.*

At this stage in the ovarian cycle, ovulation takes

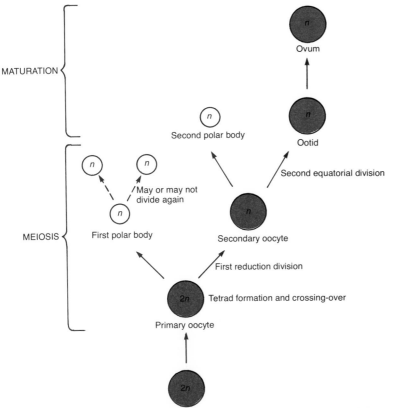

FIGURE 18-14 Oogenesis.

place. Since the secondary oocyte with its polar body and surrounding supporting cells is discharged at the time of ovulation, the "ovum" discharged is not yet mature. The discharged secondary oocyte enters the uterine tube and, if spermatozoa are present and fertilization occurs, the second divison takes place: the equatorial division. The secondary oocyte produces two cells of unequal size, both of them haploid. The larger cell is called the *ootid;* the smaller is the *second polar body.* In time, the ootid develops into an *ovum,* or mature egg.

The first polar body may undergo another division. If it does, meiosis of the primary oocyte results in a single haploid ovum and three polar bodies. In any event, the polar bodies disintegrate. Thus in the female one oogonium produces a single ovum whereas each male spermatocyte produces four spermatozoa.

UTERINE TUBES

The female body contains two **uterine,** or **fallopian, tubes,** which transport the ova from the ovaries to the uterus (see Figure 18-11). Measuring about 10 cm (4 inches) long, the tubes are positioned between the folds of the broad ligaments of the uterus. The funnel-shaped open end of each tube, called the *infundibulum,* lies close to the ovary but is not attached to it and is surrounded by a fringe of fingerlike projections called *fimbriae.* From the infundibulum the uterine tube extends inward and downward and attaches to the upper side of the uterus.

Histologically, the uterine tubes are composed of three layers (Figure 18-15). The internal *mucosa* contains ciliated columnar cells and secretory cells, which are believed to aid the nutrition of the ovum. The middle layer, the *muscularis,* is composed of a thick circular region of smooth muscle and an outer, thin, longitudinal region of smooth muscle. Wavelike contractions of the muscularis help move the ovum down into the uterus. The outer layer of the uterine tubes is a *serous membrane.*

About once a month a mature ovum ruptures from the surface of the ovary near the infundibulum of the uterine tube, a process called **ovulation.** The ovum is swept by the ciliary action of the epithelium of the infundibulum, which moves the ovum into the uterine tube. The ovum is then moved along the uterine tube by ciliary action supplemented by the wavelike contractions of the muscularis of the uterine tube. If the ovum is fertilized by a sperm cell, it usually occurs in the upper third of the uterine tubes. Fertilization may occur at any time up to 24 hours following ovulation. The ovum, fertilized or unfertilized, descends into the uterus within 7 days. Sometimes an ovum is fertilized while it is free in the pelvic cavity, and implantation

may take place on one of the pelvic viscera. Pelvic implantations usually fail because the developing fertilized ovum does not make vascular connection with the maternal blood supply. On occasion, fertilized ova fail to descend to the uterus and implant in the uterine tubes. In this instance, the pregnancy must be terminated surgically before the tube ruptures. Both pelvic and tubular implantations are referred to as *ectopic pregnancies.*

The uterine tubes are supplied by branches of the uterine and ovarian arteries. Venous return is via the uterine veins. The uterine tubes are supplied with sympathetic and parasympathetic nerve fibers from the hypogastric plexus and the pelvic splanchnic nerves. The fibers are distributed to the muscular coat of the tubes and their blood vessels.

UTERUS

The site of menstruation, implantation of a fertilized ovum, development of the fetus during pregnancy, and labor is the **uterus.** Situated between the urinary bladder and the rectum, the uterus is an inverted, pear-shaped organ (see Figures 18-10, 18-11, and 18-12). Before the first pregnancy, the adult uterus measures approximately 7.5 cm (3 inches) long, 5 cm (2 inches) wide, and 1.75 cm (1 inch) thick. Anatomical subdivisions of the uterus include the dome-shaped portion above the uterine tubes called the *fundus,* the major tapering central portion called the *body,* and the inferior narrow portion opening into the vagina called the *cervix.* Between the body and the cervix is a constricted region about 1 cm (0.5 inches) long: the *isthmus.* The interior of the body of the uterus is called the *uterine cavity,* and the interior of the narrow cervix is called the *cervical canal.* The junction of the uterine cavity with the cervical canal is the *internal os.* The *external os* is the place where the cervix opens into the vagina.

Normally the uterus is flexed between the uterine body and the cervix. In this position, the body of the uterus projects forward and slightly upward over the urinary bladder, and the cervix projects downward and backward, joining the vagina at nearly a right angle. Several structures that are either extensions of the parietal peritoneum or fibromuscular cords, referred to as ligaments, maintain the position of the uterus. The paired *broad ligaments* are double folds of parietal peritoneum attaching the uterus to either side of the pelvic cavity. Uterine blood vessels and nerves pass through the broad ligaments. The paired *uterosacral ligaments,* also peritoneal extensions, lie on either side of the rectum and connect the uterus to the sacrum. The *cardinal (lateral cervical) ligaments* extend below the bases of the broad ligaments between the pelvic wall and the cervix and vagina. These ligaments con-

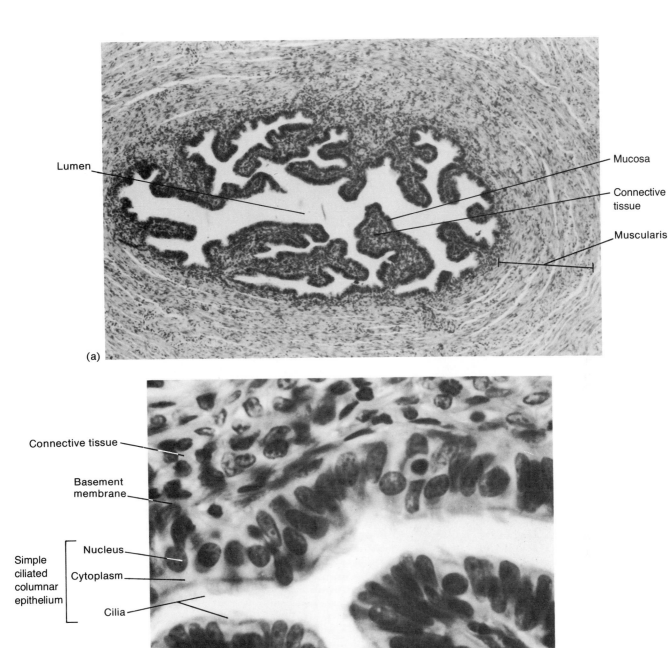

Lumen

Mucosa

Connective tissue

Muscularis

(a)

Connective tissue

Basement membrane

Simple ciliated columnar epithelium

Nucleus

Cytoplasm

Cilia

(b)

FIGURE 18-15 Histology of the uterine tube. (a) Cross section through the uterine tube at a magnification of 20×. (b) Enlarged aspect of the epithelium at a magnification of 200×. (Courtesy of Victor B. Eichler, Wichita State University.)

tain smooth muscle, uterine blood vessels, and nerves and are the chief ligaments supporting the uterus and keeping it from dropping down into the vagina. The *round ligaments* are bands of fibrous connective tissue between the layers of the broad ligament. They extend

from a point on the uterus just below the uterine tubes to a portion of the external genitalia. Although the ligaments normally maintain the position of the uterus, they also afford the uterine body with some movement. As a result, the uterus may become malposi-

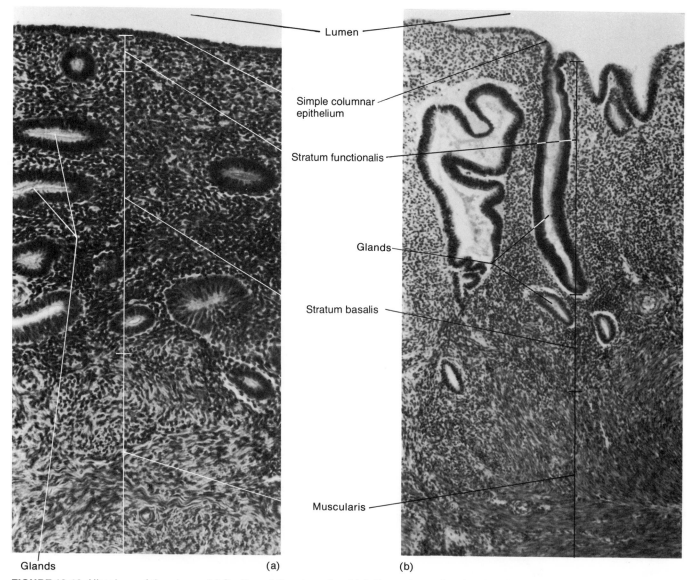

Lumen

Simple columnar epithelium

Stratum functionalis

Glands

Stratum basalis

Muscularis

Glands   (a)   (b)

FIGURE 18-16 Histology of the uterus. (a) Section of the uterus in which the endometrium has just been replaced following menstruation, shown at a magnification of 50×. (b) The endometrium during the proliferative phase of the menstrual cycle at a magnification of 20×. Note the long, straight nature of the glands. (Courtesy of Victor B. Eichler, Wichita State University.)

tioned. A backward tilting of the uterus called *retroflexion* or a forward tilting called *anteflexion* may occur.

Histologically, the uterus consists of three layers of tissue (Figure 18-16). The outer layer, part of the parietal peritoneum, is referred to as the *serous layer* or *serosa*. Laterally, the serosa becomes the broad ligament, the connective tissue of which is called the *parametrium*. Anteriorly, the serosa is reflected over the urinary bladder and forms a shallow pouch, the *vesicouterine pouch*. Posteriorly, it is reflected onto the rectum and forms a deep pouch, the *rectouterine pouch,* or *pouch of Douglas*—the lowest point in the pelvic cavity.

The middle layer of the uterus, the *myometrium,* forms the bulk of the uterine wall. This layer consists of smooth muscle fibers and is thickest in the fundus and thinnest in the cervix. During childbirth, coordinated contractions of the muscles help to expel the fetus from the body of the uterus.

The inner layer of the uterus, the *endometrium,* is a mucous membrane composed of two principal layers. The *stratum functionalis,* the layer closer to the uterine cavity, is shed during menstruation. The other layer, the *stratum basalis,* is permanent and produces a new functionalis following menstruation. The endometrium contains numerous glands.

Blood is supplied to the uterus by branches of the internal iliac artery called *uterine arteries.* Branches called *arcuate arteries* are arranged in a circular fashion underneath the serosa and give off *radial arteries* that penetrate the myometrium (Figure 18-17). Just be-

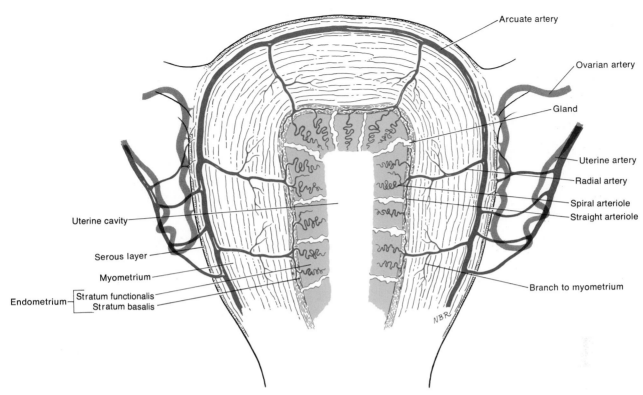

FIGURE 18-17 Blood supply to the uterus.

fore the branches enter the endometrium, they divide into two kinds of arterioles. The *straight arteriole* terminates in the basalis and supplies it with the materials necessary to regenerate the functionalis. The *spiral arteriole* penetrates the functionalis and changes markedly during the menstrual cycle. The uterus is drained by the *uterine veins*.

Sympathetic and parasympathetic fibers are supplied to the uterus via the hypogastric and pelvic plexuses. Both sets of nerves terminate on the uterine vessels. The myometrium is believed to be innervated by the sympathetic fibers alone.

ENDOCRINE RELATIONS

Gonadotropic hormones (FSH and LH) of the anterior pituitary gland initiate the menstrual cycle, ovarian cycle, and other changes associated with puberty in the female. The release of the gonadotropic hormones is controlled by regulating factors from the hypothalamus called the follicle-stimulating hormone releasing factor (FSHRF) and the luteinizing hormone releasing factor (LHRF). FSH stimulates the initial development of the ovarian follicles and the secretion of estrogens by the follicles. Another anterior pituitary hormone, the luteinizing hormone (LH), stimulates the futher development of ovarian follicles, brings about ovulation, and stimulates progesterone production by ovarian cells. The female sex hormones, estrogens and progesterone, affect the body in different ways. **Estrogens** have four main functions. First is the development and maintenance of female reproductive organs, especially the endometrium, secondary sex characteristics, and the breasts. The secondary sex characteristics include fat distribution to the breasts, abdomen, and hips, voice pitch, and hair pattern. Second, they control fluid and electrolyte balance. Third, they increase protein anabolism. Fourth, they help cause an increase in the female sex drive. High levels of estrogens in the blood inhibit the secretion of FSH by the anterior pituitary gland. This inhibition provides the basis for the action of one kind of contraceptive pill. **Progesterone,** the other female sex hormone, works with estrogens to prepare the endometrium for implantation and to prepare the mammary glands for milk secretion.

VAGINA

The **vagina** serves as a passageway for the menstrual flow. It is also the receptacle for the penis during coitus, or sexual intercourse, and the lower portion of the birth canal. It is a muscular, tubular organ lined with mucous membrane and measures about 10 cm (4 inches) in length (Figures 18-10 and 18-11). Situated between the bladder and the rectum, it is directed upward and backward where it attaches to the uterus. A recess, called the *fornix,* surrounds the vaginal attach-

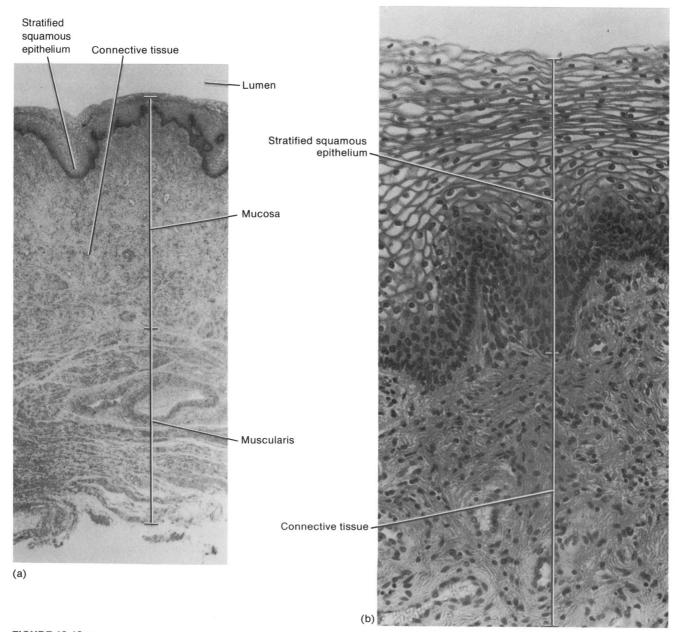

FIGURE 18-18 Histology of the vagina. (a) Section of the wall of the vagina at a magnification of 6×. (b) Enlarged aspect of the mucosa at a magnification of 50×. (Courtesy of Victor B. Eichler, Wichita State University.)

ment to the cervix. The dorsal recess, called the posterior fornix, is larger than the ventral and two lateral fornices. The fornices make possible the use of contraceptive diaphragms.

Histologically, the mucosa of the vagina consists of stratified squamous epithelium and connective tissue that lies in a series of transverse folds, the *rugae,* and is capable of a good deal of distension (Figure 18-18). The muscularis is composed of smooth muscle that can stretch considerably. This distension is important because the vagina receives the penis during sexual intercourse and serves as the lower portion of the birth

canal. At the lower end of the vaginal opening *(vaginal orifice)* is a thin fold of vascularized mucous membrane called the *hymen,* which forms a border around the orifice, partially closing it (see Figure 18-19). Sometimes the hymen completely covers the orifice, a condition called *imperforate hymen,* and surgery is required to open the orifice to permit the discharge of the menstrual flow. The mucosa of the vagina contains large amounts of glycogen that, upon decomposition, produce organic acids. These acids create a low-pH environment that retards microbial growth. However, the acidity is also injurious to sperm cells. Semen neutralizes the acidity of the vagina to ensure survival of the sperm.

VULVA

The term **vulva** or **pudendum** is a collective designation for the external genitalia of the female (Figure 18-19).

The *mons pubis (veneris),* an elevation of adipose tissue covered by coarse pubic hair, is situated over the symphysis pubis. It lies in front of the vaginal and urethral openings. From the mons pubis, two longitudinal folds of skin, the *labia majora,* extend downward and backward. The labia majora, the female homologue of the scrotum, contain an abundance of adipose tissue and sebaceous and sweat glands; they are covered by hair on their upper outer surfaces. Medial to the labia majora are two folds of skin called the *labia minora.* Unlike the labia majora, the labia minora are devoid of hair and have few sweat glands. They do, however, contain numerous sebaceous glands.

The *clitoris* is a small, cylindrical mass of erectile tissue and nerves. It is located at the anterior junction of the labia minora. A layer of skin called the *prepuce,* or *foreskin,* is formed at the point where the labia minora unite and covers the body of the clitoris. The exposed portion of the clitoris is the *glans.* The clitoris is homologous to the penis of the male in that it is capable of enlargement upon tactile stimulation and assumes a role in sexual excitement of the female.

The cleft between the labia minora is called the *vestibule.* Within the vestibule are the hymen, vaginal orifice, urethral orifice, and the openings of several ducts. The *vaginal orifice* occupies the greater portion of the vestibule and is bordered by the hymen. In front of the vaginal orifice and behind the clitoris is the *urethral orifice.* Behind and to either side of the urethral orifice are the openings of the ducts of the *lesser vestibular* or *Skene's glands.* These glands secrete mucus. On either side of the vaginal orifice itself are two small glands: the *greater vestibular* or *Bartholin's glands.* These glands open by a duct into the space between the hymen and labia minora and produce a mucoid secretion that supplements lubrication during sexual intercourse. The lesser vestibular glands are homologous to the male prostate. The greater vestibular glands are homologous to the male bulbourethral or Cowper's glands.

PERINEUM

The **perineum** is the diamond-shaped area at the lower end of the trunk between the thighs and buttocks of both males and females (Figure 18-20). It is surrounded anteriorly by the symphysis pubis, laterally by the ischial tuberosities, and posteriorly by the coccyx. A transverse line drawn between the ischial tuberosities divides the perineum into an anterior *urogenital triangle* that contains the external genitalia and

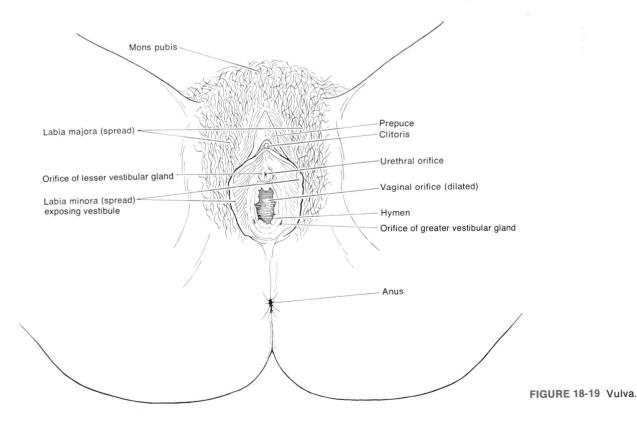

FIGURE 18-19 Vulva.

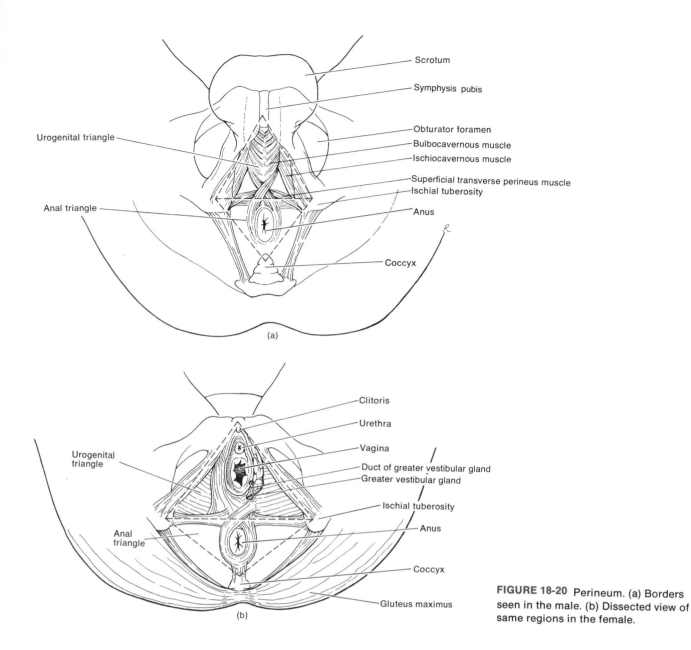

FIGURE 18-20 Perineum. (a) Borders seen in the male. (b) Dissected view of same regions in the female.

a posterior *anal triangle* that contains the anus. In the female, the region between the vagina and anus is known as the *clinical perineum*. If the vagina is too small to accommodate the head of an emerging fetus, the skin and underlying tissue of the clinical perineum tear. To avoid this, a small incision called an *episiotomy* is made in the perineal skin just prior to delivery.

MAMMARY GLANDS

The **mammary glands** are branched tubuloalveolar glands that lie over the pectoralis major muscles and are attached to them by a layer of connective tissue (Figure 18-21). Internally, each mammary gland consists of 15 to 20 *lobes,* or compartments, separated by

adipose tissue. The amount of adipose tissue determines the size of the breasts. However, breast size has nothing to do with the amount of milk produced. In each lobe are several smaller compartments, called *lobules,* composed of connective tissue in which milk-secreting cells referred to as *alveoli* are embedded (Figure 18-22). Between the lobules are strands of connective tissue called the *suspensory ligaments of Cooper.* These ligaments run between the skin and deep fascia and support the breast. Alveoli are arranged in grapelike clusters. They convey the milk into a series of *secondary tubules.* From here the milk passes into the *mammary ducts.* As the mammary ducts approach the nipple, expanded sinuses called *ampullae,* where milk may be stored, are present. The ampullae continue as *lactiferous ducts* that terminate

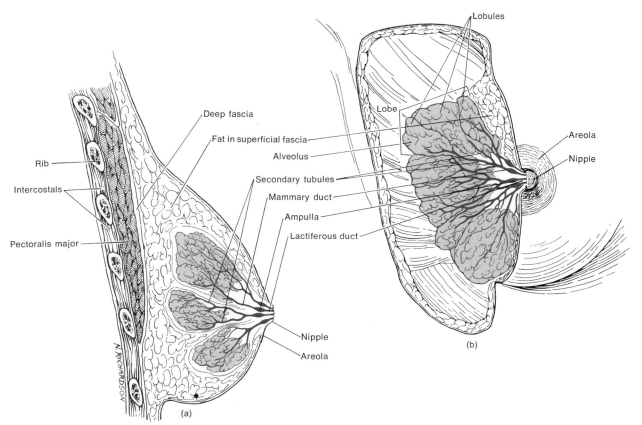

FIGURE 18-21 Mammary glands. (a) Sagittal section. (b) Front view partially sectioned.

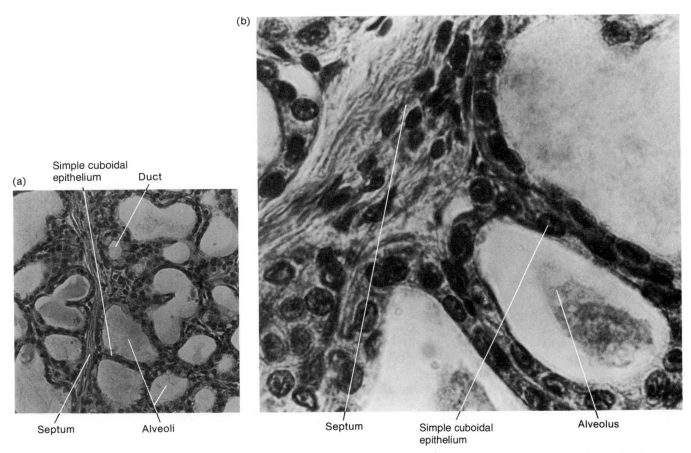

FIGURE 18-22 Histology of a lactating mammary gland. (a) Section showing numerous alveoli at a magnification of 40×. (b) Enlarged aspect at a magnification of 200×. (Courtesy of Victor B. Eichler, Wichita State University.)

in the *nipple*. Each lactiferous duct conveys milk from one of the lobes to the exterior, although some may join before reaching the surface. The circular pigmented area of skin surrounding the nipple is called the *areola*. It appears rough because it contains modified sebaceous glands.

At birth, both male and female mammary glands are undeveloped and appear as slight elevations on the chest. With the onset of puberty, the female breasts begin to develop—the mammary ducts elongate, extensive fat deposition occurs, and the areola and nipple grow and become pigmented. These changes are correlated with an increased output of estrogen by the ovary. Further mammary development occurs at sexual maturity with the onset of ovulation and the formation of the corpus luteum. During adolescence, lobules are formed and fat deposition continues, increasing the size of the glands. Although these changes are associated with estrogen and progesterone secretion by the ovaries, ovarian secretion is ultimately controlled by FSH, which is secreted in response to FSHRF by the hypothalamus.

The essential function of the mammary glands is milk secretion or lactation.

Applications to Health

VENEREAL DISEASES

The term *venereal* comes from Venus, goddess of love. **Venereal diseases** represent a group of infectious diseases that are spread primarily through sexual intercourse. After the common cold, venereal diseases are ranked as the most common communicable diseases in the United States.

Gonorrhea, more commonly known as "clap," is an infectious disease that primarily affects the mucous membrane of the urogenital tract, the rectum, and occasionally the eyes. The disease is caused by the bacterium *Neisseria gonorrhoeae*. Males usually suffer inflammation of the urethra with pus and painful urination. Fibrosis sometimes occurs in an advanced stage of gonorrhea, causing stricture of the urethra. There also may be involvement of the epididymis and prostate gland. In females, infection may occur in the urethra, vagina, and cervix, and there may be a discharge of pus. If the uterine tubes become involved, sterility and pelvic inflammation may result. Females usually present the disease without any symptoms.

Discharges from the involved mucous membranes are the source of infection, and the bacteria are transmitted by direct contact, usually sexual. If the bacteria are transmitted to the eyes of a newborn in the birth canal, blindness may result. Administration of a 1 percent silver nitrate solution or penicillin in the infant's eyes prevents infection. Penicillin is the drug of choice for the treatment of gonorrhea in adults.

Syphilis is an infectious disease caused by the bacterium *Treponema pallidum*. It also is acquired through sexual contact. The early stages of the disease primarily affect the organs most likely to have made sexual contact—the genital organs, the mouth, and the rectum. The point where the bacteria enter the body is marked by a lesion called a *chancre*. In males, it usually occurs on the penis; in females, in the vagina or cervix. The chancre heals without scarring. Following the initial infection, the bacteria enter the bloodstream and spread throughout the body. In some individuals, an active secondary stage of the disease occurs, characterized by lesions of the skin and mucous membranes and often fever. The signs of the secondary stage also go away without medical treatment. During the next several years, the disease progresses without symptoms and is said to be in a latent phase. When symptoms again appear, anywhere from 5 to 40 years after the initial infection, the person is said to be in the tertiary stage of the disease. Tertiary syphilis may involve the circulatory system, skin, bones, viscera, and the nervous system (see Chapter 14).

MALE DISORDERS

Prostate

The prostate gland is susceptible to infection, enlargement, and benign and malignant tumors. Because the prostate surrounds the urethra, any of these disorders can obstruct the flow of urine. Prolonged obstruction may result in serious changes in the bladder, ureters, and kidneys.

Acute and chronic infections of the prostate gland are common in postpubescent males, many times in association with inflammation of the urethra. In **acute prostatitis** the prostate gland becomes swollen and tender. Appropriate antibiotic therapy, bed rest, and above-normal fluid intake are effective treatment.

Chronic prostatitis is one of the most common chronic infections in men of the middle and later years. On examination, the prostate gland feels enlarged, soft, and extremely tender. The surface outline is irregular and may be hard. This disease frequently produces no symptoms, but the prostate is believed to harbor infectious microorganisms responsible for some allergic conditions, arthritis, and inflammation of nerves (neuritis), muscles (myositis), and the iris (iritis).

An **enlarged prostate** gland occurs in approximately one-third of all males over age 60. The affected gland is two to four times larger than normal. The cause is unknown, and the enlarged condition usually can be detected by rectal examination.

Tumors of the male reproductive system primarily involve the prostate gland. Carcinoma of the prostate is the second leading cause of death from cancer in men in the United States. It is responsible for approximately 19,000 deaths annually. Its incidence is related to age, race, occupation, geography, and ethnic origin. Both benign and malignant growths are common in elderly men. Both types of tumors put pressure on the urethra, making urination painful and difficult. At times, the excessive back pressure destroys kidney tissue and gives rise to an increased susceptibility to infection. Therefore, even when the tumor is benign, surgery is indicated to remove the prostate or parts of it if the growth is obstructive and perpetuates urinary tract infections.

Sexual Functional Abnormalities

Impotence is the inability of an adult male to attain or hold an erection long enough for normal intercourse. Impotence could be the result of physical abnormalities of the penis; systemic disorders such as syphilis; vascular disturbances; neurological disorders; or psychic factors such as fear of causing pregnancy, fear of venereal disease, religious inhibitions, and emotional immaturity.

Infertility, or **sterility,** is an inability to fertilize the ovum. It does not imply impotence. Male fertility requires adequate production of viable spermatozoa by the testes, unobstructed transportation of sperm through the seminal tract, and satisfactory deposition in the vagina. The tubules of the testes are sensitive to many factors—x-rays, infections, toxins, malnutrition—that may cause degenerative changes and produce male sterility.

If adequate spermatozoa production is suspected, a sperm analysis should be performed. Analysis includes measuring the volume of semen, counting the number of sperm per milliliter, evaluating sperm motility four hours after ejaculation, and determining the percentage of abnormal sperm forms (not to exceed 20 percent).

FEMALE DISORDERS

Menstrual Abnormalitites

Disorders of the female reproductive system frequently include menstrual disorders. This is hardly surprising because proper menstruation reflects not only the health of the uterus but the health of the glands that control it: the ovaries and the pituitary gland.

Amenorrhea is the absence of menstruation. If a woman has never menstruated, the condition is called *primary amenorrhea*. Primary amenorrhea can be caused by endocrine disorders, most often in the pituitary gland and hypothalamus, or by genetically caused abnormal development of the ovaries or uterus. *Secondary amenorrhea* is cessation of uterine bleeding in women who have previously menstruated. If pregnancy is ruled out, various endocrine disturbances are considered.

Dysmenorrhea is painful menstruation caused by contraction of the uterine muscles. A primary cause is believed to be inadequate progesterone, the hormone that prevents uterine contraction. It can also be caused by pelvic inflammatory disease, uterine tumors, cystic ovaries, or congenital defects.

Abnormal uterine bleeding includes menstruation of excessive duration and/or excessive amount, too frequent menstruation, intermenstrual bleeding, and postmenopausal bleeding. These abnormalities may be caused by disordered hormonal regulation, emotional factors, and systemic diseases.

Ovarian Cysts

Ovarian cysts are tumors of the ovary that contain fluid. Follicular cysts may occur in the ovaries of elderly women, in ovaries that have inflammatory diseases, and in menstruating females. They have thin walls and contain a serous albuminous material. Cysts may also arise from the corpus luteum or the endometrium. The endometrium is the inner lining of the uterus that is sloughed off in menstruation. *Endometriosis* occurs when the endometrial tissue grows outside the uterus. The tissue enters the pelvic cavity via the open uterine tubes and may be found in any of a dozen sites—on the ovaries, cervix, abdominal wall, and bladder. Causes are unknown. Endometriosis is common in women 30 to 40 years of age. Symptoms include premenstrual pain or unusual menstrual pain. The unusual pain is caused by the displaced tissue sloughing off at the same time the normal uterine endometrium is being shed during menstruation. Infertility can be a consequence. Treatment usually consists of hormone therapy or surgery. Endometriosis disappears at menopause or when the ovaries are removed.

Infertility

Female infertility, or the inability to conceive, occurs in about 10 percent of married females in the United States. Once it is established that ovulation occurs regularly, the reproductive tract is examined for functional and anatomical disorders to determine the possibility of union of the sperm and the ovum in the oviduct.

Breasts

The breasts of females are highly susceptible to cysts and tumors. Men are also susceptible to breast tumor, but certain breast cancers are 100 times more common in women.

In the female, the benign *fibroadenoma* is a common tumor of the breast. It occurs most frequently in young women. Fibroadenomas have a firm rubbery consistency and are easily moved about within the mammary tissue. The usual treatment is excision of the growth. The breast itself is not removed.

Breast cancer has the highest fatility rate of all cancers affecting women, but it is rare in men. In the female, breast cancer is rarely seen before age 30, and its occurrence rises rapidly after menopause. Breast cancer is generally not painful until it becomes quite advanced, so often it is not discovered early or, if noted, ignored. Any lump, no matter how small, should be reported to a doctor at once. If there is no evidence of *metastasis* (the spread of cancer cells to another part of the body), the treatment of choice is a *modified* or *radical mastectomy*. A radical mastectomy involves removal of the affected breast along with the underlying pectoral muscles and the axillary lymph nodes. Metastasis of cancerous cells is usually through the lymphatics or blood. Radiation treatment may follow the surgery to ensure the destruction of any stray cancer cells.

The mortality from breast cancer has not improved significantly in the last 50 years. Early detection—especially by breast self-examination and mammography—is still the most promising method to increase the survival rate.

It is estimated that 95 percent of breast cancer is first detected by the women themselves. Each month after the menstrual period the breasts should be thoroughly examined for lumps, puckering of the skin, or discharge.

Mammography is a sophisticated breast x-ray technique used to detect breast masses and determine whether they are malignant. The examination consists of two x-ray, right-angle views of each breast. Mammographic diagnosis of breast masses is 80 percent reliable. Mammographic x-ray prints are also used to guide surgeons performing mastectomies. As an aid in analyzing mammographic findings, a new x-ray technique, *xeroradiography,* is being used. In this photoelectric (rather than photochemical) method the x-ray image is reproduced on paper instead of film. Xerora-

diography provides excellent soft-tissue detail and requires less radiation than film mammography.

Modern x-ray films, xeroradiography, and special x-ray techniques have reduced the problem of mammographic radiation. Ovaries are not exposed to radiation during mammography, and the technique can be used safely on pregnant women. Most cancer experts agree that mammography should be used only after a careful clinical examination and under limited conditions. *Thermography,* a method of measuring and graphically recording heat radiation emitted by the breast, is also frequently used in conjunction with mammography. Tumors, both benign and malignant, emit more heat than nonaffected areas.

Cervical Cancer

Another common disorder of the female reproductive system is cancer of the uterine cervix. It ranks third in frequency after breast and skin cancers. **Cervical cancer** starts with a change in the shape of the cervical cells called *cervical dysplasia*. Cervical dysplasia is not a cancer in itself, but the abnormal cells tend to become malignant.

Cervical cancer, for the most part, is a venereal disease with a long incubation period. Inciting factors are not known, but herpes virus type II has recently become suspect. Smegma and the DNA of spermatozoa have also been implicated. Cancer of the cervix (except for adenocarcinoma) does not occur in celibate women, and for unknown reasons it is rare in Jewish women.

Early diagnosis of cancer of the uterus is accomplished by the *Papanicolaou test,* or "pap" smear. In this generally painless procedure, a few cells from the vaginal fornix (that part of the vagina surrounding the cervix) and the cervix are removed with a swab and examined microscopically. Malignant cells have a characteristic appearance and indicate an early stage of cancer, even before symptoms occur. Estimates indicate that the pap smear is more than 90 percent reliable in detecting cancer of the cervix. Treatment of cervical cancer may involve complete or partial removal of the uterus, called a *hysterectomy,* or radiation treatment.

KEY MEDICAL TERMS ASSOCIATED WITH THE REPRODUCTIVE SYSTEMS

Castration   Excision of the testes or ovaries.

Colposcopy (*colp* = vagina; *scopy* = look inside with a scope)   Use of a low-power binocular miscroscope to examine vaginal, exocervical, and a portion of the endocervical mucosa under magnification inside the body.

Copulation   Sexual intercourse. Coitus refers to sexual intercourse between human beings.

Leukorrhea (*leuko* = white)   A nonbloody vaginal discharge that may occur at any age and affects most women at some time.

Neoplasia (*neo* = new; *plas* = grow) A condition characterized by the presence of new growths (tumors).

Oophorectomy (*oophoron* = ovary; *ectomy* = removal) Excision of an ovary. Bilateral oophorectomy refers to the removal of both ovaries.

Pruritis (*prur* = itch) Itching.

Purulent Containing or consisting of pus.

Salpingectomy (*salping* = uterine tube) Excision of a uterine tube.

Salpingitis (*itis* = inflammation) Inflammation or infection of the uterine tube.

Smegma The secretion, consisting principally of desquamated epithelial cells, found chiefly about the external genitalia and especially under the foreskin of the male.

Vaginitis Inflammation of the vagina.

STUDY OUTLINE

Male Reproductive System

1. The scrotum supports and protects the testes and provides an appropriate temperature for the production and survival of spermatozoa.
2. The major functions of the testes are sperm production and the secretion of testosterone.
3. Ova and sperm are collectively called gametes.
4. Chromosome number is the number of chromosomes contained in each nucleated cell that is not a gamete.
5. The human chromosome number is 23 pairs—a total of 46 chromosomes (diploid).
6. Meiosis occurs in the process of producing sex cells. It causes a developing sperm or ovum to cut in half its duplicate set of chromosomes so that the mature gamete has only 23 chromosomes (haploid).
7. Spermatogenesis occurs in the testes. It is the formation of haploid spermatozoa by meiosis.
8. Oogenesis occurs in the ovaries. It is the formation of haploid ova by meiosis.
9. FSH and ICSH maintain the growth and development of the male reproductive organs.
10. Spermatozoa are conveyed from the testes to the exterior through the seminiferous tubules, straight tubules, rete testis, efferent ducts, ductus epididymis, ductus deferens, ejaculatory duct, and urethra.
11. The male urethra is divided into prostatic, membranous, and spongy portions.
12. The seminal vesicles, prostate, and Cowper's glands secrete the liquid portion of semen. They are accessory reproductive glands.
13. Semen is a mixture of sperm and secreted liquids.
14. The penis serves as the organ of copulation.

Female Reproductive System

1. The ovaries produce ova by a process called oogenesis and secrete estrogens and progesterone.
2. The uterine tubes convey ova from the ovaries to the uterus and are the sites of fertilization.
3. The normal position of the uterus is maintained by a series of ligaments.
4. The uterus is associated with menstruation, implantation of a fertilized ovum, development of the fetus, and labor.
5. FSH and LH control the ovarian cycle. Estrogens and progesterone control the menstrual cycle.
6. The vagina serves as a passageway for the menstrual flow, as the lower portion of the birth canal, and as the receptacle for the penis.
7. The vulva is a collective designation for the external genitalia of the female.
8. The perineum is a diamond-shaped area at the lower end of the trunk between the thighs and buttocks.
9. The mammary glands function in the secretion of milk.

Applications to Health

1. Venereal diseases are a group of infectious diseases spread primarily through sexual intercourse.
2. Conditions that affect the prostate are prostatitis, enlarged prostate, and tumors.
3. Impotence is the inability of the male to attain or hold an erection long enough for intercourse.
4. Infertility is the inability of a male's sperm to fertilize an ovum.
5. Menstrual disorders include amenorrhea, dysmenorrhea, and abnormal bleeding.
6. Ovarian cysts are tumors that contain fluid.
7. The mammary glands are suscepitable to benign fibroadenomas and malignant tumors. The removal of a malignant breast, pectoral muscles, and lymph nodes is called a radical mastectomy.
8. Cervical cancer can be diagnosed by a "pap" test. Complete or partial removal of the uterus is called a hysterectomy.

REVIEW QUESTIONS

1. Define reproduction. List the male and female organs of reproduction.
2. Describe the function of the scrotum in protecting the testes from temperature fluctuations. What is cryptorchidism?
3. Describe the internal structure of a testis. Where are the sperm cells made?
4. Describe the principal features of spermatogenesis.
5. Identify the principal parts of spermatozoan. List the function of each.

6. Explain the effect of FSH and ICSH on the male reproductive system.
7. Describe the physiological effects of testosterone.
8. Describe the location, structure, and histology of the ductus epididymis, ductus deferens, and ejaculatory duct.
9. Trace the course of a sperm cell through the male system of ducts from the seminiferous tubules to the urethra.
10. Explain the location of the three subdivisions of the male urethra.
11. What is the spermatic cord?
12. Briefly explain the locations and functions of the seminal vesicles, prostate gland, and Cowper's glands.
13. What is seminal fluid? What is its function?
14. Describe the structure of the penis.
15. How is the penis structurally adapted as an organ of copulation?
16. Describe the histology of the ovaries.
17. How are the ovaries held in position in the pelvic cavity? What is ovulation?
18. What is the function of the uterine tubes? Explain in terms of their histology. Define an ectopic pregnancy.
19. Diagram the principal parts of the uterus.
20. Describe the arrangement of ligaments that hold the uterus in its normal position. Explain the two major malpositions of the uterus.
21. Explain the histology of the uterus.
22. Discuss the blood supply to the uterus. Why is an abundant blood supply important?
23. List the physiological effects of estrogens and progesterone.
24. What is the function of the vagina? Describe its structure.
25. List the parts of the vulva and the functions of each part.
26. What is the perineum? Define episiotomy.
27. Describe the structure of the mammary glands. How are the breasts supported?
28. Describe the passage of milk from the areolar cells of the mammary gland to the nipple.
29. Define venereal disease. Distinguish between gonorrhea and syphilis.
30. Describe several disorders that affect the prostate gland.
31. Distinguish between impotence and infertility.
32. What are some of the causes of amenorrhea, dysmenorrhea, and abnormal uterine bleeding?
33. What are ovarian cysts? Define endometriosis.
34. Distinguish between a fibroadenoma and a malignant tumor of the breast.
35. What is a radical mastectomy? Describe the importance of mammography and thermography.
36. What is a ''pap'' smear? What is a hysterectomy?
37. Refer to the glossary of key medical terms at the end of the chapter. Be sure that you can define each term.

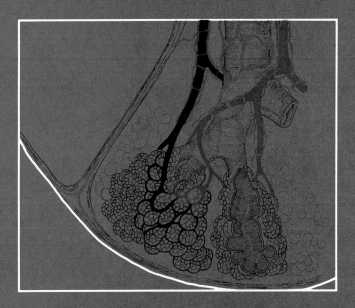

19

The Respiratory System

Cells need a continuous supply of oxygen to carry out the activities that are vital to their survival. Many of these activities release quantities of carbon dioxide. Since an excessive amount of carbon dioxide produces acid conditions that are poisonous to cells, the gas must be eliminated quickly and efficiently. The two systems that supply oxygen and eliminate carbon dioxide are the cardiovascular system and the respiratory system. The **respiratory system** consists of organs that exchange gases between the atmosphere and blood. These organs are the nose, pharynx, larynx, trachea, bronchi, and lungs (Figure 19-1). In turn, the blood transports gases between the lungs and the cells. The overall exchange of gases between the atmosphere, the blood, and the cells is **respiration.** The respiratory and cardiovascular systems participate equally in respiration. Failure of either system has the same effect on the body: rapid death of cells from oxygen starvation and disruption of homeostasis.

Organs

NOSE

The **nose** has an external portion and an internal portion inside the skull. Externally, the nose consists of a supporting framework of bone and cartilage covered with skin and lined with mucous membrane. The bridge of the nose is formed by the nasal bones, which hold it in a fixed position. Because it has a framework of pliable cartilage, the rest of the external nose is quite flexible. On the undersurface of the external nose are two openings called the *nostrils* or *external nares* (singular *naris;* see Figure 19-2). The surface anatomy of the nose may be reviewed by referring back to Exhibit 1-3.

The internal region of the nose is a large cavity in the skull that lies below the cranium and above the mouth. Anteriorly, the internal nose merges with the external nose, and posteriorly it communicates with the throat (pharynx) through two openings called the *internal nares.* Four paranasal sinuses (frontal, sphenoidal, maxillary, and ethmoidal) and the nasolacrimal ducts also open into the internal nose. The lateral walls of the internal nose are formed by the ethmoid, maxillae, and inferior conchae bones. The ethmoid forms the roof. The floor is formed by the palatine bones and the maxilla of the hard palate. Occasionally the palatine and maxillary bones fail to fuse during embryonic life, and a child is born with a crack in the bony wall that separates the internal nose from the mouth. This condition is called *cleft palate.*

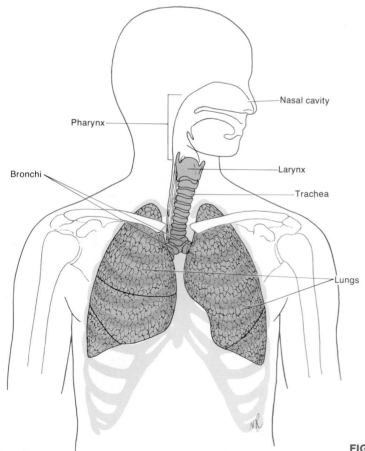

FIGURE 19-1 Organs of the respiratory system.

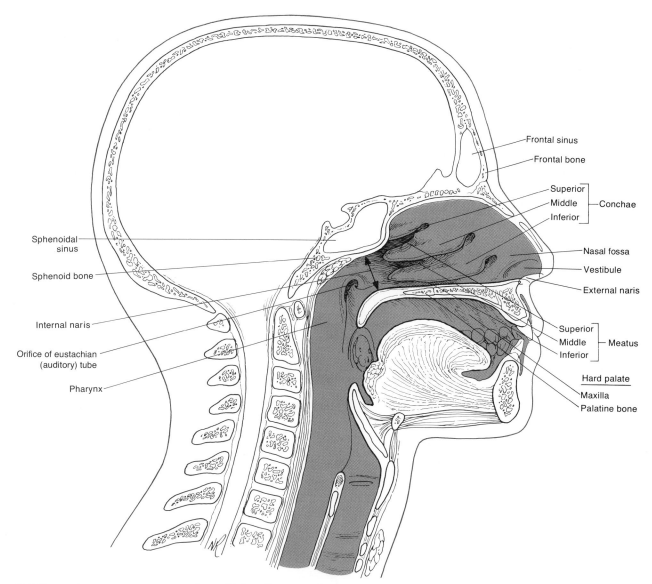

FIGURE 19-2 Left nasal fossa seen in sagittal section.

The inside of both the external and internal nose, called the *nasal cavity,* is divided into right and left chambers, the *nasal fossae,* by the vertical partition called the *nasal septum.* Cartilage is the primary material making up the anterior portion of the septum. The remainder is formed by the vomer and the perpendicular plate of the ethmoid (see Figure 6-7a). The anterior portions of the nasal fossae, which are just inside the nostrils, are called the *vestibules.* The vestibules are surrounded by cartilage as opposed to bone of the upper nasal cavity. The interior structures of the nose are specialized for three functions: incoming air is warmed, moistened, and filtered; olfactory stimuli are received; and large hollow resonating chambers are provided for speech sounds.

When air enters the nostrils, it passes first through the vestibule. The vestibule is lined by skin containing coarse hairs that filter out large dust particles. The air then passes into the rest of the fossa. Three shelves formed by projections of the superior, middle, and inferior conchae or turbinates extend out of the lateral wall of the fossa. The conchae, almost reaching the septum, subdivide each nasal fossa into a series of groovelike passageways—the *superior, middle,* and *inferior meati.* Mucous membrane lines the fossa and its shelves. The olfactory receptors lie in the membrane lining the upper portion of the fossa, also called the olfactory region. Below the olfactory region, the membrane contains pseudostratified ciliated columnar cells with many goblet cells and capillaries. As the air whirls around the turbinates and meati, it is warmed by the capillaries. Mucus secreted by the goblet cells moistens the air and traps dust particles. Drainage from the lacrimal ducts, and perhaps secretions from the paranasal sinuses, also help to moisten the air. The cilia move the resulting mucus-dust packages along to the throat so that they can be eliminated from the body. As the air passes through the top of the cavity, chemicals in the air may stimulate the olfactory receptors.

The arterial supply to the nasal cavity is principally from the sphenopalatine branch of the maxillary artery. The remainder is supplied by the ophthalmic artery. The veins of the nasal cavity drain into the sphenopalatine vein, the facial vein, and the ophthalmic vein.

The nerve supply of the nasal cavity consists of olfactory cells in the olfactory epithelium associated with the olfactory nerve (see Figure 16-7) and the

nerves of general sensation. These nerves are branches of the ophthalmic division of the trigeminal nerve and the maxillary division of the trigeminal nerve.

PHARYNX

The **pharynx,** or throat, is a tube about 13 cm (5 inches) long that starts at the internal nares and runs partway down the neck (Figure 19-3). It lies just in back of the nasal cavity and oral cavity and just in front of the cervical vertebrae. Its walls are composed of skeletal muscles and are lined with mucous membrane. The functions of the pharynx are limited to serving as a passageway for air and food and providing a resonating chamber for speech sounds.

The uppermost portion of the pharynx is called the

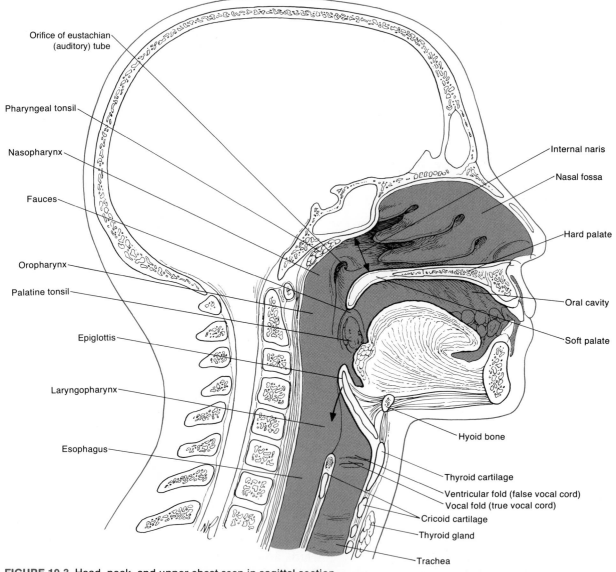

Orifice of eustachian (auditory) tube

Pharyngeal tonsil

Nasopharynx

Fauces

Oropharynx

Palatine tonsil

Epiglottis

Laryngopharynx

Esophagus

Internal naris

Nasal fossa

Hard palate

Oral cavity

Soft palate

Hyoid bone

Thyroid cartilage

Ventricular fold (false vocal cord)

Vocal fold (true vocal cord)

Cricoid cartilage

Thyroid gland

Trachea

FIGURE 19-3 Head, neck, and upper chest seen in sagittal section.

nasopharynx. This part lies behind the internal nasal cavity and extends to the plane of the soft palate. There are four openings in its walls: two internal nares plus two openings that lead into the eustachian (auditory) tubes. The posterior wall of the nasopharynx also contains the pharyngeal tonsil, or adenoid. Through the internal nares the nasopharynx exchanges air with the nasal cavities and receives the packages of dust-laden mucus. The nasopharynx has a lining of pseudostratified ciliated epithelium. Cilia in the walls of the nasopharynx move the mucus down toward the mouth. The nasopharynx also exchanges small amounts of air with the auditory canal so that the air pressure inside the middle ear equals the pressure of the atmospheric air flowing through the nose and pharynx.

The second portion of the pharynx, the *oropharynx,* lies behind the oral cavity and extends from the soft palate down to the level of the hyoid bone. It receives only one opening: the *fauces,* or opening from the mouth. It is lined by stratified squamous epithelium. This portion of the pharynx is both respiratory and digestive in function since it is a common passageway for both air and food. Two pairs of tonsils, the palatine tonsils and the lingual tonsils, are found in the oropharynx. The lingual tonsils lie at the base of the tongue (see Figure 20-5).

The lowest portion of the pharynx is called the *laryngopharynx.* The laryngopharynx extends downward from the hyoid bone and empties into the esophagus (food tube) posteriorly and into the larynx (voice box) anteriorly. Like the oropharynx, the laryngopharynx is both a respiratory and a digestive pathway in function and is lined by stratified squamous epithelium.

The arteries of the pharynx are the ascending pharyngeal, the ascending palatine branch of the facial, the descending palatine and pharyngeal branches of the maxillary, and the muscular branches of the superior thyroid artery. The veins of the pharynx drain into the pterygoid plexus and the internal jugular vein.

Most of the muscles of the pharynx are innervated by the pharyngeal plexus. This plexus is formed by the pharyngeal branches of the glossopharyngeal and vagal nerves and the superior cervical sympathetic ganglion.

LARYNX

The **larynx,** or voice box, is a short passageway that connects the pharynx with the trachea. It lies in the midline of the neck anterior to the fourth through sixth cervical vertebrae. The walls of the larynx are supported by nine pieces of cartilage. Three are single and three are paired. The three single pieces are the large thyroid cartilage and the smaller epiglottic and cricoid cartilage. Of the paired cartilages, the arytenoid carti-

lages are the most important. The paired corniculate and cuneiform cartilages are of lesser significance. The *thyroid cartilage,* or Adam's apple, consists of two fused plates that form the anterior wall of the larynx and give it its triangular shape (Figure 19-4). In males the thyroid cartilage is bigger than it is in females.

The *epiglottis* is a large, leaf-shaped piece of cartilage lying on top of the larynx. The "stem" of the epiglottis is attached to the thyroid cartilage, but the "leaf" portion is unattached and free to move up and down like a trapdoor. During swallowing, the free edge of the epiglottis forms a lid over the opening of the trachea into the larynx (glottis). In this way, the larynx is closed off and liquids and foods are routed into the esophagus and kept out of the trachea. If anything but air passes into the larynx, a cough reflex attempts to expel the material.

The *cricoid cartilage* is a ring of cartilage forming the lower walls of the larynx. It is attached to the first ring of cartilage of the trachea.

The paired *arytenoid cartilages* are pyramidal in shape and located at the superior border of the cricoid cartilage. They attach to the vocal folds and pharyngeal muscles and by their action can move the vocal cords. The *corniculate cartilages* are paired, cone-shaped cartilages. Each is located at the apex of each arytenoid cartilage. The paired *cuneiform cartilages* are rod-shaped cartilages in the mucous membrane fold that connects the epiglottis to the arytenoid cartilages.

Like the other respiratory passageways, the larynx is lined with a ciliated mucous membrane that traps dust not removed in the upper passages.

The mucous membrane of the larynx is arranged into two pairs of folds—an upper pair called the *ventricular folds* (or *false vocal cords*) and a lower pair called simply the *vocal folds* (or *true vocal cords*). The air passageway between the folds is called the *glottis.* Under the mucus membrane of the true vocal cords lie bands of elastic ligaments stretched between pieces of rigid cartilage like the strings on a guitar. Skeletal muscles of the larynx, called intrinsic muscles, are attached internally to the pieces of rigid cartilage and to the vocal folds themselves. When the muscles contract, they pull the strings of elastic ligaments tight and stretch the cords out into the air passageways so that the glottis is narrowed. If air is directed against the vocal folds, they vibrate and set up sound waves in the column of air in the pharynx, nose, and mouth. The greater the pressure of air, the louder the sound.

Pitch is controlled by the tension on the true vocal cords. If the cords are pulled taut by the muscles, they vibrate more rapidly and a higher pitch results. Lower sounds are produced by decreasing the muscular tension on the cords. Vocal cords are usually thicker and longer in males than in females and vibrate more

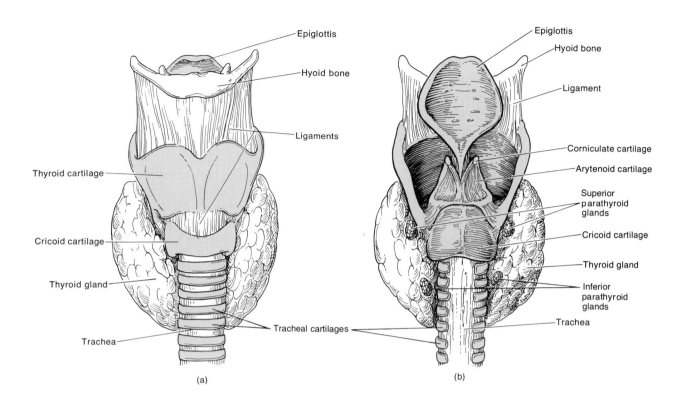

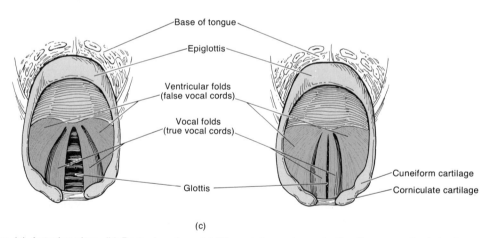

FIGURE 19-4 Larynx. (a) Anterior view. (b) Posterior view. (c) Viewed from above. In the figure on the left, the true vocal cords are relaxed. In the figure on the right, the true vocal cords are pulled taut.

slowly. Thus men have a lower range of pitch than women.

Sound originates from the vibration of the true vocal cords. But other structures are necessary for converting the sound into recognizable speech. The pharynx, mouth, nasal cavities, and paranasal sinuses all act as resonating chambers that give the voice its human and individual quality. By constricting and relaxing the muscles in the walls of the pharynx we produce the vowel sounds. Muscles of the face, tongue, and lips help us to enunciate words.

Laryngitis is an inflammation of the larynx that is most often caused by a respiratory infection or irritants such as cigarette smoke. Inflammation of the vocal folds themselves causes hoarseness or loss of voice by interfering with the contraction of the cords or by causing them to swell to the point where they cannot vibrate freely. Many long-term smokers acquire a permanent hoarseness from the damage done by chronic inflammation.

The arteries of the larynx are the superior laryngeal, inferior laryngeal, and cricothyroid. The superior and inferior laryngeal and cricothyroid veins accompany the arteries. The superior cricoid vein and cricothyroid vein empty into the superior thyroid vein, and the inferior vein empties into the inferior thyroid vein.

The nerves of the larynx are the superior and recurrent laryngeal branches of the vagus.

TRACHEA

The **trachea,** or windpipe, is a tubular passageway for air about 12 cm (4.5 inches) in length and 2.5 cm (1 inch) in diameter. It is located in front of the esophagus and extends from the larynx to the fifth thoracic vertebra, where it divides into right and left primary bronchi.

The tracheal epithelium is pseudostratified. It consists of ciliated columnar cells, goblet cells, and basal cells. The epithelium provides the same protection against dust as the membrane lining the larynx (Figure 19-5). The walls of the trachea are composed of smooth muscle and elastic connective tissue. They are encircled by a series of horizontal incomplete rings of hyaline cartilage that look like a series of letter C's stacked one on top of the other. The open parts of the C's face the esophagus posteriorly and permit it to expand into the trachea during swallowing. Transverse smooth muscle fibers, called the *trachealis muscle,* attach the open ends of the cartilage rings. The open ends of the rings of cartilage are also attached by elastic connective tissue. The solid parts of the C's provide a rigid support so the tracheal walls do not collapse inward and obstruct the air passageway.

Occasionally the respiratory passageways are unable to protect themselves from obstruction. The rings of cartilage may be accidentally crushed, or the mucous membrane may become inflamed and swell so much that it closes off the air space. Inflamed membranes also secrete a great deal of mucus that may clog the lower respiratory passageways. Or a large object may be breathed in (aspirated) while the glottis is open. In any case, the passageways must be cleared quickly. If the obstruction is above the level of the

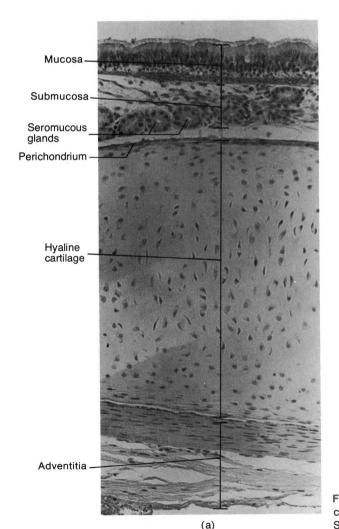

Mucosa

Submucosa

Seromucous glands

Perichondrium

Hyaline cartilage

Adventitia

(a)

FIGURE 19-5 Histology of the trachea. (a) Section of the wall of the trachea at a magnification of 40×. (Courtesy of Victor B. Eichler, Wichita State University.)

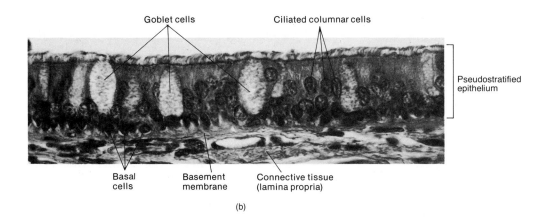

Goblet cells

Ciliated columnar cells

Pseudostratified epithelium

Basal cells

Basement membrane

Connective tissue (lamina propria)

(b)

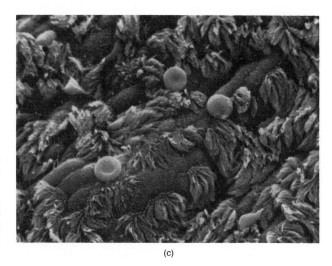

(c)

FIGURE 19-5 (Continued) Histology of the trachea. (b) Enlarged aspect of the tracheal epithelium at a magnification of 640×. (Courtesy of Edward J. Reith, from *Atlas of Descriptive Histology,* by Edward J. Reith and Michael H. Ross, Harper & Row, Publishers, Inc., New York, 1970.) (c) Scanning electron micrograph of the tracheal mucosa at a magnification of 200×. (Courtesy of Fisher Scientific Company and S.T.E.M. Laboratories, Inc., Copyright 1975.) (d) Cross section through the trachea and a portion of the esophagus.

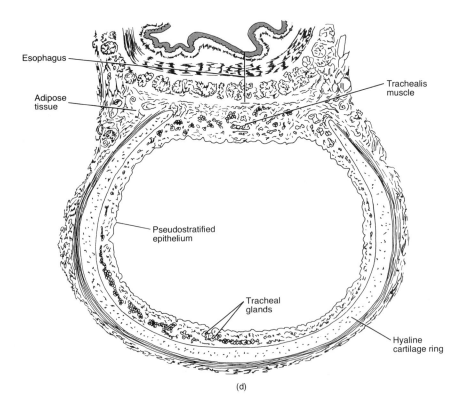

Esophagus

Adipose tissue

Trachealis muscle

Pseudostratified epithelium

Tracheal glands

Hyaline cartilage ring

(d)

chest, a *tracheostomy* may be performed. The first step in a tracheostomy is to make an incision in the neck and into the part of the trachea below the obstructed area. The patient breathes through a tube inserted through the incision. Another method is *intubation*. A tube is inserted into the mouth and passed down through the larynx and trachea. The firm walls of the tube push back any flexible obstruction, and the inside of the tube provides a passageway for air. If mucus is clogging the trachea, it can be suctioned up through the tube.

The arteries of the trachea are branches of the inferior thyroid, internal thoracic, and bronchial arteries. The veins of the trachea terminate in the inferior thyroid veins.

The smooth muscle and glands of the trachea are innervated parasympathetically via the vagus nerve, directly, and by its recurrent laryngeal branches. Sympathetic innervation is through branches from the sympathetic trunk and its ganglia.

BRONCHI

The trachea terminates in the chest by dividing into a **right primary bronchus,** which goes to the right lung, and a **left primary bronchus,** which goes to the left lung (Figure 19-6a). The right primary bronchus is more vertical, shorter, and wider than the left. As a result, foreign objects that enter the air passageways frequently lodge in it. Like the trachea, the primary bronchi contain incomplete rings of cartilage and are lined by a ciliated columnar epithelium.

The blood supply to the bronchi is via the left bronchial arteries and the right bronchial artery. The bronchial veins that drain the bronchi are the right bronchial vein which enters the azygous vein and the left bronchial vein that empties into the hemiazygous vein or the left superior intercostal vein.

Upon entering the lungs, the primary bronchi divide to form smaller bronchi—the *secondary* or *lobar bronchi,* one for each lobe of the lung. (The right lung has three lobes; the left lung has two.) The secondary bronchi continue to branch, forming still smaller bronchi, called *tertiary* or *segmental bronchi,* which divide into *bronchioles* (see Figure 19-8). Bronchioles, in turn, branch into even smaller tubes called *terminal bronchioles.* The continuous branching of the trachea into primary bronchi, secondary bronchi, bronchioles, and terminal bronchioles resembles a tree trunk with its branches and is commonly referred to as the *bronchial tree.* As the branching becomes more extensive in the bronchial tree, several structural changes may be noted. First, rings of cartilage are replaced by plates of cartilage that finally disappear in the bron-

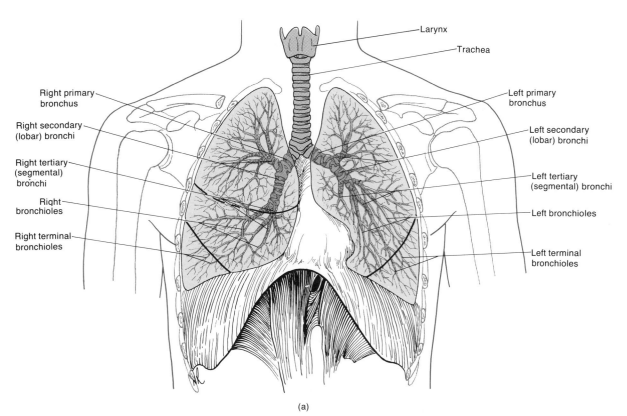

Larynx

Trachea

Right primary bronchus

Right secondary (lobar) bronchi

Right tertiary (segmental) bronchi

Right bronchioles

Right terminal bronchioles

Left primary bronchus

Left secondary (lobar) bronchi

Left tertiary (segmental) bronchi

Left bronchioles

Left terminal bronchioles

(a)

FIGURE 19-6 Air passageways to the lungs. (a) Diagrammatic representation of the bronchial tree in relationship to the lungs.

Smooth muscle　　　　　　Epithelium
　　Primary bronchus　　　　　　Cartilage

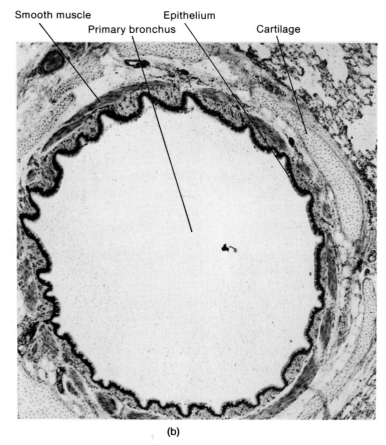

(b)

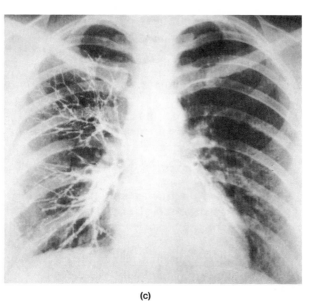

(c)

FIGURE 19-6 (Continued) Air passageways to the lungs. (b) Histology of a primary bronchial tube seen in cross section at a magnification of 40×. (Courtesy of Edward J. Reith, from *Atlas of Descriptive Histology,* by Edward J. Reith and Michael H. Ross, Harper & Row, Publishers, Inc., New York, 1970.) (c) Anteroposterior bronchogram of the lungs. (Courtesy of Lester W. Paul and John H. Juhl, *The Essentials of Roentgen Interpretation,* 3d ed., Harper & Row, Publishers, Inc., New York, 1972.)

chioles. Second, as the cartilage decreases, the amount of smooth muscle increases (Figure 19-6b). In addition, the epithelium changes from ciliated columnar to simple cuboidal in the terminal bronchioles. The fact that the walls of the bronchioles contain a great deal of smooth muscle but no cartilage is clinically significant. During an asthma attack the muscles go into spasm. Because there is no supporting cartilage, the spasms can close off the air passageways.

Bronchography is a technique for examining the bronchial tree. In this procedure, an intratracheal catheter is passed transorally or transnasally through the glottis into the trachea. Then an iodinated media is introduced, by means of gravity, into the trachea and distributed through the bronchial branches. Roentgenograms of the chest in various positions are taken and the developed film, called a *bronchogram,* provides a picture of the bronchial tree (Figure 19-6c).

LUNGS

The **lungs** are paired, cone-shaped organs lying in the thoracic cavity (Figure 19-7a). They are separated from each other by the heart and other structures in the mediastinum. Two layers of serous membrane, collectively called the *pleural membrane,* enclose and protect each lung. The outer layer is attached to the

walls of the pleural cavity and is called the *parietal pleura.* The inner layer, the *visceral pleura,* covers the lungs themselves. Between the visceral and parietal pleura is a small potential space, the *pleural cavity,* which contains a lubricating fluid secreted by the membranes (see Figure 1-4b). This fluid prevents friction between the membranes and allows them to move easily on one another during breathing. Inflammation of the pleural membrane, or *pleurisy,* causes friction during breathing that can be quite painful when the swollen membranes rub against each other.

The lungs extend from the diaphragm to a point just above the clavicles and lie against the ribs in front and back. The broad inferior portion of the lung, the *base,* is concave and fits over the convex area of the diaphragm. The narrow superior portion of the lung is termed the *apex* or *cupola.* The surface of the lung lying against the ribs, the *costal surface,* is rounded to match the curvature of the ribs. The *mediastinal (medial) surface* of each lung contains a vertical slit, the *hilum,* through which bronchi, pulmonary vessels, and nerves enter and exit. The blood vessels, bronchi, and nerves are held together by the pleura and connective tissue, and they constitute the *root* of the lung. Medially, the left lung also contains a concavity, the *cardiac notch,* in which the heart lies.

The right lung is thicker and broader than the left.

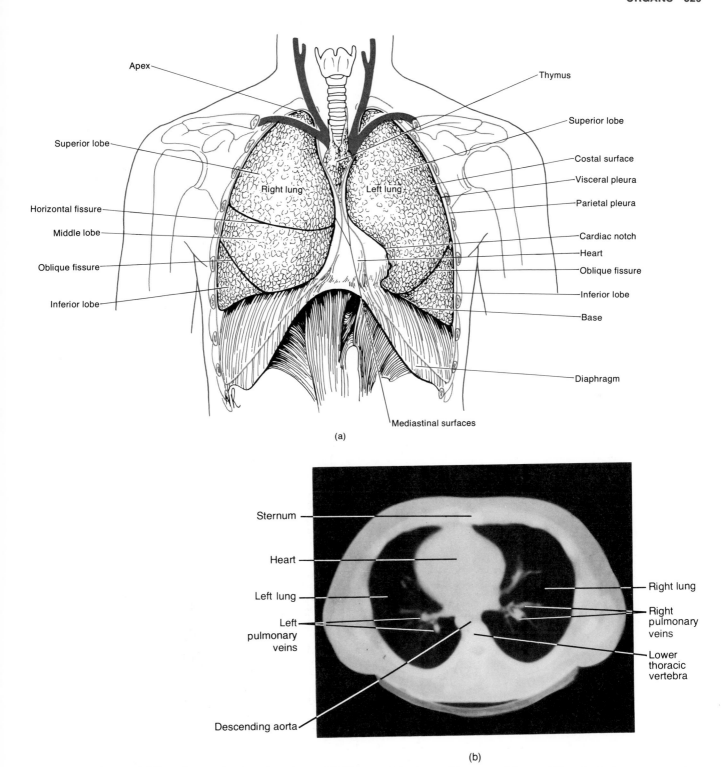

FIGURE 19-7 Lungs. (a) Coverings and external anatomy. (b) CT scan of the lungs. (Courtesy of General Electric Co.)

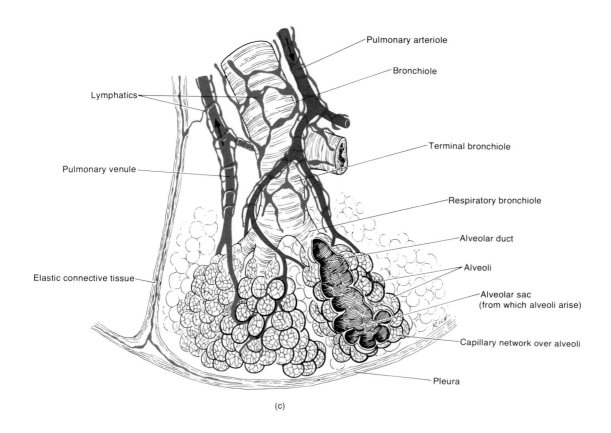

(c)

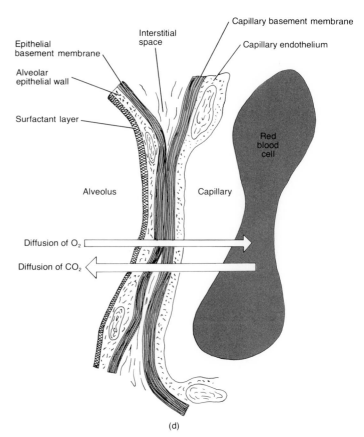

(d)

FIGURE 19-7 *(Continued)* Lungs. (c) A lobule of the lung. (d) Structure of the alveolar-capillary membrane.

It is also somewhat shorter than the left because the diaphragm is higher on the right side to accommodate the liver that lies below it. The left lung is thinner, narrower, and longer than the right.

Each lung is divided into lobes by one or more fissures. Both lungs have an *oblique fissure,* which extends downward and forward. The right lung also has a *horizontal fissure.* The oblique fissure in the left lung separates the superior from the inferior lobe. The upper part of the oblique fissure of the right lung separates the superior lobe from the inferior lobe, whereas the lower part of the oblique fissure separates the inferior lobe from the middle lobe. The horizontal fissure

of the right lung separates the superior lobe from the middle lobe.

Each lobe receives its own secondary or lobar bronchus. Thus the right primary bronchus gives rise to three lobar bronchi called the *superior, middle,* and *inferior lobar* or *secondary bronchi.* The left primary bronchus gives rise to a *superior* and an *inferior lobar* or *secondary bronchus.* Within the substance of the lung, the lobar bronchi give rise to branches that are constant in both origin and distribution. Such branches are called *tertiary* or *segmental bronchi,* and the segment of lung tissue that each supplies is called a *bronchopulmonary segment* (Figure 19-8).

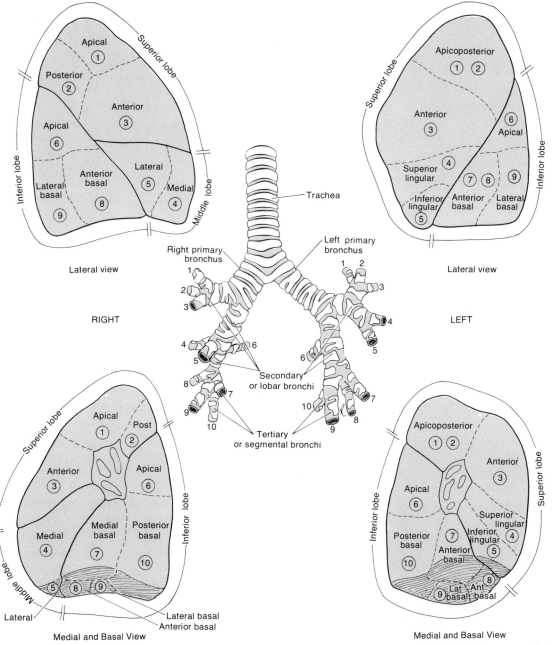

FIGURE 19-8 Bronchopulmonary segments of the lungs. The bronchial branches are shown in the center of the figure.

Smooth muscle   Alveolar sacs

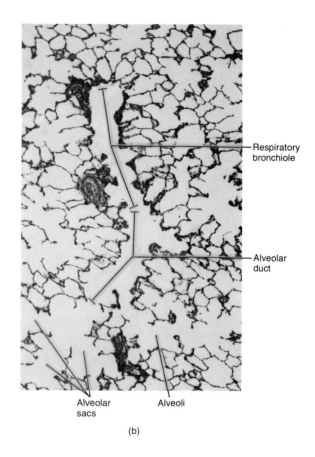

Respiratory
bronchiole

Alveolar
duct

Smooth muscle   Alveolar air spaces

Alveolar
sacs   Alveoli

(a)   (b)

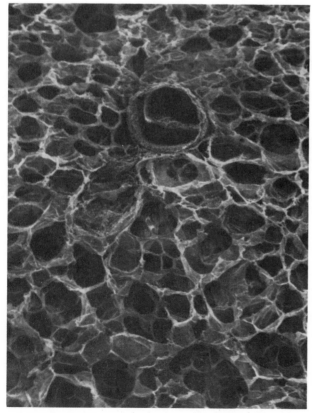

(c)

FIGURE 19-9 Histology of a respiratory bronchiole, alveolar duct, alveolar sacs, and alveoli. (a) Respiratory bronchiole at a magnification of 65×. (b) Respiratory bronchiole, alveolar duct, alveolar sacs, and alveoli at a magnification of 65×. (Courtesy of Edward J. Reith, from *Atlas of Descriptive Histology,* by Edward J. Reith and Michael H. Ross, Harper & Row, Publishers, Inc., New York, 1970.) (c) Scanning electron micrograph of alveolar sacs at a magnification of 100×. (Courtesy of Fisher Scientific Company and S.T.E.M. Laboratories, Inc., Copyright 1975.)

Each bronchopulmonary segment of the lungs is broken up into many small compartments called *lobules* (Figure 19-7c). Every lobule is wrapped in elastic connective tissue and contains a lymphatic vessel, an arteriole, a venule, and a branch from a terminal bronchiole. Terminal bronchioles subdivide into microscopic branches called *respiratory bronchioles* (Figure 19-9a, b). In the respiratory bronchioles, the epithelial lining changes from cuboidal to squamous as they become more distal. Respiratory bronchioles, in turn, subdivide into several (2 to 11) *alveolar ducts* or *atria* (Figure 19-9b).

Around the circumference of the alveolar ducts are numerous alveoli and alveolar sacs. An **alveolus** is a cup-shaped outpouching lined by epithelium and supported by a thin elastic basement membrane. **Alveolar sacs** are two or more alveoli that share a common opening (Figure 19-9b, c). The alveolar walls consist of two principal types of epithelial cells: (1) *squamous pulmonary eipthelial cells (type I alveolar cells or small alveolar cells)* and (2) *septal cells (type II alveolar cells or great alveolar cells)*. The squamous pulmonary epithelial cells are the more numerous of the two types of cells and form a continuous lining of the alveolar wall, except for occasional septal cells. Septal cells are somewhat cuboidal in shape and are dispersed among the squamous pulmonary epithelial cells. Septal cells produce a substance called *surfactant,* which lowers surface tension. Inadequate amounts of surfactant result in hyaline membrane disease (discussed at the end of the chapter). Also found within the alveolar wall are free *alveolar macrophages* or *dust cells.* They are highly phagocytic and serve to remove dust particles or other debris that gain entrance to alveolar spaces. Underneath the layer of squamous pulmonary epithelial cells is an elastic basement membrane. Over the alveoli, the arteriole and venule disperse into a capillary network (Figure 19-7c). The blood capillaries consist of a single layer of endothelial cells and a basement membrane.

The exchange of respiratory gases between the lungs and blood takes place by diffusion through the alveoli and capillary walls. This membrane, through which the respiratory gases move, is collectively known as the **alveolar-capillary (respiratory) membrane** (Figure 19-7d). It consists of: (1) a layer of squamous pulmonary epithelial cells with septal cells and free alveolar macrophages that constitute the alveolar (epithelial) wall; (2) an epithelial basement membrane underneath the alveolar wall; (3) a capillary basement membrane that is often fused to the epithelial basement membrane; and (4) the endothelial cells of the capillary. Despite the large number of layers, the alveolar-capillary membrane is only 0.004 mm thick. This is of considerable importance to the efficient diffusion of respiratory gases. Moreover, it has been estimated that the lungs contain 300 million alveoli—providing an immense surface area (70 sq m) for the exchange of gases.

The arterial supply of the lungs is derived from the pulmonary trunk. It divides into a left pulmonary artery which enters the left lung and a right pulmonary artery which enters the right lung. The venous return of the oxygenated blood is by way of the pulmonary veins, typically two in number on each side—the right and left superior and inferior pulmonary veins. All four veins drain into the left atrium.

The nerve supply of the lungs is derived from the autonomic nervous system. Parasympathetic fibers come from the vagus nerve and sympathetic fibers are derived from the second, third, and fourth thoracic ganglia.

Respiration

The principal purpose of **respiration** is to supply the cells of the body with oxygen and remove the carbon dioxide produced by cellular activities. Three basic processes are involved. The first process is *ventilation,* or breathing—the movement of air between the atmosphere and the lungs. The second and third processes involve the exchange of gases within the body. *External respiration* is the exchange of gases between the lungs and blood. *Internal respiration* is the exchange of gases between the blood and the cells.

VENTILATION

Ventilation or breathing is the process by which atmospheric gases are drawn down into the lungs and waste gases that have diffused into the lungs are expelled back up through the respiratory passageways. Air flows between the atmosphere and lungs for the same reason that blood flows through the body—a pressure gradient exists. When we make inspiratory movements, the pressure inside the lungs is less than the air pressure in the atmosphere and air flows into the lungs. When we make expiratory movements, the pressure inside the lungs is greater than the pressure in the atmosphere and air flows out of the lungs.

Inspiration

Breathing in is called **inspiration** or inhalation. Just before each inspiration the air pressure inside the lungs equals the pressure of the atmosphere, which is about 760 mm Hg at standard conditions. For air to flow into the lungs, the pressure inside the lungs must become lower than the pressure in the atmosphere. This condition is achieved by increasing the volume of the

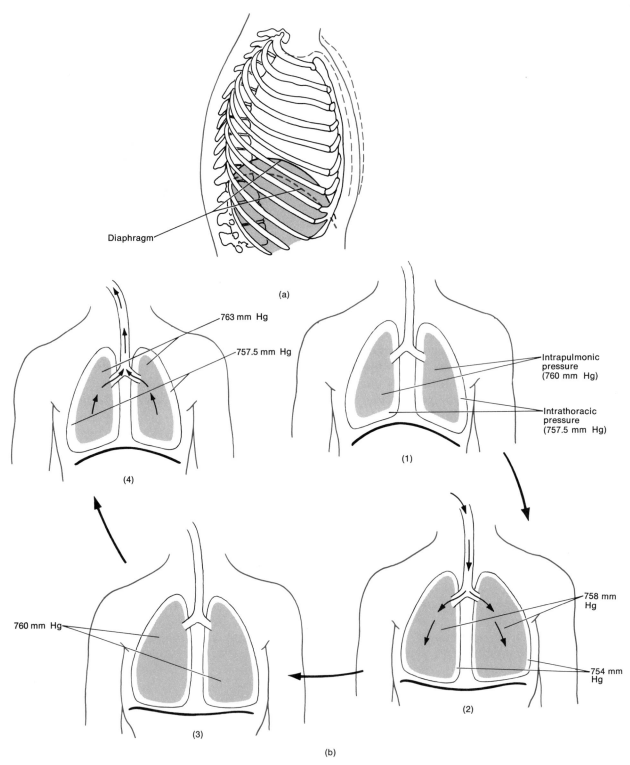

Diaphragm

(a)

763 mm Hg

757.5 mm Hg

Intrapulmonic pressure (760 mm Hg)

Intrathoracic pressure (757.5 mm Hg)

(4)

(1)

758 mm Hg

754 mm Hg

760 mm Hg

(3)

(2)

(b)

FIGURE 19-10 Breathing. (a) Changes in size of rib cage. Black indicates relaxed cage. Color indicates shape of rib cage during inspiration. (b) Pressure changes. (1) Lungs and pleural cavity just before inspiration. (2) Chest expanded and intrathoracic pressure decreased; lungs pulled outward and intrapulmonic pressure decreased. Air moves into lungs until intrapulmonic pressure equals atmospheric pressure (3). (4) Chest relaxes, intrathoracic pressure rises, and lungs snap inward. Intrapulmonic pressure raised, forcing air out until intrapulmonic pressure equals atmospheric pressure (1).

lungs. As you have observed, perhaps unknowingly, the pressure of a gas in a closed container is inversely proportional to the volume of its container. If the size of a closed container is increased, the pressure of the air inside the container decreases. If the size of the container is decreased, then the pressure inside it increases.

The first step toward increasing lung volume involves contraction of the respiratory muscles—the diaphragm and external intercostal muscles (Figure 19-10). The diaphragm is the sheet of skeletal muscle that forms the floor of the thoracic cavity. As it contracts it moves downward, thereby increasing the depth of the thoracic cavity. At the same time, the external intercostal muscles contract, pulling the ribs upward and turning them slightly so the sternum is pushed forward. In this way the circumference of the thoracic cavity also is increased.

The overall increase in the size of the thoracic cavity causes its pressure, called *intrathoracic* or *intrapleural pressure,* to fall far below the pressure of the air inside the lungs. Consequently, the walls of the lungs are sucked outward by the partial vacuum. Expansion of the lungs is aided by the pleural membranes. The parietal pleura lining the chest cavity tends to stick to the visceral pleura around the lungs and to pull the visceral pleura with it.

When the volume of the lungs increases, the pressure inside the lungs, called the *intrapulmonic* or *intraalveolar pressure,* drops from 760 to 758 mm Hg. A pressure gradient is thus established between the atmosphere and the alveoli. Air rushes from the atmosphere into the lungs, and an inspiration takes place. Inspiration is frequently referred to as an active process because it is initiated by muscle contraction.

Expiration

Breathing out, called **expiration** or exhalation, is also achieved by a pressure gradient. But this time the gradient is reversed so that the pressure in the lungs is greater than the pressure of the atmosphere. Expiration starts when the respiratory muscles relax and the size of the chest cavity decreases in depth and circumference. As the intrathoracic pressure returns to its preinspiration level, the walls of the lungs are no longer sucked out. The elastic basement membranes of the alveoli and very elastic fibers in bronchioles and alveolar ducts recoil into their relaxed shape, and lung volume decreases. Intrapulmonic pressure increases, and air moves from the area of higher pressure (the alveoli) to the area of lower pressure (the atmosphere). Expiration during rest is basically a passive process since no muscular contraction is required. However, the internal intercostals do aid in expiration, especially during exercise.

Figure 19-10b shows that the intrathoracic pressure is always a little less than the pressure inside the lungs or in the atmosphere. The pleural cavities are sealed off from the outside environment and cannot equalize their pressure with that of the atmosphere. Nor can the diaphragm and rib cage move inward enough to bring the intrathoracic pressure up to atmospheric pressure. Actually, maintenance of a low intrathoracic pressure is vital to the functioning of the lungs. The alveoli are so elastic that at the end of an expiration they attempt to recoil inward and collapse on themselves like the walls of a deflated balloon. Such a collapse, called *atelectasis,* which would obstruct the movement of air, is prevented by the slightly lower pressure in the pleural cavities that keeps the alveoli slightly inflated.

Another factor preventing the collapse of alveoli is the presence of a phospholipid produced by the alveolar cells. This substance, called *surfactant,* decreases surface tension in the lungs. That is, it forms a thin lining over the alveoli and prevents them from sticking together following expiration. Thus, as alveoli become smaller, for example following expiration, the tendency of alveoli to collapse is minimized because the surface tension does not increase.

AIR VOLUMES EXCHANGED

In clinical practice the word *respiration* means one inspiration plus one expiration. The average healthy adult has 14 to 18 respirations a minute. During each respiration the lungs exchange volumes of air with the atmosphere. A lower than normal exchange volume is usually a sign of pulmonary malfunction. The apparatus commonly used to measure the amount of air exchanged during breathing is referred to as a **respirometer** or **spirometer.**

A respirometer consists of a weighted drum inverted over a chamber of water. The drum usually contains oxygen or air. A tube connects the air-filled chamber with the subject's mouth. During inspiration, air is removed from the chamber, the drum sinks, and an upward deflection is recorded by the stylus on the graph paper on the kymograph (rotating drum). During expiration, air is added, the drum rises, and a downward deflection is recorded. The record is called a *spirogram* (Figure 19-11). Spirometric studies measure lung capacities and rates and depths of ventilation for diagnostic purposes. Spirometry is usually indicated for individuals who exhibit labored breathing. It is also used in the diagnosis of respiratory disorders such as emphysema and bronchial asthma.

During normal quiet breathing, also called *eupnea,* about 500 ml of air moves into the respiratory passageways with each inspiration. The same amount moves out with each expiration. This volume of air inspired

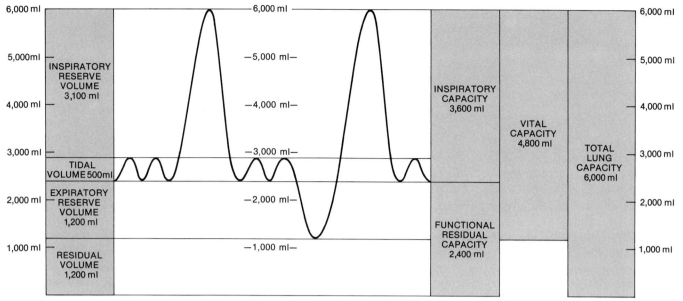

FIGURE 19-11 Spirogram of lung volumes and capacities.

(or expired) is called *tidal volume* (Figure 19-11). Actually, only about 350 ml of the tidal volume reaches the alveoli. The other 150 ml remains in the dead spaces of the nose, pharynx, larynx, trachea, and bronchi and is known as *dead air.*

By taking a very deep breath, we can inspire a good deal more than 500 ml. This excess inhaled air, called the *inspiratory reserve volume,* averages 3100 ml above the 500 ml of tidal volume. Thus the respiratory system can pull in as much as 3600 ml of air. If we inhale normally and then exhale as forcibly as possible, we should be able to push out 1200 ml of air in addition to the 500 ml tidal volume. This extra 1200 ml is called the *expiratory reserve volume.* Even after the expiratory reserve volume is expelled, a good deal of air still remains in the lungs because the lower intrathoracic pressure keeps the alveoli slightly inflated. This air, the *residual volume,* amounts to about 1200 ml. Opening the thoracic cavity allows the intrathoracic pressure to equal the atmospheric pressure, forcing out the residual volume. The air remaining is called the *minimal volume.*

Minimal volume provides a medical and legal tool for determining whether a baby was born dead or died after birth. The presence of minimal volume can be demonstrated by placing a piece of lung in water and watching it float. Fetal lungs contain no air, and so the lung of a stillborn will not float in water.

Lung capacity can be calculated by combining various lung volumes. *Inspiratory capacity,* the total inspiratory ability of the lungs, is the sum of tidal volume plus inspiratory reserve volume (3600 ml). *Functional residual capacity* is the sum of residual volume plus expiratory reserve volume (2400 ml). *Vital capacity* is the sum of inspiratory reserve volume, tidal volume, and expiratory reserve volume (4800 ml). Finally, *total lung capacity* is the sum of all volumes (6000 ml).

EXCHANGE OF RESPIRATORY GASES

As soon as the lungs fill with air, oxygen diffuses from the alveoli to the blood, through the interstitial fluid, and finally to the cells. Carbon dioxide diffuses in just the opposite direction—from the cells, through interstitial fluid to the blood, and to the alveoli.

CONTROL OF RESPIRATION

Respiration is controlled by several mechanisms that help the body maintain homeostasis.

Nervous Control

The size of the thorax is affected by the action of the respiratory muscles. These muscles contract and relax in turn as a result of nerve impulses transmitted to them from centers in the brain. The area from which nerve impulses are sent to respiratory muscles is located in the reticular formation of the brain stem and is referred to as the **respiratory center.** The respiratory center consists of a widely dispersed group of neurons and is functionally divided into three areas. These are: (1) the *medullary rhythmicity area* in the medulla, (2) the *apneustic area* in the pons, and (3) the *pneumotaxic area,* also in the pons (Figure 19-12a).

The function of the medullary rhythmicity area is to control the basic rhythm of respiration. In the normal resting state, inspiration usually lasts for about two

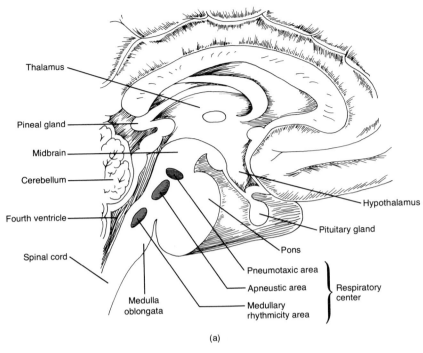

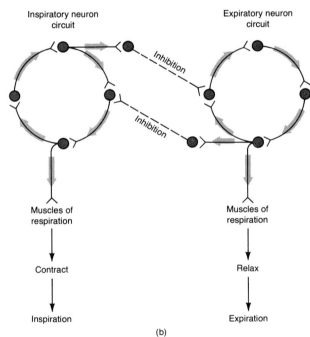

FIGURE 19-12 Respiratory center. (a) Location of the areas of the respiratory center. (b) Proposed function of the medullary rhythmicity area in controlling the basic rhythm of respiration.

seconds and expiration for about three seconds. This is the basic rhythm of respiration. Within the medullary rhythmicity area are found both inspiratory and expiratory neurons arranged in circuits (Figure 19-12b). Let us first consider the proposed role of the inspiratory neurons in respiration. When one of the neurons in the circuit becomes excited, it excites the next one, which, in turn, excites the next, and so on. In this way, the nerve impulses travel around the circuit from one neuron to the next. This continues for about two seconds, until the neurons become fatigued. However,

as some nerve impulses travel around the circuit, other impulses travel from the circuit to the muscles of respiration—the diaphragm and intercostal muscles. These impulses reach the diaphragm by the phrenic nerves and the intercostal muscles by the intercostal nerves. When the impulses reach the respiratory muscles, the muscles contract and inspiration occurs.

In addition to sending impulses to the muscles of respiration, the inspiratory center also sends inhibitory impulses to the expiratory neurons. These impulses inhibit the circuit of expiratory neurons. As long as the inspiratory neurons maintain nerve impulses through the circuit, expiration is inhibited. But, when the inspiratory neurons fatigue, the expiratory neurons are no longer inhibited. This causes the expiratory neurons to become excited and nerve impulses travel around the circuit of expiratory neurons from one neuron to another. This continues for about three seconds until the expiratory neurons fatigue. While the nerve impulses are maintained by the circuit of expiratory neurons, some impulses are transmitted to the respiratory muscles and others inhibit the inspiratory neurons. The impulses to the respiratory muscles, via the phrenic and intercostal nerves, cause the muscles to relax and expiration occurs. When the expiratory neurons fatigue, the inspiratory neurons are no longer inhibited, and the inspiratory neurons become excited once again. Thus, the basic rhythm of respiration is established by circuits of inspiratory and expiratory neurons that alternately act upon the muscles of respiration and mutually inhibit each other.

Although the medullary rhythmicity area controls the basic rhythm of respiration, other parts of the nervous system can modify the rhythm. For example, in experiments where impulses from the apneustic and pneumotaxic areas are prevented from reaching the medullary rhythmicity area, respirations consist of short inspirations and prolonged expirations. Stimulation of the apneustic area alone sends impulses to the medullary rhythmicity area that result in forceful, prolonged inspirations and weak, short expirations. When the pneumotaxic area is stimulated, impulses to the medullary rhythmicity area result in the cessation of apneustic breathing. Stimulation of the pneumotaxic area can also alter the rate of respiration.

The respiratory center has connections with the cerebral cortex, which means we can voluntarily alter our pattern of breathing. We can even refuse to breathe at all for a short time. Voluntary control is protective because it enables us to prevent water or irritating gases from entering the lungs. The ability to stop breathing is limited by the buildup of CO_2 in the blood, however. When the pCO_2 increases to a certain level, the inspiratory center is stimulated, impulses are sent to inspiratory muscles, and breathing resumes whether or not the person wishes. It is impossible for people to kill themselves by holding their breath.

Chemical and Pressure Stimuli

Certain chemical stimuli determine how fast we breathe. These stimuli may act directly on the respiratory center or on chemoreceptors located in the aortic and carotid bodies. When chemoreceptors are stimulated, impulses are sent to the respiratory center and the rate of respiration is increased. Probably the main chemical stimulus that alters respirations is CO_2. Although it is convenient to speak of CO_2 as the most important chemical stimulus for regulating the rate of respiration, it is actually hydrogen (H^+) ions that assume this role. For example, CO_2 in the blood combines with water (H_2O) to form carbonic acid (H_2CO_3) as follows: $CO_2 + H_2O \rightarrow H_2CO_3$. But, the carbonic acid quickly breaks down into H^+ ions and bicarbonate ions (HCO_3)$^-$ ions: $H_2CO_3 \rightarrow H^+ + HCO_3^-$. Any increase in CO_2 will cause an increase in H^+ ions and any decrease in CO_2 will cause a decrease in H^+ ions. In effect, it is the H^+ ions that alter the rate of respiration rather than the CO_2 molecules. Although the following discussion refers to the levels of CO_2 and their effect on respiration, keep in mind that it is really the H^+ ions that cause the effects.

Under normal circumstances, the partial pressure of CO_2 (pCO_2) in arterial blood is 40 mm Hg. If there is even a slight increase in pCO_2—a condition called *hypercapnia*—chemoreceptors in the medulla and in the carotid and aortic bodies are stimulated. Stimulation of the chemoreceptors causes the inspiratory center to become highly active, and the rate of respiration increases. This increased rate, *hyperventilation*, allows the body to expel more CO_2 until the pCO_2 is lowered to normal. Now let us consider the opposite situation. If arterial pCO_2 is lower than 40 mm Hg, the receptors are not stimulated and stimulatory impulses are not sent to the inspiratory center. Consequently, the center sets its own moderate pace until CO_2 accumulates and the pCO_2 rises to 40 mm Hg. A slow rate of respiration is called *hypoventilation*.

The oxygen receptors are sensitive only to large drops in the pO_2. If arterial pO_2 falls from a normal of 100 mm Hg to 70 mm Hg, the oxygen receptors become stimulated and send impulses to the inspiratory center. But if the pO_2 falls much below 70 mm Hg, the cells of the respiratory center suffer oxygen starvation and do not respond well to any chemical receptors. They send fewer impulses to the respiratory muscles, and the respiration rate decreases or breathing ceases altogether.

The carotid and aortic bodies also contain pressoreceptors (baroreceptors) that are stimulated by a rise in blood pressure. Although these pressoreceptors are concerned mainly with the control of circulation, they help control respiration. For example, a sudden rise in blood pressure decreases the rate of respiration, and a drop in blood pressure brings about an increase in the respiratory rate. Other factors that affect respiration are:

1. A sudden cold stimulus such as plunging into cold water causes a temporary cessation of breathing called *apnea*.

2. A sudden, severe pain brings about apnea, but a prolonged pain triggers the general adaptation syndrome and increases respiration rate.

3. Stretching the anal sphincter muscle increases the respiratory rate. This technique is sometimes employed to stimulate respiration during emergencies.

4. Irritation of the pharynx or larynx by touch or chemicals brings about an immediate cessation of breathing followed by coughing.

MODIFIED RESPIRATORY MOVEMENTS

Respirations provide human beings with methods for expressing emotions such as laughing, yawning, sighing, and sobbing. Moreover, respiratory air can be used to expel foreign matter from the upper air passages through actions such as sneezing and coughing. Some of the modified respiratory movements that express emotion or clear the air passageways are listed in Exhibit 19-1. All these movements are reflexes, but some of them also can be initiated voluntarily.

FOOD CHOKING

Food choking (cafe coronary), the sixth leading cause of accidental death, is often mistaken for myocardial infarction. However, it can be recognized easily. The victims cannot speak or breathe; they may become panic stricken and run from the room. They become pale, then deeply cyanotic, and collapse. Without intervention, death occurs in 4–5 minutes.

The food or other obstructing object causing asphyxiation may lodge in the back of the throat or enter the trachea to occlude the airway. Tracheotomy, even when a physician performs it, can be hazardous in a nonclinical setting. An instrument for removing food from the back of the throat has been developed but is seldom at hand in the presenting emergency. However, there is a first aid procedure that does not require special instruments and can be performed by any informed layman. This procedure is called the **Heimlich maneuver.**

In all probability, food choking occurs during inspiration, which causes the bolus of food to be sucked against the opening into the larynx. At the time of the accident, the lungs are therefore expanded. Even during normal expiration, however, some tidal air (500 ml) and the entire expiratory reserve volume (1200 ml) are present in the lungs.

Pressing one's fist upward into the epigastric region of the abdomen elevates the diaphragm suddenly—compressing the lungs in the rib cage and increasing the air pressure in the bronchial tree. This action forces air out through the trachea and will eject the food (or object) blocking the airway. The action can

EXHIBIT 19-1
MODIFIED RESPIRATORY MOVEMENTS

MOVEMENT	COMMENT
Coughing	Preceded by a long-drawn and deep inspiration that is followed by a complete closure of the glottis—resulting strong expiration suddenly pushes glottis open and sends a blast of air through the upper respiratory passages. Stimulus for this reflex act could be a foreign body lodged in the larynx, trachea, or epiglottis.
Sneezing	Spasmodic contraction of muscles of expiration forcefully expels air through the nose and mouth. Stimulus may be an irritation of the nasal mucosa.
Sighing	A deep and long-drawn inspiration immediately followed by a shorter but forceful expiration.
Yawning	A deep inspiration through the widely opened mouth producing an exaggerated depression of the lower jaw. May be stimulated by drowsiness or fatigue, but precise stimulus-receptor cause is unknown.
Sobbing	Starts with a series of convulsive inspirations. Glottis closes earlier than normal after each inspiration so only a little air enters the lungs with each inspiration. Immediately followed by a single prolonged expiration.
Crying	An inspiration followed by many short convulsive expirations. Glottis remains open during the entire time, and the vocal cords vibrate. Accompanied by characteristic facial expressions.
Laughing	Involves the same basic movements as crying, but the rhythm of the movements and the facial expressions usually differ from those of crying. Laughing and crying are sometimes indistinguishable.
Hiccough	Spasmodic contraction of the diaphragm followed by a spasmodic closure of the glottis. Produces a sharp inspiratory sound. Stimulus is usually irritation of the sensory nerve endings of the digestive tract.

be simulated by inserting a cork in a compressible plastic bottle and then squeezing it suddenly—the cork flies out because of the increased pressure. Figure 19-13 shows how to administer the Heimlich maneuver.

Applications to Health

HAY FEVER

An allergic reaction to the proteins contained in foreign substances such as plant pollens, dust, and certain foods is called **hay fever.** Allergic reactions are a special antigen-antibody response that initiate either a

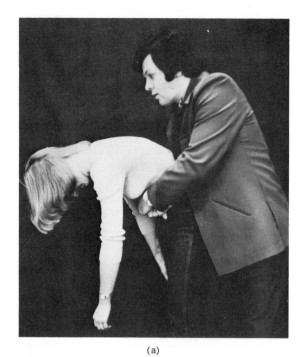
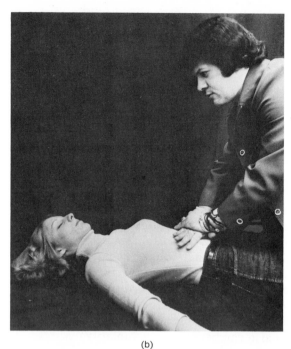

(a)   (b)

FIGURE 19-13 Heimlich maneuver. The principle employed in this antichoke first aid method is the application of force to compress the air in the lungs to expel objects in the air passageways. (a) If the victim is upright, stand behind the person and place both arms around the waist just above the belt line. Grasp your right wrist with your left hand and allow the victim's head, arms, and upper torso to fall forward. Next, the victim's abdomen is rapidly and strongly compressed. If another person is present, he or she can assist by removing the ejected object from the victim's mouth. (b) If the victim is recumbent, the rescuer may rapidly and strongly force the victim's diaphragm upward by applying pressure as shown. (Courtesy of Donald Castellaro and Deborah Massimi.)

localized or a systemic inflammatory response. In hay fever the response is localized in the respiratory membranes. The membranes become inflamed, and a watery fluid drains from the eyes and nose.

BRONCHIAL ASTHMA

Bronchial asthma is a reaction, usually allergic, characterized by attacks of wheezing and difficult breathing. Attacks are brought on by spasms of the smooth muscles that lie in the walls of the smaller bronchi and bronchioles, causing the passageways to close partially. The patient has trouble exhaling, and the alveoli may remain inflated during expiration. Usually the mucous membranes that line the respiratory passageways become irritated and secrete excessive amounts of mucus that may clog the bronchi and bronchioles and worsen the attack. About three out of four asthma victims are allergic to edible or airborne substances as common as wheat or dust. Others are sensitive to the proteins of harmless bacteria that inhabit the sinuses, nose, and throat. Asthma might also have a psychosomatic origin.

EMPHYSEMA

In **emphysema,** the alveolar walls lose their elasticity and remain filled with air during expiration. The name

means "blown up" or "full of air." Reduced forced expiratory volume is the first symptom. Later, alveoli in other areas of the lungs are damaged. Many alveoli may merge to form larger air sacs with a reduced overall volume. The lungs become permanently inflated because they have lost elasticity. To adjust to the increased lung size, the size of the chest cage increases. The patient has to work to exhale. Oxygen diffusion does not occur as easily across the damaged alveolar membrane, blood oxygen level is somewhat lowered, and any mild exercise that raises the oxygen requirements of the cells leaves the patient breathless. As the disease progresses, the alveoli are replaced with thick fibrous connective tissue. Even carbon dioxide does not diffuse easily through this fibrous tissue. If the blood cannot buffer all the hydrogen ions that accumulate, the blood pH drops or unusually high amounts of carbon dioxide may dissolve in the plasma. High carbon dioxide levels produce acid conditions that are toxic to brain cells. Consequently, the inspiratory center becomes less active and respiration rate slows down, further aggravating the problem. The compressed and damaged capillaries around the deteriorating alveoli may no longer be able to receive blood. As a result, a backup pressure increases in the pulmonary trunk and the right ventricle overworks as it attempts to force blood through the remaining capillaries.

Emphysema is generally caused by a long-term irritation. Air pollution, occupational exposure to industrial dust, and cigarette smoke are the most common irritants. Chronic bronchial asthma also may produce alveolar damage. Cases of emphysema are becoming more and more frequent in the United States. The irony is that the disease can be prevented and the progressive deterioration can be stopped by eliminating the harmful stimuli.

PNEUMONIA

The term **pneumonia** means an acute infection or inflammation of the alveoli. In this disease the alveolar sacs fill up with fluid and dead white blood cells, reducing the amount of air space in the lungs. (Remember that one of the cardinal signs of inflammation is edema.) Oxygen has difficulty diffusing through the inflamed alveoli, and the blood pO_2 may be drastically reduced. Blood pCO_2 usually remains normal because carbon dioxide always diffuses through the alveoli more easily than oxygen does. If all the alveoli of a lobe are inflamed, the pneumonia is called *lobar pneumonia.* If only parts of the lobe are involved, it is called *lobular,* or *segmental, pneumonia.* If both the alveoli and the bronchial tubes are included, it is called *bronchopneumonia.*

The most common cause of pneumonia is the pneumococcus bacterium, but other bacteria or a fungus may be the source of trouble. Viral pneumonia is caused by several viruses, including the influenza virus.

TUBERCULOSIS

The bacterium called *Mycobacterium tuberculosis* produces an inflammation called **tuberculosis.** Tuberculosis most often affects the lungs and the pleura. The bacteria destroy parts of the lung tissue, and the tissue is replaced by fibrous connective tissue. Because the connective tissue is inelastic and relatively thick, the affected areas of the lungs do not recoil during expiration, and larger amounts of air are retained. Gases no longer diffuse easily through the fibrous tissue.

Tuberculosis bacteria are spread by inhalation. Although they can withstand exposure to many disinfectants, they die quickly in sunlight. This is why tuberculosis is sometimes associated with crowded, poorly lit housing conditions. Many drugs are successful in treating tuberculosis. Rest, sunlight, and good diet are vital parts of treatment.

HYALINE MEMBRANE DISEASE (HMD)

Sometimes called glassy-lung disease or infant respiratory distress syndrome (RDS), **hyaline membrane disease (HMD)** is responsible for approximately 20,000 newborn infant deaths per year. Before birth, the respiratory passages are filled with fluid. Part of this fluid is amniotic fluid inhaled during respiratory movements in utero. The remainder is produced by the submucosal glands and the goblet cells of the respiratory epithelium.

At birth, this fluid-filled airway must become an air-filled airway, and the collapsed primitive alveoli (terminal sacs) must expand and function in gas exchange. The success of this transition depends largely on pulmonary surfactant—a mixture of lipoproteins that lowers surface tension in the fluid layer lining the primitive alveoli once air enters the lungs. Surfactant is present in the fetus's lungs as early as the twenty-third week. By 28 to 32 weeks, however, the amount of surfactant is great enough to prevent alveolar collapse during breathing. Surfactant is produced continuously by septal alveolar cells. The presence of surfactant can be detected by amniocentesis.

Although in a normal, full-term infant the second and subsequent breaths require less respiratory effort than the first, breathing is not completely normal until about 40 minutes after birth. The entire lung is not inflated fully with the first one or two breaths. In fact, for the first seven to ten days, small areas of the lungs may remain uninflated.

In the newborn whose lungs are deficient in surfactant, the effort required for the first breath is essentially the same as that required in normal newborns. However, the surface tension of the alveolar fluid is seven to fourteen times higher than the surface tension of alveolar fluid with a monomolecular layer of surfactant. Consequently, during expiration after the first inspiration, the surface tension of the alveoli increases as the alveoli deflate. The alveoli collapse almost to their original uninflated state.

Idiopathic RDS usually appears within a few hours after birth. Affected infants show difficult and labored breathing with withdrawal of the intercostal and subcostal spaces. Death may occur soon after onset of respiratory difficulty or may be delayed for a few days, although many infants survive. At autopsy, the lungs are underinflated and areas of atelectasis are prominent. (In fact, the lungs are so airless they sink in water.) If the infant survives for at least a few hours after developing respiratory distress, the alveoli are often filled with a fluid of high-protein content that resembles a hyaline (or glassy) membrane. RDS occurs frequently in premature infants and also in infants of diabetic mothers, particularly if the diabetes is untreated or poorly controlled.

A new treatment currently being developed called PEEP—positive end expiratory pressure—could reverse the mortality rate from 90 percent deaths to 90 percent survival. This treatment consists of passing a tube through the air passage to the top of the lungs to provide needed oxygen-rich air at continuous pressures of up to 14 mm Hg. Continuous pressure keeps the baby's alveoli open and available for gas exchange.

SMOKING

As part of ordinary breathing, many irritating substances are inhaled. Almost all pollutants, including inhaled smoke, have an irritating effect on the bronchial tubes and lungs and may be regarded as stresses or irritating stimuli.

Close examination of the epithelium of a bronchial tube reveals three kinds of cells (Figure 19-14). The surface cells are columnar cells that contain cilia. At intervals between the ciliated columnar cells are the mucus-secreting goblet cells. The bottom of the epithelium normally contains a row of basal cells above the basement membrane. The basal cells divide continuously, replacing the ciliated columnar epithelium as they wear down and are sloughed off. The bronchial epithelium is important clinically because a common lung cancer, **bronchogenic carcinoma,** starts in the walls of the bronchi.

The constant irritation by inhaled smoke and pollutants causes an enlargement of the goblet cells of the bronchial epithelium. They respond by secreting excessive mucus. The basal cells also respond to the stress by undergoing cell division so fast that the basal cells push into the area occupied by the goblet and columnar cells. As many as 20 rows of basal cells may be produced. Many researchers believe that if the stress is removed at this point, the epithelium can return to normal.

If the stress persists, more and more mucus is secreted and the cilia become less effective. As a result, mucus is not carried toward the throat but remains trapped in the bronchial tubes. The individual then develops a "smoker's cough." Moreover, the constant irritation from the pollutant slowly destroys the alveoli, which are replaced with thick, inelastic connective tissue. Mucus that has accumulated becomes trapped in the air sacs. Millions of sacs rupture—reducing the diffusion surface for the exchange of oxygen and carbon dioxide. The individual has now developed emphysema. If the stress is removed at this point, there is little chance for improvement. Alveolar tissue that has been destroyed cannot be repaired. But removal of the stress can stop further destruction of lung tissue.

If the stress continues, the emphysema gets progressively worse, and the basal cells of the bronchial tubes continue to divide and break through the base-

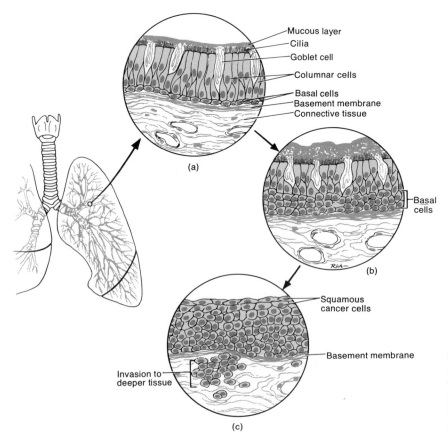

Mucous layer
Cilia
Goblet cell
Columnar cells
Basal cells
Basement membrane
Connective tissue

(a)

Basal cells

(b)

Squamous cancer cells

Basement membrane

Invasion to deeper tissue

(c)

FIGURE 19-14 Effects of smoking on the respiratory epithelium. (a) Microscopic view of the normal epithelium of a bronchial tube. (b) Initial response of the bronchial epithelium to irritation by pollutants. (c) Advanced response of the bronchial epithelium.

ment membrane. At this point the stage is set for bronchogenic carcinoma. Columnar and goblet cells disappear and may be replaced with squamous cancer cells. If this happens, the malignant growth spreads throughout the lung and may block a bronchial tube. If the obstruction occurs in a large bronchial tube, very little oxygen enters the lung and disease-producing bacteria thrive on the mucoid secretions. In the end, the patient may develop emphysema, carcinoma, and a host of infectious diseases. Treatment involves surgical removal of the diseased lung. However, metastasis of the growth through the lymphatic or blood system may result in new growths in other parts of the body such as the brain and liver.

Other factors may be associated with lung cancer. For instance, breast, stomach, and prostate malignancies can metastasize to the lungs. People who apparently have not been exposed to pollutants do occasionally develop bronchogenic carcinoma. However, the occurrence of bronchogenic carcinoma is probably over 20 times higher in heavy cigarette smokers than it is in nonsmokers.

KEY MEDICAL TERMS ASSOCIATED WITH THE RESPIRATORY SYSTEM

Apnea Absence of ventilatory movements.

Asphyxia Oxygen starvation due to low atmospheric oxygen or interference with ventilation, external respiration, or internal respiration.

Atelectasis (*asis* = state or condition) A collapsed lung or portion of a lung.

Bronchitis (*bronch* = bronchus, trachea) Inflammation of the bronchi and bronchioles.

Cheyne-Stokes respiration Irregular breathing beginning with shallow breaths that increase in depth and rapidity, then decrease and cease altogether for 15–20 seconds. The cycle repeats itself again and again. Cheyne-Stokes is normal in infants. It is also often seen just before death from pulmonary, cerebral, cardiac, and kidney disease and is referred to as the "death rattle."

Diphtheria An acute bacterial infection that causes the mucous membranes of the pharynx, nasopharynx, and larynx to enlarge and become leathery. Enlarged membranes may obstruct airways and cause death from asphyxiation.

Dyspnea (*dys* = painful, difficult) Labored or difficult breathing (short-winded).

Eupnea (*eu* = good, normal) Normal quiet breathing.

Hypoxia Reduction in oxygen supply to cells.

Influenza Viral infection that causes inflammation of respiratory mucous membranes as well as fever.

Orthopnea Inability to breathe in a horizontal position.

Pneumothorax (*pneumo* = lung) Air in pleural space causing collapse of the lung. Most common cause is surgical opening of chest during heart surgery, making intrathoracic pressure equal atmospheric pressure.

Pulmonary edema Excess amounts of interstitial fluid in the lungs producing cough and dyspnea. Common in failure of the left side of the heart.

Pulmonary embolism Presence of a blood clot or other foreign substance in a pulmonary arterial vessel stopping circulation to a part of the lungs.

Rales Sounds sometimes heard in the lungs that resemble bubbling or rattling. May be caused by air or an abnormal secretion in the lungs.

Respirator Metal chamber that entombs chest; also called "iron lung." Used to produce inspiration and expiration in patient with paralyzed respiratory muscles. Pressure inside chamber is rhythmically alternated to suck out and push in chest walls.

STUDY OUTLINE

Respiratory Organs
1. Respiratory organs include the nose, pharynx, larynx, trachea, bronchi, and lungs.
2. They act with the cardiovascular system to supply oxygen to and remove carbon dioxide from body cells.

Nose
1. The external portion is made of cartilage and skin and lined with mucous membrane; openings to exterior are external nares.
2. The internal portion communicates with pharynx through internal nares and communicates with paranasal sinuses.
3. The nose is divided into cavities by a septum. Anterior portions of the cavities are called the vestibules.
4. The nose is adapted for the warming, moistening, and filtering of air and olfaction; it also assists in speech.

Pharynx
1. The pharynx, or throat, is a muscular tube lined by mucous membrane.
2. Anatomic regions are nasopharynx, oropharynx, and laryngopharynx.
3. The nasopharynx functions in respiration. The oropharynx and laryngopharynx function in digestion and respiration.

Larynx
1. The larynx is a passageway that connects the pharynx with the trachea.
2. Prominent cartilages are the thyroid, or Adam's apple, the epiglottis, which prevents food from entering the larynx, the cricoid, which connects the larynx and trachea, the arytenoid, the corniculate, and the cuneiform cartilages.

3. The larynx contains true vocal cords that produce sound. Taut cords produce high pitches, and relaxed cords produce low pitches.

Trachea

1. The trachea extends from the larynx to the primary bronchi.
2. It is composed of smooth muscle and C-shaped rings of cartilage and is lined with ciliated mucous membrane.

Bronchi

1. The bronchial tree consists of primary bronchi, secondary bronchi, bronchioles, and terminal bronchioles. Walls of bronchi contain rings of cartilage; walls of bronchioles do not.
2. A developed picture of the bronchial tree is called a bronchogram.

Lungs

1. Lungs are paired organs in the thoracic cavity. They are enclosed by the pleural membrane (parietal pleura is outer layer; visceral pleura is inner layer).
2. The right lung has three lobes separated by two fissures; the left lung has two lobes separated by one fissure and a depression, the cardiac notch.
3. The secondary bronchi give rise to branches called segmental bronchi which supply segments of lung tissue called bronchopulmonary segments.
4. Each bronchopulmonary segment consists of lobules, which contain lymphatics, arterioles, venules, terminal bronchioles, respiratory bronchioles, alveolar ducts, alveolar sacs, and alveoli.
5. Gas exchange occurs across alveolar-capillary membranes.

Respiration

1. Inspiration occurs when intrapulmonic pressure falls below atmospheric pressure. Contraction of the diaphragm and intercostal increases the size of the thorax and decreases the intrathoracic pressure. Decreased intrathoracic pressure causes a decreased intrapulmonic pressure.
2. Expiration occurs when intrapulmonic pressure is higher than atmospheric pressure. Relaxation of diaphragm and intercostals increases intrathoracic pressure, which causes an increased intrapulmonic pressure.
3. Among the air volumes exchanged in ventilation are tidal, inspiratory reserve, expiratory reserve, residual, and minimal volumes.
4. Nervous control is regulated by the respiratory centers in medulla and pons, which control the rhythm of respiration.
5. Coughing, sneezing, sighing, yawning, sobbing, crying, laughing, and hiccoughing involve modified respiratory movements.

Applications to Health

1. Hay fever is an allergic reaction of respiratory membranes.
2. Bronchial asthma occurs when spasms of smooth muscle in bronchial tubes result in partial closure of air passageways, inflammation, inflated alveoli, and excess mucus production.
3. Emphysema is characterized by deterioration of alveoli leading to loss of their elasticity. Symptoms are reduced expiratory volume, inflated lungs, and enlarged chest.
4. Pneumonia is an acute inflammation or infection of alveoli.
5. Tuberculosis is an inflammation of pleura and lungs produced by a specific bacterium.
6. Hyaline membrane disease is an infant disorder in which surfactant is lacking and alveolar ducts and alveoli have a glassy appearance.
7. Smoking
   a. Pollutants, including smoke, act as stresses on the epithelium of the bronchi and lungs. Constant irritation results in excessive secretion and mucus and rapid division of bronchial basal cells.
   b. Additional irritation may cause retention of mucus in bronchioles, loss of elasticity of alveoli, and less surface area for gaseous exchange.
   c. In the final stages, bronchial epithelial cells may be replaced by cancer cells. The growth may block a bronchial tube and spread throughout the lung and other body tissues.

REVIEW QUESTIONS

1. What organs constitute the respiratory system? What function do the respiratory and cardiovascular systems have in common?
2. Describe the structures of the external and internal nose and describe their functions in filtering, warming, and moistening air.
3. What is the pharynx? Differentiate the three regions of the pharynx, and indicate their roles in respiration.
4. Where is the larynx located? Describe the positions and functions of the laryngeal cartilages. How does the larynx function in voice production?
5. Describe the location and structure of the trachea. What is a tracheostomy? Intubation?
6. What is the bronchial tree? Describe its structure. What is a bronchogram?
7. Where are the lungs located? Distinguish the parietal pleura from the visceral pleura. What is pleurisy?
8. Define each of the following parts of a lung: base, apex, costal surface, medial surface, hilum, root, and cardiac notch.
9. What is a lobule of the lung? Describe its composition and function in respiration. What is a bronchopulmonary segment?
10. What are the basic differences between ventilation, external respiration, and internal respiration?
11. Discuss the basic steps involved in inspiration and ex-

piration. Be sure to include values for all pressures involved.

12. What is a spirometer? Define the various lung volumes and capacities.

13. How does the medullary rhythmicity area function in controlling respiration? What is the role of the apneustic area and the pneumotaxic area in controlling respiration?

14. Define the various kinds of modified respiratory movements.

15. Describe how the Heimlich maneuver is performed.

16. For each of the following, list the outstanding clinical symptoms: hay fever, bronchial asthma, emphysema, pneumonia, tuberculosis, hyaline membrane disease, and bronchogenic carcinoma.

17. Discuss the stages in the destruction of respiratory epithelium as a result of continued irritation by pollutants.

18. Refer to the glossary of key medical terms associated with the respiratory system. Be sure that you can define each term.

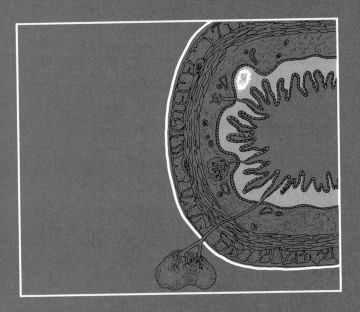

20

The Digestive System

STUDENT OBJECTIVES

- Define digestion as a chemical and mechanical process.
- Identify the organs of the gastrointestinal tract and the accessory organs of digestion.
- Describe the structure of the wall of the gastrointestinal tract.
- Define the mesentery, mesocolon, falciform ligament, lesser omentum, and greater omentum.
- Describe the structure of the mouth and its role in mechanical digestion.
- Identify the location and histology of the salivary glands and define the composition and function of saliva.
- Identify the parts of a typical tooth and compare deciduous and permanent dentitions.
- Discuss the sequence of events involved in swallowing.
- Describe the location, anatomy, and histology of the stomach and compare mechanical and chemical digestion.
- Describe the location, structure, and histology of the pancreas.
- Define the position, structure, and histology of the liver.
- Describe the structure and histology of the gallbladder.
- Describe those structural features of the small intestine that adapt it for digestion and absorption.
- Describe those digestive activities of the small intestine by which carbohydrates, proteins, and fats are reduced to their final products.
- Describe the mechanical movements of the small intestine and define absorption.
- Describe those structural features of the large intestine that adapt it for absorption and feces formation and elimination and describe the mechanical movements of the large intestine.
- Describe the processes involved in feces formation and discuss the mechanisms involved in defecation.

- List the causes and symptoms of dental caries and periodontal disease.
- Contrast between the location and effects of gastric and duodenal ulcers.
- Describe the causes and dangers of peritonitis.
- Define cirrhosis as a disorder of an accessory organ of digestion.
- Describe the location and identification of tumors of the gastrointestinal tract.
- Define key medical terms associated with the digestive system.

We all know that food is vital to life. Food is required for the chemical reactions that occur in every cell—both those that synthesize new enzymes, cell structures, bone, and all the other components of the body and those that release the energy needed for the building processes. However, the vast majority of foods we eat are simply too large to pass through the plasma membranes of the cells. Therefore, chemical and mechanical **digestion** must occur first.

Chemical and Mechanical Digestion

Chemical digestion is a series of catabolic reactions that break down the large carbohydrate, lipid, and protein molecules which we eat into molecules that are usable by body cells. These products of digestion are small enough to pass through the walls of the digestive organs, into the blood and lymph capillaries, and eventually into the body's cells. **Mechanical digestion** consists of various movements that aid chemical digestion. Food must be pulverized by the teeth before it

can be swallowed. Then the smooth muscles of the stomach and small intestine churn the food so it is thoroughly mixed with the enzymes that catalyze the reactions.

The digestive system therefore prepares food for consumption by the cells. It does this through five basic activities:

1. Ingestion, or eating, which is taking food into the body

2. Peristalsis, the movement of food along the digestive tract

3. Mechanical and chemical digestion

4. Absorption, the passage of digested food from the digestive tract into the cardiovascular and lymphatic systems for distribution to cells

5. Defecation, the elimination of indigestible substances from the body

General Organization

The organs of digestion are traditionally divided into two main groups. First is the **gastrointestinal (GI) tract**

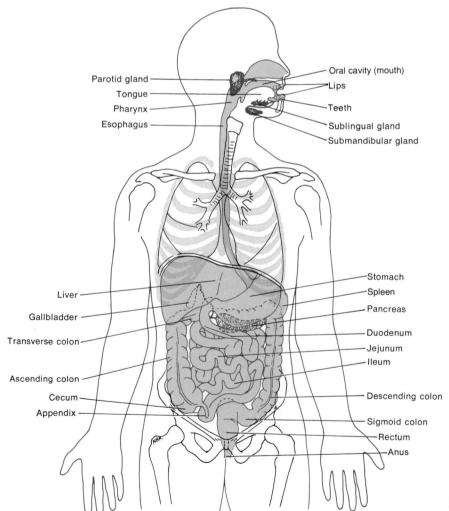

Parotid gland
Tongue
Pharynx
Esophagus

Oral cavity (mouth)
Lips
Teeth
Sublingual gland
Submandibular gland

Liver
Gallbladder
Transverse colon
Ascending colon
Cecum
Appendix

Stomach
Spleen
Pancreas
Duodenum
Jejunum
Ileum
Descending colon
Sigmoid colon
Rectum
Anus

FIGURE 20-1 Organs of the digestive system and related structures.

or **alimentary canal,** a continuous tube running through the ventral body cavity and extending from the mouth to the anus (Figure 20-1). The relationship of the digestive organs to the nine regions of the abdominopelvic cavity may be reviewed in Figure 1-5. The length of a tract taken from a cadaver is about 9 m (30 ft). In a living person it is somewhat shorter because the muscles in its walls are in a state of tone. Organs composing the gastrointestinal tract include the mouth, pharynx, esophagus, stomach, small intestine, and large intestine. The GI tract contains the food from the time it is eaten until it is digested and prepared for elimination. Muscular contractions in the walls of the GI tract break down the food physically by churning it. Secretions produced by cells along the GI tract break down the food chemically.

The second group of organs composing the digestive system consists of the **accessory organs**—the teeth, tongue, salivary glands, liver, gallbladder, pancreas, and appendix. Teeth are cemented to bone, protrude into the GI tract, and aid in the physical breakdown of food. The other accessory organs except for the tongue lie totally outside the tract and produce or store secretions that aid in the chemical breakdown of food. These secretions are released into the tract through ducts.

GENERAL HISTOLOGY

The walls of the GI tract, especially from the esophagus to the anal canal, have the same basic arrangement of tissues. The four tunics or coats of the tract from the inside out are the mucosa, submucosa, muscularis, and serosa or adventitia (Figure 20-2).

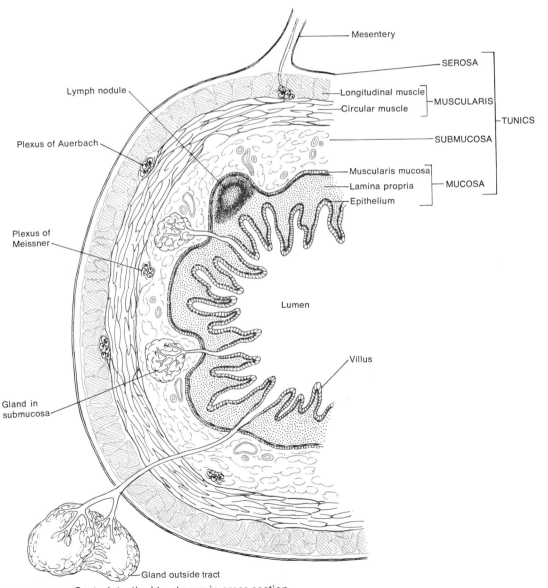

FIGURE 20-2 Gastrointestinal tract seen in cross section.

The **tunica mucosa,** or inner lining of the tract, is a mucous membrane attached to a thin layer of visceral muscle. Two layers compose the membrane: a *lining epithelium,* which is in direct contact with the contents of the GI tract, and an underlying layer of connective tissue called the *lamina propria.* Under the lamina propria are two thin layers of visceral muscle called the *muscularis mucosa.*

The epithelial layer is composed of nonkeratinized cells that are stratified in the mouth and esophagus but are simple throughout the rest of the tract. The functions of the stratified epithelium are protection and secretion. The functions of the simple epithelium are secretion and absorption. However, the lack of keratin allows some absorption to occur in all parts of the tract.

The lamina propria is made of loose connective tissue containing many blood and lymph vessels and scattered lymph nodules. This layer supports the epithelium, binds it to the muscularis mucosa, and provides it with a blood and lymph supply. The blood and lymph vessels are the avenues by which nutrients in the tract reach the other tissues of the body. The lymph tissue also protects against disease. Remember that the GI tract is in contact with the outside environment and contains food which often carries harmful bacteria. Unlike the skin, the mucous membrane of the tract is not protected from bacterial entry by keratin. The lamina propria also contains glandular epithelium that secretes products necessary for chemical digestion.

The muscularis mucosa contains visceral muscle fibers that throw the mucous membrane of the intestine into small folds which increase the digestive and absorptive area. With one exception, which will be described later, the other three coats of the intestine contain no glandular epithelium.

The **tunica submucosa** consists of dense connective tissue binding the tunica mucosa to the tunica muscularis. It is highly vascular and contains a portion of the *plexus of Meissner,* which is part of the autonomic nerve supply to the muscularis mucosa.

The **tunica muscularis** of the mouth, pharynx, and esophagus consists in part of skeletal muscle that produces voluntary swallowing. Throughout the rest of the tract, the muscularis consists of smooth muscle that is generally found in two sheets: an inner ring of circular fibers and an outer sheet of longitudinal fibers. Contractions of the smooth muscles help to break down food physically, mix it with digestive secretions, and propel it through the tract. The muscularis also contains the major nerve supply to the alimentary tract—the *plexus of Auerbach,* which consists of fibers from both autonomic divisions.

The **tunica serosa,** the outermost layer of the canal, is a serous membrane composed of connective tissue and epithelium. This covering, also called the visceral peritoneum, is worth discussing in detail.

Peritoneum

The **peritoneum** is the largest serous membrane of the body. Serous membranes are also associated with the heart (pericardium), lungs (pleura), and other thoracic organs. Serous membranes consist of a layer of simple squamous epithelium (called mesothelium) and an underlying supporting layer of connective tissue. The *parietal peritoneum* lines the walls of the abdominal cavity. The *visceral peritoneum* covers some of the organs and constitutes their serosa. The space between the parietal and visceral portions of the peritoneum is called the *peritoneal cavity.*

Unlike the two other serous membranes of the body, the pericardium and pleura, the peritoneum contains large folds that weave in between the viscera. The folds bind the organs to each other and to the walls of the cavity and contain the blood and lymph vessels and the nerves that supply the abdominal organs. One extension of the peritoneum is called the **mesentery** and is an outward fold of the serous coat of the small intestine (Figure 20-3). Attached to the posterior abdominal wall is the tip of the fold. The mesentery binds the small intestine to the wall. A similar fold of parietal peritoneum, called the **mesocolon,** binds the large intestine to the posterior body wall. It also carries blood vessels and lymphatics to the intestines.

Other important peritoneal folds are the falciform ligament, the lesser omentum, and the greater omentum. The **falciform ligament** attaches the liver to the anterior abdominal wall and diaphragm. The **lesser omentum** arises as two folds in the serosa of the stomach and duodenum suspending the stomach and duodenum from the liver. The **greater omentum** is a large fold in the serosa of the stomach that hangs down like an apron over the intestines. It then passes up to part of the large intestine (the transverse colon), wraps itself around it, and finally attaches to the parietal peritoneum of the posterior wall of the abdominal cavity. Because the greater omentum contains large quantities of adipose tissue, it commonly is called the "fatty apron." The greater omentum contains numerous lymph nodes. If an infection occurs in the intestine, plasma cells formed in these lymph nodes combat the infection and help prevent it from spreading to the peritoneum. Inflammation of the peritoneum (*peritonitis*) is a serious condition because the peritoneal membranes are continuous with each other, enabling the infection to spread to all the organs in the cavity.

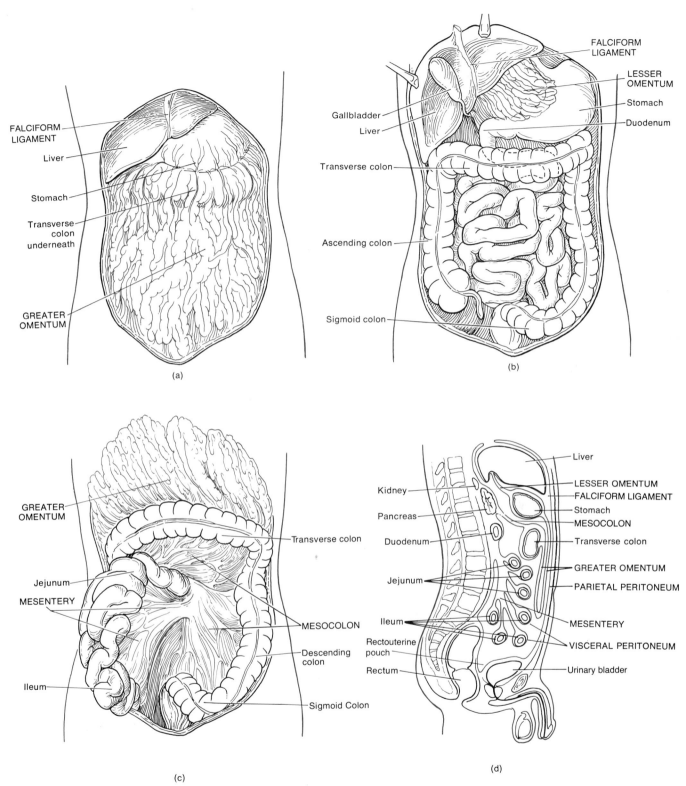

FIGURE 20-3 Extensions of the peritoneum. (a) Greater omentum. (b) Lesser omentum. The liver and gallbladder have been lifted. (c) Mesentery. The greater omentum has been lifted. (d) Sagittal section through the abdomen and pelvis indicating the relationship of the peritoneal extensions to each other.

Organs

MOUTH OR ORAL CAVITY

The **mouth,** also referred to as the **oral** or **buccal cavity,** is formed by the cheeks, hard and soft palates, and tongue (Figure 20-4). Forming the lateral walls of the oral cavity are the cheeks—muscular structures covered on the outside by skin and lined by stratified squamous nonkeratinized epithelium. The anterior portions of the cheeks terminate in the superior and inferior lips. The lips are fleshy folds surrounding the orifice of the mouth. They are covered on the outside by skin and on the inside by a mucous membrane. The transition zone where the two kinds of covering tissue meet is called the *vermilion*—this portion of the lips is not keratinized and the color of the blood in the underlying blood vessels is visible through the transparent surface layer of the vermilion. The inner surface of each lip is attached to its corresponding gum by a midline fold of mucous membrane called the *labial frenulum*. The orbicularis oris muscle and connective tissue lie between the external integumentary covering and the internal mucosal lining. During chewing the cheeks and lips help to keep food between the upper and lower teeth. They also assist in speech.

The *vestibule* of the oral cavity is bounded externally by the cheeks and lips and internally by the gums and teeth. The *oral cavity proper* extends from the vestibule to the *fauces,* the opening between the oral cavity and the pharynx or throat.

The **hard palate**—the anterior portion of the roof and the mouth—is formed by the maxillae and palatine bones and is lined by mucous membrane. The **soft palate** forms the posterior portion of the roof of the mouth. It is an arch-shaped muscular partition between the oropharynx and nasopharynx and is lined by mucous membrane. Hanging from the middle of the lower border of the soft palate is a fingerlike muscular process called the *uvula*. On either side of the base of the uvula are two muscular folds that run down the lateral side of the soft palate. Anteriorly, the *glossopalatine arch (anterior pillar)* runs downward, laterally, and forward to the side of the base of the tongue. Posteriorly, the *pharyngopalatine arch (posterior pil-*

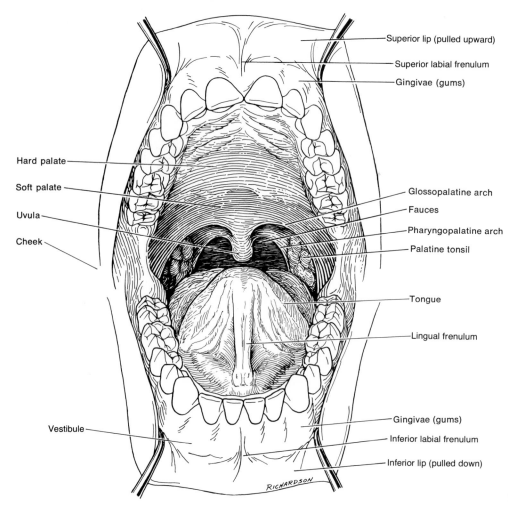

Superior lip (pulled upward)

Superior labial frenulum

Gingivae (gums)

Hard palate

Soft palate

Uvula

Cheek

Glossopalatine arch

Fauces

Pharyngopalatine arch

Palatine tonsil

Tongue

Lingual frenulum

Gingivae (gums)

Inferior labial frenulum

Inferior lip (pulled down)

Vestibule

RICHARDSON

FIGURE 20-4 Oral cavity.

lar) projects downward, laterally, and backward to the side of the pharynx. The palatine tonsils are situated between the arches, and the lingual tonsils are situated at the base of the tongue. At the posterior border of the soft palate, the mouth opens into the oropharynx through the fauces.

The **tongue,** together with its associated muscles, forms the floor of the oral cavity. It is composed of skeletal muscle covered with mucous membrane (Figure 20-5). The tongue is divided into symmetrical lateral halves by a median septum which extends its entire length and attached inferiorly to the hyoid bone. Each half of the tongue consists of an identical complement of muscles called extrinsic and intrinsic muscles. The extrinsic muscles of the tongue originate outside the tongue and insert into it. Among the extrinsic muscles are the hyoglossus, chondroglossus, genioglossus, styloglossus, and palatoglossus. (See Figure 10-7). They move the tongue from side to side and in and out and maneuver food for chewing and swallowing. They also form the floor of the mouth and hold the tongue in position. The intrinsic muscles originate and insert within the tongue, and they alter the shape of the tongue for speech and swallowing. The intrinsic muscles include the longitudinalis superior, longitudinalis inferior, transversus linguae, and verticalis linguae (Figure 20-5c). The *lingual frenulum,* a fold of

mucous membrane in the midline of the undersurface of the tongue, aids in limiting the movement of the tongue posteriorly. If the lingual frenulum is too short, tongue movements are restricted, speech is faulty, and the person is said to be "tongue-tied." These functional problems can be corrected by cutting the lingual frenulum.

The upper surface and sides of the tongue are covered with **papillae,** projections of the lamina propria covered with epithelium. Taste buds are located in some papillae. *Filiform papillae* are conical projections distributed in parallel rows over the anterior two-thirds of the tongue and contain no taste buds (Figure 20-5b). *Fungiform papillae* are mushroomlike elevations distributed among the filiform papillae and are more numerous near the tip of the tongue. They appear as red dots on the surface of the tongue, and most of them contain taste buds. *Circumvallate papillae* are arranged in the form of an inverted V on the posterior surface of the tongue, and all contain taste buds. Note the taste zones of the tongue in Figure 20-5a.

The tongue has a rich blood supply. The lingual artery is the main blood supply, and its veins eventually drain into the internal jugular vein. Most of the muscles of the tongue are supplied by the hypoglossal nerve. Taste sensations from the tongue are conveyed by the facial and glossopharyngeal nerves.

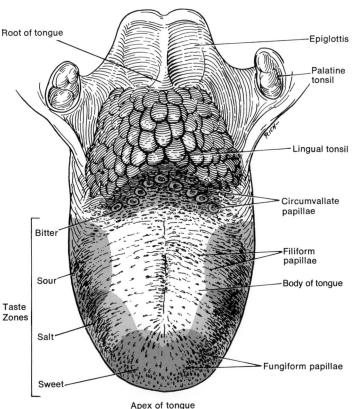

Root of tongue

Epiglottis

Palatine tonsil

Lingual tonsil

Circumvallate papillae

Filiform papillae

Body of tongue

Fungiform papillae

Taste Zones

Bitter

Sour

Salt

Sweet

Apex of tongue

(a)

FIGURE 20-5 The tongue. (a) Locations of the papillae and the four taste zones.

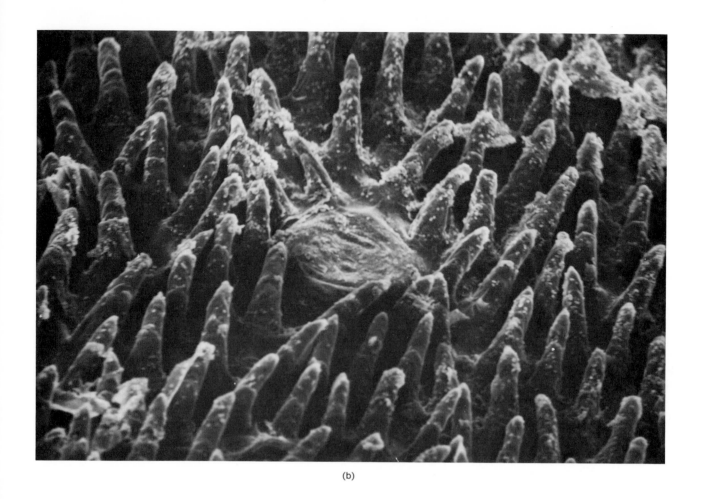

(b)

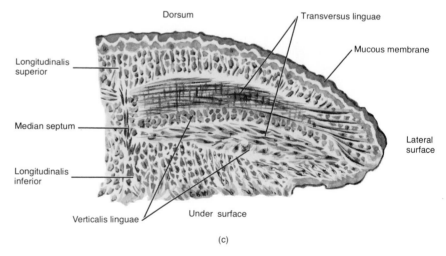

FIGURE 20-5 (Continued) The tongue. (b)
Scanning electron micrograph of filiform
papillae at a magnification of 100×. (Courtesy
of Fisher Scientific Company and S.T.E.M.
Laboratories, Inc., Copyright 1975.) (c) Frontal
section through the left side of the tongue
showing its intrinsic muscles.

SALIVARY GLANDS

Saliva is a fluid that is continuously secreted by glands in or near the mouth. Ordinarily, just enough saliva is secreted to keep the mucous membranes of the mouth moist. But when food enters the mouth, secretion increases so the saliva can lubricate, dissolve, and chemically break down the food. The mucous membrane lining the mouth contains many small glands, the *buccal glands,* that secrete small amounts of saliva. However, the major portion of saliva is secreted by the **salivary glands,** which lie outside the mouth and pour their contents into ducts that empty into the oral cavity. There are three pairs of salivary glands: the parotid, submandibular (submaxillary), and sublingual glands (Figure 20-6). The *parotid glands* are located under and in front of the ears between the skin and the masseter muscle. Each secretes into the oral cavity vestibule via a duct, called *Stensen's duct,* that pierces the buccinator muscle to open into the vestibule opposite the upper second molar tooth. The *submandibular glands* are found beneath the base of the tongue in the posterior part of the floor of the mouth. Their ducts *(Wharton's ducts)* run superficially under the mucosa on either side of the midline of the floor of the mouth and enter the oral cavity proper just behind

the central incisors. The *sublingual glands* are anterior to the submandibular glands, and their ducts open into the floor of the mouth in the oral cavity proper. The parotid glands are compound tubuloacinar glands, whereas the submandibular and sublinguals are compound acinar glands (Figure 20-7).

The parotid gland receives its blood supply from branches of the external carotid artery and is drained by vessels that are tributaries of the external jugular vein. The submandibular gland is supplied by branches of the external maxillary and lingual arteries and drained by tributaries of the deep facial and lingual veins, and the sublingual gland is supplied by the sublingual branch of the lingual artery and the submental branch of the facial artery and is drained by tributaries of the sublingual and submental veins.

The salivary glands receive both sympathetic and parasympathetic innervation. The sympathetic fibers form plexuses on the blood vessels that supply the glands and serve as vasoconstrictors. The parotid gland receives sympathetic fibers from the plexus on the external carotid artery, whereas the submandibular and sublingual glands receive sympathetic fibers that contribute to the sympathetic plexus and accompany the facial artery to the glands. The parasympathetic fibers of the glands consist of secretomotor fibers to the glands.

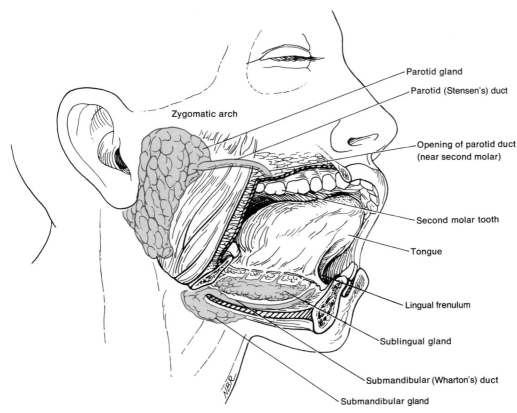

Zygomatic arch

Parotid gland
Parotid (Stensen's) duct
Opening of parotid duct (near second molar)
Second molar tooth
Tongue
Lingual frenulum
Sublingual gland
Submandibular (Wharton's) duct
Submandibular gland

FIGURE 20-6 Salivary glands.

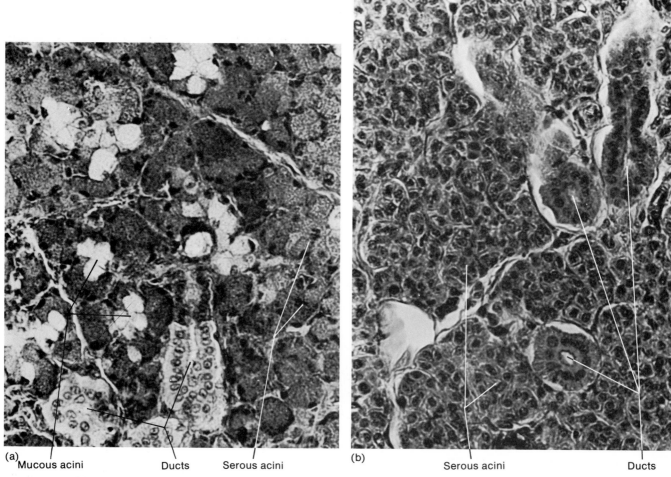

(a) Mucous acini   Ducts   Serous acini   (b)   Serous acini   Ducts

FIGURE 20-7 Histology of the salivary glands. (a) Submandibular gland showing serous acini and mucous acini at a magnification of 90×. (b) Parotid gland showing serous acini at a magnification of 90×. (Courtesy of Victor B. Eichler, Wichita State University.)

Saliva

The fluids secreted by the buccal glands and the three pairs of salivary glands constitute **saliva.** Amounts of saliva secreted daily vary considerably but range from 1000 to 1500 ml. Chemically, saliva is 99.5 percent water and 0.5 percent solutes. Among the solutes are salts—chlorides, bicarbonates, and phosphates of sodium and potassium. Some dissolved gases and various organic substances including urea and uric acid, serum albumin and globulin, mucin, the bacteriolytic enzyme lysozyme, and the digestive enzyme amylase are also present.

The water in saliva provides a medium for dissolving foods so they can be tasted and digestive reactions can take place. The chlorides in the saliva activate the amylase. The bicarbonates and phosphates buffer chemicals that enter the mouth and keep the saliva at a slightly acidic pH of 6.35 to 6.85. Urea and uric acid are found in saliva because the saliva-producing glands (like the sweat glands of the skin) help the body to get rid of wastes. Mucin is a protein that forms mucus when dissolved in water. Mucus lubricates the food so it can be easily turned in the mouth, formed into a ball or bolus, and swallowed. The enzyme lysozyme de-

stroys bacteria, thereby protecting the mucous membrane from infection and the teeth from decay.

Digestion

Depending on the cells the gland contains, each saliva-producing gland supplies different ingredients to saliva. The parotids contain cells that secrete a thin watery serous liquid containing the enzyme salivary amylase. The submandibular glands contain cells similar to those found in the parotids plus some mucous cells. Therefore, they secrete a fluid that is thickened with mucus but still contains quite a bit of enzyme. The sublingual glands contain mostly mucous cells, so they secrete a much thicker fluid that contributes only a small amount of enzyme to the saliva.

The enzyme salivary amylase initiates the breakdown of polysaccharides (carbohydrates)—this is the only chemical digestion that occurs in the mouth. Carbohydrates are starches and sugars and are classified as either monosaccharides, disaccharides, or polysaccharides. Monosaccharides are small molecules containing several carbon, hydrogen, and oxygen atoms—an example is glucose. Disaccharides consist of two monosaccharides linked together; polysaccharides are chains of three or more monosaccharides. The vast majority of carbohydrates we eat are polysaccharides. Since only monosaccharides can be absorbed into the bloodstream, ingested disaccharides and polysaccharides must be broken down. The function of *salivary amylase* is to break the chemical bonds between some of the monosaccharides that make up the polysaccharides. In this way, the enzyme breaks the long-chain polysaccharides into shorter polysaccharides called *dextrins*. Given sufficient time, salivary amylase also can break down the dextrins into the disaccharide maltose. Food usually is swallowed too quickly for more than 3 to 5 percent of the carbohydrates to be reduced to disaccharides in the mouth. However, salivary amylase in the swallowed food continues to act on polysaccharides for another 15 to 30 minutes in the stomach before the stomach acids eventually inactivate it. Saliva continues to be secreted heavily some time after food is swallowed. This flow of saliva washes out the mouth and dilutes and buffers the chemical remnants of irritating substances.

TEETH

The **teeth** or **dentes,** are located in sockets of the alveolar processes of the mandible and maxillae. The alveolar processes are covered by the *gingivae* (gums), which extend slightly into each socket (Figure 20-8) forming the gingival sulcus. The sockets are lined by the *periodontal ligament,* which consists of dense fibrous connective tissue and is attached to the socket walls and the cemental surface of the roots. Thus it anchors the teeth in position and also acts as a shock absorber to dissipate the forces of chewing.

A typical tooth consists of three principal portions. The *crown* is the portion above the level of the gums. The *root* consists of one to three projections embedded in the socket. The *cervix* is the constricted junction line of the crown and the root.

Teeth are composed primarily of *dentin,* a bonelike substance that gives the tooth its basic shape and rigidity. The dentin encloses a space that is referred to as the *pulp cavity.* It lies in the crown and is filled with *pulp,* a connective tissue containing blood vessels,

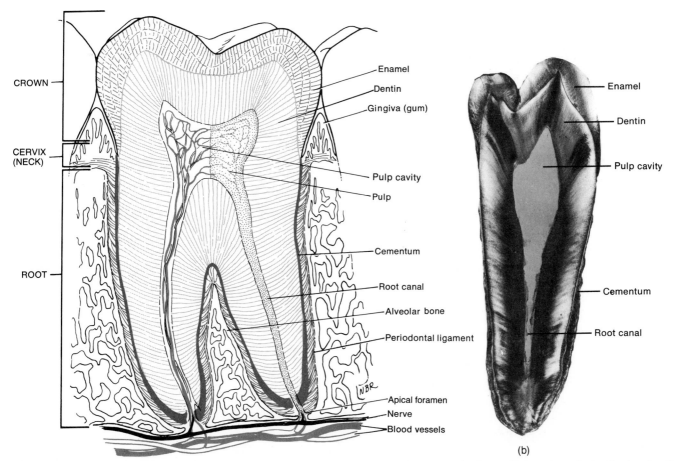

(b)

FIGURE 20-8 Parts of a typical tooth. (a) Sagittal section through a molar. (b) Photograph of a sagittal section of a tooth showing the relationship of the pulp cavity to the root canal. (Courtesy of Carolina Biological Supply Company.)

nerves, and lymphatics. From the pulp cavity there are narrow extensions that run through the root of the tooth and are called *root canals*. Each root canal has an opening at its base, the *apical foramen*. Through the foramen enter blood vessels bearing nourishment, lymphatics affording protection, and nerves providing sensation. The dentin of the crown is covered by *enamel* that consists primarily of calcium phosphate and calcium carbonate. Enamel is the hardest substance in the body and protects the tooth from the wear of chewing. It is also a barrier against acids that easily dissolve the dentin. The dentin of the root is covered by *cementum,* another bonelike substance. *Pyorrhea* is inflammation of the periodontal ligament and adjacent gums. A prolonged, severe case of pyorrhea can weaken the periodontal ligament, erode the alveolar bone, and thereby loosen the tooth.

Because of the curvature of the dental arches, it is necessary to use terms other than anterior, posterior, medial, and lateral in describing the surfaces of the teeth. Accordingly, the following directional terms are used: *Labial* refers to the surface of a tooth in contact with or directed toward the lips. *Buccal* refers to the surface in contact with or directed toward the cheeks.

Lingual is restricted to the teeth of the lower jaw and refers to the surface directed toward the tongue. *Palatal*, on the other hand, is restricted to the teeth of the upper jaw and refers to the surface directed toward the palate. The term *mesial* is used to designate the anterior or medial side of the tooth relative to its position in the dental arch. *Distal* refers to the posterior or lateral side of the tooth relative to its position in the dental arch. Essentially, mesial and distal refer to the sides of adjacent teeth that are in contact with each other. Finally, the term *occlusal* refers to the biting surface of a tooth.

Dentitions

Everyone has two **dentitions,** or sets of teeth. The first of these—the *deciduous teeth,* milk teeth, or baby teeth—begin to erupt at about six months of age, and one pair appears at about each month thereafter until all 20 are present. Figure 20-9a illustrates the deciduous teeth. The incisors, which are closest to the midline, are chisel-shaped and adapted for cutting into food. Next to the incisors, moving posteriorly, are the canines or cuspids, which have a pointed surface

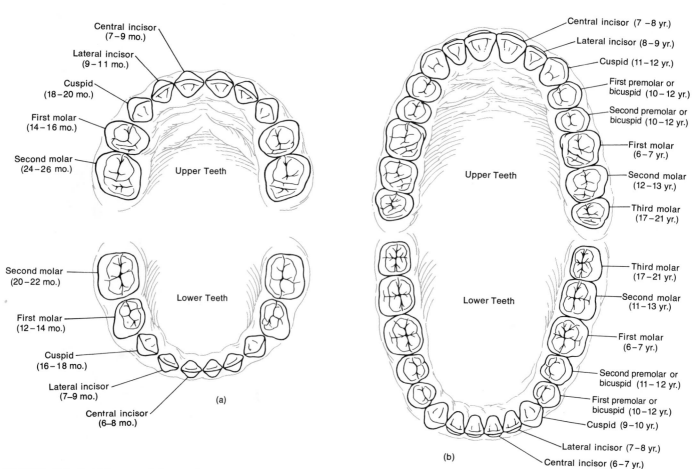

FIGURE 20-9 Dentitions and times of eruptions. The times of eruptions are indicated in parentheses. (a) Deciduous dentition. (b) Permanent dentition.

called the cusp. Canines are used to tear and shred food. The incisors and canines have only one root apiece. Behind them lie the first and second primary molars, which have four cusps and two or three roots. Upper molars have three roots; lower molars have two roots. The molars crush and grind food.

All the deciduous teeth are lost—generally between 6 and 12 years of age—and are replaced by the *permanent dentition* (Figure 20-9b). The permanent dentition contains 32 teeth that appear between the age of 6 and adulthood. It resembles the deciduous dentition with the following exceptions: the deciduous molars are replaced with premolars or bicuspids that have two cusps and one root (upper first bicuspids have two roots and are used for crushing and grinding). The permanent molars erupt into the mouth behind the bicuspids. They do not replace any primary teeth and erupt as the jaw grows to accommodate them—the first molars at age 6, the second molars at age 12, the third molars after age 18. The human jaw is becoming increasingly smaller and often does not afford enough room behind the second molars for the eruption of the third molars or wisdom teeth. In this case, the third molars remain embedded in the alveolar bone and are said to be "impacted." Most often they cause pressure and pain and must be surgically removed. In some individuals, however, third molars may be dwarfed in size or may not develop at all.

The arteries that supply blood to the teeth are distributed to the pulp cavity and surrounding periodontal membrane. The upper incisors and canines are supplied by anterior superior alveolar branches of the infraorbital artery; the upper premolars and molars are supplied by the posterior superior alveolar branches of the internal maxillary artery; the lower incisors and canines are supplied by the incisive branches of the inferior alveolar artery; and the lower premolars and molars are supplied by the dental branches of the inferior alveolar artery.

The teeth receive sensory fibers from branches of the maxillary and mandibular divisions of the trigeminal (V) nerve—the upper teeth from branches of the maxillary division and the lower teeth from branches of the mandibular division.

Teeth and Digestion

Through chewing, or *mastication,* the teeth pulverize food and mix it with saliva. As a result, the food is reduced to a soft, flexible mass, called a *bolus,* that is easily swallowed.

DEGLUTITION

Swallowing, or **deglutition,** moves food from the mouth to the stomach. Swallowing starts with the bolus on the upper side of the tongue. Then the tip of the tongue rises and presses against the palate (Figure 20-10). The bolus slides to the back of the mouth and

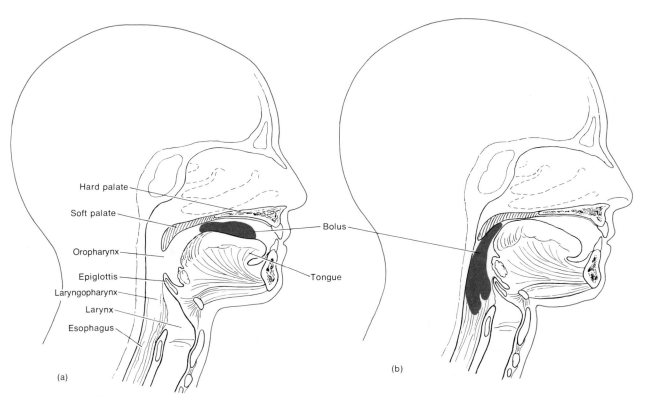

Hard palate
Soft palate
Oropharynx
Epiglottis
Laryngopharynx
Larynx
Esophagus
(a)

Bolus
Tongue
(b)

FIGURE 20-10 Deglutition. (a) Position of structures prior to swallowing. (b) During swallowing, the tongue rises against the palate, the nose is closed off, the larynx rises, the epiglottis seals off the larynx, and the bolus is passed into the esophagus.

is pulled through the fauces by muscles that lie in the oropharynx. During this period the respiratory passageways close and breathing is temporarily interrupted. The soft palate and uvula move upward to close off the nasopharynx, and the larynx is pulled forward and upward under the tongue. As the larynx rises, it meets the epiglottis, which seals off the glottis. The movement of the larynx also pulls the vocal cords together, further sealing off the respiratory tract, and widens the opening between the laryngopharynx and esophagus. The bolus passes through the laryngopharynx and enters the esophagus in one second. The respiratory passageways then reopen and breathing resumes.

Esophagus

The **esophagus,** the third organ involved in deglutition, is a muscular, collapsible tube that lies behind the trachea. It is about 23 to 25 cm (10 inches) long and be-

gins at the end of the laryngopharynx, passes through the mediastinum in front of the vertebral column, pierces the diaphragm through an opening called the esophageal hiatus, and terminates in the upper portion of the stomach.

The *mucosa* of the esophagus consists of stratified squamous nonkeratinized epithelium, lamina propria, and a muscularis mucosae (Figure 20-11). The *submucosa* contains connective tissue and blood vessels. The *muscularis* of the upper third is striated, the middle third is striated and smooth, and the lower third is smooth. The outer layer is known as the *tunica adventitia* since it contains no serosa.

The esophagus does not produce digestive enzymes and does not carry on absorption. It does, however, secrete mucus and transport food to the stomach.

The arteries of the esophagus are derived from the inferior thyroid, thoracic aorta, intercostal arteries, phrenic, and left gastric arteries and, it is drained by the adjacent veins. Innervation of the esophagus is by

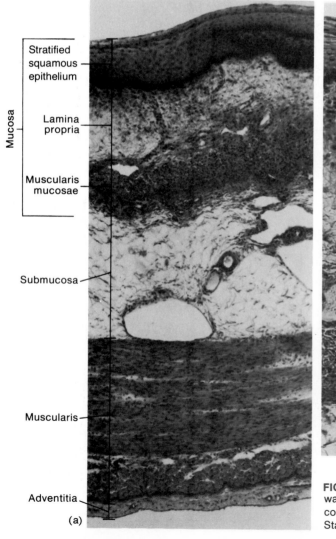

Mucosa
Stratified squamous epithelium
Lamina propria
Muscularis mucosae

Submucosa

Muscularis

Adventitia

(a)

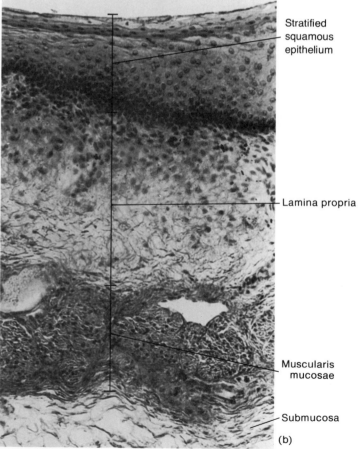

Stratified squamous epithelium

Lamina propria

Muscularis mucosae

Submucosa

(b)

FIGURE 20-11 Histology of the esophagus. (a) Section through the wall of the esophagus at a magnification of 20×. (b) Details of the mucosa at a magnification of 40×. (Courtesy of Victor B. Eichler, Wichita State University.)

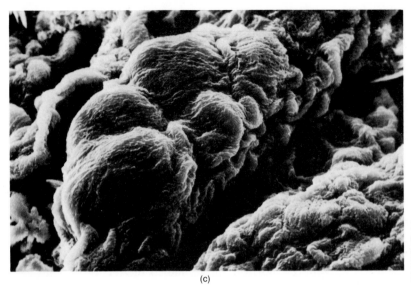

(c)

FIGURE 20-11 (Continued) Histology of the esophagus. (c) Scanning electron micrograph of the esophageal mucosa at a magnification of 2000×. (Courtesy of Fisher Scientific Company and S.T.E.M. Laboratories, Inc. Copyright 1975)

laryngeal nerves, the cervical sympathetic chain, and vagi.

Food is pushed through the esophagus by muscular movements called **peristalsis** (Figure 20-12). Peristalsis is a function of the tunica muscularis. In the section of the esophagus lying just above and around the top of the bolus, the circular muscle fibers contract. The contraction constricts the esophageal wall and squeezes

the bolus downward. Meanwhile, longitudinal fibers lying around the bottom of the bolus and just below it also contract. Contraction of the longitudinal fibers shortens this lower section, pushing its walls outward so it can receive the bolus. The contractions are repeated in a wave that moves down the esophagus, pushing the food toward the stomach. Passage of the bolus is further facilitated by glands secreting mucus.

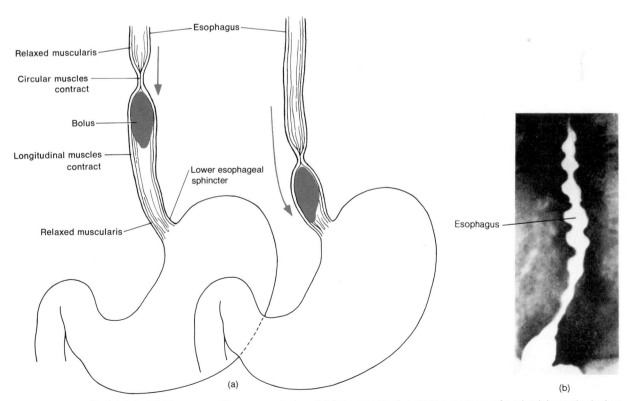

FIGURE 20-12 Peristalsis. (a) Diagrammatic representation. (b) Anteroposterior roentgenogram of peristalsis made during fluoroscopic examination while a patient was swallowing a barium "meal." (Courtesy of Lester W. Paul and John H. Juhl, *The Essentials of Roentgen Interpretation,* 3d ed., Harper & Row, Publishers, Inc., New York, 1972.)

The passage of solid or semisolid food from the mouth to the stomach takes four to eight seconds. Very soft foods and liquids pass through in about one second.

Just above the level of the diaphragm, the esophagus is slightly narrowed. This narrowing has been attributed to a physiological sphincter in the inferior part of the esophagus known as the *lower esophageal* or *gastroesophageal sphincter*. A **sphincter** is an opening that has a thick circle of muscle around it. The lower esophageal sphincter relaxes during swallowing and thus aids the passage of the bolus from the esophagus into the stomach. The movement of the diaphragm against the stomach during breathing presses on the stomach and helps prevent the regurgitation of gastric contents from the stomach to the esophagus.

STOMACH

The **stomach** is a J-shaped enlargement of the GI tract directly under the diaphragm in the epigastric, umbilical, and left hypochondriac regions of the abdomen

(see Figure 1-5c). The superior portion of the stomach is a continuation of the esophagus. The inferior portion empties into the duodenum, the first part of the small intestine. Within each individual, the position and size of the stomach vary continually. For instance, the diaphragm pushes the stomach downward with each inspiration and pulls it upward with each expiration. Empty, it is about the size of a large sausage, but the stomach can stretch itself to accommodate large amounts of food.

Anatomy

The stomach is divided into four areas: cardia, fundus, body, and pylorus (Figure 20-13). The *cardia* surrounds the lower esophageal sphincter, and the rounded portion above and to the left of the cardia is the *fundus*. Below the fundus, the large central portion of the stomach is called the *body*. The narrow, inferior region is the *pylorus*. The concave medial border of the stomach is called the *lesser curvature*, and the convex

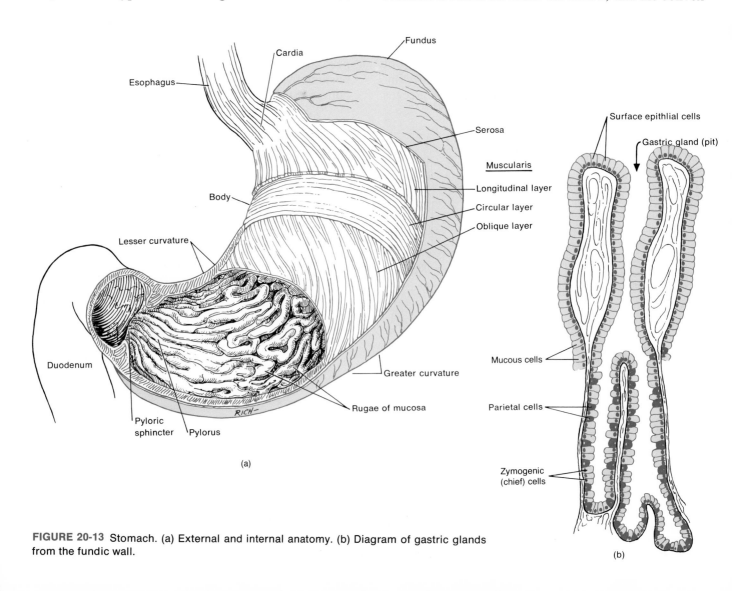

FIGURE 20-13 Stomach. (a) External and internal anatomy. (b) Diagram of gastric glands from the fundic wall.

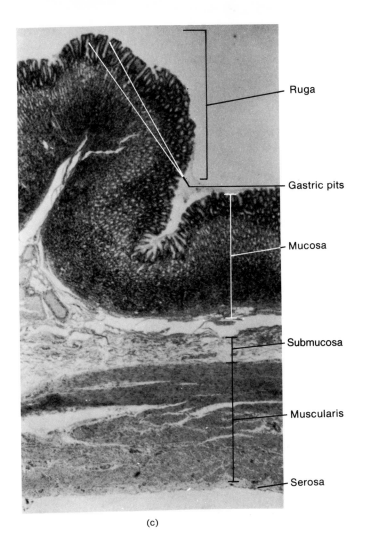

Ruga

Gastric pits

Mucosa

Submucosa

Muscularis

Serosa

(c)

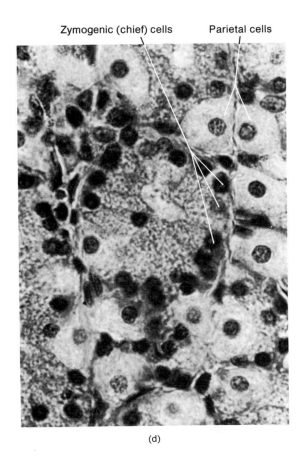

Zymogenic (chief) cells   Parietal cells

(d)

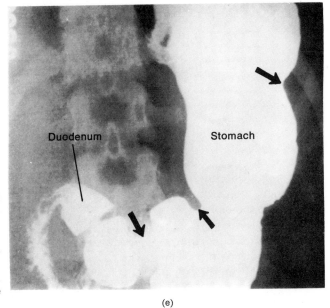

Duodenum   Stomach

(e)

FIGURE 20-13 (*Continued*) Stomach. (c) Histology of the fundic wall at a magnification of 6×. (d) Enlarged aspect of parietal and zymogenic cells at a magnification of 160×. (Courtesy of Victor B. Eichler, Wichita State University.) (e) Anteroposterior projection of a normal stomach. Note the peristaltic waves indicated by the arrows. (Courtesy of Lester W. Paul and John H. Juhl, *The Essentials of Roentgen Interpretation,* 3d ed., Harper & Row, Publishers, Inc., New York, 1972.) See Figure 20-24e for a scanning electron micrograph of the mucosa of the stomach.

lateral border is the *greater curvature*. The pylorus communicates with the duodenum of the small intestine via a sphincter called the *pyloric valve*.

Two abnormalities of the pyloric valve can occur in infants. *Pylorospasm* is characterized by failure of the muscle fibers encircling the opening to relax normally. Ingested food does not pass easily from the stomach to the small intestine, the stomach becomes overly full, and the infant vomits frequently to relieve the

pressure. Pylorospasm is treated by adrenergic drugs that relax the muscle fibers of the valve. *Pyloric stenosis* is a narrowing of the pyloric valve caused by a tumorlike mass that apparently is formed by enlargement of the circular muscle fibers. It must be surgically corrected.

The stomach wall is composed of the same four basic layers as the rest of the alimentary canal, with certain modifications. When the stomach is empty, the mucosa lies in large folds that can be seen with the naked eye. These folds are called *rugae*. As the stomach fills and distends, the rugae gradually smooth out and disappear. Microscopic inspection of the mucosa reveals a layer of simple columnar epithelium containing many narrow openings that extend down into the lamina propria. These pits—*gastric glands*—are lined with three kinds of secreting cells: zymogenic, parietal, and mucous. The *zymogenic,* or *chief cells,* secrete the principal gastric enzyme called pepsinogen. Hydrochloric acid, which activates the pepsinogen, is produced by the *parietal cells.* The *mucous cells* secrete mucus and the intrinsic factor—a substance involved in the absorption of vitamin B_{12}. Secretions of the gastric glands are called *gastric juice.*

The submucosa of the stomach is composed of loose areolar connective tissue, and it connects the mucosa to the muscularis. The muscularis, unlike that in other areas of the alimentary canal, has three layers of smooth muscle: an outer longitudinal layer, a middle circular layer, and an inner oblique layer. This arrangement of fibers allows the stomach to contract in a variety of ways to churn food, break it into small particles, mix it with gastric juice, and pass it to the duodenum. The serosa covering the stomach is part of the visceral peritoneum. At the lesser curvature the two layers of the visceral peritoneum come together and extend upward to the liver as the lesser omentum. At the greater curvature, the visceral peritoneum continues downward as the greater omentum hanging over the intestines.

The arterial supply of the stomach is derived from the celiac artery. The right and left gastric arteries form an anastomosing arch along the lesser curvature, and the right and left gastroepiploic arteries form a similar arch on the greater curvature. Short gastric arteries supply the fundus. The veins of the same name accompany the arteries and drain, directly or indirectly, into the hepatic portal vein.

The vagi convey parasympathetic fibers to the stomach. These fibers form synapses within the plexus of Meissner in the submucosa. The sympathetic nerves arise from the celiac ganglia.

Digestion

Several minutes after food enters the stomach, gentle, rippling, peristaltic movements called *mixing waves* pass over the stomach every 15 to 25 seconds. These waves macerate food, mix it with the secretions of the digestive glands, and reduce it to a thin liquid called *chyme.* Few mixing waves are observed in the fundus, which is primarily a storage area. Foods may remain in the fundus for an hour or more without becoming mixed with gastric juice. During this time, salivary digestion continues.

The principal chemical activity of the stomach is to begin the digestion of proteins. In the adult, digestion is achieved primarily through the enzyme *pepsin.* Pepsin breaks certain peptide bonds between the amino acids making up proteins. Thus a protein chain of many amino acids is broken down into fragments of amino acids—long fragments called *proteoses;* somewhat shorter ones are *peptones.* Pepsin is most effective in the very acidic environment of the stomach (pH of 1). It becomes inactive in an alkaline environment. What keeps pepsin from digesting the protein in stomach cells along with the food? First of all, pepsin is secreted in an inactive form called *pepsinogen,* so it cannot digest the proteins in the zymogenic cells that produce it. When pepsinogen comes in contact with the hydrochloric acid secreted by the parietal cells, it is converted to active pepsin. Once pepsin has been activated, the stomach cells are protected by mucus. The mucus coats the mucosa and forms a barrier between the gastric juice and the cells. Sometimes the mucus fails to do its job, and the pepsin and hydrochloric acid eat a hole in the stomach wall known as a *gastric ulcer.*

Another enzyme of the stomach is *gastric lipase.* Gastric lipase splits the butterfat molecules found in milk. The enzyme operates best at a pH of 5 to 6 and has a limited role in the adult stomach. Adults rely exclusively on an enzyme found in the small intestine to digest fats.

As digestion proceeds in the stomach, more vigorous peristaltic waves begin at about the middle of the stomach, pass downward, reach the pyloric valve, and sometimes go into the duodenum. The movement of chyme from the stomach into the duodenum depends on a pressure gradient between the two organs. When the pressure in the stomach (intragastric pressure) is greater than that in the duodenum (intraduodenal pressure), chyme is forced into the duodenum. Peristaltic waves are largely responsible for increased intragastric pressure. It is estimated that 2–5 ml of chyme are passed into the duodenum with each peristaltic wave. When intraduodenal pressure exceeds intragastric pressure, the pyloric valve closes and prevents the regurgitation of chyme from the duodenum to the stomach. The stomach empties all its contents into the duodenum two to six hours after ingestion. Food rich in carbohydrate leaves the stomach in a few hours. Protein foods are somewhat slower, and emptying is slowest after a meal containing large amounts of fat. The stomach wall is impermeable to the passage of most materials into the blood, so most substances are not absorbed until they reach the small intestine. However, the stomach does participate in the absorption of some water and salts, certain drugs, and alcohol.

The next step in the breakdown of food is digestion

in the small intestine. Chemical digestion in the small intestine depends not only on its own secretions but on activities of three organs outside the alimentary canal: the pancreas, liver, and gallbladder.

PANCREAS

The **pancreas** is a soft, oblong tubuloacinar gland about 12.5 cm (6 inches) long and 2.5 cm (1 inch) thick. It lies posterior to the greater curvature of the stomach and is connected by a duct (sometimes two) to the duodenum (Figure 20-14). The pancreas is divided into a head, body, and tail. The *head* is the expanded portion near the C-shaped curve of the duodenum. Moving superiorly and to the left of the head are the centrally located *body* and the terminal tapering *tail*. The pancreas is made up of small clusters of glandular epithelial cells. Some clusters, called *islets of Langerhans,* form the endocrine portions of the pancreas and consist of alpha and beta cells that secrete glucagon and insulin. The other masses of cells, called *acini,* are the exocrine portions of the organ (see Figure 17-14). Secreting cells of the acini release a mixture of digestive enzymes called *pancreatic juice,* which is dumped into

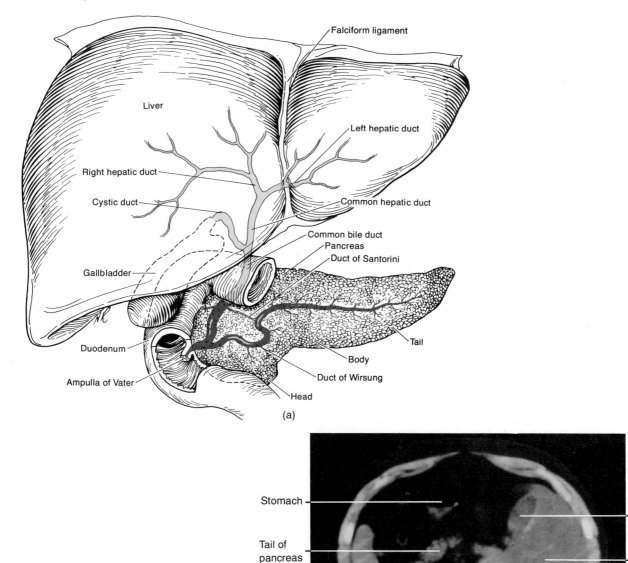

FIGURE 20-14 Pancreas. (a) Relations to liver, gallbladder, and duodenum. (b) CT scan. (Courtesy of General Electric Co.)

small ducts attached to the acini. Pancreatic juice eventually leaves the pancreas through a large main tube called the *pancreatic duct,* or *duct of Wirsung.* In most people the pancreatic duct unites with the common bile duct from the liver and gallbladder and enters the duodenum in a small, raised area called the *ampulla of Vater.* An accessory duct, the *duct of Santorini,* may also lead from the pancreas and empty into the duodenum about 2.5 cm (1 inch) about the ampulla of Vater.

The arterial supply of the pancreas is from the splenic artery through its pancreatic branches and from the superior mesenteric and celiac arteries by way of the inferior and superior pancreatoduodenal arteries. The veins, in general, correspond to the arteries. Venous blood reaches the hepatic portal vein by means of the splenic and superior mesenteric veins.

The nerves to the pancreas are branches of the celiac plexus that accompany the arteries entering the gland. The glandular innervation is from the parasympathetic division of the autonomic nervous system, and the innervation of the blood vessels is from the sympathetic division of the autonomic nervous system.

The functions of the pancreas are twofold. The acini secrete enzymes that digest food in the small intestine, and the alpha and beta cells secrete glucagon and insulin, which control the fate of digested and absorbed carbohydrates.

LIVER

The **liver** performs so many vital functions that we cannot live without it. Among these are the following:

1. The liver manufactures the anticoagulant heparin and most of the other plasma proteins.

2. The reticuloendothelial cells of the liver phagocytose worn-out red blood cells and some bacteria.

3. Liver cells contain enzymes that either break down poisons or transform them into less harmful compounds. When amino acids are burned for energy, for example, they leave behind toxic nitrogenous wastes that are converted to urea by the liver cells. Moderate amounts of urea are harmless to the body and are easily excreted by the kidneys and sweat glands.

4. Newly absorbed nutrients are collected in the liver. It can change any excess monosaccharides into glycogen or fat, both of which can be stored. In addition it can transform glycogen, fat, and protein into glucose and vice versa, depending on the body's needs.

5. The liver stores glycogen, copper, iron, and vitamins A, D, E, and K. It also stores some poisons that cannot be broken down and excreted. (High levels of DDT are found in the livers of animals, including humans, who eat sprayed fruits and vegetables.)

6. The liver manufactures bile, which is used in the small intestine for the emulsification and absorption of fats.

The liver weighs about 1.4 kg (4 lb) in the average adult and is located under the diaphragm. It occupies most of the right hypochondrium and part of the epigastrium of the abdomen. The liver is covered largely by peritoneum and completely by a dense connective tissue layer that lies beneath the peritoneum. Anatomically, the liver is divided into two principal lobes— the **right lobe** and the **left lobe**—separated by the *falciform ligament* (Figure 20-15). The right lobe, besides the main lobe, also has associated with it an inferior *quadrate lobe* and a posterior *caudate lobe.* The falciform ligament is a reflection of the parietal peritoneum, which extends from the undersurface of the diaphragm to the superior surface of the liver as visceral peritoneum where it separates the two principal lobes of the liver. In the free border of the falciform ligament is the *ligamentum teres (round ligament).* It extends from the liver to the umbilicus. The ligamentum teres is a fibrous cord homologous to the umbilical vein of the fetus (see Figure 11-24).

The lobes of the liver are made up of numerous functional units called *lobules,* which may be seen under a microscope. A lobule consists of cords of *hepatic* (liver) *cells* arranged in a radial pattern around a *central vein.* Between the cords are endothelial-lined spaces called *sinusoids* through which blood passes. The sinusoids are also partly lined with phagocytic cells, termed *Kupffer cells,* that destroy worn-out white and red blood cells and bacteria.

Unlike the other products of the liver, bile normally is not secreted into the bloodstream. It is manufactured by the hepatic cells and secreted into *bile capillaries* or *canaliculi* that empty into small ducts. These small ducts eventually merge to form the larger *right* and *left hepatic ducts,* which unite to leave the liver as the *common hepatic duct.* Further on, the common hepatic duct joins the *cystic duct* from the gallbladder. The two tubes become the *common bile duct,* which empties into the duodenum at the ampulla of Vater. The *sphincter of Oddi* is a valve in the common bile duct. When the small intestine is empty the sphincter closes, and the backed-up bile overflows into the cystic duct to the gallbladder, where it is stored.

The liver receives a double supply of blood. From the hepatic artery it obtains oxygenated blood, and from the hepatic portal vein it receives deoxygenated blood containing newly absorbed nutrients. Branches of both the hepatic artery and the hepatic portal vein carry the blood into the sinusoids of the lobules, where oxygen, most of the nutrients, and certain poisons are

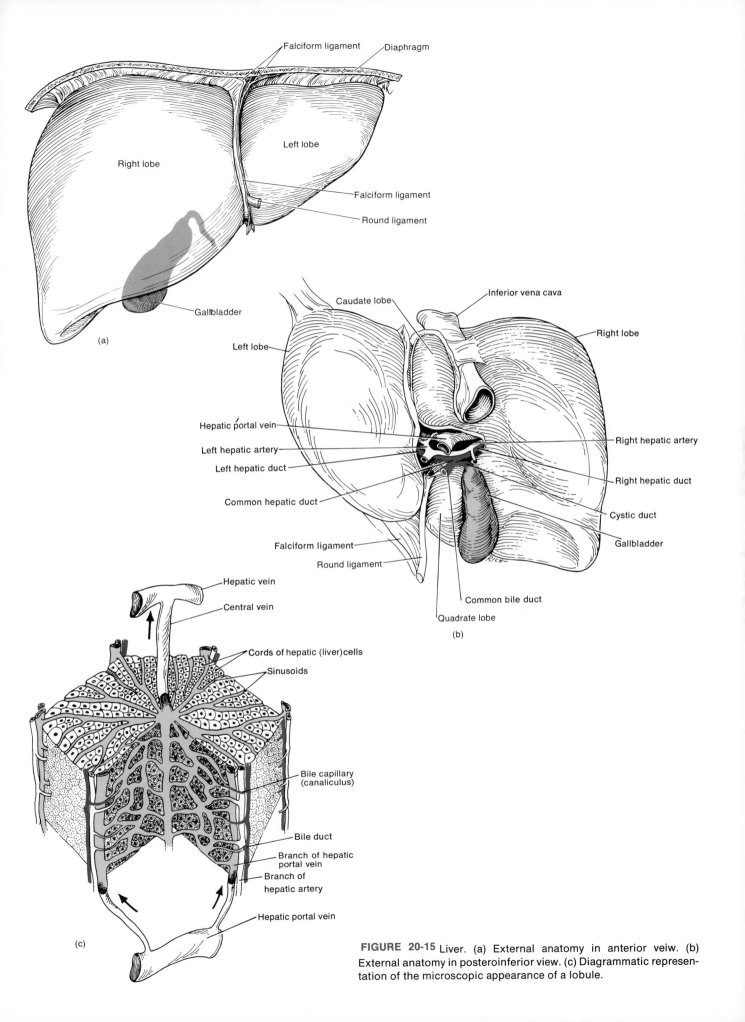

Falciform ligament — **Diaphragm**

Left lobe

Right lobe

Falciform ligament

Round ligament

Gallbladder

(a)

Caudate lobe — **Inferior vena cava**

Left lobe — **Right lobe**

Hepatic portal vein

Left hepatic artery

Left hepatic duct

Common hepatic duct

Right hepatic artery

Right hepatic duct

Cystic duct

Gallbladder

Falciform ligament

Round ligament

Common bile duct

Quadrate lobe

(b)

Hepatic vein

Central vein

Cords of hepatic (liver) cells

Sinusoids

Bile capillary (canaliculus)

Bile duct

Branch of hepatic portal vein

Branch of hepatic artery

Hepatic portal vein

(c)

FIGURE 20-15 Liver. (a) External anatomy in anterior veiw. (b) External anatomy in posteroinferior view. (c) Diagrammatic representation of the microscopic appearance of a lobule.

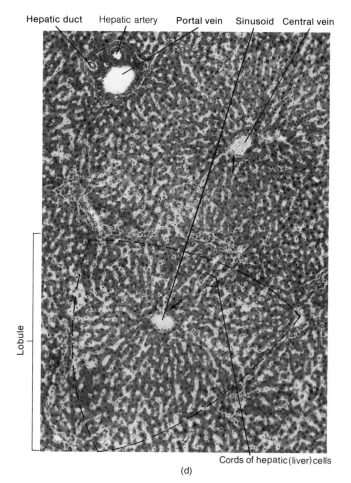

Hepatic duct   Hepatic artery   Portal vein   Sinusoid   Central vein

Lobule

Cords of hepatic (liver) cells

(d)

FIGURE 20-15 (Continued) Liver. (d) Photomicrograph of a lob-ule at a magnification of 65×. The broken line indicates the boundary of a lobule and the arrows indicate sinusoids empty-ing into the central vein. (Courtesy of Edward J. Reith, from *Atlas of Descriptive Histology,* by Edward J. Reith and Michael H. Ross, Harper & Row, Publishers, Inc., New York, 1970.)

extracted by the hepatic cells. Nutrients are stored or used to make new materials. The poisons are stored and detoxified. Products manufactured by the hepatic cells and nutrients needed by other cells are secreted back into the blood. The blood then drains into the central vein and eventually passes into a hepatic vein.

The nerve supply to the liver consists of vagal pre-ganglionic parasympathetic fibers and postganglionic sympathetic fibers from the celiac ganglia.

GALLBLADDER

The **gallbladder** is a sac located along the underside of the liver (Figure 20-16). Its inner walls consist of a mu-cous membrane arranged in rugae resembling those of the stomach. When the gallbladder fills with bile, the rugae allow it to expand to the size and shape of a pear. The middle, muscular coat of the wall consists of smooth muscle fibers. Contraction of these fibers

ejects the bile into the cystic duct. The outer coat is the visceral peritoneum. Figure 20-17 presents a cross section of the abdomen at the level of the pancreas. Note the relationship of the digestive organs to one another.

SMALL INTESTINE

The major portions of digestion and absorption occur in a long tube called the **small intestine.** The small in-testine begins at the pyloric valve of the stomach, coils through the central and lower part of the abdominal cavity, and eventually opens into the large intestine. It averages 2.5 cm (1 inch) in diameter and about 6.35 m (21 ft) in length.

Anatomy

The small intestine is divided into three segments: duo-denum, jejunum, and ileum. The *duodenum,* the broadest part, originates at the pyloric valve of the stomach and extends about 25 cm (10 inches) until it merges with the jejunum. The *jejunum* is about 2.5 m (8 ft) long and extends to the ileum. The final portion of the small intestine, the *ileum,* measures about 3.6 m (12 ft) and joins the large intestine at the *ileocecal valve.* A roentgenogram of the normal small intestine is shown in Figure 20-18a.

The wall of the small intestine is composed of the same four tunics or coats that make up most of the GI tract. However, both the mucosa and the submucosa are modified to allow the small intestine to complete the processes of digestion and absorption. The mucosa contains many pits lined with glandular epithelium. These pits—the *intestinal glands,* or *crypts of Lieber-kühn*—secrete the intestinal digestive enzymes (Fig-ure 20-18e). The submucosa of the duodenum contains *Brunner's glands,* which secrete mucus to protect the walls of the small intestine from the action of the en-zymes.

Since almost all the absorption of nutrients occurs in the small intestine, its wall is specially adapted for this function. The epithelium covering and lining the mucosa consists of simple columnar epithelium. Some of the epithelial cells have been transformed to goblet cells, which secrete additional mucus. The rest contain *microvilli*—fingerlike projections of the plasma mem-brane. Larger amounts of digested nutrients diffuse into the intestinal wall because the microvilli increase the surface area of the plasma membrane.

The mucosa lies in a series of *villi*—projections 0.5 to 1 mm high giving the intestinal mucosa its velvety appearance (Figure 20-18e). The enormous number of villi (4 to 5 million) vastly increases the surface area of the epithelium available for the epithelial cells special-izing in absorption. Each villus has a core of lamina propria, the connective tissue layer of the mucosa.

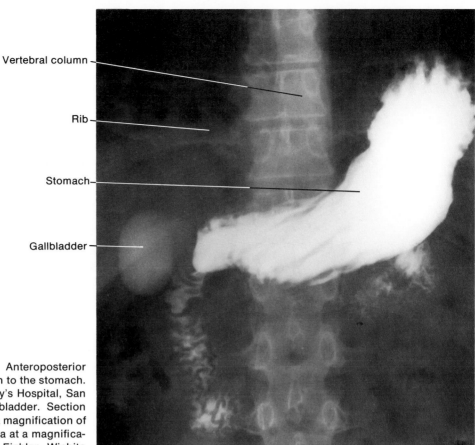

(a)

FIGURE 20-16 The gallbladder. (a) Anteroposterior projection of the gallbladder in relation to the stomach. (Courtesy of John C. Bennett, St. Mary's Hospital, San Francisco.) (b) Histology of the gallbladder. Section through the wall of the gallbladder at a magnification of 20×. (c) Enlarged aspect of the mucosa at a magnification of 100×. (Courtesy of Victor B. Eichler, Wichita State University.)

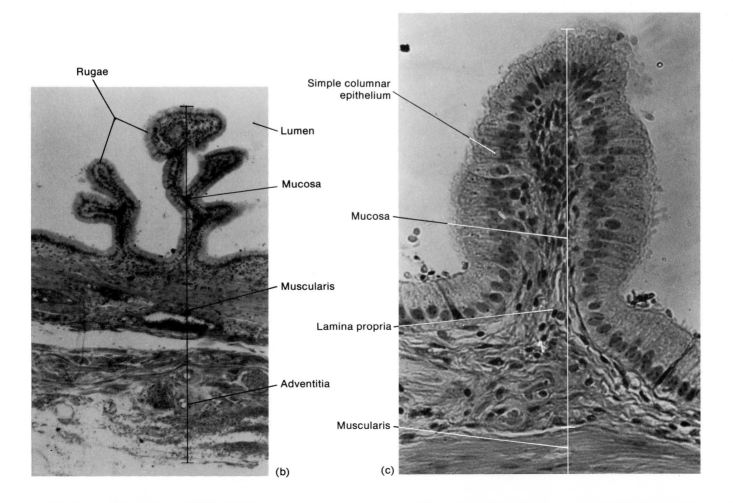

(b)

(c)

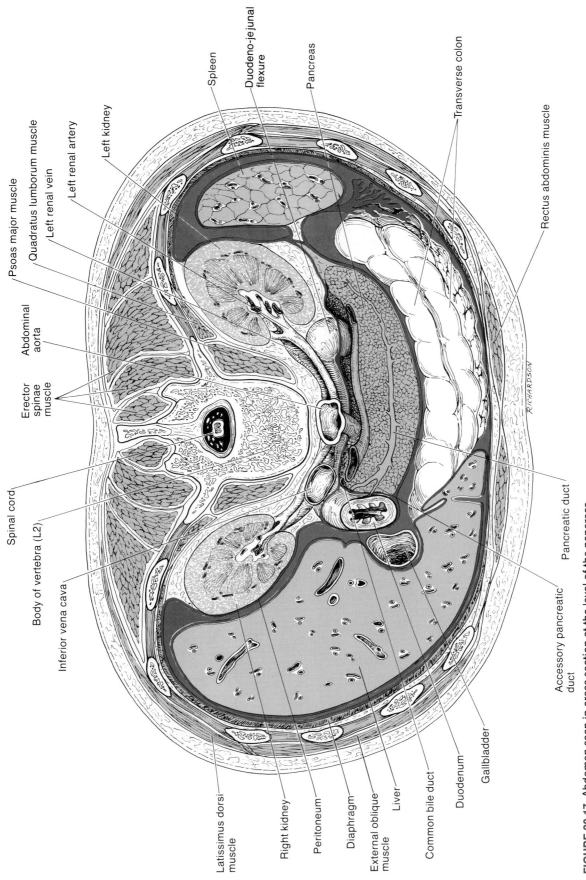

FIGURE 20-17 Abdomen seen in cross section at the level of the pancreas.

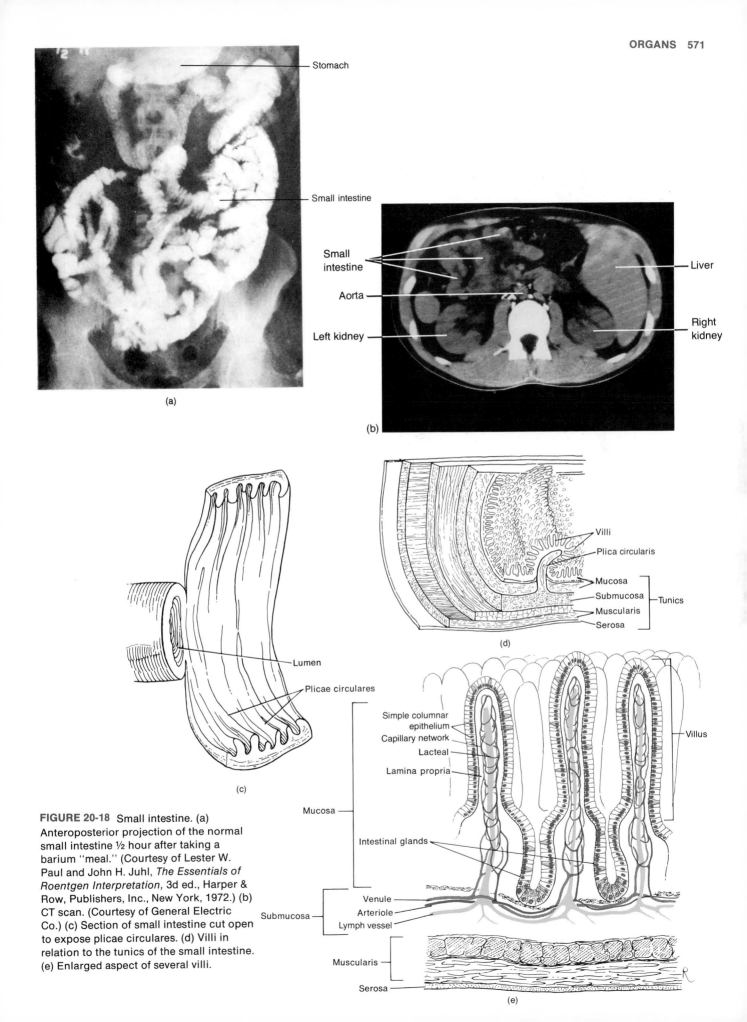

FIGURE 20-18 Small intestine. (a) Anteroposterior projection of the normal small intestine ½ hour after taking a barium "meal." (Courtesy of Lester W. Paul and John H. Juhl, *The Essentials of Roentgen Interpretation,* 3d ed., Harper & Row, Publishers, Inc., New York, 1972.) (b) CT scan. (Courtesy of General Electric Co.) (c) Section of small intestine cut open to expose plicae circulares. (d) Villi in relation to the tunics of the small intestine. (e) Enlarged aspect of several villi.

Labels in figure:
Stomach
Small intestine
Small intestine
Aorta
Left kidney
Liver
Right kidney
(a)
(b)
Lumen
Plicae circulares
(c)
Villi
Plica circularis
Mucosa
Submucosa
Muscularis
Serosa
Tunics
(d)
Simple columnar epithelium
Capillary network
Lacteal
Lamina propria
Mucosa
Intestinal glands
Venule
Arteriole
Lymph vessel
Submucosa
Muscularis
Serosa
Villus
(e)

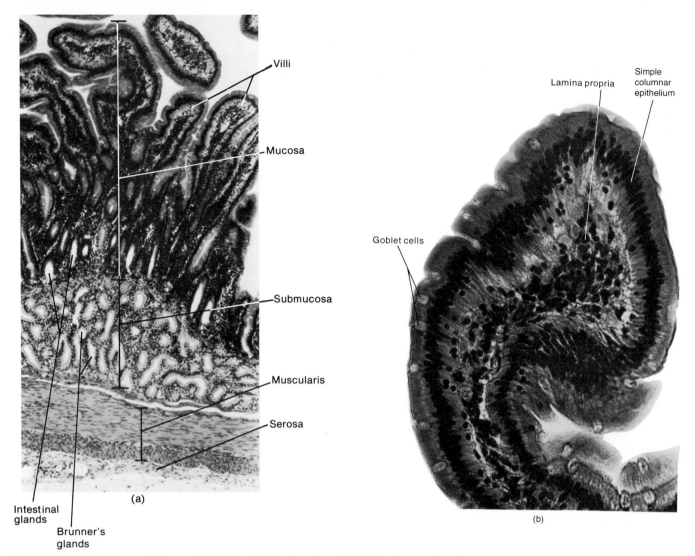

FIGURE 20-19 Histology of the small intestine. (a) Section through the duodenum showing various layers at a magnification of 15×. (b) Enlarged aspect of a villus at a magnification of 80×.

Embedded in this connective tissue are an arteriole, a venule, a capillary network, and a *lacteal* (lymphatic vessel). Nutrients that diffuse through the epithelial cells which cover the villus are able to pass through the capillary walls and the lacteal and enter the blood.

In addition to the microvilli and villi, a third set of projections called *plicae circulares* further increases the surface area for absorption. The plicae are permanent deep folds in the mucosa and submucosa (Figure 20-18c). Some of the folds extend all the way around the intestine, and others extend only part way around.

The muscularis of the small intestine consists of two layers of smooth muscle. The outer, thinner layer contains longitudinally arranged fibers. The inner, thicker layer contains circularly arranged fibers. Except for a major portion of the duodenum, the serosa (or visceral peritoneum) completely covers the small intestine. The histological aspects of the small intestine are shown in Figure 20-19.

There is an abundance of lymphatic tissue in the walls of the small intestine. Single lymph nodules, called *solitary lymph nodules,* are most numerous in the lower part of the ileum. Aggregated lymph nodules, referred to as *Peyer's patches,* are also most numerous in the ileum.

The arterial blood supply of the small intestine is from the superior mesenteric artery and the gastroduodenal artery, coming indirectly from the celiac artery. Blood is returned by way of the superior mesenteric vein which, with the splenic vein, forms the hepatic portal vein.

The nerves to the small intestine are supplied by the superior mesenteric plexus. The branches of the plexus contain postganglionic sympathetic fibers, preganglionic parasympathetic fibers, and afferent fibers. The afferent fibers are both vagal and of spinal nerves. In the wall of the small intestine are two autonomic plexuses, the plexus of Auerbach between the muscular layers, and the plexus of Meissner in the submucosa. The nerve fibers are derived chiefly from the sympathetic division of the autonomic nervous system and partly from the vagus.

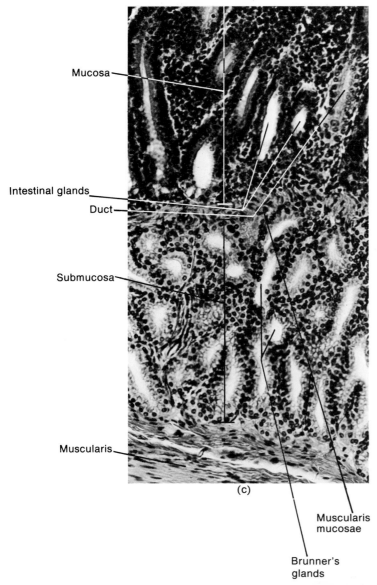

Mucosa

Intestinal glands

Duct

Submucosa

Muscularis

(c)

Muscularis
mucosae

Brunner's
glands

FIGURE 20-19 (*Continued*) Histology of the small intestine. (c) Enlarged aspect of intestinal glands at a magnification of 40×. (Courtesy of Victor B. Eichler, Wichita State University.) (d) Scanning electron micrograph of villi. (Courtesy of Carroll H. Weiss, Camera M.D. Studios.)

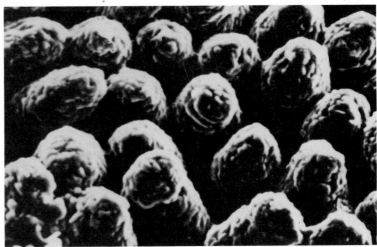

(d)

Chemical Digestion

The digestion of carbohydrates, proteins, and lipids in the small intestine requires secretions from the pancreas, liver, and intestinal glands.

● *Secretions* Each day the liver secretes 800 to 1000 ml (almost 1 qt) of the yellow, brownish, or olive-green liquid called *bile.* Bile consists of water, bile salts, bile acids, a number of lipids, and two pigments called biliverdin and bilirubin. Bile is partially an excretory product and partially a digestive secretion. When red blood cells are broken down, iron, globin, and bilirubin are released. The iron and globin are recycled, but the bilirubin is excreted into the bile ducts. Bilirubin eventually is broken down in the intestines, and its breakdown products give feces their color. If the liver is unable to remove bilirubin from the blood or if the bile ducts are obstructed, large amounts of bilirubin circulate through the bloodstream and collect in other tissues, giving the skin and eyes a yellow color called *jaundice.* Other substances found in bile aid in the digestion of fats and are required for their absorption.

Each day the pancreas produces 1200 to 1500 ml (about 1 to 1½ qt) of a clear, colorless liquid called *pancreatic juice.* Pancreatic juice consists mostly of water, some salts, sodium bicarbonate, and enzymes. The sodium bicarbonate gives pancreatic juice a slightly alkaline pH (7.1 to 8.2) that stops the action of pepsin from the stomach and creates the proper environment for the enzymes in the small intestine. The enzymes in pancreatic juice include a carbohydrate-digesting enzyme, several protein-digesting enzymes, and the only active fat-digesting enzyme in the adult body.

The intestinal juice, or *succus entericus,* is a clear yellow fluid secreted in amounts of about 2 to 3 liters (2 to 3 qt) a day. It has a pH of 7.6, which is slightly alkaline, and contains water, mucus, and enzymes that complete the digestion of carbohydrates and proteins.

● *Digestive Process* When chyme reaches the small intestine, the carbohydrates and proteins have been only partly digested and are not ready for absorption. Lipid digestion has not even begun. Digestion in the small intestine continues as follows:

1. Carbohydrates. In the mouth, some polysaccharides are broken down into dextrins containing several monosaccharide units. Even though the action of salivary amylase may continue in the stomach, few polysaccharides are reduced to disaccharides by the time chyme leaves the stomach. *Pancreatic amylase,* an enzyme in pancreatic juice, breaks dextrins into the disaccharide maltose. Sucrose and lactose, two other

disaccharides, are ingested. Next, three enzymes in the intestinal juice digest the disaccharides into monosaccharides. *Maltase* splits maltose into two molecules of glucose. *Sucrase* breaks sucrose into a molecule of glucose and a molecule of fructose. *Lactase* digests lactose into a molecule of glucose and a molecule of galactose. This process completes the digestion of carbohydrates.

2. Proteins. Protein digestion starts in the stomach, where most of the proteins are fragmented by the action of pepsin into short chains of amino acids called peptones and proteoses. Enzymes found in pancreatic juice continue the digestion. *Trypsin* digests any intact proteins into peptones and proteoses, breaks the peptones and proteoses into dipeptides (containing only two amino acids), and breaks some of the dipetides into single amino acids. *Chymotrypsin* duplicates trypsin's activities. *Carboxypeptidase,* the third enzyme, reduces whole or partly digested proteins to amino acids. To prevent these enzymes from digesting the proteins in the cells of the pancreas, they are secreted in inactive forms—trypsin as *trypsinogen,* activated by an intestinal enzyme called *enterokinase;* chymotrypsin as *chymotrypsinogen,* activated in the small intestine by trypsin; and carboxypeptidase as *procarboxypeptidase,* also activated in the small intestine by trypsin. Protein digestion is completed by several intestinal enzymes grouped together under the name *erepsin,* which converts all the remaining dipeptides into single amino acids. Single amino acids can be absorbed.

3. Lipids. In an adult, almost all lipid digestion occurs in the small intestine. The first step in the process is the *emulsification* of neutral fats (triglycerides), which is a function of bile. Neutral fats, or just simply fats, are the most abundant lipids in the diet. They are called triglycerides because they consist of a molecule of glycerol and three molecules of fatty acid. Bile salts break the globules of fat into droplets (emulsification) so the fat-splitting enzyme can get at the lipid molecules. In the second step, *pancreatic lipase,* an enzyme found in pancreatic juice, hydrolyzes each fat molecule into fatty acids, glycerol, and glycerides, end products of fat digestion. Glycerides consist of glycerol with one or two fatty acids still attached and are known as monoglycerides and diglycerides, respectively.

Mechanical Digestion

In the small intestine, three distinct movements occur as a result of contractions of the longitudinal and circular muscles. These movements are rhythmic segmentation, pendular movements, and propulsive peristalsis. Rhythmic segmentation and pendular movements are strictly localized contractions in areas containing food. The two movements mix the chyme with the digestive juices and bring every particle of food into contact with the mucosa for absorption. They do not push the intestinal contents along the tract. *Rhythmic segmentation* starts with the contractions of circular muscle fibers in a portion of the intestine, an action that constricts the intestine into segments. Next muscle fibers that encircle the middle of each segment also contract, dividing each segment into two. Finally, the fibers that contracted first relax, and each small segment unites with an adjoining small segment so that large segments are reformed. This sequence of events is repeated 12 to 16 times a minute, sloshing the chyme back and forth. *Pendular movements* consist of alternating contractions and relaxations of the longitudinal muscles. The contractions cause a portion of the intestine to shorten and lengthen, spilling the chyme back and forth.

The third movement, *propulsive peristalsis*, propels the chyme onward through the intestinal tract. Peristaltic movement in the intestine is similar to that in the esophagus. In the intestine, these waves may be as slow as 5 cm (2 inches)/minute or as fast as 50 cm (20 inches)/second.

Absorption

All the chemical and mechanical phases of digestion from the mouth down through the small intestine are directed toward changing foods into forms that can diffuse through the epithelial cells lining the mucosa into the underlying blood and lymph vessels. The diffusible forms are monosaccharides (glucose, fructose, and galactose), amino acids, fatty acids, glycerol, and glycerides. Passage of these digested nutrients from the alimentary canal into the blood or lymph is called **absorption.**

About 90 percent of all absorption takes place throughout the length of the small intestine. The other 10 percent occurs in the stomach and large intestine. Absorption of materials in the small intestine occurs specifically through the villi and depends on diffusion, facilitated diffusion, osmosis, and active transport (see Chapter 2). Monosaccharides and amino acids are absorbed into the blood capillaries of the villi and transported in the bloodstream to the liver via the hepatic portal system. Fatty acids, glycerol, and glycerides do not enter the bloodstream immediately. They cluster together and become surrounded by bile salts to form water-soluble particles called *micelles.* These are absorbed into the intestinal epithelial cell. Here they enter the smooth endoplasmic reticulum where triglycerides are resynthesized. The triglycerides, along with small quantities of phospholipids and cholesterol, are

then organized into protein-coated lipid droplets called *chylomicrons.* The protein coat keeps the chylomicrons suspended and prevents them from sticking to each other or to the walls of the lymphatics or blood vessels. Small chylomicrons leave the intestinal cells and enter the blood capillaries in the villus. Larger chylomicrons enter the lacteal in the villus and are transported by way of lymphatic vessels to the thoracic duct and enter the cardiovascular system at the left subclavian vein. Finally they arrive at the liver through the hepatic artery.

Most of the products of carbohydrate, protein, and lipid digestion are processed by the liver before they are delivered to the other cells of the body. Large amounts of water, electrolytes, mineral salts, and some vitamins also are absorbed in the small intestine.

In summary, then, the principal chemical activity of the small intestine is to digest all foods into forms that are usable by body cells. Any undigested materials that are left behind are processed in the large intestine.

LARGE INTESTINE

The overall functions of the large intestine are the completion of absorption, the manufacture of certain vitamins, the formation of feces, and the expulsion of feces from the body.

Anatomy

The **large intestine** is about 1.5 m (5 ft) in length and averages 6.5 cm (2.5 inches) in diameter. It extends from the ileum to the anus and is attached to the posterior abdominal wall by its *mesocolon* of visceral peritoneum. Structurally, the large intestine is divided into four principal regions: cecum, colon, rectum, and anal canal (Figure 20-20).

The opening from the ileum into the large intestine is guarded by a fold of mucous membrane called the *ileocecal valve.* This structure allows materials from the small intestine to pass into the large intestine. Hanging below the ileocecal valve is the *cecum,* a blind pouch abour 6 cm (2 or 3 inches) long. Attached to the cecum is a twisted, coiled tube, measuring about 8 cm (3 inches) in length, called the *vermiform appendix* (*vermis* = worm). The visceral peritoneum of the appendix, called the *mesoappendix,* attaches the appendix to the inferior part of the ileum and adjacent part of the posterior abdominal wall. Inflammation of the appendix is called *appendicitis.*

The open end of the cecum merges with a long tube called the *colon.* The colon is divided into ascending, transverse, descending, and sigmoid portions. The *ascending colon* ascends on the right side of the abdomen, reaches the undersurface of the liver, and turns

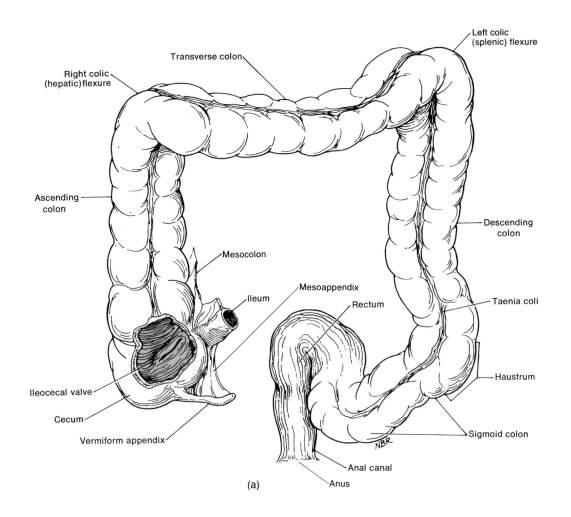

(a)

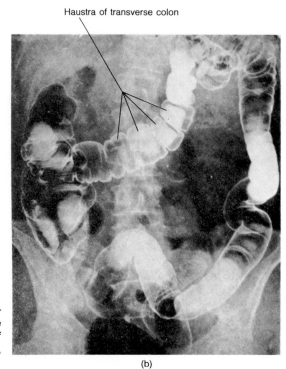

(b)

FIGURE 20-20 The large intestine. (a) Anatomy of the large intestine. (b) Anteroposterior projection of the large intestine in which several haustra are clearly visible. (Courtesy of Lester W. Paul and John H. Juhl, *The Essentials of Roentgen Interpretation,* 3d ed., Harper & Row, Publishers, Inc., New York, 1972.)

abruptly to the left. Here it forms the *right colic (hepatic) flexure*. The colon continues across the abdomen to the left side as the *transverse colon*. It curves beneath the lower end of the spleen on the left side as the *left colic (splenic) flexure* and passes downward to the level of the iliac crest as the *descending colon*. The *sigmoid colon* begins at the left iliac crest, projects inward to the midline, and terminates as the rectum at about the level of the third sacral vertebra.

The *rectum*, the last 20 cm (7 to 8 inches) of GI tract, lies anterior to the sacrum and coccyx. The terminal 2 to 3 cm of the rectum is called the *anal canal* (Figure 20-21). Internally, the mucous membrane of the anal canal is arranged in longitudinal folds called *anal columns* that contain a network of arteries and veins. Inflammation and enlargement of the anal veins is known as *hemorrhoids* or *piles*. The opening of the anal canal to the exterior is called the *anus*. It is guarded by an internal sphincter of smooth muscle and an external sphincter of skeletal muscle. Normally the anus is closed except during the elimination of the wastes of digestion.

The wall of the large intestine differs from that of the small intestine in several respects. No villi or permanent circular folds are found in the mucosa, which does, however, contain simple columnar epithelium with numerous goblet cells (Figure 20-22). These cells secrete mucus that lubricates the colonic contents as they pass through the colon. Solitary lymph nodes also are found in the mucosa. The submucosa of the large intestine is similar to that found in the rest of the alimentary canal. The muscularis consists of an external layer of longitudinal muscles and an internal layer of circular muscles. Unlike other parts of the digestive tract, the longitudinal muscles do not form a continuous sheet around the wall but are broken up into three flat bands called *taeniae coli*. Each band runs the length of the large intestine. Tonic contractions of the bands gather the colon into a series of pouches called *haustra*, which give the colon its puckered appearance. The serosa of the large intestine is part of the visceral peritoneum.

The arterial supply of the cecum and colon is derived from branches of the superior mesenteric and inferior mesenteric arteries. The venous return is by way of the superior and inferior mesenteric veins and ultimately to the hepatic portal vein and into the liver. The arterial supply of the rectum and anal canal is derived from the superior, middle, and inferior rectal arteries. The rectal veins correspond to the rectal arteries.

The nerves to the large intestine consist of sympa-

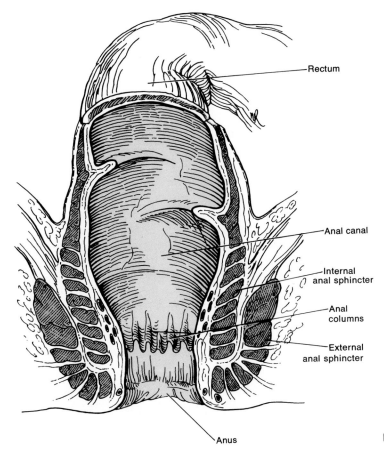

Rectum

Anal canal

Internal anal sphincter

Anal columns

External anal sphincter

Anus

FIGURE 20-21 Anal canal seen in longitudinal section.

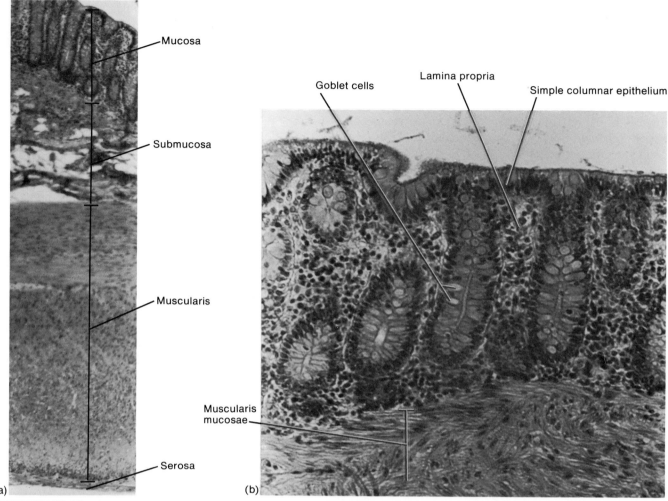

FIGURE 20-22 Histology of the large intestine. (a) Section through the wall of the large intestine at a magnification of 20×. (b) Enlarged aspect of the mucosa of the large intestine at a magnification of 50×. (Courtesy of Victor B. Eichler, Wichita State University.)

thetic, parasympathetic, and afferent components. The sympathetic innervation is derived from the celiac, superior and inferior mesenteric, and internal iliac plexuses. The fibers reach the plexuses by way of the thoracic and lumbar splanchnic nerves. The parasympathetic innervation is derived from the vagus and pelvic splanchnic nerves.

Activities

The principal activities of the large intestine are concerned with mechanical movements, absorption, and the formation and elimination of feces.

● *Movements* Movements of the colon begin when substances enter through the ileocecal valve. Since chyme moves through the small intestine at a fairly constant rate, the time required for a meal to pass into the colon is determined by gastric evacuation time. As food passes through the ileocecal valve, it fills the cecum and accumulates in the ascending colon.

One movement characteristic of the large intestine is *haustral churning*. In this process the haustra remain relaxed and distended while they fill up. When the distension reaches a certain point, the walls contract and squeeze the contents into the next haustrum. Peristalsis also occurs, although at a slower rate than in other portions of the tract (3 to 12 contractions per minute). A final type of movement is *mass peristalsis*, a strong peristaltic wave that begins in about the middle of the transverse colon and drives the colonic contents into the rectum. Food in the stomach initiates this reflex action in the colon. Thus mass peristalsis usually takes place three or four times a day, during a meal or immediately after.

● *Absorption and Feces Formation* By the time the intestinal contents arrive at the large intestine, digestion and absorption are almost complete. When the chyme has been in the large intestine 3 to 10 hours, it has become solid or semisolid as a result of absorption and is now known as *feces*. Chemically, feces consist of water, inorganic salts, epithelial cells from the mucosa of the alimentary canal, bacteria, products of bacterial decomposition, and undigested parts of food not attacked by bacteria. Mucus is secreted by the glands of the large intestine, but no enzymes are secreted.

Chyme is prepared for elimination in the large intestine by the action of bacteria. These bacteria ferment any remaining carbohydrates and release hydrogen, carbon dioxide, and methane gas. They also convert remaining proteins to amino acids and break down the amino acids into simpler substances: indole, skatole, hydrogen sulfide, and fatty acids. Some of the indole and skatole is carried off in the feces and contributes to its odor. The rest is absorbed. Bacteria also decompose bilirubin, the breakdown product of red blood cells that is excreted in bile, to simpler pigments which give feces their brown color. Intestinal bacteria also aid in the synthesis of several vitamins needed for normal metabolism, including some B vitamins and vitamin K.

Although most water absorption occurs in the small intestine, the large intestine absorbs enough to make it an important organ in maintaining the body's water balance. Intestinal water absorption is greatest in the cecum and ascending colon. The large intestine also absorbs inorganic solutes plus some products of bacterial action, including vitamins and large amounts of indole and skatole. The indole and skatole are transported to the liver, where they are converted to less toxic compounds and excreted in the urine.

● *Defecation*   Mass peristaltic movement pushes fecal material into the rectum. The resulting distension of the rectal walls stimulates pressure-sensitive receptors, initiating a reflex for **defecation,** which is emptying of the rectum. Contraction of the longitudinal rectal mucles shortens the rectum, thereby increasing the pressure inside it. The pressure forces the sphincters open, and the feces are expelled through the anus. Voluntary contractions of the diaphragm and abdominal muscles aid defecation by increasing the pressure inside the abdomen, which pushes the walls of the sigmoid colon and rectum inward. If defecation does not occur, the feces remain in the rectum until the next wave of mass peristalsis again stimulates the pressure-sensitive receptors, creating the desire to defecate.

Applications to Health

DENTAL CARIES

Dental caries, or tooth decay, involve a gradual demineralization (softening) of the enamel and dentin (Figure 20-23a). If this condition remains untreated, various

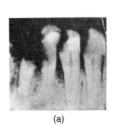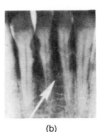

(a)   (b)

FIGURE 20-23 Diseases of the teeth. (a) Anteroposterior projection of dental caries. (b) Anteroposterior projection of periodontitis in which the alveolar bone (arrow) has been destroyed. (Courtesy of Lester W. Paul and John H. Juhl, *The Essentials of Roentgen Interpretation,* 3d ed., Harper & Row, Publishers, Inc., New York, 1972.)

microorganisms may invade the pulp, causing inflammation and infection with subsequent death (necrosis) of the dental pulp and abscess of the alveolar bone surrounding the root's apex. Such teeth are treated by root canal therapy.

The process of dental caries is initiated when bacteria act on carbohydrates deposited on the tooth, giving off acids which demineralize the enamel. Microbes that digest carbohydrates include two bacteria, *Lactobacillus acidophilus* and *Streptococcus mutans.* Research suggests that the streptococci break down carbohydrates into *dental plaque,* a polysaccharide that adheres to the tooth surface. When other bacteria digest the plaque, acid is produced. Saliva cannot reach the tooth surface to buffer the acid because the plaque covers the teeth.

PERIODONTAL DISEASES

Periondontal disease is a collective term for a variety of conditions characterized by inflammation and degeneration of the gingivae, alveolar bone, periodontal ligament, and cementum. The initial symptoms are enlargement and inflammation of the soft tissue and bleeding gums. Without treatment, the soft tissue may deteriorate and the alveolar bone may be resorbed, causing loosening of the teeth and receding of the gums (Figure 20-23b).

Periodontal diseases are frequently caused by local irritants, such as bacteria, impacted food, and cigarette smoke, or by a poor "bite." The latter may put a strain on the tissues supporting the teeth. Periodontal diseases may also be caused by allergies, vitamin deficiencies, and a number of systemic disorders, especially those that affect bone (blood dyscrasias), connective tissue, or circulation.

PEPTIC ULCERS

An **ulcer** is a craterlike lesion in a membrane. Ulcers that develop in areas of the alimentary canal exposed

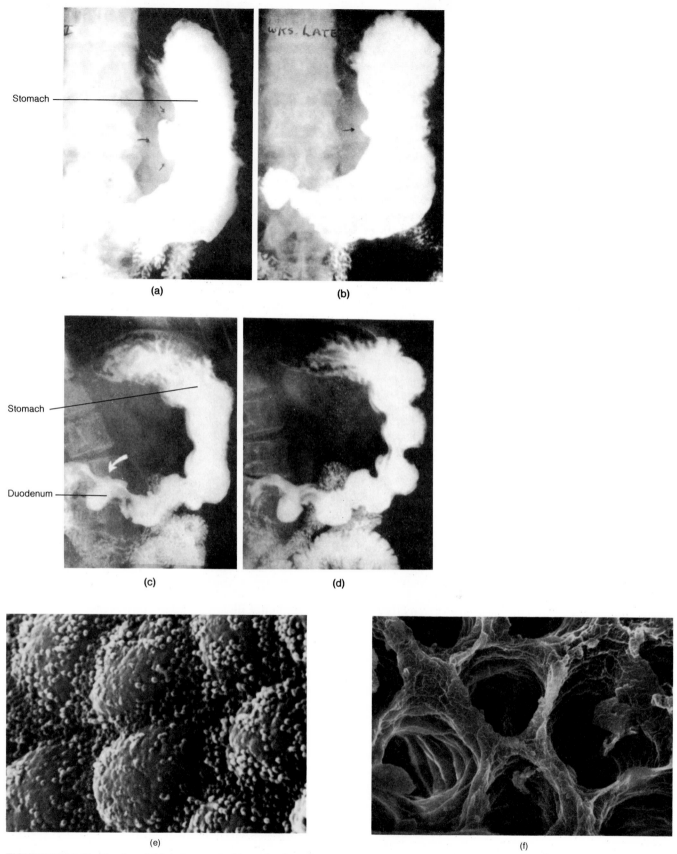

FIGURE 20-24 Peptic ulcers. (a) Anteroposterior projection of gastric ulcers (arrows) at the time of diagnosis. (b) The same ulcer three weeks after treatment. (c) Anteroposterior projection of a duodenal ulcer (arrow) prior to treatment. (d) The same ulcer after treatment. (Courtesy of Lester W. Paul and John H. Juhl, *The Essentials of Roentgen Interpretation,* 3d ed., Harper & Row, Publishers, Inc., New York, 1972.) (e) Scanning electron micrograph of the normal stomach mucosa at a magnification of 5000×. (f) Scanning electron micrograph of ulceration of the stomach mucosa at a magnification of 1000×. (Courtesy of Fisher Scientific Company and S.T.E.M. Laboratories, Inc., Copyright 1975.)

to acid gastric juice are called *peptic ulcers*. Peptic ulcers occasionally develop in the lower end of the esophagus. However, most of them occur on the lesser curvature of the stomach, in which case they are called *gastric ulcers*, or in the first part of the duodenum, where they are called *duodenal ulcers* (Figure 20-24).

Hypersecretion of acid gastric juice seems to be the immediate cause of duodenal ulcers. In gastric ulcer patients, because the stomach walls are highly adapted to resist gastric juice through their secretion of mucus, the cause may be hyposecretion of mucus. Hypersecretion of pepsin also may contribute to ulcer formation.

Among the factors believed to stimulate an increase in acid secretion are emotions, certain foods or medications (alcohol, coffee, aspirin) and overstimulation of the vagus nerve. Normally, the mucous membrane lining the stomach and duodenal walls resist the secretions of hydrochloric acid and pepsin. In some people, however, this resistance breaks down and an ulcer develops.

The danger inherent in ulcers is the erosion of the muscular portion of the wall of the stomach or duodenum. This erosion could damage blood vessels and produce fatal hemorrhage. If an ulcer erodes all the way through the wall, the condition is called *perforation*. Perforation allows bacteria and partially digested food to pass into the peritoneal cavity, producing peritonitis.

PERITONITIS

Peritonitis is an acute inflammation of the serous membrane lining the abdominal cavity and covering the abdominal viscera. One possible cause is contamination of the peritoneum by pathologic bacteria from the external environment. This contamination could result from accidental or surgical wounds in the abdominal wall or from perforation of organs exposed to the outside environment. Another possible cause is perforation of the walls of organs that contain bacteria or chemicals which are normally beneficial to the organ but are toxic to the peritoneum. For example, the large intestine contains colonies of bacteria that live on undigested nutrients and break them down so they can be eliminated more easily. But if the bacteria enter the peritoneal cavity, they attack the cells of the peritoneum for food and produce acute infection. As another example, the normal bacteria of the female reproductive tract protect the tract by giving off acid wastes that produce an acid environment unfavorable to many yeasts, protozoa, and bacteria which might otherwise attack the tract. However, these acid-producing bacteria are harmful to the peritoneum. A third cause may be chemical irritation. The peritoneum does not have

any natural barriers that keep it from being irritated or digested by chemical substances such as bile and digestive enzymes. However, it does contain a great deal of lymphatic tissue and can fight infection fantastically well. The danger stems from the fact that the peritoneum is in contact with most of the abdominal organs. If the infection gets out of hand, it may destroy vital organs and bring on death. For these reasons, perforation of the alimentary canal from an ulcer or perforation of the uterus from an incompetent abortion are considered serious. If a surgeon plans to do extensive surgery on the colon, the patient may be given high doses of antibiotics for several days preceding surgery to kill intestinal bacteria and reduce the risk of peritoneal contamination.

CIRRHOSIS

Cirrhosis is a chronic disease of the liver in which the parenchymal (functional) liver cells are replaced by fibrous connective tissue, a process called *stromal repair*. Often there is a lot of replacement with adipose connective tissue as well. The liver has a high ability for parenchymal regeneration, so stromal repair occurs whenever a parenchymal cell is killed or cells are damaged continuously for a long time. These conditions could be caused by *hepatitis* (inflammation of the liver), certain chemicals that destroy liver cells, parasites that infect the liver, and alcoholism.

TUMORS

Both benign and malignant **tumors** can occur in all parts of the gastrointestinal tract. The benign growths are much more common, but malignant tumors are responsible for 30 percent of all deaths from cancer in the United States (Figure 20-25a). For early diagnosis, complete routine examinations are necessary. Cancers of the mouth usually are detected through routine dental checkups.

A regular physical checkup should include rectal examination. Fifty percent of all rectal carcinomas are within reach of the finger, and 75 percent of all colonic carcinomas can be seen with the sigmoidoscope (Figure 20-25b). Both the fiberoptic sigmoidoscope and the more recent fiberoptic endoscope are flexible tubular instruments composed of a light and many tiny glass fibers. They allow visualization, magnification, biopsy, electrosurgery, and even photography of the entire length of the gastrointestinal tract. The greatest contribution of colonoscopy may be identification and removal of malignant polyps of the colon (gastric polypectomy) before invasion of the bowel wall or lymphatic metastasis. It has proved to be a safe and effective treatment that avoids the significant expense, risk,

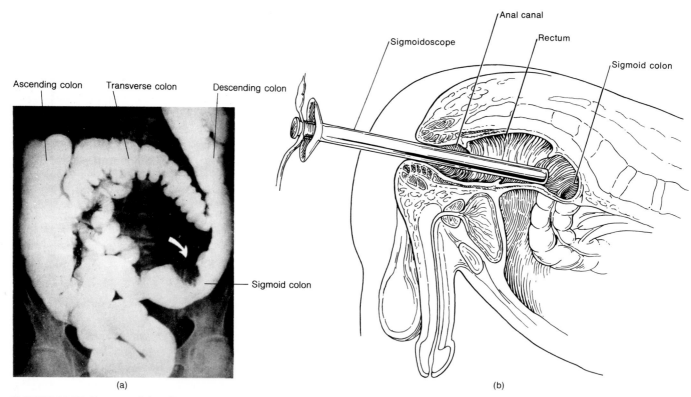

FIGURE 20-25 Tumors of the alimentary canal. (a) Anteroposterior projection of carcinoma (arrow) of the sigmoid colon. (Courtesy of Lester W. Paul and John H. Juhl, *The Essentials of Roentgen Interpretation,* 3d ed., Harper & Row, Publishers, Inc., New York, 1972.) (b) Detection of carcinomas by use of the sigmoidoscope.

and discomfort of major surgery. Colonoscopy may be the greatest advance yet toward lowering the death rate from cancer of the colon. Unfortunately, this type of cancer has shown a considerable increase in incidence over the last 20 years.

Another test in a routine examination for intestinal disorders is the filling of the gastrointestinal tract with barium, which is either swallowed or given in an enema. Barium, a mineral, shows up on x-rays the same way that calcium appears in bone x-rays. Tumors as well as ulcers can be diagnosed this way. The only definitive treatment of gastrointestinal carcinomas, if they cannot be removed using the endoscope, is surgery.

KEY MEDICAL TERMS ASSOCIATED WITH THE DIGESTIVE SYSTEM

Cholecystitis (*chole* = bile, gall)   Inflammation of the gallbladder that often leads to infection. Some cases are caused by obstruction of the cystic duct with bile stones. Stagnating bile salts irritate the mucosa. Dead mucosal cells provide medium for bacteria.

Cholelithiases   Calculi composed of bile salts, lecithin, and cholesterol that are formed in the gallbladder or its ducts and are also called *gallstones*.

Colitis   An inflammation of the colon and rectum. Inflammation of the mucosa reduces absorption of water and salts, producing watery, bloody feces, and —in severe cases—dehydration and salt depletion. Irritated muscularis spasms produce cramps.

Colonoscopy   The endoscopic examination of the entire colon.

Colostomy   The cutting of the colon and bringing the proximal end to the surface of the abdomen to serve as a substitute anus. Feces are eliminated through the upper end. A temporary colostomy may be done to allow a badly inflamed colon to rest and heal. Later the two halves may be rejoined and the abdominal opening closed. If the rectum is removed for malignancy, the colostomy provides a permanent outlet for feces.

Constipation   Infrequent or difficult defecation.

Diarrhea   Frequent defecation of liquid feces.

Flatus   Excessive amount of air (gas) in the stomach or intestine, usually expelled through the anus. If the gas is expelled through the mouth, it is called *belching* (burping). Flatus may result from gas released during the breakdown of foods in the stomach or from swallowing air or gas-containing substances such as carbonated drinks.

Heartburn   A burning sensation in the region of the esophagus and stomach. It may result from regurgitation of gastric contents into the lower end of the esophagus or from distention stemming from causes such as the retention of regurgitated food and gastric contents in the lower esophagus.

Hepatitis (*hepato* = liver)   A liver inflammation. It may be caused by organisms such as viruses, bacteria, and protozoa or by the absorption of materials, such as carbon tetrachloride and certain anesthetics and drugs, that are toxic to liver cells.

Hernia   Protrusion of an organ or part of an organ through a membrane or through the wall of a cavity, usually the abdominal cavity. *Diaphragmatic hernia* is the protrusion of the lower esophagus, stomach, or intestine into the thoracic cavity through the hole in the diaphragm that allows passage of the esophagus. *Umbilical hernia* is the protrusion of abdominal organs through the naval area of the abdominal wall. *Inguinal hernia* is the protrusion of the hernial sac containing the intestine into the inguinal opening. It may extend into the scrotal compartment, causing strangulation of the herniated part.

Lesion   Any pathological or traumatic discontinuity of tissue or loss of function of a part.

Mumps   Viral disease causing painful inflammation and enlargement of the salivary glands, particularly the parotids. In adults, the sex glands and pancreas may be involved. Inflammation of testes may cause male sterility.

Nausea   Discomfort preceding vomiting. Possibly, it is caused by distention or irritation of the gastrointestinal tract, most commonly the stomach.

Pancreatitis   Inflammation of the pancreas. The pancreas secretes active trypsin instead of trypsinogen, and the trypsin digests the pancreatic cells and blood vessels.

Vomiting   Expulsion of stomach (and sometimes duodenal) contents through the mouth by reverse peristalsis. The abdominal muscle walls forcibly empty the stomach.

STUDY OUTLINE

Chemical and Mechanical Digestion
1. Chemical digestion is a series of catabolic reactions that break down the large carbohydrate, lipid, and protein molecules of food into molecules that are usable by body cells.
2. Mechanical digestion consists of movements that aid chemical digestion.

General Organization
1. The organs of digestion are usually divided into two main groups.
2. First is the gastrointestinal (GI) tract, or alimentary canal, a continuous tube running through the ventral body cavity from the mouth to the anus.
3. The second group consists of the accessory organs: teeth, tongue, salivary glands, liver, gallbladder, pancreas, appendix.

General Histology
1. The basic arrangement of tissues in the alimentary canal from the inside outward is the mucosa, submucosa, muscularis, and serosa.

Peritoneum
1. Extensions of the peritoneum include the mesentery, lesser omentum, greater omentum, falciform ligament, and mesocolon.

Organs
Mouth or Oral Cavity
1. The mouth is formed by the cheeks, palates, lips, and tongue, which aid mechanical digestion.
2. The tongue, together with its associated muscles, forms the floor of the oral cavity. It is composed of skeletal muscle covered with mucous membrane.
3. The upper surface and sides of the tongue are covered with papillae. Some papillae contain taste buds.

Salivary Glands
1. The major portion of saliva is secreted by the salivary glands, which lie outside the mouth and pour their contents into ducts that empty into the oral cavity.
2. There are three pairs of salivary glands: the parotid, submandibular (submaxillary), and sublingual glands.
3. The salivary glands produce saliva that lubricates food and starts the chemical digestion of carbohydrates.

Teeth
1. The teeth, or dentes, project into the mouth and are adapted for mechanical digestion.
2. A typical tooth consists of three principal portions: crown, root, and cervix.
3. Teeth are composed primarily of dentin covered by enamel—the hardest substance in the body.

Deglutition
1. Both pharynx and esophagus assume a role in deglutition, or swallowing.
2. When a bolus is swallowed, the respiratory tract is sealed off and the bolus moves into the esophagus.
3. Peristaltic movements of the esophagus pass the bolus into the stomach.

Stomach
1. The stomach begins at the bottom of the esophagus and ends at the pyloric valve.
2. Adaptations of the stomach for digestion include rugae that permit distension; glands that produce mucus, hydrochloric acid, and enzymes; and a three-layered muscularis for efficient mechanical movement.
3. Nervous and hormonal mechanisms initiate the secretion of gastric juice.
4. Proteins are chemically digested into peptones and proteoses through the action of pepsin in the stomach.
5. The stomach also stores food, produces the intrinsic factor, and carries on some absorption.

Pancreas, Liver, Gallbladder

1. Pancreatic acini produce enzymes that enter the duodenum via the pancreatic duct. Pancreatic enzymes digest proteins, carbohydrates, and fats.
2. Cells of the liver produce bile, which is needed to emulsify fats.
3. Bile is stored in the gallbladder and passed into the duodenum via the common bile duct.

Small Intestine

1. This organ extends from the pyloric valve to the ileocecal valve.
2. It is very highly adapted for digestion and absorption. Its glands produce enzymes and mucus, and its wall contains microvilli, villi, and plicae circulares.
3. The enzymes of the small intestine digest carbohydrates, proteins, and fats into the end products of digestion: monosaccharides, amino acids, fatty acids, and glycerol.
4. The entrance of chyme into the small intestine stimulates the secretion of several hormones that coordinate the secretion and release of bile, pancreatic juice, and intestinal juice and inhibit gastric activity.
5. Mechanical digestion in the small intestine involves rhythmic segmentation, pendular movements, and propulsive peristalsis.
6. Absorption is the passage of the end products of digestion from the alimentary canal into the blood or lymph.
7. Absorption in the small intestine occurs through the villi. Monosaccharides and amino acids pass into the blood capillaries, small aggregations (chylomicrons) of fatty acids and glycerol pass into the blood capillaries, and large chylomicrons enter the lacteal.

Large Intestine

1. This organ extends from the ileocecal valve to the anus.
2. Mechanical movements of the large intestine include haustral churning, peristalsis, and mass peristalsis.
3. The large intestine functions in the synthesis of several vitamins and in water absorption, leading to feces formation.
4. The elimination of feces from the large intestine is called defecation. Defecation is a reflex action aided by voluntary contractions of the diaphragm and abdominal muscles.

Applications to Health

1. Dental caries are started by acid-producing bacteria.
2. Periodontal diseases are characterized by inflammation and/or degeneration of gingivae, alveolar bone, peridontal membrane, and cementum.
3. Peptic ulcers are craterlike lesions that develop in the mucous membrane of the alimentary canal in areas exposed to gastric juice.
4. Peritonitis is inflammation of the peritoneum.
5. In cirrhosis, parenchymal cells of the liver are replaced by fibrous connective tissue.
6. Tumors may be detected by sigmoidoscope and barium x-rays.

REVIEW QUESTIONS

1. Define digestion. Distinguish between chemical and mechanical digestion.
2. Identify the organs of the gastrointestinal tract in sequence. How does the gastrointestinal tract differ from the accessory organs of digestion?
3. Describe the structure of each of the four coats of the gastrointestinal tract.
4. What is the peritoneum? Describe the location and function of the mesentery, mesocolon, falciform ligament, lesser omentum, the greater omentum.
5. What structures form the oral cavity? How does each of the structures contribute to digestion?
6. Make a simple diagram of the tongue. Indicate the location of the papillae and the four taste zones.
7. Describe the location of the salivary glands and their ducts. What are buccal glands?
8. Contrast the histology of the salivary glands.
9. Describe the composition of saliva and the role of each of its components in digestion.
10. What are the principal portions of a typical tooth? What are the functions of each of the parts?
11. Compare deciduous and permanent dentitions with regard to numbers of teeth and times of eruption.
12. Contrast the functions of incisors, cuspids, premolars, and molars. What is pyorrhea?
13. What is a bolus? How is it formed?
14. Define deglutition. List the sequence of events that are involved in passing a bolus from the mouth to the stomach.
15. Describe the location of the stomach. List and briefly explain the anatomic features of the stomach.
16. What is the importance of rugae, zymogenic cells, parietal cells, and mucous cells in the stomach?
17. What is a gastric ulcer? How is it formed?
18. Where is the pancreas located? Describe the duct system by which the pancreas is connected to the duodenum.
19. What are pancreatic acini? Contrast their functions with those of the islets of Langerhans.
20. Where is the liver located? What are the principal functions of the liver?
21. Draw a labeled diagram of a liver lobule.
22. How is blood supplied to and drained from the liver?
23. Once bile has been formed by the liver, how is it collected and transported to the gallbladder for storage?
24. Where is the gallbladder located? How is the gallbladder connected to the duodenum?
25. What are the subdivisions of the small intestine? How are the coats of the small intestine adapted for digestion and absorption?
26. Describe the movements that occur in the small intestine.

27. Define absorption. How are the end products of carbohydrate and protein digestion absorbed? How are the end products of fat digestion absorbed?
28. What routes are taken by absorbed nutrients to reach the liver?
29. What are the principal subdivisions of the large intestine? How does the muscularis of the large intestine differ from that of the rest of the digestive tract?
30. Describe the mechanical movements that occur in the large intestine.
31. Explain the activities of the large intestine that change chyme into feces.
32. Define defecation. How does defecation occur?
33. Define dental caries. How are they started? What is dental plaque? What are three preventive measures that can be taken against dental caries?
34. Define periodontal disease, and describe the best method of prevention.
35. What is a peptic ulcer? Distinguish between gastric and duodenal ulcers. Describe some of the suspected causes of ulcers. What is perforation?
36. Define peritonitis. Explain some possible causes. Why is peritonitis a potentially dangerous condition?
37. Define cirrhosis.
38. How are tumors of the alimentary canal detected?
39. Refer to the glossary of key medical terms associated with the digestive system and be sure that you can define each term.

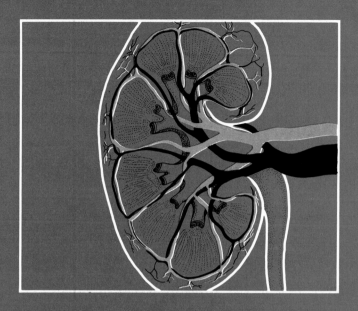

21

The Urinary System

he metabolism of nutrients results in the production of wastes by body cells, including carbon dioxide and excess water and heat. Protein catabolism produces toxic nitrogenous wastes such as ammonia and urea. In addition, many of the essential ions such as sodium, chloride, sulfate, phosphate, and hydrogen tend to accumulate in the body. All the toxic materials and the excess essential materials must be eliminated.

The primary function of the **urinary system** is to keep the body in homeostasis by controlling the concentration and volume of blood. It does so by removing and restoring selected amounts of water and solutes. It also excretes selected amounts of various wastes. Two kidneys, two ureters, one urinary bladder, and a single urethra make up the system (Figure 21-1). The kidneys regulate the concentration and volume of the blood and removes wastes from the blood in the form of urine. Urine is excreted from each kidney through its ureter and is stored in the urinary blad-

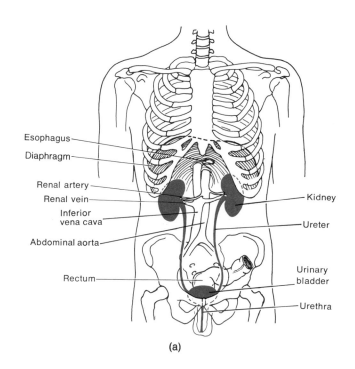

(a)

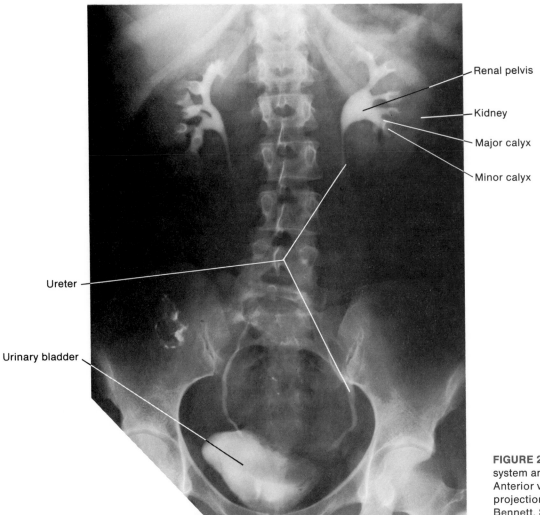

(b)

FIGURE 21-1 Organs of the urinary system and related structures. (a) Anterior view. (b) Anteroposterior projection. (Courtesy of John C. Bennett, St. Mary's Hospital, San Francisco.)

der until it is expelled from the body through the urethra. Other systems that aid in waste elimination are the respiratory, integumentary, and digestive systems.

Kidneys

The paired **kidneys** are reddish organs that resemble kidney beans in shape. The are found just above the waist between the parietal peritoneum and the posterior wall of the abdomen. Since they are external to the peritoneal lining of the abdominal cavity, their placement is described as *retroperitoneal* (see Figure 20-17). Relative to the vertebral column, the kidneys are located between the levels of the last thoracic and third lumbar vertebrae. The right kidney is slightly lower than the left because of the large area occupied by the liver.

EXTERNAL ANATOMY

The average adult kidney measures about 11.25 cm (4 inches) long, 5.0 to 7.5 cm (2 to 3 inches) wide, and 2.5 cm (1 inch) thick. Its concave medial border faces the vertebral column. Near the center of the concave border is a notch called the *hilum* through which the ureter leaves the kidney. Blood and lymph vessels and nerves also enter and exit the kidney through the hilum. The hilum is the entrance to a cavity in the kidney called the *renal sinus.*

Three layers of tissue surround each kidney. The innermost layer, the *renal capsule,* is a smooth, transparent, fibrous membrane that can easily be stripped off the kidney and is continuous with the outer coat of the ureter at the hilum. It serves as a barrier against trauma and the spread of infection to the kidney. The second layer, the *adipose capsule,* is a mass of fatty tissue surrounding the renal capsule. It also protects the kidney from trauma and holds it firmly in place in the abdominal cavity. The outermost layer, the *renal fascia,* is a thin layer of fibrous connective tissue that anchors the kidneys to their surrounding structures and to the abdominal wall. Some individuals, especially thin people in whom either the adipose capsule or renal facia is deficient, may develop *ptosis* (dropping) of one or both kidneys. Ptosis is dangerous because it may cause kinking of the ureter with reflux of urine and retrograde pressure. Ptosis of the kidneys below the rib cage also makes these organs susceptible to blows and penetrating injuries.

INTERNAL ANATOMY

A coronal (frontal) section through a kidney reveals an outer, reddish area called the *cortex* and an inner, reddish brown region called the *medulla* (Figure 21-2).

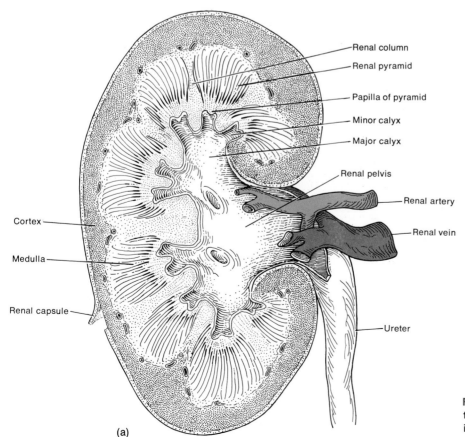

Renal column
Renal pyramid
Papilla of pyramid
Minor calyx
Major calyx
Renal pelvis
Renal artery
Renal vein
Ureter
Cortex
Medulla
Renal capsule

(a)

FIGURE 21-2 Kidney. (a) Coronal section through the right kidney illustrating gross internal anatomy.

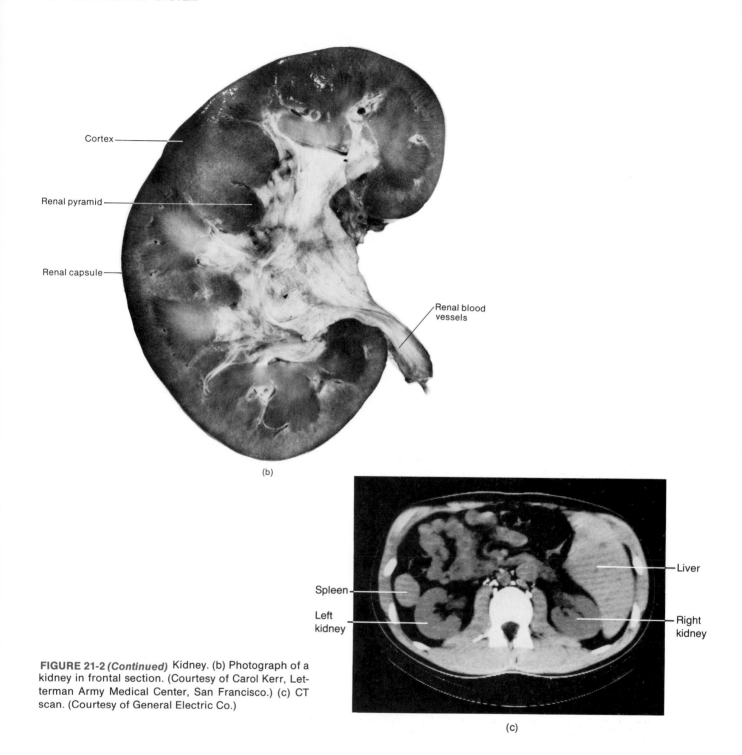

Cortex

Renal pyramid

Renal capsule

Renal blood vessels

(b)

Liver

Spleen

Left kidney

Right kidney

(c)

FIGURE 21-2 *(Continued)* Kidney. (b) Photograph of a kidney in frontal section. (Courtesy of Carol Kerr, Letterman Army Medical Center, San Francisco.) (c) CT scan. (Courtesy of General Electric Co.)

The cortex is arbitrarily divided into an outer cortical zone and an inner juxtamedullary zone. Likewise, the medulla is divided into an outer zone (one-third) and an inner zone (two-thirds). Within the medulla are 8 to 18 striated, triangular structures termed *renal,* or *medullary, pyramids.* The striated appearance is due to the presence of straight tubules and blood vessels. The bases of the pyramids face the cortical area, and their apices, called *renal papillae,* are directed toward

the center of the kidney. The cortex is the smooth-textured area extending from the renal capsule to the bases of the pyramids and into the spaces between them. The cortical substance between the renal pyramids forms the *renal columns.* Together the cortex and renal pyramids constitute the parenchyma of the kidney. Structurally, the parenchyma of each kidney consists of approximately one million microscopic units called nephrons, collecting ducts, and their associated vascular supply. Nephrons are the functional units of the kidney. They help form the urine and regulate blood composition.

In the renal sinus of the kidney is a large cavity called the *renal pelvis*. The edge of the pelvis contains cuplike extensions called the *major* and *minor calyces*. There are 2 or 3 major calyces and 7 to 13 minor calyces. Each minor calyx collects urine from collecting ducts of the pyramids. From the major calyces, the urine drains into the pelvis and out through the ureter.

NEPHRON

The physiological unit of the kidney is the **nephron** (Figure 21-3). Essentially, a nephron is a *renal tubule* and its vascular component. A nephron begins as a double-walled globe, called *Bowman's (glomerular) capsule*, lying in the cortex of the kidney. The inner layer or wall of the capsule, known as the visceral layer, consists of epithelial cells called podocytes. The visceral layer surrounds a capillary network called the

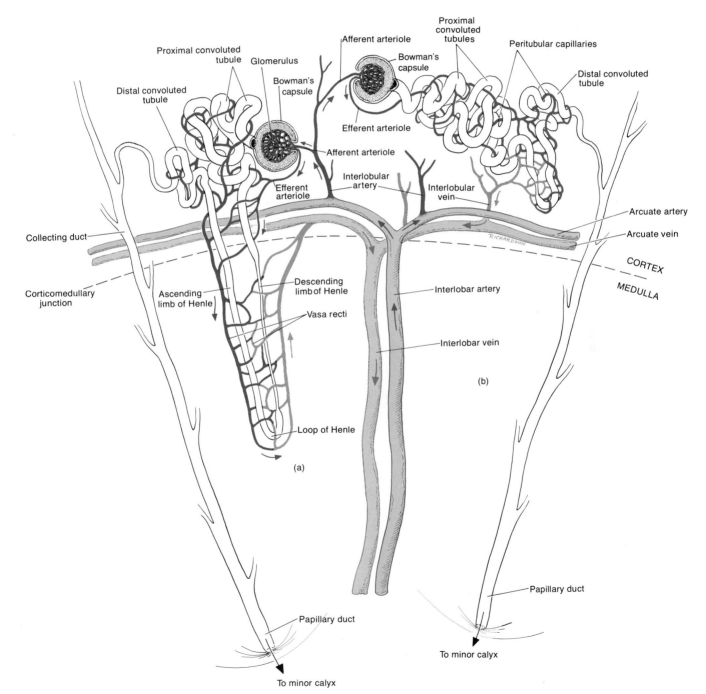

FIGURE 21-3 Nephrons. (a) Juxtamedullary nephron. (b) Cortical nephron.

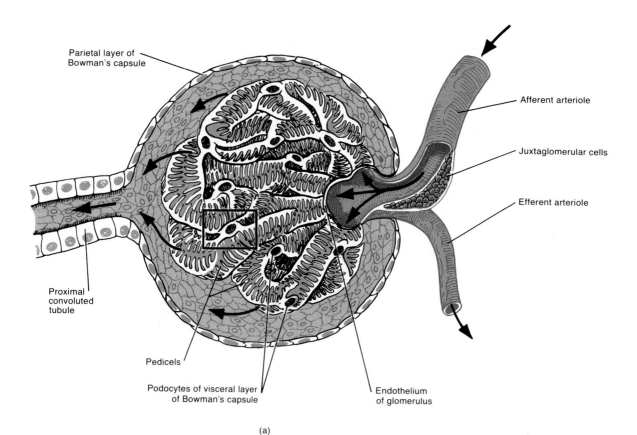

Parietal layer of
Bowman's capsule

Afferent arteriole

Juxtaglomerular cells

Efferent arteriole

Proximal
convoluted
tubule

Pedicels

Podocytes of visceral layer
of Bowman's capsule

Endothelium
of glomerulus

(a)

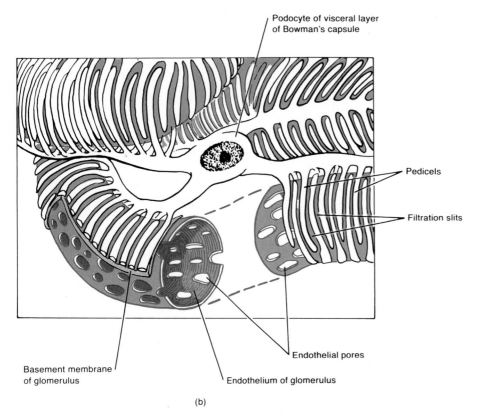

Podocyte of visceral layer
of Bowman's capsule

Pedicels

Filtration slits

Basement membrane
of glomerulus

Endothelial pores

Endothelium of glomerulus

(b)

FIGURE 21-4 Endothelial-capsular membrane. (a) Overview of a renal corpuscle. (b) Enlarged aspect of a portion of the endothelial-capsular membrane.

glomerulus. A space separates the inner wall from the outer one known as the parietal layer. The parietal layer is composed of simple squamous epithelium. Collectively, Bowman's capsule and the enclosed glomerulus constitute a *renal corpuscle.*

The visceral layer of Bowman's capsule and the endothelium of the glomerulus form an **endothelial-capsular membrane.** This membrane consists of the following parts in order in which substances filtered by the kidney must pass through (Figure 21-4).

1. Endothelium of the glomerulus. This single layer of endothelial cells has completely open pores averaging 500–1000 Å in diameter.

2. Basement membrane of the glomerulus. This extracellular membrane lies beneath the endothelium and contains no pores. It consists of fibrils in a glycoprotein matrix. It serves as the dialyzing (filtering) membrane.

3. Epithelium of the visceral layer of Bowman's capsule. These epithelial cells, because of their peculiar shape, are called *podocytes.* The podocytes contain footlike structures called *pedicels.* The pedicels are arranged parallel to the circumference of the glomerulus and cover the basement membrane except for spaces between them called *filtration slits* or *slip pores.*

The endothelial-capsular membrane filters water and solutes in the blood. Large molecules, such as proteins, and the formed elements in blood do not normally pass through it. The substances that are filtered pass into the space between the visceral and parietal layers of Bowman's capsule. The filtered fluid moves into the renal tubule.

Bowman's capsule opens into the *proximal convoluted tubule,* which also lies in the cortex. Convoluted means the tubule is coiled rather than straight; proximal signifies that this portion of the tubule originates at the Bowman's capsule. The wall of the proximal convoluted tubule consists of cuboidal epithelium with microvilli. These cytoplasmic extensions, like those of the small intestine, increase the surface area of reabsorption and secretion.

The next section of the tubule, the *descending limb of Henle,* dips into the medulla. It consists of squamous epithelium. The tubule then bends into a U-shaped structure called the *loop of Henle.* As the tubule straightens, it increases in diameter and ascends toward the cortex as the *ascending limb of Henle,* which consists of cuboidal and low columnar epithelium. In the cortex the tubule again becomes convoluted. Because of its distance from the point of origin at the Bowman's capsule, this section is referred to as the *distal convoluted tubule.* The cells of the distal tubule, like those of the proximal tubule, are cuboidal. Unlike the cells of the proximal tubule, however, the cells of the distal tubule have few microvilli. The distal tubule terminates by merging with a straight *collecting duct.* In the medulla, the collecting ducts receive the distal tubules of several nephrons, pass through the renal pyramids, and open into the calyces of the pelvis through a number of large *papillary ducts.* Cells of the collecting ducts are cuboidal; those of the papillary ducts are columnar.

Nephrons are frequently classified into two kinds. A *cortical nephron* usually has its glomerulus in the outer cortical zone, and the remainder of the nephron rarely penetrates the medulla. A *juxtamedullary nephron* usually has its glomerulus close to the corticomedullary junction, and other parts of the nephron penetrate deeply into the medulla (Figure 21-3). The histology of the various components of a nephron and its glomerulus is shown in Figure 21-5.

BLOOD AND NERVE SUPPLY

The nephrons are largely responsible for removing wastes from the blood and regulating its fluid and electrolyte content. Thus they are abundantly supplied with blood vessels. The right and left *renal arteries* transport about one-fourth the total cardiac output to the kidneys (Figure 21-6). Approximately 1200 ml passes through the kidneys every minute. Before or immediately after entering through the hilum, the renal artery divides into several branches that enter the parenchyma and pass as the *interlobar arteries* between the renal pyramids in the renal columns. At the base of the pyramids, the interlobar arteries arch between the medulla and cortex and become known as the *arcuate arteries.* Divisions of the arcuate arteries produce a series of *interlobular arteries* which enter the cortex and divide into *afferent arterioles* (Figure 21-3). One afferent arteriole is distributed to each glomerular capsule, where the arteriole divides into a tangled capillary network termed the *glomerulus.* The glomerular capillaries then reunite to form an *efferent arteriole,* leading away from the capsule, that is smaller in diameter than the afferent arteriole. This variation in diameter helps raise the glomerular pressure. The afferent-efferent arteriole situation is unique because blood usually flows out of capillaries into venules and not into other arterioles. Each efferent arteriole of a cortical nephron divides to form a network of capillaries, called the *peritubular capillaries,* around the convoluted tubules. The efferent arteriole of a juxtamedullary nephron also forms peritubular capillaries. In addition, it forms long loops of thin-walled vessels called *vasa recti* that dip down alongside the loop of Henle into the medullary region of the papilla. The peritubular capillaries eventually reunite to form *interlobular veins.* The blood then drains through the *arcuate veins* to the *interlobar veins* running between

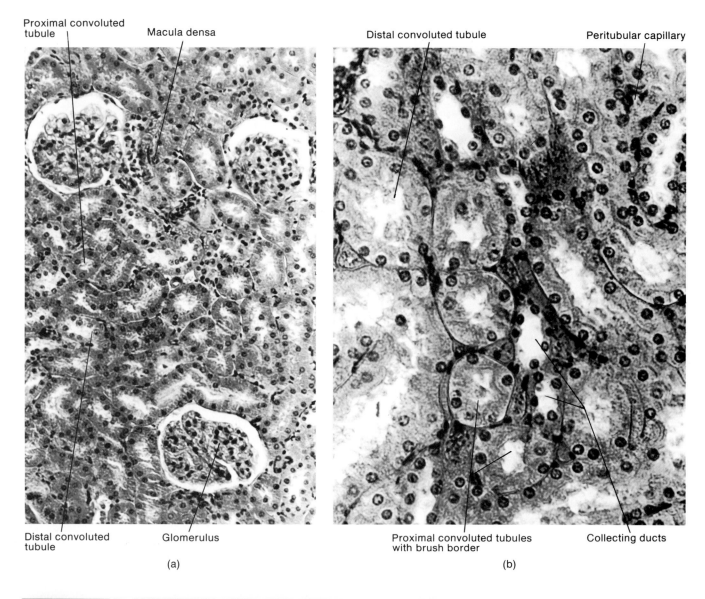

Proximal convoluted tubule

Macula densa

Distal convoluted tubule

Peritubular capillary

Distal convoluted tubule

Glomerulus

Proximal convoluted tubules with brush border

Collecting ducts

(a)

(b)

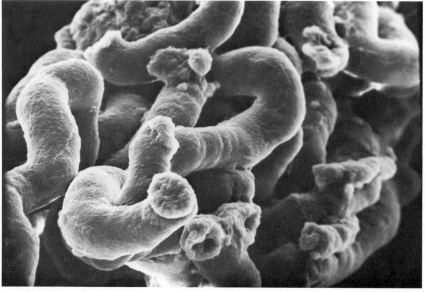

(c)

FIGURE 21-5 Histology of a nephron and the glomerulus. (a) Nephron and glomerulus at a magnification of 100×. (b) Enlarged aspect of tubules, collecting duct, and peritubular capillaries at a magnification of 100×. (Courtesy of Victor B. Eichler, Wichita State University.) (c) Scanning electron micrograph of renal tubules at a magnification of 500×. (Courtesy of Fisher Scientific Company and S.T.E.M. Laboratories, Inc., Copyright 1975.)

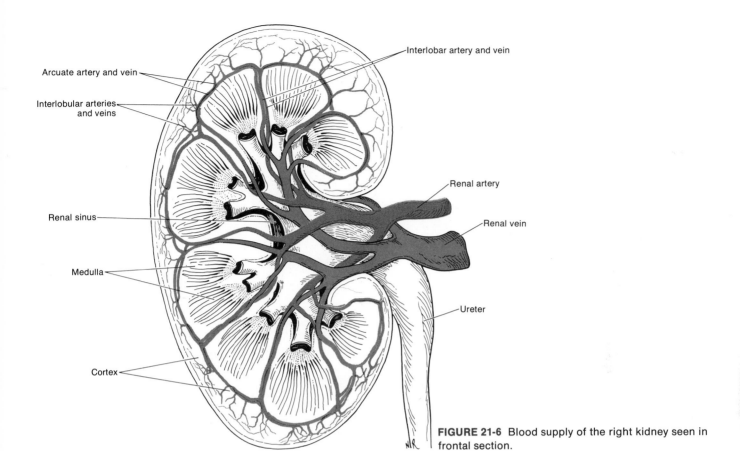

Interlobar artery and vein

Arcuate artery and vein

Interlobular arteries and veins

Renal artery

Renal sinus

Renal vein

Medulla

Ureter

Cortex

FIGURE 21-6 Blood supply of the right kidney seen in frontal section.

the pyramids and leaves the kidney through a single *renal vein* that exists at the hilum. The vasa recta pass blood into the interlobular veins. From here, it goes to the arcuate veins, the interlobar veins, and then into the renal vein.

The nerve supply to the kidneys is derived from the *renal plexus* of the autonomic system. Nerves from the plexus accompany the renal arteries and their branches and are distributed to the vessels. Because the nerves are vasomotor, they regulate the circulation of blood in the kidney by regulating the diameters of the small blood vessels.

JUXTAGLOMERULAR APPARATUS

As an afferent arteriole approaches the renal corpuscle, the smooth muscle cells of the tunica media become modified. The nuclei become rounded (instead of elongated), and their cytoplasm contains granules (instead of myofibrils). Such modified cells are called *juxtaglomerular cells*. The cells of the distal convoluted tubule adjacent to the afferent arteriole become considerably narrower. Collectively, these cells are known as the *macula densa*. Together with the modified cells of the afferent arteriole they constitute the **juxtaglomerular apparatus** (Figure 21-7).

PHYSIOLOGY

The major work of the urinary system is done by the nephrons. The other parts of the system are primarily passageways and storage areas. Nephrons carry out three important functions. They control blood concentration and volume by removing selected amounts of water and solutes. They help regulate blood pH. And they remove toxic wastes from the blood. As the nephrons go about these activities, they remove many materials from the blood, return the ones that the body requires, and eliminate the remainder. The eliminated materials are collectively called *urine*. The entire volume of blood in the body is filtered by the kidneys 60 times a day.

Ureters

Once urine is formed by the nephrons and collecting ducts, it drains through papillary ducts into the calyces surrounding the renal papillae. The minor calyces join to become the major calyces that unite to become the renal pelvis. From the pelvis, the urine drains into the ureters and is carried by peristalsis to the urinary bladder. From the urinary bladder, the urine is discharged from the body through the single urethra.

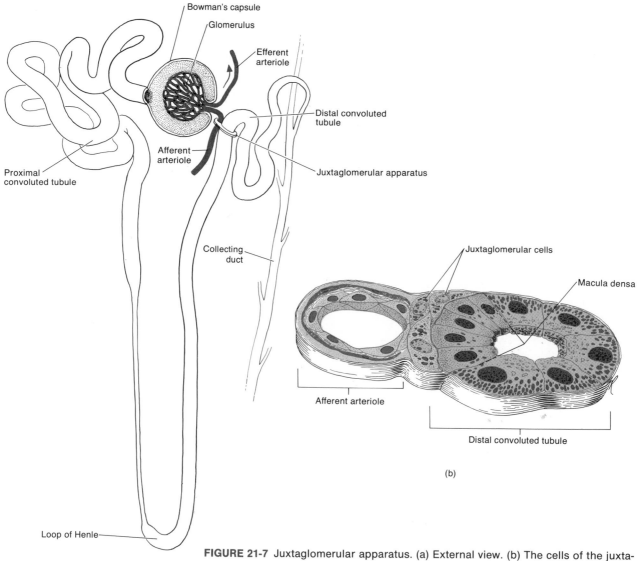

Bowman's capsule

Glomerulus

Efferent arteriole

Distal convoluted tubule

Afferent arteriole

Juxtaglomerular apparatus

Proximal convoluted tubule

Juxtaglomerular cells

Macula densa

Collecting duct

Afferent arteriole

Distal convoluted tubule

Loop of Henle

(a)

(b)

FIGURE 21-7 Juxtaglomerular apparatus. (a) External view. (b) The cells of the juxtaglomerular appatatus seen in cross section.

STRUCTURE

The body has two **ureters**—one for each kidney. Each ureter is an extension of the pelvis of the kidney and runs 25 to 30 cm (10 to 12 inches) to the urinary bladder (see Figure 21-1). As the ureters descend, their thick walls increase in diameter, but at their widest point they measure less than 1.7 cm (0.5 inches) in diameter. Like the kidneys, the ureters are retroperitoneal in placement. The ureters enter the urinary bladder at the superior lateral angle of its base. Since there is no valve or sphincter at the openings of the ureters into the urinary bladder, it is quite possible for cystitis (bladder inflammation) to develop into kidney infection. Three coats of tissue form the walls of the ureters (Figure 21-8). A lining of mucous membrane, the *mucosa,* with transitional epithelium is the inner coat.

The solute concentration and pH of urine differ drastically from the internal environment of the cells that form the walls of the ureters. Mucus secreted by the mucosa prevents the cells from coming in contact with urine. Throughout most of the length of the ureters, the second or middle coat, the *muscularis,* is composed of inner longitudinal and outer circular layers of smooth muscle. The muscularis of the proximal third of the ureters also contains a layer of outer longitudinal muscle. Peristalsis is the major function of the muscularis. The third, or external, coat of the ureters is a *fibrous coat*. Extensions of the fibrous coat anchor the ureters in place.

The arterial supply of the ureters is from the renal, testicular or ovarian, common iliac, and inferior vesical arteries. The veins terminate in the corresponding trunks. The ureters are innervated by the renal and vesical plexuses.

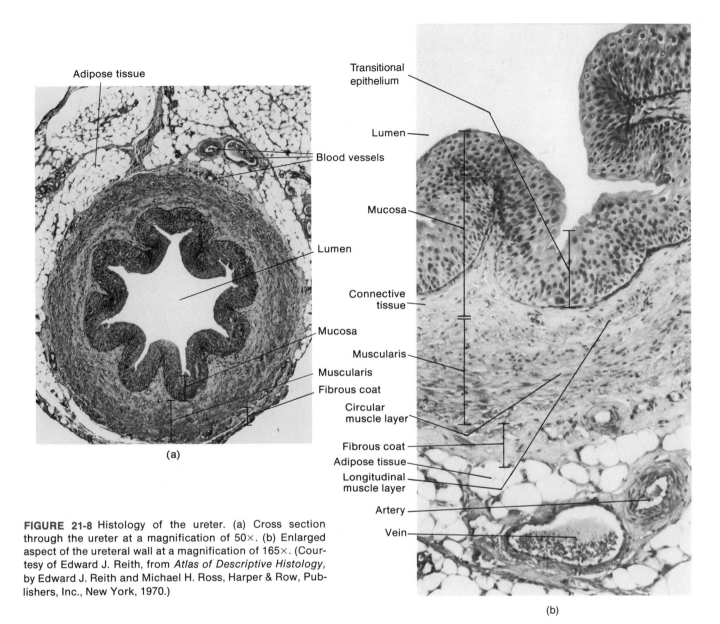

FIGURE 21-8 Histology of the ureter. (a) Cross section through the ureter at a magnification of 50×. (b) Enlarged aspect of the ureteral wall at a magnification of 165×. (Courtesy of Edward J. Reith, from *Atlas of Descriptive Histology,* by Edward J. Reith and Michael H. Ross, Harper & Row, Publishers, Inc., New York, 1970.)

PHYSIOLOGY

The principal function of the ureters is to transport urine from the renal pelvis into the urinary bladder. Urine is carried through the ureters primarily by peristaltic contractions of the muscular walls of the ureters, but hydrostatic pressure and gravity also contribute. Peristaltic waves pass from the kidney to the urinary bladder, varying in rate from one to five per minute depending on the amount of urine formation.

Urinary Bladder

The **urinary bladder** (Figure 21-9) is a hollow muscular organ situated in the pelvic cavity posterior to the symphysis pubis. In the male it is directly anterior to the rectum. In the female it is anterior to the vagina and inferior to the uterus. It is a freely movable organ held in position by folds of the peritoneum. The shape of the urinary bladder depends on how much urine it contains. When empty it looks like a deflated balloon. It becomes spherical when slightly distended. As urine volume increases, it becomes pear-shaped and rises into the abdominal cavity.

STRUCTURE

At the base of the urinary bladder is a small triangular area—the *trigone*—that points anteriorly. The opening to the urethra is found in the apex of this triangle. At the two points that form the base, the ureters drain into the urinary bladder. It is easily identified because the mucosa is firmly bound to the muscularis so that the trigone is always smooth.

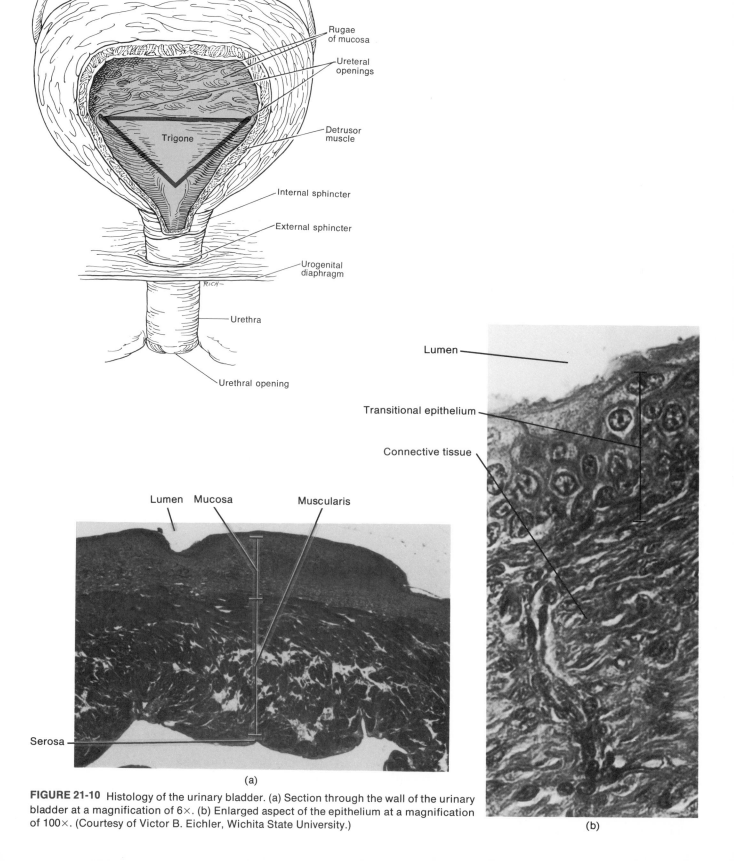

Ureters

FIGURE 21-9 Urinary bladder and female urethra.

Serosa

Rugae
of mucosa

Ureteral
openings

Detrusor
muscle

Trigone

Internal sphincter

External sphincter

Urogenital
diaphragm

RICH—

Urethra

Urethral opening

Lumen

Transitional epithelium

Connective tissue

Lumen Mucosa   Muscularis

Serosa

(a)

FIGURE 21-10 Histology of the urinary bladder. (a) Section through the wall of the urinary
bladder at a magnification of 6×. (b) Enlarged aspect of the epithelium at a magnification
of 100×. (Courtesy of Victor B. Eichler, Wichita State University.)

(b)

Four coats make up the walls of the urinary bladder (Figure 21-10). The mucosa, the innermost coat, is a mucous membrane containing transitional epithelium. Transitional epithelium is able to stretch—a marked advantage for an organ that must continually inflate and deflate. Stretchability is further enhanced by the rugae (folds in the mucosa) that appear when the urinary bladder is empty. The second coat, the submucosa, is a layer of dense connective tissue that connects the mucous and muscular coats. The third coat—a muscular one called the *detrusor muscle*—consists of three layers: inner longitudinal, middle circular, and outer longitudinal muscles. In the area around the opening to the urethra, the circular fibers form an *internal sphincter* muscle. Below the internal sphincter is the *external sphincter,* which is composed of skeletal muscle. The outermost coat, the serous coat, is formed by the peritoneum and covers only the superior surface of the organ.

The arteries of the urinary bladder are the superior vesical, the middle vesical, and the inferior vesical. The veins from the urinary bladder pass to the internal iliac trunk. The nerves are derived partly from the hypogastric sympathetic plexus and partly from the second and third sacral nerves. The fibers from the sacral nerves constitute the nervi erigentes.

PHYSIOLOGY

Urine is expelled from the urinary bladder by an act called *micturition,* commonly known as urination or voiding. This response is brought about by a combination of involuntary and voluntary nervous impulses. The average capacity of the urinary bladder is 700 to 800 ml. When the amount of urine in the urinary bladder exceeds 200 to 400 ml, stretch receptors in the urinary bladder wall transmit impulses to the lower portion of the spinal cord. These impulses initiate a conscious desire to expel urine and an unconscious reflex referred to as the *micturition reflex.* Parasympathetic impulses transmitted from the spinal cord reach the urinary bladder wall and internal urethral sphincter, bringing about contraction of the detrusor muscle of the urinary bladder and relaxation of the internal sphincter. Then the conscious portion of the brain sends impulses to the external sphincter, the sphincter relaxes, and urination takes place. Although emptying of the urinary bladder is controlled by reflex, it may be initiated voluntarily and may be started or stopped at will because of cerebral control of the external sphincter.

A lack of voluntary control over micturition is referred to as *incontinence.* In infants about two years old and under, incontinence is normal because neurons to the external sphincter muscle are not completely developed. Infants void whenever the urinary bladder is sufficiently distended to arouse a reflex stimulus. Proper training overcomes incontinence if the latter is not caused by emotional stress or irritation of the urinary bladder.

Involuntary micturition in the adult may occur as a result of unconsciousness, injury to the spinal nerves controlling the urinary bladder, irritation due to abnormal constituents in urine, disease of the urinary bladder, and emotional stress due to an inability of the detrusor muscle to relax.

Retention, a failure to void urine, may be due to an obstruction in the urethra or neck of the urinary bladder, nervous contraction of the urethra, or lack of sensation to urinate. Far more serious than retention is *suppression,* or *anuria*—failure of the kidneys to secrete urine. It usually occurs when blood plasma is prevented from reaching the glomerulus as a result of inflammation of the glomeruli. Anuria also may be caused by low filtration pressure.

Urethra

The **urethra** is a small tube leading from the floor of the urinary bladder to the exterior of the body (see Figure 21-9). In females, it lies directly behind the symphysis pubis and is embedded in the anterior wall of the vagina. Its undilated diameter is about 6 mm (0.25 inches), and its length is approximately 3.8 cm (1.5 inches). The female urethra is directed obliquely downward and forward. The opening of the urethra to the exterior, the *urinary meatus,* is located between the clitoris and vaginal opening.

In males, the urethra is about 20 cm (8 inches) long, and it follows a different course from that of the female urethra. Immediately below the urinary bladder it runs vertically through the prostate gland, then pierces the urogenital diaphragm, and finally penetrates the penis and takes a curved course through its body (see Figures 18-1 and 18-9).

STRUCTURE

The walls of the female urethra consist of three coats: an inner mucous coat that is continuous externally with that of the vulva, an intermediate thin layer of spongy tissue containing a plexus of veins, and an outer muscular coat that is continuous with that of the urinary bladder and consists of circularly arranged fibers of smooth muscle. The male urethra is composed of two membranes. An inner mucous membrane is continuous with the mucous membrane of the urinary bladder. An outer submucous tissue connects the urethra with the structures through which it passes.

PHYSIOLOGY

Since the urethra is the terminal portion of the urinary system, it serves as the passageway for discharging urine from the body. The male urethra also serves as the duct through which reproductive fluid (semen) is discharged from the body.

Applications to Health

HEMODIALYSIS THERAPY

If the kidneys are so impaired by disease or injury that they are unable to excrete nitrogenous wastes and regulate pH and electolyte concentration of the plasma, the blood must be filtered by an artificial device. Such filtering of the blood is called *hemodialysis. Dialysis* means using a semipermeable membrane to separate large nondiffusible particles from smaller diffusible ones. One of the best-known devices for accomplishing dialysis is the kidney machine (Figure 21-11). A tube connects it with the patient's radial artery. The blood is pumped from the artery and through the tubes to one side of a semipermeable cellophane membrane. The other side of the membrane is continually washed with an artificial solution called the dialyzing solution. The blood that passes through the artificial kidney is treated with an anticoagulant. Only about 500 ml of the patient's blood is in the machine at a time. This volume is easily compensated for by vasoconstriction and increased cardiac output.

All substances (including wastes) in the blood except protein molecules and blood cells can diffuse back and forth across the semipermeable membrane. The electrolyte level of the plasma is controlled by keeping the dialyzing solution electrolytes at the same concentration found in normal plasma. Any excess plasma electrolytes move down the concentration gradient and into the dialyzing solution. If the plasma electrolyte level is normal, it is in equilibrium with the

dialyzing solution and no electrolytes are gained or lost. Since the dialyzing solution contains no wastes, substances such as urea move down the concentration gradient and into the dialyzing solution. Thus wastes are removed and normal electrolyte balance is maintained.

A great advantage of the kidney machine is that nutrition can be bolstered by placing large quantities of glucose in the dialyzing solution. While the blood gives up its wastes, the glucose diffuses into the blood.

There are obvious drawbacks to the artificial kidney, however. Anticoagulants must be added to the blood during dialysis, and a large amount of the patient's blood must flow through this apparatus to make it work. To date, no artificial kidney has been capable of becoming a permanent implant.

ABNORMAL URINE CONSTITUENTS

If the body's chemical processes are not operating efficiently, traces of substances not normally present may appear in the urine. Normal constituents may appear in abnormal amounts. Analyzing the physical and chemical properties of urine often provides information that aids diagnosis. Such an analysis is called a **urinalysis.**

Albumin

Protein albumin is one of the things a technician looks for when performing a urinalysis. Albumin is a normal constituent of plasma, but it usually does not appear in urine because the particles are too large to pass through the pores in the capillary walls. The presence of albumin in the urine—*albuminuria*—indicates an increase in the permeability of the glomerular membrane. Conditions that lead to albuminuria include injury to the glomerular membrane as a result of disease, increased blood pressure, and irritation of kidney cells by substances such as bacterial toxins, ether, or heavy metals. Other proteins, such as globulin and fibrinogen, may also appear in the urine under certain conditions.

Glucose

The presence of sugar in the urine is termed *glycosuria.* Normal urine contains such small amounts of glucose that clinically it may be considered absent. The most common cause of glycosuria is a high blood sugar level. Glucose is filtered into the Bowman's capsule. Later, in the proximal convoluted tubules, the tubule cells actively transport the glucose back into the blood. However, the number of glucose carrier molecules is limited. If you ingest more carbohydrates

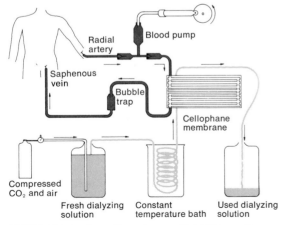

FIGURE 21-11 Operation of an artificial kidney. The blood route is indicated in color. The route of the dialyzing solution is indicated in gray.

than your body can convert to glycogen or fat, more sugar is filtered into the Bowman's capsule than can be removed by the carriers. This condition, called *temporary* or *alimentary glycosuria,* is not considered pathological. Another nonpathological cause is emotional stress. Stress can cause excessive amounts of epinephrine to be secreted. Epinephrine stimulates the breakdown of glycogen and the liberation of glucose from the liver. A pathological glycosuria results from diabetes mellitus. In this case, there is a frequent or continuous elimination of glucose because the pancreas fails to produce sufficient insulin. When glycosuria occurs with a normal blood sugar level, the problem lies in failure of the kidney tubular cells to reabsorb glucose.

Erythrocytes

The appearance of red blood cells in the urine is called *hematuria.* Hematuria generally indicates a pathological condition. One cause is acute inflammation of the urinary organs as a result of disease or irritation from stones. Whenever blood is found in the urine, additional tests are performed to ascertain the part of the urinary tract that is bleeding. One should also make sure the sample was not contaminated with menstrual blood from the vagina.

Leucocytes

The presence of leucocytes and other components of pus in the urine, referred to as *pyuria,* indicates infection in the kidney or other urinary organs. Again the source of the pus must be located, and care should be taken that the urine was not contaminated.

Ketone Bodies

Ketone, or acetone, bodies appear in normal urine in small amounts. Their appearance in high quantities, a condition called *ketosis,* or *acetonuria,* may indicate abnormalities. It may be caused by diabetes mellitus, starvation, or simply too little carbohydrate in the diet. Whatever the cause, excessive quantities of fatty acids are oxidized in the liver and the ketone bodies are filtered from the plasma into the Bowman's capsule.

Casts

Microscopic examination of urine may reveal *casts*— tiny masses of material that have hardened and assumed the shape of the lumens of the tubules, later flushed out of the tubules by a buildup of filtrate behind them. Casts are named after the substances that compose them or for their appearance. There are white-blood-cell casts, red-blood-cell casts, epithelial casts that contain cells from the walls of the tubes, granular casts that contain decomposed cells which form granules, and fatty casts from cells that have become fatty.

Calculi

Occasionally, the salts found in urine may solidify into insoluble stones called *calculi.* They may be formed in any portion of the urinary tract from the kidney tubules to the external opening. Conditions leading to calculi formation include the ingestion of excessive mineral salts, a decrease in the amount of water, abnormally alkaline or acid urine, and overactivity of the parathyroid glands.

Urinary Tract Disorders

PHYSICAL

Floating kidney (ptosis) occurs when the kidney is no longer held in place securely by the adjacent organs or its covering of fat and slips from its normal position. Pain occurs if the ureter is twisted. Such an abnormal orientation may also obstruct the flow of urine.

CHEMICAL

If urine becomes too concentrated, chemicals normally dissolved in it may crystallize out, forming **kidney stones (renal calculi).** Common constituents of stones are uric acid, calcium oxalate, and calcium phosphate. The stones usually form in the pelvis of the kidney, where they cause pain, hematuria, and pyuria. A staghorn calculus may fill the entire collecting system; a dendritic stone involves some but not all of the calyces. Severe pain occurs when a stone passes through a ureter and stretches its walls. Ureteral stones are seldom completely obstructive because they are usually needle-shaped and urine can flow around them.

Gout is a hereditary condition associated with an excessively high level of uric acid in the blood. When the purine-type nucleic acids are catabolized, a certain amount of uric acid is produced as a waste. Some people, however, seem to produce excessive amounts of uric acid, and others seem to have trouble excreting normal amounts. In either case, uric acid accumulates in the body and tends to solidify into crystals that are deposited in the joints and kidney tissue. Gout is aggravated by excessive use of diuretics, dehydration, and starvation.

INFECTIOUS

Glomerulonephritis or **Bright's disease** is an inflammation of the kidney that involves the glomeruli. One of the most common causes of glomerulonephritis is an allergic reaction to the toxins given off by streptococci bacteria that have recently infected another part of the body, especially the throat. The glomeruli become so inflamed, swollen, and engorged with blood that the glomerular membranes become highly permeable and allow blood cells and proteins to enter the filtrate. Thus the urine contains many erythrocytes and much protein. The glomeruli may be permanently changed, leading to chronic renal disease and renal failure.

Pyelitis is an inflammation of the kidney pelvis and its calyces. **Pyelonephritis** is the interstitial inflammation of one or both kidneys. It usually involves both the parenchyma and the renal pelvis and is due to bacterial invasion from the middle and lower urinary tracts or the bloodstream.

Cystitis is an inflammation of the urinary bladder involving principally the mucosa and submucosa. It may be caused by bacterial infection, chemicals, or mechanical injury.

KEY MEDICAL TERMS ASSOCIATED WITH THE URINARY SYSTEM

Cystoscope (*cyst* = bladder, *scope* = to view) Instrument used for examination of the urinary bladder.

Dysuria (*dys* = painful) Painful urination.

Nephrosis (*neph* = kidney) Any disease of the kidney but usually one that is degenerative.

Oliguria (*olig* = scanty) Scanty urine.

Polyuria (*poly* = many, much) Excessive amounts of urine.

Stricture Narrowing of the lumen of a canal or hollow organ, as the ureter or urethra.

Uremia (*emia* = condition of blood) Toxic levels of urea in the blood resulting from severe malfunction of the kidneys.

STUDY OUTLINE

Kidneys
1. The primary function of the urinary system is to regulate the concentration and volume of blood by removing and restoring selected amounts of water and solutes. It also excretes wastes.
2. The organs of the urinary system are the kidneys, ureters, urinary bladder, and urethra.
3. The kidneys are retroperitoneal and consist internally of a cortex, medulla, pyramids, papillae, columns, calyces, and a pelvis.
4. The functional unit of the kidney is the nephron.
5. The extensive blood flow through the kidney begins in the renal artery and terminates in the renal vein.

Ureters, Urinary Bladder, and Urethra
1. The paired ureters convey urine from the kidneys to the urinary bladder, mostly by peristaltic contractions.
2. The urinary bladder stores urine. The expulsion of urine from the bladder is called micturition.
3. A lack of control over micturition is called incontinence, failure to void urine is referred to as retention, and inability of the kidneys to produce urine is called suppression.

4. The urethra extends from the floor of the urinary bladder to the exterior and discharges urine from the body.

Hemodialysis
1. Filtering blood through an artificial device is called hemodialysis.
2. The kidney machine filters the blood of wastes and adds nutrients.

Applications to Health
1. Abnormal conditions diagnosed through urinalysis include albuminuria, glycosuria, hematuria, pyuria, ketosis, casts, and calculi.
2. Ptosis or floating kidney occurs when the kidney slips from its normal position.
3. Renal calculi are kidney stones.
4. Gout is a high level of uric acid in the blood.
5. Glomerulonephritis is an inflammation of the glomeruli of the kidney.
6. Pyelitis is an inflammation of the kidney and calyces.
7. Cystitis is an inflammation of the urinary bladder.

REVIEW QUESTIONS

1. What are the functions of the urinary system? What organs compose the system?

2. Describe the location of the kidneys. Why are they said to be retroperitoneal?

3. Prepare a labeled diagram that illustrates the principal external and internal features of the kidney.
4. What is a nephron? List and describe the parts of a nephron from the Bowman's capsule to the collecting duct.
5. How are nephrons supplied with blood?
6. Describe the structure, histology, and function of the ureters.
7. Describe the location, histology, and function of the urinary bladder.
8. How is the urinary bladder adapted to its storage function? What is micturition?

9. Contrast the causes of incontinence, retention, and suppression.
10. Compare the position of the urethra in the male and female. What is the function of the urethra?
11. What is hemodialysis? Briefly describe the operation of an artificial kidney.
12. Define each of the following: albuminuria, glycosuria, hematuria, pyuria, ketosis, casts, and calculi.
13. Define each of the following: ptosis, kidney stones, gout, glomerulonephritis, pyelitis, and cystitis.
14. Refer to the glossary of key medical terms for this chapter. Be sure that you can define each term.

Selected Readings

Barr, M. L., *The Human Nervous System,* 2d ed. New York: Harper & Row, 1974.

Basmajian, J. V., *Grant's Method of Anatomy,* 9th ed. Baltimore: Williams & Wilkins, 1975.

Bloom, W., and D. W. Fawcett, *A Textbook of Histology,* 10th ed. Philadelphia: Saunders, 1975.

Carpenter, M. B., *Human Neuroanatomy,* 7th ed. Baltimore: Williams & Wilkins, 1976.

Christensen, J. B., and I. R. Telford, *Synopsis of Gross Anatomy,* 3d ed. New York: Harper & Row, 1978.

Clemente, C. D., *Anatomy: A Regional Atlas of the Human Body.* Philadelphia: Lea & Febiger, 1975.

Cunningham's Textbook of Anatomy, 11th ed. Edited by G. J. Romanes. London: Oxford University Press, 1972.

Dawson, H. L., *Basic Human Anatomy,* 2d ed. New York: Appleton-Century-Crofts, 1974.

DiFiore, M. S. H., *An Atlas of Human Histology,* 4th ed. Philadelphia: Lea & Febiger, 1974.

Ellis, H., *Clinical Anatomy,* 5th ed. Philadelphia: Lippincott, 1971.

Gardner, E. D., D. J. Gray, and R. O'Rahilly, *Anatomy,* 4th ed. Philadelphia: Saunders, 1975.

Grant, J. C. B., *Grant's Atlas of Anatomy,* 6th ed. Baltimore: Williams & Wilkins, 1972.

Gray, H., *Anatomy of the Human Body,* 29th ed. Edited by Charles Mayo Goss. Philadelphia: Lea & Febiger, 1973.

Hamilton, W. J., and H. W. Mossman, *Hamilton, Boyd, and Mossman's Human Embryology,* 4th ed. Cambridge: Heffer & Sons, 1972.

Hamilton, W. J., G. Simon, and S. G. I. Hamilton, *Surface and Radiological Anatomy,* 5th ed. Cambridge: Heffer & Sons, 1971.

Hollinshead, W. H., *Textbook of Anatomy,* 3rd ed. New York: Harper & Row, 1974.

Kieffer, S. A., and E. R. Heitzman, *An Atlas of Cross-Sectional Anatomy.* New York: Harper & Row, 1979.

Lachman, E., *Case Studies in Anatomy,* 2d ed. New York: Oxford University Press, 1971.

Langman, J., *Medical Embryology,* 3d ed. Baltimore: Williams & Wilkins, 1975.

Langman, J., and M. W. Woerdeman, *Atlas of Medical Anatomy.* Philadelphia: Saunders, 1978.

Lockhart, R. D., G. F. Hamilton, F. W. Fyfe, *Anatomy of the Human Body,* 2d ed. Philadelphia: Lippincott, 1969.

Lockhart, R. D., *Living Anatomy,* 7th ed. London: Faber & Faber, 1974.

McMinn, R. M. H., and R. T. Hutchings, *Color Atlas of Human Anatomy.* Chicago: Year Book Medical Pub., 1977.

Netter, F. H., *CIBA Collection of Medical Illustrations,* Vols. 1–6. Summit, NJ: CIBA, 1962–1974.

Pansky, B., and E. L. House, *Review of Gross Anatomy,* 3d ed. New York: Mac Millan, 1975.

Patt, D. I., and G. R. Patt, *Comparative Vertebrate Histology.* New York: Harper & Row, 1969.

Pernkopf, E., *Atlas of Topographical and Applied Human Anatomy.* H. Ferner, ed.; H. Monsen, trans. Philadelphia: Saunders, Vol. 1, 1963; Vol. 2, 1964.

Ranson, S. W., and S. L. Clark, *Anatomy of the Nervous System, Its Development and Function,* 10th ed. Philadelphia: Saunders, 1959.

Reith, E. J., and M. N. Ross, *Atlas of Descriptive Histology,* 3d ed. New York: Harper & Row, 1977.

Royce, J., *Surface Anatomy.* Philadelphia: Davis, 1965.

Sobotta/Hammersen, *Histology.* Philadelphia: Lea & Febiger, 1976.

Woodburne, R. T., *Essentials of Human Anatomy,* 6th ed. New York: Oxford University Press, 1978.

Yokochi, C., *Photographic Anatomy of the Human Body.* Baltimore: University Park Press, 1971.

Glossary

Pronunciation appears in parentheses immediately following the word or phrase. The strongest accented syllable appears in capital letters; for example. ab-DO-men. If there is a secondary accent, it is noted by a double quotation mark (''); for example, ad''ĕ-no-hi-POF-ĭ-sis.

In a syllable ending with a consonant, the vowel preceding the consonant is pronounced with the short sound; for example, ab-DUK-shun.

ă as in ăt, ĕ as in mĕt, ĭ as in bĭt,
ŏ as in nŏt, ŭ as in bŭd

In a syllable ending with a vowel or if a vowel stands alone as a syllable, the vowel is pronounced with the long sound; for example, ab-DO-men and A-nal.

ā as in māke, ē as in bē, ī as in īvy,
ō as in pōle, ū as in pūre

If the pronunciation of a vowel does not follow these rules, then an accent mark appears above the vowel: ˘ for a short sound or ¯ for a long sound.

Other phonetics are as follows: a as in father is written ah, and the diphthong oi is written oy.

Abdomen (ab-DO-men) The area between the diaphragm and pelvis.

Abduction (ab-DUK-shun) Movement away from the axis or midline of the body or one of its parts.

Abortion (ab-OR-shun) The premature loss or removal of the embryo or nonviable fetus; any failure in the normal process of developing or maturing.

Abscess (AB-ses) A localized collection of pus and liquefied tissue in a cavity.

Absorption (ab-SŌRP-shun) The taking up of liquids by solids or of gases by solids or liquids.

Accommodation (ak-kom-mo-DA-shun) A change in the shape of the eye lens so that vision is more acute; an adjustment of the eye lens for various distances; focusing.

Accretion (ah-KRE-shun) A mass of material that has accumulated in a space or cavity; the adhesion of parts.

Acetabulum (as''i-TAB-yu-lum) The rounded cavity on the external surface of the innominate bone that receives the head of the femur.

Acetone (AS-e-tōne) **bodies** Organic compounds that may be found in excessive amounts in the urine and blood of diabetics. Found whenever too much fat in proportion to

carbohydrate is being oxidized. Also called *ketone bodies* (KĒ-tōne).

Acetylcholine (a sēt-il-KO-lēn) A chemical transmitter substance that is liberated at synapses in the central nervous system. It stimulates skeletal muscle contraction.

Achilles (ah-KIL-ēz) **reflex** Extension of foot and contraction of calf muscles following a tap upon tendon of Achilles.

Achilles tendon The tendon of the soleus and gastrocnemius muscles at the back of the heel.

Achlorhydria (a-klōr-HI-dre-ah) Absence of hydrochloric acid in the gastric juice.

Acid (AS-id) A proton donor; excess hydrogen ions producing a pH less than 7.

Acinus (AS-ĭ-nus) A small sac; a grapelike dilatation.

Acoustic (ah-KOOS-tik) Pertaining to sound or the sense of hearing.

Acromion (ah-KRŌ-mē-on) The lateral triangular projection of the spine of the scapula, forming the point of the shoulder and articulating with the clavicle.

Actin (AK-tin) One of the contractile proteins in muscle fiber; another is myosin.

Action potential A wave of negativity along a conducting neuron; a nerve impulse.

Active transport The movement of substances across cell membranes, against a concentration gradient, requiring the expenditure of energy.

Actomyosin (ak-to-MI-o-sin) The combination of actin and myosin in a muscle cell.

Acuity (ak-U-i-te) Clearness, or sharpness, usually of vision.

Acute (ak-ŪT) Having rapid onset, severe symptoms, and a short course; not chronic.

Adam's apple The laryngeal prominence of the thyroid cartilage.

Adaptation (ad-ap-TA-shun) The adjustment of the pupil of the eye to light variations.

Addison's (AD-i-sonz) **disease** Disorder due to deficiency in the secretion of adrenocortical hormones.

Adduction (ad-DUK-shun) Movement toward the axis or midline of the body or one of its parts.

Adenohypophysis (ad''ĕ-no-hi-POF-ĭ-sis) The anterior, glandular portion of the pituitary gland or hypophysis.

Adenoids (AD-i-noyds) The pharyngeal tonsils.

Adenosine triphosphate or **ATP** (ah-DEN-o-sen tri-FOS-fāt) A compound containing sugar, adenine, nitrogen, and three phosphoric acids; the breakdown of ATP provides the energy for cellular work.

Adhesion (ad-HE-zhun) Abnormal joining of parts to each other.

Adrenal cortex (ad-RE-nal KOR-teks) The outer portion of an adrenal gland; the cortex is divided into three zones, each of which has a different cellular arrangement and secretes different hormones.

Adrenal glands Two glands, one superior to each kidney; also called the suprarenal glands.

Adrenal medulla (mĕ-DUL-ah) The inner portion of an adrenal gland, consisting of cells which secrete epinephrine and norepinephrine in response to the stimulation of preganglionic sympathetic neurons.

Adrenalin (ad-REN-ah-lin) Proprietary name for epinephrine; one of the active secretions of the medulla of the adrenal gland.

Adrenergic fiber (ad''ren-ER-jik) A postganglionic sympathetic neuron which, when stimulated, releases epinephrine (adrenalin) or norepinephrine (noradrenalin) at its termination.

Adrenocorticotropic (ad-rēn-o-kŏrtĭ-ko-TROP-ik) **hormone (ACTH)** Hormone produced by the anterior pituitary gland, which influences the cortex of the adrenal glands.

Adsorption (ad-SORP-shun) Process whereby a gas or a dissolved substance becomes concentrated at the surface of a solid or at the interfaces of a colloid system.

Adventitia (ad-ven-TISH-yah) The outermost covering of a structure or organ.

Afferent arteriole (AF-er-ent ar-TE-re-ōl) A blood vessel of a kidney that breaks up into the capillary network called a glomerulus; there is one afferent arteriole for each glomerulus.

Afferent neuron (NU-ron) A nerve cell that carries an impulse toward the central nervous system.

Agglutination (ag-glu-tin-A-shun) Clumping of microorganisms, blood corpuscles or particles; an immunity response; an antigen-antibody reaction.

Agglutinin (ag-GLU-tin-in) A specific principle or antibody in blood serum of an animal affected with a microbic disease that is capable of causing the clumping of bacteria, blood corpuscles, or particles.

Agglutinogen (ag-GLU-tin-o-jen) A genetically determined antigen located on the surface of erythrocytes. These proteins are responsible for the two major blood group classifications: the ABO group and the Rh system.

Agnosia (ag-NO-ze-ah) A loss of the ability to recognize the meaning of stimuli from the various senses (visual, auditory, touch).

Agonist (AG-o-nist) Literally, ''prime mover.'' The muscle directly engaged in the contraction that produces a desired movement; it is opposed by the antagonist or relaxed muscle.

Albinism (AL-bin-ism) Abnormal, nonpathological, partial or total absence of pigment in skin, hair, and eyes.

Albumin (al-BU-min) A protein substance found in nearly every animal or plant tissue and fluid.

Albuminuria (al-bu''min-UR-ēa) Presence of albumin in the urine.

Aldosterone (al-DOS-te-rōne) Powerful salt-retaining hormone of the adrenal cortex.

Alimentary (al-i-MENT-ar-e) Pertaining to nutrition.

Alkaline (AL-ka-lin) Containing more hydroxyl than hydrogen ions and producing a pH of more than 7.

Allantois (al-AN-to-is) A kind of elongated bladder between the chorion and amnion of the fetus.

Allergic (ah-LER-jik) Pertaining to or sensitive to an allergen.

All-or-none principle In muscle physiology: muscle fibers of a motor unit contract to their fullest extent or not at all. In neuron physiology: if a stimulus is strong enough to initiate an action potential, an impulse is transmitted along the entire neuron at a constant and maximum strength.

Alveolar sacs (al-VE-o-lar) A collection or cluster of alveoli opening into a central atrium or chamber.

Alveolus (al-VE-o-lus) A small hollow or cavity; a single air cell.

Amenorrhea (a-men-o-RE-ah) Absence or suppression of menstruation.

Amino (am-E-no) **acids** Any one of a class of organic compounds occurring naturally in plant and animal tissues and forming the chief constituents of protein. About 22 different ones are known.

Amniocentesis (am''ne-o-sen-TE-sis) Removal of amniotic fluid by inserting a needle transabdominally into the amniotic cavity.

Amnion (AM-ne-on) The inner of the fetal membranes, a thin transparent sac that holds the fetus suspended in amniotic fluid. Also called the ''bag of waters.''

Amorphous (a-MOR-fus) Without definite shape or differentiation in structure; pertains to solids without crystalline structure.

Amphiarthroses (am''fe-ar-THRO-sēs) Joints or articulations midway between diarthroses and synarthroses, in which the articulating bony surfaces are separated by an elastic substance to which both are attached, so that the mobility is slight, but may be exerted in all directions.

Ampulla (am-PU-lah) A saclike dilatation of a canal.

Ampulla of Vater (VAH-ter) A small, raised area in the duodenum where the combined common bile duct and main pancreatic duct empty into the duodenum.

Amyl nitrate (A-mil NI-trate) An organic compound that produces dilatation of the blood vessels when inhaled. Used in attacks of angina pectoris.

Anabolism (ah-NA-bo-lizm) The building up of a body substance.

Anal canal (A-nal) The terminal 2 or 3 cm of the rectum that open to the exterior at the anus.

Anal column A longitudinal fold in the mucous membrane of the anal canal that contains a network of arteries and veins.

Analgesia (an-al-JE-ze-ah) Absence of normal sense of pain.

Anal triangle The subdivision of the male or female perineum that contains the anus.

Anaphase (AN-ă-fāz) The third stage of mitosis when the chromatids that have separated at the centromeres move to opposite poles.

Anaphylactic (an''a-fil-AC-tik) Pertaining to increasing susceptibility to any foreign protein introduced into the body; decreasing immunity.

Anastomosis (ah-nas-to-MO-sis) A communication between either blood vessels, lymphatics, or nerves; an end-to-end union or joining together.

Anatomical position (an''ah-TOM-i-kal) The body is erect, facing the observer, with the arms at the sides and the palms of the hands facing forward.

Anatomy (ah-NAT-o-me) The structure, or the study of the structure, of the body and the relationship of its parts to one another; also called morphology.

Androgen (AN-dro-jen) Substance producing or stimulating male characteristics, such as the male hormone.

Anemia (ah-NEM-e-ah) A condition in which there is a decreased number of erythrocytes, or the erythrocytes contain a decreased amount of hemoglobin.

Aneurysm (AN-u-rizm) A saclike enlargement of a blood vessel, usually an artery, that occurs when the blood vessel wall becomes weakened.

Angina pectoris (AN-ji-nah PEK-to-ris) An agonizing pain in the chest that may or may not involve heart or artery disease.

Angiography (an-je-OG-rah-fe) The injection of a contrast medium into the common carotid or vertebral artery in order to make the cerebral blood vessels visible in x-rays. Angiography may detect brain tumors with specific vascular patterns.

Anisocytosis (an''eso-si-TO-sis) An abnormal condition in which there is a lack of uniformity in the size of red blood cells.

Ankylose (ANG-ke-lōs) Immobilization of a joint by pathological or surgical process.

Anomaly (ah-NOM-ah-le) An abnormality that may be a developmental (congenital) defect; a variant from the usual standard.

Anorexia (an-o-REK-se-ah) Loss of appetite; anorexia ner-

vosa is a nervous condition in which a person may become seriously weakened from lack of food.

Anoxia (an-OK-se-ah) Deficiency of oxygen.

Antagonist (an-TAG-o-nist) A muscle that exerts an action opposite to that of the prime mover or agonist.

Antepartum (an-te-PAR-tum) Before delivery of the child; occurring (to the mother) before childbirth.

Anterior or **ventral** (an-TE-re-or) Toward the front; opposite of posterior or dorsal.

Anterior root The structure composed of axons of motor or efferent fibers that emerges from the anterior aspect of the spinal cord and extends laterally to join a posterior root, forming a spinal nerve; there are 31 pairs of anterior or motor or ventral roots.

Antibody (AN-ti-bod-e) A specific substance produced by an animal or person that produces immunity to the presence of an antigen.

Antidiuretic (an-ti-di-ur-ET-ik) Substance that inhibits urine formation.

Antigen (AN-tĭ-jen) Any substance that when introduced into the tissues or blood induces the formation of antibodies or reacts with them.

Antrum (AN-trum) Any nearly closed cavity or chamber, especially one within a bone, such as a sinus.

Anulus fibrosus (AN-u-lus fi-BRO-sus) A ring of fibrous tissue and fibrocartilage that encircles the pulpy substance (nucleus pulposus) of an intervertebral disc.

Anuria (an-U-re-ah) Absence of urine formation.

Anus (A-nus) The distal end and outlet of the rectum.

Aorta (a-OR-tah) The main systemic trunk of the arterial system of the body, emerging from the left ventricle of the heart.

Aperture (AP-er-chur) An opening or orifice.

Apex (A-peks) The pointed end of a conical structure.

Aphasia (a-FA-sze-ah) Loss of ability to express oneself properly through speech or loss of verbal comprehension.

Apocrine (AP-o-krin) Pertaining to cells that lose part of their cytoplasm while secreting.

Apocrine gland A type of gland in which the secretory products gather at the free end of the secreting cell and are pinched off, along with some of the cytoplasm, to become the secretion, as in the mammary glands.

Aponeurosis (ap''o-nu-RO-sis) A white, flattened, sheetlike tendon that connects a muscle to the part that it moves, or functions as a sheath enclosing a muscle.

Appendage (ah-PEN-dij) A part or thing attached to the body.

Appositional (exogenous) growth Growth due to surface deposition of material as in the growth in diameter of cartilage and bone.

Aqueduct (AK-we-duct) A canal or passage, especially for the conduction of a liquid.

Aqueous humor (AK-we-us) The watery fluid that fills the anterior and posterior chambers of the anterior cavity of the eye.

Arachnoid (ar-AK-noyd) The middle of the three coverings (meninges) of the brain and spinal cord, so named because of its weblike or spidery structure.

Arachnoid villi (VIL-e) Berrylike tufts of arachnoid that protrude into the superior sagittal sinus and through which the cerebrospinal fluid enters the blood stream; arachnoid granulations.

Arbor vitae (AR-bor VI-te)   The treelike appearance of the white-matter tracts of the cerebellum when seen in mid-sagittal section.

Arch of aorta (a-OR-tah)   The most superior portion of the aorta, lying between the ascending and descending segments of the aorta; the brachiocephalic, left common carotid, and left subclavian arteries are the three branches of the arch of the aorta.

Areflexia (ah''re-FLEK-se-ah)   Absence of reflexes.

Areola (ah-RE-o-lah)   Any tiny space in a tissue; the pigmented ring around the nipple of the breast.

Areolar (ah-RE-o-lar)   A type of connective tissue.

Arm   The portion of the upper extremity from the shoulder to the elbow.

Arrector pili (ah-REK-tor PI-le)   Smooth muscles attached to hairs; contraction of the muscles pulls the hairs into a more vertical position, resulting in "goose bumps."

Arrhythmia (ah-RITH-me-ah)   Irregular heart action causing absence of rhythm.

Arteriogram (ar-TE-re-o-gram)   A roentgenogram of an artery, obtained by injecting radiopaque substances into the blood and then taking the roentgenogram.

Arteriole (ar-TE-re-ōl)   A very small arterial branch.

Artery (AR-ter-e)   A blood vessel carrying blood away from the heart.

Arthritis (ar-THRI-tis)   Inflammation of a joint.

Arthrology (ar-THROL-o-je)   The scientific study or description of joints.

Arthrosis (ar-THRO-sis)   A joint of articulation.

Articular cartilage (ar-TIK-ular KAR-tĭ-lij)   The gristle or white elastic substance attached to articular bone surfaces.

Articulate (ar-TIK-u-lāt)   To join together as a joint to permit motion between parts.

Articulation (ar-tik''u-LA-shun)   A joint.

Arytenoid (ar-IT-en-oyd)   Resembling a ladle.

Arytenoid cartilages (ar-IT-en-oyd KAR-tĭ-lijs)   A pair of small cartilages of the larynx that articulates with the cricoid cartilage.

Ascending colon (KO-lon)   The portion of the large intestine that passes upward from the cecum to the lower edge of the liver where it bends at the hepatic flexure to become the transverse colon.

Ascites (as-SI-tēz)   Serous fluid in the peritoneal cavity.

Asphyxia (as-FIX-e-ah)   Unconsciousness due to interference with the oxygen supply of the blood.

Aspirate (AS-pir-āte)   To remove by suction.

Association area   A portion of the cerebral cortex connected by many motor and sensory fibers to other parts of the cortex; the association areas are concerned with motor patterns, memory, concepts of word-hearing and word-seeing, reasoning, will, judgment, and personality traits.

Association neuron (NU-ron)   A nerve cell lying completely within the central nervous system that carries impulses from sensory neurons to motor neurons; internuncial or connecting or central neuron.

Astereognosis (as-ter''e-og-NO-sis)   Inability to recognize objects or forms by touch.

Asthenia (as-THE-ne-ah)   Lack or loss of strength; debility.

Astigmatism (ah-STIG-mah-tizm)   An irregularity of the lens or cornea of the eye, causing the image to be out of focus, producing faulty vision.

Astrocyte (AS-tro-sīt)   A neuroglial cell having a star shape.

Ataxia (ah-TAK-se-ah)   The lack of muscular coordination; lack of precision.

Atherosclerosis (ath''er-o-sklĕ-RO-sis)   A disease involving mostly the lining of large arteries, in which yellow patches of fat are deposited, forming plaques that decrease the size of the lumen.

Atresia (ah-TRE-ze-ah)   The abnormal closure of a passage, or the absence of a normal body opening.

Atrioventricular bundle (a''tre-o-ven-TRIK-u-lar)   The portion of the conduction system of the heart that begins at the atrioventricular node and passes through the cardiac skeleton separating the atria and the ventricles, then runs a short distance down the interventricular septum before splitting into right and left bundle branches; the bundle of His or the AV bundle.

Atrioventricular node   The portion of the conduction system of the heart made up of a compact mass of conducting cells located near the orifice of the coronary sinus in the right atrial wall; AV node or node of Tawara.

Atrioventricular valve   A structure made up of membranous flaps or cusps that allows blood to flow in one direction only, from an atrium into a ventricle; AV valve.

Atrium (A-tre-um)   Either of the two superior cavities of the heart that act as receiving chambers; a chamber at the end of an alveolar duct of the lungs that connects with a variable number of alveolar sacs.

Atrophy (AT-ro-fe)   A wasting away or decrease in size of a part, due to a failure or abnormality of nutrition.

Aura (AW-rah)   A feeling or sensation that precedes an epileptic seizure or any paroxysmal attack (like those of bronchial asthma).

Auricle (AW-rĭ-kul)   An ear-shaped appendage of each atrium of the heart; the trumpet or pinna of the ear.

Auscultation (aws-kul-TA-shun)   Examination by listening to sounds in the body.

Autonomic ganglion (aw''to-NOM-ik GANG-gle-on)   Clusters of sympathetic or parasympathetic cell bodies located outside the central nervous system.

Autonomic nervous system   Visceral efferent neurons, both sympathetic and parasympathetic, that transmit impulses from the central nervous system to smooth muscle, cardiac muscle, and glands; so named because this portion of the nervous system was thought to be self-governing or spontaneous.

Autonomic plexus (PLEK-sus)   An extensive network of sympathetic and parasympathetic fibers; the cardiac, celiac, and pelvic plexuses are located in the thorax, abdomen, and pelvis, respectively.

Autopsy (AW-top-se)   The examination of the body after death; a postmortem study of the corpse. Also called necropsy.

Axilla (ak-SIL-ah)   The small hollow beneath the arm where it joins the body at the shoulders; the armpit.

Axon (AK-son)   The process of a nerve cell that carries an impulse away from the cell body.

Azygos (AZ-i-gos) An anatomical structure that is not paired; occurring singly.

Back The posterior part of the body; the dorsum.

Ball-and-socket joint A synovial joint in which the rounded surface of one bone moves within a cup-shaped depression or fossa of another bone; the shoulder or hip joint.

Baroreceptor (bar-o-re-SEP-tor) Receptor stimulated by pressure change; also called a pressoreceptor.

Bartholin's (BAR-to-linz) **glands** Two small mucous glands located one on each side of the vaginal opening.

Basal ganglia (BA-sal GANG-gle-ah) Clusters of cell bodies that make up the central gray matter of the cerebral hemispheres; they include the caudate nucleus, the lentiform nucleus, the claustrum, and the amygdaloid body.

Basal (BA-zel) **metabolic** (met-ah-BOL-ik) **rate (BMR)** The rate of metabolism measured under standard or basal conditions.

Base The broadest part of a pyramidal structure; a nonacid, or a proton acceptor, characterized by excess of OH⁻ ions and a pH greater than 7; a ring-shaped organic molecule, containing nitrogen, which is one of the components of a nucleotide; for example, adenine, guanine, cytosine, thymine, and uracil.

Basement membrane A sheet of extracellular material underlying the basal surface of epithelial cells; the basal lamina.

Basilar membrane (BAS-ĭ-lar) A membrane in the cochlea of the inner ear that separates the cochlear duct from the scala tympani; the organ of Corti rests on the basilar membrane.

Basophil (BA-so-fil) A polymorphonuclear leukocyte with a pale nucleus, and cytoplasm containing large granules that stain a deep blue, normally constituting about 0.5 percent of the white blood cells.

Belly The abdomen; the prominent, fleshy part of a skeletal muscle; gaster.

Benign (be-NĪN) Not malignant.

Bicipital (bi-CIP-i-tal) Having two heads, as in muscle.

Bifurcate (bi-FUR-kāt) Having two branches or divisions; forked.

Bilateral (bi-LAT-er-al) Pertaining to two sides of the body.

Bile (bīl) A secretion of the liver.

Biliary (BIL-e-a-re) Relating to bile, the gallbladder, or the bile ducts.

Bilirubin (bil-i-ROO-bin) The orange or yellowish pigment in bile.

Biliverdin (bil-i-VER-din) A greenish pigment in bile formed in the oxidation of bilirubin.

Biopsy (BI-op-se) Removal of tissue or other material from the living body for examination, usually microscopic.

Blepharism (BLEF-ah-rizm) Spasm of the eyelids; continuous blinking.

Blind spot Special area in the retina at the end of the optic nerve in which there are no light receptor cells.

Blood The fluid that circulates through the heart, arteries, capillaries, and veins, and that constitutes the chief means of transport within the body.

Blood-brain barrier A special mechanism that prevents the passage of materials from the blood to the cerebrospinal fluid and brain; astrocytes, which in most parts of the brain are situated between neurons and capillaries, play a role in this barrier.

Body cavity A space within the body that contains various internal organs.

Bolus (BO-lus) A soft, rounded mass, usually food, that is swallowed.

Bony labyrinth (LAB-ĭ-rinth) A series of cavities within the petrous portion of the temporal bone, or the vestibule, cochlea, and semicircular canals of the inner ear.

Bowman's capsule (KAP-sul) A double-walled globe at the proximal end of a nephron that encloses the glomerulus.

Brachial plexus (BRA-ke-al PLEK-sus) A network of nerve fibers of the anterior rami of spinal nerves C5, C6, C7, C8, and T1; the nerves that emerge from the brachial plexus supply the upper extremity.

Bradycardia (brā-dē-KAR-dē-ah) Slow heart rate.

Brain A mass of nerve tissue located in the cranial cavity.

Brain stem The portion of the brain immediately superior to the spinal cord, made up of the medulla oblongata, pons, and midbrain.

Broad ligament A double fold of parietal peritoneum attaching the uterus to the side of the pelvic cavity.

Bronchi (BRONG-ki) The two divisions of the trachea.

Bronchial tree (BRONG-ke-al) The bronchi and their branching structures.

Bronchiectasis (brong"-ke-EK-tah-sis) A chronic disorder in which there is a loss of the normal elastic tissue and expansion of lung air passages. Characterized by difficult breathing, coughing, expectoration of pus, and foul breath.

Bronchioles (BRONG-ke-ōlz) The smaller divisions of the bronchi.

Bronchogenic carcinoma (brong"ko-JEN-ik kar"sĭ-NO-mah) Cancer originating in the bronchi.

Bronchogram (BRONG-ko-gram) A roentgenogram of the lungs and bronchi.

Bronchopulmonary segment (brong"ko-PUL-mo-ner"e) One of the smaller divisions of a lobe of a lung supplied by its own branch of a bronchus.

Bronchoscope (BRONG-ko-skōp) An instrument used to examine the interior of the large tubes (bronchi) of the lungs.

Bronchus (BRONG-kus) One of the two large branches of the trachea.

Brunner's glands Glands in the submucosa of the duodenum that secrete intestinal juice.

Buccal cavity (BUK-al) The mouth or oral cavity.

Buffer (BU-fer) A substance that tends to preserve a certain hydrogen-ion concentration by adding acids or bases.

Bulbourethral glands (BUL-bo-u-RE-thral) A pair of glands, also known as Cowper's glands, located inferior to the prostate on either side of the urethra that secrete an alkaline fluid into the cavernous urethra.

Bundle branch One of the two branches of the bundle of His or the atrioventricular bundle, made up of specialized muscle fibers that transmit electrical impulses to the ventricles.

Bunion (BUN-yun) Inflammation and thickening of the bursa of the joint at the base of the big toe.

Burn An injury caused by fire, steam, chemicals, electricity, lightning, or the ultraviolet rays of the sun.

Bursa (BUR-sah) A fluid-filled, saclike cavity in connective tissue situated at points of friction or pressure, usually about joints.

Bursitis (bur-SI-tis) Inflammation of a bursa.

Buttocks (BUT-oks) The two fleshy masses on the posterior aspect of the lower trunk, formed by the gluteal muscles.

Cachexia (kah-KEK-sē-ah) A state of ill health, malnutrition, and wasting.

Calcification (kal''sĭ-fĭ-KA-shun) The process by which calcium salts are deposited in the matrix of bone or cartilage.

Calcify (KAL-si-fi) To harden by deposits of calcium salts.

Calcitonin (kal-si-TŌN-in) Hormone secreted by the parafollicular cells of the thyroid gland.

Calculus (KAL-ku-lus) A stone formed within the body, as in the gallbladder, kidney, or urinary bladder.

Callus (KAL-lus) A growth of new bone tissue in and around a fractured area, ultimately replaced by mature bone; an acquired, localized thickening of the stratum corneum, as a result of continued physical trauma.

Calorie (KAL-o-re) A unit of heat. The small calorie is the standard unit and is the amount of heat necessary to raise 1 g of water 1°C. The large Calorie (kilocalorie) is used in metabolic and nutrition studies; it is the amount of heat necessary to raise 1 kg of water 1°C.

Calyx (KA-liks) A cup-shaped tube extending from the renal pelvis (a renal portion of the ureter) that encircles renal papillae.

Canal (kan-AL) A narrow tube, channel, or passageway.

Canaliculus (kan''al-IK-u-lus) A small channel or canal; a microscopic channel in bone connecting lucunae; a short duct that conveys tears from the region of the medial canthus of the eye to the lacrimal sac.

Canal of Schlemm (shlem) A circular venous sinus located at the junction of the sclera and the cornea by which aqueous humor drains from the anterior chamber of the eyeball into the blood.

Cancellous (KAN-sel-us) Having a reticular or latticework structure, as in spongy tissue of bone.

Cancer (KAN-ser) A malignant tumor of epithelial origin tending to infiltrate and give rise to new growths or metastases; also called *carcinoma* (kar-si-NŌ-mah).

Canthus (KAN-thus) The angle formed by the joining of the eyelids at either corner of the eye.

Capillary (KAP-ĭ-ler-e) A tiny blood vessel connecting an arteriole and a venule; the smallest of the lymphatic vessels.

Carbohydrate (car-bo-HĪD-rāt) An organic compound containing carbon, hydrogen, and oxygen in a particular amount and arrangement and comprised of sugar subunits.

Carbuncle (KAR-bun-kal) A hard, pus-filled, and painful inflammation involving the skin and underlying connective tissues. Similar to boils but larger and more deeply rooted with several openings.

Cardiac notch (KAR-de-ak) An angular notch in the anterior border of the left lung.

Cardinal ligament (KAR-dĭ-nal LIG-ah-ment) A ligament of the uterus, extending laterally from the cervix and vagina as a continuation of the broad ligament.

Caries (KA-ri-ēz) Decay of a tooth or bone.

Carotene (KAR-o-teen) A yellow pigment in carrots, tomatoes, egg yolk, and other substances that can be converted by the body into vitamin A.

Carotid (kah-ROT-id) The main artery in the neck extending to the head.

Carpus (KAR-pus) The wrist; a collective term for the eight bones that make up the wrist.

Cartilage (KAR-tĭ-lij) A specialized connective tissue forming part of the skeleton; gristle.

Cartilaginous joint (kar''tĭ-LAJ-ĭ-nus) A joint without a joint cavity where the articulating bones are held tightly together by cartilage, allowing little or no movement.

Cast A solid mold of a part; can originate from different areas of the body and be composed of various materials.

Castration (kas-TRA-shun) The removal of the testes or ovaries.

Cataract (KAT-ah-rakt) Loss of transparency of the crystalline lens of the eye, or its capsule, or both.

Catheter (KATH-ĭ-ter) A tube that can be inserted into a body cavity through a canal to remove fluids such as urine or blood; also a tube inserted into a blood vessel through which opaque materials are injected into blood vessels or the heart for x-ray visualization.

Cauda equina (KAW-dă e-KWI-nah) A taillike collection of roots of spinal nerves at the inferior end of the spinal canal.

Caudal (KAW-dal) Pertaining to any taillike structure; inferior in position.

Cecum (SE-kum) A blind pouch at the junction of the ileum of the small intestine with the ascending colon, and to which the appendix is attached.

Celiac (SE-li-ak) Pertaining to the abdomen.

Celiac plexus (SE-le-ak PLEK-sus) A large mass of ganglia and nerve fibers located at the level of the upper part of the first lumbar vertebra; the solar plexus.

Cell The basic, living, structural and functional unit of all organisms; the smallest structure capable of performing all the activities vital to life.

Cell inclusion A lifeless, often temporary, constituent in the cytoplasm of a cell as opposed to an organelle.

Cellulitis (sel-u-LI-tis) Inflammation of cellular or connective tissue, especially the subcutaneous tissue.

Cementum (se-MEN-tum) Calcified tissue covering the root of a tooth.

Center An area in the brain where a particular function is localized.

Center of ossification (os''ĭ-fĭ-KA-shun) An area in the cartilage model of a future bone where the cartilage cells hypertrophy, secrete enzymes that result in the calcification of their matrix resulting in the death of the cartilage cells, followed by the invasion of the area by osteoblasts that then lay down bone.

Central canal A microscopic tube running the length of the spinal cord in the gray commissure.

Central fovea (FO-ve-ah) A cuplike depression in the center of the macula lutea of the retina, containing cones only; the area of clearest vision.

Central nervous system The brain and spinal cord.

Centrosome (SEN-tro-sōm) A rather dense area of cytoplasm, near the nucleus of a cell, containing a pair of centrioles.

Cephalic (se-FA-lik) Pertaining to the head; superior in position.

Cerebellar peduncle (ser-ĕ-BEL-ar PE-dung-kl) A bundle of nerve fibers connecting the cerebellum with the brain stem.

Cerebellum (ser-ĕ-BEL-um) The portion of the hindbrain lying posterior to the medulla and pons, concerned with coordination of movements.

Cerebral aqueduct (SER-ĕ-bral AK-we-dukt) A channel through the midbrain connecting the third and fourth ventricles and containing cerebrospinal fluid; the aqueduct of Sylvius.

Cerebral peduncles (PE-dung-kls) A pair of nerve fiber bundles located on the ventral surface of the midbrain, conducting impulses between the pons and the cerebral hemispheres.

Cerebrospinal fluid (SER-ĕ-bro-SPI-nal) A fluid produced in the choroid plexuses of the ventricles of the brain.

Cerebrum (SER-ĕ-brum) The two hemispheres of the forebrain, making up the largest part of the brain.

Cerumen (se-RU-men) Ear wax.

Cervical (SER-vĭ-kal) Pertaining to the neck.

Cervical ganglion (GANG-glĭ-on) A cluster of nerve cell bodies of postganglionic sympathetic neurons, located in the neck, near the vertebral column.

Cervical plexus (PLEK-sus) A network of neuron fibers formed by the anterior rami of the first four cervical nerves.

Cervix (SER-viks) The neck or a necklike portion of an organ.

Chalazion (kah-LA-ze-on) A small tumor of the eyelid.

Chemotherapy (kēm″o-THER-ă-pe) The treatment of illness or disease by chemicals.

Chiasm (KI-azm) A crossing; especially the crossing of the optic nerve fibers.

Choana (ko-AN-a) Funnel-shaped, like the posterior openings of the nasal fossa.

Cholecystectomy (ko″-le-sis-TEK-to-me) Surgical removal of the gallbladder.

Cholesterol (ko-LES-ter-ol) An organic sterol found in many parts of the body.

Cholinergic (kōl-in-ER-jik) Applied to nerve endings of the nervous system that liberate acetylcholine at a synapse.

Cholinergic fiber (kōl″in-ER-jik) An axon of a neuron that liberates acetylcholine at a synapse.

Cholinesterase (kōl″in-ES-ter-ās) An enzyme that hydrolyzes acetylcholine.

Chondrocyte (KON-dro-sīt) A cartilage cell.

Chordae tendineae (KOR-de ten-DIN-e-e) Fine strings of tendinous tissue connecting the papillary muscles of the ventricles of the heart with the atrioventricular valves.

Chorion (KŌR-e-on) The outermost fetal membrane; serves a protective and nutritive function.

Choroid (KO-royd) The vascular coat of the eye, located between the sclera and the retina.

Choroid plexus (PLEK-sus) A vascular structure located in the roof of each of the four ventricles of the brain; produces cerebrospinal fluid.

Chromaffin cells (kro-MAF-in) Cells that have an affinity for chrome salts, due in part to the presence of the precursors of the chemical transmitter epinephrine; also called pheochrome cells and are found, among other places, in the adrenal medulla.

Chromatin (KRO-mah-tin) The threadlike mass of the genetic material consisting principally of DNA, which is present in the nucleus of a nondividing or interphase cell.

Chromatolysis (kro″mah-TOL-ĭ-sis) Disappearance of the Nissl bodies of a neuron cell body as a result of trauma to the cell; disintegration of the chromatin of a cell nucleus.

Chromosome (KRO-mo-sōm) One of the 46 small, dark-staining bodies that appear in the nucleus of a human diploid cell during cell division.

Chronic (KRO-nik) Long-term; applied to a disease that is not acute.

Chyle (kīl) The milky fluid found in the lacteals of the small intestine after digestion.

Chyme (kīm) The semifluid mixture of partly digested food and digestive secretions found in the stomach and small intestine during digestion of a meal.

Cicatrix (SIK-ah-triks) A scar left by a healed wound.

Ciliary (SIL-e-ār-e) Pertaining to any hairlike processes; an eyelid, an eyelash.

Ciliary body (SIL-ĭ-ar-e) A part of the middle tunic of the eyeball that, along with the choroid and the iris, make up the vascular layer; the ciliary body includes the ciliary muscle and the ciliary processes.

Ciliary ganglion (GANG-glĭ-on) A very small parasympathetic ganglion whose preganglionic fibers come from the oculomotor nerve and whose postganglionic fibers carry impulses to the ciliary muscle and the sphincter muscle of the iris.

Cilium (SIL-ĭ-um) A hair or hairlike process; cilia projecting from cells can be used to move the entire cell or can move substances along the surface of the cell.

Circle of Willis A ring of arteries forming an anastomosis at the base of the brain between the internal carotid and vertebral arteries and arteries supplying the brain.

Circumcision (ser-kum-SIZH-un) Removal of the foreskin, the fold over the glans penis.

Circumduction (sir″kum-DUK-shun) A movement at a synovial joint in which the distal end of a bone moves in a circle while the proximal end remains relatively stable.

Circumvallate papilla (sir″kum-VAL-āt pah-PIL-ah) The largest of the elevations on the upper surface of the tongue; they are circular and arranged in an inverted V-shaped row at the posterior portion of the tongue.

Cirrhosis (si-RO-sis) A liver disorder in which the parenchymal cells are destroyed and replaced by connective tissue.

Clitoris (KLIT-or-is) An erectile organ of the female homologous to the male penis.

Coccygeal (kok-SIJ-e-al) Pertaining to or located in the region of the coccyx, the caudal extremity of the vertebral column.

Coccygeal plexus (PLEK-sus) A network of nerves formed by the anterior rami of the coccygeal and the fourth and fifth sacral nerves; fibers from the plexus supply the skin in the region of the coccyx.

Cochlea (KŌK-le-ah) A winding cone-shaped tube forming a portion of the inner ear and containing the organ of hearing.

Cochlear duct (KOK-le-ar) The membranous cochlea consisting of a spirally arranged tube enclosed in the bony cochlea and lying along its outer wall.

Coitus (KO-ĭ-tus) Sexual intercourse; also coition or copulation.

Collagen (KOL-ah-jen) A protein that is the main organic constituent of connective tissue.

Colostomy (ko-LOS-to-me) The surgical creation of a new opening from the colon to the body surface.

Colostrum (ko-LOS-trum) The first milk secreted at the end of pregnancy.

Colposcope (KOL-po-skōp) An instrument used to examine the vagina; a vaginal speculum.

Coma (KO-mah) Profound unconsciousness from which one cannot be roused.

Common bile duct A tube formed by the union of the common hepatic duct and the cystic duct that empties bile into the duodenum at the ampulla of Vater.

Compact (dense) bone Bone tissue with no apparent spaces in which the layers or lamellae are fitted tightly together; compact bone is found immediately deep to the periosteum and external to spongy bone.

Computed tomography (CT) scanning Radiographic procedure in which an x-ray source moves in an arc around the part of the body being scanned; result is a cross-sectional picture of an area of the body.

Conductivity The ability to transmit an impulse.

Condyle (KON-dīl) A rounded projection or process at the end of a bone that forms an articulation or joint.

Cone The light-sensitive receptor cell in the retina concerned with color vision.

Congenital (kon-JEN-ĭ-tal) Present at the time of birth.

Conjunctiva (kon''junk-TI-vah) The delicate membrane covering the eyeball and lining the eyelids.

Conjunctivitis (kon-junk''tiv-I-tis) Inflammation of the delicate membrane covering the eyeball and lining the eyelids.

Connective tissue The most abundant of the four tissue types in the body, performing the functions of binding and supporting; there are relatively few cells and a great deal of intercellular substance.

Contractility (kon''trak-TIL-ĭ-te) The ability to shorten.

Contralateral (kon''trah-LAT-er-al) Situated on, or pertaining to, the opposite side.

Conus medullaris (KO-nus med''u-LAR-is) The terminal tapering portion of the spinal cord.

Convergence (kon-VER-jens) The coordinated movement of the eyes so that both focus on a single set of objects.

Convoluted (CON-vo-lu-ted) Rolled together or coiled.

Convolution (kon''vo-LU-shun) An elevation of the gray matter of the cerebral cortex formed by the folding of the tissue upon itself.

Cornea (KOR-ne-ah) The clear, transparent, anterior covering of the eyeball; the anterior portion of the outermost tunic of the eyeball.

Coronary circulation (KOR-o-nă-re) The pathway followed by the blood from the ascending aorta through the blood vessels supplying the heart and returning to the right atrium.

Coronary sinus (SI-nus) A wide venous channel on the posterior surface of the heart that collects the blood from the coronary circulation and returns it to the right atrium.

Corpora quadrigemina (KOR-por-ah kwad-rĭ-JEM-ĭ-nah) Four small elevations on the dorsal region of the midbrain concerned with visual and auditory functions.

Corpus (KOR-pus) The principal part of any organ; any mass or body.

Corpus callosum (KOR-pus kal-LO-sum) An arch of white matter located at the bottom of the longitudinal fissure composed of myelinated fibers connecting the right and left hemispheres of the cerebrum.

Corpus striatum (stri-A-tum) An area in the interior of each cerebral hemisphere composed of the caudate and lentiform nuclei of the basal ganglia and white matter of the internal capsule, arranged in a striated manner.

Cortex (KOR-teks) An outer layer of an organ; the convoluted layer of gray matter covering each cerebral hemisphere.

Costal (KOS-tal) Pertaining to a rib.

Costal cartilage (KOS-tal KAR-tĭ-lij) A strip of hyaline cartilage by which each of the first through seventh ribs is connected directly to the sternum.

Cramp A spasmodic, especially a tonic, contraction of one or many muscles; usually painful.

Cranial nerve (KRA-ne-al) One of twelve pairs of nerves that leaves the brain, passes through formina in the skull, to supply the head, neck, and part of the trunk; each is designated both with Roman numerals and with names.

Craniosacral outflow (kra''ne-o-SA-kral) The parasympathetic division of the autonomic system, with cell bodies of preganglionic neurons located in nuclei in the brain stem and in the lateral gray matter of the sacral portion of the spinal cord.

Craniotomy (kra-ne-OT-o-me) Any operation on the skull, as for surgery on the brain or decompression of the fetal head in difficult labor.

Cranium (KRA-ne-um) The bones of the skull that enclose and protect the brain and the organs of sight, hearing, and balance.

Cremaster muscle (kre-MAS-ter) A thin layer of skeletal muscle that arises from the internal oblique abdominal muscle and extends as a series of loops down the spermatic cord to attach to the tunica vaginalis of the scrotum; it draws the testis up toward the superficial inguinal ring.

Crest A projection, peak, or ridge—as the iliac crest, the thickened, expanded superior border of the ilium.

Cretinism (KRE-tin-izm) Severe congenital thyroid deficiency during childhood leading to physical and mental retardation.

Cricoid cartilage (KRI-koyd KAR-tĭ-lij) A circle of cartilage shaped like a signet ring forming the inferior and posterior portion of the larynx.

Crista (KRIS-tah) A crest, as in the crista galli, a triangular process projecting superiorly from the cribriform plate of the ethmoid bone.

Cryosurgery (CRI-o-ser-jer-e) The destruction of tissue by application of extreme cold, as in the destruction of lesions in the thalamus for the treatment of Parkinsonism.

Crypts of Lieberkuhn (kripts of LI-ber-kun) Simple tubular glands that open onto the surface of the intestinal mucosa; these intestinal glands secrete digestive enzymes.

Cupula (KU-pu-lah) A mass of gelatinous material covering the hair cells of a crista, a receptor in the ampulla of a semicircular canal stimulated when the head moves.

Curvature (KUR-vah-tūr) A nonangular deviation of a straight line, as in the greater and lesser curvatures of the stomach; abnormal curvatures of the vertebral column include kyphosis, lordosis, and scoliosis.

Cutaneous (ku-TA-ne-us) Pertaining to the skin.

Cyanosis (si-an-O-sis) Slightly bluish or dark purple discoloration of the skin and the mucous membrane due to an oxygen deficiency.

Cystic duct (SIS-tik) The duct that transports bile from the gallbladder to the common bile duct.

Cystitis (sis-TI-tis) Inflammation of the urinary bladder.

Cystoscope (SIS-to-skōp) An instrument used to examine the inside of the urinary bladder.

Cytokinesis (si″to-kĭ-NE-sis) The division of the cytoplasm of a cell following mitosis (nuclear division).

Cytology (si-TOL-o-je) The study of cells.

Cytopenia (si-to-PE-ne-ah) Reduction or lack of cellular elements in the circulating blood.

Cytoplasm (SI-to-plazm) The protoplasm of a cell, excluding the nucleus.

Dartos (DAR-tos) Smooth muscle fibers in the scrotum.

Debility (de-BIL-i-te) Weakness of tonicity in functions or organs of the body.

Deciduous (de-SID-u-us) Anything that is cast off at maturity, especially the first set of teeth.

Decubitus (de-KYOO-be-tus) The lying-down position; a decubitus ulcer is caused by pressure when a patient is confined to bed for a long period of time.

Decussation of pyramids (de″kus-SA-shun) The crossing of the fibers of the pyramids of the medulla from one side to the other.

Deep fascia (FASH-e-ah) A sheet of connective tissue wrapped around a muscle to hold it in place.

Defecation (def-ě-KA-shun) The elimination of wastes and undigested food as feces from the rectum.

Deglutition (deg″lu-TĬ-shun) The act of swallowing.

Dehydration (de-hīd-RA-shun) A condition due to excessive water loss from the body or its parts.

Deleterious (del-e-TĒR-e-us) Harmful; noxious.

Demineralization (de-min″er-al-ĭ-ZA-shun) The loss of calcium and phosphorus from bones.

Dendrite (DEN-drīt) A nerve cell process carrying an impulse toward the cell body.

Dens (denz) Tooth.

Denticulate ligament (den-TIK-u-lāt LIG-ah-ment) A band of pia mater extending the entire length of the spinal cord on each side between the dorsal and ventral spinal nerve roots.

Dentin (DEN-tin) The main substance of a tooth, covered by enamel on the exposed part of the tooth and by cementum on the root of the tooth.

Dentition (den-TĬ-shun) The number, shape, and arrangement of teeth; the eruption of teeth (also called teething).

Deoxyribonucleic acid (de-ok″sĭ-ri″bo-nu-KLA-ik) A nucleic acid that transmits the genetic code from generation to generation and regulates protein synthesis; DNA is a double helix composed of units called nucleotides, each of which contains deoxyribose sugar, a phosphate group, and a nitrogen base.

Dermatome (DER-mah-tōm) An area of skin supplied by sensory fibers of a single dorsal root; an instrument for incising the skin or for cutting thin transplants of skin.

Dermis (DER-mis) A layer of dense connective tissue lying deep to the epidermis; the true skin or corium.

Descending colon (KO-lon) The part of the large intestine descending from the level of the lower end of the spleen on the left side to the level of the left iliac crest.

Detritus (de-TRI-tus) Any broken-down or degenerative tissue or carious matter.

Detrusor muscle (de-TRU-sor) The smooth muscle coat or tunic of the urinary bladder.

Developmental anatomy The study of development from the fertilized egg to the adult form; the branch of anatomy called embryology is generally restricted to the study of development from the fertilized egg to the period just before birth.

Diabetes insipidus (di″ah-BE-tēz in-SIP-ĭ-dus) A metabolic disease resulting from a hyposecretion of antidiuretic hormone from the posterior pituitary gland, associated with excretion of large amounts of urine.

Diabetes mellitus (MEL-ĭ-tus) A metabolic disease resulting from a hyposecretion of insulin from the pancreas, associated with elevated blood sugar, sugar in the urine, excessive urination, and increased thirst.

Diagnosis (di-ag-NO-sis) Recognition of disease states from symptoms, inspection, palpation, posture, reflexes, general appearance, abnormalities, and other means.

Dialysis (di-AL-ĭ-sis) The process of separating crystalloids (smaller particles) from colloids (larger particles) by the difference in their rates of diffusion through a semipermeable membrane.

Diapedesis (di″ah-pě-DE-sis) The passage of blood cells through intact blood vessel walls.

Diaphragm (DI-a-fram) Any partition that separates one area from another, especially the dome-shaped skeletal muscle between the thoracic and abdominal cavities.

Diaphysis (di-AF-ĭ-sis) The shaft of a long bone, or the portion of a long bone between the epiphyses.

Diarthrosis (di-ar-THRO-sis) An articulation in which opposing bones move freely, as in a hinge joint; a synovial joint.

Diastole (di-AS-to-le) The relaxing dilatation period of the heart muscle, especially of the ventricles.

Diencephalon (di″en-SEF-ah-lon) The "tween" brain, located between the cerebral hemispheres and the midbrain, consisting of the hypothalamus, thalamus, metathalamus, and epithalamus.

Differentiation (dif″er-en-she-A-shun) Acquirement of specific functions different from those of the original general type.

Diffusion (dĭ-FU-zhun) A passive process in which there is a net or greater movement of molecules or ions from a region of high concentration to a region of low concentration until equilibrium is reached.

Digestion (dī-JES-chun) The mechanical and chemical breakdown of food to simple molecules that can be absorbed by the body.

Digitalis (dij-i-TAL-is) The dried leaf of foxglove used in the treatment of heart disease.

Dilate (DI-lāt) To expand or swell.

Diplopia (dip-LO-pe-ah) Double vision.

Dissect (DI-sekt) To separate tissues and parts of a cadaver (corpse) or an organ for anatomical study.

Distal (DIS-tal) Away from the attachment of a limb to the trunk; the opposite of proximal.

Diuretic (di-ūr-ET-ik) Any agent that increases the secretion or lack of absorption of urine.

Diurnal (di-UR-nal) Daily.

Diverticulum (di-ver-TIK-u-lum) A sac or pouch in the walls of a canal or organ, especially in the colon.

Dorsal (DOR-sal) Pertaining to the back.

Dorsal ramus (DOR-sal RA-mus) A branch of a spinal nerve containing motor and sensory fibers supplying the muscles, skin, and bones of the posterior part of the head, neck, and trunk.

Dorsiflexion (dor''sĭ-FLEK-shun) Flexion of the foot at the ankle.

Dropsy (DROP-se) A condition rather than a disease: abnormal accumulation of water in the tissues and cavities.

Duct of Santorini (san''tor-E-ne) An accessory duct of the pancreas that empties into the duodenum about 2.5 cm superior to the ampulla of Vater.

Ductus arteriosus (DUK-tus ar-ter-ĭ-O-sus) A fetal blood vessel connecting the pulmonary trunk with the aorta.

Ductus deferens (DEF-er-ens) The vas deferens or seminal duct that conducts spermatozoa from the epididymis to the ejaculatory duct.

Ductus venosus (ven-O-sus) A fetal blood vessel connecting the umbilical vein and the inferior vena cava.

Duodenum (du''o-DE-num) The first or proximal portion of the small intestine, originating at the pyloric valve of the stomach and extending about 25 cm until it merges with the jejunum.

Dura mater (DU-rah MA-ter) Literally, "hard or tough mother"; the outermost meninx.

Dysentery (DIS-en-ter-e) A painful disorder due to intestinal inflammation and accompanied by frequent, loose, bloody stools.

Dysfunction (dis-FUNK-shun) Absence of complete normal function.

Dysmenorrhea (dis''men-o-RE-ah) Painful or difficult menstruation.

Dystrophia (dis-TRO-fe-ah) Progressive weakening of a muscle.

Ectopic (ek-TOP-ik) Out of the normal location.

Edema (ĕ-DE-mah) An abnormal accumulation of fluid in the body tissues.

Effector (ĕ-FEK-tor) The organ of the body, either a muscle or a gland, that responds to a motor neuron impulse.

Efferent arteriole (EF-er-ent ar-TE-re-ōl) A vessel of the renal vascular system that transports blood from the glomerulus to the peritubular capillary.

Efferent ducts A series of coiled tubes that transport spermatozoa from the rete testis to the epididymis.

Efferent neuron (NU-ron) A motor neuron that conveys impulses from the brain and spinal cord to effectors which may be either muscles or glands.

Effusion (ef-U-zhun) The escape of fluid from the lymphatics or blood vessels into a cavity or into tissues.

Ejaculatory duct (e-JAK-u-lah-tor''e) A tube that transports spermatozoa from the ductus deferens (seminal duct) to the prostatic urethra.

Elasticity (e-las-TIS-ĭ-te) The ability of muscle to return to its original shape after contraction or extension.

Elastin (e-LAS-tin) A substance found in elastic fibers of connective tissue that gives elasticity to the skin and to the tissues that form the walls of blood vessels.

Elbow The joint between the upper arm and the forearm.

Electrocardiogram (e-lek''tro-KAR-de-o-gram) A recording of the electrical changes that accompany the cardiac cycle.

Electroencephalogram (e-lek''tro-en-SEF-ah-lo-gram) A recording of the electrical impulses of the brain.

Electrolyte (e-LEK-tro-līt) Any compound that separates into ions when dissolved in water.

Ellipsoidal joint (e-lip-SOY-dal) A synovial or diarthrotic joint structured so that an oval-shaped condyle of one bone fits into an elliptical cavity of another bone, permitting side-to-side and back-and-forth movements, as at the joint at the wrist between the radius and carpals.

Embolism (EM-bo-lizm) Obstruction or closure of a vessel by a transported blood clot, a mass of bacteria, or other foreign material.

Embryo (EM-bre-o) The young of any organism in an early stage of development; in humans between the second and eighth weeks inclusively.

Embryology (em''bre-OL-o-je) The study of development from the fertilized egg through the eighth week of utero.

Emesis (EM-e-sis) Vomiting.

Emphysema (em''fi-SE-mah) A swelling or inflation of air passages with resulting stagnation of air in parts of the lungs; loss of elasticity in the alveoli.

Emulsification (e-mul''si-fi-CA-shun) The dispersion of large fat globules in the intestine to smaller uniformly distributed particles.

Enamel (e-NAM-el) The very hard white substance covering the crown of a tooth.

End bulb of Krause (krows) The cutaneous receptor for the sensation of cold.

End organ of Ruffini (roo-FE-ne) A cutaneous receptor for heat.

Endocardium (en-do-KAR-dĭ-um) A thin layer of endothelium and smooth muscle that lines the inside of the heart and covers the valves and tendons that hold the valves open.

Endochondral ossification (en''do-KON-dral os''ĭ-fi-KA-shun) The replacement of cartilage by bone.

Endocrine gland (EN-do-krin) A gland that secretes a hormone directly into the blood.

Endogenous (en-DOJ-en-us) Growing from or beginning within the organism.

Endometrium (en''do-ME-trĭ-um) The mucous membrane lining the uterus.

Endomysium (en''do-MIZ-ĭ-um) Connective tissue within a bundle or fascicle of muscle cells that separates the muscle cells from one another.

Endoneurium (en''do-NU-rĭ-um) Connective tissue within a bundle or fascicle of nerve fibers that separates and supports the nerve fibers.

Endoplasmic reticulum (en''do-PLAZ-mik re-TIK-u-lum) A network of canals running through the cytoplasm of a cell; in the areas where ribosomes are attached to the outer surface of the ER, it is referred to as granular or rough reticulum; portions of the ER that have no ribosomes are called agranular or smooth reticulum.

Endoscope (EN-do-skōp) An instrument used to look inside hollow organs such as the stomach (gastroscope) or urinary bladder (cystoscope).

Endosteum (en-DOS-te-um) The membrane that lines the medullary cavity of bones.

Endothelium (en''do-THE-lĭ-um) The layer of simple squamous epithelium that lines the cavities of the heart and blood and lymphatic vessels.

Enuresis (en-ūr-E-sis) Involuntary discharge of urine, complete or partial, after age 3.

Eosinophil (e''o-SIN-o-fil) A white blood cell readily stained by eosin, normally constituting about 2–4 percent of the white blood cells.

Epicardium (ep''ĭ-KAR-dĭ-um) The thin outer layer of the heart, also called the visceral pericardium.

Epicondyle (ep''ĭ-KON-dīl) A projection of a bone on or above a condyle.

Epidermis (ep''ĭ-DERM-is) The outermost layer of skin, composed of stratified squamous epithelium.

Epididymis (ep''ĭ-DID-ĭ-mus) A highly coiled tube located along the posterior border of the testis that receives spermatozoa from the efferent ducts and stores them until they are passed along to the ductus deferens.

Epidural space (ep''ĭ-DU-ral) A space between the spinal dura mater and the vertebral canal, containing loose connective tissue and a plexus of veins.

Epiglottis (ep''ĭ-GLOT-is) A large, leaf-shaped piece of cartilage lying on top of the larynx, with its ''stem'' attached to the thyroid cartilage and its ''leaf'' portion unattached and free to move up and down to cover the glottis.

Epimysium (ep''ĭ-MIZ-ĭ-um) A fibrous sheath of connective tissue wrapped around a muscle.

Epineurium (ep''ĭ-NU-rĭ-um) A sheath of connective tissue wrapped around a nerve.

Epiphyseal line (ep''ĭ-FIZ-e-al) The remnant of the epiphyseal plate in a long bone.

Epiphyseal plate A layer of cartilage between the epiphysis and diaphysis of a long bone that allows the bone to increase in length until early adulthood.

Epiphysis (ĕ-PIF-ĭ-sis) The end of a long bone, usually larger in diameter than the shaft (the diaphysis). The plural form is epiphyses.

Episiotomy (ĕ-piz''e-OT-o-me) Incision of the clinical perineum at the end of the second stage of labor to avoid tearing the perineum.

Epistaxis (ep-e-STAKS-is) Hemorrhage from the nose; nosebleed.

Epithelial tissue (ep''ĭ-THE-lĭ-al) The tissue that forms the outer part of the skin, lines blood vessels, hollow organs, and passages that lead to the outside of the body.

Eponychium (ep''o-NIK-ĭ-um) The thin layer of stratum corneum of the skin that overlaps the lunula of the nail.

Eructation (e-ruk-TA-shun) The forceful expulsion of gas from the stomach; belching.

Erythema (er-ĭ-THE-mah) Skin redness usually caused by engorgement of the capillaries in the lower layers of the skin.

Erythrocyte (ĕ-RITH-ro-sīt) A red blood cell or corpuscle.

Erythropoiesis (e-rith''ro-po-E-sis) The production of red blood cells.

Esophagus (ĕ-SOF-ah-gus) A hollow muscular tube connecting the pharynx and the stomach.

Estrogen (ES-tro-jen) Any substance that induces estrogenic activity or stimulates the development of secondary female characteristics; female hormones.

Etiology (e''te-OL-o-je) The study of the causes of disease, including theories of origin and what organisms, if any, are involved.

Euphoria (u-FŌR-e-ah) A subjectively pleasant feeling of well-being marked by confidence and assurance.

Eustachian tube (u-STA-ke-an) A narrow channel connecting the middle ear with the nasopharynx; the auditory tube.

Euthanasia (u-than-A-ze-ah) The proposed practice of ending a life in case of incurable disease.

Eversion (e-VER-zhun) Turning outward.

Exacerbation (eks-as-er-BA-shun) An increase in the severity of symptoms or of disease.

Excrement (EKS-kre-ment) Material cast out from the body as waste, especially fecal matter.

Excretion (eks-KRE-shun) The elimination of waste products from the body. It can refer to the expulsion of any matter—whether from a single cell or from the entire body—or to the matter excreted.

Exocrine gland (EK-so-krin) One that secretes its product into a duct or tube that empties out at the surface of the covering and lining epithelium.

Exogenous (ex-OJ-en-us) Originating outside an organ or part.

Exophthalmos (ek''sof-THAL-mos) An abnormal protrusion or bulging of the eyeball.

Expiration (ek''spĭ-RA-shun) Breathing out, or exhalation.

Extensibility (ek-sten''sĭ-BIL-ĭ-te) The ability to be stretched when pulled.

Extension (ek-STEN-shun) An increase in the anterior angle between two bones, except in extension of the knee and toes, in which the posterior angle is involved; restoring a body part to its anatomical position after flexion.

External or **superficial** (su''per-FISH-al) Located on or near the surface.

External auditory meatus (AW-dĭ-tor-e me-A-tus) A canal in the temporal bone that leads to the middle ear.

External ear The outer ear, consisting of the pinna, the external auditory canal, and the tympanic membrane or eardrum.

External nares (NA-rēs) The external nostrils, or the openings into the nasal cavity on the exterior of the body.

Exteroceptor (ek''ster-o-SEP-tor) A sense organ adapted for the reception of stimuli from outside the body.

Extravasation (eks-trah-va-SA-shun) The process of escaping from a vessel into the tissues, especially blood, lymph, or serum.

Extrinsic (ek-STRIN-sik) Of external origin.

Eyebrow The hairy ridge above the eye.

Exudate (EKS-u-dat) Escaping fluid or semifluid material that oozes from a space that may contain serum, pus, and cellular debris.

Face The anterior aspect of the head.

Facet (FAS-et) A smooth surface for articulation.

Facilitated diffusion (fah-SIL-ĭ-ta-ted dĭ-FU-zuhn) A passive process that moves a substance down its own concentration gradient but that is carrier-mediated.

Falciform ligament (FAL-sĭ-form LIG-ah-ment) A sheet of parietal peritoneum that separates the two principal lobes of the liver; the ligamentum teres, or remnant of the umbilical vein, lies within its fold.

False vocal folds A pair of folds of the mucous membrane of the larynx superior to the true vocal cords.

Falx cerebelli (falks ser''ĕ-BEL-le) A small triangular process of the dura mater attached to the occipital bone in the posterior cranial fossa and projecting inward between the two cerebellar hemispheres.

Falx cerebri (SER-ĕ-bre) A fold of the dura mater extending down into the longitudinal fissure between the two cerebral hemispheres.

Fascia (FASH-e-ah) A fibrous membrane covering, supporting, and separating muscles.

Fasciculi (fah-SIK-u-le) Small bundles or clusters, especially of nerves or muscle fibers.

Fauces (FAW-sēz) The passageway from the mouth to the pharynx.

Febrile (FEB-ril) Feverish; pertaining to a fever.

Feces (FE-sēz) Material discharged from the bowel that is made up of bacteria, secretions, and food residue; also called stool.

Fenestration (fen-es-TRA-shun) Surgical opening made into the labyrinth of the ear for some conditions of deafness.

Fetal circulation (FE-tal) The circulatory system of the fetus, including the placenta and special blood vessels involved in the exchange of materials between fetus and mother.

Fetus (FE-tus) The latter stages of the developing young of an animal. In human beings, the child in utero from the third month to birth.

Fibrillation (fi-brĭ-LA-shun) Irregular twitching of individual muscle cells (fibers) or small groups of muscle fibers preventing effective action by an organ or muscle.

Fibroblast (FI-bro-blast) A flat, long connective tissue cell that forms the fibroelastic tissues of the body.

Fibrocyte (FI-bro-sīt) A mature fibroblast, found in the fibroelastic connective tissues of the body.

Fibrosis (fi-BRO-sis) Abnormal formation of fibrous tissue.

Fibrous joint (FI-brus) A joint that allows little or no movement, such as a suture and syndesmosis.

Fibrous tunic (TU-nik) The outer coat of the eyeball, made up of the posterior sclera and the anterior cornea.

Fight-or-flight response The effect of the stimulation of the sympathetic division of the autonomic nervous system, particularly a marked increase in heart rate, blood pressure, oxygen consumption, respiration, and a feeling of tenseness; the defense-alarm reaction.

Filiform papilla (FIL-ĭ-form pah-PIL-ah) One of the many conical projections distributed in parallel rows over the anterior two-thirds of the tongue and containing no taste buds.

Filtration (fil-TRA-shun) A passive process involving the movement of solvents and dissolved substances across a semipermeable membrane by mechanical pressure, from an area of higher pressure to an area of lower pressure.

Filum terminale (FI-lum tur-min-AL-e) A thread of fibrous tissue continuous with the pia mater that extends from the terminal end of the spinal cord (conus medullaris) to the first segment of the coccyx.

Fimbriae (FIM-bre-e) A fringe of fingerlike projections surrounding the open end (infundibulum) of a uterine or fallopian tube.

Fissure (FISH-er) A groove, fold, or slit that may be normal or abnormal.

Fistula (FIS-tu-lah) An abnormal passage between two organs or between an organ cavity and the outside.

Flaccid (FLAK-sid) Relaxed, flabby, or soft; lacking muscle tone.

Flagellum (fla-JEL-um) A hairlike, motile process on the extremity of a cell; in particular, the tail of a spermatozoan that propels it along its way.

Flatfoot A condition that results if the ligaments and tendons of the arches of the foot are weakened and the height in the longitudinal arch decreases or "falls."

Flatus (FLA-tus) Gas or air in the digestive tract; commonly used to denote passage of gas rectally.

Flexion (FLEK-shun) A folding movement in which there is a decrease in the angle between two bones anteriorly except in flexion of the knee and toes, in which the bones are approximated posteriorly.

Fluoroscope (floo-O-ro-skōp) An instrument for visual observation of the body (heart, bowels) by means of x-ray.

Follicle (FOL-i-kul) A small secretory sac or cavity.

Fontanel (fon-tah-NEL) A soft spot in a baby's skull; a membrane-covered spot where bone formation has not yet occurred.

Foot The terminal part of the lower extremity.

Foramen (for-A-men) A passage or opening; a communication between two cavities of an organ, or a hole in a bone for passage of vessels or nerves.

Foramen ovale (o-VAL-e) An opening in the fetal heart in the septum between the right and left atria; a hole in the greater wing of the sphenoid bone that transmits the mandibular branch of the trigeminal nerve (V).

Forearm (FŌR-arm) The part of the upper extremity between the elbow and the wrist.

Fornix (FŌR-niks) A tract in the brain made up of association fibers, connecting the hippocampus with the mammillary bodies; a recess around the cervix of the uterus where it protrudes into the vagina.

Fossa (FOS-ah) A depressed area or a shallow depression.

Fourth ventricle (VEN-tri-kl) A cavity within the brain lying between the cerebellum and the medulla and pons.

Fovea (FŌV-e-ah) A pit or cuplike depression.

Fracture (FRAK-chūr) Any break in a bone.

Frenulum (FREN-u-lum)   A small fold of mucous membrane that connects two parts and limits movement.

Frontal or **coronal** (ko-RO-nal)   A plane that runs vertical to the ground and divides the body into anterior and posterior portions.

Fulminate (FUL-min-āte)   To occur suddenly with great intensity.

Fundus (FUN-dus)   The part of a hollow organ farthest from the opening.

Fungiform papilla (FUN-jĭ-form pah-PIL-ah)   A mushroomlike elevation on the upper surface of the tongue appearing as a red dot; most fungiform papillae contain taste buds.

Furuncle (FU-rung-kal)   A boil; painful nodule caused by bacteria. It is usually due to infection and inflammation of a hair follicle or oil gland.

Gallbladder (GAWL-blad-er)   A sac attached to the underside of the liver that serves as a storage place for bile.

Gamete (GAM-ēt)   A male or female reproductive cell; the spermatozoan or ovum.

Ganglion (GANG-glĭ-on)   A cluster of nerve cell bodies located outside the central nervous system.

Gangrene (GANG-grēn)   Death of tissue accompanied by bacterial invasion and putrefaction; usually due to blood vessel obstruction.

Gastrointestinal tract (gas''tro-in-TES-tĭ-nal)   The portion of the digestive tract from the cardia of the stomach to the anus.

Gavage (gah-VAHZH)   Feeding through a tube passed through the esophagus and into the stomach.

Gene (jēn)   One of the biological units of heredity; an ultramicroscopic, self-reproducing DNA particle located in a definite position on a particular chromosome.

Genitalia (jen-i-TA-le-ah)   Reproductive organs.

Gestation (jes-TA-shun)   The period of intrauterine fetal development.

Gingiva (jin-JIV-ah)   The fleshy structure covering a tooth-bearing border of the maxilla and the mandible; the gum.

Gingivitis (jin-je-VI-tis)   Inflammation of the gums.

Gland   A secretory structure.

Glans penis (glanz PE-nis)   The slightly enlarged region at the distal end of the penis.

Glaucoma (glaw-KO-mah)   An eye disorder in which there is increased pressure due to an excess of fluid within the eye.

Gliding joint   A synovial or diarthrotic joint having articulating surfaces that are usually flat, permitting only side-to-side and back-and-forth movements, as between carpal bones, tarsal bones, and the scapula and clavicle.

Glomerulus (glo-MER-u-lus)   A rounded mass of nerves or blood vessels, especially the microscopic tuft of capillaries that is surrounded by the Bowman's capsule of each kidney tubule.

Glottis (GLOT-is)   The air passageway between the vocal folds in the larynx.

Glucosuria (gloo-ko-SU-re-ah)   Abnormal amount of sugar in the urine.

Goblet cell   A goblet-shaped unicellular gland that secretes mucus.

Goiter (GOY-ter)   An enlargement of the thyroid gland.

Golgi complex (GOL-je)   An organelle in the cytoplasm of cells consisting of four to eight flattened channels, stacked upon one another with expanded areas at their ends.

Golgi tendon organ   A receptor found chiefly near the junction of tendons and muscles.

Gonad (GO-nad)   A term referring to the female sex glands (ovaries) and the male sex glands (testes).

Gonadocorticoids (go-NAD-o-KOR-tĭ-koyds)   Male and female sex hormones secreted by the adrenal cortex.

Gradient (GRA-de-ent)   A slope or gradation in the body; difference in concentration or electrical charges across a semipermeable membrane.

Gray commissure (KOM-ĭ-shur)   A narrow strip of gray matter connecting the two lateral gray masses within the spinal cord.

Gray matter   Areas in the central nervous system consisting of nonmyelinated nerve tissue.

Gray ramus communicans (RA-mus kom-U-nik-ans)   A short nerve containing postganglionic sympathetic fibers; the cell bodies of the fibers are in a sympathetic chain ganglion, and the nonmyelinated axons run by way of the gray ramus to a spinal nerve and then to the periphery to supply smooth muscle in blood vessels, arrector pili muscles, and sweat glands.

Greater omentum (o-MEN-tum)   A large fold in the serosa of the stomach that hangs down like an apron over the front of the intestines.

Greater vestibular glands (ves-TIB-u-lar)   A pair of glands on either side of the vaginal orifice that open by a duct into the space between the hymen and the labia minora (also called Bartholin's glands); these glands are homologous to the male Cowper's or bulbourethral glands.

Groin (groyn)   The depression between the thigh and the trunk; the inguinal region.

Groove or **sulcus** (SUL-kus)   A shallow downfold of the gray matter of the cerebral cortex; a furrow on the surface of a bone that accommodates a soft structure such as a blood vessel, nerve, or tendon.

Gross anatomy   The branch of anatomy that deals with structures that can be studied without using a microscope.

Gyrus (JĪ-rus)   One of the tortuous elevations (convolutions) of the cerebral cortex region of the brain. Plural form is gyri.

Hair   A threadlike structure produced by the specialized epidermal structure developing from a papilla sunk in the dermis.

Hand   The terminal portion of an upper extremity, including the carpus, metacarpus, and phalanges.

Hard palate (PAL-at)   The anterior portion of the roof of the mouth, formed by the maxillae and palatine bones and lined by mucous membrane.

Haustra (HOS-trah)   Pouches or sacculations of the colon.

Haversian canal (ha-VER-shan)   A circular channel running longitudinally in the center of a Haversian system of mature compact bone, containing blood and lymph vessels and nerves.

Haversian system or **osteon**   A Haversian canal with its concentrically arranged lamellae, lacunae, osteocytes, and canaliculi constituting the basic unit of structure in adult compact bone.

Head The superior part of a human being, cephalic to the neck; the superior or proximal part of a structure.

Heart A hollow muscular organ lying slightly to the left of the midline of the chest that pumps the blood through the cardiovascular system.

Heart block An arrhythmia of the heart in which the atria and ventricles contract independently because of a blocking of electrical impulses through the heart at a critical point in the conduction system

Hematology (he''mah-TOL-o-je) The study of the blood.

Hematoma (hem''-ah-TO-mah) A tumor or swelling filled with blood.

Hematopoiesis (hem''ah-to-poy-E-sis) The process by which blood cells are formed.

Hematuria (hem-at-U-re-ah) Blood in the urine.

Hemiballismus (hem''ĭ-bah-LIZ-mus) Violent muscular restlessness of half of the body, especially of the upper extremity.

Hemocytometer (hēm-o-si-TOM-e-ter) An instrument used in counting blood cells.

Hemodialysis (he''mo-di-AL-ĭ-sis) Filtering of the blood by means of an artificial device so that certain substances are removed from the blood as a result of the difference in rates of their diffusion through a semipermeable membrane while the blood is being circulated outside the body.

Hemoglobin (he''mo-GLO-bin) A red pigment in erythrocytes constituting about 33 percent of the cell volume, involved in the transport of oxygen and carbon dioxide.

Hemolysis (he-MOL-i-sis) The rupture of red blood cells with release of hemoglobin into the plasma.

Hemophilia (he-mo-FĒL-e-ah) A hereditary blood disorder where there is a deficient production of certain factors involved in blood clotting, resulting in excessive bleeding into joints, deep tissues, and elsewhere.

Hemorrhoids (HEM-o-royds) Dilated or varicosed blood vessels (usually veins) in the anal region; also called piles.

Hemostat (HE-mo-stat) An agent or instrument used to prevent the flow or escape of blood.

Hepatic duct (hē-PAT-ik) A duct that receives bile from the bile capillaries; small hepatic ducts merge to form the larger right and left hepatic ducts that unite to leave the liver as the common hepatic duct.

Hepatic portal circulation (POR-tal) The flow of blood from the digestive organs to the liver before returning to the heart.

Hernia (HER-ne-ah) The protrusion or projection of an organ or part of an organ through the wall of the cavity containing it.

Hilus (HI-lus) An area, depression, or pit where blood vessels and nerves enter or leave an organ; hilum.

Hinge joint A synovial or diarthrotic joint characterized by a convex surface of one bone that fits into a concave surface of another bone, such as the elbow, knee, ankle, and interphalangeal joints.

Histology (his-TOL-o-je) Microscopic study of the structure of tissues.

Holocrine gland (HOL-o-krin) The type of glandular secretion in which the entire secreting cell, along with its accumulated secretions, makes up the secretory product of the gland, as in the sebaceous glands.

Homogeneous (ho-mo-JEN-e-us) Having similar or the same consistency and composition throughout.

Homologous (ho-MOL-o-gus) Similar in structure and origin, but not necessarily in function.

Hordeolum (hor-DE-o-lum) Inflammation of a sebaceous gland of the eyelid; a sty.

Horizontal or **transverse** A plane that runs parallel to the ground and divides the body into superior and inferior portions.

Hormone (HŌR-mŏn) A secretion of an endocrine or ductless gland secreted directly into the blood for transport.

Horn Principal area of gray matter in the spinal cord.

Hyaluronidase (hi''al-u-RON-i-dāse) An enzyme that breaks down hyaluronic acid, increasing the permeability of connective tissues by dissolving the substances that hold body cells together.

Hydrocele (HI-dro-sēl) A fluid-containing sac or tumor. Specifically, a collection of fluid formed in the space along the spermatic cord and in the scrotum.

Hydrophobia (hi''dro-FO-be-ah) Rabies. Characterized by severe muscle spasms when attempting to drink water. An abnormal fear of water.

Hydrostatic (hi-dro-STAT-ik) Pertaining to the pressure of liquids in equilibrium and that exerted on liquids.

Hymen (HI-men) A thin fold of vascularized mucous membrane at the inferior end of the vaginal opening.

Hypercapnia (hi-per-KAP-ne-ah) An abnormal amount of carbon dioxide in the blood.

Hyperemia (hi-per-E-me-ah) An excess of blood in an area or part of the body.

Hyperextension (hi''per-ek-STEN-shun) Continuation of extension beyond the anatomic position, as in bending the head backward.

Hypermetropia (hi''per-me-TRO-pe-ah) Farsightedness; hyperopia.

Hyperplasia (hi-per-PLA-ze-ah) An abnormal increase in the number of normal cells in a tissue or organ, increasing its size.

Hypersecretion (hi''per-se-KRE-shun) Overactivity of glands resulting in excessive secretion.

Hypertension (hi''per-TEN-shun) High blood pressure.

Hypertonic (hi''per-TON-ik) Having an osmotic pressure greater than that of the solution with which it is compared.

Hypertrophy (hi-PER-tro-fe) An excessive enlargement or overgrowth of an organ or part.

Hyponychium (hi''po-NIK-ĭ-um) The thickened epidermis beneath the free distal end of the nail of a digit.

Hyposecretion (hi''po-se-KRE-shun) Underactivity of glands resulting in diminished secretion.

Hypothalamic–hypophyseal tract (hi''po-thal-AM-ik–hi''po-FIZ-e-al) A bundle of nerve processes made up of fibers that have their cell bodies in the hypothalamus but release their neurosecretions in the posterior pituitary gland or neurohypophysis.

Hypothalamus (hi''po-THAL-ā-mus) A portion of the forebrain, lying beneath the thalamus and forming the floor and part of the walls of the third ventricle.

Hypotonic (hi''po-TON-ik) Having an osmotic pressure lower than that of the solution with which it is compared.

Hypoxia (hī-POKS-ē-ah) Lack of adequate oxygen; also anoxia.

Ileocecal valve (il''e-o-SE-kal) A fold of mucous membrane that guards the opening from the ileum into the large intestine.

Ileum (IL-e-um) The distal or terminal portion of the small intestine, extending between the jejunum of the small intestine and the large intestine.

Immunity (im-U-ni-te) The state of being resistant to injury, particularly by poisons, foreign proteins, and invading parasites.

Implantation (im-plan-TA-shun) The insertion of a tissue or a part into a part of the body. The attachment of the preembryonic ball of cells (blastula) into the lining of the uterus.

Impotence (IM-po-tens) Weakness; inability to copulate; failure to maintain an erection.

Inclusion A secretion or storage area in the cytoplasm of a cell.

Incontinence (in-KON-tin-ens) Inability to retain urine, semen, or feces, through loss of sphincter control.

Incus (ING-kus) The middle of the three ossicles of the middle ear, so named because of its anvil shape; also called the anvil.

Infarction (in-FARK-shun) The presence of a localized area of necrotic tissue, produced by inadequate oxygenation of the tissue.

Inferior or **caudal** (KAW-dal) Away from the head, or toward the lower part of a structure.

Infundibulum (in''fun-DIB-u-lum) The stalklike structure that attaches the pituitary gland or hypophysis to the hypothalamus of the brain.

Insertion The manner or place of attachment of a muscle to the bone that it moves.

In situ (in SI-tu) In position.

Inspiration (in''spi-RA-shun) The drawing of air into the lungs.

Insula (IN-su-lah) A triangular area of cerebral cortex that lies deep within the lateral cerebral fissure, under the parietal, frontal, and temporal lobes, and cannot be seen in an external view of the brain; the island or isle of Reil.

Integument (in-TEG-u-ment) A covering, especially the skin.

Intercalated disc (in-TER-kah-la-ted) A transverse thickening of the sarcolemma that separates one cardiac cell from another.

Intercellular (in-ter-SEL-ū-lar) Between the cells of a structure.

Intercellular substance (in-ter-SEL-u-lar) The matrix of connective tissue that largely determines the quality of the tissue.

Intercostal nerve (in''ter-KOS-tal) A nerve supplying a muscle located between the ribs.

Internal capsule (KAP-sul) A thick sheet of white matter made up of myelinated fibers connecting various parts of the cerebral cortex and lying between the thalamus and the caudate and lentiform nuclei of the basal ganglia.

Internal or **deep** Away from the surface of the body.

Internal ear The inner ear or labyrinth, lying inside the temporal bone, containing the organs of hearing and balance.

Internal nares (NA-rēs) The two openings posterior to the nasal cavities opening into the nasopharynx; the choanae.

Interphase (IN-ter-fāz) The period during its life cycle when a cell is carrying on every life process except division; the stage between two mitotic divisions.

Interstitial cells of Leydig (in''ter-STĬ-shal . . . LI-dig) Secretory cells that secrete testosterone, located in the connective tissue between seminiferous tubules in a mature testis.

Interstitial fluid The fluid filling the microscopic spaces between the cells of tissues.

Interstitial (endogenous) growth Growth from within, as in the growth of cartilage.

Interventricular foramen (in''ter-ven-TRIK-u-lar) A narrow, oval opening through which the lateral ventricles of the brain communicate with the third ventricle; also called the foramen of Monro.

Intervertebral disc (in''ter-VER-tĕ-bral) A pad of fibrocartilage located between the bodies of two vertebrae.

Intracellular (in-tra-SEL-u-lar) Within cells.

Intramembranous ossification (in''tra-mem-BRA-nus os''sĭ-fĭ-KA-shun) The method of bone formation in which the bone is formed directly in membranous tissue.

Intrinsic (in-TRIN-sik) Of internal origin; for example, the intrinsic factor, a mucoprotein formed by the gastric mucosa that is necessary for the absorption of an extrinsic factor called vitamin B_{12}.

Intubation (in-tu-BA-shun) Insertion of a tube into the larynx through the glottis for entrance of air or to dilate a stricture.

Intussusception (in''tus-sus-SEP-shun) The infolding (invagination) of one part of the intestine within another segment.

In utero (in U-ter-o) Within the uterus.

Invagination (in-vaj-in-A-shun) The pushing of the wall of a cavity into the cavity itself.

Inversion (in-VER-zhun) The movement of the sole inward at the ankle joint.

In vitro (in VIT-ro) In a glass, as in a test tube.

In vivo (in VIV-o) In the living body.

Ipsilateral (ip-se-LAT-er-al) On the same side, affecting the same side of the body.

Irritability (ir''rĭ-tah-BIL-ĭ-te) The ability to receive and respond to stimuli.

Ischemia (is-KE-me-ah) A lack of sufficient blood to a part, because of obstruction of the circulation to it.

Islets of Langerhans (I-lets of LAHNG-er-hans) Clusters of endocrine gland cells in the pancreas that secrete insulin and glucagon.

Isotonic (i-so-TON-ik) Having equal tension or tone; the existence of equality of osmotic pressure between two different solutions or between two elements in a solution.

Isthmus (IS-mus) A narrow strip of tissue or a narrow passage connecting two larger parts; for example, the band of tissue connecting the two lobes of the thyroid gland.

Jaundice (JAWN-dis) A condition characterized by yellowness of skin, white of eyes, mucous membranes, and body fluids.

Jejunum (je-JU-num) The second portion of the small intestine, located between the duodenum and the ileum.

Joint capsule (KAP-sūl) A sleevelike, fibrous cuff lined with a synovial membrane that encases the articulating bones of a diarthrotic or synovial joint.

Joint kinesthetic receptor (kin''es-THET-ik) A proprioceptive receptor located in a joint, stimulated by joint movement.

Keratin (KER-ă-tin) A special insoluble protein found in the hair, nails, and other horny tissues of the epidermis.

Kidney (KID-ne) One of the paired reddish organs located in the lumbar region that secrete urine.

Kinesthesia (kin-es-THE-sze-ah) Ability to perceive extent, direction, or weight of movement; muscle sense.

Kinesiology (kin-e''-se-OL-o-je) The study of the movement of body parts.

Knee A hinge joint located between the thigh and the leg of the lower extremity.

Krebs cycle The citric acid cycle; a series of energy-yielding steps in the catabolism of carbohydrates.

Kupffer cells (KUP-fer) Phagocytic cells that partly line the sinusoids of the liver.

Kyphosis (ki-FO-sis) An increased curvature of the chest, giving a hunchback appearance.

Labia majora (LA-be-ah ma-JOR-ah) Two longitudinal folds of skin extending downward and backward from the mons pubis of the female.

Labia minora (min-OR-ah) Two small folds of skin lying between the labia majora of the female.

Labial frenulum (LA-be-al FREN-u-lum) A medial fold of mucous membrane between the inner surface of the lip and the gums.

Labium (LA-be-um) A lip; liplike. Plural form is labia.

Labyrinth (LAB-i-rinth) Intricate communicating passageways, especially the internal ear.

Lacrimal (LAK-rim-al) Pertaining to tears.

Lacrimal canal (LAK-rĭ-mal) A duct, one in each eyelid, commencing at the punctum at the medial margin of an eyelid and conveying the tears medially into the nasolacrimal sac.

Lacrimal glands Secretory cells located at the superior lateral portion of both orbits that secrete tears into excretory ducts that open onto the surface of the conjunctiva.

Lacrimal sac The superior expanded portion of the nasolacrimal duct that receives the tears from a lacrimal canal.

Lactation (lac-TA-shun) The period of milk release; suckling in mammals.

Lacteal (LAK-te-al) Related to milk; one of the many intestinal lymph vessels that take up fat from digested food.

Lacuna (lă-KU-nah) A small, hollow space, such as that found in bones, in which lie the osteoblasts. The plural form is lacunae.

Lamella (lah-MEL-ah) One of the concentric rings surrounding the Haversian canal in a Haversian system of mature compact bone.

Lamina (LAM-in-ah) A thin, flat layer or membrane, as the flattened part of either side of the arch of a vertebra. The plural form is laminae.

Lamina propria (PRO-pre-ah) The connective tissue layer of a mucous membrane.

Large intestine The portion of the digestive tract extending from the ileum of the small intestine to the anus, divided structurally into the cecum, colon, rectum, and anal canal.

Laryngopharynx (lah-rin''go-FAR-ingks) The inferior portion of the pharynx, extending downward from the level of the hyoid bone to empty posteriorly into the esophagus and anteriorly into the larynx.

Laryngoscope (lar-INJ-o-skōp) An instrument for examining the larynx.

Larynx (LAR-ingks) The voice box, a short passageway that connects the pharynx with the trachea.

Lateral (LAT-er-al) Farther from the midline of the body; toward the side.

Lateral apertures (AP-er-chūrz) Two of the three openings in the roof of the fourth ventricle through which cerebrospinal fluid enters the subarachnoid space of the brain and spinal cord; also known as the foramina of Luschka.

Lateral ventricle (VEN-trĭ-kl) A cavity within a cerebral hemisphere that communicates with the lateral ventricle in the other cerebral hemisphere and with the third ventricle by way of the interventricular foramen; cerebrospinal fluid flows through the ventricular system.

Leg The part of the lower extremity between the knee and the ankle.

Lens A transparent organ lying posterior to the pupil and iris of the eyeball and anterior to the vitreous humor.

Lesion (LE-zhun) Any local diseased change in tissue formation.

Lesser omentum (o-MEN-tum) A fold of the peritoneum that extends from the liver to the lesser curvature of the stomach and the commencement of the duodenum.

Lesser vestibular glands (ves-TIB-u-lar) Mucous-secreting glands that have ducts that open on either side of the urethral orifice in the vestibule of the female; also known as Skene's glands; these glands are female homologues of the male prostate gland.

Leucocyte (LU-ko-sīt) A white blood cell; also leukocyte.

Leukemia (lu-KE-me-ah) A malignant disease of the tissues in the bone marrow, spleen, and lymph nodes, characterized by an uncontrolled, greatly accelerated production of white blood cells; also known as cancer of the blood.

Leukocytosis (lu-ko-si-TO-sis) An increase in the number of white blood cells, characteristic of many infections and other disorders.

Leukopenia (lu-ko-PE-ne-ah) A decrease in the number of white blood cells below 5000 per cubic millimeter.

Leukoplakia (lu-ko-PLA-ke-ah) A disorder in which there are white patches in the mucous membranes of the tongue, gums, and cheeks.

Ligament (LIG-ah-ment) Collagenous connective tissue, with numerous collagen molecules arranged parallel to one another, that attaches muscles to bones.

Limbic system A portion of the forebrain, sometimes

termed the visceral brain, concerned with various aspects of emotion and behavior, that includes the limbic lobe (hippocampus and associated areas of gray matter plus the cingulate gyrus), certain parts of the temporal and frontal cortex, some thalamic and hypothalamic nuclei, and parts of the basal ganglia.

Line A mark, narrow ridge, stripe, or streak; often an imaginary line connecting different anatomical landmarks, as the imaginary line from the superior border of the pubic bone to the sacral promontory that separates the abdominopelvic cavity into the abdominal and pelvic subdivisions.

Lingual frenulum (LIN-gwal FREN-u-lum) A fold of mucous membrane that connects the tongue to the floor of the mouth.

Lipoma (li-PO-mah) A fatty tissue tumor, usually benign.

Lobe (lōbe) A curve or rounded projection.

Local Pertaining to or restricted to one spot or part.

Lordosis (lor-DO-sis) Abnormal anterior convexity of the lumbar spine; swayback.

Lower extremity An inferior limb, including the thigh, leg, and foot.

Lumbar (LUM-bar) The region of the back and side between the ribs and pelvis; loins.

Lumbar plexus (PLEK-sus) A network formed by the anterior branches of spinal nerves L1 through L4.

Lumen (LU-men) The space within an artery, vein, intestine, or tube.

Lung One of the two main organs of respiration, lying on either side of the heart in the thoracic cavity.

Lunula (LU-nu-lah) The moon-shaped white area at the base of a nail.

Lymph (limf) Tissue fluid confined in lymphatic vessels and flowing through the lymphatic system to be returned to the blood.

Lymph node An oval- or bean-shaped structure located along the lymphatic vessels.

Lymphangiogram (lim-FAN-je-o-gram) A film produced by roentgenography in which lymphatic vessels and lymph organs are filled with an opaque substance in order to be filmed.

Lymphatic (lim-FAT-ik) Pertaining to lymph or referring to one of the vessels that collects lymph from the tissues and carries it to the blood.

Lymphocyte (LIM-fo-sīt) An agranular white blood cell that fights infection by producing antibodies, normally constituting about 20–25 percent of the white blood cells.

Lysosome (LI-so-sōm) An organelle in the cytoplasm of a cell, enclosed in a single membrane, and containing powerful digestive enzymes.

Macrophage or **histiocyte** (MAK-ro-fāj, HIS-te-o-sīt) A large phagocytic cell of the reticuloendothelial system common in loose connective tissue.

Macula (MAK-u-lah) A discolored spot or a colored area.

Macula lutea (MAK-u-lah LU-te-ah) A yellow spot situated in the exact center of the posterior retina and containing the fovea centralis, the area of keenest vision.

Malaise (ma-LĀYZ) Discomfort, uneasiness, indisposition, often indicative of infection.

Malignancy (mah-LIG-nan-se) A cancerous growth.

Malleus (MAL-e-us) The largest of the three ossicles of the middle ear; the one attached to the inner side of the eardrum; also called the hammer.

Mammary gland (MAM-ar-e) Gland of the female that secretes milk for the nourishment of the young.

Marrow (MAR-o) Soft, spongelike material in the cavities of bone; red marrow produces blood cells; yellow marrow, formed mainly of fatty tissue, has no blood-producing function.

Mast cell A cell found in loose connective tissue along blood vessels that produces heparin, an anticoagulant; the name given to a basophil after it has left the bloodstream and entered the tissues.

Mastication (mas''tĭ-KA-shun) Chewing.

Meatus (me-A-tus) A passageway or opening, especially the external portion of a canal.

Medial (ME-de-al) Nearer the midline of the body.

Medial lemniscus (lem-NIS-kus) A flat band of myelinated nerve fibers extending through the medulla, pons, and midbrain and terminating in the thalamus on the opposite side; second-order sensory neurons in this tract transmit impulses for discriminating touch and pressure sensations.

Median (ME-de-an) A vertical plane dividing the body into right and left halves; situated in the middle.

Median aperture (AP-er-chūr) One of the three openings in the roof of the fourth ventricle through which cerebrospinal fluid enters the subarachnoid space of the brain and cord; also called the foramen of Magendie.

Mediastinum (me''de-ah-STI-num) A space between the pleurae of the lungs that extends from the sternum to the backbone.

Medulla oblongata (mĕ-DUL-lah ob''lon-GAH-tah) A continuation of the upper portion of the spinal cord forming the most inferior portion of the brain stem; it lies just above the level of the foramen magnum and extends upward to the lower portion of the pons.

Medullary (marrow) cavity (MED-u-lar-rī) The space within the diaphysis of a long bone that contains fatty yellow marrow.

Meibomian gland (mi-BO-mĭ-an) A sebaceous gland embedded in the tarsal plates of each eyelid, with a duct that opens onto the edge of the eyelid; its secretion lubricates the margins of the lids and prevents the overflow of tears.

Meissner's corpuscle (MĪS-ners KOR-pusl) An encapsulated sensory receptor found in the papillae of the skin, sensitive to light touch; makes two-point discrimination possible.

Melanin (MEL-an-in) The dark pigment found in some parts of the body, such as the skin.

Melanocyte (MEL-ah-no-sīt'') A pigmented cell located in the deepest layer of the epidermis that synthesizes melanin, a black pigment.

Melanoma (mel-an-O-mah) A tumor containing melanin, usually dark.

Membrane The combination of an epithelial layer and an underlying connective tissue layer.

Membranous labyrinth (mem-BRA-nus LAB-ĭ-rinth) The portion of the labyrinth of the inner ear that is located inside the bony labyrinth and separated from it by the perilymph; it is made up of the membranous semicircular canals, the saccule and utricle, and the cochlear duct.

Meninges (mĕ-NIN-jēz)   Three membranes covering the brain and spinal cord, called the dura mater, arachnoid, and pia mater.

Menisci (mĕ-NIS-ke)   Crescent-shaped fibrocartilages in the knee joint; also called the semilunar cartilages.

Menopause (MEN-o-pawz)   The termination of the menstrual cycles.

Merkel's disc (MER-kls)   An encapsulated, cutaneous receptor for touch attached to deeper layers of epidermal cells.

Merocrine gland (MER-o-krin)   A secretory cell that remains intact throughout the process of formation and discharge of the secretory product, as in the salivary and pancreatic glands.

Mesentery (MES-en-ter''e)   A fold of peritoneum attaching the small intestine to the posterior abdominal wall.

Mesocolon (mez''o-KO-lon)   A fold of peritoneum attaching the colon to the posterior abdominal wall.

Mesothelium (mez''o-THE-li-um)   The simple squamous epithelium that lines the serous cavities.

Mesovarium (mez''o-VAR-ĭ-um)   A short fold of peritoneum that attaches an ovary to the broad ligament of the uterus.

Metabolism (me-TAB-o-lizm)   The physical and chemical changes or processes by which living substance is maintained, producing energy for the use of the organism.

Metacarpus (met''ah-KAR-pus)   A collective term for the five bones that make up the palm of the hand.

Metaphase   The second stage in mitosis in which chromatid pairs line up on the equatorial plane of the spindle fibers.

Metastasis (mĕ-TAS-tah-sis)   The transfer of disease from one organ or part of the body to another part that is not connected with it.

Metatarsus (met''ah-TAR-sus)   A collective term for the five bones located in the foot between the tarsals and the phalanges.

Microcephalus (mi-kro-SEF-ah-lus)   A fetus or individual with an abnormally small head.

Micrometer (mi-KRO-me-ter)   One one-millionth of a meter or 1/1000 of a millimeter (1/25,000 of an inch); formerly a micron.

Microtubules   Small protein tubules that afford cell shape and stiffness and form cilia, flagella, and spindle fibers.

Microvilli (mi''kro-VIL-e)   Microscopic fingerlike projections of the cell membranes of certain cells; the brush border of light microscopy.

Micturition (mik-tu-RĬ-shun)   The act of expelling urine from the urinary bladder; urination or voiding.

Midbrain   The part of the brain between the pons and the forebrain.

Middle ear   A small, epithelial-lined cavity hollowed out of the temporal bone, separated from the external ear by the eardrum and from the internal ear by a thin bony partition containing the oval and round windows; extending across the middle ear are the three auditory ossicles. Also called the tympanic cavity.

Midsagittal (mid-SAJ-ĭ-tal)   A plane through the midline of the body, running vertical to the ground and dividing the body into equal right and left sides; midline or median.

Mitochondrion (mi''to-KON-drĭ-on)   The powerhouse of the cell; a double-membraned organelle that plays a central role in the production of ATP.

Mitosis (mi-TO-sis)   The orderly division of the nucleus of a cell that ensures that each new daughter nucleus has the same number and kind of chromosomes as the original parent nucleus; the process of mitosis includes the replication of chromosomes and the distribution of the two sets of chromosomes into two separate and equal nuclei.

Mittelschmerz (MIT-el-shmerz)   Pain in the abdominopelvic region that supposedly indicates the release of an egg from the ovary.

Modiolus (mo-DI-o-lus)   The central pillar or column of the cochlea.

Monocyte (MON-o-sīt)   The larger of the agranular white blood cells that combats inflammation and infection by phagocytosis; normally constitutes about 3–8 percent of the white blood cells.

Mons pubis (monz PU-bis)   The rounded, fatty prominence over the symphysis pubis, covered by coarse pubic hair.

Morbid (MOR-bid)   Diseased; pertaining to disease.

Morphology   See Anatomy.

Motor area   The region of the cerebral cortex that governs muscular movement, particularly the precentral gyrus of the frontal lobe.

Motor unit   A motor neuron, together with the muscle cells it stimulates.

Mucin (MU-sin)   A protein found in mucus.

Mucosa (mu-KO-sah)   A mucous membrane.

Mucous cell (MU-kus)   A unicellular gland that secretes mucus; also called a goblet cell.

Mucous membrane   A membrane that lines a body cavity that opens to the exterior; also called the mucosa.

Mucus (MU-kus)   The thick fluid secretion of the mucous glands and mucous membranes.

Muscle   Unless otherwise specified, the term "muscle" implies a skeletal muscle, an organ specialized for contraction, composed of striated muscle cells supported by connective tissue, attached to a bone by a tendon or an aponeurosis, and stimulated by a somatic efferent neuron.

Muscularis (mus''ku-LA-ris)   A muscular layer or tunic of an organ.

Muscularis mucosa (mu-KO-sah)   A thin layer of smooth muscle cells located in the outermost layer of the mucosa of the alimentary canal, underlying the lamina propria of the mucosa.

Muscular tissue   Tissue responsible for movement through contraction.

Myelin sheath (MI-ĕ-lin)   A white, lipid, segmented covering, formed by Schwann cells, around the axons and dendrites of many peripheral neurons.

Myocardium (mi''o-KAR-dĭ-um)   The layer of the heart made up of cardiac muscle that comprises the bulk of the heart and lies between the epicardium and the endocardium.

Myofibril (mi''o-FI-bril)   A threadlike structure, running longitudinally through a muscle cell, consisting mainly of molecules of the protein actin and the protein myosin.

Myology (mi-OL-o-je)   The science or study of the muscles and their parts.

Myometrium (mi"o-ME-trĭ-um)   The smooth muscle coat or tunic of the uterus.

Myopia (mi-O-pe-ha)   Defect in vision so that objects can be seen distinctly only when very close to the eyes; nearsightedness.

Myosin (MI-o-sin)   The protein that makes up the thick filaments of myofibrils of muscle cells.

Nail   A horny plate, composed largely of keratin, that develops from the epidermis of the skin to form a protective covering on the dorsal surface of the distal phalanges of the fingers and toes.

Nail matrix (MA-triks)   The part of the nail beneath the body and root from which the nail is produced.

Narcosis (nar-KO-sis)   Unconscious state due to narcotics.

Nasal cavity (NA-zal)   A mucosa-lined cavity on either side of the nasal septum that opens onto the face at an external nostril or naris and into the nasopharynx at an internal nostril or naris.

Nasal septum (SEP-tum)   A vertical partition composed of bone and cartilage, covered with a mucous membrane, separating the right and left nasal cavities.

Nasolacrimal duct (na"zo-LAK-rĭ-mal)   A canal that transports the lacrimal secretion from the nasolacrimal sac into the nose.

Nasopharynx (na"zo-FAR-ingks)   The uppermost portion of the pharynx, lying posterior to the nose and extending down to the soft palate.

Nebulization (ne"būl-i-ZA-shun)   Treatment with spray method.

Neck   The part of the body connecting the head and the trunk, or a constricted portion of an organ such as the neck of a femur or the neck of the uterus.

Necrosis (ne-KRO-sis)   Death of a cell or group of cells as a result of disease or injury.

Neonatal (ne-o-NA-tal)   Pertaining to the first four weeks after birth.

Neoplasm (NE-o-plazm)   A mass of new, abnormal tissue; a tumor.

Nephritis (ne-FRI-tis)   Kidney inflammation.

Nephron (NEF-ron)   The physiological unit of a kidney, made up of a glomerular capsule, proximal convoluted tubule, descending limb of Henle, loop of Henle, ascending limb of Henle, and distal convoluted tubule.

Nerve   A cordlike bundle of nerve fibers and their associated connective tissue coursing together outside the central nervous system.

Nerve impulse   An action potential or a wave of negativity that sweeps along the outside of the membrane of a neuron.

Nervous tissue   Tissue that initiates and transmits nerve impulses to coordinate homeostasis.

Neurilemma (nu"rĭ-LEM-ah)   A delicate sheath around a nerve fiber composed of the cytoplasm and nucleus and enclosing cell membrane of a Schwann cell; there may or may not be a layer of myelin between the neurilemma and the axis cylinder (axon or dendrite) of the neuron.

Neuroeffector junction (nu"ro-ĕ-FEK-tor)   A synapse between an autonomic fiber and a visceral effector.

Neurofibril (nu"ro-FI-bril)   One of the delicate threads that form a complicated network in the cytoplasm of the cell body and processes of a neuron.

Neuroglia (nu-ro-GLE-ah)   Cells of the nervous system that are specialized to perform the functions of connective tissue; the neuroglia of the central nervous system are the astrocytes, oligodendrocytes, and microglia; neuroglia of the peripheral nervous system include the Schwann cells and the ganglion satellite cells.

Neurohypophysis (nu"ro-hi-POF-ĭ-sis)   The posterior lobe of the pituitary gland.

Neuromuscular junction (nu"ro-MUS-ku-lar)   The area of contact or synapse between a neuron and a muscle fiber.

Neuromuscular spindle   An encapsulated receptor in a skeletal muscle, consisting of specialized muscle cells and nerve endings, stimulated by change in length or tension of muscle cells; a proprioceptor.

Neuron (NU-ron)   A nerve cell, consisting of a cell body, dendrites, and an axon.

Neurosyphilis (nu"ro-SIF-ĭ-lis)   Syphilis of the nervous system; for example, tabes dorsalis.

Neutrophil (NU-tro-fil)   A granular, polymorphonuclear white blood cell, actively phagocytic; normally constituting about 60–70 percent of the white blood cells.

Nipple   A round or cone-shaped projection at the tip of the breast, or any similarly shaped structure.

Nissl bodies (NIS-l)   Rough endoplasmic reticulum in the cell bodies of neurons.

Node of Ranvier (ron-ve-A)   A space, along a myelinated nerve fiber, between the individual Schwann cells that form the myelin sheath and the neurilemma.

Nucleus (NU-kle-us)   A spherical or oval organelle of a cell that contains the hereditary factors of the cell, called genes; a cluster of nerve cell bodies in the central nervous system.

Nucleus pulposus (pul-PO-sus)   A soft, pulpy, highly elastic substance in the center of an intervertebral disc, a remnant of the notochord.

Nystagmus (nis-TAG-mus)   Constant, involuntary, rhythmic movement of the eyeballs; horizontal, rotary, or vertical.

Occlusion (o-KLU-zhun)   The act of closure or state of being closed.

Odontoid (o-DON-toyd)   Toothlike.

Olfactory (ōl-FAK-to-re)   Pertaining to smell.

Olfactory bulb (ōl-FAK-to-re)   A mass of gray matter at the termination of an olfactory nerve, lying beneath the frontal lobe of the cerebrum on either side of the crista galli of the ethmoid bone.

Olfactory cell   A bipolar neuron with its cell body lying between supporting cells located in the mucous membrane lining the upper portion of each nasal cavity.

Olfactory tract   A bundle of axons that extends from the olfactory bulb posteriorly to the olfactory portion of the cortex.

Oligospermia (ol-i-go-SPER-me-ah)   A deficiency of spermatozoa in the semen.

Olive   A prominent oval mass on each lateral surface of the superior part of the medulla.

Oogenesis (o-o-JEN-e-sis)   Formation and development of the ovum.

Ophthalmic (of-THAL-mik) Pertaining to the eye.

Optic chiasma (OP-tik ki-AZ-mah) A crossing point of the optic nerves, anterior to the pituitary gland.

Optic disc A small area of the retina containing openings through which the fibers of the ganglion neurons emerge as the optic nerve; the blind spot.

Optic tract A bundle of axons that transmits impulses from the retina of the eye between the optic chiasma and the thalamus.

Ora serrata (O-rah ser-AH-tah) The jagged, anterior margin of the retina, near the ciliary body, where the nervous portion of the retina ends.

Orbit (OR-bit) The bony pyramid-shaped cavity of the skull that holds the eyeball.

Organ A group of two or more different kinds of tissue that performs a particular function.

Organ of Corti (KOR-te) The spiral organ or organ of hearing consisting of supporting cells and hair cells that rest on the basilar membrane and extend into the endolymph of the cochlear duct; the receptor for hearing.

Organelle (or-gan-EL) A tiny, specific particle of living material present in most cells and serving a specific function.

Organism (OR-gah-nizm) A total living form; one individual.

Orgasm (OR-gazm) A state of highly emotional excitement; especially that which occurs at the climax of sexual intercourse.

Orifice (OR-i-fis) Any aperture or opening.

Origin (OR-ĭ-jin) The place of attachment of a muscle to the more stationary bone, or the end opposite the insertion.

Oropharynx (or''o-FAR-ingks) The second portion of the pharynx, lying posterior to the mouth and extending from the soft palate down to the hyoid bone.

Osmosis (os-MO-sis) The net movement of water molecules through a semipermeable membrane from an area of high water concentration to an area of lower water concentration.

Osseous tissue (OS-se-us) Bone tissue.

Ossicle (OS-i-kul) Any small bone; as the three tiny bones of the ear.

Ossification (os''ĭ-fĭ-KA-shun) Formation of bone substance.

Osteoblast (OS-te-o-blast'') A bone-forming cell.

Osteoclast (OS-te-o-clast'') A large, multinuclear cell that destroys or resorbs bone tissue.

Osteocyte (OS-te-o-sīt'') A mature osteoblast that has lost its ability to produce new bone tissue.

Osteology (os''te-OL-o-je) The study of bones.

Osteomalacia (os''te-o-mah-LA-she-ah) Demineralization of bones due to vitamin D deficiency.

Osteomyelitis (os''te-o-mi-ĭ-LI-tis) Inflammation of bone marrow, or of the bone and the marrow.

Osteoporosis (os''te-o-pōr-O-sis) Increased porosity of bone.

Ostium (OS-te-um) Any small opening; especially entrance into a hollow organ or canal.

Otic (O-tic) Pertaining to the ear.

Otic ganglion (O-tik GANG-glĭ-on) A ganglion of the parasympathetic subdivision of the autonomic nervous system, made up of cell bodies of postganglionic parasympathetics to the parotid gland.

Otolith (O-to-lith) A particle of calcium carbonate embedded in the gelatinous layer that coats the hair cells of the sensory receptor (the macula) in the saccule and utricle of the inner ear.

Oval window A small opening between the middle and inner ear into which the footplate of the stapes fits; also called the fenestra vestibuli.

Ovarian follicle (o-VAR-ĭ-an FOL-ĭ-kl) A general name for an ovum in any stage of development, along with its surrounding group of epithelial cells.

Ovarian ligament (LIG-ah-ment) A rounded cord of connective tissue that attaches the ovary to the uterus.

Ovary (O-var-e) The female gonad in which the ova are formed.

Ovulation (o-vu-LA-shun) The discharge of a mature egg cell (ovum) from the follicle of the ovary.

Ovum (O-vum) The female reproductive or germ cell; an egg cell.

Pacinian corpuscle (pă-SIN-ĭ-an KOR-pusl) An encapsulated pressure receptor found in the deep subcutaneous tissues, in tissues that lie under mucous membranes, in serous membranes of the abdominal cavity, around joints and tendons, and in some viscera.

Paget's disease (PAJ-ets) A disorder characterized by an irregular thickening and softening of the bones; the cause is unknown, but the bone-producing osteoblasts and the bone-destroying osteoclasts apparently become uncoordinated.

Palate (PAL-at) The horizontal structure separating the mouth and the nasal cavity; the roof of the mouth.

Palliative (PAL-e-ah-tiv) Serving to relieve or alleviate without curing.

Palpate (PAL-pāt) To examine by touch; to feel.

Palpebra (PAL-pĕ-brah) An eyelid.

Pancreas (PAN-kre-as) A soft, oblong-shaped organ lying along the greater curvature of the stomach and connected by a duct to the duodenum; it is both exocrine (secreting pancreatic juice) and endocrine (secreting insulin and glucagon).

Pancreatic duct (pan''kre-AT-ik) A single large tube that drains pancreatic juice into the duodenum at the ampulla of Vater; also called the duct of Wirsung; sometimes an accessory duct, the duct of Santorini, is also present and empties into the duodenum above the ampulla of Vater.

Papillae (pah-PIL-e) Small projections or elevations.

Papillary muscle (PAP-ĭ-lar-e) Muscular columns located on the inner surface of the ventricles from which the chordae tendineae extend to attach to the cusps of the atrioventricular valves.

Paranasal sinus (par''ah-NA-zal SI-nus) A mucous-lined air cavity in a skull bone that communicates with the nasal cavity; paranasal sinuses are located in the frontal, maxillary, ethmoid, and sphenoid bones.

Parasympathetic (par''ah-sim-pah-THET-ik) One of the two subdivisions of the autonomic nervous system, having cell bodies of preganglionic neurons in nuclei in the brain stem and in the lateral gray matter of the sacral portion of the spinal cord and concerned with activities that restore and conserve body energy; the craniosacral division.

Parathyroids (par''ah-THI-roydz) The four small endocrine glands embedded on the posterior surfaces of the lateral lobes of the thyroid; their secretory product is parathormone, or parathyroid hormone.

Parenchyma (par-EN-ki-mah) The essential parts of any organ concerned with its function.

Parenteral (par-EN-ter-al) Situated or occurring outside the intestines; as by a subcutaneous method.

Paries (PA-rēs) The enveloping wall of any structure, especially hollow organs.

Parietal (pah-RI-ĕ-tal) Pertaining to the walls of a cavity.

Parietal cell One of the three kinds of secreting cells of the gastric glands, the one that produces hydrochloric acid.

Parietal pleura (PLU-rah) The outer layer of the serous pleural membrane that encloses and protects the lung, the layer that is attached to the walls of the pleural cavity.

Parotid gland (pah-ROT-id) One of the paired salivary glands located inferior and anterior to the ears connected to the oral cavity via a duct that opens into the inside of the cheek opposite the upper second molar tooth.

Paroxysm (PAR-oks-sizm) A sudden periodic attack or recurrence of symptoms of a disease.

Parturition (par-tu-RISH-un) Act of giving birth to young; childbirth, delivery.

Pathogenesis (path''-o-JEN-e-sis) The development of disease or a morbid or pathological state.

Pathological (path''o-LOJ-ĭ-kal) Pertaining to or caused by disease.

Pathological anatomy The study of structural changes caused by disease.

Pectoral (PEK-to-ral) Pertaining to the chest or breast.

Pedicle (PED-ĭ-kl) The part of the vertebral or neural arch of a vertebra that connects the body with a lamina.

Pelvic splanchnic nerves (PEL-vik SPLANGK-nik) Preganglionic parasympathetic fibers from the levels of S2, S3, and S4 that supply the urinary bladder, reproductive organs, and the descending and sigmoid colon and rectum.

Pelvis The basinlike structure formed by the two pelvic bones, the sacrum, and the coccyx; the expanded, proximal portion of the ureter, lying within the kidney and into which the major calyces open.

Penis (PE-nis) The male copulary organ, used to introduce spermatozoa into the female vagina.

Pericardium (per''ĭ-KAR-dĭ-um) A loose-fitting serous membrane that encloses the heart, consisting of an outer fibrous layer and an internal serous layer containing a parietal layer that lines the inside of the fibrous pericardium, and a visceral layer, also known as the epicardium.

Perichondrium (per''ĭ-KON-drĭ-um) A connective tissue that covers cartilage.

Perikaryon (per''ĭ-KAR-ĭ-on) The cell body of a neuron.

Perimysium (per''ĭ-MIZ-ĭ-um) Connective tissue around a bundle or fascicle of muscle cells.

Perineum (per''ĭ-NE-um) The pelvic floor; the space between the anus and the scrotum in the male, and the anus and the vulva in the female.

Perineurium (per''ĭ-NU-rĭ-um) Connective tissue around a bundle or fascicle of nerve fibers.

Periodontal membrane (per''ĭ-o-DON-tal) The periosteum lining the sockets for the teeth in the alveolar processes of the mandible and maxillae.

Periosteum (per''ĭ-OS-te-um) A dense, white, fibrous membrane covering the surface of a bone, except at the articular surface.

Peripheral nervous system (per-IF-er-al) The part of the nervous system that lies outside the central nervous system—nerves and ganglia.

Periphery (pe-RIF-er-e) Outer part or surface of the body; part away from the center.

Peristalsis (per''ĭ-STAL-sis) A wave of contraction along a hollow muscular tube, followed by relaxation.

Peritoneum (per''ĭ-to-NE-um) The serous membrane lining the abdominal cavity and covering the abdominal organs and some pelvic organs.

Peritonitis (per''ĭ-to-NI-tis) Inflammation of the peritoneum, the membranous coat lining the abdominal cavity and covering the viscera.

Peyer's patches (PI-urs) Clusters of lymph nodules on the walls of the ileum.

pH The symbol commonly used in expressing hydrogen ion concentration.

Phagocytosis (fag''o-si-TO-sis) Literally, "cell eating"; the engulfing of solid particles by living cells.

Phalanges (fah-LAN-jēz) Bones of a finger or toe.

Pharynx (FAR-ingks) The throat, a tube that starts at the internal nares and runs partway down the neck where it opens into the esophagus posteriorly and into the larynx anteriorly.

Phlebotomy (fleb-OT-o-me) The cutting of a vein to allow the escape of blood.

Physiology (fiz''e-OL-o-je) The study of the functions of body parts.

Pia mater (PE-a MA-ter) Literally, "tender mother"; the innermost meninx, the transparent layer of connective tissue that adheres to the surfaces of the brain and cord and contains blood vessels.

Pilonidal (pi-lo-NI-dal) Containing hairs resembling a tuft inside a cyst or sinus.

Pineal gland (PIN-e-al) The cone-shaped gland located in the roof of the third ventricle; also called the epiphysis cerebri.

Pinna (PIN-nah) The projecting part of the external ear; the auricle.

Pinocytosis (pin''o-si-TO-sis) Literally, "cell drinking"; the engulfing of liquid material by living cells.

Pituitary gland (pit-U-ĭ-tar-e) The hypophysis, nicknamed the "master gland," a small endocrine gland lying in the sella turcica of the sphenoid bone and attached to the hypothalamus by the infundibulum.

Pivot joint (PIV-ot) A synovial or diarthrotic joint in which a rounded, pointed, or conical surface of one bone articulates with a shallow depression of another bone, as in the joint between the atlas and axis and between the proximal ends of the radius and ulna.

Plantar flexion (PLAN-tar FLEK-shun) Extension of the foot at the ankle joint.

Plasma (PLAZ-mah)   The extracellular fluid found in blood vessels.

Plasma cell   A cell of loose connective tissue that gives rise to antibodies; the name given to a lymphocyte after it leaves the circulatory system and becomes a connective tissue cell.

Pleura (PLUR-ah)   The serous membrane that enfolds the lungs and lines the walls of the chest and diaphragm.

Plexus (PLEK-sus)   A network of nerves, veins, or lymphatic vessels.

Plexus of Auerbach (OW-ur-bok)   A network of nerve fibers from both autonomic divisions located in the muscularis coat or tunic of the small intestine; also called the myenteric plexus.

Plexus of Meissner (MĪS-ner)   A network of autonomic nerve fibers located in the outer portion of the submucous layer or tunic of the small intestine.

Plica circularis (PLI-ca sur-ku-LAR-is)   A permanent, deep, transverse fold in the mucosa and submucosa of the small intestine that increases the surface area for absorption.

Pons varolii (ponz var-O-le-i)   The portion of the brainstem that forms a ''bridge'' between the medulla and the midbrain, anterior to the cerebellum.

Posterior or **dorsal** (pos-TĔR-ĭ-or)   Nearer to or at the back of the body.

Posterior root   The structure composed of afferent (sensory) fibers lying between a spinal nerve and the dorsolateral aspect of the spinal cord; also called the dorsal root or sensory root.

Posterior root ganglion   A group of cell bodies of sensory (afferent) neurons and their supporting cells located along the posterior root of a spinal nerve; also called a dorsal or sensory root ganglion.

Postganglionic neuron (pōst''gang-glī-ON-ik NU-ron)   The second visceral efferent neuron in an autonomic pathway, having its cell body and dendrites located in an autonomic ganglion and its unmyelinated axon ending at cardiac muscle, smooth muscle, or a gland.

Postpartum (pōst-PAR-tum)   After parturition; occurring after the delivery of a baby.

Preganglionic neuron (pre''gang-glī-ON-ik)   The first visceral efferent neuron in an autonomic pathway, with its cell body and dendrites in the brain or spinal cord and its myelinated axon ending at an autonomic ganglion where it synapses with a postganglionic neuron.

Premonitory (pre-MON-i-to-re)   Giving previous warning; as premonitory symptoms.

Prepuce (PRE-pus)   The foreskin; a fold of loosely fitting skin covering the glans of the penis or clitoris.

Presbyopia (prez-be-O-pe-ah)   Defect of vision due to advancing age; loss of elasticity of the lens of the eye.

Pressor (PRES-or)   Stimulating the activity of a function, especially vasomotor, usually accompanied by an increase in blood pressure.

Prevertebral ganglion (pre-VER-tĕ-bral GANG-glī-on)   A cluster of cell bodies of postganglionic sympathetic neurons lying anterior to the vertebral column and close to the large abdominal artery from which its name is derived; also called a collateral ganglion; examples are the superior mesenteric, inferior mesenteric, and celiac ganglia.

Primordial (pri-MŌR-de-al)   Existing first; especially primordial egg cells in the ovary.

Proctology (prok-TOL-o-je)   The branch of medicine that treats the rectum and its disorders.

Progeny (PROJ-e-ne)   Refers to offspring or descendants.

Prognosis (prog-NO-sis)   A forecast of the probable results of a disorder; the outlook for recovery.

Prolapse (PRO-laps)   A dropping or falling down of an organ, especially the uterus or rectum.

Proliferation (pro-lif''er-A-shun)   Rapid and repeated reproduction of new parts, especially cells.

Pronation (pro-NA-shun)   A movement of the flexed forearm in which the palm of the hand is turned backward (posteriorly).

Prophase (PRO-fāz)   The first stage in mitosis.

Proprioceptor (pro''pre-o-SEP-tor)   A receptor located in a muscle, tendon, or joint that provides information about body position and movements.

Prostate gland (PROS-tāt)   A muscular and glandular doughnut-shaped organ inferior to the urinary bladder that surrounds the upper portion of the male urethra.

Prosthesis (PROS-the-sis)   Replacement of a missing part by an artificial substitute.

Protraction (pro-TRAK-shun)   The movement of the mandible or clavicle forward on a plane parallel with the ground.

Protuberance (pro-TU-ber-ans)   A part that is prominent beyond a surface, like a knob.

Proximal (PROK-sĭ-mal)   Nearer the attachment of a limb to the trunk.

Psychosomatic (si-ko-so-MĂ-tik)   Pertaining to the relationship between mind and body.

Pterygopalatine ganglion (ter''ĭ-go-PAL-ah-tĭn GANG-glī-on)   A cluster of cell bodies of parasympathetic postganglionic neurons ending at the lacrimal and nasal glands.

Ptosis (TO-sis)   A condition of dropping or downward displacement of an organ or body structure, for example, the kidney or the upper eyelid.

Puberty (PU-ber-te)   Period of life at which the reproductive organs become functional.

Pulmonary (PUL-mōn-ary)   Concerning or affected by the lungs.

Pulmonary circulation   The flow of deoxygenated blood from the right ventricle to the lungs and the return of oxygenated blood from the lungs to the left atrium.

Pulp cavity   A cavity within the crown and neck of a tooth, filled with pulp, a connective tissue containing blood vessels, nerves, and lymphatics.

Pulse (puls)   Throbbing caused by the regular recoil and alternate expansion of an artery; the periodic thrust felt over arteries in time with the heartbeat.

Pupil   The black hole in the center of the iris, the area through which light enters the posterior cavity of the eyeball.

Purkinje fibers (pur-KIN-je)   Muscle fibers in the subendocardial tissue of the heart specialized for conducting an impulse to the myocardium; part of the conduction system of the heart.

Pus The liquid product of inflammation containing leucocytes, or their remains, and debris of dead cells.

Pyloric sphincter (pi-LOR-ik SFINGK-ter) A thickened ring of smooth muscle through which the pylorus of the stomach communicates with the duodenum; also called the pyloric valve.

Pyorrhea (pi-o-RE-ah) A discharge or flow of pus, especially from the tooth sockets and the tissues of the gums.

Pyramids (PIR-ah-midz) Pointed or cone-shaped structures; two roughly triangular structures on the ventral side of the medulla composed of the largest motor tracts that run from the cerebral cortex to the spinal cord; triangular-shaped structures in the renal medulla composed of the straight segments of renal tubules.

Pyrexia (pi-REK-se-ah) A condition in which the temperature is above normal.

Radiographic anatomy Diagnostic branch of anatomy that includes the use of x-rays.

Ramus (RA-mus) A branch, especially a nerve or blood vessel. Plural form is rami.

Raphe (RA-fe) A seam or ridge.

Receptor (re-SEP-tor) A sense organ that receives stimuli from the environment; the distal end of a dendrite of a sensory neuron.

Rectouterine pouch (rek''to-U-ter-in) A pocket formed by the parietal peritoneum as it moves posteriorly from the surface of the uterus and is reflected onto the rectum; the lowest point in the pelvic cavity; also called the pouch or cul de sac of Douglas.

Rectum (REK-tum) The last 20 cm of the gastrointestinal tract, from the sigmoid colon to the anus.

Recumbent (re-KUM-bent) Lying down.

Red nucleus A cluster of cell bodies in the midbrain, occupying a large portion of the tegmentum, and sending fibers into the rubroreticular and rubrospinal tracts.

Reflex arc The most basic conduction pathway through the nervous system, connecting a receptor and an effector and consisting of a receptor, a sensory neuron, a center in the central nervous system for a synapse, a motor neuron, and an effector.

Regimen (REJ-i-men) A strictly regulated scheme of diet, exercise, or activity designed to achieve certain ends.

Regional anatomy The division of anatomy dealing with a specific region of the body, such as the head, neck, chest, or abdomen.

Regurgitation (re-gur-ji-TA-shun) Return of solids or fluids to the mouth from the stomach; flowing backward of blood through incompletely closed heart valves.

Relapse (re-LAPS) The return of a disease weeks or months after its apparent cessation.

Renal (RE-nal) Pertaining to the kidney.

Renal pelvis (RE-nal) A cavity in the center of the kidney formed by the expanded, proximal portion of the ureter, lying within the kidney, and into which the major calyces open.

Renal pyramid A triangular-shaped structure in the renal medulla composed of the straight segments of renal tubules.

Respiration (res-pi-RA-shun) The exchange of gases between the atmosphere and the cells.

Resuscitation (re-sus-i-TA-shun) Act of bringing a person back to full consciousness.

Rete testis (RE-te TES-tis) A network of ducts in the center of the testis formed by the straight tubules.

Reticular formation (re-TIK-u-lar) A network of small groups of nerve cells scattered among bundles of fibers beginning in the medulla as a continuation of the spinal cord and extending upward through the central part of the brain stem.

Reticulin (re-TIK-u-lin) A fiber of small diameter in the matrix of connective tissue that branches to form a netlike supporting framework around fat cells, nerve fibers, muscle cells, and blood vessels; also called a reticular fiber.

Reticulum (ri-TIK-u-lum) A network.

Retina (RET-i-nah) The inner coat of the eyeball, lying only in the posterior portion of the eye and consisting of nervous tissue and a pigmented layer comprised of epithelial cells lying in contact with the choroid.

Retraction (re-TRAK-shun) The movement of a protracted part of the body backward on a plane parallel to the ground, as in pulling the lower jaw back in line with the upper jaw.

Retroperitoneal (ret''ro-per-i-to-NE-al) Behind the peritoneum.

Retroversion (re-tro-VER-zhun) A turning backward of an entire organ, especially the uterus.

Rhodopsin (ro-DOP-sin) A photo-sensitive, reddish-purple pigment in rod cells of the retina, consisting of the protein scotopsin plus retinene; visual purple.

Ribosomes (RI-bo-sōms) Organelles in the cytoplasm of cells, composed of ribosomal RNA, that synthesize proteins; nicknamed the "protein factories."

Rickets (RIK-ets) A disease of metabolism affecting children, caused by a vitamin D deficiency in which normal ossification does not take place, often resulting in deformities.

Right lymphatic duct (lim-FAT-ik) A vessel of the lymphatic system that drains lymph from the upper right side of the body and empties it into the right subclavian vein.

Rod A visual receptor in the retina of the eye that is specialized for vision in dim light.

Roentgen (RENT-gen) The international unit of radiation; a standard quantity of x or gamma radiation.

Roentgenogram (RENT-gen-o-gram'') A film exposed to x-rays.

Root canal A narrow extension of the pulp cavity lying within the root of a tooth.

Root hair plexus (PLEK-sus) A network of dendrites arranged around the root of a hair as free or naked nerve endings that are stimulated when a hair shaft is moved.

Rotation (ro-TA-shun) Moving a bone around its own axis, with no other movement permitted.

Round ligament (LIG-ah-ment) A band of fibrous connective tissue enclosed between the folds of the broad ligament of the uterus; it emerges from a point on the uterus just below the uterine tube, extends laterally along the pelvic wall, and penetrates the abdominal wall through the deep inguinal ring to end in the labia majora.

Round window A small opening between the middle and inner ear, directly below the oval window, covered by the second tympanic membrane; also called the fenestra cochleae.

Rugae (RU-je) Temporary large folds that appear in the mucosa of an empty hollow organ that gradually smooth out and disappear as the organ is distended; for example, in the mucosa of the stomach and vagina.

Saccule (SAK-ūl) The lower and smaller of the two chambers in the membranous labyrinth inside the vestibule of the inner ear containing a receptor organ for equilibrium.

Sacral (SA-kral) Pertaining to the sacrum.

Sacral plexus (PLEK-sus) A network formed by the anterior branches of spinal nerves L4 through S3.

Sacral promontory (PROM-on-tor-e) The superior surface of the body of the first sacral vertebra that projects anteriorly into the pelvic cavity; a line from the sacral promontory to the superior border of the symphysis pubis divides the abdominal and pelvic cavities.

Saddle joint A synovial or diarthrotic joint in which the articular surfaces of both of the bones are saddle-shaped or concave in one direction and convex in the other direction, as in the joint between the trapezium and the metacarpal of the thumb.

Sagittal (SAJ-ĭ-tal) A plane that runs vertically to the ground and divides the body into unequal left and right portions; also called parasagittal.

Salivary gland (SAL-ĭ-ver-e) The three pairs of glands that lie outside the mouth and pour their secretory product (called saliva) into ducts that empty into the oral cavity; the parotid, submandibular, and sublingual glands.

Salpingitis (sal-pin-JI-tis) Inflammation of the uterine (fallopian) tube or of the auditory (eustachian) tube.

Sarcolemma (sar″ko-LEM-mah) The cell membrane of a muscle cell, especially of a skeletal muscle cell.

Sarcoma (sar-KO-mah) A connective tissue tumor, often highly malignant.

Sarcomere (SAR-ko-mēr) A contractile unit in a striated muscle cell extending from one Z-line to the next Z-line.

Sarcoplasm (SAR-ko-plazm) The cytoplasm of a striated muscle cell.

Scala tympani (SKA-lah TIM-pan-e) The lower spiral-shaped channel of the bony cochlea, filled with perilymph.

Scala vestibuli (ves-TIB-u-le) The upper spiral-shaped channel of the bony cochlea, filled with perilymph.

Schwann cell (shwon) A glial cell of the peripheral nervous system that forms the myelin sheath and neurilemma of a nerve fiber by spiraling around a nerve fiber in a jelly-roll fashion.

Sciatica (si-AT-ik-ah) Inflammation and pain along the sciatic nerve; felt at the back of the thigh running down the inside of the leg.

Sclera (SKLE-rah) The white coat of fibrous tissue that forms the outer protective covering over the eyeball except in the area of the anterior cornea; the posterior portion of the fibrous tunic.

Sclerosis (skle-RO-sis) A hardening with loss of elasticity of the tissues.

Scoliosis (sko″le-O-sis) An abnormal curvature sideways (laterally) from the normal vertical line of the spine.

Scrotum (SKRO-tum) A skin-covered pouch that contains the testes and their accessory organs.

Sebaceous gland (se-BA-shus) An exocrine gland in the dermis of the skin, almost always associated with a hair follicle, that secretes sebum.

Sebum (SE-bum) A fatty secretion of the sebaceous glands of the skin.

Segmental bronchi (seg-MEN-tal BRONG-ki) Branches of the secondary bronchi that lead into the segmental subdivisions of the lobes of the lungs.

Sella turcica (SEL-ah TUR-si-kah) A saddlelike depression on the middle upper surface of the sphenoid bone enclosing the pituitary gland.

Semen (SE-men) A fluid discharge at ejaculation by a male that consists of a mixture of spermatozoa and the secretions of the seminal vesicles, the prostate gland, and the bulbourethral glands.

Semicircular canals Three bony channels projecting upward and posteriorly from the vestibule of the inner ear, filled with perilymph, in which lie the membranous semicircular canals filled with endolymph; they contain receptors for equilibrium.

Semicircular ducts The membranous semicircular canals filled with endolymph and floating in the perilymph of the bony semicircular canals.

Semilunar valve (sem″i-LU-nar) A valve guarding the entrance into the aorta or the pulmonary trunk from a ventricle of the heart.

Seminal vesicles (SEM-ĭ-nal VES-ĭ-klz) A pair of convoluted pouchlike structures lying posterior and inferior to the urinary bladder, anterior to the rectum, that secrete their component of semen into the ejaculatory ducts.

Seminiferous tubule (sem″ĭ-NĬ-fer-us TU-būl) A tightly coiled duct, located in a lobule of the testis, where spermatozoa are produced.

Semipermeable Having pores of a size that will permit some but not all substances to diffuse through a membrane; differentially permeable.

Senescence (sen-ES-ens) The process of growing old; the period of old age.

Senility (se-NIL-i-te) A loss of mental ability in old people. The state of being old.

Sensation A state of awareness of external or internal conditions of the body.

Sensory area A region of the cerebral cortex concerned with the interpretation of sensory impulses.

Sepsis (SEP-sis) A morbid condition resulting from the presence of pathogenic bacteria and their products.

Septum (SEP-tum) A wall dividing two cavities.

Serosa (ser-O-sah) Any serous membrane; the outermost layer or tunic of an organ formed by a serous membrane, the membrane that lines the pleural, pericardial, and peritoneal cavities.

Serous membrane (SIR-us) An epithelial membrane (a combination of an epithelial and a connective tissue layer) that lines body cavities that do not open to the exterior and that covers the organs that lie within these cavities; each consists of two parts: a parietal portion and a visceral portion.

Sertoli cells (ser-TO-le) "Nurse" cells located in seminiferous tubules embedded between the developing spermatozoa that produce secretions for the supplying of nutrients to the spermatozoa.

Sesamoid bone (SES-ah-moyd) A bone that forms in a tendon and near a joint; for example, the patella.

Shoulder A synovial or diarthrotic joint where the humerus joins the scapula.

Sigmoid (SIG-moyd) Shaped like the Greek letter sigma (S).

Sigmoid colon (SIG-moyd KO-lon) The S-shaped portion of the large intestine that begins at the level of the left iliac crest, projects inward to the midline, and terminates at the rectum at about the level of the third sacral vertebra.

Sinuatrial node (si"nu-A-tre-al) A compact mass of cardiac muscle cells specialized for conduction, located in the right atrium beneath the opening of the superior vena cava; also called the sinoatrial node, SA node, or pacemaker.

Sinus (SI-nus) A hollow in a bone or other tissue; a channel for blood; any cavity having a narrow opening.

Sinusoid (SI-nus-oyd) A blood space in certain organs, as the liver or spleen.

Sliding-filament hypothesis The most commonly accepted explanation for muscle contraction in which actin and myosin move into interdigitation with each other, decreasing the length of the sarcomeres.

Slipped disc The popular name for a rupture of an intervertebral disc so that the nucleus pulposus protrudes into the vertebral cavity; also called a ruptured or herniated disc.

Small intestine A long tube of the gastrointestinal tract that begins at the pyloric valve of the stomach, coils through the central and lower part of the abdominal cavity, and ends at the large intestine; the small intestine is divided into three segments: duodenum, jejunum, and ileum.

Soft palate (PAL-at) The posterior portion of the roof of the mouth, extending posteriorly from the palatine bones and ending at the uvula; it is a muscular partition lined with mucous membrane.

Somatic nervous system (so-MAT-ik) The portion of the peripheral nervous system made up of the somatic efferent fibers that run between the central nervous system and the skeletal muscles and skin.

Somatic reflex (RE-fleks) A reflex arc in which the effector is a skeletal muscle.

Spasm (spazm) An involuntary, convulsive, muscular contraction.

Spermatic cord (sper-MAT-ik) A supporting structure of the male reproductive system, extending from a testis to the deep inguinal ring, that includes the ductus deferens, arteries, veins, lymphatics, nerves, cremaster muscle, and connective tissue.

Spermatogenesis (sper"mah-to-JEN-ĭ-sis) The formation and development of spermatozoa.

Spermatozoan (sper"mah-to-ZO-an) A mature male germ cell or sperm cell.

Spermicidal (sper-mi-SI-dal) An agent that kills spermatozoa.

Sphincter (SFINK-ter) A circular muscle constricting an orifice.

Sphincter of Oddi (SFINGK-ter of O-de) A circular muscle at the opening of the common bile and main pancreatic ducts in the duodenum that protrudes as a mass of tissue called the ampulla of Vater.

Sphygmomanometer (sfig"mo-man-OM-e-ter) An instrument for measuring arterial blood pressure.

Spina bifida (SPI-nah BIF-ĭ-dah) A defect of the vertebral column in which the two halves of the neural arch of a vertebra fail to fuse in midline.

Spinal nerve (SPI-nal) One of the 31 pairs of nerves that originates on the spinal cord.

Spinal puncture Withdrawal of some of the cerebrospinal fluid from the subarachnoid space in the lumbar region.

Spinous process or **spine** (SPI-nus) A sharp or thornlike process or projection; a sharp ridge running diagonally across the posterior surface of the scapula.

Spirometer (spi-ROM-et-er) An apparatus used to measure air capacity of the lungs.

Spongy or **cancellous bone** (SPUN-je or KAN-sel-lus) Bone tissue containing many large spaces filled with marrow.

Sputum (SPU-tum) Substance ejected from the mouth containing saliva and mucus.

Squamous (SKWA-mus) Scalelike.

Stapes (STA-pēz) The innermost of the three auditory ossicles of the middle ear, the one that fits into the oval window; also called the stirrup.

Stasis (STA-sis) Stagnation or halt of normal flow of fluids, as blood, urine, or of the intestinal mechanism.

Stereocilia (ste"re-o-SIL-e-ah) Groups of extremely long, slender, nonmotile microvilli projecting from epithelial cells lining the epididymis.

Sterility (ster-IL-it-e) Infertility; absence of reproductive power.

Stomach The J-shaped enlargement of the gastrointestinal tract directly under the diaphragm in the epigastric, umbilical, and left hypochondriac regions of the abdomen, between the esophagus and the small intestine.

Straight tubules (TU-bŭlz) The ducts in a testis that lead from the convoluted seminiferous tubules to the rete testis.

Stratum (STRA-tum) A layer.

Stratum basalis (ba-SAL-is) The outer layer of the endometrium, next to the myometrium, that is maintained during menstruation and gestation and produces a new functionalis following menstruation or parturition.

Stratum functionalis (funk"shun-AL-is) The inner layer of the endometrium, the layer next to the uterine cavity, that is shed during menstruation and that forms the maternal portion of the placenta during gestation.

Stricture (STRIK-tur) A local contraction of a tubular structure.

Stroma (STRO-mah) The tissue that forms the ground substance, foundation, or framework of an organ, as opposed to its functional parts.

Subarachnoid space (sub"ah-RAK-noyd) A wide space between the arachnoid and the pia mater that surrounds the brain and spinal cord and through which cerebrospinal fluid circulates.

Subcutaneous layer or **superficial fascia** (sub''ku-TA-ne-us *or* su''per-FISH-al FASH-e-ah) A continuous sheet of fibrous connective tissue between the dermis of the skin and the deep fascia of the muscles.

Subdural space (sub-DU-ral) A space between the dura mater and the arachnoid of the brain and spinal cord that contains a small amount of fluid.

Sublingual gland (sub-LING-gwal) One of a pair of salivary glands situated in the floor of the mouth under the mucous membrane and to the side of the lingual frenulum, with a duct that opens into the floor of the mouth; the smallest of the three pairs of salivary glands.

Submandibular ganglion (sub-man-DIB-u-lar GANG-lĭ-on) A cluster of cell bodies of postganglionic parasympathetic neurons, located above the submandibular gland with fibers ending at the submandibular and sublingual salivary glands and other small salivary glands in the floor of the mouth.

Submandibular gland One of a pair of salivary glands found beneath the base of the tongue under the mucous membrane in the posterior part of the floor of the mouth, posterior to the sublingual glands, with a duct situated to the side of the lingual frenulum; also called the submaxillary gland.

Submucosa (sub-mu-KO-sah) A layer of connective tissue located beneath a mucous membrane, as in the alimentary canal where a submucosa binds the mucosa to the muscularis tunic, and in the urinary bladder where a submucosa connects the mucosa to the muscularis tunic.

Subserous fascia (sub-SE-rus FASH-e-ah) A layer of connective tissue internal to the deep fascia, lying between the deep fascia and the serous membrane that lines the body cavities.

Sulcus (SUL-kus) A groove or depression between parts; especially a fissure between the convolutions of the brain; the plural form is sucli.

Superficial fascia (soo''per-FISH-al FASH-e-ah) A continuous sheet of fibrous connective tissue between the dermis of the skin and the deep fascia of the muscles; also called the subcutaneous layer.

Superior or **cephalic** (su-PĒR-e-or *or* sĕ-FAL-ik) Toward the head; toward the upper part of a structure.

Supination (su''pĭ-NA-shun) A movement of the forearm in which the palm of the hand is turned forward (anteriorly).

Suppuration (sup-u-RA-shun) The process of pus formation.

Surface anatomy The study of the structures that can be identified from the outside of the body.

Surfactant (sur-FAK-tant) A substance formed in the lungs that helps to keep the small air sacs expanded by reducing surface tension.

Susceptibility (sŭ-sep''-tĭ-BIL-ĭ-te) Lack of resistance of a body to the deleterious or other effects of an agent such as pathogenic microorganisms.

Suspensory ligament (sus-PEN-so-re LIG-ah-ment) A fold of peritoneum extending laterally from the surface of the ovary to the pelvic wall.

Suture (SU-chur) A type of fibrous joint found between bones of the skull where the bones are very slightly sep-

arated by a thin layer of fibrous tissue, with the result that the joints are immovable.

Sweat gland A gland widely distributed through the skin, particularly in the skin of the palms, soles, armpits, and forehead, with the secretory portion lying in the subcutaneous tissue and the excretory duct projecting upward through the dermis and epidermis to open in a pore at the surface.

Sympathetic (sim''pah-THET-ik) One of the two subdivisions of the autonomic nervous system, having cell bodies of preganglionic neurons in the lateral gray columns of the thoracic segment and first two or three lumbar segments of the spinal cord; primarily concerned with processes involving the expenditure of energy; the thoracolumbar division.

Sympathetic trunk ganglion A cluster of cell bodies of postganglionic sympathetic neurons lateral to the vertebral column, close to the body of a vertebra. These ganglia extend downward through the neck, thorax, and abdomen to the coccyx, on both sides of the vertebral column and are connected to one another to form a chain on each side of the vertebral column; also called lateral or sympathetic chain or vertebral chain ganglia.

Symphysis (SIM-fĭ-sis) A slightly movable cartilaginous joint in which the material connecting two bones is a broad, flat disc of fibrocartilage, as between the bodies of vertebrae and between the anterior surfaces of the pelvic bones.

Symphysis pubis (PU-bis) A slightly movable cartilaginous joint between the anterior surfaces of the pelvic bones.

Symptom (SIMP-tum) A specific recognizable abnormality.

Synapse (SIN-aps) A small gap that serves as the functional junction between two neurons, where a nerve impulse is conducted from one neuron to another by a neurohumor.

Synarthrosis (sin''ar-THRO-sis) An immovable joint, such as a suture, synchondrosis, and synostosis.

Synchondrosis (sin''kon-DRO-sis) An immovable cartilaginous joint in which the connecting material is hyaline cartilage; a synchondrosis may be temporary, as in the epiphyseal plate, or permanent, as between the ribs and the sternum.

Syncytium (sin-SISH-ĭ-um) A multinucleated mass of protoplasm produced by the merging of cells.

Syndesmosis (sin''des-MO-sis) A fibrous joint in which two bones are united by dense fibrous tissue, producing a slightly movable joint such as the distal articulation between the tibia and the fibula.

Syndrome (SIN-drōm) A group of abnormalities that occur together in a characteristic pattern; the complete picture of a disease.

Synergist (SIN-er-jist) A group of muscles that assists the prime mover or agonist by reducing undesired action or unnecessary movement; also called a fixator.

Synostosis (sin''os-TO-sis) An immovable joint in which the connecting tissue between two bones is bone, as in the epiphyseal line that replaces the cartilaginous epiphyseal plate of a growing bone.

Synovial cavity (sin-O-ve-al) The space between the articulating bones of a synovial or diarthrotic joint, filled with synovial fluid.

Synovial joint A fully movable or diarthrotic joint in which a joint or synovial cavity is present between the two articulating bones.

Synovial membrane   The inner of the two layers of the articular capsule of a synovial joint, composed of loose connective tissue covered with epithelium that secretes synovial fluid into the joint cavity.

System   An association of organs that have a common function.

Systemic (sis-TEM-ik)   Affecting the whole body; generalized.

Systemic anatomy   The study of particular systems of the body, such as the system of nerves, spinal cord, and brain, or the system of heart, blood vessels, and blood.

Systemic circulation   All of the circulatory routes taken by oxygenated blood that leaves the left ventricle through the aorta and returns to the right atrium, carrying oxygenated blood to all of the organs of the body.

Systole (SIS-to-le)   Heart muscle contraction, especially that of the ventricles.

T tubule (TU-būl)   In a muscle cell, an invagination of the sarcolemma that runs transversely through the fiber perpendicular to the sarcoplasmic reticulum.

Tactile (TAK-til)   Pertaining to the sense of touch.

Taeniae coli (TE-nĭ-e KO-li)   Three flat bands of smooth muscle tissue, formed by concentrating the longitudinal muscle layer into bands instead of a sheet, that run the length of the large intestine and are easily seen in a gross specimen.

Target organ   The organ or group of organs affected by a particular hormone.

Tarsal plate (TAR-sal)   A thin, elongated sheet of connective tissue, one in each eyelid, giving the eyelid form and support; the aponeurosis of the levator palpebrae superioris is attached to the tarsal plate of the superior eyelid.

Tarsus (TAR-sus)   A collective designation for the seven bones of the ankle.

Tectorial membrane (tek-TO-re-al)   A gelatinous membrane projecting over and in contact with the hair cells of the organ of Corti in the cochlear duct.

Telophase (TEL-o-fāz)   The final stage of mitosis in which the nuclei become established.

Tendinitis (ten''din-I-tis)   A disorder involving the inflammation of a tendon and synovial membrane at a joint.

Tendon (TEN-don)   A white fibrous cord of dense, regular connective tissue that attaches muscle to bone.

Tentorium cerebelli (ten-TO-re-um ser''ĕ-BEL-e)   A transverse shelf of dura mater that forms a partition between the occipital part of the cerebral hemispheres and the cerebellum and that covers the cerebellum.

Terminal ganglion (TER-min-al GANG-lĭ-on)   A cluster of cell bodies of postganglionic parasympathetic neurons either lying very close to the visceral effectors or located within the walls of the visceral effectors supplied by the postganglionic fibers.

Testosterone (tes-TOS-ter-ōn)   A hormone secreted by the interstitial cells (of Leydig) of a mature testis.

Tetany (TET-an-e)   A nervous condition characterized by intermittent or continuous tonic muscular contractions of the extremities.

Thalamus (THAL-ah-mus)   A large, oval structure, located above the midbrain, consisting of two masses of gray matter covered by a thin layer of white matter.

Therapy (THER-a-pe)   The treatment of disease or disorder.

Thigh   The portion of the lower extremity between the hip and the knee.

Third ventricle (VEN-trĭ-kl)   A slitlike cavity between the right and left halves of the thalamus and between the lateral ventricles.

Thoracic (thor-Ă-sik)   Pertaining to the chest.

Thoracic duct   A lymphatic vessel that begins as a dilation called the cisterna chyli and receives lymph from the left side of the head, neck, and chest, and the left arm, and the entire body below the ribs, finally emptying the lymph into the left subclavian vein; also called the left lymphatic duct.

Thoracolumbar outflow (tho''rah-ko-LUM-bar)   The sympathetic division of the autonomic nervous system, so called because the preganglionic sympathetic neurons have their cell bodies in the lateral gray columns of the thoracic segment and first two or three lumbar segments of the spinal cord.

Thorax (THO-raks)   The chest.

Thrombocyte (THROM-bo-sīt)   A blood platelet, a fragment of a megakaryocyte, that plays a role in the chain of reactions that results in blood clotting.

Thrombophlebitis (throm''bo-flĭ-BI-tis)   A disorder in which inflammation of a vein wall is followed by the formation of a blood clot (thrombus).

Thymectomy (thy-MEK-to-me)   Surgical removal of the thymus.

Thymus gland (THI-mus)   A bilobed organ, located in the upper mediastinum posterior to the sternum and between the lungs, that plays a role in the immunity mechanism of the body.

Thyroid cartilage (THI-royd KAR-tĭ-lij)   The largest single cartilage of the larynx, consisting of two fused plates that form the anterior wall of the larynx; also called the Adam's apple.

Tinnitus (tin-I-tus)   A ringing or tinkling sound in the ears.

Tissue   A group of similar cells, and their intercellular substance, joined together to perform a specific function.

Tongue   A large skeletal muscle on the floor of the oral cavity.

Tonsil (TON-sil)   A mass of lymphoid tissue embedded in mucous membrane.

Toxic (TOK-sik)   Pertaining to poison; poisonous.

Trabeculae (tra-BEK-u-le)   Fibrous cords of connective tissue, serving as supporting fiber by forming a septum extending into an organ from its walls or capsule; an anastomosing network of lamellae of spongy bone, with the spaces filled with red marrow.

Trachea (TRA-ke-ah)   The windpipe.

Tract   A bundle of nerve fibers in the central nervous system.

Transection (tran-SEK-shun)   A cross cut.

Transplantation (trans-plan-TA-shun)   The transfer or implantation of body tissue from one part of the body to another or from one person to another.

Transverse (trans-VERS)   Lying across; crosswise.

Transverse colon (trans-VERS KO-lon)   The portion of the large intestine extending across the abdomen from the hepatic flexure to the splenic flexure.

Transverse fissure (FISH-er)   The deep cleft that separates the cerebrum from the cerebellum.

Trauma (TRAU-mah)   An injury or wound that may be produced by external force or by shock, as in psychic trauma.

Triad (TRI-ad)   A complex of three units in a muscle cell composed of a T tubule and the segments of sarcoplasmic reticulum on both sides of it.

Trigone (TRI-gōn)   A triangular area in the base of the urinary bladder between the ureteral and urethral orifices where the mucous membrane is firmly attached to the muscular coat and is always smooth.

Trochanter (tro-KAN-ter)   A large projection on the femur that serves as a point of muscle attachment.

True vocal cords   A pair of folds of the mucous membrane of the larynx enclosing two strong bands of fibrous tissue, the vocal ligaments, that vibrate to make sounds during speaking; the true vocal cords are inferior to the false vocal folds.

Trunk   The part of the body to which the upper and lower extremities are attached.

Tubercle (TU-ber-kl)   A small elevation.

Tuberosity (tu''bĕ-ROS-ĭ-te)   An elevation or protuberence.

Tumor (TU-mor)   A swelling or enlargement.

Tunica albuginea (TU-nĭ-kah al''bu-JIN-e-ah)   A tough, whitish layer of fibrous tissue around an organ; for example, covering the testis.

Tunica externa (eks-TER-nah)   The white, fibrous outer coat of an artery or vein.

Tunica interna (in-TER-nah)   The inner coat of an artery or vein, consisting of a lining of endothelium and its supporting layer of connective tissue; also called the tunica intima.

Tunica media (ME-de-ah)   The middle coat of an artery or vein composed of smooth muscle and elastic fibers.

Tympanic antrum (tim-PAN-ik AN-trum)   An air space in the posterior wall of the middle ear that leads into the mastoid air cells or sinus.

Tympanic membrane   The eardrum.

Ulcer (UL-ser)   An open lesion upon the skin or mucous membrane of the body, with loss of substance and necrosis of the tissue.

Umbilicus (um-BIL-i-kus)   A small scar on the abdomen that marks the former attachment of the umbilical cord to the fetus; the navel.

Upper extremity   The appendage attached at the shoulder, consisting of an arm, a forearm, and a hand.

Uremia (u-RE-me-ah)   A toxic condition from urea and other waste products in the blood.

Ureter (u-RE-ter)   The duct that conveys urine from the kidney to the urinary bladder.

Urethra (u-RE-thrah)   The duct that conveys urine from the urinary bladder to the exterior of the body.

Urinary bladder (U-rĭ-ner-e)   A hollow, muscular organ situated in the pelvic cavity posterior to the symphysis pubis.

Urogenital triangle (u''ro-JEN-ĭ-tal)   The region of the pelvic floor below the symphysis pubis, bounded by the symphysis pubis and the ischial tuberosities and containing the external genitalia.

Urticaria (ur-ti-KA-re-ah)   A skin reaction to certain foods, drugs, or other substances to which a person may be allergic; hives.

Uterine tubes (U-ter-in)   Ducts that transport ova from the ovary to the uterus; also called the fallopian tubes or the oviducts.

Uterosacral ligament (u''ter-o-SA-kral LIG-ah-ment)   A fibrous band of tissue extending from the cervix of the uterus laterally to attach to the sacrum.

Uterovesical pouch (u''ter-o-VES-ĭ-kal)   A shallow pouch formed by the reflection of the peritoneum from the anterior surface of the uterus, at the junction of the cervix and the body, to the posterior surface of the urinary bladder.

Uterus (U-ter-us)   An inverted, pear-shaped, hollow organ situated between the urinary bladder and the rectum in the female pelvis; the womb.

Utricle (U-trĭ-kl)   The larger of the two divisions of the membranous labyrinth located inside the vestibule of the inner ear.

Uvula (U-vu-lah)   A soft, fleshy mass, especially the V-shaped pendant part, descending from the soft palate.

Vagina (vah-JI-nah)   A muscular, tubular organ that leads from the uterus to the vestibule, situated between the urinary bladder and the rectum of the female.

Varicocele (VAR-i-ko-sēl)   A twisted vein; especially the accumulation of blood in the veins of the spermatic cord.

Varicose vein (VAR-ĭ-kōs)   A swollen or distended vein, especially in the legs, due to the failure of the valves to prevent the backflow of blood.

Vas   A vessel or duct.

Vasa vasorum (VA-sah va-SO-rum)   Blood vessels supplying nutrients to the larger arteries and veins.

Vascular (VAS-ku-lar)   Pertaining to or containing many vessels.

Vascular tunic (TU-nik)   The middle layer of the eyeball, composed of the choroid, the ciliary body, and the iris.

Vasectomy (vah-SEK-to-me)   A means of sterilization of males in which a portion of each ductus deferens is removed.

Vein   A blood vessel that conveys blood from the tissues back to the heart.

Venae cavae (VE-ne KA-ve)   The two large veins that open into the right atrium, returning to the heart all of the deoxygenated blood from the systemic circulation except from the coronary circulation.

Ventilation (ven''tĭ-LA-shun)   Breathing, or the process by which atmospheric gases are drawn down into the lungs and waste gases that have diffused into the lungs are expelled back up through the respiratory passageways.

Ventral (VEN-tral)   Pertaining to the anterior or front side of the body; opposite of dorsal.

Ventral ramus (VEN-tral RA-mus)   The anterior branch of a spinal nerve, containing sensory and motor fibers to the muscles and skin of the anterior surface of the head, neck, trunk, and the extremities.

Ventricle (VEN-trĭ-kl) One of the two lower chambers of the heart; a cavity within the brain, containing cerebrospinal fluid.

Venule (VEN-ul) A small vein.

Vermiform appendix (VER-mĭ-form ah-PEN-diks) A twisted, coiled tube attached to the cecum.

Vermis (VER-mis) The central constricted area of the cerebellum that separates the two cerebellar hemispheres.

Vertebral or **spinal canal** (VER-tĕ-bral *or* SPI-nal) The cavity formed by the vertebral foramina of all vertebrae together; the inferior portion of the dorsal cavity, containing the spinal cord.

Vertigo (VUR-tĭ-go) Sensation of dizziness.

Vesicle (VESĭ-kal) A small bladder or sac containing liquid.

Vestibular membrane (ves-TIB-u-lar) The membrane that separates the cochlear duct from the scala vestibuli; also called Reissner's membrane.

Vestibule (VES-tib-ūl) A small space or cavity at the beginning of a canal; especially the inner ear, larynx, mouth, nose, vagina.

Villus (VIL-us) A minute, fingerlike projection of the intestinal mucosa into the lumen of the intestine that increases the surface area for absorption.

Viscera (VIS-er-ah) Organs within body cavities, especially within the abdomen; the singular form is viscus.

Visceral (VIS-er-al) Pertaining to the covering of an organ.

Visceral autonomic reflex (aw"to-NOM-ik RE-fleks) A quick, involuntary response in which the impulse travels over visceral efferent neurons to smooth muscle, cardiac muscle, or a gland.

Visceral effector (ĕ-FEK-tor) Cardiac muscle, smooth muscle, and glandular epithelium.

Visceral pleura (PLU-rah) The inner layer of the serous membrane that covers the lungs.

Visceroceptor (vis"er-o-SEP-tor) A sensory receptor, located in blood vessels and viscera, that picks up information about the internal environment.

Viscosity (vis-KOS-i-te) The state of being sticky or thick.

Vitreous humor (VIT-re-us HU-mor) A soft, jellylike substance that fills the posterior cavity of the eyeball, lying between the lens and the retina.

Volkmann's canal (FOLK-mahns) A minute passageway by means of which blood vessels and nerves from the periosteum penetrate into compact bone.

Vulva (VUL-vah) The external genitalia of the female; also called the pudendum.

Wallerian degeneration (wah-LER-ĭ-an) The changes that occur in a cut nerve cell proximal to the site of injury, first described by Waller, making it possible to follow peripheral nerve fibers into the central nervous system; the changes that occur in the proximal portion of the cut neuron include fragmentation of the fiber retrograde to the first node of Ranvier, phagocytosis of the myelin sheath, the Nissl reaction or chromatolysis in which the Nissl substance is dispersed in the cytoplasm, and the displacement of the nucleus to the periphery of the cell.

White matter Aggregations or bundles of myelinated axons located in the brain and spinal cord.

White pulp The portion of the spleen composed of ovoid masses of lymphoid tissue called lymph follicles that contain germinal centers and where lymphocytes are produced.

White ramus communicans (RA-mus com-MU-nĭ-kans) The portion of a preganglionic sympathetic nerve fiber that branches away from the anterior ramus of a spinal nerve to enter the nearest sympathetic trunk ganglion.

Wormian or **sutural bone** A small bone located within a suture or immovable joint of certain cranial bones.

Zona fasciculata (ZO-nah fah-sik"u-LAH-tah) The middle zone of the adrenal cortex that consists of cells arranged in long, straight cords and that secretes glucocorticoid hormones.

Zona glomerulosa (glo-mer"u-LO-sah) The outer zone of the adrenal cortex, directly under the connective tissue covering, that consists of cells arranged in arched loops or round balls and that secretes mineralocorticoids.

Zona reticularis (ret-ik"u-LAR-is) The inner zone of the adrenal cortex, consisting of cords of branching cells that secrete sex hormones, chiefly androgens.

Zymogenic cell (zi"mo-JEN-ik) A cell that secretes enzymes; for example, the chief cells of the gastric glands that secrete pepsinogen.

Index